# Communications in Computer and Information Science 1793

Rationale

The CCIS series is devoted to the publication of proceedings of computer science conferences. Its aim is to efficiently disseminate original research results in informatics in printed and electronic form. While the focus is on publication of peer-reviewed full papers presenting mature work, inclusion of reviewed short papers reporting on work in progress is welcome, too. Besides globally relevant meetings with internationally representative program committees guaranteeing a strict peer-reviewing and paper selection process, conferences run by societies or of high regional or national relevance are also considered for publication.

Topics

The topical scope of CCIS spans the entire spectrum of informatics ranging from foundational topics in the theory of computing to information and communications science and technology and a broad variety of interdisciplinary application fields.

Information for Volume Editors and Authors

Publication in CCIS is free of charge. No royalties are paid, however, we offer registered conference participants temporary free access to the online version of the conference proceedings on SpringerLink (http://link.springer.com) by means of an http referrer from the conference website and/or a number of complimentary printed copies, as specified in the official acceptance email of the event.

CCIS proceedings can be published in time for distribution at conferences or as post-proceedings, and delivered in the form of printed books and/or electronically as USBs and/or e-content licenses for accessing proceedings at SpringerLink. Furthermore, CCIS proceedings are included in the CCIS electronic book series hosted in the SpringerLink digital library at http://link.springer.com/bookseries/7899. Conferences publishing in CCIS are allowed to use Online Conference Service (OCS) for managing the whole proceedings lifecycle (from submission and reviewing to preparing for publication) free of charge.

Publication process

The language of publication is exclusively English. Authors publishing in CCIS have to sign the Springer CCIS copyright transfer form, however, they are free to use their material published in CCIS for substantially changed, more elaborate subsequent publications elsewhere. For the preparation of the camera-ready papers/files, authors have to strictly adhere to the Springer CCIS Authors' Instructions and are strongly encouraged to use the CCIS LaTeX style files or templates.

Abstracting/Indexing

CCIS is abstracted/indexed in DBLP, Google Scholar, EI-Compendex, Mathematical Reviews, SCImago, Scopus. CCIS volumes are also submitted for the inclusion in ISI Proceedings.

How to start

To start the evaluation of your proposal for inclusion in the CCIS series, please send an e-mail to ccis@springer.com.

Mohammad Tanveer · Sonali Agarwal ·
Seiichi Ozawa · Asif Ekbal · Adam Jatowt
Editors

# Neural Information Processing

## 29th International Conference, ICONIP 2022
## Virtual Event, November 22–26, 2022
## Proceedings, Part VI

*Editors*
Mohammad Tanveer
Indian Institute of Technology Indore
Indore, India

Sonali Agarwal
Indian Institute of Information Technology - Allahabad
Prayagraj, India

Seiichi Ozawa
Kobe University
Kobe, Japan

Asif Ekbal
Indian Institute of Technology Patna
Patna, India

Adam Jatowt
University of Innsbruck
Innsbruck, Austria

ISSN 1865-0929 ISSN 1865-0937 (electronic)
Communications in Computer and Information Science
ISBN 978-981-99-1644-3 ISBN 978-981-99-1645-0 (eBook)
https://doi.org/10.1007/978-981-99-1645-0



This Springer imprint is published by the registered company Springer Nature Singapore Pte Ltd.
The registered company address is: 152 Beach Road, #21-01/04 Gateway East, Singapore 189721, Singapore

# Preface

Welcome to the proceedings of the 29th International Conference on Neural Information Processing (ICONIP 2022) of the Asia-Pacific Neural Network Society (APNNS), held virtually from Indore, India, during November 22–26, 2022.

The mission of the Asia-Pacific Neural Network Society is to promote active interactions among researchers, scientists, and industry professionals who are working in neural networks and related fields in the Asia-Pacific region. APNNS has Governing Board Members from 13 countries/regions – Australia, China, Hong Kong, India, Japan, Malaysia, New Zealand, Singapore, South Korea, Qatar, Taiwan, Thailand, and Turkey. The society's flagship annual conference is the International Conference of Neural Information Processing (ICONIP).

The ICONIP conference aims to provide a leading international forum for researchers, scientists, and industry professionals who are working in neuroscience, neural networks, deep learning, and related fields to share their new ideas, progress, and achievements. Due to the current situation regarding the pandemic and international travel, ICONIP 2022, which was planned to be held in New Delhi, India, was organized as a fully virtual conference.

The proceedings of ICONIP 2022 consists of a multi-volume set in LNCS and CCIS, which includes 146 and 213 papers, respectively, selected from 1003 submissions reflecting the increasingly high quality of research in neural networks and related areas. The conference focused on four main areas, i.e., "Theory and Algorithms," "Cognitive Neurosciences," "Human Centered Computing," and "Applications." The conference also had special sessions in 12 niche areas, namely

1 International Workshop on Artificial Intelligence and Cyber Security (AICS)
2. Computationally Intelligent Techniques in Processing and Analysis of Neuronal Information (PANI)
3. Learning with Fewer Labels in Medical Computing (FMC)
4. Computational Intelligence for Biomedical Image Analysis (BIA)
5 Optimized AI Models with Interpretability, Security, and Uncertainty Estimation in Healthcare (OAI)
6. Advances in Deep Learning for Biometrics and Forensics (ADBF)
7. Machine Learning for Decision-Making in Healthcare: Challenges and Opportunities (MDH)
8. Reliable, Robust and Secure Machine Learning Algorithms (RRS)
9. Evolutionary Machine Learning Technologies in Healthcare (EMLH)
10 High Performance Computing Based Scalable Machine Learning Techniques for Big Data and Their Applications (HPCML)
11. Intelligent Transportation Analytics (ITA)
12. Deep Learning and Security Techniques for Secure Video Processing (DLST)

Our great appreciation goes to the Program Committee members and the reviewers who devoted their time and effort to our rigorous peer-review process. Their insightful reviews and timely feedback ensured the high quality of the papers accepted for publication.

The submitted papers in the main conference and special sessions were reviewed following the same process, and we ensured that every paper has at least two high-quality single-blind reviews. The PC Chairs discussed the reviews of every paper very meticulously before making a final decision. Finally, thank you to all the authors of papers, presenters, and participants, which made the conference a grand success. Your support and engagement made it all worthwhile.

December 2022

Mohammad Tanveer
Sonali Agarwal
Seiichi Ozawa
Asif Ekbal
Adam Jatowt

# Organization

## Program Committee

### General Chairs

| | |
|---|---|
| M. Tanveer | Indian Institute of Technology Indore, India |
| Sonali Agarwal | IIIT Allahabad, India |
| Seiichi Ozawa | Kobe University, Japan |

### Honorary Chairs

| | |
|---|---|
| Jonathan Chan | King Mongkut's University of Technology Thonburi, Thailand |
| P. N. Suganthan | Nanyang Technological University, Singapore |

### Program Chairs

| | |
|---|---|
| Asif Ekbal | Indian Institute of Technology Patna, India |
| Adam Jatowt | University of Innsbruck, Austria |

### Technical Chairs

| | |
|---|---|
| Shandar Ahmad | JNU, India |
| Derong Liu | University of Chicago, USA |

### Special Session Chairs

| | |
|---|---|
| Kai Qin | Swinburne University of Technology, Australia |
| Kaizhu Huang | Duke Kunshan University, China |
| Amit Kumar Singh | NIT Patna, India |

### Tutorial Chairs

| | |
|---|---|
| Swagatam Das | ISI Kolkata, India |
| Partha Pratim Roy | IIT Roorkee, India |

**Finance Chairs**

| | |
|---|---|
| Shekhar Verma | Indian Institute of Information Technology Allahabad, India |
| Hayaru Shouno | University of Electro-Communications, Japan |
| R. B. Pachori | IIT Indore, India |

**Publicity Chairs**

| | |
|---|---|
| Jerry Chun-Wei Lin | Western Norway University of Applied Sciences, Norway |
| Chandan Gautam | A*STAR, Singapore |

**Publication Chairs**

| | |
|---|---|
| Deepak Ranjan Nayak | MNIT Jaipur, India |
| Tripti Goel | NIT Silchar, India |

**Sponsorship Chairs**

| | |
|---|---|
| Asoke K. Talukder | NIT Surathkal, India |
| Vrijendra Singh | IIIT Allahabad, India |

**Website Chairs**

| | |
|---|---|
| M. Arshad | IIT Indore, India |
| Navjot Singh | IIIT Allahabad, India |

**Local Arrangement Chairs**

| | |
|---|---|
| Pallavi Somvanshi | JNU, India |
| Yogendra Meena | University of Delhi, India |
| M. Javed | IIIT Allahabad, India |
| Vinay Kumar Gupta | IIT Indore, India |
| Iqbal Hasan | National Informatics Centre, Ministry of Electronics and Information Technology, India |

**Regional Liaison Committee**

| | |
|---|---|
| Sansanee Auephanwiriyakul | Chiang Mai University, Thailand |
| Nia Kurnianingsih | Politeknik Negeri Semarang, Indonesia |

| | |
|---|---|
| Md Rafiqul Islam | University of Technology Sydney, Australia |
| Bharat Richhariya | IISc Bangalore, India |
| Sanjay Kumar Sonbhadra | Shiksha 'O' Anusandhan, India |
| Mufti Mahmud | Nottingham Trent University, UK |
| Francesco Piccialli | University of Naples Federico II, Italy |

**Program Committee**

| | |
|---|---|
| Balamurali A. R. | IITB-Monash Research Academy, India |
| Ibrahim A. Hameed | Norwegian University of Science and Technology (NTNU), Norway |
| Fazly Salleh Abas | Multimedia University, Malaysia |
| Prabath Abeysekara | RMIT University, Australia |
| Adamu Abubakar Ibrahim | International Islamic University, Malaysia |
| Muhammad Abulaish | South Asian University, India |
| Saptakatha Adak | Philips, India |
| Abhijit Adhikary | King's College, London, UK |
| Hasin Afzal Ahmed | Gauhati University, India |
| Rohit Agarwal | UiT The Arctic University of Norway, Norway |
| A. K. Agarwal | Sharda University, India |
| Fenty Eka Muzayyana Agustin | UIN Syarif Hidayatullah Jakarta, Indonesia |
| Gulfam Ahamad | BGSB University, India |
| Farhad Ahamed | Kent Institute, Australia |
| Zishan Ahmad | Indian Institute of Technology Patna, India |
| Mohammad Faizal Ahmad Fauzi | Multimedia University, Malaysia |
| Mudasir Ahmadganaie | Indian Institute of Technology Indore, India |
| Hasin Afzal Ahmed | Gauhati University, India |
| Sangtae Ahn | Kyungpook National University, South Korea |
| Md. Shad Akhtar | Indraprastha Institute of Information Technology, Delhi, India |
| Abdulrazak Yahya Saleh Alhababi | University of Malaysia, Sarawak, Malaysia |
| Ahmed Alharbi | RMIT University, Australia |
| Irfan Ali | Aligarh Muslim University, India |
| Ali Anaissi | CSIRO, Australia |
| Ashish Anand | Indian Institute of Technology, Guwahati, India |
| C. Anantaram | Indraprastha Institute of Information Technology and Tata Consultancy Services Ltd., India |
| Nur Afny C. Andryani | Universiti Teknologi Petronas, Malaysia |
| Marco Anisetti | Università degli Studi di Milano, Italy |
| Mohd Zeeshan Ansari | Jamia Millia Islamia, India |
| J. Anuradha | VIT, India |
| Ramakrishna Appicharla | Indian Institute of Technology Patna, India |

| | |
|---|---|
| V. N. Manjunath Aradhya | JSS Science and Technology University, India |
| Sunil Aryal | Deakin University, Australia |
| Muhammad Awais | COMSATS University Islamabad, Wah Campus, Pakistan |
| Mubasher Baig | National University of Computer and Emerging Sciences (NUCES) Lahore, Pakistan |
| Sudhansu Bala Das | NIT Rourkela, India |
| Rakesh Balabantaray | International Institute of Information Technology Bhubaneswar, India |
| Sang-Woo Ban | Dongguk University, South Korea |
| Tao Ban | National Institute of Information and Communications Technology, Japan |
| Dibyanayan Bandyopadhyay | Indian Institute of Technology, Patna, India |
| Somnath Banerjee | University of Tartu, Estonia |
| Debajyoty Banik | Kalinga Institute of Industrial Technology, India |
| Mohamad Hardyman Barawi | Universiti Malaysia, Sarawak, Malaysia |
| Mahmoud Barhamgi | Claude Bernard Lyon 1 University, France |
| Kingshuk Basak | Indian Institute of Technology Patna, India |
| Elhadj Benkhelifa | Staffordshire University, UK |
| Sudip Bhattacharya | Bhilai Institute of Technology Durg, India |
| Monowar H Bhuyan | Umeå University, Sweden |
| Xu Bin | Northwestern Polytechnical University, China |
| Shafaatunnur Binti Hasan | UTM, Malaysia |
| David Bong | Universiti Malaysia Sarawak, Malaysia |
| Larbi Boubchir | University of Paris, France |
| Himanshu Buckchash | UiT The Arctic University of Norway, Norway |
| George Cabral | Federal Rural University of Pernambuco, Brazil |
| Michael Carl | Kent State University, USA |
| Dalia Chakrabarty | Brunel University London, UK |
| Deepayan Chakraborty | IIT Kharagpur, India |
| Tanmoy Chakraborty | IIT Delhi, India |
| Rapeeporn Chamchong | Mahasarakham University, Thailand |
| Ram Chandra Barik | C. V. Raman Global University, India |
| Chandrahas | Indian Institute of Science, Bangalore, India |
| Ming-Ching Chang | University at Albany - SUNY, USA |
| Shivam Chaudhary | Indian Institute of Technology Gandhinagar, India |
| Dushyant Singh Chauhan | Indian Institute of Technology Patna, India |
| Manisha Chawla | Amazon Inc., India |
| Shreya Chawla | Australian National University, Australia |
| Chun-Hao Chen | National Kaohsiung University of Science and Technology, Taiwan |
| Gang Chen | Victoria University of Wellington, New Zealand |

| | |
|---|---|
| He Chen | Hebei University of Technology, China |
| Hongxu Chen | University of Queensland, Australia |
| J. Chen | Dalian University of Technology, China |
| Jianhui Chen | Beijing University of Technology, China |
| Junxin Chen | Dalian University of Technology, China |
| Junyi Chen | City University of Hong Kong, China |
| Junying Chen | South China University of Technology, China |
| Lisi Chen | Hong Kong Baptist University, China |
| Mulin Chen | Northwestern Polytechnical University, China |
| Xiaocong Chen | University of New South Wales, Australia |
| Xiaofeng Chen | Chongqing Jiaotong University, China |
| Zhuangbin Chen | The Chinese University of Hong Kong, China |
| Long Cheng | Institute of Automation, China |
| Qingrong Cheng | Fudan University, China |
| Ruting Cheng | George Washington University, USA |
| Girija Chetty | University of Canberra, Australia |
| Manoj Chinnakotla | Microsoft R&D Pvt. Ltd., India |
| Andrew Chiou | CQ University, Australia |
| Sung-Bae Cho | Yonsei University, South Korea |
| Kupsze Choi | The Hong Kong Polytechnic University, China |
| Phatthanaphong Chomphuwiset | Mahasarakham University, Thailand |
| Fengyu Cong | Dalian University of Technology, China |
| Jose Alfredo Ferreira Costa | UFRN, Brazil |
| Ruxandra Liana Costea | Polytechnic University of Bucharest, Romania |
| Raphaël Couturier | University of Franche-Comte, France |
| Zhenyu Cui | Peking University, China |
| Zhihong Cui | Shandong University, China |
| Juan D. Velasquez | University of Chile, Chile |
| Rukshima Dabare | Murdoch University, Australia |
| Cherifi Dalila | University of Boumerdes, Algeria |
| Minh-Son Dao | National Institute of Information and Communications Technology, Japan |
| Tedjo Darmanto | STMIK AMIK Bandung, Indonesia |
| Debasmit Das | IIT Roorkee, India |
| Dipankar Das | Jadavpur University, India |
| Niladri Sekhar Dash | Indian Statistical Institute, Kolkata, India |
| Satya Ranjan Dash | KIIT University, India |
| Shubhajit Datta | Indian Institute of Technology, Kharagpur, India |
| Alok Debnath | Trinity College Dublin, Ireland |
| Amir Dehsarvi | Ludwig Maximilian University of Munich, Germany |
| Hangyu Deng | Waseda University, Japan |

| | |
|---|---|
| Mingcong Deng | Tokyo University of Agriculture and Technology, Japan |
| Zhaohong Deng | Jiangnan University, China |
| V. Susheela Devi | Indian Institute of Science, Bangalore, India |
| M. M. Dhabu | VNIT Nagpur, India |
| Dhimas Arief Dharmawan | Universitas Indonesia, Indonesia |
| Khaldoon Dhou | Texas A&M University Central Texas, USA |
| Gihan Dias | University of Moratuwa, Sri Lanka |
| Nat Dilokthanakul | Vidyasirimedhi Institute of Science and Technology, Thailand |
| Tai Dinh | Kyoto College of Graduate Studies for Informatics, Japan |
| Gaurav Dixit | Indian Institute of Technology Roorkee, India |
| Youcef Djenouri | SINTEF Digital, Norway |
| Hai Dong | RMIT University, Australia |
| Shichao Dong | Ping An Insurance Group, China |
| Mohit Dua | NIT Kurukshetra, India |
| Yijun Duan | Kyoto University, Japan |
| Shiv Ram Dubey | Indian Institute of Information Technology, Allahabad, India |
| Piotr Duda | Institute of Computational Intelligence/Czestochowa University of Technology, Poland |
| Sri Harsha Dumpala | Dalhousie University and Vector Institute, Canada |
| Hridoy Sankar Dutta | University of Cambridge, UK |
| Indranil Dutta | Jadavpur University, India |
| Pratik Dutta | Indian Institute of Technology Patna, India |
| Rudresh Dwivedi | Netaji Subhas University of Technology, India |
| Heba El-Fiqi | UNSW Canberra, Australia |
| Felix Engel | Leibniz Information Centre for Science and Technology (TIB), Germany |
| Akshay Fajge | Indian Institute of Technology Patna, India |
| Yuchun Fang | Shanghai University, China |
| Mohd Fazil | JMI, India |
| Zhengyang Feng | Shanghai Jiao Tong University, China |
| Zunlei Feng | Zhejiang University, China |
| Mauajama Firdaus | University of Alberta, Canada |
| Devi Fitrianah | Bina Nusantara University, Indonesia |
| Philippe Fournierviger | Shenzhen University, China |
| Wai-Keung Fung | Cardiff Metropolitan University, UK |
| Baban Gain | Indian Institute of Technology, Patna, India |
| Claudio Gallicchio | University of Pisa, Italy |
| Yongsheng Gao | Griffith University, Australia |

| | |
|---|---|
| Yunjun Gao | Zhejiang University, China |
| Vicente García Díaz | University of Oviedo, Spain |
| Arpit Garg | University of Adelaide, Australia |
| Chandan Gautam | I2R, A*STAR, Singapore |
| Yaswanth Gavini | University of Hyderabad, India |
| Tom Gedeon | Australian National University, Australia |
| Iuliana Georgescu | University of Bucharest, Romania |
| Deepanway Ghosal | Indian Institute of Technology Patna, India |
| Arjun Ghosh | National Institute of Technology Durgapur, India |
| Sanjukta Ghosh | IIT (BHU) Varanasi, India |
| Soumitra Ghosh | Indian Institute of Technology Patna, India |
| Pranav Goel | Bloomberg L.P., India |
| Tripti Goel | National Institute of Technology Silchar, India |
| Kah Ong Michael Goh | Multimedia University, Malaysia |
| Kam Meng Goh | Tunku Abdul Rahman University of Management and Technology, Malaysia |
| Iqbal Gondal | RMIT University, Australia |
| Puneet Goyal | Indian Institute of Technology Ropar, India |
| Vishal Goyal | Punjabi University Patiala, India |
| Xiaotong Gu | University of Tasmania, Australia |
| Radha Krishna Guntur | VNRVJIET, India |
| Li Guo | University of Macau, China |
| Ping Guo | Beijing Normal University, China |
| Yu Guo | Xi'an Jiaotong University, China |
| Akshansh Gupta | CSIR-Central Electronics Engineering Research Institute, India |
| Deepak Gupta | National Library of Medicine, National Institutes of Health (NIH), USA |
| Deepak Gupta | NIT Arunachal Pradesh, India |
| Kamal Gupta | NIT Patna, India |
| Kapil Gupta | PDPM IIITDM, Jabalpur, India |
| Komal Gupta | IIT Patna, India |
| Christophe Guyeux | University of Franche-Comte, France |
| Katsuyuki Hagiwara | Mie University, Japan |
| Soyeon Han | University of Sydney, Australia |
| Palak Handa | IGDTUW, India |
| Rahmadya Handayanto | Universitas Islam 45 Bekasi, Indonesia |
| Ahteshamul Haq | Aligarh Muslim University, India |
| Muhammad Haris | Universitas Nusa Mandiri, Indonesia |
| Harith Al-Sahaf | Victoria University of Wellington, New Zealand |
| Md Rakibul Hasan | BRAC University, Bangladesh |
| Mohammed Hasanuzzaman | ADAPT Centre, Ireland |

| | |
|---|---|
| Takako Hashimoto | Chiba University of Commerce, Japan |
| Bipan Hazarika | Gauhati University, India |
| Huiguang He | Institute of Automation, Chinese Academy of Sciences, China |
| Wei He | University of Science and Technology Beijing, China |
| Xinwei He | University of Illinois Urbana-Champaign, USA |
| Enna Hirata | Kobe University, Japan |
| Akira Hirose | University of Tokyo, Japan |
| Katsuhiro Honda | Osaka Metropolitan University, Japan |
| Huy Hongnguyen | National Institute of Informatics, Japan |
| Wai Lam Hoo | University of Malaya, Malaysia |
| Shih Hsiung Lee | National Cheng Kung University, Taiwan |
| Jiankun Hu | UNSW@ADFA, Australia |
| Yanyan Hu | University of Science and Technology Beijing, China |
| Chaoran Huang | UNSW Sydney, Australia |
| He Huang | Soochow University, Taiwan |
| Ko-Wei Huang | National Kaohsiung University of Science and Technology, Taiwan |
| Shudong Huang | Sichuan University, China |
| Chih-Chieh Hung | National Chung Hsing University, Taiwan |
| Mohamed Ibn Khedher | IRT-SystemX, France |
| David Iclanzan | Sapientia Hungarian University of Transylvania, Romania |
| Cosimo Ieracitano | University "Mediterranea" of Reggio Calabria, Italy |
| Kazushi Ikeda | Nara Institute of Science and Technology, Japan |
| Hiroaki Inoue | Kobe University, Japan |
| Teijiro Isokawa | University of Hyogo, Japan |
| Kokila Jagadeesh | Indian Institute of Information Technology, Allahabad, India |
| Mukesh Jain | Jawaharlal Nehru University, India |
| Fuad Jamour | AWS, USA |
| Mohd. Javed | Indian Institute of Information Technology, Allahabad, India |
| Balasubramaniam Jayaram | Indian Institute of Technology Hyderabad, India |
| Jin-Tsong Jeng | National Formosa University, Taiwan |
| Sungmoon Jeong | Kyungpook National University Hospital, South Korea |
| Yizhang Jiang | Jiangnan University, China |
| Ferdinjoe Johnjoseph | Thai-Nichi Institute of Technology, Thailand |
| Alireza Jolfaei | Federation University, Australia |

| | |
|---|---|
| Nagendra Kumar | IIT Indore, India |
| Pranaw Kumar | Centre for Development of Advanced Computing (CDAC) Mumbai, India |
| Puneet Kumar | Jawaharlal Nehru University, India |
| Raja Kumar | Taylor's University, Malaysia |
| Sachin Kumar | University of Delhi, India |
| Sandeep Kumar | IIT Patna, India |
| Sanjaya Kumar Panda | National Institute of Technology, Warangal, India |
| Chouhan Kumar Rath | National Institute of Technology, Durgapur, India |
| Sovan Kumar Sahoo | Indian Institute of Technology Patna, India |
| Anil Kumar Singh | IIT (BHU) Varanasi, India |
| Vikash Kumar Singh | VIT-AP University, India |
| Sanjay Kumar Sonbhadra | ITER, SoA, Odisha, India |
| Gitanjali Kumari | Indian Institute of Technology Patna, India |
| Rina Kumari | KIIT, India |
| Amit Kumarsingh | National Institute of Technology Patna, India |
| Sanjay Kumarsonbhadra | SSITM, India |
| Vishesh Kumar Tanwar | Missouri University of Science and Technology, USA |
| Bibekananda Kundu | CDAC Kolkata, India |
| Yoshimitsu Kuroki | Kurume National College of Technology, Japan |
| Susumu Kuroyanagi | Nagoya Institute of Technology, Japan |
| Retno Kusumaningrum | Universitas Diponegoro, Indonesia |
| Dwina Kuswardani | Institut Teknologi PLN, Indonesia |
| Stephen Kwok | Murdoch University, Australia |
| Hamid Laga | Murdoch University, Australia |
| Edmund Lai | Auckland University of Technology, New Zealand |
| Weng Kin Lai | Tunku Abdul Rahman University of Management & Technology (TAR UMT), Malaysia |
| Kittichai Lavangnananda | King Mongkut's University of Technology Thonburi (KMUTT), Thailand |
| Anwesha Law | Indian Statistical Institute, India |
| Thao Le | Deakin University, Australia |
| Xinyi Le | Shanghai Jiao Tong University, China |
| Dong-Gyu Lee | Kyungpook National University, South Korea |
| Eui Chul Lee | Sangmyung University, South Korea |
| Minho Lee | Kyungpook National University, South Korea |
| Shih Hsiung Lee | National Kaohsiung University of Science and Technology, Taiwan |
| Gurpreet Lehal | Punjabi University, India |
| Jiahuan Lei | Meituan-Dianping Group, China |

| | |
|---|---|
| Pui Huang Leong | Tunku Abdul Rahman University of Management and Technology, Malaysia |
| Chi Sing Leung | City University of Hong Kong, China |
| Man-Fai Leung | Anglia Ruskin University, UK |
| Bing-Zhao Li | Beijing Institute of Technology, China |
| Gang Li | Deakin University, Australia |
| Jiawei Li | Tsinghua University, China |
| Mengmeng Li | Zhengzhou University, China |
| Xiangtao Li | Jilin University, China |
| Yang Li | East China Normal University, China |
| Yantao Li | Chongqing University, China |
| Yaxin Li | Michigan State University, USA |
| Yiming Li | Tsinghua University, China |
| Yuankai Li | University of Science and Technology of China, China |
| Yun Li | Nanjing University of Posts and Telecommunications, China |
| Zhipeng Li | Tsinghua University, China |
| Hualou Liang | Drexel University, USA |
| Xiao Liang | Nankai University, China |
| Hao Liao | Shenzhen University, China |
| Alan Wee-Chung Liew | Griffith University, Australia |
| Chern Hong Lim | Monash University Malaysia, Malaysia |
| Kok Lim Yau | Universiti Tunku Abdul Rahman (UTAR), Malaysia |
| Chin-Teng Lin | UTS, Australia |
| Jerry Chun-Wei Lin | Western Norway University of Applied Sciences, Norway |
| Jiecong Lin | City University of Hong Kong, China |
| Dugang Liu | Shenzhen University, China |
| Feng Liu | Stevens Institute of Technology, USA |
| Hongtao Liu | Du Xiaoman Financial, China |
| Ju Liu | Shandong University, China |
| Linjing Liu | City University of Hong Kong, China |
| Weifeng Liu | China University of Petroleum (East China), China |
| Wenqiang Liu | Hong Kong Polytechnic University, China |
| Xin Liu | National Institute of Advanced Industrial Science and Technology (AIST), Japan |
| Yang Liu | Harbin Institute of Technology, China |
| Zhi-Yong Liu | Institute of Automation, Chinese Academy of Sciences, China |
| Zongying Liu | Dalian Maritime University, China |

| | |
|---|---|
| Jaime Lloret | Universitat Politècnica de València, Spain |
| Sye Loong Keoh | University of Glasgow, Singapore, Singapore |
| Hongtao Lu | Shanghai Jiao Tong University, China |
| Wenlian Lu | Fudan University, China |
| Xuequan Lu | Deakin University, Australia |
| Xiao Luo | UCLA, USA |
| Guozheng Ma | Shenzhen International Graduate School, Tsinghua University, China |
| Qianli Ma | South China University of Technology, China |
| Wanli Ma | University of Canberra, Australia |
| Muhammad Anwar Ma'sum | Universitas Indonesia, Indonesia |
| Michele Magno | University of Bologna, Italy |
| Sainik Kumar Mahata | JU, India |
| Shalni Mahato | Indian Institute of Information Technology (IIIT) Ranchi, India |
| Adnan Mahmood | Macquarie University, Australia |
| Mohammed Mahmoud | October University for Modern Sciences & Arts - MSA University, Egypt |
| Mufti Mahmud | University of Padova, Italy |
| Krishanu Maity | Indian Institute of Technology Patna, India |
| Mamta | IIT Patna, India |
| Aprinaldi Mantau | Kyushu Institute of Technology, Japan |
| Mohsen Marjani | Taylor's University, Malaysia |
| Sanparith Marukatat | NECTEC, Thailand |
| José María Luna | Universidad de Córdoba, Spain |
| Archana Mathur | Nitte Meenakshi Institute of Technology, India |
| Patrick McAllister | Ulster University, UK |
| Piotr Milczarski | Lodz University of Technology, Poland |
| Kshitij Mishra | IIT Patna, India |
| Pruthwik Mishra | IIIT-Hyderabad, India |
| Santosh Mishra | Indian Institute of Technology Patna, India |
| Sajib Mistry | Curtin University, Australia |
| Sayantan Mitra | Accenture Labs, India |
| Vinay Kumar Mittal | Neti International Research Center, India |
| Daisuke Miyamoto | University of Tokyo, Japan |
| Kazuteru Miyazaki | National Institution for Academic Degrees and Quality Enhancement of Higher Education, Japan |
| U. Mmodibbo | Modibbo Adama University Yola, Nigeria |
| Aditya Mogadala | Saarland University, Germany |
| Reem Mohamed | Mansoura University, Egypt |
| Muhammad Syafiq Mohd Pozi | Universiti Utara Malaysia, Malaysia |

| | |
|---|---|
| Anirban Mondal | University of Tokyo, Japan |
| Anupam Mondal | Jadavpur University, India |
| Supriyo Mondal | ZBW - Leibniz Information Centre for Economics, Germany |
| J. Manuel Moreno | Universitat Politècnica de Catalunya, Spain |
| Francisco J. Moreno-Barea | Universidad de Málaga, Spain |
| Sakchai Muangsrinoon | Walailak University, Thailand |
| Siti Anizah Muhamed | Politeknik Sultan Salahuddin Abdul Aziz Shah, Malaysia |
| Samrat Mukherjee | Indian Institute of Technology, Patna, India |
| Siddhartha Mukherjee | Samsung R&D Institute India, Bangalore, India |
| Dharmalingam Muthusamy | Bharathiar University, India |
| Abhijith Athreya Mysore Gopinath | Pennsylvania State University, USA |
| Harikrishnan N. B. | BITS Pilani K K Birla Goa Campus, India |
| Usman Naseem | University of Sydney, Australia |
| Deepak Nayak | Malaviya National Institute of Technology, Jaipur, India |
| Hamada Nayel | Benha University, Egypt |
| Usman Nazir | Lahore University of Management Sciences, Pakistan |
| Vasudevan Nedumpozhimana | TU Dublin, Ireland |
| Atul Negi | University of Hyderabad, India |
| Aneta Neumann | University of Adelaide, Australia |
| Hea Choon Ngo | Universiti Teknikal Malaysia Melaka, Malaysia |
| Dang Nguyen | University of Canberra, Australia |
| Duy Khuong Nguyen | FPT Software Ltd., FPT Group, Vietnam |
| Hoang D. Nguyen | University College Cork, Ireland |
| Hong Huy Nguyen | National Institute of Informatics, Japan |
| Tam Nguyen | Leibniz University Hannover, Germany |
| Thanh-Son Nguyen | Agency for Science, Technology and Research (A*STAR), Singapore |
| Vu-Linh Nguyen | Eindhoven University of Technology, Netherlands |
| Nick Nikzad | Griffith University, Australia |
| Boda Ning | Swinburne University of Technology, Australia |
| Haruhiko Nishimura | University of Hyogo, Japan |
| Kishorjit Nongmeikapam | Indian Institute of Information Technology (IIIT) Manipur, India |
| Aleksandra Nowak | Jagiellonian University, Poland |
| Stavros Ntalampiras | University of Milan, Italy |
| Anupiya Nugaliyadde | Sri Lanka Institute of Information Technology, Sri Lanka |

| | |
|---|---|
| Anto Satriyo Nugroho | Agency for Assessment & Application of Technology, Indonesia |
| Aparajita Ojha | PDPM IIITDM Jabalpur, India |
| Akeem Olowolayemo | International Islamic University Malaysia, Malaysia |
| Toshiaki Omori | Kobe University, Japan |
| Shih Yin Ooi | Multimedia University, Malaysia |
| Sidali Ouadfeul | Algerian Petroleum Institute, Algeria |
| Samir Ouchani | CESI Lineact, France |
| Srinivas P. Y. K. L. | IIIT Sri City, India |
| Neelamadhab Padhy | GIET University, India |
| Worapat Paireekreng | Dhurakij Pundit University, Thailand |
| Partha Pakray | National Institute of Technology Silchar, India |
| Santanu Pal | Wipro Limited, India |
| Bin Pan | Nankai University, China |
| Rrubaa Panchendrarajan | Sri Lanka Institute of Information Technology, Sri Lanka |
| Pankaj Pandey | Indian Institute of Technology, Gandhinagar, India |
| Lie Meng Pang | Southern University of Science and Technology, China |
| Sweta Panigrahi | National Institute of Technology Warangal, India |
| T. Pant | IIIT Allahabad, India |
| Shantipriya Parida | Idiap Research Institute, Switzerland |
| Hyeyoung Park | Kyungpook National University, South Korea |
| Md Aslam Parwez | Jamia Millia Islamia, India |
| Leandro Pasa | Federal University of Technology - Parana (UTFPR), Brazil |
| Kitsuchart Pasupa | King Mongkut's Institute of Technology Ladkrabang, Thailand |
| Debanjan Pathak | Kalinga Institute of Industrial Technology (KIIT), India |
| Vyom Pathak | University of Florida, USA |
| Sangameshwar Patil | TCS Research, India |
| Bidyut Kr. Patra | IIT (BHU) Varanasi, India |
| Dipanjyoti Paul | Indian Institute of Technology Patna, India |
| Sayanta Paul | Ola, India |
| Sachin Pawar | Tata Consultancy Services Ltd., India |
| Pornntiwa Pawara | Mahasarakham University, Thailand |
| Yong Peng | Hangzhou Dianzi University, China |
| Yusuf Perwej | Ambalika Institute of Management and Technology (AIMT), India |
| Olutomilayo Olayemi Petinrin | City University of Hong Kong, China |
| Arpan Phukan | Indian Institute of Technology Patna, India |

| | |
|---|---|
| Nabin Sharma | University of Technology Sydney, Australia |
| Raksha Sharma | IIT Bombay, India |
| Sourabh Sharma | Avantika University, India |
| Suraj Sharma | International Institute of Information Technology Bhubaneswar, India |
| Ravi Shekhar | Queen Mary University of London, UK |
| Michael Sheng | Macquarie University, Australia |
| Yin Sheng | Huazhong University of Science and Technology, China |
| Yongpan Sheng | Southwest University, China |
| Liu Shenglan | Dalian University of Technology, China |
| Tomohiro Shibata | Kyushu Institute of Technology, Japan |
| Iksoo Shin | University of Science & Technology, China |
| Mohd Fairuz Shiratuddin | Murdoch University, Australia |
| Hayaru Shouno | University of Electro-Communications, Japan |
| Sanyam Shukla | MANIT, Bhopal, India |
| Udom Silparcha | KMUTT, Thailand |
| Apoorva Singh | Indian Institute of Technology Patna, India |
| Divya Singh | Central University of Bihar, India |
| Gitanjali Singh | Indian Institute of Technology Patna, India |
| Gopendra Singh | Indian Institute of Technology Patna, India |
| K. P. Singh | IIIT Allahabad, India |
| Navjot Singh | IIIT Allahabad, India |
| Om Singh | NIT Patna, India |
| Pardeep Singh | Jawaharlal Nehru University, India |
| Rajiv Singh | Banasthali Vidyapith, India |
| Sandhya Singh | Indian Institute of Technology Bombay, India |
| Smriti Singh | IIT Bombay, India |
| Narinder Singhpunn | Mayo Clinic, Arizona, USA |
| Saaveethya Sivakumar | Curtin University, Malaysia |
| Ferdous Sohel | Murdoch University, Australia |
| Chattrakul Sombattheera | Mahasarakham University, Thailand |
| Lei Song | Unitec Institute of Technology, New Zealand |
| Linqi Song | City University of Hong Kong, China |
| Yuhua Song | University of Science and Technology Beijing, China |
| Gautam Srivastava | Brandon University, Canada |
| Rajeev Srivastava | Banaras Hindu University (IT-BHU), Varanasi, India |
| Jérémie Sublime | ISEP - Institut Supérieur d'Électronique de Paris, France |
| P. N. Suganthan | Nanyang Technological University, Singapore |

| | |
|---|---|
| Alex To | University of Sydney, Australia |
| Stefania Tomasiello | University of Tartu, Estonia |
| Anh Duong Trinh | Technological University Dublin, Ireland |
| Enkhtur Tsogbaatar | Mongolian University of Science and Technology, Mongolia |
| Enmei Tu | Shanghai Jiao Tong University, China |
| Eiji Uchino | Yamaguchi University, Japan |
| Prajna Upadhyay | IIT Delhi, India |
| Sahand Vahidnia | University of New South Wales, Australia |
| Ashwini Vaidya | IIT Delhi, India |
| Deeksha Varshney | Indian Institute of Technology, Patna, India |
| Sowmini Devi Veeramachaneni | Mahindra University, India |
| Samudra Vijaya | Koneru Lakshmaiah Education Foundation, India |
| Surbhi Vijh | JSS Academy of Technical Education, Noida, India |
| Nhi N. Y. Vo | University of Technology Sydney, Australia |
| Xuan-Son Vu | Umeå University, Sweden |
| Anil Kumar Vuppala | IIIT Hyderabad, India |
| Nobuhiko Wagatsuma | Toho University, Japan |
| Feng Wan | University of Macau, China |
| Bingshu Wang | Northwestern Polytechnical University Taicang Campus, China |
| Dianhui Wang | La Trobe University, Australia |
| Ding Wang | Beijing University of Technology, China |
| Guanjin Wang | Murdoch University, Australia |
| Jiasen Wang | City University of Hong Kong, China |
| Lei Wang | Beihang University, China |
| Libo Wang | Xiamen University of Technology, China |
| Meng Wang | Southeast University, China |
| Qiu-Feng Wang | Xi'an Jiaotong-Liverpool University, China |
| Sheng Wang | Henan University, China |
| Weiqun Wang | Institute of Automation, Chinese Academy of Sciences, China |
| Wentao Wang | Michigan State University, USA |
| Yongyu Wang | Michigan Technological University, USA |
| Zhijin Wang | Jimei University, China |
| Bunthit Watanapa | KMUTT-SIT, Thailand |
| Yanling Wei | TU Berlin, Germany |
| Guanghui Wen | RMIT University, Australia |
| Ari Wibisono | Universitas Indonesia, Indonesia |
| Adi Wibowo | Diponegoro University, Indonesia |
| Ka-Chun Wong | City University of Hong Kong, China |

| | |
|---|---|
| Kevin Wong | Murdoch University, Australia |
| Raymond Wong | Universiti Malaya, Malaysia |
| Kuntpong Woraratpanya | King Mongkut's Institute of Technology Ladkrabang (KMITL), Thailand |
| Marcin Woźniak | Silesian University of Technology, Poland |
| Chengwei Wu | Harbin Institute of Technology, China |
| Jing Wu | Shanghai Jiao Tong University, China |
| Weibin Wu | Sun Yat-sen University, China |
| Hongbing Xia | Beijing Normal University, China |
| Tao Xiang | Chongqing University, China |
| Qiang Xiao | Huazhong University of Science and Technology, China |
| Guandong Xu | University of Technology Sydney, Australia |
| Qing Xu | Tianjin University, China |
| Yifan Xu | Huazhong University of Science and Technology, China |
| Junyu Xuan | University of Technology Sydney, Australia |
| Hui Xue | Southeast University, China |
| Saumitra Yadav | IIIT-Hyderabad, India |
| Shekhar Yadav | Madan Mohan Malaviya University of Technology, India |
| Sweta Yadav | University of Illinois at Chicago, USA |
| Tarun Yadav | Defence Research and Development Organisation, India |
| Shankai Yan | Hainan University, China |
| Feidiao Yang | Microsoft, China |
| Gang Yang | Renmin University of China, China |
| Haiqin Yang | International Digital Economy Academy, China |
| Jianyi Yang | Shandong University, China |
| Jinfu Yang | BJUT, China |
| Minghao Yang | Institute of Automation, Chinese Academy of Sciences, China |
| Shaofu Yang | Southeast University, China |
| Wachira Yangyuen | Rajamangala University of Technology Srivijaya, Thailand |
| Xinye Yi | Guilin University of Electronic Technology, China |
| Hang Yu | Shanghai University, China |
| Wen Yu | Cinvestav, Mexico |
| Wenxin Yu | Southwest University of Science and Technology, China |
| Zhaoyuan Yu | Nanjing Normal University, China |
| Ye Yuan | Xi'an Jiaotong University, China |
| Xiaodong Yue | Shanghai University, China |

| | |
|---|---|
| Aizan Zafar | Indian Institute of Technology Patna, India |
| Jichuan Zeng | Bytedance, China |
| Jie Zhang | Newcastle University, UK |
| Shixiong Zhang | Xidian University, China |
| Tianlin Zhang | University of Manchester, UK |
| Mingbo Zhao | Donghua University, China |
| Shenglin Zhao | Zhejiang University, China |
| Guoqiang Zhong | Ocean University of China, China |
| Jinghui Zhong | South China University of Technology, China |
| Bo Zhou | Southwest University, China |
| Yucheng Zhou | University of Technology Sydney, Australia |
| Dengya Zhu | Curtin University, Australia |
| Xuanying Zhu | ANU, Australia |
| Hua Zuo | University of Technology Sydney, Australia |

## Additional Reviewers

Acharya, Rajul
Afrin, Mahbuba
Alsuhaibani, Abdullah
Amarnath
Appicharla, Ramakrishna
Arora, Ridhi
Azar, Joseph
Bai, Weiwei
Bao, Xiwen
Barawi, Mohamad Hardyman
Bhat, Mohammad Idrees Bhat
Cai, Taotao
Cao, Feiqi
Chakraborty, Bodhi
Chang, Yu-Cheng
Chen
Chen, Jianpeng
Chen, Yong
Chhipa, Priyank
Cho, Joshua
Chongyang, Chen
Cuenat, Stéphane
Dang, Lili
Das Chakladar, Debashis
Das, Kishalay
Dey, Monalisa
Doborjeh, Maryam
Dong, Zhuben
Dutta, Subhabrata
Dybala, Pawel
El Achkar, Charbel
Feng, Zhengyang
Galkowski, Tomasz
Garg, Arpit
Ghobakhlou, Akbar
Ghosh, Soumitra
Guo, Hui
Gupta, Ankur
Gupta, Deepak
Gupta, Megha
Han, Yanyang
Han, Yiyan
Hang, Bin
Harshit
He, Silu
Hua, Ning
Huang, Meng
Huang, Rongting
Huang, Xiuyu
Hussain, Zawar
Imran, Javed
Islam, Md Rafiqul

Jain, Samir
Jia, Mei
Jiang, Jincen
Jiang, Xiao
Jiangyu, Wang
Jiaxin, Lou
Jiaxu, Hou
Jinzhou, Bao
Ju, Wei
Kasyap, Harsh
Katai, Zoltan
Keserwani, Prateek
Khan, Asif
Khan, Muhammad Fawad Akbar
Khari, Manju
Kheiri, Kiana
Kirk, Nathan
Kiyani, Arslan
Kolya, Anup Kumar
Krdzavac, Nenad
Kumar, Lov
Kumar, Mukesh
Kumar, Puneet
Kumar, Rahul
Kumar, Sunil
Lan, Meng
Lavangnananda, Kittichai
Li, Qian
Li, Xiaoou
Li, Xin
Li, Xinjia
Liang, Mengnan
Liang, Shuai
Liquan, Li
Liu, Boyang
Liu, Chang
Liu, Feng
Liu, Linjing
Liu, Xinglan
Liu, Xinling
Liu, Zhe
Lotey, Taveena
Ma, Bing
Ma, Zeyu
Madanian, Samaneh
Mahata, Sainik Kumar
Mahmud, Md. Redowan
Man, Jingtao
Meena, Kunj Bihari
Mishra, Pragnyaban
Mistry, Sajib
Modibbo, Umar Muhammad
Na, Na
Nag Choudhury, Somenath
Nampalle, Kishore
Nandi, Palash
Neupane, Dhiraj
Nigam, Nitika
Nigam, Swati
Ning, Jianbo
Oumer, Jehad
Pandey, Abhineet Kumar
Pandey, Sandeep
Paramita, Adi Suryaputra
Paul, Apurba
Petinrin, Olutomilayo Olayemi
Phan Trong, Dat
Pradana, Muhamad Hilmil Muchtar Aditya
Pundhir, Anshul
Rahman, Sheikh Shah Mohammad Motiur
Rai, Sawan
Rajesh, Bulla
Rajput, Amitesh Singh
Rao, Raghunandan K. R.
Rathore, Santosh Singh
Ray, Payel
Roy, Satyaki
Saini, Nikhil
Saki, Mahdi
Salimath, Nagesh
Sang, Haiwei
Shao, Jian
Sharma, Anshul
Sharma, Shivam
Shi, Jichen
Shi, Jun
Shi, Kaize
Shi, Li
Singh, Nagendra Pratap
Singh, Pritpal

Singh, Rituraj
Singh, Shrey
Singh, Tribhuvan
Song, Meilun
Song, Yuhua
Soni, Bharat
Stommel, Martin
Su, Yanchi
Sun, Xiaoxuan
Suryodiningrat, Satrio Pradono
Swarnkar, Mayank
Tammewar, Aniruddha
Tan, Xiaosu
Tanoni, Giulia
Tanwar, Vishesh
Tao, Yuwen
To, Alex
Tran, Khuong
Varshney, Ayush
Vo, Anh-Khoa
Vuppala, Anil
Wang, Hui
Wang, Kai
Wang, Rui
Wang, Xia
Wang, Yansong
Wang, Yuan
Wang, Yunhe
Watanapa, Saowaluk
Wenqian, Fan
Xia, Hongbing
Xie, Weidun
Xiong, Wenxin
Xu, Zhehao
Xu, Zhikun
Yan, Bosheng
Yang, Haoran
Yang, Jie
Yang, Xin
Yansui, Song
Yu, Cunzhe
Yu, Zhuohan
Zandavi, Seid Miad
Zeng, Longbin
Zhang, Jane
Zhang, Ruolan
Zhang, Ziqi
Zhao, Chen
Zhou, Xinxin
Zhou, Zihang
Zhu, Liao
Zhu, Linghui

# Contents – Part VI

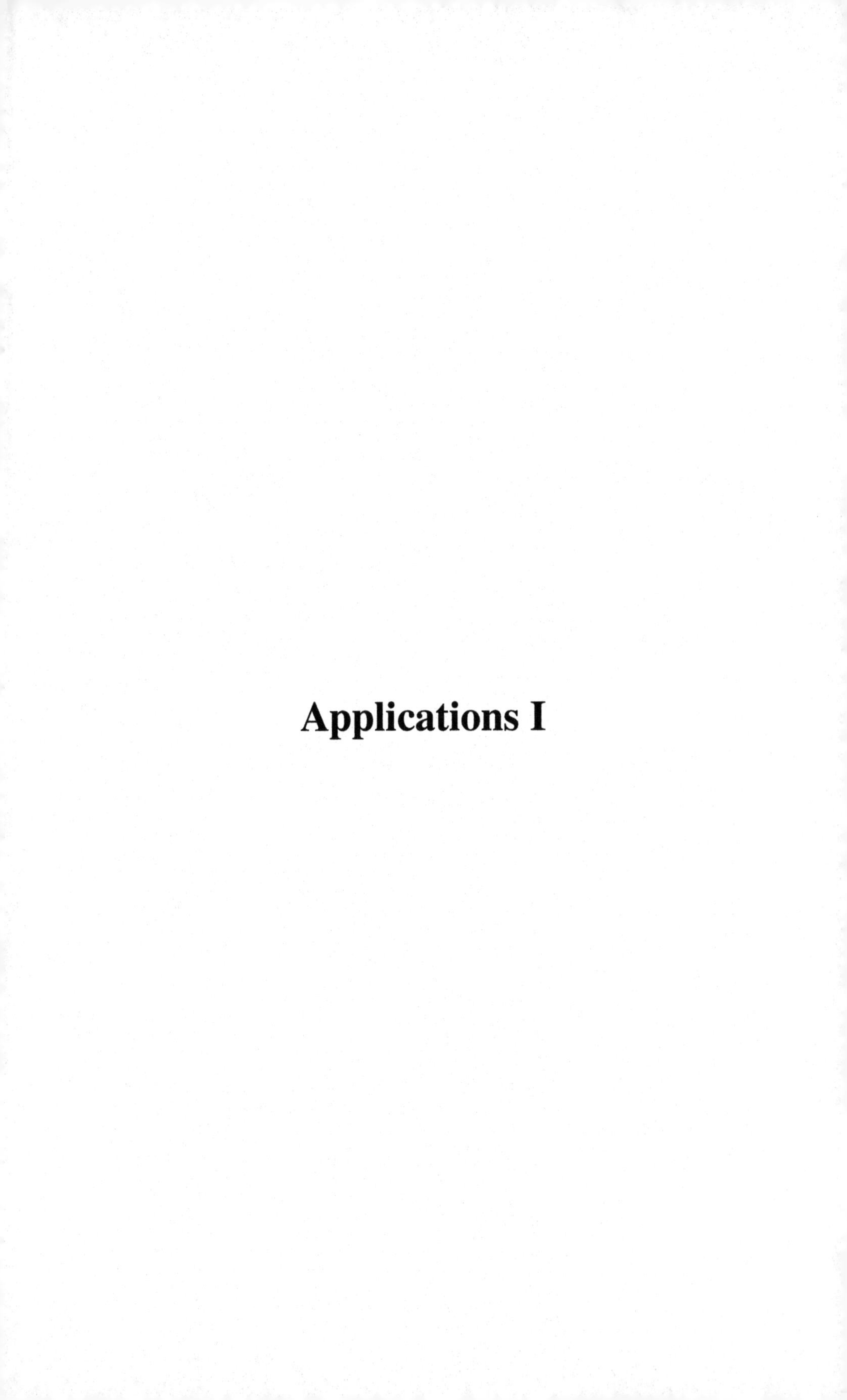

# Applications I

# Transfer Learning Based Long Short-Term Memory Network for Financial Time Series Forecasting

Ruibin Lin, Dabin Zhang(✉), Liwen Ling, Junjie Huang, and Guotao Cai

College of Mathematics and Informatics, South China Agricultural University, Guangzhou 510642, China
zdbff@aliyun.com

**Abstract.** Long Short-Term Memory (LSTM) neural network is widely used to deal with various temporal modelling problems, including financial Time Series Forecasting (TSF) task. However, accurate forecasting of financial time series remains a difficult problem due to its implicit complex information and lack of labeled training data. To alleviate the limitation of overfitting caused by insufficient clean data, a new approach using LSTM based on transfer learning is proposed in our study for financial TSF task, termed as ADA-LSTM for short. Concretely, we not only implement a typical Adversarial Domain Adaptation architecture, but also tactfully introduce a smoothed formulation of Dynamic Time Warping (soft-DTW) in adversarial training phase to measure the shape loss during the transfer of sequence knowledge. Compared to many existing methods of selecting potential source domain during transfer learning in TSF, in our study, appropriate source dataset is selected from a novel perspective using temporal causal discovery method via transfer entropy instead of using statistical similarity across different time series. The feasibility and effectiveness of ADA-LSTM are validated by the empirical experiments conducting on different financial datasets.

**Keywords:** Financial time series forecasting · Deep learning · Transfer learning · Temporal causal discovery

## 1 Introduction

Financial market is a typical chaotic system composed of a large number of related markets, which has always been an important subject in the economic system. Forecasting time series data from financial markets accurately is of great decision-making significance for investors. Modeling traditional TSF problem mainly relies on econometric statistic models, including Auto-Regression (AR), Auto-Regressive Integrated Moving Average (ARIMA) and Exponential Smoothing (ETS), etc. These models are highly interpretable and have been applied in many practical fields. However, they require the data to have the characteristics of normality, stationarity, low noise, etc., and the modeling process for financial data usually has limited conditions. Unfortunately, the update


M. Tanveer et al. (Eds.): ICONIP 2022, CCIS 1793, pp. 3–13, 2023.
https://doi.org/10.1007/978-981-99-1645-0_1

speed of financial time series is fast and the implicit relationship is complex. These traditional statistic models are difficult to capture nonlinear components in the case of low signal-to-noise environment.

Most existing financial TSF researches [1,2] assume that the training data and test data conform to the same distribution and there is sufficient labeled data for model training. But in many practical financial markets, time series often change greatly over time, resulting in a big different distribution between the old data and the new recent data. In addition, re-collecting and labeling a large amount of training data to rebuild the model is very expensive and requires a lot of manpower and material resources. To alleviate the mentioned data dilemma, we hope to leverage the knowledge from old data to finish new financial prediction tasks based on the idea of transfer learning [3–5].

The purpose of transfer learning is to effectively transfer knowledge between different but related domains or tasks, which can utilize prior knowledge previously learned from source domain to help learn better on target domain. It has been proved to be effective in many practical applications [6,7] like images, text, and sound, etc. Recently, many studies about TSF researches using transfer learning have been proposed. Ye [8] proposes a hybrid algorithm abbreviated as TrEnOS-ELMK. It aims to make the most of, rather than discard, the adequate long-ago data, and constructs an algorithm framework for transfer learning in time series forecasting. Gu [9] proposes MS-AMTL algorithm integrating multi-source transfer learning, active learning and metric learning paradigms for time series prediction. It aims to transfer knowledge in a relatively long-time span effectively. Gu [10] also proposes MultiSrcTL algorithm and AcMultiSrcTL algorithm to overcome the problem of negative transfer due to single-source TSF transfer. Ye [11] proposes DTr-CNN which embeds the transferring phase into the feature learning process. It incorporates DTW and Jensen-Shannon (JS) divergence to quantify the similarity between source time series and target times.

In particular, there are some significant differences between this paper and the previous researches. The main contributions of our research can be concisely summarized as follows:

- We propose a novel deep transfer learning method to enhance forecasting performance across different financial datasets, named ADA-LSTM. Figure 1 intuitively shows the diagram of it.

- We not only implement a domain adaptation architecture that tactfully introduces soft-DTW in the adversarial training phase, but also provide a new perspective using temporal causality discovery via transfer entropy for source domain selection.

## 2 Basic Methodology

Over the past few years, deep learning models have achieved almost state-of-the-art performance in many machine learning fields. Traditional deep forward neural networks assume that each input vector is independent, so it is difficult to process

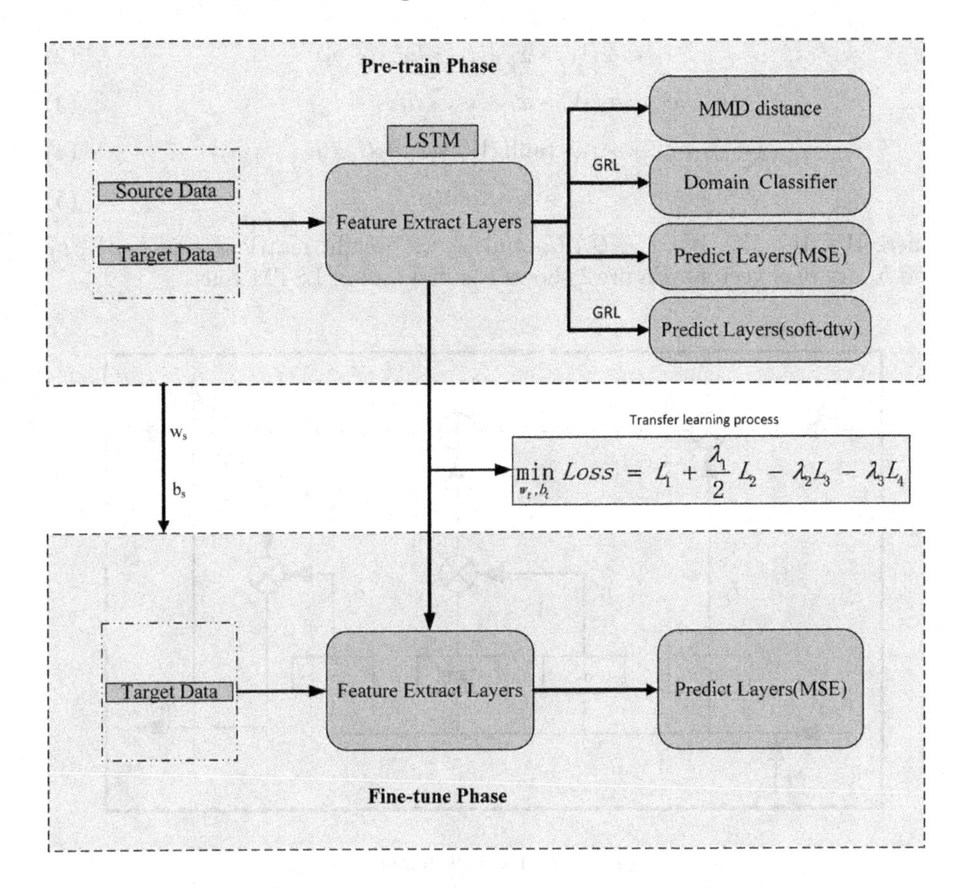


**Fig. 1.** The overall diagram of ADA-LSTM

sequence modeling. To meet the practical requirements for temporal scenarios, various types of Recurrent Neural Network (RNN) have been proposed.

LSTM [12] is a special RNN model, which is mainly used to solve the problems of gradient disappearance and gradient explosion in long sequence training. It has shown excellent non-stationary modelling ability and robustness for financial time series [13,14]. Compared with ordinary RNN, LSTM performs better in longer-sequence modelling. Concretely, it regulates the flow of information through a separate memory cell and a cross-network gating mechanism, which is controlled by three gates, called forget gate, input gate, and output gate. This mechanism determines what information is kept, for how long, and when it is read from a memory location.

The calculations for the forget gate ($f_t$), the input gate ($i_t$), the output gate ($o_t$), the cell state ($c_t$) and the hidden state ($h_t$) are performed using the following formulas:

$$f_t = \sigma \left( W_f * x_t + U_f * h_{t-1} + b_f \right) \tag{1}$$

$$i_t = \sigma\left(W_i * x_t + U_i * h_{t-1} + b_i\right) \tag{2}$$

$$o_t = \sigma\left(W_o * x_t + U_o * h_{t-1} + b_o\right) \tag{3}$$

$$c_t = f_t * c_{t-1} + i_t * \tanh\left(W_c * x_t + U_c * h_{t-1} + b_c\right) \tag{4}$$

$$h_t = o_t * \tanh\left(c_t\right) \tag{5}$$

where $W_f$, $W_i$, $W_o$, $W_c$, $U_f$, $U_i$, $U_o$ and $U_c$ are weight matrices, and $b_f$, $b_i$, $b_f$ and $b_c$ are bias vectors. Figure 2 shows the diagram of LSTM cells.

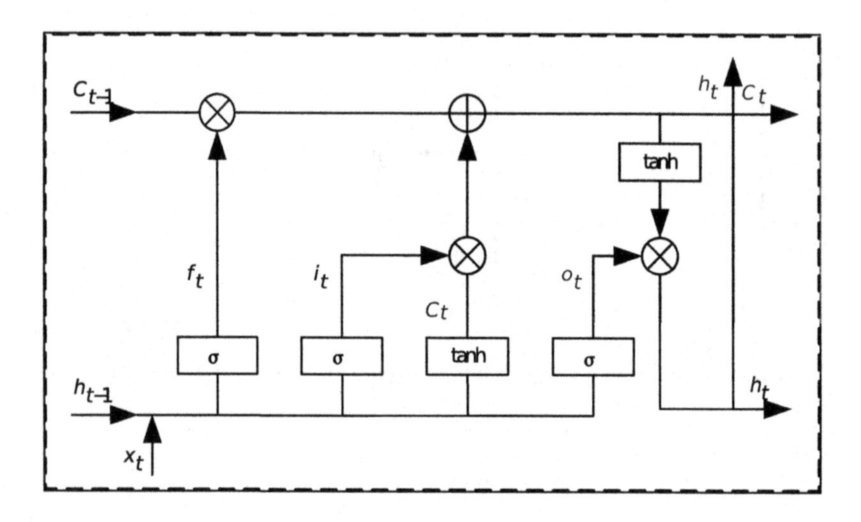


**Fig. 2.** LSTM cell diagram

## 3 Proposed Method

### 3.1 Selection of Potential Source Dataset

Transfer learning aims to use the prior knowledge from source domain to guide the training on target domain. Therefore, selecting source domain correctly will make important impact on forecasting performance on target domain. Although there are several related financial datasets, their association with the target dataset is often unclear. Selecting source domain blindly may lead to negative transfer and affect forecasting performance. To alleviate the tedious procedure of selecting source domain, much research related to this field has been developed recently. In [9–11], a distance-based measure and a feature-based measure are respectively used to compute the similarity of series and select time series with the greatest similarity as source domain.

Here, we provide a novel perspective for the selection of potential source domain in TSF transfer learning task. Schreiber [15] defined the concept of

Transfer Entropy (TE) for discovering causality in steady-state time series, as follows:

$$TE_{Y \to X} = \sum_{x_{n+\tau}, x_n, y_n} p(x_{n+\tau}, x_n, y_n) \log \left( \frac{p(x_{n+\tau} \mid x_n, y_n)}{p(x_{n+\tau} \mid x_n)} \right) \tag{6}$$

where $Y$ is cause and $X$ is effects. Temporal causality means that one variable is helpful for forecasting the future value of another variable. In our case, we compute the average TE from each variable of source dataset to the target variable of target dataset and select the source dataset with the maximum TE as source domain. The results are shown in Table 1.

**Table 1.** Average TE between potential source datasets and target dataset

| | DJI | HSI | NCI | SHI | SZI | SP |
|---|---|---|---|---|---|---|
| DJI | | 0.192533 | **0.303757** | 0.265046 | 0.248814 | 0.204539 |
| HSI | 0.301333 | | 0.223934 | **0.338623** | 0.313138 | 0.300018 |
| NCI | 0.244414 | 0.181682 | | **0.261499** | 0.254737 | 0.213614 |
| SHI | **0.412704** | 0.315008 | 0.356929 | | 0.250056 | 0.407840 |
| SZI | **0.406966** | 0.358996 | 0.341988 | 0.279806 | | 0.388393 |
| SP | 0.204768 | 0.191986 | 0.234973 | **0.274989** | 0.273868 | |

### 3.2 Explanation of ADA-LSTM

To alleviate the pressure of insufficient labeled data, some transfer learning methods such as domain adaptation are applied to our proposed method.

By reviewing previous studies [16], Domain Adversarial Neural Network (DANN) is one of the most widely used methods for domain adaptation. Adversarial training helps the feature extraction layers to produce more robust features, i.e., the discrepancy between the output vectors from feature extract modules of source domain and target domain is reduced.

What's more, Maximum Mean Discrepancy (MMD) [17] is one of the commonly used metrics which aims to align the distribution of feature space, and we add it to our proposed method as one part of the hybrid loss function to ensure that the extracted features from source domain and target domain are approximatively similar.

While in the proposed ADA-LSTM algorithm, we first stack two LSTM layers as feature extract layers, two full connect layers as predict layers, and then add a domain classifier following the feature extract layers. The loss functions in the pre-train phase are designed for the alignment between source and target distribution, which is expounded as follows:

$$\min_{w_t, b_t} \; Loss = L_1 + \frac{\lambda_1}{2} L_2 - \lambda_2 L_3 - \lambda_3 L_4 \tag{7}$$

$$L_1 = \sum_{i=1}^{N_s} L_s\left(f\left(x_i\right)\right) \tag{8}$$

$$L_2 = MMD = \left\| \frac{1}{N_s} \sum_{i=1}^{N_s} f\left(x_i\right) - \frac{1}{N_t} \sum_{j=1}^{N_t} f\left(x_j\right) \right\|^2 \tag{9}$$

$$L_3 = \sum_{i=1}^{N_s+N_t} L_d\left(G_d\left(R_\lambda\left(f\left(x_i\right)\right)\right)\right) \tag{10}$$

$$L_4 = soft\text{–}DTW_\gamma(R_\lambda(f(x_{s,pred})), R_\lambda(f(x_{t,true}))) \tag{11}$$

According to Eqs. (7)–(11), the hybrid loss function is made up of four parts. $N_s$ and $N_t$ represent the source labeled data and target unlabeled data, respectively. $L_1$ represents the training error using Mean Square Error (MSE) loss function. $f(x_i)$ represents the feature extract layers. $L_2$ represents the MMD distance between source and target distribution. $L_3$ represents the domain classifier which aims to maximize the classification loss between source and target domain. Note that the negative sign in $L_{loss}$ indicate the Gradient Reversal Layer (GRL) $R_\lambda$. Moverover, soft-DTW loss [18] is introduced tactfully in $L_4$ to measure the similarity of shape between the predictions of source time series $x_{s,pred}$ and the ground truth of target time series $x_{t,true}$. During the pre-train phase, we expect that the prediction results can be closer to the ground truth of training set on source domain while the knowledge of shape similarity can be closer to the ground truth of training set on target domain by adversarial training. Its introduction transfer shape knowledge of time series and enhances the generalization of the model.

After pre-training, ADA-LSTM is fine-tuned on the target labeled dataset, using only MSE as a loss function.

# 4 Experiments

## 4.1 Experimental Datasets

In this work, we analyze the forecasting performance of ADA-LSTM on one group of financial indexes from different financial markets, including Shanghai Composite Index (SHI), Shenzhen Component Index (SZI), DowJones Industrial Average Index (DJI), S&P500 (SP), Hang Seng Index (HSI) of Hong Kong and NASDAQ Composite Index (NCI). All datasets are day-grained and cover the period from January 5, 2010, to December 31, 2021. For each dataset, 80% for training and the rest for testing. Four variables including open price, highest price, lowest price, and close price are used to modeling. Then, Vector Auto-Regressive (VAR) model is used to select the lag period according to BIC criterion and time series are converted into the input format that is suitable for the supervised learning problem according to rolling window strategy. The lag of each dataset is shown in Table 2.

**Table 2.** The lag of each dataset

| | DJI | HSI | NCI | SHI | SZI | SP |
|---|---|---|---|---|---|---|
| BIC | 36.42 | 38.98 | 29.35 | 23.61 | 35.33 | 18.88 |
| Lag | 10 | 7 | 8 | 8 | 8 | 10 |

## 4.2 Experimental Settings

In our experiments, the parameters of ADA-LSTM are set as $\lambda_1 = \lambda_2 = 1, \lambda_3 = 0.1$. Moreover, we stack two LSTM layers and take the output of all time steps as the input of the final fully connected layer. The best hidden size of each layer is chosen from {16, 32, 64} according to prediction error. During the training process, the learning rate is adjusted adaptively according to the training loss. The initial learning rate is set as 0.1, and the minimum learning rate is set as 0.0000001. To fully train the model and avoid the randomization, the number of iterations is set to 10,000 and the model is trained five times to get the average error results.

To better compare the performance of different models, we use Max-Min scale method to adjust the dimension of variables. Meanwhile, as the evaluation criterion, root mean square error (RMSE) and symmetric mean absolute percentage (sMAPE) are computed as follows:

$$RMSE = \sqrt{\frac{1}{N_{test}} \sum_{i=1}^{N_{test}} \left(y_{i,pred} - y_{i,target}\right)^2} i = 1, 2, \ldots, N_{test} \tag{12}$$

$$sMAPE = \frac{100\%}{N_{test}} \sum_{i=1}^{N_{test}} \frac{\left|y_{i,pred} - y_{i,target}\right|}{\left(\left|y_{i,pred} + y_{i,target}\right|\right)/2} \quad i = 1, 2, \ldots, N_{test} \tag{13}$$

## 4.3 Empirical Results

As described in Table 1, we select source domain according to causality. Each column in Table 3 and Table 4 contains the error for a single task, e.g., column DJI(NCI) means the forecasting task for DJI with the selected source domain NCI. Similar notations have the same meanings in the following tables.

By analyzing Table 3 and Table 4, on six datasets, ADA-LSTM gets the lowest error result five times. TL-LSTM which use typical transfer learning strategy performs the second best. These show that using transfer learning strategy is better than training from scratch. Gated Recurrent Units (GRU) performs the third best on the whole. Temporal Convolutional Network (TCN) wins once on DJI, but among results, it performs unstably due to its complex network structure in oversimplified data. Obviously, traditional machine learning models such as Support Vector Regression (SVR) and Random Forest (RF) are not competitive with deep neural networks. Besides, the prediction results of ADA-LSTM for each dataset are shown in Fig. 3. The result of ADA-LSTM is almost coincident with the ground truth. Most of the time series datasets used in this paper are not stationary. Experimental results verify the modelling ability of the proposed model under non-stationary conditions and demonstrate the end-to-end characteristics.

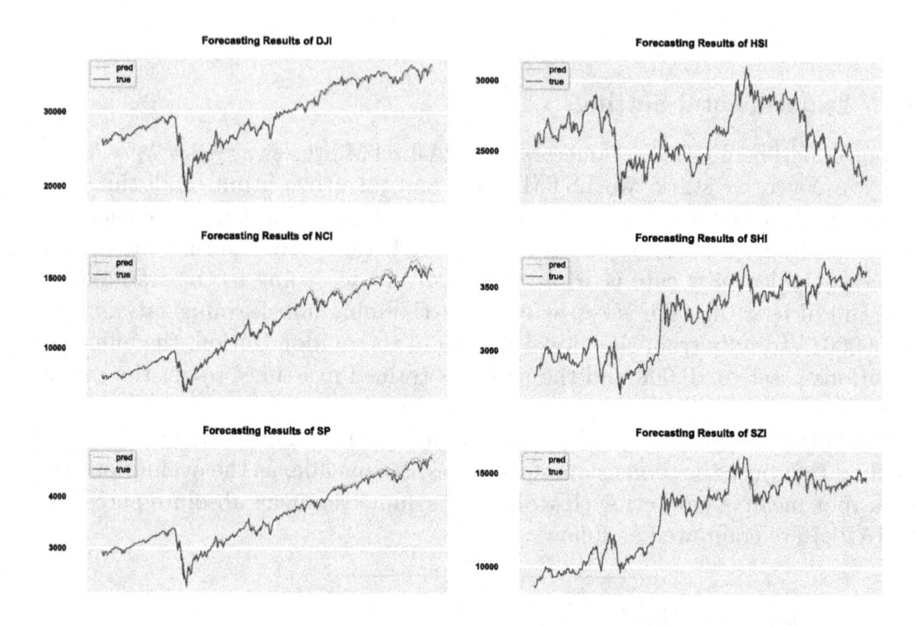


**Fig. 3.** Prediction result of ADA-LSTM

**Table 3.** Average RMSE of different models on corresponding target dataset

| | DJI(NCI) | HSI(SHI) | NCI(SHI) | SHI(DJI) | SZI(DJI) | SP(SHI) |
|---|---|---|---|---|---|---|
| LSTM | 1370.83 | 343.52 | 1928.23 | 51.01 | 244.22 | 282.81 |
| TL-LSTM | 815.03 | 338.22 | 269.14 | 33.69 | 183.33 | 150.00 |
| GRU | 1655.49 | 408.23 | 1200.66 | 43.83 | 199.98 | 565.08 |
| TCN | **395.42** | 2733.93 | 4686.87 | 539.97 | 186.20 | 1678.71 |
| SVR | 5740.81 | 532.68 | 5117.48 | 104.97 | 378.65 | 756.07 |
| RF | 4730.45 | 341.18 | 4269.79 | 37.61 | 240.26 | 867.19 |
| ADA-LSTM | 433.88 | **329.51** | **244.56** | **32.79** | **180.40** | **53.92** |

**Table 4.** Average sMAPE% of different models on corresponding target dataset

| | DJI(NCI) | HSI(SHI) | NCI(SHI) | SHI(DJI) | SZI(DJI) | SP(SHI) |
|---|---|---|---|---|---|---|
| LSTM | 3.62 | 0.99 | 12.32 | 1.16 | 1.44 | 5.67 |
| TL-LSTM | 2.06 | 0.97 | 1.78 | 0.76 | 1.06 | 3.31 |
| GRU | 4.49 | 1.21 | 7.32 | 1.01 | 1.16 | 10.85 |
| TCN | **0.95** | 8.45 | 36.97 | 14.72 | 1.10 | 53.11 |
| SVR | 16.93 | 1.61 | 43.49 | 2.97 | 2.64 | 19.62 |
| RF | 11.55 | 0.98 | 31.48 | 0.86 | 1.37 | 18.46 |
| ADA-LSTM | 1.08 | **0.94** | **1.48** | **0.73** | **1.05** | **1.07** |

### 4.4 DM Test for Results

DM test [19] is used to compare the prediction effectiveness between two models. For hypothesis testing, the null hypothesis indicates that there is no difference between the prediction errors of the two models, and the alternative hypothesis indicates that the prediction errors of the two models are different. To save space, we only compare ADA-LSTM with the model TL-LSTM with the second-best performance.

All results of the DM test are reported in Table 5. It can be seen that the prediction performance of ADA-LSTM is better than that of TL-LSTM with more than 95% confidence level on DJI, HSI, NCI, and SZI. On SHI, ADA-LSTM also has a confidence level of over 90% on SHI. The above results demonstrate the competitive performance of ADA-LSTM. On SP, the prediction performance of the two models is not significant, which may be related to the high volatility of SP.

**Table 5.** The results of DM test

| | DJI(NCI) | HSI(SHI) | NCI(SHI) | SHI(DJI) | SZI(DJI) | SP(SHI) |
|---|---|---|---|---|---|---|
| DM statistic | −15.79 | −2.51 | −2.19 | −1.88 | −19.85 | −1.46 |
| P value | 0.00 | 0.01 | 0.03 | 0.06 | 0.00 | 0.14 |

## 5 Conclusion

In order to alleviate the overfitting dilemma of low signal-to-noise ratio of financial data and insufficient labeled samples, motivated by outstanding performance of transfer learning, we propose ADA-LSTM model based on a novel transfer learning strategy for financial TSF task. ADA-LSTM combines the typical adversarial domain adaption architecture and introduces soft-DTW to measure the loss of shape knowledge in the process of adversarial training. In addition, temporal causal discovery via transfer entropy is introduced into our approach: a new perspective for selecting source domains.

There is still some room for improvement in our work. The effect of source domain selection based on causality is not compared with that based on statistical similarity. We will expand this work in the future.

**Acknowledgments.** This work was supported by the National Natural Science Foundation of China (71971089, 72001083) and Natural Science Foundation of Guangdong Province (No. 2022A1515011612)

## References

1. Sezer, O.B., Gudelek, U., Ozbayoglu, M.: Financial time series forecasting with deep learning: a systematic literature review: 2005–2019. Appl. Soft Comput. **90**, 106181 (2020)
2. Mohan, B.H.: Krishna: a review of two decades of deep learning hybrids for financial time series prediction. Int. J. Emerging Technol. **10**(1), 324–331 (2019)
3. Tan, C., Sun, F., Kong, T., et al.: A Survey on Deep Transfer Learning. ARTIFICIAL NEURAL NETWORKS AND MACHINE LEARNING (ICANN). PT III, 270–279 (2018)
4. Pan, S.J., Qiang, Y.: A survey on transfer learning. IEEE Trans. Knowl. Data Eng. **22**(10), 1345–1359 (2010)
5. Jiang, J., Shu, Y., Wang, J., et al.: Transferability in Deep Learning: A Survey. arXiv:2201.05867 (2022)
6. Amaral, T., Silva, L.M., Alexandre, L.A., Kandaswamy, C., de Sá, J.M., Santos, J.M.: Transfer learning using rotated image data to improve deep neural network performance. In: Campilho, A., Kamel, M. (eds.) ICIAR 2014. LNCS, vol. 8814, pp. 290–300. Springer, Cham (2014). https://doi.org/10.1007/978-3-319-11758-4_32
7. Vu, N.T., Imseng, D., Povey, D., et al.: Multilingual deep neural network based acoustic modelling for rapid language adaptation. In: 2014 IEEE International Conference on Acoustics, Speech and Signal Processing (ICASSP), pp. 7639–7643. IEEE(2014). https://doi.org/10.1109/ICASSP.2014.6855086
8. Ye, R., Dai, Q.: A novel transfer learning framework for time series forecasting. Knowl.-Based Syst. **156**(sep.15), 74–99 (2018)
9. Gu, Q., Dai, Q., Yu, H., et al.: Integrating multi-source transfer learning, active learning and metric learning paradigms for time series prediction. Appl. Soft Comput. **109**(3), 107583- (2021)
10. Gu, Q., Dai, Q.: A novel active multi-source transfer learning algorithm for time series forecasting. Appl. Intell. **51**(2), 1–25 (2021)
11. Ye, R., Dai, Q.: Implementing transfer learning across different datasets for time series forecasting. Pattern Recogn. **109**, 107617 (2020)
12. Hochreiter, S., Schmidhuber, J.: Long short-term memory. Neural Comput. **9**(8), 1735–1780 (1997)
13. Zhang, Y., Yan, B., Aasma, M.: A novel deep learning framework: prediction and analysis of financial time series using CEEMD and LSTM. Expert Systems with Applications. 113609 (2020)
14. Niu, T., Wang, J., Lu, H., et al.: Developing a deep learning framework with two-stage feature selection for multivariate financial time series forecasting. Expert Syst. Appl. **148**, 113237 (2020)

15. Schreiber, T.: Measuring information transfer. Phys. Rev. Lett. **85**(2), 461–464 (2000)
16. Ganin, Y., Ustinova, E., Ajakan, H., et al.: Domain-adversarial training of neural networks. J. Mach. Learn. Res. **17**(1), 2096–2030 (2016)
17. Smola, A., Gretton, A., Le, S., et al.: A hilbert space embedding for distributions. Discovery Sci. **4754**, 13–31 (2007)
18. Cuturi, M., Blondel, M.: Soft-DTW: a differentiable loss function for time-series. In: International Conference on Machine Learning, pp. 894–903 (2017)
19. Mariano, D.R.S.: Comparing predictive accuracy. J. Bus. Econ. Stat. **20**(1), 134–144 (2002)

# ScriptNet: A Two Stream CNN for Script Identification in Camera-Based Document Images

Minzhen Deng[1], Hui Ma[2], Li Liu[1](✉), Taorong Qiu[1], Yue Lu[3], and Ching Y. Suen[4]

[1] School of Mathematics and Computer Sciences, Nanchang University, Nanchang 330031, China
liuli_033@163.com

[2] Xi'an Microelectronics Technology Institute, Xi'an 710065, China

[3] School of Communication and Electronic Engineering, East China Normal University, Shanghai 200241, China

[4] Centre for Pattern Recognition and Machine Intelligence, Concordia University, Montreal H3G 1M8, Canada

**Abstract.** Script identification is an essential part of a document image analysis system, since documents written in different scripts may undergo different processing methods. In this paper, we address the issue of script identification in camera-based document images, which is challenging since the camera-based document images are often subject to perspective distortions, uneven illuminations, etc. We propose a novel network called ScriptNet that is composed of two streams: spatial stream and visual stream. The spatial stream captures the spatial dependencies within the image, while the visual stream describes the appearance of the image. The two streams are then fused in the network, which can be trained in an end-to-end manner. Extensive experiments demonstrate the effectiveness of the proposed approach. The two streams have been shown to be complementary to each other. An accuracy of 99.1% has been achieved by our proposed network, which compares favourably with state-of-the-art methods. Besides, the proposed network achieves promising results even when it is trained with non-camera-based document images and tested on camera-based document images.

**Keywords:** Script identification · camera-based document images · ScriptNet · spatial stream · visual stream

## 1 Introduction

Script identification plays a vital role in document image analysis systems, since it is usually a precursor to downstream stages, e.g. Optical Character Recognition (OCR) [1]. It is often the case that document images written in different scripts entail different processing methods. We are mainly concerned with script identification in camera-based document images in this paper. As smartphones are becoming ubiquitous nowadays, the number of camera-based document images has increased substantially. Manual identification of the script is a


M. Tanveer et al. (Eds.): ICONIP 2022, CCIS 1793, pp. 14–25, 2023.
https://doi.org/10.1007/978-981-99-1645-0_2

tedious and time-consuming task. Therefore, it is of great significance to propose an automatic script identification method.

However, script identification is a nontrival task, because different scripts usually share common characters. For example, Chinese and Japanese have a large portion of characters in common. In this way, script identification can be considered as a fine-grained classification problem. Besides, the camera-based document images usually suffer from perspective distortions, uneven illumination, and blur, etc., making script identification in camera-based document images even more challenging.

A lot of works have been devoted to script identification in the literature [2], which can be broadly classified into two categories: handcrafted feature-based methods and deep learning-based methods. With respect to handcrafted feature-based methods, various types of features have been employed. Hangarge et al. [3] employ directional Discrete Cosine Transform for script identification. In [4], three types of features involving LBP (Local Binary Pattern), HOG (Histogram of Oriented Gradients) and Gradient Local Auto-Correlation are compared for script identification. Besides, they test two different classifiers: SVM (Support Vector Machines) and ANN (Artificial Neural Networks). Likewise, LBP is also employed in [5] for script identification. Shivakumara et al. [6] propose a novel feature named Gradient-Angular-Features to identify script.

With the popularity of deep learning in many fields [7,8], it has also been employed to solve the problem of script identification. The Convolutional Neural Network (CNN) and Recurrent Neural Network (RNN) are brought into one end-to-end trainable network for script identification in [9]. Cheng et al. [10] employ patches extracted from the image for script identification. They present a novel module named PA (Patch Aggregator), which can be integrated in a convolutional neural network. In [11], an attention-based multi-scale neural network is proposed. To be more precise, a Res2Net is utilized to obtain a multi-scale feature map. Afterwards, a soft channel-wise attention model is further employed to boost the local sensitivity of the feature map. In order to improve the efficiency of the classifier, the fully-connected layer is replaced with a global maximum pooling layer. A multi-modal deep network is proposed in [12] to solve the indic handwritten script identification problem, which takes both offline and online modality of the data as input to make use of the information from both modalities jointly. Ghosh et al. [13] propose a lightweight CNN architecture called LWSINet to address the issue of script identification. An end-to-end multi-task deep neural network is proposed in [14] for simultaneous script identification and keyword spotting. Bhunia et al. [15] propose a novel method which extracts local and global features using CNN-LSTM framework and weighs them dynamically for script identification.

The script identification methods mentioned above have been designed either for scanned document images [3,5,12,14] or videos [4,6,13,15] or scene texts [9–11,15]. To the best of our knowledge, only a few works have been devoted to script identification in camera-based document images [16,17], all of which employ handcrafted features. For instance, Li and Tan [16] represent a document image by a frequency vector of affine invariant signatures of characters, and then identify the script of the image by comparing the vector with some predefined script templates. Yet, the handcrafted features are usually designed for particular scripts and are lack of generalization ability.

In this paper, we propose a novel network called ScriptNet to address the issue of script identification in camera-based document images. The network is composed of two streams: spatial stream and visual stream. Regarding the spatial stream, it combines a CNN model with BiLSTM [18] to fully exploit the spatial dependencies within text. For the visual stream, different abstraction levels of features are extracted from the image. The early fusion approach is then adopted to fuse the two streams. The network can be trained in an end-to-end manner. Extensive experiments show that the two streams in the proposed network are complementary. The promising experimental results demonstrate the effectiveness and validity of the proposed method.

## 2 Proposed Methodology

The overall structure of the proposed ScriptNet is shown in Fig. 1, which consists of two streams: spatial stream and visual stream. The details of each stream will be elaborated in this section.

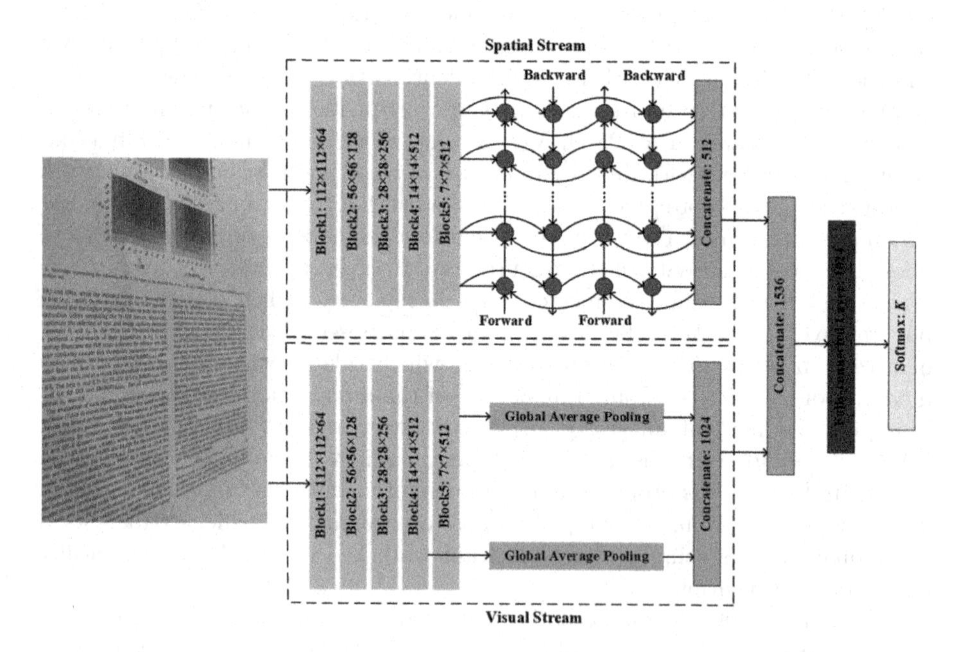


**Fig. 1.** Architecture of our proposed ScriptNet.

### 2.1 Spatial Stream

In this stream, we first employ a CNN model to generate image representation. Various famous CNN models such as AlexNet [19], VGG-16 [20] and EfficientNet [21] have been compared. VGG-16 is shown to outperform the other models and is thus employed in our work. The fully-connected layers are discarded and the last convolutional layer is considered as the representation of the image.

As indicated in [9], the spatial dependencies in the image can provide important clues for script identification. In order to capture the spatial dependencies within the image, we employ LSTM [18], which is an improved variant of RNN. LSTM overcomes the issues of vanishing gradient and gradient explosion that are often seen in RNN and is thus good at capturing long-term dependencies. Compared with RNN, LSTM is extended with memory cells. Besides, three gates are introduced in LSTM: input gate, output gate and forget gate. The input gate determines the amount of information in input that should be stored in a hidden state. The output gate decides how much information in the hidden state should be output for the current time step. The forget gate is employed to govern what information to drop from the hidden state. The basic LSTM cell is illustrated in Fig. 2.

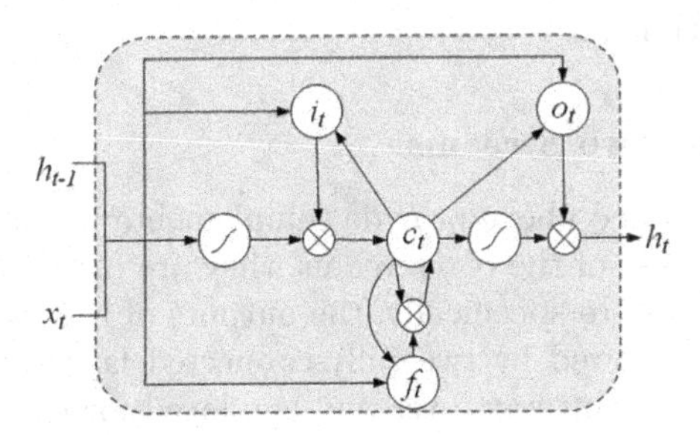


**Fig. 2.** The basic LSTM cell.

Given the input data $x_t$ for each time step $t$, the hidden value $h_t$ of an LSTM cell is computed as follows:

$$i_t = \sigma(W_{xi}x_t + W_{hi}h_{t-1} + W_{ci}c_{t-1} + b_i) \tag{1}$$

$$f_t = \sigma(W_{xf}x_t + W_{hf}h_{t-1} + W_{cf}c_{t-1} + b_f) \tag{2}$$

$$c_t = f_t c_{t-1} + i_t \tanh(W_{xc}x_t + W_{hc}h_{t-1} + b_c) \tag{3}$$

$$o_t = \sigma(W_{xo}x_t + W_{ho}h_{t-1} + W_{co}c_t + b_o) \tag{4}$$

$$h_t = o_t \tanh(c_t) \tag{5}$$

where $\sigma$ is the sigmoid function and tanh is the hyperbolic tangent activation function. $i, f, o$ and $c$ represent the input gate, forget gate, output gate and cell activation vector, respectively. $h$ denotes the hidden vector and $b$ is the bias vector. $W$ is the weight matrix, whose meaning can be easily inferred from the subscripts. For example, $W_{hi}$ represents the hidden-input gate weight matrix.

In this study, we exploit BiLSTM which benefits from capturing the spatial dependencies in two directions: from left to right and from right to left. We feed the extracted image representation, viz. the last convolutional layer of VGG-16 as stated above, to BiLSTM. Following [9], we stack two BiLSTM layers as shown in Fig. 1.

### 2.2 Visual Stream

For the visual stream, it aims to describe the appearance of the image. Since CNN can provide a hierarchical abstraction of the image, it is employed in this stream. Because of the effectiveness demonstrated in preliminary experiments, VGG-16 is adopted again. As different layers in CNN model describe the image with different levels of abstraction, with the first layers depicting the basic structures of the image and the deeper layers describing more complex structures and semantic information, it is helpful to exploit multiple layers to describe the image. To this end, we insert a global average pooling layer after each of the last $M$ convolutional layers of VGG-16. Subsequently, the outputs of the $M$ global average pooling layers are concatenated. We set $M = 2$ in our study according to preliminary experiments.

### 2.3 Fusion of the Two Streams

The two streams as stated above provide complementary information about the image. To make best use of the two streams, they are fused and the early fusion approach is adopted. More specifically, the outputs of the two streams are concatenated, which is followed by two fully-connected layers with the last fully-connected layer being a softmax layer and employed for classification. To train our proposed two stream network, the cross-entropy loss [22] is employed, which is defined as:

$$L_{\mathrm{CE}} = -\frac{1}{N}\sum_{i=1}^{N}\sum_{k=1}^{K} y_{i,k} \log p_{i,k} \tag{6}$$

where $N$ is the training set size and $K$ denotes the number of possible classes. $p_{i,k}$ represents the predicted probability that the $i$th training sample belongs to the $k$th class. $y_{i,k} = 1$ if the groundtruth label of the $i$th training sample is $k$; otherwise, $y_{i,k} = 0$.

## 3 Experiments

Since there are no public datasets for script identification in camera-based document images, we collect a dataset, the details of which are given below. In an attempt to validate the proposed approach, we have thoroughly evaluated it on the dataset and compared the proposed approach with state-of-the-art methods.

### 3.1 Dataset

We have collected a dataset which is composed of camera-based document images in four popular scripts: Chinese, Japanese, English and German. To generate camera-based document images, we first download some free books in the above four scripts. The books are all in PDF format. Afterwards, we print out the books and capture them using a cellphone camera (QuidWay Honor8X JSN-AL00a) at different angles between the cellphone and the imaging plane. Table 1 details the number of images for each script. In Fig. 3, we illustrate some sample images from the dataset.

**Table 1.** Number of images for each script in the dataset.

| | English | Chinese | Japanese | German |
|---|---|---|---|---|
| Number | 1,980 | 1,829 | 1,914 | 1,678 |

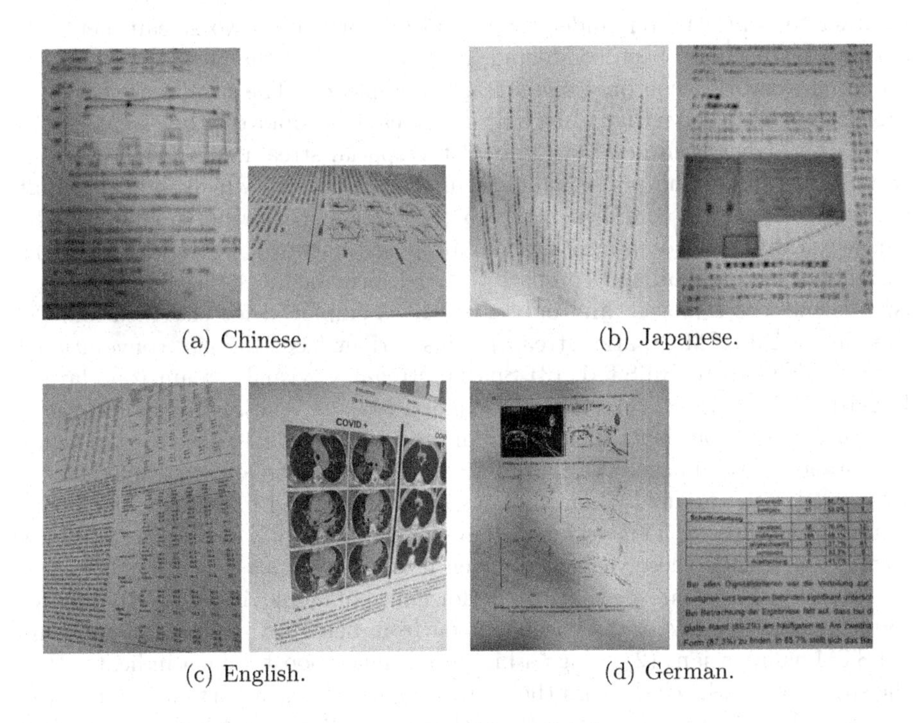


(a) Chinese. (b) Japanese.

(c) English. (d) German.

**Fig. 3.** Sample images from the dataset.

This newly created dataset is extremely challenging, since the camera-based document images suffer from diverse perspective distortions, uneven illuminations, blur, etc., as shown in Fig. 3. In addition, there are a lot of common characters between Chinese and Japanese as well as between English and German, making script identification on this dataset a nontrivial task.

We further split the dataset into training set, validation set and test set with a ratio of 6 : 2 : 2. The classification accuracy is employed as the evaluation metric. In detail, ten random dataset splits have been made and the average classification accuracy is reported.

### 3.2 Implementation Details

The images fed to the network are resized to $224 \times 224$. The network is trained with stochastic gradient descent [23], with the size of minibatch set as 32. The learning rate is initialized as 0.001, and is then reduced by a factor of 10 when the loss on the validation set plateaus. The momentum is set at 0.9. The network is trained for 15 epochs. We employ NVIDIA quadro p4000 GPU in our experiment.

### 3.3 Ablation Study

In order to gain a better understanding of our proposed two stream network, ablation studies have been conducted. In detail, we test the performance of the proposed network when only one stream is employed. The results are given in Fig. 4. One can observe that when only one stream is employed, the visual stream achieves a higher accuracy than that of the spatial stream. Yet, when the two streams are fused, the performance is further improved, which indicates that the two streams are complementary to each other. To further investigate the efficacy of different components in our proposed network, we consider various variants of our method and compare their performance: (1) Baseline: All extra components are removed, and only VGG-16 is employed for classification. (2) Baseline + BiLSTM (Spatial stream): This variant feeds the last convolutional layer of VGG-16 to BiLSTM. (3) Spatial stream + visual stream (one layer): Regarding this variant, two streams are employed, viz. spatial stream and visual stream. Yet, only one layer is employed in the visual stream to describe the visual appearance of the image. (4) Spatial stream + visual stream (multiple layers): Compared with variant (3), multiple layers are employed to describe the image in the visual stream. This variant is our proposed ScriptNet. The performance comparison among these variants is given in Fig. 5.

From Fig. 5, we can see that the performance of employing VGG-16 only for script identification as in the baseline is far from being satisfactory. Introducing BiLSTM as in variant (2) brings a big performance boost, which indicates that the spatial dependencies within the text is of paramount importance in identifying the script. Comparing variant (3) with variant (2), the performance is further improved, suggesting that visual stream and the spatial stream are complementary to each other. Since different layers in CNN model describe the image with different levels of abstraction, it is beneficial to exploit multiple layers to depict the image in the visual stream as in variant (4).

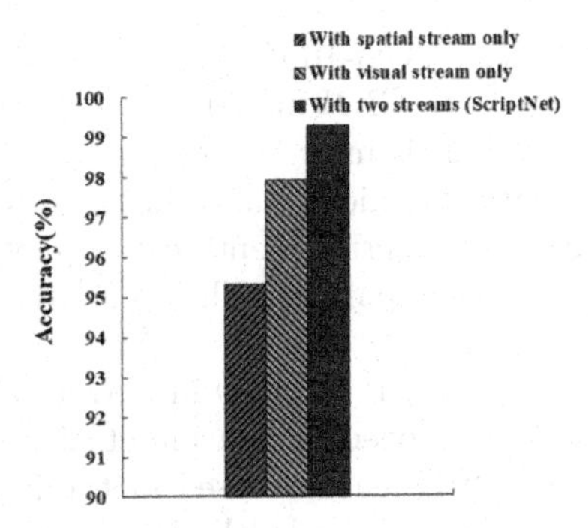


**Fig. 4.** Performance comparison between the proposed network and that when only one stream is considered.

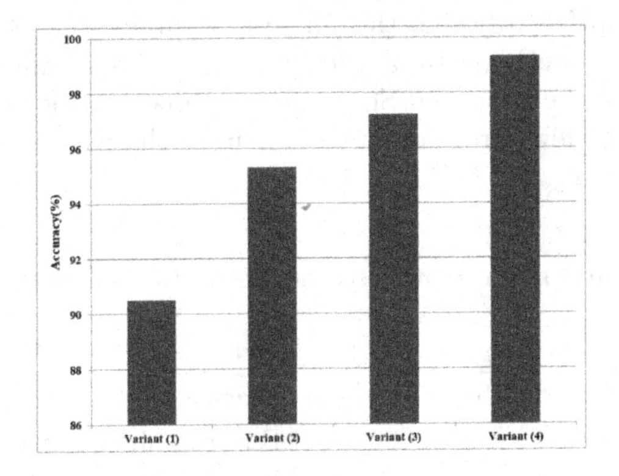


**Fig. 5.** Performance comparison among the four variants.

### 3.4 Comparison with Literature

In order to validate the effectiveness of our proposed ScriptNet, we compare it with some state-of-the-art script identification methods. Compared with camera-based document images, it is often the case that the non-camera-based document images are easier to obtain. So it would be desirable if a method can identify the script of camera-based document images while only employing non-camera-based document images for training. To this end, we also test the performance of different methods when the training set is replaced with the same number of non-camera-based document images.

**Training with Camera-Based Document Images.** Both handcrafted feature-based and deep learning-based methods have been employed for comparison. Regarding handcrafted feature-based methods, those proposed by Sharma et al. [4] and Ferrer et al. [5] are employed, which identify the script of images based on HOG and LBP features, respectively. The SVM classifier is then employed for classification. For deep learning-based methods, we first employ two baseline CNN models, viz. VGG-16 [20] and AlexNet [19], for performance comparison. Since these baseline CNN models have achieved promising results in generic image classification, it is interesting to find out how they perform for the task of script identification, which is also an image classification problem. Besides, the deep learning-based script identification method proposed by Mei et al. [9] is also employed for comparison. The comparison results are given in Table 2.

According to the table, the LBP features in Ferrer et al.'s method are preferred to the HOG features employed in Sharma et al.'s method. Yet, the performance of the two handcrafted feature-based methods is far from being satisfactory. Regarding CNN-based methods, the baseline VGG-16 model demonstrates great potential in solving the script identification problem compared with AlexNet. Mei et al.'s method greatly outperforms the two baseline CNN models. Our proposed ScriptNet obtains the highest accuracy, viz. 99.1%, which surpasses all the other methods by a wide margin. Besides, we demonstrate the confusion matrix of the proposed ScriptNet as shown in Fig. 6, from which one can observe that the few errors mainly come from the misclassification between Chinese and Japanese.

**Table 2.** Comparison of the proposed method with state-of-the-art methods.

| | Methods | Accuracy(%) |
|---|---|---|
| Handcrafted feature-based methods | Sharma et al. [4] | 86.3 |
| | Ferrer et al. [5] | 89.7 |
| CNN-based methods | AlexNet [19] | 87.6 |
| | VGG-16 [20] | 90.8 |
| | Mei et al. [9] | 94.4 |
| | ScriptNet | **99.1** |

**Training with Non-Camera-Based Document Images.** In order to test the performance of different methods when trained with non-camera-based document images, we replace the training set with the same number of non-camera-based document images. In greater details, the non-camera-based document images come from the same book pages that compose the training set. Instead of printing out these pages and then capturing them using the cellphone as stated in Sect. 3.1, the pages are directly converted into images using Adobe Acrobat. The performance comparisons among different methods are shown in Fig. 7. We can see that the performance of all the methods deteriorate when trained with non-camera-based document images. An extremely sharp performance decrease is observed for the two handcrafted feature-based methods, which indicates the poor generalization ability of these methods. Yet, our proposed ScriptNet still achieves promising results, far exceeding the other methods.

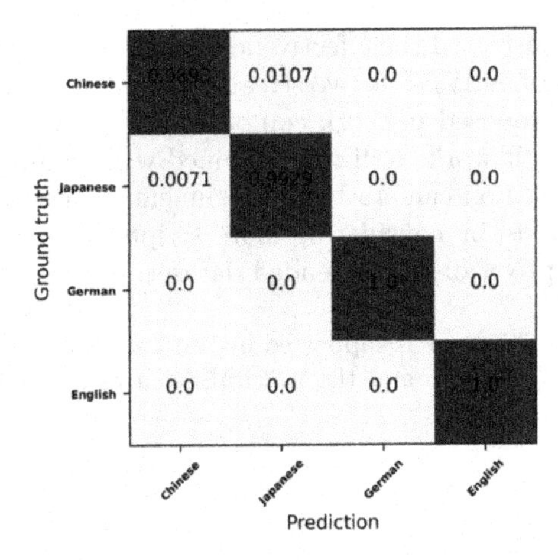


**Fig. 6.** Confusion matrix of the proposed ScriptNet.

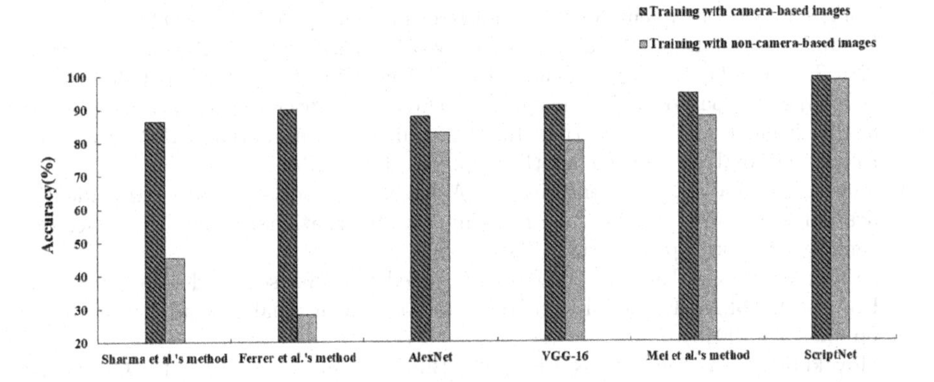


**Fig. 7.** Performance comparison among different methods when trained with non-camera-based images.

## 4 Conclusion

The number of camera-based document images has increased dramatically due the popularity of smartphones. We address the issue of script identification in camera-based document images in this paper, which often acts as an a prerequisite in document image analysis systems. We propose a novel network called ScriptNet that consists of two streams: spatial stream and visual stream. The two streams describe the image from different aspects. More specifically, the spatial stream captures the spatial dependencies within the text by combining a CNN model with BiLSTM, while the visual stream extracts different abstraction levels of features to depict the image appearance. The early fusion approach is then adopted to fuse the two streams. The network can be trained in an end-to-end manner. Extensive experiments on a dataset which is composed of four popular scripts have demonstrated the effectiveness of the proposed ScriptNet, which achieves an accuracy of 99.1%. The two streams have shown to be complementary to each other. The proposed network compares favorably with state-of-the-art methods. In addition, it works well when trained with non-camera-based document images and tested on camera-based document images. In future work, we will extend the datasets by considering more scripts and test the performance of the proposed ScriptNet on the extended dataset.

**Acknowledgments.** This work is supported by National Natural Science Foundation of China under Grant 61603256 and the Natural Sciences and Engineering Research Council of Canada.

## References

1. Randika, A., Ray, N., Xiao, X., Latimer, A.: Unknown-box approximation to improve optical character recognition performance. In: Proceedings of International Conference on Document Analysis and Recognition, pp. 481–496 (2021)
2. Ubul, K., Tursun, G., Aysa, A., Impedovo, D., Pirlo, G., Yibulayin, T.: Script identification of multi-script documents: a survey. IEEE Access **5**, 6546–6559 (2017)
3. Hangarge, M., Santosh, K., Pardeshi, R.: Directional discrete cosine transform for handwritten script identification. In: Proceedings of International Conference on Document Analysis and Recognition, pp. 344–348 (2013)
4. Sharma, N., Pal, U., Blumenstein, M.: A study on word-level multi-script identification from video frames. In: Proceedings of International Joint Conference on Neural Networks, pp. 1827–1833 (2014)
5. Ferrer, M.A., Morales, A., Pal, U.: LBP based line-wise script identification. In: Proceedings of International Conference on Document Analysis and Recognition, pp. 369–373 (2013)
6. Shivakumara, P., Sharma, N., Pal, U., Blumenstein, M., Tan, C.L.: Gradient-angular-features for word-wise video script identification. In: Proceedings of International Conference on Pattern Recognition, pp. 3098–3103 (2014)
7. Dong, S., Wang, P., Abbas, K.: A survey on deep learning and its applications. Comput. Sci. Rev. **40**, 100379 (2021)
8. Vaquero, L., Brea, V.M., Mucientes, M.: Tracking more than 100 arbitrary objects at 25 FPS through deep learning. Pattern Recogn. **121**, 108205 (2022)
9. Mei, J., Dai, L., Shi, B., Bai, X.: Scene text script identification with convolutional recurrent neural networks. In: Proceedings of International Conference on Pattern Recognition, pp. 4053–4058 (2016)
10. Cheng, C., Huang, Q., Bai, X., Feng, B., Liu, W.: Patch aggregator for scene text script identification. In: Proceedings of International Conference on Document Analysis and Recognition, pp. 1077–1083 (2019)
11. Ma, M., Wang, Q.F., Huang, S., Huang, S., Goulermas, Y., Huang, K.: Residual attention-based multi-scale script identification in scene text images. Neurocomputing **421**, 222–233 (2021)
12. Bhunia, A.K., Mukherjee, S., Sain, A., Bhunia, A.K., Roy, P.P., Pal, U.: Indic handwritten script identification using offline-online multi-modal deep network. Inf. Fusion **57**, 1–14 (2020)

13. Ghosh, M., Mukherjee, H., Obaidullah, S.M., Santosh, K., Das, N., Roy, K.: LWSINet: a deep learning-based approach towards video script identification. Multimedia Tools Appl. **80**(19), 29095–29128 (2021)
14. Cheikhrouhou, A., Kessentini, Y., Kanoun, S.: Multi-task learning for simultaneous script identification and keyword spotting in document images. Pattern Recogn. **113**, 107832 (2021)
15. Bhunia, A.K., Konwer, A., Bhunia, A.K., Bhowmick, A., Roy, P.P., Pal, U.: Script identification in natural scene image and video frames using an attention based Convolutional-LSTM network. Pattern Recogn. **85**, 172–184 (2019)
16. Li, L., Tan, C.L.: Script identification of camera-based images. In: Proceedings of International Conference on Pattern Recognition, pp. 1–4 (2008)
17. Dhandra, B., Mallappa, S., Mukarambi, G.: Script identification of camera based bilingual document images using SFTA features. Int. J. Technol. Human Interact. **15**(4), 1–12 (2019)
18. Dileep, P., et al.: An automatic heart disease prediction using cluster-based bidirectional LSTM (C-BiLSTM) algorithm. Neural Comput. Appl. **35**, 1–14 (2022). https://doi.org/10.1007/s00521-022-07064-0
19. Krizhevsky, A., Sutskever, I., Hinton, G.E.: ImageNet classification with deep convolutional neural networks. Adv. Neural. Inf. Process. Syst. **25**, 1097–1105 (2012)
20. Simonyan, K., Zisserman, A.: Very deep convolutional networks for large-scale image recognition. In: Proceedings of International Conference on Learning Representations (2015)
21. Tan, M., Le, Q.: Efficientnet: Rethinking model scaling for convolutional neural networks. In: Proceedings of International Conference on Machine Learning, pp. 6105–6114 (2019)
22. Zhang, J., Zhao, L., Zeng, J., Qin, P., Wang, Y., Yu, X.: Deep MRI glioma segmentation via multiple guidances and hybrid enhanced-gradient cross-entropy loss. Expert Syst. Appl. **196**, 116608 (2022)
23. Lou, Z., Zhu, W., Wu, W.B.: Beyond sub-gaussian noises: Sharp concentration analysis for stochastic gradient descent. J. Mach. Learn. Res. **23**, 1–22 (2022)

# Projected Entangled Pair State Tensor Network for Colour Image and Video Completion

Rongfeng Huang[1,2], Shifang Liu[1,2], Xinyin Zhang[1,2], Yang Liu[1,2], and Yonghua Zhao[1(✉)]

[1] Computer Network Information Center, Chinese Academy of Sciences, Beijing, China
yhzhao@sccas.cn
[2] University of Chinese Academy of Sciences, Beijing, China

**Abstract.** Tensor decompositions, such as the CP, Tucker, tensor train, and tensor ring decomposition, have yielded many promising results in science and engineering. However, a more general tensor model known as the projected entangled pair state (PEPS) tensor network has not been widely considered for colour image and video processing, although it has long been studied in quantum physics. In this study, we constructed the relationship between the generalized tensor unfolding matrices and the PEPS ranks. Furthermore, we employed the PEPS tensor network for the high-order tensor completion problem and developed an efficient gradient-based optimisation algorithm to find the latent factors of the incomplete tensor, which we used to fill the missing entries of the tensor. Comparing the proposed method with state-of-the-art methods, experimental results for colour image and video completion confirm the effectiveness of the proposed methods.

**Keywords:** Projected entangled pair states · Tensor decomposition · Tensor completion · Gradient-based optimization

## 1 Introduction

Many kinds of real-world data are represented in high dimensional form. For example, colour images are represented by third-order tensors (*height* × *width* × *RGB channels*) and videos composed of colour images are represented by fourth-order tensors (*height* × *width* × *RGB channels* × *time*). However, tensor data from real applications are commonly involved in missing values owing to measurement errors or other non-human factors. Tensor completion (TC) is a problem of recovering tensor data with missing values from the partially observed entries of the tensor. As colour images and videos are perfect examples of third-order and fourth-order tensors, their completion can be formulated as tensor completion problems.

This work is supported by National Key Research and Development Program of China (2017YFB0202202) and Strategic Priority Research Program of Chinese Academy of Sciences (XDC05000000).


M. Tanveer et al. (Eds.): ICONIP 2022, CCIS 1793, pp. 26–38, 2023.
https://doi.org/10.1007/978-981-99-1645-0_3

Tensor decompositions (TDs) have received much attention in a great number of fields, such as signal processing [2], image recovery [3], video surveillance [4]. The most classic TDs are the CP decomposition [5] and Tucker decomposition [6]. Although the number of parameters scales linearly with tensor dimension, finding optimal CP rank is already an NP-hard problem. Tucker decomposition sustains the "curse of dimensionality". To address these drawbacks, tensor train (TT) decomposition [7] and tensor ring (TR) decomposition [8] have been developed. A common limitation of these two decompositions is that connections are built only between two adjacent modes, resulting in limited tensor dimension interactions and the inability to capture sufficient latent factors. Suffering from the weakness discussed, many existing tensor completion algorithms based on these TDs cannot handle certain missing patterns, such as high missing ratios.

As a generalisation of TT decomposition in two dimensions, the projected entangled pair state (PEPS) tensor network constructs enough connections between different modes without sustaining the "curse of dimensionality" and has achieved great success in quantum physics [9]. However, it has not received much attention in the TC field. To this end, we demonstrate the application of the PEPS tensor network in the TC field.

The major contributions of this paper can be summarized as follows: 1) We constructed the relationship between the generalized tensor unfolding matrices and the PEPS ranks. 2) We employ the PEPS tensor network to the TC problem and develop an efficient gradient-based algorithm to solve it. 3) We conduct experiments on colour image and video datasets with different missing patterns to validate our method's strength.

## 2 Related Works

Tensor completion approaches can be mainly categorized into the "rank minimization based" approach and the "tensor decomposition based" approach.

**Rank Minimization Based Approaches:** Rank minimization based approaches formulate the convex surrogate models of low-rank tensors. The most representative surrogate is the nuclear norm. Based on different definitions of tensor rank, various nuclear norm regularized algorithms have been proposed. Based on Tucker rank, Liu et al. [10] constructed the nuclear norm by summing the nuclear norm of model-$n$ unfolding matrices. A drawback is that model-$n$ unfolding matrices are highly unbalanced matrices and only capable of capturing the correlation between one mode and the others. To overcome this weakness, Bengua et al. [11] proposed the TT nuclear norm for the TT model, which is better to capture the global correlation of a tensor. Yuan et al. [12] based on the TR model and formulated the convex surrogate by replacing the TR rank with the rank of TR factors. Zhang et al. [13] obtained the tensor nuclear norm by the convex relaxation of tensor tubal rank [14]. Rank minimization-based approaches can find the rank of the recovered tensor from the observed entries automatically. However, multiple costly singular value decomposition (SVD) operations on the unfolding matrices of the tensor are necessary.

**Decomposition Based Approaches:** Instead of finding the low-rank tensor directly, "tensor decomposition based" approaches formulate TC as a tensor decomposition problem with predefined tensor ranks, and finds the latent factors of the incomplete tensor by solving the optimal tensor decomposition, then the latent factors are used to fill the missing entries. Different tensor decomposition models result in various "tensor decomposition based" approaches. Acar et al. [15] and Karatzoglou et al. [16] proposed CP-based and Tucker-based tensor completion algorithms, respectively. Zheng et al. [17] proposed a fully-connected tensor network decomposition based tensor completion algorithm. Due to the limitations of the employed tensor decomposition model as discussed in Sect. 1, the above three "tensor decomposition based" algorithms face challenges when the tensor order is high. To this end, Yuan et al. [18,19] and Wang et al. [20,21] proposed weighted TT and TR optimisation models to find the latent factors, respectively. The performance of decomposition based approaches is rather sensitive to rank selection. However, specifying the optimal rank beforehand is very challenging because the optimal rank is generally data dependent.

Apart from the two kinds of approaches, there are also some other supplementary tensor completion approaches. For instance, Zhao et al. [22] and Long et al. [23] considered the probabilistic framework with a fully Bayesian treatment and constructed Bayesian probabilistic models for CP and TR decompositions respectively. Similar to "rank minimization based" approaches, the ranks of the decomposition model do not need to be specified beforehand, because they can be caught by Bayesian inference.

## 3 PEPS Tensor Network

The PEPS tensor network is illustrated graphically using a two-dimensional tensor network, as shown in Fig. 1(a). Width and height are useful parameters to describe the PEPS tensor network and are denoted by $W$ and $H$, respectively. For simplicity, we denote the PEPS model by $PEPS\left(\{\mathcal{G}^{i,j}\}_{i=1,j=1}^{W,H}\right)$. According to the number of connected tensors, there are three types of tensors: interior tensors, four tensors on corners, and other tensors on the boundary, which have four, two, and three connected tensors, respectively. The dimensions of the two connected indices must be equal; thus, we only need to assign a rank for each edge connecting the two tensors. We use a vector of size $W(H-1)+H(W-1)$ to store the ranks of all edges and call this vector the PEPS ranks. Without a loss of generality, we assume that all edge ranks are equal to $R$. Owing to the complicated structure and connectivity, designing algorithms based on the PEPS model is tedious.

## 4 Tensor Completion with PEPS Tensor Network

The mode-$n$ matricisation (unfolding) of tensor $\mathcal{X} \in \mathbb{R}^{I_1 \times I_2 \times \cdots \times I_N}$ is denoted by $\mathcal{X}_{(n)} \in \mathbb{R}^{I_n \times \underline{I_1 \cdots I_{n-1} I_{n+1} \cdots I_N}}$. Given a vector $\mathbf{n}$, whose elements are permutations

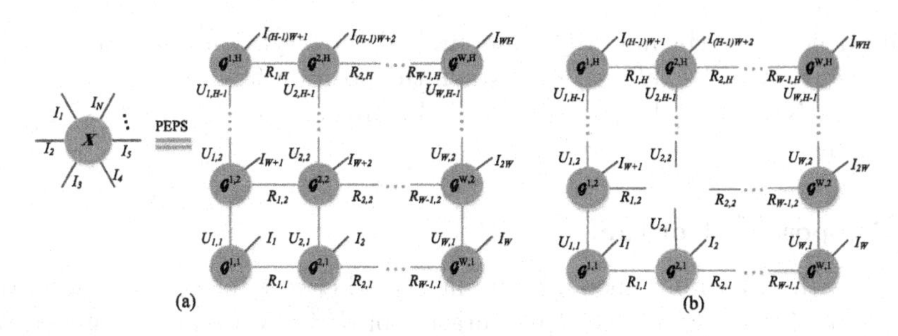


**Fig. 1.** (a) Graphical representation PEPS tensor network. (b) Graphical representation of $\mathcal{G}^{\overline{i,j}}$.

of $1, 2, \cdots, N$, the transposition of $\mathcal{X}$ based on $\mathbf{n}$ is also a $N$th order tensor $\mathcal{X}' \in R^{I_{n_1} \times I_{n_2} \times \cdots \times I_{n_N}}$. The elements of $\mathcal{X}'$ and $\mathcal{X}$ satisfy $x'_{i_{n_1} i_{n_2} \cdots i_{n_N}} = x_{i_1 i_2 \cdots i_N}$. Furthermore, the generalized tensor unfolding of $\mathcal{X}$ based on $\mathbf{n}$ is defined as a matrix $\mathcal{X}_{[n_1,\cdots,n_M;n_{M+1},\cdots,n_N]} = reshape(\mathcal{X}', [I_{n_1}, \cdots, I_{n_M}; I_{n_{M+1}} \cdots I_{n_N}])$.

The relationship between the generalized tensor unfolding matrices and the PEPS ranks can be constructed.

**Theorem 1.** *Given a tensor $\mathcal{X}$ of size $I_1 \times I_2 \times \cdots \times I_N$, suppose it can be expressed using a PEPS model $PEPS\left(\{\mathcal{G}^{i,j}\}_{i=1,j=1}^{W,H}\right)$, the following inequalities can be verified:*

$$rank(\mathcal{X}_{[1,\cdots,i,\cdots,(H-1)W+1,\cdots,(H-1)W+i;\; i+1,\cdots,W,\cdots,(H-1)W+i+1,\cdots,WH]}) \leq \prod_{k=1}^{H} R_{i,k}, \tag{1}$$

*and*

$$rank(\mathcal{X}_{[1,\cdots,W,\cdots,(j-1)W+1,\cdots,jW;\; jW+1,\cdots,(j+1)W,\cdots,(H-1)W+1,\cdots,WH]}) \leq \prod_{k=1}^{W} U_{k,j}. \tag{2}$$

The proof is not difficult and omitted for lack of space. As shown in (1) and (2), the PEPS tensor network uses multiple PEPS ranks to bound the ranks of generalized tensor unfolding matrices, thus PEPS ranks are usually far less than TT and TR ranks, which indicates that the PEPS model would be better than the other two models for low-rank tensor data.

Given an incomplete tensor $\mathcal{X} \in R^{I_1 \times I_2 \times \cdots \times I_N}$ with observed locations $\Omega$, the TC problem by the PEPS tensor network can be expressed as follows:

$$\min_{\{\mathcal{G}^{i,j}\}_{i=1,j=1}^{W,H}} \frac{1}{2} \left\| \mathcal{W} * \left(\mathcal{X} - PEPS(\{\mathcal{G}^{i,j}\}_{i,j=1}^{W,H})\right) \right\|_F^2 \tag{3}$$

where $*$ denotes the Hadamard product and $\mathcal{W}$ is a weight tensor of the same size as $\mathcal{X}$. The entry of $\mathcal{W}$ values 1 only when the position belongs to $\Omega$, otherwise, it values 0. Note that the sizes of the PEPS model and PEPS ranks are both predefined.

### 4.1 Proposed Algorithm

Let $\mathcal{G}^{\overline{i,j}}$ be the tensor obtained by merging the other core tensors, except for $\mathcal{G}^{i,j}$. Figure 1(b) shows a graphical representation of $\mathcal{G}^{\overline{2,2}}$. Two types of indices exist in $\mathcal{G}^{\overline{i,j}}$. One is the result of removing $\mathcal{G}^{i,j}$, and the other inherits from other core tensors. We further merge the two types of indices and obtain a matrix denoted as $\mathcal{G}^{\overline{i,j}}_{<4>}$.

**Theorem 2.** *Suppose PEPS* $\left(\{\mathcal{G}^{i,j}\}_{i=1,j=1}^{W,H}\right)$ *is a PEPS tensor network of* $\mathcal{X}$*, the mode-*$((j-1)W+i)$ *unfolding matrix of* $\mathcal{X}$ *can be written as follows:*

$$\mathcal{X}_{((j-1)W+i)} = \mathcal{G}^{i,j}_{(5)}\mathcal{G}^{\overline{i,j}}_{<4>} \tag{4}$$

*where* $\mathcal{G}^{i,j}_{(5)}$ *is the mode-5 matricisation of* $\mathcal{G}^{i,j}$.

*Proof.* As shown in Fig. 1(a), we have

$$\begin{aligned}
x_{i_1 i_2 \cdots i_N} = & \sum_{r_{i,j-1},u_{i,j-1},r_{i,j},u_{i,j}} \Big\{\mathcal{G}^{i,j}(r_{i-1,j},u_{i,j-1},r_{i,j},u_{i,j},i_{(j-1)W+i})\cdot \\
& \Big\{\sum_{u_{1,1},u_{1,2}\cdots,u_{1,H-1}} \cdots \sum_{u_{i,1},\cdots,u_{i,j-2},u_{i,j+1},\cdots,u_{i,H-1}} \cdots \sum_{u_{W,1},u_{W,2}\cdots,u_{W,H-1}} \\
& \sum_{r_{1,1},r_{2,1},\cdots,r_{W-1,1}} \cdots \sum_{r_{1,j},\cdots,r_{i-2,j},r_{i+1,j}\cdots,r_{W-1,j}} \cdots \sum_{r_{1,H},r_{2,H}\cdots,r_{W-1,H}} \\
& \{\mathcal{G}^{1,1}(1,1,r_{1,1},u_{1,1},i_1)\mathcal{G}^{2,1}(r_{1,1},1,r_{2,1},u_{2,1},i_2)\cdots\mathcal{G}^{W,1}(r_{W-1,1},1,1,u_{W-1,1},i_W) \\
& \cdots\mathcal{G}^{1,j}(1,u_{1,j-1},r_{1,j},u_{1,j},i_{(j-1)W+1})\mathcal{G}^{2,j}(r_{1,j},u_{2,j-1},r_{2,j},u_{2,j},i_{(j-1)W+2})\cdots \\
& \mathcal{G}^{i-1,j}(r_{i-2,j},u_{i-1,j-1},r_{i-1,j},u_{i-1,j},i_{(j-1)W+i-1})\cdot \\
& \mathcal{G}^{i+1,j}(r_{i,j},u_{i+1,j-1},r_{i+1,j},u_{i+1,j},i_{(j-1)W+i+1})\cdots \\
& \mathcal{G}^{W,j}(r_{W-1,j},u_{W,j-1},1,u_{W,j},i_{jW})\cdots\mathcal{G}^{1,H}(1,u_{1,H-1},r_{1,H},1,i_{(H-1)W+1}) \\
& \mathcal{G}^{2,H}(r_{1,H},u_{2,H-1},r_{2,H},1,i_{(H-1)W+2})\cdots\mathcal{G}^{W,H}(r_{W-1,H},u_{W,H-1},1,1,i_N)\}\Big\}\Big\}
\end{aligned}$$

Hence, the mode-$((j-1)W+i)$ unfolding matrix can be rewritten as

$$\begin{aligned}
& \mathcal{X}_{((j-1)W+i)}(i_{(j-1)W+i},\underline{i_1 i_2\cdots i_{(j-1)W+i-1},i_{(j-1)W+i+1}i_{(j-1)W+i+2}\cdots i_N}) \\
& = \sum_{r_{i-1,j},u_{i,j-1},r_{i,j},u_{i,j}} \Big\{\mathcal{G}^{i,j}_{(5)}(i_{(j-1)W+i},\underline{r_{i,j-1}u_{i,j-1}r_{i,j}u_{i,j}})\cdot \\
& \mathcal{G}^{\overline{i,j}}_{<4>}(\underline{r_{i-1,j}u_{i,j-1}r_{i,j}u_{i,j}},\underline{i_1 i_2\cdots i_{(j-1)W+i-1},i_{(j-1)W+i+1}i_{(j-1)W+i+2}\cdots i_N})\Big\}.
\end{aligned}$$

*which is the product of two matrices. Other mode-n unfolding matrices can be obtained with the same procedure, which indicates the correctness of (4).*

Based on theorem 2, the objective function of the TC problem in (3) can be expressed as follows:

$$f\left(\{\mathcal{G}^{i,j}\}_{i,j=1}^{W,H}\right) = \frac{1}{2}\left\|\mathcal{W}_{((j-1)W+i)} * \left(\mathcal{X}_{((j-1)W+i)} - \mathcal{G}_{(5)}^{i,j}\mathcal{G}_{<4>}^{\overline{i,j}}\right)\right\|_F^2 \tag{5}$$

Subsequently, the partial derivatives of the objective function (5) with respect to $\mathcal{G}_{(5)}^{i,j}$ is calculated as follows:

$$\frac{\partial f}{\partial \mathcal{G}_{(5)}^{i,j}} = \mathcal{W}_{((j-1)W+i)} * \left(\mathcal{G}_{(5)}^{i,j}\mathcal{G}_{<4>}^{\overline{i,j}} - \mathcal{X}_{((j-1)W+i)}\right)\left(\mathcal{G}_{<4>}^{\overline{i,j}}\right)' \tag{6}$$

After the objective function values and partial derivatives are obtained, the objective function can be optimised using various optimisation algorithms such as the steepest descent algorithm, stochastic gradient descent algorithm, and nonlinear conjugate gradient methods. Finally, we propose the PEPS-TC algorithm to determine the PEPS tensor network for the incomplete tensor $\mathcal{X}$. Algorithm 1 describes the algorithm in detail. The cores are initialised randomly. The iterations are repeated until certain stopping conditions are satisfied. Common stopping conditions include little or no improvement in the objective function and a predefined maximum number of iterations.

**Algorithm 1:** Tensor completion algorithm with PEPS tensor network (PEPS-TC)

**Input**: incomplete tensor $\mathcal{X}$, weight tensor $\mathcal{W}$, the size of PEPS model $W, H$, and the PEPS ranks $\mathbf{r}$.

**Output**: core tensors $\{\mathcal{G}^{i,j}\}_{i=1,j=1}^{W,H}$.

1 Initialise the core tensors $\{\mathcal{G}^{i,j}\}_{i=1,j=1}^{W,H}$.

2 **while** *the stopping conditions are not satisfied* **do**

3 | The objective function values are computed according to (5).

4 | The gradients are computed according to (6).

5 | We update $\{\mathcal{G}^{i,j}\}_{i=1,j=1}^{W,H}$ using the gradient descent method.

6 **end**

## 4.2 Complexity Analysis

For an $N$th-order incomplete tensor $\mathcal{X} \in \mathbb{R}^{I\times I\times\cdots\times I}$, we analyse the storage complexity and computational complexity of the proposed PEPS-TC algorithm by simply setting all PEPS ranks to the same value $R$ and let the size of the PEPS model be $(W, H)$.

**Storage Complexity.** The costly quantities to be stored in the PEPS-TC algorithm are $\mathcal{G}^{\overline{i,j}}_{<4>}$ and $\mathcal{G}^{i,j}_{(5)}\mathcal{G}^{\overline{i,j}}_{<4>}$. Therefore, the storage complexity is $\mathrm{O}(R^4 I^{N-1} + I^N)$.

**Computational Complexity.** The computational cost lies in three parts: 1) calculating $\mathcal{G}^{\overline{i,j}}_{<4>}$, 2) calculating $\mathcal{G}^{i,j}_{(5)}\mathcal{G}^{\overline{i,j}}_{<4>}$, and 3) calculating $\frac{\partial f}{\partial \mathcal{G}^{i,j}_{(5)}}$, which costs $\mathrm{O}\big(R^{W+3}I^{N-1}\big)$, $\mathrm{O}\big(R^4 I^N\big)$, and $\mathrm{O}\big(R^4 I^N\big)$, respectively. Therefore, the entire computational complexity at each iteration in the PEPS-TC algorithm is $max\left(\mathrm{O}\left(R^{W+3}I^{N-1}\right), \mathrm{O}\left(R^4 I^N\right)\right)$.

## 5 Numerical Experiments

All experiments were performed on a Linux workstation with a 2.4 GHz Intel(R) Xeon(R) CPU and 372 GB memory. We compared our method with state-of-the-art tensor completion algorithms, including BCPF [22], TT-WOPT [18], SiLRTC-TT [11], TR-WOPT [19], TR-VBI [23], T-SVD [13], and HiLRTC [10]. To ensure a fair comparison with other algorithms, all the hyperparameters of the compared algorithms were set according to the recommended settings in the corresponding studies to obtain the best results. Three commonly used quantities, relative squared error (RSE), peak signal-to-noise ratio (PSNR), and structural similarity index measure (SSIM), are used to evaluate and compare the performance of various algorithms. A smaller RSE value and larger PSNR and SSIM values indicate better performance.

For the optimisation algorithms of PEPS-TC, the poblano toolbox [24] was used for the underlying nonlinear optimisation. Moreover, we used two stopping conditions: the number of iterations reaches 500, and the objective function value between two iterations satisfies $|f_t - f_{t-1}| < tol$, where $f_t$ is the objective function value of the $t$th iteration and $tol = 10^{-10}$. The optimisation algorithms are stopped when one of the stopping conditions is satisfied.

### 5.1 Synthetic Data

In this experiment, we considered the completion problem of an 8th-order tensor of the size $4 \times 4 \times 4 \times 4 \times 4 \times 4 \times 4 \times 4$. The tensor was generated by a $4 \times 2$ PEPS model. For simplicity, we set all PEPS ranks are equals to four. Every entry in the core tensors of the PEPS model was sampled from the standard Gaussian distribution $\mathcal{N}(0, 1)$ independently.

Figure 2 shows the experimental results of the RSE values with respect to $mr$ for PEPS-TC, TT-WOPT, and TR-WOPT. The missing ratio of the experimental data tensors ranges from 0.1 to 0.99. PEPS-TC shows the lowest recovery error compared with the other algorithms for almost all test cases, and the $RSE$ drops to $10^{-8}$ for a missing ratio not greater than 0.8. The large RSE values of the TT-WOPT and TR-WOPT at every missing ratio indicate that the TT and TR models cannot complete this kind of tensor data.

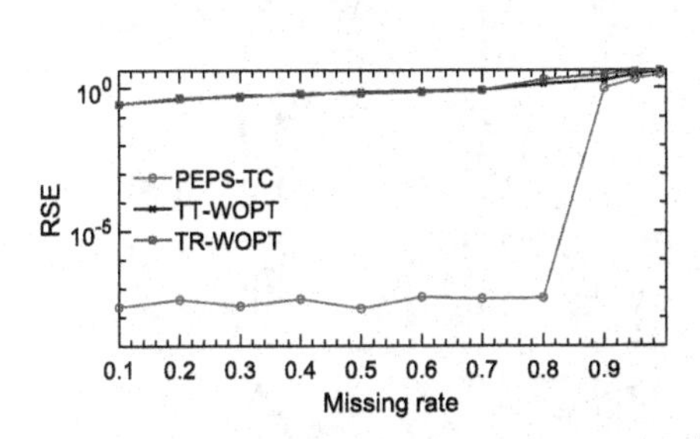


**Fig. 2.** Completion results for the synthetic tensor via PEPS-TC, TT-WOPT, and TR-WOPT under different data missing rates.

**Fig. 3.** Benchmark RGB images used in numerical experiments.

### 5.2 Image Completion

In this section, we consider the completion of the colour images. The benchmark colour images are shown in Fig. 3. All benchmark images are represented by third-order tensors of size $256 \times 256 \times 3$. Reshaping low-order tensors into high-order tensors has shown improved performance in terms of completion [11]. Therefore, the 3rd-order colour images are reshaped directly into 9th-order tensors of size $4 \times 4 \times 4 \times 4 \times 4 \times 4 \times 4 \times 4 \times 3$ for high-order tensor completion algorithms such as TT-WOPT, TR-WOPT, TR-VBI, and SiLRTC-TT. However, for BCPT, T-SVD, and HiLRTC, the original 3rd-order tensor was used. In the case of PEPS-TC, 8th-order tensors of size $4 \times 4 \times 4 \times 4 \times 4 \times 4 \times 4 \times 12$ are adopted so that a $4 \times 2$ PEPS model can be constructed. All tested algorithms' ranks and hyperparameters were tuned to achieve the best possible performance.

The reconstructed images of different approaches are shown in Fig. 4. Detailed RSE, PSNR, and SSIM values corresponding to Fig. 4 are reported in Table 1. The first four test patterns are structural missing, and the last four belong to random missing. The distribution of missing pixels for the former is more regular.

In the case of the structural missing, the visual results of BCPF, T-SVD, HaLRTC, and SiLRTC-TT are not good for the line missing and grid missing. A lot of black lines or grids are observed clearly, which indicates that these methods did not fill lost pixels properly. Nevertheless, compared with the original colour images, no significant difference is observed in the reconstructed images of the other four methods, including the proposed method. All methods obtained considerable visual results for block missing and test covering. In terms of RSE, PSNR, and SSIM values, the proposed method outperformed the other completion methods, particularly in the line missing testcase, where the $PSNR$ of our method was up to 28.35, however, the largest of $PSNR$ of other methods was just 26.05.

In the case of randomly missing, most algorithms failed to solve the completion task when the random missing ratio was 0.95 and 0.99. In particular,

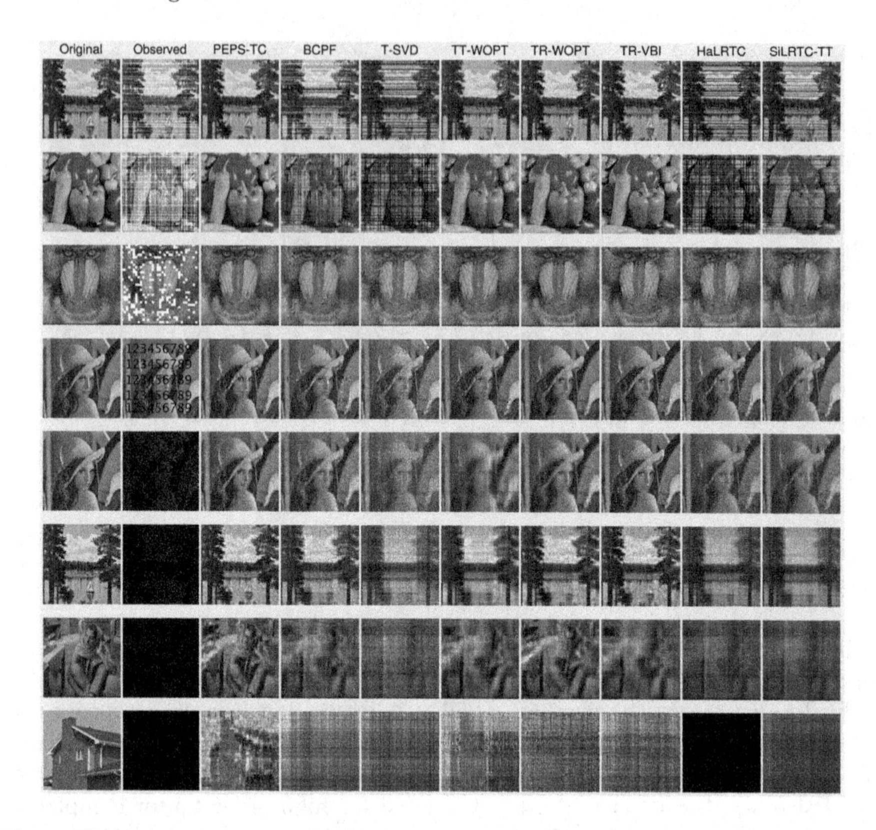


**Fig. 4.** Visual comparison results of completed images for eight algorithms under eight image missing situations. From left to right are the original images, missing images, and completed results of the eight algorithms, respectively. From top to bottom, the visual completion results for the eight algorithms under different missing patterns (that is, line missing, grid missing, block missing, text covering, 0.8 random missing, 0.9 random missing, 0.95 random missing, and 0.99 random missing).

when the missing ratio was 0.99, only the proposed algorithm obtained considerable results. As far as RSE, PSNR, and SSIM are concerned, the completed results of all algorithms decreased as the missing ratio increased, and the proposed algorithm outperformed other algorithms in most patterns except for the random missing with $mr = 0.8$, where TR-VBI provides the highest RSE and PSNR values ($RSE = 0.094, PSNR = 25.64$), however, the proposed algorithm performed best on SSIM ($SSIM = 0.741$). The reason for this exception is that TR-VBI easily catches the low-rank structure of the image for low missing ratios. However, as the missing ratio of the image gets higher and most of the information is lost, finding the low-rank structure of the image is very challenging, therefore, the performance of the algorithm drops quickly for $mr \geq 0.9$. This demonstrates the superiority of using PEPS-TC over the other algorithms in high missing ratio colour image completion problems.

**Table 1.** Performance comparison using RSE, PSNR, and SSIM for benchmark images.

| | | PEPS-TC | BCPF | T-SVD | TT-WOPT | TR-WOPT | TR-VBI | HaLRTC | SiLRTC-TT |
|---|---|---|---|---|---|---|---|---|---|
| Sailboat | RSE | **0.067** | 1.331 | 0.506 | 0.105 | 0.087 | 0.096 | 0.422 | 0.257 |
| with line | PSNR | **28.35** | 2.390 | 10.80 | 24.46 | 26.05 | 25.23 | 12.37 | 16.68 |
| missing | SSIM | **0.894** | 0.301 | 0.269 | 0.812 | 0.852 | 0.753 | 0.384 | 0.555 |
| Peppers | RSE | **0.085** | 0.448 | 0.604 | 0.150 | 0.112 | 0.089 | 0.554 | 0.279 |
| with grid | PSNR | **26.60** | 12.18 | 9.580 | 21.67 | 24.25 | 26.24 | 10.33 | 16.29 |
| missing | SSIM | **0.847** | 0.210 | 0.131 | 0.666 | 0.749 | 0.767 | 0.181 | 0.403 |
| Baboon | RSE | **0.096** | 0.167 | 0.330 | 0.110 | **0.096** | 0.148 | 0.108 | 0.102 |
| with block | PSNR | **25.75** | 20.89 | 15.00 | 24.54 | 25.74 | 21.95 | 24.67 | 25.18 |
| missing | SSIM | **0.862** | 0.562 | 0.521 | 0.851 | **0.862** | 0.666 | 0.849 | 0.857 |
| Lena | RSE | **0.067** | 0.124 | 0.314 | 0.093 | 0.069 | 0.081 | 0.081 | 0.074 |
| with text | PSNR | **28.62** | 23.23 | 15.17 | 25.77 | 28.29 | 26.98 | 26.98 | 27.70 |
| | SSIM | **0.885** | 0.700 | 0.386 | 0.830 | 0.868 | 0.773 | 0.846 | 0.869 |
| Lena | RSE | 0.097 | 0.131 | 0.288 | 0.164 | 0.124 | **0.094** | 0.130 | 0.128 |
| with 0.8 | PSNR | 25.41 | 22.79 | 15.94 | 20.83 | 23.26 | **25.64** | 22.84 | 22.98 |
| missing | SSIM | **0.741** | 0.588 | 0.254 | 0.554 | 0.633 | 0.684 | 0.646 | 0.682 |
| Sailboat | RSE | **0.159** | 0.211 | 0.330 | 0.220 | 0.187 | 0.170 | 0.236 | 0.239 |
| with 0.9 | PSNR | **20.82** | 18.38 | 14.51 | 18.02 | 19.43 | 20.27 | 17.43 | 17.31 |
| missing | SSIM | **0.562** | 0.397 | 0.220 | 0.443 | 0.479 | 0.501 | 0.422 | 0.462 |
| Barbara | RSE | **0.190** | 0.294 | 0.408 | 0.240 | 0.208 | 0.262 | 0.340 | 0.326 |
| with 0.95 | PSNR | **20.68** | 16.87 | 14.02 | 18.62 | 19.88 | 17.88 | 15.62 | 15.98 |
| missing | SSIM | **0.520** | 0.257 | 0.154 | 0.382 | 0.446 | 0.365 | 0.302 | 0.363 |
| House | RSE | **0.243** | 0.319 | 0.403 | 0.419 | 0.683 | 0.418 | 0.995 | 0.463 |
| with 0.99 | PSNR | **16.59** | 14.22 | 12.20 | 11.86 | 7.620 | 11.87 | 4.350 | 10.99 |
| missing | SSIM | **0.283** | 0.123 | 0.078 | 0.092 | 0.036 | 0.062 | 0.005 | 0.055 |

### 5.3 Video Completion

In this experiment, we considered a video-recovery task. We used the GunShot video as our dataset. It was downloaded from Youtube[1] with 85 frames, and each frame was considered as a $100 \times 260 \times 3$ image. Thus, the video was a 4th-order tensor of size $100 \times 260 \times 3 \times 85$. For ease of computation, we chose the first 50 frames and downsampled each frame to get a 4th-order tensor of size $50 \times 130 \times 3 \times 50$. Furthermore, the 4th-order tensor is reshaped to an 8th-order tensor of size $5 \times 5 \times 2 \times 13 \times 5 \times 3 \times 3 \times 50$ for better completion performance. Because only the third-order tensor is supported for T-SVD, The *RGB channels* mode is merged with the *time* mode to form a third-order tensor of size $50 \times 130 \times 150$.

Figure 5 show two frames selected from the recovery video for $mr = 0.95$. The RSE values for the entire video and RSE, PSNR, and SSIM values for selected frames are presented in Table 2. As shown in Fig. 5, T-SVD and HaLRTC are much less than other methods, which also can be confirmed by the data in Table 2 with T-SVD and HaLRTC having $RSE = 0.478$ and $RSE = 0.689$, respectively. What's more, the proposed one gives the best result of $RSE = 0.146$, whereas

[1] https://youtu.be/7y9apnbI6GA.

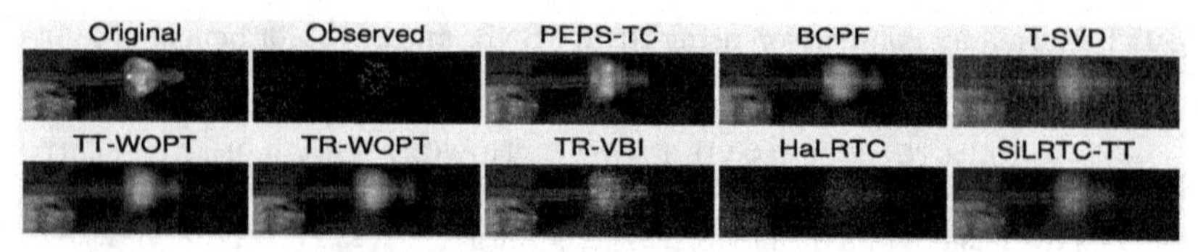


(a) Visual comparison of the 10th frame.

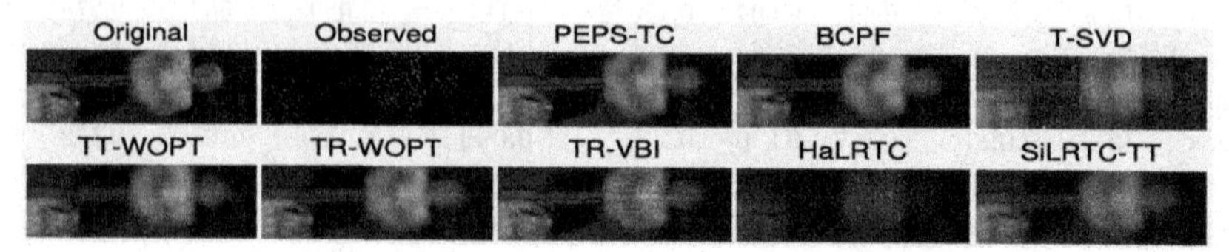


(b) Visual comparison of the 40th frame.

**Fig. 5.** Visual comparison of the 10th and 40th frames for GunShot.

**Table 2.** Performance comparison using RSE, PSNR and SSIM for GunShot.

| | | PEPS-TC | BCPF | T-SVD | TT-WOPT | TR-WOPT | TR-VBI | HaLRTC | SiLRTC-TT |
|---|---|---|---|---|---|---|---|---|---|
| Entire video | RSE | **0.146** | 0.191 | 0.478 | 0.196 | 0.173 | 0.289 | 0.689 | 0.243 |
| 10th Frame | RSE | **0.206** | 0.226 | 0.552 | 0.234 | 0.220 | 0.328 | 0.732 | 0.310 |
| | PSNR | **27.86** | 27.03 | 19.29 | 26.75 | 27.29 | 23.80 | 16.84 | 24.30 |
| | SSIM | **0.859** | 0.786 | 0.305 | 0.778 | 0.786 | 0.744 | 0.593 | 0.812 |
| 40th Frame | RSE | **0.114** | 0.159 | 0.414 | 0.173 | 0.147 | 0.249 | 0.657 | 0.204 |
| | PSNR | **29.11** | 26.26 | 17.93 | 25.52 | 26.93 | 22.34 | 13.92 | 24.08 |
| | SSIM | **0.893** | 0.821 | 0.375 | 0.776 | 0.825 | 0.759 | 0.510 | 0.814 |

the next best result of TR-WOPT had only $RSE = 0.173$. The RSE, PSNR, and SSIM values for selected frames also show the superiority of our method.

## 6 Conclusion

In this study, we present a TC algorithm based on the PEPS tensor network, which is a non-trivial extension of TC algorithms based on the TT and TR decompositions. The evaluation of various datasets, including synthetic data, colour images, and Gunshot video, demonstrates a significant improvement in PEPS completion compared with CP, Tucker, TT, and TR completion. The weakness of our work is that the PEPS ranks need to be predefined, where the optimal PEPS ranks are hard very to find. In the future, we will study how to determine the PEPS ranks automatically.

## References


1. Stoudenmire, E.M., Schwab, D.J.: Supervised learning with quantum-inspired tensor networks. arXiv preprint arXiv:1605.05775 (2017)
2. Cichocki, A., Mandic, D., Lathauwer, L.: Tensor decompositions for signal processing applications: from two-way to multiway component analysis. IEEE Signal Process. Mag. **32**(2), 145–163 (2015)
3. Yang, J., et al.: Remote sensing images destriping using unidirectional hybrid total variation and nonconvex low-rank regularization. J. Comput. Appl. Math. **363**, 124–144 (2020)
4. Ratre, A., Pankajakshan, V.: Tucker tensor decomposition-based tracking and Gaussian mixture model for anomaly localisation and detection in surveillance videos. IET Comput. Vision **12**(6), 933–940 (2018)
5. Carroll, J., Chang, J.: Analysis of individual differences in multidimensional scaling via an $N$-way generalization of "Eckart-Young decomposition." Psychometrika **35**(3), 283–319 (1970)
6. Tucker, L.: Some mathematical notes on three-mode factor analysis. Psychometrika **31**(3), 279–311 (1966)
7. Oseledets, I.: Tensor-train decomposition. SIAM J. Sci. Comput. **33**(5), 2295–2317 (2011)
8. Zhao, Q., et al.: Tensor ring decomposition. arXiv preprint arXiv:1606.05535 (2016)
9. Orús, R.: A practical introduction to tensor networks: matrix product states and projected entangled pair states. Ann. Phys. **349**(10), 117–158 (2014)
10. Liu, J., et al.: Tensor completion for estimating missing values in visual data. IEEE Trans. Pattern Anal. Mach. Intell. **35**(1), 208–220 (2013)
11. Bengua, J., et al.: Efficient tensor completion for color image and video recovery: low-rank tensor train. IEEE Trans. Image Process. **26**(5), 2466–2479 (2017)
12. Yuan, L., et al.: Tensor ring decomposition with rank minimization on latent space: an efficient approach for tensor completion. In: Proceedings of the AAAI Conference on Artificial Intelligence, pp. 9151–9158 (2019)
13. Zhang, Z., et al.: Novel methods for multilinear data completion and denoising based on tensor-SVD. In: Proceedings of the IEEE Conference on Computer Vision and Pattern Recognition, pp. 3842–3849 (2014)
14. Kilmer, M., et al.: Third-order tensors as operators on matrices: a theoretical and computational framework with applications in imaging. SIAM J. Matrix Anal. Appl. **34**(1), 148–172 (2013)
15. Acar, E., et al.: Scalable tensor factorizations for incomplete data. Chemom. Intell. Lab. Syst. **106**(1), 41–56 (2011)
16. Karatzoglou, A., et al.: Multiverse recommendation: n-dimensional tensor factorization for context-aware collaborative filtering. In: Proceedings of the Fourth ACM Conference on Recommender Systems, pp. 79–86 (2010)
17. Zheng, Y., et al. : Fully-connected tensor network decomposition and its application to higher-order tensor completion. In: Thirty-Fifth AAAI Conference on Artificial Intelligence, pp. 11071–11078 (2021)
18. Yuan, L., Zhao, Q., Cao, J.: Completion of high order tensor data with missing entries via tensor-train decomposition. In: The 24th international Conference on Neural Information Processing (ICONIP), pp. 222–229 (2017)
19. Yuan, L., et al.: Higher-dimension tensor completion via low-rank tensor ring decomposition. In: Asia-Pacific Signal and Information Processing Association Annual Summit and Conference (APSIPA ASC), pp. 1071–1076 (2018)

20. Wang, W., Aggarwal, V., Aeron, S.: Tensor completion by alternating minimization under the tensor train (TT) model. arXiv preprint arXiv:1609.05587 (2016)
21. Wang, W., Aggarwal, V., Aeron, S.: Efficient low rank tensor ring completion. In: IEEE International Conference on Computer Vision (ICCV), pp. 5697–5705 (2017)
22. Zhao, Q., Zhang, L., Cichocki, A.: Bayesian CP factorization of incomplete tensors with automatic rank determination. IEEE Trans. Pattern Anal. Mach. Intell. **37**(9), 1751–1763 (2015)
23. Long, Z., et al.: Bayesian low rank tensor ring for image recovery. IEEE Trans. Image Process. **30**(5), 3568–3580 (2021)
24. Poblano Homepage. https://github.com/sandialabs/poblano_toolbox

# Artificial Neural Networks for Downbeat Estimation and Varying Tempo Induction in Music Signals

Sarah Nadi(✉) and Jianguo Yao

Shanghai Jiao Tong University, Shanghai, China
{sarahnadi4,jianguo.yao}@sjtu.edu.cn

**Abstract.** The human capability to feel the musical downbeat and perceive the beats within a unit of time is intuitive and insensitive to the varying tempo of music audio signals. Yet, this mechanism is not straightforward for automated systems and requires further scientific depiction. As automatic music analysis is a crucial step for music structure discovery and music recommendation, downbeat tracking and varying tempo induction have been persistent challenges in the music information retrieval field. This paper introduces an architecture based on bidirectional long-short-term memory artificial recurrent neural networks to distinguish downbeat instants, supported by a dynamic Bayesian network to jointly infer the tempo estimation and correct the estimated downbeat locations according to the optimal solution. The proposed system outperforms existing algorithms with an achieved F1 score of 88.0%.

**Keywords:** Music Information Retrieval · Audio Signal Processing · Artificial Neural Networks · Dynamic Bayesian Networks

## 1 Introduction

Artificial intelligence is gaining tremendous momentum across diverse industries; it is one of the crucial promoters of industrial growth and a fundamental factor in evolving technological advancements such as natural language processing, automated speech recognition, visual recognition, and robotics. Even though a considerable effort was invested in such artificial intelligence implementations; researchers did not intensely carry out other significant areas, namely, the music information retrieval field. We, as humans, can naturally synchronize our brain's electrical activity and movements to a distinct musical beat. Yet, a machine's understanding and interpretation of the musical metrical structure, namely the patterns of beats and downbeats (Fig. 1), is still a crucial issue in music information retrieval. The beat is the most distinguishable unit division in music. It incorporates a sequence of systematic time instants that spontaneously appeal to human responses like head nodding or foot tapping when harmonizing with the regular succession of musical beats. The recurrence rate of these beats denotes


M. Tanveer et al. (Eds.): ICONIP 2022, CCIS 1793, pp. 39–51, 2023.
https://doi.org/10.1007/978-981-99-1645-0_4

the tempo of the music piece, which is measured in beats per minute (BPM). In music theory, the same number of beats are structured into groups defined as measures. The initial beat of every measure indicates the chiefly emphasized note is considered a downbeat. Downbeats are mainly employed to assist musicians and composers in studying and exploring music tracks. Specifying beats, downbeats and defining the precise tempo of music pieces opens new opportunities for all varieties of music application purposes, for instance, defining the genre of songs [23,36], analyzing the rhythmic structure and dance style [19,32], and identifying the similarity between songs [18,26]. Detected beats and downbeats can as well be used in music audio transcription [1,4], chord recognition [28,33–35], and music recommendation [25,31].

Manual identification of downbeat locations is a high-cost and time-consuming process. It can also differ from one annotator to another depending on their own perception and intuition [6,12]. The purpose of downbeat tracking and tempo induction systems is to automatically mark the time instants of all the beats and downbeats present in music audio signals, along with estimating the tempo. Thus, tackling down (beat) tracking using fully automated techniques, along with achieving a reasonable and optimal accuracy score, is a challenge of critical significance. The authors of the work described in [6] performed the beat detection process first, then identified downbeats by finding the spectral difference between band-limited beat synchronous analysis frames, considering that low-frequency bands are perceptually more critical. The system of Krebs et al. [21] utilized probabilistic state-space observation models to infer structural elements from rhythmic patterns in the audio data. Support Vector Machines (SVMs) methods were also adopted to identify downbeats in a semi-automated mechanism [20]. Meanwhile, Bruno et al. [8] adopted a fully automated system that employs a deterministic time-warping procedure based on convolutional neural networks learning from rhythmic patterns independently of tempo changes. In recent works on the subject, researchers started to focus on developing systems that could deal with various types of music. For example, the authors of [3] presented a source separation module that separates the non-percussive and percussive elements of the input audio, then performs beat and downbeat tracking independently before combining the results via a learnable fusion process. Despite research efforts, various approaches have restrictions. They either rely on hand-crafted beat annotations that are not available for the majority of music recordings [6,12], or can only be applied to limited music styles [21]. From our comprehension of the strengths and drawbacks of previous approaches, in this paper, we implement a fully automated intelligent system that can match spontaneous human foot tapping or head nodding ability while listening to musical pieces by detecting the beats/downbeats and tempo in a wide range of musical styles and genres. The system is based on bidirectional long-short-term memory artificial recurrent neural networks used to distinguish downbeat instants, followed by a dynamic Bayesian network to jointly infer the tempo estimation and correct the estimated downbeat positions according to the optimal global solution. The following is how the remaining of the paper is structured: In Sect. 2, we give a detailed explanation of the proposed architecture. Section 3 summarizes the experimental results and presents a performance comparison to other downbeat tracking systems. Finally, Sect. 4 is dedicated for conclusion and future work.

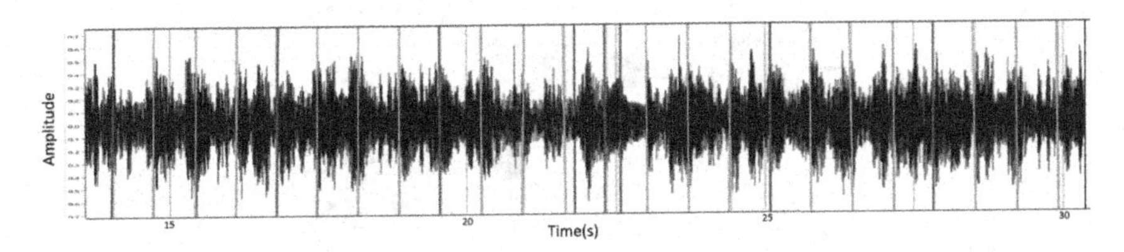


**Fig. 1.** Illustration of a typical downbeat annotation, displaying the beats (green lines) and downbeats (red lines) time instants. (Color figure online)

## 2 Methods

### 2.1 Recurrent Neural Networks

During the past decades, neural networks have been widely used for machine learning applications. A Multilayer perceptron (MLP) is a primitive approach to establishing a feed-forward neural network (FNN). MLP is composed of at least three layers with non-linear activation functions of neurons. The network output values resulting from the last hidden layer are assembled in the output nodes when the input data is forwarded through the hidden layers. A recurrent neural network (RNN) is formed if the hidden layers allow cyclic associations. RNNs are distinguished by their ability to memorize past values, although they practically encounter the limitation of the vanishing gradient or exponential expansion over time. In [10], an alternative approach named "Long short-term memory (LSTM)" is presented to tackle this issue.

As shown in Fig. 2, every LSTM cell owns a recurrent link with a weight value of 1.0 that enables the cell to function as memory. The information in the memory cell is managed by the input, output, and forget gates linked to other neurons. The usage of LSTM cells allows the network to access the preliminary input values. If future factors of the input data are additionally required to regulate the output, either an arranged time window of the input is added, or a suspension is made between the input data and the output targets. Even though both options have drawbacks, they either augment the input size or shift the input values and output targets from one to the other. Bidirectional recurrent neural networks (BRNN) represent a sophisticated resolution to this matter. It duplicates the hidden layers and feeds the input values in reverse temporal order. This process empowers the network with the ability to access prior input values simultaneously and envision future context. A bidirectional long-short-term memory (BLSTM) is constituted when bidirectional recurrent neural networks are jointly used with LSTM cells.

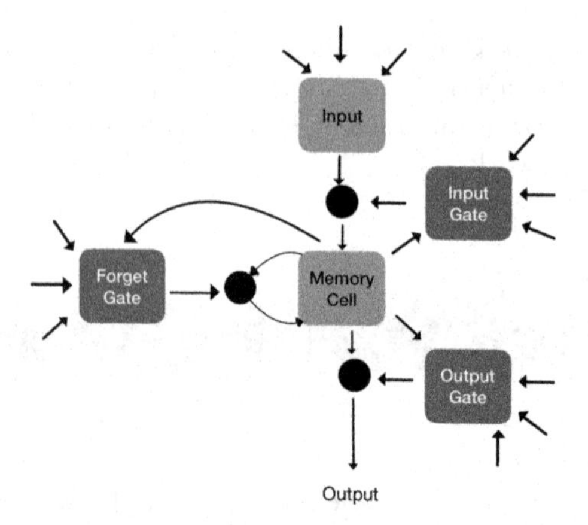


**Fig. 2.** Long Short-Term Memory (LSTM) block with memory cell.

BLSTM has the characteristic of modeling any kind of temporal factors in the input data.

## 2.2 Algorithm Description

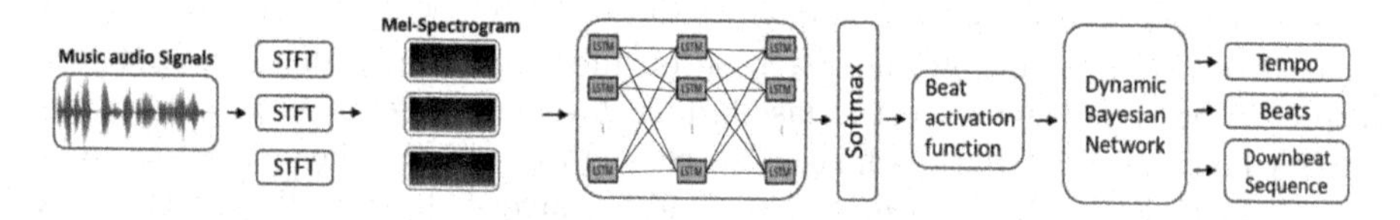


**Fig. 3.** Overview of the downbeat tracking and varying tempo induction algorithm workflow

The algorithm for downbeat tracking and tempo estimation in musical audio data is described in this section and illustrated in Fig. 3. First, the audio data signals are processed by three short-time Fourier transforms (STFTs) built-in parallel with different window widths. The Mel Spectrogram, which is a depiction for better viewing the various pressure strengths present in a sound, is the output of the first stage. The BLSTM artificial neural networks are fed the acquired magnitude spectrogram. Next, the SoftMax layer generates a beat activation function by modeling the beat and downbeat classes. Finally, as a post-processing phase, a dynamic Bayesian network is used in conjunction with the neural network to govern the optimal downbeat sequence and induct its tempo.

**Feature Extraction.** Before being transformed to a time-frequency representation with the short-time Fourier transform (STFT) [10], the audio signal is divided into intersecting frames and weighted with windows of 23.2 ms, 46.4 ms, and 92.8 ms in length (STFT window sizes of $N = 1024$, $N = 2048$ and $N = 4096$). Every two subsequent frames are separated by 10 milliseconds, which creates a frame rate of 100 frames per second (fps). We employ three diverse magnitude spectrograms with STFT to allow the network to collect exact characteristics both in time and frequency at a signal sample rate value of 44.1 kHz.

### Artificial Neural Network

*Network Topology.* The artificial neural network takes the logarithmic preprocessed signals in the form of Mel-spectrograms according to the window sizes of 23 ms, 46 ms, and 92 ms, respectively, along with their conforming difference of median. The network has three fully connected bidirectional recurrent layers. Every layer is composed of 25 LSTM units. The subsequent three-unit Softmax layer classifies every frame into a beat, downbeat, or non-beat classes by computing the probability of a frame $i$ defining a beat, downbeat or non-beat position. Lastly, it outputs 3 activation functions $b_i$, $d_i$ and $no_i$.

*Network Training.* We train the neural network on the datasets cited in Table 1. We divide the datasets into training and validation sets by forming eight non-conflicting but equivalently distributed subsets of identical size with 8-fold cross-validation. The weights and biases of the network are chosen arbitrarily within a range of $[-0.1, 0.1]$. The training is done with stochastic gradient descent to reduce the cross-entropy loss, adopting a learning rate of $10^{-5}$. To avoid overfitting, we endorse early stopping if no improvement in the validation set is perceived.

**Dynamic Bayesian Network.** In order to achieve a high level of responsiveness to varying tempo in music audio signals, we feed the neural network model's output into a dynamic Bayesian network (DBN) as observations for the simultaneous induction of downbeat sequence phase and tempo value. The DBN excels at handling unclear RNN observations and determining their global optimum state sequence. The state-space [22] is employed to simulate a complete bar with a random number of beats. We do not consent to meter alterations throughout a piece of music, and we have the ability to model distinct meters using separate state spaces. On the validation set, all DBN parameters are tweaked to optimize the obtained downbeat tracking performance.

*State Space.* A state $s(\psi, \pi, \omega)$ is a DBN state space composed of three hidden space variables:

- `Bar position:` $\psi \in \{ 1 .. \psi \}$
- `Tempo state:` $\pi \in \{ 1 .. \pi \}$
- `Time signature:` $\omega \in \{ 3/4 , 4/4 \}$

The states crossing downbeat points $\psi = 1$ and beat positions originate from the downbeat and beat states $D$ and $B$ respectively. The bar boundary of a tempo $\pi$ is relational to its conforming beat period $1/\ \pi$.

By allocating the tempo states logarithmically over the beat intervals, the state space size can be significantly minimized without influencing the performance.

*Transition Model.* The transition model $P(x_k|x_{k-1})$ is degraded into a distribution for the bar position $\psi$ and the tempo state $\pi$ hidden variables by:

$$P(x_k|x_{k-1}) = P(\psi_k|\psi_{k-1}, \pi_{k-1})...P(\pi_k|\pi_{k-1}) \tag{1}$$

The first factor is:

$$P(\psi_k|\psi_{k-1}, \pi_{k-1}) = \mathbb{J}_x \tag{2}$$

where $\mathbb{J}_x$ represents an indicator whose value is one if: $\psi_k = (\psi_{k-1} + \pi_{k-1} - 1)$ mod $(M+1)$ and zero otherwise. The bar position becomes cyclic with the modulo operator. Afterwards, the secondary factor $P(\pi_k|\pi_{k-1})$ is executed with respect to: $\pi_{min} \leq \pi_k \leq \pi_{max}$ The three tempo transitions performed by the pointer are shown in Eq. (3)

$$P(\pi_k|\pi_{k-1}) = \begin{cases} 1 - p_\pi, \pi_k = \pi_{k-1} \\ p_\pi/2, \pi_k = \pi_{k-1} + 1 \\ p_\pi/2, \pi_k = \pi_{k-1} - 1 \end{cases} \tag{3}$$

Else, $P(\pi_k|\pi_{k-1}) = 0$ .

*Observation Model.* The observation model stands for predicting both beats and downbeat locations. As the artificial neural network in the prior phase of the system architecture outputs the activation functions $b$ and $d$ that are in the range of [0, 1], the activations are subsequently transformed accordingly to fit the observation distributions of the dynamic Bayesian network. The conversion to the state-conditional observations is shown as follows:

$$P(o_k|s_{k-1}) = \begin{cases} d_k, s_k \in D \\ b_k, s_k \in B \\ n_k/(\theta_o - 1), otherwise \end{cases} \tag{4}$$

Noting that $B$ and $D$ are beats and downbeats states, respectively.

The observation $\theta_o \in [\frac{\psi}{\psi-1}, \psi]$ is a parameter that determines how much of the beat/downbeat interval is deemed beat/downbeat and non-beat positions within one beat/downbeat time-frame.

*Initial State Distribution and Inference.* Any previous information about the hidden states, like the meter and tempo distributions, can be integrated into the initial state distribution. We adopt a uniform distribution across all states for reasons of simplification and better generalization. As for the inference, we aim to acquire the sequence of hidden states that optimizes the posterior probability of the hidden states given the observations. To this end, the popular Viterbi Algorithm is used [30].

Finally, we get the set of time frames $k$ that were used to generate the beats and downbeats sequences. $\mathbb{B}$ and $\mathbb{D}$:

$$\mathbb{B} = \{k : s_k^* \in B\} \tag{5}$$

$$\mathbb{D} = \{k : s_k^* \in D\} \tag{6}$$

### 2.3 Evaluation Metrics and Procedure

To evaluate our system's performance, we utilize the standard metrics F1-measure which is reported in percent, in accordance with other downbeat tracking publications. The metric is calculated as:

$$F1 = \frac{2RP}{R+P} \tag{7}$$

With the recall $R$ representing the ratio of appropriately defined downbeats and the overall number of annotated downbeats in the specified window, and $P$ depicting the ratio of accurately detected downbeats within the window and all the estimated downbeats. Since annotations are occasionally absent or not always precise, we apply a tolerance window value of 70 ms and ignore the first 5 s and last 3 s of the music audio. Finally, in terms of fairness to systems in which the training was not performed on all datasets, we utilize the leave-one-dataset-out technique described in [24], in which we employ all datasets for training and exclude one holdout dataset for testing in each iteration. The network is trained with 90% of the training dataset and the hyper parameters are set with the remaining 10%.

### 2.4 Data Sets

We evaluate our system's performance on a variety of real world music audio recording datasets. (Table 1) While mostly focused on western music, the files span a broad range of genres and styles, including Samba, Cha Cha, Tango, Rumba, Slow Waltz, Viennese Waltz, Jive, Quickstep, as well as bass, drum, hardcore, jungle, Japanese and western English popular tunes. The files contain around 1600 audio files ranging from 30 s to 10 min in length, totaling about 45 h of music. Some datasets concentrate on a single musical genre or musician, while others cover the entire musical spectrum. Using many datasets creates a diversity in annotation along with a variety in content. We would note that tracking downbeats can be unclear, and different annotators might have different interpretations. As a result, it is common to use tolerance windows framing every downbeat location and consider the downbeats falling within these windows to be correctly annotated. In earlier assessments of downbeat tracking systems, tolerance windows were either established in absolute time, with a tolerance window of ±70 ms, or in relative time, with a tolerance window of ±20 ms. As for this matter, there is still no ultimate agreement on these windows' definition or their appropriate size.

**Table 1.** Datasets used for training and evaluation

| Dataset | Files | Length |
|---|---|---|
| Hainsworth [16] | 222 | 3 h 19 m |
| Beatles [5] | 180 | 8 h 09 m |
| Ballroom [15] | 685 | 5 h 57 m |
| RWC Popular [14] | 100 | 6 h 47 m |
| Rock [7] | 200 | 12 h 53 m |
| Robbie Williams [9] | 65 | 4 h 31 m |
| HJDB [17] | 235 | 3 h 19 m |

# 3 Results and Discussion

## 3.1 Ablation Study

**Impact of the DBN Stage.** To gain a better understanding of the systems behavior and evaluate the significance of the DBN stage, we implemented a baseline system ANN that employs the basic peak detection technique instead of DBN to report downbeat positions whenever the subsequent ANN activations surpass a certain threshold. A threshold value of 0.2 was used as it achieved optimal results on the validation set. Table 2 shows the empirical accuracy results of the proposed system ANNDBN and the baseline ANN across all datasets mentioned in Table 1. As can be observed, the ANN coupled with the DBN notably out-performs the ANN baseline. Remarkably, the performance of the ANNDBN varies between datasets. It typically outperforms on datasets covering Western music with regular-tempo, such as Ballroom, HJDB, Robbie Williams, and the Beatles. Whereas the performance is relatively lower on the rest of the datasets that include non-western music with inconsistent tempo and time signature.

**Influence of the Bar Length Selection.** The performance is moderately decreased if the DBN is restricted to bars with three or four beats exclusively. Highest performance is attained when both 4/4 and 3/4 time signatures within the music audio are simultaneously modeled. When bar lengths up to eight beats per bar are executed by the DBN, the performance marginally reduces while maintaining a fairly high accuracy (Table 3). This shows the framework's ability to automatically detect the right bar length.

**Table 2.** F1-measure mean of downbeat tracking across all data sets used.

| System | Ballroom | Beatles | Hainsworth | RWC popular | Rock | Robbie Williams | HJDB | Mean |
|---|---|---|---|---|---|---|---|---|
| ANN | 78.4 | 73.1 | 64.2 | 55.1 | 82.9 | 91.7 | 81.0 | 75.2 |
| ANNDBN | 96.8 | 89.8 | 68.9 | 85.1 | 85.3 | 94.9 | 95.2 | 88.0 |

**Table 3.** Bar lengths modelled by the DBN [beats per bar]

| [ Beats per Bar ] | F-measure |
|---|---|
| 3 | 0.480 |
| 4 | 0.742 |
| 3,4 | 0.880 |
| 3...5 | 0.879 |
| 3...6 | 0.854 |
| 3...7 | 0.840 |
| 3...8 | 0.820 |

### 3.2 Comparative Analysis

**Downbeat Tracking.** In this section, we present a comparative analysis where we assess our system, named in this paper ANNDBN, to the downbeat tracking systems of Sebastian Bock [2], Durand et al. [11], Fuentes et al. [13], Papadopoulos [27], Peeters et al. [29] and Krebs [22] (Table 4).

Both systems presented in [13] and [22] evaluated their works using a leave-one-dataset-out technique proposed in [21] using all of the aforementioned sets listed in Table 1. It is worth mentioning that a data augmentation approach was used for the training phase introduced in [2] to rise the network's exposure to a broader range of tempo, as well as beat and downbeat information. Besides, the authors treated the task as a multi-label classification where downbeat detection is processed as a distinct binary classification instead of jointly modeling beats and downbeats. As for the work presented by Durand [11], various sorts of features based upon rhythm, bass, melody and harmony content were fed to a convolutional neural network adopting a leave-one-dataset technique.

Performance results are illustrated in Table 4. The shown results demonstrate that the ANNDBN system outperforms all systems on Ballroom and Beatles datasets, with an inclusive improvement of 5.2% over the second top system. The adoption of a bidirectional long short-term memory artificial recurrent neural network for downbeat instants distinction supported by a dynamic Bayesian network for a joint inference of tempo estimation and correction of estimated downbeat locations appear to be proper in this circumstance. Conversely, in some other music datasets, such as the Hainsworth set, where estimating downbeats is more complex, the total boost in performance was substantially lower at 3.3% points than the best performing system.

One potential explanation is that our advanced feature extraction and model training, complemented with the DBN phase, performed well on excerpts when downbeat cues were more tough to obtain, and gave the system the ability to choose the appropriate bar length efficiently by itself. Unfortunately, it is still not achievable to accomplish a modular comparison due to the variety of the training and evaluation data-sets used in different works. Thus, we only assess the overall competitiveness of the end-to-end system and leave the extensible comparison for future works.

**Table 4.** F-measure results for published downbeat tracking systems.

| F-measure | Ballroom | Beatles | Hainsworth | RWC popular |
|---|---|---|---|---|
| ANNDBN | 96.8 | 89.8 | 68.9 | 85.1 |
| Sebastian Böck [2] | 91.6 | 37.0 | 72.2 | - |
| Durand et al. [11] | 79.7 | 84.7 | 66.4 | 87.6 |
| Fuentes et al. [13] | 83.0 | 86.0 | 67.0 | 91.0 |
| Papadopoulos [27] | 50.0 | 65.3 | 44.2 | 75.8 |
| Peeters et al. [29] | 45.5 | 53.3 | 42.3 | 69.8 |
| Krebs [22] | 52.5 | 72.1 | 51.7 | 72.1 |

**Tempo Induction.** As previously stated, the dynamic Bayesian network was used with the aim of modeling measures with the exact number of beats within a bar for optimal performance. The presented ANNDBN system is set up to function with 3/4 and 4/4 time signatures by default. Although, considering that dynamic Bayesian network parameters are not learned, parameters can be altered during the runtime to adapt to any time signature and varying tempo. We conducted an experiment in which we confined the DBN to modeling just bars fixed to a length value of 3 or 4 beats, or bar lengths of up to 8 beats, to study the system's capacity to automatically pick which bar length to choose. As a result, the f-measure of the downbeat tracking achieved high performance in a wide varying temporal range.

## 4 Conclusion and Future Work

In this paper, we used a bidirectional long short-term memory artificial recurrent neural network in aggregation with a dynamic Bayesian network to provide a method for detecting downbeats positions within music audio signals and inducting the tempo. The artificial neural network based on BLSTM is in charge of modeling the musical piece's metrical structure at numerous interconnected levels and classifying each audio frame as a beat or downbeat. By simultaneously inferring the sequence's meter, tempo, and phase, the dynamic

Bayesian network then post-processes the artificial neural network's probability functions obtained from the Softmax layer to align the beats and downbeats to the overall optimal solution. The system demonstrates outstanding performance in downbeat tracking in a wide range of musical genres and styles. It attains this end by learning essential elements directly from audio signals rather than relying on hand-crafted features such as rhythmic patterns or harmonic changes. Despite the relatively good performance achieved on the datasets in hand, it is not assured that the proposed system has decoded the means to unambiguously infer the tempo, beat, and downbeat in exceptionally complex musical compositions. Thus, future works may focus on reformulating inference mechanisms adopted to adjust the last system's output. Upcoming works may also address the issue of performing varying time signatures present in music audio signals to achieve more optimal results on challenging music data.

## References

1. Benetos, E., Dixon, S., Duan, Z., Ewert, S.: Automatic music transcription: an overview. IEEE Signal Process. Mag. **36**(1), 20–30 (2018)
2. Böck, S., Davies, M.E.: Deconstruct, analyse, reconstruct: how to improve tempo, beat, and downbeat estimation. In: Proceedings of the 21st International Society for Music Information Retrieval Conference (ISMIR), Montreal, QC, Canada, pp. 12–16 (2020)
3. Chiu, C.Y., Su, A.W.Y., Yang, Y.H.: Drum-aware ensemble architecture for improved joint musical beat and downbeat tracking. IEEE Signal Process. Lett. **28**, 1100–1104 (2021)
4. Cogliati, A., Duan, Z., Wohlberg, B.: Context-dependent piano music transcription with convolutional sparse coding. IEEE/ACM Trans. Audio Speech Lang. Process. **24**(12), 2218–2230 (2016)
5. Davies, M.E., Degara, N., Plumbley, M.D.: Evaluation methods for musical audio beat tracking algorithms. Queen Mary University of London, Centre for Digital Music, Technical report C4DM-TR-09-06 (2009)
6. Davies, M.E., Plumbley, M.D.: A spectral difference approach to downbeat extraction in musical audio. In: 2006 14th European Signal Processing Conference, pp. 1–4. IEEE (2006)
7. De Clercq, T., Temperley, D.: A corpus analysis of rock harmony. Pop. Music **30**(1), 47–70 (2011)
8. Di Giorgi, B., Mauch, M., Levy, M.: Downbeat tracking with tempo-invariant convolutional neural networks. arXiv preprint arXiv:2102.02282 (2021)
9. Di Giorgi, B., Zanoni, M., Sarti, A., Tubaro, S.: Automatic chord recognition based on the probabilistic modeling of diatonic modal harmony. In: nDS'13; Proceedings of the 8th International Workshop on Multidimensional Systems, pp. 1–6. VDE (2013)
10. Durak, L., Arikan, O.: Short-time fourier transform: two fundamental properties and an optimal implementation. IEEE Trans. Signal Process. **51**(5), 1231–1242 (2003)
11. Durand, S., Bello, J.P., David, B., Richard, G.: Robust downbeat tracking using an ensemble of convolutional networks. IEEE/ACM Trans. Audio Speech Lang. Process. **25**(1), 76–89 (2016)

12. Durand, S., David, B., Richard, G.: Enhancing downbeat detection when facing different music styles. In: 2014 IEEE International Conference on Acoustics, Speech and Signal Processing (ICASSP), pp. 3132–3136. IEEE (2014)
13. Fuentes, M., McFee, B., Crayencour, H., Essid, S., Bello, J.: Analysis of common design choices in deep learning systems for downbeat tracking. In: The 19th International Society for Music Information Retrieval Conference (2018)
14. Goto, M., Hashiguchi, H., Nishimura, T., Oka, R.: Rwc music database: popular, classical and jazz music databases. In: Ismir, vol. 2, pp. 287–288 (2002)
15. Gouyon, F., Klapuri, A., Dixon, S., Alonso, M., Tzanetakis, G., Uhle, C., Cano, P.: An experimental comparison of audio tempo induction algorithms. IEEE Trans. Audio Speech Lang. Process. **14**(5), 1832–1844 (2006)
16. Hainsworth, S.W., Macleod, M.D.: Particle filtering applied to musical tempo tracking. EURASIP J. Adv. Sig. Process. **2004**(15), 1–11 (2004)
17. Hockman, J., Davies, M.E., Fujinaga, I.: One in the jungle: Downbeat detection in hardcore, jungle, and drum and bass. In: ISMIR, pp. 169–174 (2012)
18. Holzapfel, A., Stylianou, Y.: Rhythmic similarity of music based on dynamic periodicity warping. In: 2008 IEEE International Conference on Acoustics, Speech and Signal Processing, pp. 2217–2220. IEEE (2008)
19. Huang, Y.S., Yang, Y.H.: Pop music transformer: Beat-based modeling and generation of expressive pop piano compositions. In: Proceedings of the 28th ACM International Conference on Multimedia, pp. 1180–1188 (2020)
20. Jehan, T.: Downbeat prediction by listening and learning. In: IEEE Workshop on Applications of Signal Processing to Audio and Acoustics, 2005, pp. 267–270. IEEE (2005)
21. Krebs, F., Böck, S., Widmer, G.: Rhythmic pattern modeling for beat and downbeat tracking in musical audio. In: Ismir, pp. 227–232. Citeseer (2013)
22. Krebs, F., Böck, S., Widmer, G.: An efficient state-space model for joint tempo and meter tracking. In: ISMIR, pp. 72–78 (2015)
23. Lidy, T., Rauber, A.: Evaluation of feature extractors and psycho-acoustic transformations for music genre classification. In: ISMIR, pp. 34–41 (2005)
24. Livshin, A., Rodex, X.: The importance of cross database evaluation in sound classification. In: ISMIR 2003, p. 1 (2003)
25. Logan, B.: Music recommendation from song sets. In: ISMIR, pp. 425–428 (2004)
26. Moritz, M., Heard, M., Kim, H.W., Lee, Y.S.: Invariance of edit-distance to tempo in rhythm similarity. Psychology of Music p. 0305735620971030 (2020)
27. Papadopoulos, H., Peeters, G.: Joint estimation of chords and downbeats from an audio signal. IEEE Trans. Audio Speech Lang. Process. **19**(1), 138–152 (2010)
28. Park, J., Choi, K., Jeon, S., Kim, D., Park, J.: A bi-directional transformer for musical chord recognition. arXiv preprint arXiv:1907.02698 (2019)
29. Peeters, G., Papadopoulos, H.: Simultaneous beat and downbeat-tracking using a probabilistic framework: theory and large-scale evaluation. IEEE Trans. Audio Speech Lang. Process. **19**(6), 1754–1769 (2010)
30. Rabiner, L.R.: A tutorial on hidden Markov models and selected applications in speech recognition. Proc. IEEE **77**(2), 257–286 (1989)
31. Schedl, M.: Deep learning in music recommendation systems. Frontiers in Applied Mathematics and Statistics **5**, 44 (2019)
32. Schuller, B., Eyben, F., Rigoll, G.: Tango or waltz?: putting ballroom dance style into tempo detection. EURASIP J. Audio Speech Music Process. **2008**, 1–12 (2008)
33. Sigtia, S., Boulanger-Lewandowski, N., Dixon, S.: Audio chord recognition with a hybrid recurrent neural network. In: ISMIR, pp. 127–133 (2015)

34. Ullrich, K., Schlüter, J., Grill, T.: Boundary detection in music structure analysis using convolutional neural networks. In: ISMIR, pp. 417–422 (2014)
35. Wang, J.C., Smith, J.B., Chen, J., Song, X., Wang, Y.: Supervised chorus detection for popular music using convolutional neural network and multi-task learning. In: ICASSP 2021–2021 IEEE International Conference on Acoustics, Speech and Signal Processing (ICASSP), pp. 566–570. IEEE (2021)
36. Yu, Y., Luo, S., Liu, S., Qiao, H., Liu, Y., Feng, L.: Deep attention based music genre classification. Neurocomputing **372**, 84–91 (2020)

# FedSpam: Privacy Preserving SMS Spam Prediction

Jiten Sidhpura(✉), Parshwa Shah, Rudresh Veerkhare, and Anand Godbole

Sardar Patel Institute of Technology, Mumbai, India
{jiten.sidhpura,parshwa.shah,
rudresh.veerkhare,anand_godbole}@spit.ac.in

**Abstract.** SMSes are a great way to communicate via short text. But, some people take advantage of this service by spamming innocent people. Deep learning models need more SMS data to be adaptive and accurate. But, SMS data being sensitive, should not leave the device premises. Therefore, we have proposed a federated learning approach in our research study. Initially, a distilBERT model having validation accuracy of 98% is transported to mobile clients. Mobile clients train this local model via the SMSes received and send their local model weights to server for aggregation. The process is done iteratively making the model robust and resistant to latest spam techniques. Model prediction analysis is done at server side using global model to check which words in message influence spam and ham. On-device training experiment is conducted on a client and it is observed that the losses of the global model converge after every iteration.

**Keywords:** Spam Detection · Federated Learning · Distil-BERT · SHAP Analysis · Android

## 1 Introduction

Short Message Service (SMS) is one of the popular modes of communication. This service is provided by telecom operators and used by people and organizations. These organizations can be banks, governments, companies, startups, etc. Sensitive details like authentication OTPs, bank balance, verification codes, and booking details are shared via SMS which becomes a loophole in the system. Spammers can exploit this vulnerability by spamming and trapping innocent users using cheap tricks. Novice users blindly believe in the fake and lucrative rewards and benefits given by spammers and fall into the trap.

According to [1], there are more than 6 billion smartphone users currently, and will be 7 billion by 2025. This will allow spammers to bulk SMS spam on multiple smartphones with the hope that significant people get fooled. Spammers usually attract users by giving them fake rewards, lottery results, giveaways, and promises of some free and affordable services. They ask users for PINs and

J. Sidhpura, P. Shah, R. Veerkhare—Contributed equally to this work.


M. Tanveer et al. (Eds.): ICONIP 2022, CCIS 1793, pp. 52–63, 2023.
https://doi.org/10.1007/978-981-99-1645-0_5

passwords by gaining their trust. They also send hyperlinks that redirect them to some phishing page or it downloads some malware or spyware on the phone. In this way, they get unauthorized access to users' sensitive data. Users ignore these threats due to a lack of awareness and repent later. Thus, there is a clear need for an intelligent automated deep learning-based spam classifier that alerts mobile users about potential spam and ham messages.

Keeping in mind, the SMS data confidentiality, it will be more vulnerable if SMS data is sent to a centralized server to get spam classification. Also, bombarding the server with requests for each SMS by billion users at a time is not feasible as the server may crash. Deep learning models become robust if they are trained on large data. Smart spammers have multiple tricks to spam users. So, a system needs to be created that is adaptive to new spam techniques and at the same time classifies SMS spam on-device, thus respecting the privacy of users' sensitive SMS data. Federated Learning i.e. on-device privacy-preserving learning approach can solve all problems of SMS data privacy, server latency, and model robustness and adaptability.

The objective of this research is to apply federated learning to classify SMS spam messages. We have used a federated learning approach provided by the Flower framework to classify SMS spam on-device. The distilBERT deep learning model is used to classify whether the SMS is spam or ham. Initially, a global model is sent to each mobile device from a centralized server. This global model gets trained on the local mobile's SMS data on-device so every device will have different model weights. After these local models get trained, they will send their model weights and not the SMS data to a centralized server periodically at a certain interval. The centralized server will aggregate the weights received from each local model and send the final aggregated global model to each device. This happens iteratively and the model becomes more and more adaptive and robust and users' privacy is also not breached. Model prediction analysis is also done on the server-side to determine keywords in spam SMSes.

The structure of the paper is as follows: Sect. 2 describes the literature survey, where we discussed existing approaches. In Sect. 3, we have proposed our methodology. In Sect. 4 we have explained our experiments. In Sect. 5, we have presented our results. Finally, in Sect. 6, we conclude our paper.

## 2 Literature Survey

Sandhya Mishra et al. [2] proposed a Smishing (SMS Phishing) detection system and evaluated it on real time datasets and they also performed a case study on Paytm smishing scam. Their proposed system consisted of Domain Checking Phase and SMS Classification Phase. Only the SMS containing URLs are fed into the domain checking phase where they verify the authenticity of the URL. Features such as domain name and query parameters are extracted and searched on Google. If the domain name is in top 5 results they marked them as legitimate else they carried second-level domain checking. In second-level domain checking, source code of a website along with all URLS in that source code are also extracted. Domain names of all these URLS are compared and if they belong to the same domain then they are marked legitimate else it is passed to SMS Classification phase. In the final phase misspelled words, leet words, symbols, special characters, and smishing keywords are extracted. Leet words are the words in which some characters of the genuine word are replaced by digits and symbols to make it appear like a genuine word. Finally a neural network and classifications algorithms such as Naive Bayes, Decision Tree and Random Forests were used for the final prediction. Neural network gave the best performance among all and gave a 97.93% accuracy score.

M.Rubin Julis et al. [3] followed the Natural Language Processing (NLP) approaches for spam classification. They used the SMS Spam Corpus v.0.1 to apply their proposed methodology. They applied stemming on their dataset to change over the similar words to their base word format so that the number of unique words does not become too large. Eg: works, working would be converted to work. Spam SMS have the same looking words (homoglyphs) to trick the user to believe that it is a legitimate message. They converted these words into their genuine word format for better insights. Various classical machine learning algorithms such as Logistic Regression, Naive Bayes Classifier, K Nearest Neighbours, Support Vector Machines and Decision Tree Classifier were used. To compare the efficiency of models they also computed the prediction time. Support Vector Machine got the best accuracy score (98%) among all the models. Naive Bayes was fastest in terms of prediction and since it gave nearly the same results as the Support vector machine it was most effective.

Wael Hassan Gomaa et al. [4] proposed the impact of deep learning on SMS Spam filtering. They used the dataset from the UCI repository with 5574 English language emails containing 4827 non-spam and 747 spam emails. The dataset was converted to semantic word vectors using the Glove model. They used 6 machine learning classifiers and 8 deep learning techniques. The maximum accuracy they got from machine learning classifiers was 96.86% given by Gradient Boosted Trees. In deep learning, the maximum accuracy was given by RDML which is 99.26% accuracy.

X. Liu et al. [5] have compared various machine learning approaches for SMS spam detection. Neural network-based approaches performed better compared to classical machine learning algorithms. Further, the authors proposed a seq2seq transformers multi-head attention model for spam classification and it has outperformed other methods. To encode textual data, authors have used Glove embeddings for neural networks and TF-IDF representation for machine learning algorithms. SMS Spam Collection v1 and UtkMl's Twitter datasets are used for training and evaluation purposes.

Sergio Rojas-Galeano [6] aims to investigate the possible advantages of language models that are sensitive to the context of words. First, they performed a baseline analysis of a large database of spam messages using Google's BERT. Then, build a thesaurus of the vocabulary that contained these messages. The author has performed a Mad-lib attack experiment where he modified each message with different rates of substitution. The classic models maintained a 93%

balanced accuracy (BA) while the BERT model scored a 96%. The performance of the TFIDF and BoW encoders was also affected by the attacks. These results indicate that BERT models could provide a better alternative to address these types of attacks.

H. Brendan McMahan et al. [11] introduced a distributed machine learning technique that keeps the data for the model training on the edge device such as mobile and develops a global model combined with the knowledge gained from each client by training models at edge devices. This new methodology is termed Federated Learning by the authors. The decentralized approach respects the privacy of the user-sensitive data and sends only the model weights to a centralized server through the network. They introduced the FederatedAveraging algorithm to combine the results of the local models gained after applying the Stochastic Gradient Descent (SGD) algorithm to the local models of all the clients. This algorithm executes on the server-side and performs model weights averaging. Their algorithm reduces the communication costs for transferring weights from clients to the centralized server by introducing more independent clients and performing extra computation on devices. Mobiles handling these computations is feasible since training dataset size is generally small, and mobile processors are modern and advanced. Later they conducted two experiments, one image classification task to identify the images that will be viewed frequently. The second task was to develop language models for predicting the next word, complete responses to messages. They achieved a test set accuracy of 99% for the image classification task with the help of the CNN model on the MNIST dataset with 1, 10, 20, 50, and 100 clients. In case of the language models, they used the LSTM model on a dataset created from The Complete Works of William Shakespeare with a total of 1146 clients and achieved a score of 54%. These results show the performance of federated learning and how one can use it to solve other data-sensitive problems with it.

Federated Learning has shown promising results in machine learning where the private-sensitive data of the user comes into the picture. However, implementing federated learning systems has many challenges due to scalability. Due to these difficulties involved while developing federated systems, researchers have to rely on simulation results. Daniel J. Beutel et al. [15] introduced Flower, a federated learning open source framework that helps researchers test their federated systems in the real world on edge devices such as mobile. Their framework supports many deep learning frameworks and supports model training on many heterogeneous devices. The framework was capable to support a total of 15M clients with the help of a couple of GPUs only.

Andrew Hard et al. [13] used a federated learning mechanism to train a recurrent neural network to train a federated learning system to predict the next word for a sentence from Google's keyboard application, GBoard. The authors of the paper chose federated learning because most of the datasets available generally have a different distribution than chatting between users. Their model was trained using Tensorflow Framework and used Tensorflow Lite for prediction on mobile devices. Client devices needed to have a vocabulary of 10000 words, and after compression, a model of size 1.4 MB was placed in the mobile phones. The GBoard application gave three suggestions for the next word, so they used Top-3 recall and Top-1 recall as their evaluation metrics. Results of the federated learning approach were better than the server approach for real chat data.

## 3 Methodology

### 3.1 Dataset

We have merged three datasets which amount to 39792 rows in our proposed methodology. In total, we have 21768 ham messages and 18024 spam messages. In each dataset, we have two columns i.e. message body and label whether the message is spam or ham.

**Enron Spam Dataset:** Enron Spam dataset is a dataset created by [8]. There are a total of 33716 records and 4 columns in the dataset. The fields in this email dataset are the subject, message, spam/ham label, and date of the email. We have used message and label fields for our research and also discarded 371 rows as they were unlabelled. Finally, we have 33345 messages, out of which 16852 are spam and 16493 are ham.

**UCI SMS Spam Collection Data Set:** UCI SMS Spam collection dataset is a publicly available dataset made by [9] containing 5572 messages and 2 columns. The 2 columns are message and label. There are a total of 747 spam messages and 4825 ham messages.

**British English SMS Corpora:** British English SMS Corpora is a publicly available dataset made by [10] containing 875 messages with each message with a label i.e. spam or ham. It contains 425 spam messages and 450 ham messages.

Data preprocessing like removing special characters and punctuations are done on each dataset and then merged to create a final dataset.

### 3.2 DistilBERT

The DistilBERT [7] model is a distilled version of BERT. DistilBERT is a small, fast, cheap, and light Transformer model trained by distilling BERT base. It has 40% fewer parameters than bert-base-uncased and runs 60% faster while preserving over 95% of BERT's performances as measured on the GLUE language understanding benchmark. This Model is specifically calibrated for edge devices or under constrained computation conditions. Given this, DistilBERT is a suitable candidate for federated SPAM detection.

### 3.3 Federated Learning

In machine learning systems data is stored in a centralized format to train and improve machine learning models. Also, the input data is sent from devices over the network to compute inference. Sometimes the input data for the model is sensitive to the user, and in such scenarios, we can use Federated Learning [11]. Federated Learning is a modern distributed machine learning technique that utilizes sensitive data of the user present on mobile devices to train machine learning models directly on mobile phones. Clients then send the model weights and not the data over the network to a server. With the help of the FedAvg [11] algorithm, a weighted average of all the model parameters from various clients gets computed. Google has applied this technique in Google Assistant [12] and Google keyboard (Gboard) [13]. We have used this same Federated Learning technology to predict whether SMS received by the user is spam or ham.

Each part of our Federated Machine Learning system is described in following subsections:

**Initial Global Model Training:** For initial global model training, we have fine-tuned the DistilBERT model for our task of SMS Spam detection. The input message is processed using the DistilBERT tokenizer and converted into input_ids and attention_masks. Then the classifier network as shown in Fig. 1 after DistilBERT consists of 3 Dense Layers each having a Dropout with a rate of 0.2. 90% of the combined data is used for training and rest 10% for validation.

**Model Training and Inference on Edge Devices** - For training the Deep Learning model on an Edge device like android phones, our deep learning model has to be exported in multiple parts.

**Base** - This is the fine-tuned DistilBERT model, which would not get trained on the user's device, it will be only used to extract bottleneck features, and these features are used as an input to the head.

**Head** - This is the classifier part of the model, which comes after the base model. The Head will get trained on the user's device.

**Scheduling of on Device Model Training:** Training model on mobile devices is a resource-intensive task, and the computing power of mobiles is also limited. We should not train a model when the device is operated to avoid a bad user experience, has a low battery, or has no internet connection. Due to these reasons, we need to train our model after the above constraints are followed, and to accomplish it we have used the WorkManager API. We have used this API to schedule training of models on recent SMS data available on the user's device once every 15 days.

**Federated Cycle:** The server will wait for a specific number of clients to start the federated learning iterations. Once the server has connections from the required number of clients, the local model training on clients initiates, and

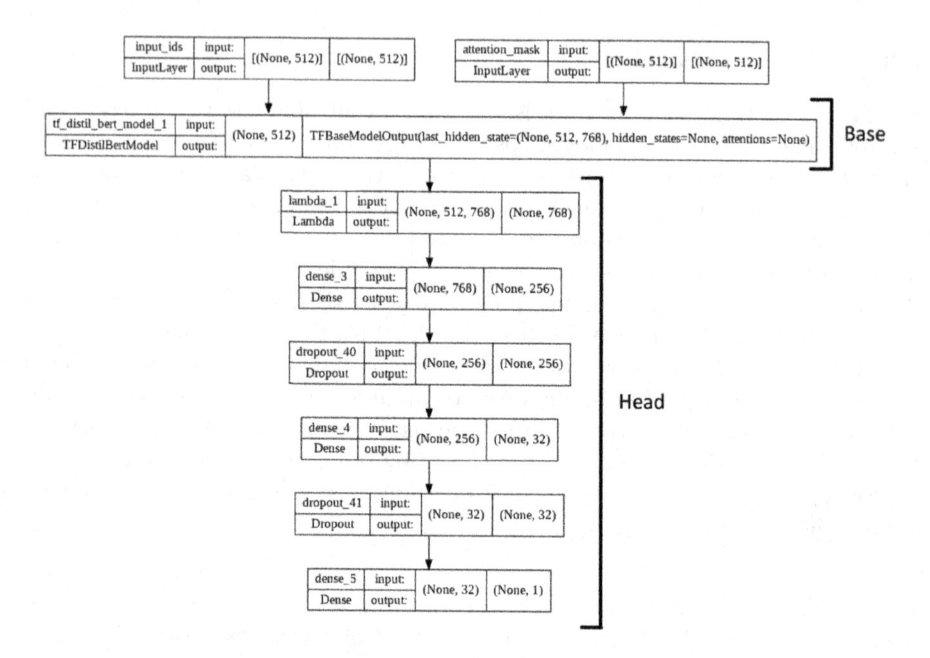


**Fig. 1.** Deep Learning Model Architecture

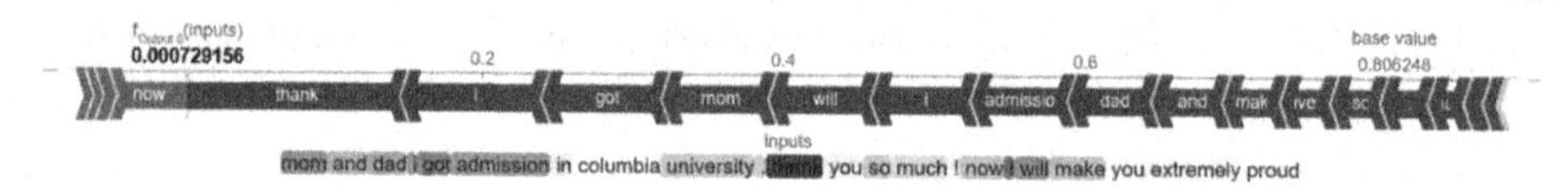


**Fig. 2.** Shap Text Plot for a Ham Message (Color figure online)

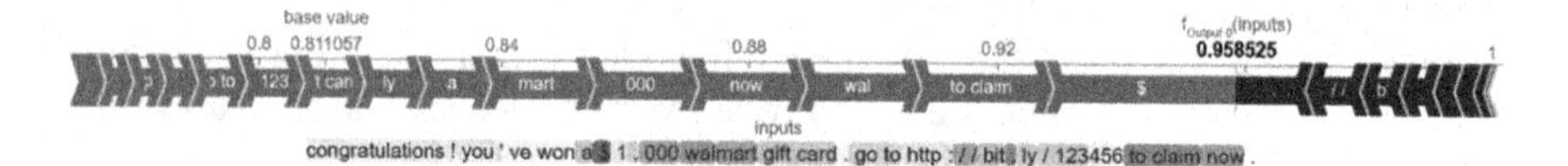


**Fig. 3.** Shap Text Plot for a Spam Message (Color figure online)

the weights are sent securely to the server. Later a global model is created from an aggregation of local model weights. This global model has knowledge of all the clients and is sent back to all the clients. In this way, every client gets the knowledge from all other clients without losing their privacy. We carry out analysis of the global model to understand its performance. This analysis is explained in detail in Sect. 4.2.

### 3.4 System Flow

The FedSpam system flow is demonstrated in Fig. 4. The steps mentioned in the flow diagram are explained below:

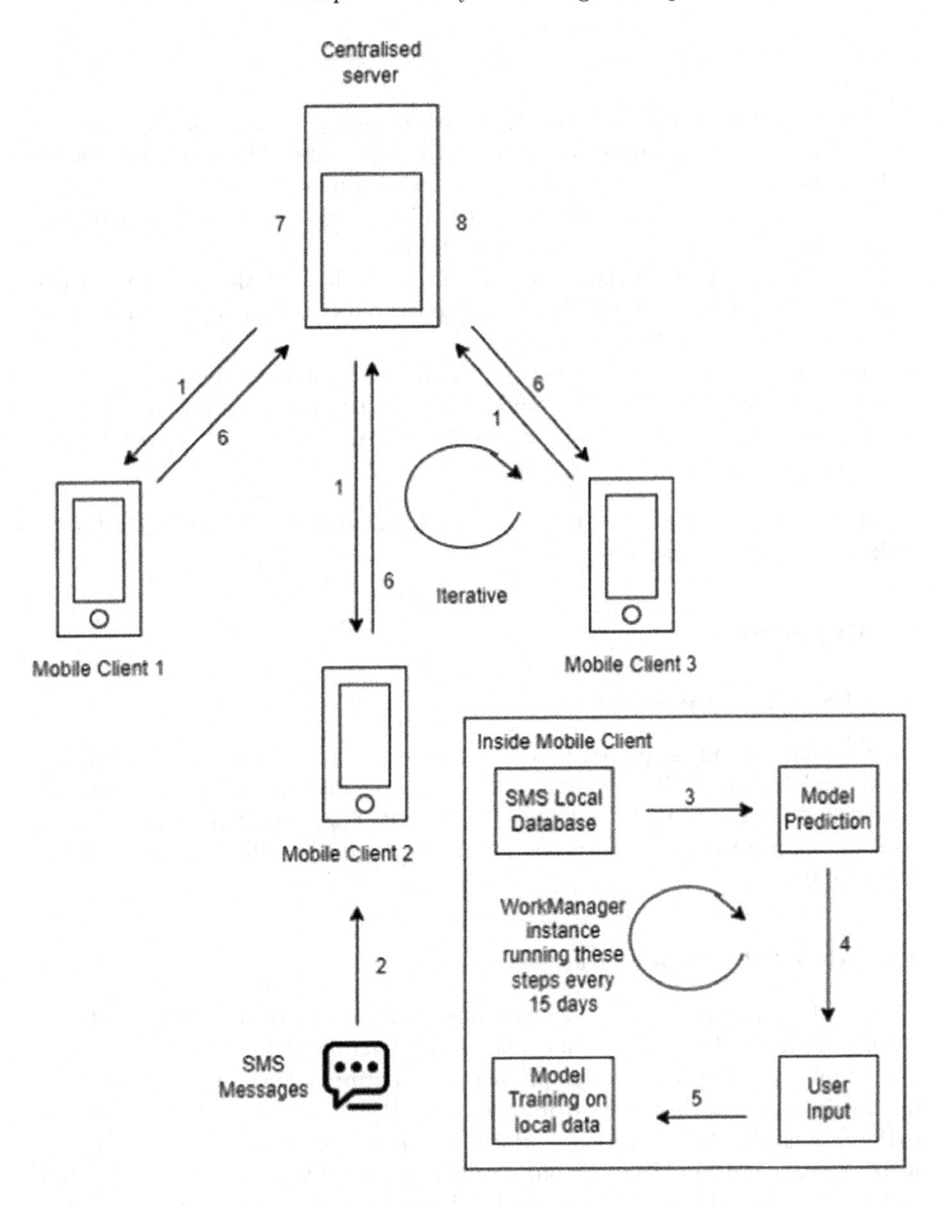


**Fig. 4.** FedSpam System Flow Diagram

1. Initially, the centralized server will send the global model to all local mobile clients.
2. Mobile clients receive SMS messages from companies, operators, friends, etc.
3. All SMSes received during the sprint of the last 15 days stored in the mobile's local database are fetched and sent to the local model.
4. The local model will predict whether the messages are spam or ham and assign a label and store it in a local database.
5. Now, the user is given the option to rectify the label if the local model predicted wrong and writes the final label to the database. Then, the local model will train on these SMS data.
6. The local model weights are sent to the server for aggregation.
7. Secure Aggregation of all local model weights to make a global model.
8. Model Prediction Analysis of the global model to check which keywords influence messages towards spam or ham.

All these federated learning steps are scheduled periodically every 15 days to make model robust and adaptive.

# 4 Experiment

## 4.1 Experimental Setup

For training the initial global model, we used Google Colab with 12.69GB of RAM, and 107GB of ROM, having TPU runtime. Packages used are tensorflow 2.8.0, transformers 4.17.0, and flwr [15] 0.18.0. For federated learning setup, multiple android devices are used, all having android SDK version $\geq$ 28 and RAM $\geq$ 4GB.

## 4.2 Model Prediction Analysis

Machine Learning these days plays an important role for many businesses today. Businesses not only require a machine learning model with great accuracy but also require justification for the prediction. The importance of the model prediction justification is equal to its accuracy. We have used SHapley Additive exPlanations (SHAP) [14] for the global model to see which words or group of words present in the SMS made our model predict either the spam or ham class. In both Figs. 2 and 3, words that are marked with red color push the prediction towards class 1 (spam), and those marked with blue push the prediction towards class 0 (ham). From Fig. 2, it is clear that many words are blue, and their contributions towards prediction are much heavier than the few words marked in red. In the case of Fig. 3, many words are red, and their contributions are more dominant than those marked as blue.

### 4.3 On Device Resources Analysis

We used Xiaomi's Redmi Note 5 Pro (launched back in February 2018) to understand the CPU and RAM consumption of our federated learning mobile app. Since we are training machine learning models on mobile phones, it becomes very important to analyze the CPU usage and RAM consumption to understand the impact of the application on mobile devices. We have used Android Studio's CPU Profiler to monitor the CPU and RAM consumption. After experimenting, we found that the CPU usage on average is around 12%, and in some cases, it can rise to 25% because of data loading and model training operations. Similarly, we found RAM consumption to be around 400 MB. With these requirements also, the application was working smoothly and did not cause any performance issues on the device. Due to the advancements in mobile processors, our proposed application will become more and more efficient on modern mobile devices.

### 4.4 Evaluation Metrics

We have used Categorical Cross Entropy mentioned in Eq. 1 as the loss function for this classification problem

$$-\sum_{1}^{n} y_i . \log \widehat{y_i} \tag{1}$$

where n is the output size.

$$\sigma(z_i) = \frac{e^{z_i}}{\sum_{j=1}^{K} e^{z_j}} \quad for\ i = 1, 2, \ldots, K \tag{2}$$

where K is the number of classes. The last layer activation function is the softmax activation function mentioned in Eq. 2. The optimizer we have used is Adam.

## 5 Results

Initial Global Model - We fine-tuned the DistilBERT on the combined dataset. There we achieved 98% accuracy on the validation set.

Figure 5 depicts how the federated loss of 2 clients converged after ten epochs. With just 32 samples used for training on both devices, the loss of the global model decreased. In the real world, we can use thousands of clients and then perform model weight averaging with the help of the FedAveraging algorithm.

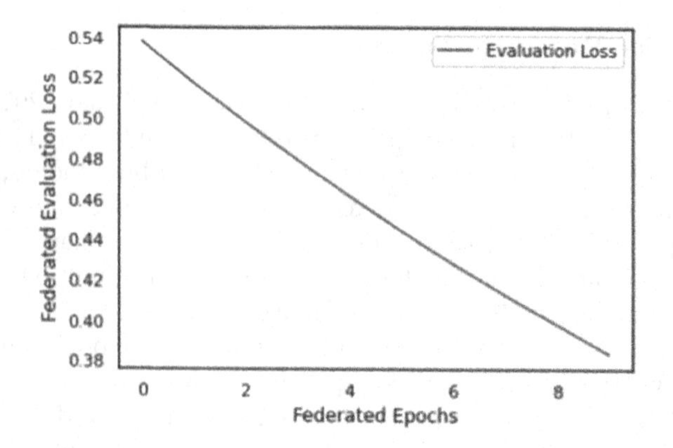


**Fig. 5.** Federated Loss Convergence

## 6 Conclusion and Future Scope

SMS Spamming is a problem billions of mobile users face. Spammers bulk spam a large number of people. Innocent people fall into the trap set by spammers by attracting them with exciting rewards and lotteries. In our research study, we have proposed a federated learning based on-device SMS Spam classification approach. We have used the DistilBERT model to classify our SMS. The validation accuracy we achieved is 98%. The SMS received on the user's mobile are sensitive so they will never leave the device premises. Only, the model will train on the user's SMS and its weights will be transferred to the centralized server. As a part of the future scope, we would like to extend this project to predict SMS spam in vernacular languages.

## References

1. HOW MANY SMARTPHONES ARE IN THE WORLD? https://www.bankmycell.com/blog/how-many-phones-are-in-the-world. Accessed 19 April 2022
2. Mishra, S., Soni, D.: DSmishSMS-A System to Detect Smishing SMS. Neural Comput. Appl. (2021)
3. Julis, M., Alagesan, S.: Spam detection in sms using machine learning through text mining. Int. J. Sci. Technol. Res. **9**, 498–503 (2020)
4. Gomaa, W.: The impact of deep learning techniques on SMS spam filtering. Int. J. Adv. Comput. Sci. Appl. **11** (2020). https://doi.org/10.14569/IJACSA.2020.0110167
5. Liu, X., Lu, H., Nayak, A.: A spam transformer model for SMS spam detection. IEEE Access **9**, 80253–80263 (2021). https://doi.org/10.1109/ACCESS.2021.3081479
6. Rojas-Galeano, S.: Using BERT Encoding to Tackle the Mad-lib Attack in SMS Spam Detection (2021)
7. Sanh, V., Debut, L., Chaumond, J., Wolf, T.: DistilBERT, a distilled version of BERT: smaller, faster, cheaper and lighter (2019)

8. Metsis, V., Androutsopoulos, I., Paliouras, G.: Spam filtering with Naive Bayes - which Naive Bayes? In: Proceedings of the 3rd Conference on Email and Anti-Spam (CEAS 2006), Mountain View, CA, USA (2006)
9. Almeida, T.A., Gómez Hidalgo, J.M., Yamakami, A.: Contributions to the study of SMS spam filtering: new collection and results. In: Proceedings of the 2011 ACM Symposium on Document Engineering (DOCENG 2011), Mountain View, CA, USA (2011)
10. Nuruzzaman, M.T., Lee, C., Choi, D.: Independent and personal SMS spam filtering. In: 2011 IEEE 11th International Conference on Computer and Information Technology, pp. 429–435 (2011). https://doi.org/10.1109/CIT.2011.23
11. McMahan, B., et al.: Communication-efficient learning of deep networks from decentralized data. Artificial intelligence and statistics. PMLR (2017)
12. Your voice & audio data stays private while Google Assistant improves. https://support.google.com/assistant/answer/10176224?hl=en#zippy=. Accessed 19 April 2022
13. Hard, A., et al.: Federated Learning for Mobile Keyboard Prediction (2018)
14. Lundberg, S.M., Lee, S.I.: A unified approach to interpreting model predictions. In: Advances in Neural Information Processing Systems, 30 (2017)
15. Beutel, D.J., et al.: Flower: a friendly federated learning research framework. arXiv preprint arXiv:2007.14390 (2020)

# Point Cloud Completion with Difference-Aware Point Voting

Lihua Lu[1,2,3], Ruyang Li[1,2], Hui Wei[1,2(✉)], Yaqian Zhao[1,2], Rengang Li[1,2], and Binqiang Wang[1,2]

[1] Inspur Electronic Information Industry Co., Ltd., Jinan, China
{lulihua,weihui}@inspur.com
[2] Inspur (Beijing) Electronic Information Industry Co., Ltd., Beijing, China
[3] Shandong Massive Information Technology Research Institute, Jinan, China

**Abstract.** Previous methods usually recover the complete point cloud according to the shared geometric information between the known and unknown shapes, which neglects that the partially known points and the points to be inferred may be different referring to both spatial and semantic aspects. To address this problem, we propose a model with Difference-aware Point Voting to reason about the differences between the known and unknown point clouds. Before voting, we devise a Multi-scale Feature Extraction block to filter out low-quality observed points obtaining the seed points, and learn each seed point with multi-scale features facilitating the voting process. By voting, we fill in the gaps between the known and unknown shapes in feature space, and infer the features to represent the missing shapes. Further, we propose a hierarchical point decoder to progressively refine the voting process under the guidance of the geometric commonalities shared by the observed and missing parts. The decoder finally generates the new points that can be taken as the center points for generating missing regions. Quantitative and visual results on PCN and ShapeNet-55 datasets show that our model outperforms the state-of-the-art methods.

**Keywords:** Point Cloud Completion · Difference-Aware Point Voting · Multi-scale Feature Extraction · Deep Hough Voting

## 1 Introduction

Point clouds captured by 3D scanners or depth cameras are widely-used in 3D computer vision, such as 3D object detection [1,28], instance reconstruction [4, 10,11] and point cloud classification and segmentation [2,7]. However, raw point clouds are usually sparse and incomplete due to occlusion, specular surface, poor sensor resolution, etc. Therefore, point cloud completion which aims to reconstruct a complete shape from its partial observations is deadly in need.

Existing researches [5,17,25,27,30] usually take an encoder-decoder pipeline to generate a complete point cloud. The pioneering FoldNet [27] adopts a graph-based encoder to learn features from the input points, and then proposes a


M. Tanveer et al. (Eds.): ICONIP 2022, CCIS 1793, pp. 64–75, 2023.
https://doi.org/10.1007/978-981-99-1645-0_6

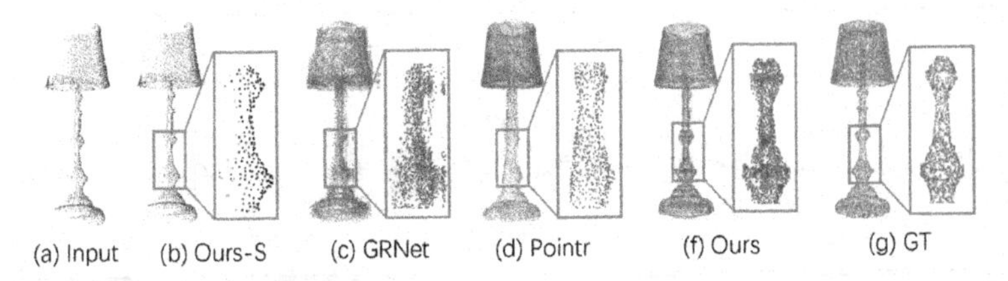


**Fig. 1.** Visual completion results of our model and other methods. "Ours-S" shows the complete shape at the coarse level predicted by our model (the points in missing regions are depicted in red). Our model can complete the point cloud with more realistic and detailed geometric structures (as shown in the orange boxes). (Color figure online)

folding-based decoder deforming a canonical 2D grid onto the underlying 3D object surface of a point cloud. Recent methods [19,29–31] improve the completion performance by designing elaborate decoders to predict the complete shape in a coarse-to-fine way. For example, Pointr [29] employs a two-stage generation architecture, in which a decoder firstly generates a coarse point cloud and then expands the number of points for a dense point cloud. These methods generally recover the complete point cloud according to the common geometric information like the global shape feature between the known and unknown shapes. However, this neglects that the partially observed points and the points in missing regions can be different in both geometrics and semantics.

Deep Hough Voting has been successfully applied in the 3D object detection task [1,14,26], which aims to localize objects in a 3D scene by estimating the 3D bounding boxes of objects from point clouds. When directly predicting bounding boxes from point clouds, the major challenge is that a 3D object centroid can be far from any surface point. Typical methods like VoteNet [1] and VENet [26] adopt deep Hough voting to address this challenge, and generate new points that lie close to the centers of 3D objects from the input surface points. Inspired by these methods, we propose a Difference-aware Point Voting (DPV) block to bridge the gap between the observed and missing shapes. Rather than directly rebuilding the complete shape from the partial point cloud, the DPV block reasons about the differences between the known and unknown shapes in feature space, making our model focus on inferring the geometric structures of missing shapes (depicted in Fig. 1). By voting, we essentially generate the new features representing the missing geometrics.

In addition, many researches [27,29,30] have proved geometric commonalities like the global shape feature shared by the observed and missing shapes are the foundations of rebuilding the complete shape. Therefore, we propose a hierarchical point decoder to progressively refine the votes and predict the missing points. It initially generates the votes guided by the common shape feature, which can be iteratively lifted and detailed by the geometric details conveyed from the incomplete point cloud using multiple DPV blocks. The decoder finally infers the missing points at a coarse level (depicted in Fig. 1 (b)), which serves as the center points to generate the dense point cloud for the missing regions.

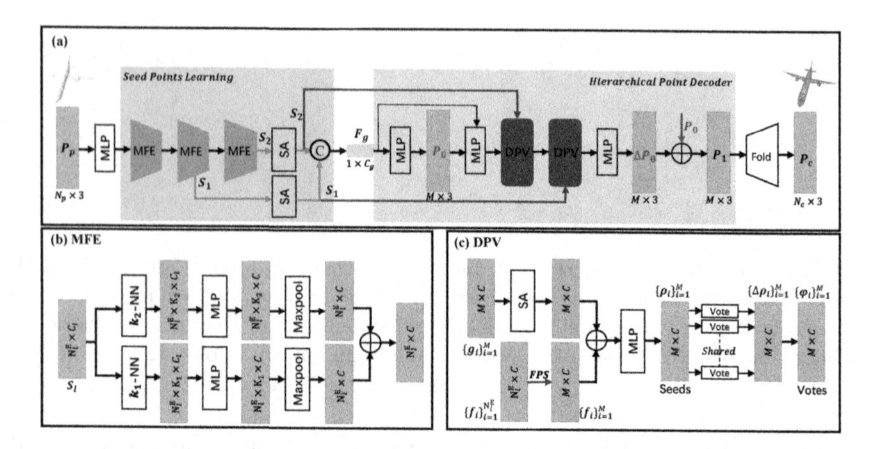


**Fig. 2.** (a) The overall architecture of our model. (b) The details of the Multi-scale Feature Extraction (MFE) block. (c) The details of the Difference-aware Point Voting (DPV) block. © and ⊕ respectively denote concatenation and element-wise summation operations. "SA" and "FPS" represent the self-attention mechanism and farthest point sampling.

To support the voting mechanism, we also design a Multi-scale Feature Extraction (MFE) block to learn the seed points. Our MFE block can extract multi-scale point features by incorporating the neighborhood information at multiple scales, and then refine the features via a self-attention operation. The block offers each seed point with both low-level and high-level semantics, facilitating the DPV block to learn the semantic and geometric features of missing points. As portrayed in Fig. 2, we adopt multiple MFE blocks to progressively filter out the low-quality observed points and learn multiple sets of seed points. Every MFE block endows each seed point with the multi-scale features. The subsequent hierarchical point decoder contains multiple DPV blocks. Each DPV block takes the corresponding seed points as the inputs, and outputs the votes for the missing shapes. Specifically, each seed point independently votes for one new point in the missing regions.

In summary, our contributions are three-fold: (1) We propose a difference-aware point voting block to reason about the differences between the known and unknown shapes. It can generate features for the missing shapes directly from the observed points by voting. (2) Before voting, we devise a multi-scale feature extraction block to filter out low-quality observed points obtaining the seed points, and learn each seed point with multi-scale features supporting the voting process. (3) We propose a hierarchical point decoder, which can refine the votes using the common geometric structures shared by the known and unknown shapes. It finally predicts new points in the missing shapes. Experimental results demonstrate our model can outperform state-of-the-art methods, and recover the complete point cloud of both observed and missing regions.

## 2 Related Work

### 2.1 Point Cloud Completion

Point cloud completion aims to recover the complete shape from its raw and partial observation. Previous notable studies [17,20,27,30] follow an encoder-decoder architecture, which predicts a complete point cloud from the highly-encoded global shape representation. However, a single shape representation loses local geometric details, which can hardly be recovered during the decoder process. Some follow-up works [22,31] learn local features to capture geometric details of the input point cloud. Recently, many researches [5,19,24,29] dedicate to generate the complete point cloud in a coarse-to-fine pipeline. Pointr [29] firstly generates a coarse point cloud using a transformer encoder-decoder, which is detailed to a dense point cloud by a lifting module. The coarse-to-fine decoders can achieve better performance since they can impose more constraints on the point generation process. Nevertheless, existing methods usually lose sight of the differences between the observed and missing regions. In this paper, we introduce the deep Hough voting and propose a Different-aware Point Voting block to explicitly learn the differences between the partially observed and missing regions.

### 2.2 The Voting Mechanism

Hough voting has been applied to many tasks, such as 6D pose estimation [6], global registration [8], semantic and instance segmentation [12]. Recently, inspired by Hough voting in 2D object detection [9], many researches [1,13,14,26] adapt deep Hough voting to 3D object detection. VoteNet [1] implements the voting mechanism in a deep learning framework for 3D object detection. It selects a set of seed points from the input point cloud, and generates votes for object proposals. By voting, these methods can bridge the gap between the center points and surface points of objects, and generate new points that lie close to the centers of 3D objects from the input surface points. Illuminated by the success of the deep Hough voting mechanism in 3D object detection, we introduce it to point cloud completion.

## 3 Method

### 3.1 Overview

The overall architecture of our model is depicted in Fig. 2(a). Given the partially observed point cloud $P_p$ with $N_p$ points, multiple MFE blocks followed by the self-attention (SA) mechanism [18] are stacked to hierarchically learn the seed points with multi-scale features. The outputs of these last two blocks are used to compute the shared geometric feature $F_g$ with size $1 \times C_g$ between the observed and missing regions. Secondly, the hierarchical point decoder uses $F_g$ to generate an initial point cloud $P_0$ with $M$ points for the missing regions. Then two DPV blocks are employed to progressively utilize the last two sets of seed points and generate votes to refine $P_0$, acquiring the point cloud $P_1$ for the missing regions at a coarse level. Finally, the FoldNet [27] takes $P_1$ as the center points to infer the dense point cloud for the missing regions, which is merged with the input $P_p$ to generate the complete point cloud $P_c$ with $N_c$ points.

### 3.2 Seed Points Learning

Before voting, we propose the multi-scale feature extraction (MFE) block to learn high-performance seed points. Different to the classical PointNet [15], PointNet++ [16] and EdgeConv [21], our MFE block can offer each point with multi-scale features. As depicted in Fig. 2 (a), we alternately stack three MFE blocks and two self-attention layers to learn multiple sets of seed points. Firstly, we adopt three MFE blocks to hierarchically get rid of useless observed points and obtain seed points with multi-scale features. Each MFE block performs the farthest point sampling operation to choose a subset $S_l$ with size $N_l^E \times C_l$. As depicted in Fig. 2(b), for each point $p_i \in \mathbb{R}^{C_l}$ in $S_l$, both a "small" $k_1$-nearest and a "large" $k_2$-nearest neighborhoods (denoted as $k_1$-$NN$ and $k_2$-$NN$) are selected from $S_l$ according to the Euclidean distance of point coordinates. For each neighborhood $k\text{-}NN(p_i) = \{p_j, 1 \leq j \leq k\}$, $k = \{k_1, k_2\}$, the Multi-Layer Perceptions (MLP) and max-pooling operations are conducted to compute the new feature $p_i{}' \in \mathbb{R}^C$ for each point as follows:

$$p_i{}' = MLP(\max_j \{MLP([p_i; p_j - p_i])\}). \tag{1}$$

The features at both small and large scales are added together to finally update the seed points with multi-scale features at the low level.

Secondly, we pick the last two MFE blocks, where each block is followed by a self-attention layer to lift the multi-scale features. The self-attention layer integrates long-range correlations among points, and lifts the features of each seed point with no-local geometric contexts. Later, we concatenate the last two seed sets $S_1$ and $S_2$, and operate a max-pooling operation to obtain the shared shape feature $F_g$ among the observed and missing regions. Concretely, $S_1$ and $S_2$ respectively have sizes of $N_1^E \times C$ and $N_2^E \times C$ for further voting operations. In our experiments, we set $N_1^E = 512$, $N_2^E = 128$, $C = 384$ to achieve a high computational efficiency.

### 3.3 Hierarchical Point Decoder

We propose a hierarchical point decoder to predict the missing shapes by fully utilizing the shared information between the observed and missing regions. As depicted in Fig. 2(a), the shared shape feature $F_g$ is firstly fed into a MLP followed by a reshape operation to produce an initial point cloud $P_0 = \{x_i\}_{i=1}^M$ with $x_i \in \mathbb{R}^3$ for missing regions. In addition, to exploit the differences between the known and unknown shapes in feature space, we propose the DPV block to generate $M$ votes, where each vote has a high dimensional feature vector for inferring the center points in missing regions. Two DPV blocks are adopted to iteratively generate votes to update features of each point in $P_0$.

Given the set of seed points $S_l$ with $N_l^E$ points, where each point has the feature $f_i \in \mathbb{R}^C$, it's intuitive to directly use the seed feature $f_i$ to generate the vote. However, this voting mode neglects prior information supplied by the outputs $g_i \in \mathbb{R}^C$ of the last DPV block. To deal with this problem, we take both $f_i$ and $g_i$ as the feature of each seed point (depicted in Fig. 2(c)). Specifically, our DPV block first adopts a self-attention layer ($SA$) to advance the supplementary feature $g_i$. Then $M$ points are sampled from the subset $S_l$ via a farthest point sampling layer, where each point feature $f_i$ is further merged with the lifted supplementary feature $SA(g_i)$ to form the feature $\rho_i$ using a linear layer ($Linear$). The point voting operation ($Vote$) implemented by MLPs utilizes the merged seed feature $\rho_i$ to compute the feature offset $\Delta\rho_i \in \mathbb{R}^C$ such that the vote $\varphi_i$ has $\varphi_i = \rho_i + \Delta\rho_i$, where

$$\Delta\rho_i = Vote\left\{Linear([f_i; SA(g_i)])\right\}. \tag{2}$$

The DPV block is shared by all points in seed set $S_l$, which takes each seed point as the input to generate one vote independently. For the first DPV block, we take the initial point cloud $P_0$ as the prior information. Virtually, we concatenate the point coordinates in $P_0$ with the repeated global features, and feed them into an MLP to learn the features $G = \{g_i\}_{i=1}^M$.

Each feature $\varphi_i \in \mathbb{R}^C$ in the outputs of the second DPV block is fed into an additional MLP to compute the Euclidean space offset $\Delta x_i \in \mathbb{R}^3$. Finally, the initial point cloud $P_0$ is refined to $P_1 = \{x_i + \Delta x_i\}_{i=1}^M$ for the missing regions, which are taken as the center points to generate the dense point cloud. To control the voting process, we combine the inferred $P_1$ with the input point cloud $P_p$ to form the coarse but complete point cloud $P_s$.

We take $P_1$ as the center points to recover detailed local shapes for the missing regions, and generate the dense and complete point cloud via FoldNet [27]. The inferred points are merged with the input points $P_p$ to form the final complete point cloud $P_c$. We adopt Chamfer Distance as our loss function. Given the ground-truth point cloud $P_t$, the predicted sparse point cloud $P_s$ and the predicted dense point cloud $P_c$, our final loss function is the sum of these two losses $\mathcal{L} = \mathcal{L}_s + \mathcal{L}_c$, where

$$\mathcal{L}_s = \frac{1}{|P_s|}\sum_{p\in P_s}\min_{g\in P_t}\|p-g\| + \frac{1}{|P_t|}\sum_{g\in P_t}\min_{p\in P_s}\|g-p\|, \tag{3}$$

$$\mathcal{L}_c = \frac{1}{|P_c|}\sum_{p\in P_c}\min_{g\in P_t}\|p-g\| + \frac{1}{|P_t|}\sum_{g\in P_t}\min_{p\in P_c}\|g-p\|. \tag{4}$$

# 4 Experiments

## 4.1 Datasets and Implementation Details

**PCN Dataset.** The PCN dataset [30] contains 30,974 3D CAD models derived from 8 categories. Partial point clouds are obtained by back-projecting the 2.5D depth images into 3D. Each ground truth point cloud contains 16,384 points, which are uniformly sampled on the mesh surfaces. We use the same train/val/test splits on this dataset as previous work [25,29].

**ShapeNet-55 Dataset.** The ShapeNet-55 dataset [29] is a challenging dataset, which contains 55 object categories from diverse viewpoints, different incomplete patterns and various incompleteness levels. 41,952 models are chosen for training and 10,518 models for testing. Each ground truth contains 8,192 points. During training, a viewpoint is randomly chosen and $n$ furthest points are removed from this viewpoint. From the remaining point cloud, 2,048 points are downsampled as the input. $n$ is randomly selected from 25% to 75%. For a fair comparison, we keep the standard protocol on this dataset as [29].

**Implementation Details.** Our model is trained on two RTX 3090Ti GPUs with a batch size of 48 and optimized by an AdamW optimizer. During the training process, the hierarchical point decoder firstly generates $M = 224$ points for the missing regions. $k_1$ and $k_2$ are set to be 8 and 16. In our experiments, we adopt the mean Chamfer Distance (CD) as the evaluation metric. The CD-$\mathcal{L}1$ and CD-$\mathcal{L}2$ denote the distances in $\mathcal{L}1$-norm and $\mathcal{L}2$-norm. For the ShapeNet-55 dataset, we employ the same experimental setting in Pointr [29]. During training, we randomly choose a viewpoint and remove the $n$ furthest points from the viewpoint. And then, from the remaining point cloud, we downsample 2,048 points as the input partial point cloud. $n$ is randomly selected from 25% to 75% of the complete point cloud, and 2,048 to 6,144 points are sampled to denote different incomplete levels. During evaluation, $n$ is set to 25%, 50% or 75% of the whole point cloud, and 2,048, 4,096 or 6,144 points are sampled to represent three incomplete degrees: simple, moderate and hard. Besides, 8 viewpoints are fixed for evaluation.

## 4.2 Comparisons with State-of-the-Art Methods

**Evaluation on PCN Dataset.** Table 1 shows the comparative results on PCN dataset. Compared with other methods, our model achieves the best performance. PCN [30] follows a coarse-to-fine pipeline, which firstly generates a coarse point cloud via MLPs and then refines it by the folding operation, providing a baseline for point cloud completion. Pointr [29] relies on transformer to complete the point cloud, and achieve the state-of-the-art results. But, it ignores to use geometric details of the observed points during the point generation. Differently, our model employs the deep Hough voting to build a DPV block, which can reason about the differences between observed and missing points. Multiple DPV

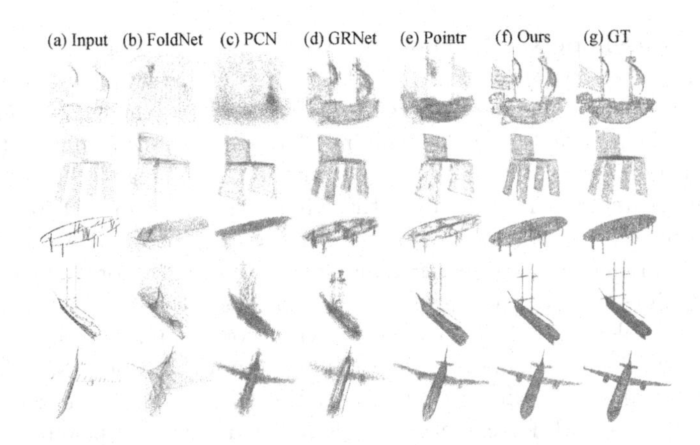


**Fig. 3.** Visual comparison results on PCN dataset. Our model can generate the complete point cloud with smoother and more detailed geometrics.

blocks can fully exploit geometric information of the observed shapes and finally reduce the average CD by 1.2. In addition, our model outperforms all counterparts on 8 categories, which proves the generalization ability of our model across different categories.

Figure 3 depicts the visual results, which demonstrates our model can predict the complete point cloud with higher fidelity. In the first line, both FoldNet [27] and PCN [30] fail to predict the complex watercraft, and the latest Pointr [29] misses the clear shape. But, our model can generate the complete shape with detailed bows and sails of the watercraft. Furthermore, the third line demonstrates our model can maintain the observed regions like table legs and predict smoother table surfaces, while other methods, including GRNet [25], Pointr [29], fail to predict the missing regions.

**Table 1.** Completion results on PCN dataset in terms of $\mathcal{L}1$ Chamfer Distance $\times 10^3$. Lower is better.

| Methods | Average | Plane | Cabinet | Car | Chair | Lamp | Sofa | Table | Watercraft |
|---|---|---|---|---|---|---|---|---|---|
| FoldNet [27] | 14.31 | 9.49 | 15.80 | 12.61 | 15.55 | 16.41 | 15.97 | 13.65 | 14.99 |
| TopNet [17] | 12.15 | 7.61 | 13.31 | 10.90 | 13.82 | 14.44 | 14.78 | 11.22 | 11.12 |
| AtlasNet [3] | 10.85 | 6.37 | 11.94 | 10.10 | 12.06 | 12.37 | 12.99 | 10.33 | 10.61 |
| PCN [30] | 9.64 | 5.50 | 22.70 | 10.63 | 8.70 | 11.00 | 11.34 | 11.68 | 8.59 |
| GRNet [25] | 8.83 | 6.45 | 10.37 | 9.45 | 9.41 | 7.96 | 10.51 | 8.44 | 8.04 |
| PMP-Net [23] | 8.73 | 5.65 | 11.24 | 9.64 | 9.51 | 6.95 | 10.83 | 8.72 | 7.25 |
| CDN [19] | 8.51 | 4.79 | 9.97 | 8.31 | 9.49 | 8.94 | 10.69 | 7.81 | 8.05 |
| Pointr [29] | 8.38 | 4.75 | 10.47 | 8.68 | 9.39 | 7.75 | 10.93 | 7.78 | 7.29 |
| Ours | **7.18** | **4.03** | **9.36** | **8.26** | **7.71** | **5.90** | **9.45** | **6.52** | **6.20** |

**Evaluation on ShapeNet-55 Dataset.** We report 10 categories out of 55 categories on ShapeNet-55 dataset in Table 2 for convenience. Our model is evaluated under three incomplete degrees, where 25%, 50% or 75% of the whole point cloud are respectively removed, and the remaining points are downsampled to $2,048$ as the input. Our model outperforms other methods referring to the average CD loss (CD-Avg) and the F-Score over the whole dataset, and CD losses under three incomplete degrees: simple, moderate and hard (CD-S, CD-M and CD-H). It proves our model can address challenging scenarios with more categories, different viewpoints and diverse incomplete patterns. Besides, the first five columns display the completion results of categories with more than $2,500$ samples in the training samples, and the following five columns report categories with less than 80 samples. Compared with the second-best Pointr [29], our model gives better completion results both for categories with sufficient and insufficient samples.

**Table 2.** Completion results on ShapeNet-55 dataset in terms of $\mathcal{L}2$ Chamfer Distance $\times 10^3$ and F-Score@1%.

| Methods | Table | Chair | Airplane | Car | Sofa | Birdhouse | Bag | Remote | Keyboard | Rocket | CD-S | CD-M | CD-H | CD-Avg | F1 |
|---|---|---|---|---|---|---|---|---|---|---|---|---|---|---|---|
| FoldNet [27] | 2.53 | 2.81 | 1.43 | 1.98 | 2.48 | 4.71 | 2.79 | 1.44 | 1.24 | 1.48 | 2.67 | 2.66 | 4.05 | 3.12 | 0.082 |
| PCN [30] | 2.13 | 2.29 | 1.02 | 1.85 | 2.06 | 4.50 | 2.86 | 1.33 | 0.89 | 1.32 | 1.94 | 1.96 | 4.08 | 2.66 | 0.133 |
| TopNet [17] | 2.21 | 2.53 | 1.14 | 2.18 | 2.36 | 4.83 | 2.93 | 1.49 | 0.95 | 1.32 | 2.26 | 2.16 | 4.3 | 2.91 | 0.126 |
| PFNet [5] | 3.95 | 4.24 | 1.81 | 2.53 | 3.34 | 6.21 | 4.96 | 2.91 | 1.29 | 2.36 | 3.83 | 3.87 | 7.97 | 5.22 | 0.339 |
| GRNet [25] | 1.63 | 1.88 | 1.02 | 1.64 | 1.72 | 2.97 | 2.06 | 1.09 | 0.89 | 1.03 | 1.35 | 1.71 | 2.85 | 1.97 | 0.238 |
| Pointr [29] | 0.90 | 1.11 | 0.52 | 1.10 | 0.90 | 2.07 | 1.12 | 0.87 | 0.43 | 0.66 | 0.73 | 1.13 | 2.04 | 1.09 | 0.464 |
| Ours | **0.89** | **1.00** | **0.48** | **0.92** | **0.84** | **2.02** | **1.00** | **0.58** | **0.39** | **0.59** | **0.60** | **0.92** | **1.82** | **0.97** | **0.482** |

The completion results of our model and Pointr [29] under the moderate incomplete degree are shown in Fig. 4, which demonstrates our model works a bit better. In the first column, the input loses many geometrics and can hardly be completed. However, our method can achieve the decent prediction of the overall and detailed structures, like sharp edges and corners.

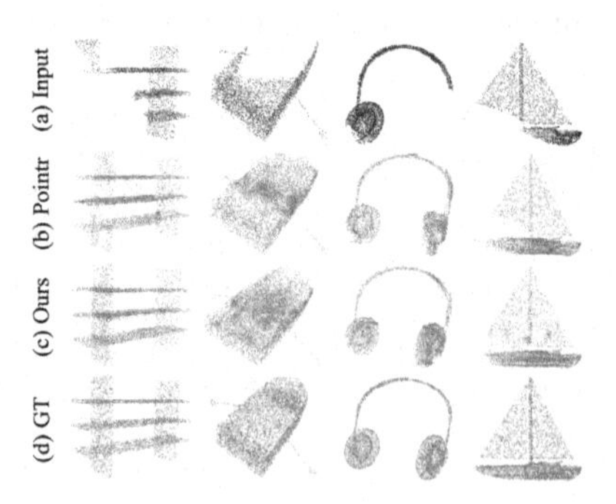


**Fig. 4.** Visual completion results on ShapeNet-55 dataset. Despite of the input point cloud losing most geometric information, our model can generate the complete shape.

### 4.3 Ablation Studies

**Analysis of Our Model.** We conduct experiments on PCN dataset to validate the effectiveness of the key parts in our model. The results are reported in Table 3. The baseline model uses three stacked PointNet [15] layers to abstract the global feature. Based on it, a coarse point cloud is predicted by MLPs, which is refined to the dense one by FoldNet [27]. We replace the PointNet layers with our MFE blocks (model A). On the contrary, we add one DPV block before FoldNet in the baseline model (model B). Both model A and B achieve better performance than the baseline, demonstrating the effectiveness of the proposed blocks. Model B reduces the average CD by 1.3, better than model A, proving the DPV block is the most contributing factor in our model. Model C is similar to our model but with only one DPV block. It is obvious two DPV blocks work better since they can fully utilize multi-level seed features to progressively refine the features of missing points.

**Table 3.** Effects of the key parts in our model on PCN dataset in terms of CD-$\mathcal{L}1$. Lower is better.

| Methods | MFE | DPV-1 | DPV-2 | CD-$\mathcal{L}1$ |
|---|---|---|---|---|
| Baseline | | | | 9.22 |
| Model A | ✓ | | | 8.96 |
| Model B | | ✓ | | 8.25 |
| Model C | ✓ | ✓ | | 7.94 |
| Ours | ✓ | ✓ | ✓ | **7.18** |

**Visual Results of the Hierarchical Point Decoder.** Our hierarchical point decoder adopts two DPV blocks to yield the coarse point cloud for missing regions, serving as the center points for the dense point cloud. Figure 5 (b)

**Fig. 5.** Visual results of the hierarchical point decoder on PCN dataset. (b) The points in missing regions predicted by the hierarchical point decoder are denoted in red. (Color figure online)

portrays the coarse point cloud consisting of the points predicted by our decoder (in red) and the input points (in black). It indicates our decoder can generate the smooth and uniform points to complete missing regions, even for the objects with large missing regions, like the sofa category in the second column.

## 5 Conclusion

In this paper, we propose the difference-aware point voting block to infer the differences between the observed and missing shapes. Before voting, the multi-scale feature extraction and lift block is proposed to learn the seed points with multi-scale features to support the voting process. The hierarchical point decoder exploits both the similarities and dissimilarities between the known and unknown shapes, and progressively generates the missing points. Experimental results on PCN and ShapeNet-55 datasets demonstrate the superiority of our model.

**Acknowledgements.** This work was supported by Shandong Provincial Natural Science Foundation under Grant ZR2021QF062.

## References

1. Ding, Z., Han, X., Niethammer, M.: VoteNet: a deep learning label fusion method for multi-atlas segmentation. In: Shen, D., et al. (eds.) MICCAI 2019. LNCS, vol. 11766, pp. 202–210. Springer, Cham (2019). https://doi.org/10.1007/978-3-030-32248-9_23
2. Fan, S., Dong, Q., Zhu, F., Lv, Y., Ye, P., Wang, F.Y.: SCF-Net: learning spatial contextual features for large-scale point cloud segmentation. In: CVPR, pp. 14504–14513 (2021)
3. Groueix, T., Fisher, M., Kim, V.G., Russell, B.C., Aubry, M.: A papier-mâché approach to learning 3D surface generation. In: CVPR, pp. 216–224 (2018)
4. Hou, J., Dai, A., Nießner, M.: RevealNet: seeing behind objects in RGB-D scans. In: CVPR, pp. 2098–2107 (2020)
5. Huang, Z., Yu, Y., Xu, J., Ni, F., Le, X.: PF-Net: point fractal network for 3D point cloud completion. In: CVPR, pp. 7662–7670 (2020)
6. Kehl, W., Milletari, F., Tombari, F., Ilic, S., Navab, N.: Deep learning of local RGB-D patches for 3D object detection and 6D pose estimation. In: Leibe, B., Matas, J., Sebe, N., Welling, M. (eds.) ECCV 2016. LNCS, vol. 9907, pp. 205–220. Springer, Cham (2016). https://doi.org/10.1007/978-3-319-46487-9_13
7. Lai, X., et al.: Stratified transformer for 3D point cloud segmentation. In: CVPR, pp. 8500–8509 (2022)
8. Lee, J., Kim, S., Cho, M., Park, J.: Deep hough voting for robust global registration. In: CVPR, pp. 15994–16003 (2021)
9. Leibe, B., Leonardis, A., Schiele, B.: Robust object detection with interleaved categorization and segmentation. IJCV **77**(1), 259–289 (2008)
10. Najibi, M., et al.: DOPS: Learning to detect 3D objects and predict their 3D shapes. In: CVPR, pp. 11913–11922 (2020)
11. Nie, Y., Hou, J., Han, X., Nießner, M.: RFD-Net: point scene understanding by semantic instance reconstruction. In: CVPR, pp. 4608–4618 (2021)

12. Novotny, D., Albanie, S., Larlus, D., Vedaldi, A.: Semi-convolutional operators for instance segmentation. In: Ferrari, V., Hebert, M., Sminchisescu, C., Weiss, Y. (eds.) ECCV 2018. LNCS, vol. 11205, pp. 89–105. Springer, Cham (2018). https://doi.org/10.1007/978-3-030-01246-5_6
13. Qi, C.R., Chen, X., Litany, O., Guibas, L.J.: ImVoteNet: boosting 3D object detection in point clouds with image votes. In: CVPR, pp. 4404–4413 (2020)
14. Qi, C.R., Litany, O., He, K., Guibas, L.J.: Deep hough voting for 3D object detection in point clouds. In: CVPR, pp. 9277–9286 (2019)
15. Qi, C.R., Su, H., Mo, K., Guibas, L.J.: PointNet: deep learning on point sets for 3D classification and segmentation. In: CVPR, pp. 652–660 (2017)
16. Qi, C.R., Yi, L., Su, H., Guibas, L.J.: PointNet++: deep hierarchical feature learning on point sets in a metric space. arXiv preprint arXiv:1706.02413 (2017)
17. Tchapmi, L.P., Kosaraju, V., Rezatofighi, H., Reid, I., Savarese, S.: TopNet: structural point cloud decoder. In: CVPR, pp. 383–392 (2019)
18. Vaswani, A., et al.: Attention is all you need. In: NeurIPS, pp. 5998–6008 (2017)
19. Wang, X., Ang Jr, M.H., Lee, G.H.: Cascaded refinement network for point cloud completion. In: CVPR, pp. 790–799 (2020)
20. Wang, Y., Tan, D.J., Navab, N., Tombari, F.: SoftPoolNet: shape descriptor for point cloud completion and classification. In: Vedaldi, A., Bischof, H., Brox, T., Frahm, J.-M. (eds.) ECCV 2020. LNCS, vol. 12348, pp. 70–85. Springer, Cham (2020). https://doi.org/10.1007/978-3-030-58580-8_5
21. Wang, Y., Sun, Y., Liu, Z., Sarma, S.E., Bronstein, M.M., Solomon, J.M.: Dynamic graph CNN for learning on point clouds. ACM Trans. Graph. **38**(5), 1–12 (2019)
22. Wen, X., Li, T., Han, Z., Liu, Y.S.: Point cloud completion by skip-attention network with hierarchical folding. In: CVPR, pp. 1939–1948 (2020)
23. Wen, X., et al.: PMP-Net: point cloud completion by learning multi-step point moving paths. In: CVPR, pp. 7443–7452 (2021)
24. Xiang, P., et al.: SnowflakeNet: point cloud completion by snowflake point deconvolution with skip-transformer. In: CVPR, pp. 5499–5509 (2021)
25. Xie, H., Yao, H., Zhou, S., Mao, J., Zhang, S., Sun, W.: GRNet: gridding residual network for dense point cloud completion. In: Vedaldi, A., Bischof, H., Brox, T., Frahm, J.-M. (eds.) ECCV 2020. LNCS, vol. 12354, pp. 365–381. Springer, Cham (2020). https://doi.org/10.1007/978-3-030-58545-7_21
26. Xie, Q., et al.: VENet: voting enhancement network for 3D object detection. In: ICCV, pp. 3712–3721 (2021)
27. Yang, Y., Feng, C., Shen, Y., Tian, D.: FoldingNet: point cloud auto-encoder via deep grid deformation. In: CVPR, pp. 206–215 (2018)
28. You, Y., et al.: Canonical voting: towards robust oriented bounding box detection in 3D scenes. In: CVPR, pp. 1193–1202 (2022)
29. Yu, X., Rao, Y., Wang, Z., Liu, Z., Lu, J., Zhou, J.: PoinTr: diverse point cloud completion with geometry-aware transformers. In: ICCV, pp. 12498–12507 (2021)
30. Yuan, W., Khot, T., Held, D., Mertz, C., Hebert, M.: PCN: point completion network. In: International Conference on 3D Vision (3DV), pp. 728–737 (2018)
31. Zhang, W., Yan, Q., Xiao, C.: Detail preserved point cloud completion via separated feature aggregation. In: Vedaldi, A., Bischof, H., Brox, T., Frahm, J.-M. (eds.) ECCV 2020. LNCS, vol. 12370, pp. 512–528. Springer, Cham (2020). https://doi.org/10.1007/978-3-030-58595-2_31

# External Knowledge and Data Augmentation Enhanced Model for Chinese Short Text Matching

Haoyang Ma[1,2(✉)] and Hongyu Guo[1]

[1] North China Institute of Computing Technology, Beijing, China
guohongyu@sina.com
[2] National University of Defense Technology, Changsha, Hunan, China
mhy99@nudt.edu.cn

**Abstract.** With the rapid development of the network, a large amount of short text data has been generated in life. Chinese short text matching is an important task of natural language processing, but it still faces challenges such as ambiguity in Chinese words and imbalanced ratio of samples in the training set. To address these problems, we propose an external knowledge and data augmentation enhanced model (EDM) for Chinese short text matching. EDM uses jieba, thulac and ltp to generate word lattices and employs HowNet as an external knowledge source for disambiguation. Additionally, dropout and EDA hybrid model is adopted for data augmentation to balance the proportion of samples in the training set. The experimental results on two Chinese short text datasets show that EDM outperforms most of the existing models. Ablation experiments also demonstrate that external knowledge and data augmentation can significantly improve the model.

**Keywords:** External Knowledge · Data Augmentation · Short Text Matching

## 1 Introduction

Short text matching (STM) is an important task in natural language processing (NLP). By inputting two sentence-level or phrase-level short texts, an algorithm or model is used to predict whether these two texts are matched. STM can be applied in language inference [1], paraphrasing questions [2], answer selection [3], and question answering [5] and other tasks.

STM focuses on calculating text similarity, and text similarity algorithms can be divided into two categories: unsupervised and supervised. Unsupervised text similarity algorithms are mainly divided into three categories [5]: character-based, statistical model-based, and ontology knowledge-based. Supervised similarity calculation models are generally dominated by deep neural network models, such as ESIM [6], BiMPM [7], DIIN [8] and so on. Thanks to the achievements of deep learning in recent years, the pre-training models led by BERT [9]


M. Tanveer et al. (Eds.): ICONIP 2022, CCIS 1793, pp. 76–87, 2023.
https://doi.org/10.1007/978-981-99-1645-0_7

have shined in the NLP fields, and achieved a significant improvement in the performance of multiple tasks.

Nonetheless, the polysemy of Chinese words may bring about great problems, which often lead to ambiguity in semantic understanding [10]. The word framed on the left has two senses/meanings: "Antique" and "Old Fogey". When matching it with another sentence, the latter sense is likely to prevail. To resolve this semantic ambiguity, we use HowNet [11] as an external knowledge source. In HowNet, words have different senses and sememes. The word "antique" contains both "past" and "stiff" sememes, and the latter in the original sentence has the same sememe as "stubborn", enabling a better judgement of whether the two texts match.

Meanwhile, although there are many public datasets available for testing, there are also problems such as imbalanced ratio of training set samples and few training samples. To solve these problems, we propose an external knowledge and data augmentation enhanced model (EDM) for Chinese short text matching. EDM uses a combination of dropout [12] and EDA [13] to process the dataset, and input the processed text pairs into jieba, thulac and ltp to generate word lattices. EDM obtains the sememes through OpenHowNet [14], uses the multi-dimensional graph attention transformer to obtain the initial semantic representation, and uses the attentive pooling and fusion layer to fuse and update words and semantics. Finally, the sentence matching is performed. The experimental results on two Chinese short text datasets show that EDM outperforms most of the existing models. Ablation experiments also demonstrate that external knowledge and data augmentation can significantly improve the model.

The contributions of our work can be summarized as:

1) We propose a new EDM model, which uses hybrid data augmentation to optimize training samples, integrates multi-granularity information and external knowledge to eliminate semantic ambiguity, and improves the accuracy of the model in Chinese short text matching tasks.
2) Extensive experimental results on two Chinese datasets demonstrate the effectiveness of the model, and ablation experiments demonstrate the improvement of model accuracy by external knowledge and data augmentation.

## 2 Research Status and Background Knowledge

### 2.1 Text Similarity Algorithms

The text similarity algorithms involve many research fields and have a wide range of applications. The bag-of-words model [15] is the most commonly used text similarity algorithm model, especially for long texts. It directly uses corpus, and can quickly process large amounts of data with simplified learning process. However, the simple bag-of-words model cannot solve the word polysemy problem, so language models based on topic analysis that count the co-occurrence probability of words are gradually developed. According to the length of the text, Gerard Salton and McGill proposed a vector space model (VSM) [16], which is a method of mapping long texts to a vector space, and calculates the similarity by counting the frequency of word occurrences and the dimensionality reduction of vectors. Most scholars focus on the study of string-based similarity algorithms, such as the LCS [17] algorithm. Semantic-based similarity algorithms generally rely on external semantic dictionaries, Chinese are generally based on HowNet [11], and English are generally based on WordNet [18]. Although the research on text similarity can be taken as the theoretical basis, due to the short length, refined content, high timeliness, non-socialization and other characteristics of short texts, the traditional text similarity algorithms have limitations in solving short text matching problems.

By virtue of the development of deep learning, recent years have seen new breakthroughs in text similarity calculation [19]. The specific steps of a text similarity calculation model using deep learning technology: first segment the input text, then use word embedding techniques to convert the segmented words into word vectors, and then map the two sentence vectors to be compared into the same vector space through the same deep neural network, and finally use classification techniques to calculate how similar the two sentences are. Word vectors are trained by word embedding techniques (the commonly used techniques are Word2Vec [20] and Glove [21]), the essence of which is to train word vectors based on co-occurrence information. This vector representation of words enhances the contextual information representation of words, but does not provide a solution to semantic representations of sentences. Therefore, it is necessary to input the word vectors into the deep neural network, i.e. the encoder, to obtain sentence-level contextual semantic information. BiLSTM-based text matching model [22] implements encoded representations of sentences using bidirectional long short-term memory neural networks. ESIM [6] employs two BiLSTMs to encode different stages, the first one is used to encode sentences, and the other is employed to fuse word alignment information between two sentences, which achieves state-of-the-art results on various matching tasks. LET [23] proposes a linguistic knowledge enhanced graph transformer for Chinese short text matching, which achieves good results on two Chinese datasets. The model proposed in this paper will be improved on the basis of LET.

## 2.2 Data Augmentation

Data augmentation is a common practice to address limited training data and disproportionate training samples. Its working principle is to generate new data with original labels based on the original data, so as to solve the above problems. This year, data augmentation has become more widely used in the NLP fields, but there are still big problems. Unlike images, text data do not have an easy way to be replaced or to generate new data without changing the original semantics. The more commonly used method is EDA [13], which consists of four simple but powerful operations: synonym replacement (SR), random insertion (RI), random swap (RS), and random deletion (RD). AEDA [24] inserts various punctuation marks into the input sequence, and can change the position of characters without changing the order of characters. Sennrich [25] proposed the use of back-translation method to improve the neural network model. The principle of back-translation is to translate the source language into other languages through translation software and then translate it back. Sosuke [26] stochastically replaced words with other words that are predicted by a bidirectional language model at the word position, and called it contextual augmentation.

## 3 Proposed Model

Given two sentences, we denote them by $C^a = \{c_1^a, c_2^a, \ldots, c_{T_a}^a\}$ and $C^b = \{c_1^b, c_2^b, \ldots, c_{T_b}^b\}$. We need method $f(C^a, C^b)$ to predict whether the senses of $C^a$ and $C^b$ are equal. For each text, we use jieba, thulac and ltp to generate a word lattice graph $G = (V, E)$. A word $w_i$ is its corresponding node $x_i \in V$ in the word lattice graph. Then we can obtain all senses through the HowNet. If there is an edge connecting nodes $x_i$ itself and all its reachable nodes in its forward and backward directions as $N_{fw}^{+-}(x_i)$ and $N_{bw}^{+-}(x_i)$. Figure 1 shows a block-level view of our model. Next, we'll explain each component in more detail.

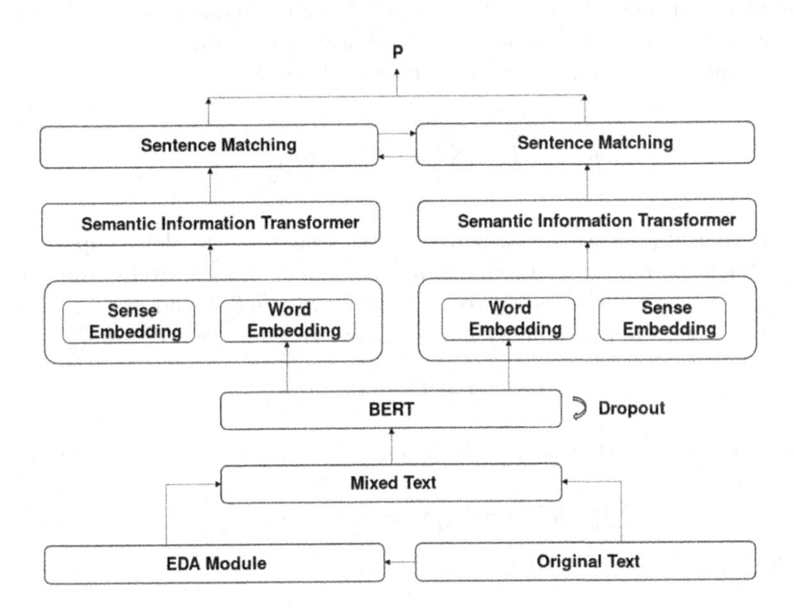


**Fig. 1.** Schematic diagram of EDM model structure.

### 3.1 Data Augmentation Model

The data augmentation model adopted consists of the EDA method module and the dropout module, and the EDA method and the dropout method are the more commonly used data augmentation methods. We get the augmented text through

EDA method, and then mix it with the original text to get the final training data. For all training data, we use the dropout method for data enhancement. This borrows from the data augmentation approach in SimCSE [12].

### 3.2 Input Model

We feed the data-augmented text pairs $C^a$ and $C^b$ into input model, and form two word lattice graphs, so there are two word lattice graphs $G^a = (V^a, E^a)$ and $G^b = (V^b, E^b)$ for each text pair, where $V$ is the set of nodes $x_i$, and $E$ is the set of edges. The word lattice graph contains all possible segmentation results, including characters and words. A word lattice is a directed acyclic graph, and each edge of the graph represents a character or word.

In 2018, Petar et al. [27] proposed a graph attention network applied to graph structured data. The set of all nodes connected to $x_i$ is denoted by $N^+(x_i)$. We use $h_i$ and $h_i^{'}$ to represent the feature vector and the new feature vector of the node $x_i$. The weight coefficient of neighboring node $x_j$ to $x_i$ can be set as $\alpha_{ij} = a(Wh_i, Wh_j)$. After calculation, the degree of relevance between $x_i$ and all neighboring nodes can be obtained. After normalization by softmax, the attention weight of $x_i$ and all neighboring nodes can be obtained. The weighted average value of updated node $h_i^l$ can be calculated as:

$$h_i^l = \sigma\left(\sum_{x_j \in \mathcal{N}^+(x_i)} \alpha_{ij}^l \cdot \left(W^l h_j^{l-1}\right)\right) \tag{1}$$

To avoid the possible limitations in the capacity to model complex dependencies, Shen et al. proposed a multi-dimensional attention mechanism [28]. For each $h_j^{l-1}$, a feature-wise score vector is first calculated and then normalized using feature-wise multi-dimensional softmax, which is denoted by $\beta$,

$$\alpha_{i,j}^l = \beta_j\left(\hat{\alpha}_{i,j}^l + f_m^l\left(h_j^{l-1}\right)\right) \tag{2}$$

where $f_m^l$ is used to estimate the contribution of each feature dimension of $h_j^{l-1}$,

$$f_m^l\left(h_j^{l-1}\right) = W_2^l \sigma\left(W_1^l h_j^{l-1} + b_1^l\right) + b_2^l \tag{3}$$

then the Eq. (1) can be revised as:

$$h_i^l = \sigma\left(\sum_{x_j \in \mathcal{N}^+(x_i)} \alpha_{ij}^l \odot \left(W^l h_j^{l-1}\right)\right) \tag{4}$$

We need to input text pairs into the BERT model to get a contextual representation of each character as $\left\{c^{CLS}, c_1^a, c_2^a, \ldots, c_{T_a}^a, c^{SEP}, c_1^b, c_2^b, \ldots, c_{T_b}^b, c^{SEP}\right\}$. Then we use a feed forward network to obtain a feature-wise score vector for each character, which is denoted by $\gamma$. After that, we can normalize it with feature-wise multi-dimensional softmax,

$$u_k = \beta_k \left(\gamma \left(c_k\right)\right) \tag{5}$$

then we can obtain contextual word embedding:

$$v_i = \sum_{k=t_1}^{t_2} u_k \odot c_k \tag{6}$$

To get sense embedding, we need to use HowNet. HowNet has a established sense and sememe architecture. As an example of the HowNet structure, the word "antique" has two senses, i.e. "antique" and "old fogey", and the "old fogey" has two sememes: "human" and "stiff". HowNet makes it easier to calculate whether two sentences match.

We generate the set of the senses as $S^{w_i} = \{S_{i,1}, S_{i,2}, \ldots, S_{i,k}\}$ for each word $w_i$, then generate the set of sememes as $O^{s_{i,k}} = \left\{O_{i,k}^1, O_{i,k}^2, \ldots, O_{i,k}^n\right\}$ for each sense. We use sememe attention over target model [29] to calculate each sememe's embedding vector $e_{i,k}^n$, then use multi-dimensional attention function to calculate each sememe's representation $o_{i,k}^n$ as:

$$o_{i,k}^{n\prime} = \beta \left(e_{i,k}^n, \left\{e_{i,k}^{n\prime} \mid o_{i,k}^{n\prime} \in O^{s_{i,k}}\right\}\right) \tag{7}$$

For the embedding of each sense, we can obtain it with attentive pooling of all its sememe representations:

$$s_{i,k} = AP\left(\left\{o_{i,k}^n \mid o_{i,k}^n \in O^{s_{i,k}}\right\}\right) \tag{8}$$

### 3.3 Semantic Information Transformer

Contextual information is now separated from semantic information, we propose a word lattice graph transformer. For word $w_i$, we use $v_i$ and $s_{i,k}$ as the original sense representation $g_{i,k}^o$. Then update them iteratively.

To update sense representation from $g_{i,k}^{l-1}$ to $g_{i,k}^l$, we need both backward information and forward information of $x_i$, then update its representation with a gated recurrent unit (GRU) [30],

$$m_{i,k}^{l,bw} = \beta\left(g_{i,k}^{l-1}, \left\{h_j^{l-1} \mid x_j \in N_{bw}^{+}\left(x_i\right)\right\}\right) \tag{9}$$

$$m_{i,k}^{l,fw} = \beta\left(g_{i,k}^{l-1}, \left\{h_j^{l-1} \mid x_j \in N_{fw}^{+}\left(x_i\right)\right\}\right) \tag{10}$$

$$g_{i,k}^{l} = GRU\left(g_{i,k}^{l-1}, m_{i,k}^{l}\right) \tag{11}$$

where $m_{i,k}^{l}=\left\{m_{i,k}^{l,bw},m_{i,k}^{l,fw}\right\}$. We use GRU to control the mix of contextual information and semantic information because $m_{i,k}^{l}$ contains contextual information merely. Then we use $g_{i,k}^{l}$ to update the word representation from $h_i^{l-1}$ to $h_i^l$. The transformer uses multi-dimensional attention to obtain the first sense of word $w_i$ from its semantic information, then updates it with GRU.

$$q_i^l=\beta\left(h_i^{l-1},\{g_{i,k}^l \mid s_{i,k}\in S^{w_i}\}\right) \tag{12}$$

$$h_i^l=GRU\left(h_i^{l-1},q_i^l\right) \tag{13}$$

### 3.4 Sentence Matching Layer

To incorporate word representation into characters, we use characters in $C^a$ as example. We generate a set $W^{c_t^a}$ that contains the words using character $c_t^a$, then use attentive pooling to get $\hat{c}_t^a$ of each character:

$$\hat{c}_t^a=AP\left(\left\{h_i^a \mid w_i^a\in W^{c_t^a}\right\}\right) \tag{14}$$

After obtaining $\hat{c}_t^a$ and $c_t^a$, we use layer normalization to get semantic information enhanced character representation $y_t^a$:

$$y_t^a=LN\left(\hat{c}_t^a+c_t^a\right) \tag{15}$$

Then for each character $c_t^a$, we can use multi-dimensional attention to obtain its aggregative information from $C^a$ and $C^b$, which are denoted by $m_t^s$ and $m_t^c$. When they are almost equal, we can know that these two sentences are matched. To compare them, we need to use multi-perspective cosine distance [7],

$$m_t^s=\beta\left(y_t^a,\{y_{t'}^a \mid c_{t'}^a\in C^a\}\right) \tag{16}$$

$$m_t^c=\beta\left(y_t^a,\{y_{t'}^a \mid c_{t'}^b\in C^b\}\right) \tag{17}$$

$$d_k=CD\left(w_k^{\cos}\odot m_t^s,w_k^{\cos}\odot m_t^c\right) \tag{18}$$

where $w_k^c os$ represents the different weights of different dimensions of a text. We can get the final character representation $d_t:=[d_1,d_2,\ldots,d_k]$ using feed forward networks, then use attentive pooling to obtain the sentence representation vector:

$$\hat{y}_t^a=\gamma\left([m_t^s,d_t]\right) \tag{19}$$

$$r^a=AP\left(\hat{y}_t^a \mid \hat{y}_t^a\in\hat{Y}^a\right) \tag{20}$$

### 3.5 Relation Classifier Layer

Finally, our model can predict the similarity between two sentences by using $r^a$, $r^b$, and $c^{CLS}$.

$$P = \gamma\left(c^{CLS}, r^a, r^b, \left|r^a - r^b\right|, r^a \odot r^b\right) \tag{21}$$

For each $\left\{C_i^a, C_i^b, y_i\right\}$ raining sample, our ultimate goal is to reduce the BCE loss:

$$\mathcal{L} = -\sum_{i=1}^{N}\left(y_i \log\left(p_i\right) + \left(1 - y_i\right)\log\left(1 - p_i\right)\right) \tag{22}$$

where $y_i \in \{0, 1\}$ is the label of the $i$-th training sample we input to the model and $p_i \in [0, 1]$ is the prediction of our model taking sentence pairs as input.

## 4 Experiments

### 4.1 Datasets and Parameter Settings

To validate the performance of EDM, we conduct extensive experiments on two Chinese datasets: AFQMC [31] and BQ corpus [32].

The AFQMC dataset is the dataset of ANT Financial ATEC: NLP Problem Similarity Calculation Competition, and it is a dataset for classification task. All data are from the actual application scenarios of Ant Financial's financial brain, that is, two sentences described by users in a given customer service are determined by algorithms to determine whether they represent the same semantics. Synonymous sentences are represented by 1, non-synonymous sentences are represented by 0, and the format is consistent with LCQMC dataset. Since the AFQMC dataset was not pre-divided into training set, test set and validation set, we divide the dataset by the ratio of 6:2:2 in this experiment. The data volumes of training set, test set and validation set are 61,486, 20,496, and 20,495, respectively, and the total number of all samples is 102,477.

The BQ dataset is a question matching dataset in the field of banking and finance. Comprising question text pairs extracted from one year of online banking system logs, it is the largest question matching dataset in the banking domain. It classifies two paragraphs of bank credit business according to whether they are semantically similar or not. 1 represents the similarity judgment while 0 represents the dissimilarity judgment. The BQ dataset contains 120,000 pieces of data in total, including 100,000 for training set, 10,000 for validation set and 10,000 for test set.

We use the BERT with 12 layers, 4 attention heads, the embedding dimension of 128 and the hidden dimension of 512. The number of graph updating layers L is 2, the dimensions of both word and sense representations are 128, and the hidden size is also set as 128. The dropout rate is 0.2 and the warmup rate is 0.1. We use Open-howNet [14] with 200 dimensions for sememe embedding.

### 4.2 Compared Systems

EDM will be compared with six baseline models, as follows:

**Text-CNN:** It is a representation-based model that utilizes Convolutional Neural Networks (CNN) to encode text to generate the representations of n-gram in the sentence, followed by softmax for match prediction [33].

**BiLSTM:** Bi-directional Long Short Term Memory (BiLSTM) is an improved version of Long Short Term Memory (LSTM), which is composed of forward and backward LSTM layers, it can better obtain bidirectional semantics [22].

**ALBERT:** Based on BERT, it uses factorized embedding and parameter sharing between layers to reduce the amount of parameters [34].

**ERNIE:** By changing the BERT model structure, ERNIE fuses knowledge and language semantic information to enhance semantic representation [35].

**BERT:** Bidirectional Encoder Representation from Transformers(BERT) is a pre-training model proposed by Google that uses a bidirectional encoding architecture [9].

**RoBERTa:** RoBERTa is an improved version of BERT that improves model performance by improving training tasks and data generation, prolonging training time, and using larger batches and more data [36].

### 4.3 Main Results

Accuracy (ACC.) and F1 score are used as the evaluation metrics. Table 1 shows the accuracy and F1 values of EDM and seven comparison models on two Chinese datasets. It can be seen that in the AFQMC dataset, EDM reaches the best level, and the accuracy is improved by 1.1% compared with the RoBERTa model. In the BQ dataset, although the accuracy of EDM is slightly lower than that of the RoBERTa model, it has a good accuracy improvement compared to the other six models. Compared with the existing data, on the two Chinese datasets, the EDM achieves an average improvement of 0.62% and 0.85% in the F1 value.

**Table 1.** Comparison of experimental results with other models.

| MODEL | AFQMC | | BQ | |
|---|---|---|---|---|
| | ACC | F1 | ACC | F1 |
| Text-CNN | 61.15 | 60.09 | 68.52 | 69.17 |
| BiLSTM | 64.68 | 54.53 | 73.51 | 72.68 |
| ALBERT | 72.55 | - | 83.34 | - |
| ERNIE | 73.83 | 73.91 | 84.67 | 84.20 |
| BERT | 73.70 | 74.12 | 84.50 | 84.00 |
| RoBERTa | 74.04 | - | **85.00** | - |
| **EDM** | **74.87** | **74.65** | 84.82 | **84.73** |

### 4.4 Analysis

**Ablation Study.** We conduct ablation experiments to demonstrate the effectiveness of data augmentation with external knowledge. We select three indicators: data augmentation (DA) and lattice graph (LG) to conduct experiments. When an ablation experiment is performed on a certain metric, other parts of the model are not changed. The results of the ablation experiments are shown in Table 2.

**Table 2.** Results of ablation experiments on the BQ dataset.

| | | ACC. | F1 |
|---|---|---|---|
| EDM | | 84.82 | 84.73 |
| -DA | | 84.48 | 84.33 |
| -LG | | 83.88 | 83.76 |
| -DA | -LG | 83.71 | 83.24 |

**Influences of Data Augmentation on Performance.** Existing datasets have the problem of unreasonable number of labels, taking the AFQMC dataset as an example. Before data augmentation, the ratio of 0 and 1 labels of AFQMC reached 4:1. By enhancing the data with the label of 1, we increase the label ratio to 2:1, which reduces the scale of labels without wasting the original data. Data augmentation can generate augmented data that are similar to the original data, and introduce a certain amount of noise, which helps prevent overfitting.

We use random swap and synonym replacement in the EDA method. As shown in Table 3, using the hybrid method can effectively improve the accuracy of the model on the basis of solving the above problems. Compared with dropout method or EDA method, the accuracy of hybrid method is improved by about 0.2% and 0.4%, respectively.

**Table 3.** Comparison of data augmentation methods.

| Method | ACC. | F1 |
|---|---|---|
| dropout | 74.72 | 74.69 |
| EDA | 74.56 | 74.51 |
| hybrid | 74.87 | 74.65 |

## 5 Conclusion

In this paper, we propose an external knowledge and data augmentation enhanced model (EDM). The proposed model is mainly applicable to Chinese short text matching, and its effectiveness has been verified on two public Chinese datasets. To address the label imbalance in the existing datasets, we use a fusion of EDA method and dropout method for data augmentation. To solve the problem of ambiguity in Chinese words, we use jieba, thulac and ltp to generate word lattice graphs, and use OpenHowNet to fuse external knowledge. Extensive experimental results show that EDM outperforms most of the existing benchmark models on Chinese datasets.

## References


1. MacCartney, B., Galley, M., Manning, C.D.: A phrase-based alignment model for natural language inference. In: Proceedings of the 2008 Conference on Empirical Methods in Natural Language Processing, pp. 802–811 (2008)
2. Dolan, B., Brockett, C.: Automatically constructing a corpus of sentential paraphrases. In: Third International Workshop on Paraphrasing (IWP2005) (2005)
3. Yang, Y., Yih, W., Meek, C.: Wikiqa: a challenge dataset for open-domain question answering. In: Proceedings of the 2015 Conference on Empirical Methods in Natural Language Processing, pp. 2013–2018 (2015)
4. Brill, E., Dumais, S., Banko, M.: An analysis of the AskMSR question-answering system. In: Proceedings of the 2002 Conference on Empirical Methods in Natural Language Processing (EMNLP 2002), pp. 257–264 (2002)
5. Gomaa, W.H., Fahmy, A.A., et al.: A survey of text similarity approaches. Int. J. Comput. Appl. **68**, 13–18 (2013)
6. Chen, Q., Zhu, X., Ling, Z., Wei, S., Jiang, H., Inkpen, D.: Enhanced LSTM for natural language inference. arXiv preprint arXiv:1609.06038 (2016)
7. Wang, Z., Hamza, W., Florian, R.: Bilateral multi-perspective matching for natural language sentences. arXiv preprint arXiv:1702.03814 (2017)
8. Gong, Y., Luo, H., Zhang, J.: Natural language inference over interaction space. arXiv preprint arXiv:1709.04348 (2017)
9. Devlin, J., Chang, M.W., Lee, K., Toutanova, K.: BERT: Pre-training of Deep Bidirectional Transformers for Language Understanding (2018)
10. Xu, J., Liu, J., Zhang, L., Li, Z., Chen, H.: Improve Chinese word embeddings by exploiting internal structure. In: Proceedings of the 2016 Conference of the North American Chapter of the Association for Computational Linguistics: Human Language Technologies, pp. 1041–1050 (2016)
11. Dong, Z., Dong, Q.: HowNet-a hybrid language and knowledge resource. In: Proceedings of the International Conference on Natural Language Processing and Knowledge Engineering, 2003, pp. 820–824. IEEE (2003)
12. Gao, T., Yao, X., Chen, D.: Simcse: simple contrastive learning of sentence embeddings. arXiv preprint arXiv:2104.08821 (2021)
13. Wei, J., Zou, K.: Eda: Easy data augmentation techniques for boosting performance on text classification tasks. arXiv preprint arXiv:1901.11196 (2019)
14. Qi, F., Yang, C., Liu, Z., Dong, Q., Sun, M., Dong, Z.: Openhownet: an open sememe-based lexical knowledge base. arXiv preprint arXiv:1901.09957 (2019)
15. Zhang, Y., Jin, R., Zhou, Z.-H.: Understanding bag-of-words model: a statistical framework. Int. J. Mach. Learn. Cybern. **1**, 43–52 (2010)
16. Blei, D.M., Ng, A.Y., Jordan, M.I.: Latent Dirichlet allocation. J. Mach. Learn. Res. **3**, 993–1022 (2003)

17. Mohler, M., Mihalcea, R.: Text-to-text semantic similarity for automatic short answer grading. In: Proceedings of the 12th Conference of the European Chapter of the ACL (EACL 2009), pp. 567–575 (2009)
18. Fang, H., Tao, T., Zhai, C.: Diagnostic evaluation of information retrieval models. ACM Trans. Inf. Syst. (TOIS) **29**, 1–42 (2011)
19. Lan, W., Xu, W.: Neural network models for paraphrase identification, semantic textual similarity, natural language inference, and question answering. In: Proceedings of the 27th International Conference on Computational Linguistics, pp. 3890–3902 (2018)
20. Mikolov, T., Chen, K., Corrado, G., Dean, J.: Efficient estimation of word representations in vector space. Computer Science (2013)
21. Pennington, J., Socher, R., Manning, C.D.: Glove: global vectors for word representation. In: Proceedings of the 2014 Conference on Empirical Methods in Natural Language Processing (EMNLP), pp. 1532–1543 (2014)
22. Mueller, J., Thyagarajan, A.: Siamese recurrent architectures for learning sentence similarity. In: Proceedings of the AAAI Conference on Artificial Intelligence (2016)
23. Lyu, B., Chen, L., Zhu, S., Yu, K.: Let: linguistic knowledge enhanced graph transformer for Chinese short text matching. arXiv preprint arXiv:2102.12671 (2021)
24. Karimi, A., Rossi, L., Prati, A.: AEDA: An Easier Data Augmentation Technique for Text Classification. arXiv preprint arXiv:2108.13230 (2021)
25. Sennrich, R., Haddow, B., Birch, A.: Improving neural machine translation models with monolingual data. arXiv preprint arXiv:1511.06709 (2015)
26. Kobayashi, S.: Contextual augmentation: Data augmentation by words with paradigmatic relations. arXiv preprint arXiv:1805.06201 (2018)
27. Veličković, P., Cucurull, G., Casanova, A., Romero, A., Lio, P., Bengio, Y.: Graph attention networks. arXiv preprint arXiv:1710.10903. (2017)
28. Shen, T., Zhou, T., Long, G., Jiang, J., Pan, S., Zhang, C.: Disan: directional self-attention network for RNN/CNN-free language understanding. In: Proceedings of the AAAI Conference on Artificial Intelligence (2018)
29. Niu, Y., Xie, R., Liu, Z., Sun, M.: Improved word representation learning with sememes. In: Proceedings of the 55th Annual Meeting of the Association for Computational Linguistics (Volume 1: Long Papers), pp. 2049–2058 (2017)
30. Caruana, R.: Learning many related tasks at the same time with backpropagation. In: Advances in Neural Information Processing Systems, pp. 657–664 (1995)
31. Xu, L., Zhang, X., Dong, Q.: CLUECorpus2020: a large-scale Chinese corpus for pre-training language model. arXiv preprint arXiv:2003.01355 (2020)
32. Chen, J., Chen, Q., Liu, X., Yang, H., Lu, D., Tang, B.: The bq corpus: a large-scale domain-specific Chinese corpus for sentence semantic equivalence identification. In: Proceedings of the 2018 Conference on Empirical Methods in Natural Language Processing, pp. 4946–4951 (2018)
33. Chen, Y.: Convolutional neural network for sentence classification (2015)
34. Lan, Z., Chen, M., Goodman, S., Gimpel, K., Sharma, P., Soricut, R.: Albert: a lite bert for self-supervised learning of language representations. arXiv pre-print arXiv:1909.11942 (2019)
35. Zhang, Z., Han, X., Liu, Z., Jiang, X., Sun, M., Liu, Q.: ERNIE: enhanced language representation with informative entities. arXiv preprint arXiv:1905.07129 (2019)
36. Liu, Y., et al.: Roberta: a robustly optimized bert pretraining approach. arXiv preprint arXiv:1907.11692 (2019)

# Improving Oracle Bone Characters Recognition via A CycleGAN-Based Data Augmentation Method

Wei Wang[1], Ting Zhang[1(✉)], Yiwen Zhao[1], Xinxin Jin[1], Harold Mouchere[2], and Xinguo Yu[1]

[1] Faculty of Artificial Intelligence in Education, Central China Normal University, Wuhan, China
ting.zhang@ccnu.edu.cn

[2] Nantes Université, École Centrale Nantes, LS2N, UMR 6004, 44000 Nantes, France

**Abstract.** Oracle bone inscription is the earliest writing system in China which contains rich information about the history of Shang dynasty. Automatically recognizing oracle bone characters is of great significance since it could promote the research on history, philology and archaeology. The proposed solutions for oracle bone characters recognition are mainly based on machine learning or deep learning algorithms which rely on a large number of supervised training data. However, the existing dataset suffers from the problem of severe class imbalance. In this work, we propose a CycleGAN-based data augmentation method to overcome the limitation. Via learning the mapping between the glyph images data domain and the real samples data domain, CycleGAN could generate oracle character images of high-quality. The quality is evaluated using the quantitative measure. Totally, 185362 samples are generated which could serve as a complementary to the existing dataset. With these generated samples, the state of the art results of recognition task on OBC306 are improved greatly in terms of mean accuracy and total accuracy.

**Keywords:** Characters recognition · Class imbalance · Data augmentation · CycleGAN

## 1 Introduction

Oracle bone inscriptions, usually inscribed on turtle shells and animal bones, are the oldest hieroglyphs in China. They record the development of civilization of late Shang dynasty. The study of oracle bone inscriptions could promote the research on the origin of Chinese characters, as well as the research on the history of Shang

This work is supported by National Natural Science Foundation of China (No. 62007014), China Post doctoral Science Foundation (No. 2019M652678) and the Fundamental Research Funds for the Central Universities(No. CCNU20ZT019).


M. Tanveer et al. (Eds.): ICONIP 2022, CCIS 1793, pp. 88–100, 2023.
https://doi.org/10.1007/978-981-99-1645-0_8

dynasty. Since firstly discovered in the city of Anyang, China, around 160000 bones or shells (see Fig. 1) were unearthed gradually from which about 4500 classes of characters were found however more than 2000 characters remain not deciphered so far. Currently, the recognition and decipherment of oracle bone characters majorly rely on the manual method by domain experts with high expertise which to a certain degree blocks the development of this field. Thus, developing automatic recognition methods is in need and of great significance.

**Fig. 1.** Examples of unearthed oracle bone inscriptions.

The state of the art methods for oracle bone characters recognition are mainly based on deep learning algorithms. This task is more difficult than other handwritten characters recognition tasks since it exhibits several challenges. (1) Large intra-class variance. As oracle bone characters were carved by different ancients from multiple regions over several historic periods without much standardization [1], the same category could have several variants. Practically in China, characters were unified for the first time in Qin dynasty, after Shang dynasty. (2) Class imbalance. Due to the scarcity of the unearthed bones or shells, few instances were found for some characters. The distribution of the instances with respect to each category is quite imbalanced, following the long-tail distribution [8]. (3) Low quality. Some inscriptions were damaged to some degree during the long burial period or the excavation process. Accordingly, there exist some noisy or broken character instances in the collected dataset [8].

Several oracle bone character datasets have been built up to now for the research aim such as Oracle-20K [7] (20039 instances belonging to 261 categories), Oracle-AYNU [6] (39062 instances belonging to 2583 categories) and OBC306 [8] (309551 instances belonging to 306 categories) among which OBC306 has the largest scale and is closer to the real case as it uses the published oracle prints as the raw materials. As stated previously, the existing oracle datasets are quite imbalanced due to the scarcity of the unearthed materials. To overcome this limitation, in this work we propose a data augmentation method for oracle characters using Cycle-Consistent Adversarial Network (CycleGAN) [12]. CycleGAN is trained to learn the mapping between the glyph images and the real samples to develop the capability of generating oracle images of high-quality.

The main contributions of this paper are as follows:

- We propose a feasible solution to generate oracle character images using CycleGAN by enabling it to learn the mapping between the glyph images

collection and the real samples collection. Totally, 185362 samples are generated which will be released later as a supplementary to the existing dataset for boosting the research in this field.
- To our knowledge, it is the first attempt to use the GAN-based methods to generate oracle character images.
- With the generated samples, the state of the art results of recognition task on OBC306 are greatly improved in terms of mean accuracy and total accuracy.

The rest of this paper is organized as follows. The related works are reviewed shortly in Sect. 2. Section 3 introduces the proposed method. The experiments and results are discussed in Sect. 4. Section 5 concludes the work and proposes several perspectives for future work.

## 2 Related Works

### 2.1 Oracle Bone Characters Recognition

The research on oracle bone characters recognition is quite limited. Before the deep learning era, the proposed methods [2–4] for oracles recognition mainly follows the traditional pipeline, being pre-processing, feature extraction and classification. The accuracy of the models relies much on features extracted. Researchers usually design features representing oracles based on graph theory and topology theory.

The idea proposed in [2] regards an oracle character as a non-directed graph and uses the topological features for classification. The method in [3] firstly transforms the inscriptions to labelled graphs, then extracts the quasi-adjacency matrix of the graphs as features, finally perform the recognition using an algorithm based on graph isomorphism. [4] takes the skeleton and line points as features which later are fed into a two-stage recognition process. Even though these manually designed features were reported to work well on specific datasets, they have a low generalization capability.

Owing to the advantage of deep convolutional neural networks (DCNNs) [5], feature representations could be learned from data automatically. These feature representations are robust, and have strong generalization capability. Guo et al. [7] develop a novel hierarchical representation which combines a Gabor-related low-level representation and a sparse-encoder-related mid-level representation for oracle bone characters. This hierarchical representation serves as a complementary to CNN-based models. One point should be noted that [7] discards oracle characters with less than 25 instances which means they skip from the problem of class imbalance. To overcome the limitations, [6] propose a deep metric learning based classification model which uses a CNN to map oracle images into an Euclidean space and then carries out the classification by the Nearest Neighbour (NN) rule. This method achieves the state of the art results on the public dataset Oracle-20K and Oracle-AYNU. Almost at the same time, a large-scale oracle bone character dataset was released which consists of 309551 samples. However, this dataset also suffers the problem of class imbalance by nature. To solve the imbalance problem, [19] proposed a mix-up augmentation strategy to synthesize samples in the feature space by leveraging information from both majority and minority classes. Experimental results verify the effectiveness of the proposed method. However, [19] synthesize samples in the feature space which makes it difficult to evaluate the quality of the synthesized samples explicitly. Different from work [19], we propose to generate samples directly for minority classes with GAN-based methods.

### 2.2 Generative Adversarial Networks (GANs)

GAN [9,10] was firstly proposed by Ian GoodFellow in 2014 and since then it has quickly become a hot research topic. GAN consists of two models, being a generator and a discriminator, which can be implemented by any form of differentiable system such as neural network. The generator aims to capture the distribution of real samples for new data generation while the discriminator struggles to discriminate generated data from the real data as accurately as possible [11]. GANs are widely applied in the field of image processing and computer vision, such as image super-resolution, image synthesis and manipulation, etc. Many interesting variants of GANs have been proposed since 2014.

Conditional GANs (cGANs) [13] extend GANs to a conditional model by conditioning the generator and discriminator on some extra information. With cGANs, samples can be generated conditioning on class labels, texts, etc. In [14], cGAN is used for image-to-image translation. The developed software is named as pix2pix. The generator aims to learn a mapping from the input image (instead of random noise) to the generated image. pix2pix successfully applied cGANs for image-to-image translation however the need for the aligned image pairs hinders its generalisation.

The limitation of pix2pix is well solved by Cycle-consistent GANs (CycleGAN) [12] which could learn to translate an image from one domain to another without paired examples. By exploiting the property that translation should be cycle consistent, CycleGAN introduce a cycle consistency loss in addition to the typical adversarial loss.

Even though the limitation of pix2pix seems not to be a hinder in the context of our task and a paired image-to-image translation network can be trained to generate higher quality images, our task benefits a lot from the unpaired image-to-image training. Because being different from other image generation tasks which pursue only the high quality of images, our main goal is to improve the recognition accuracy, and it's mostly influenced by the diversity of the data. CycleGAN [12] tries to learn to the mapping between two domain collections without aligned image pairs. This working mechanism is beneficial to the oracle character images generation task especially for minority classes which could utilize rich stroke variety information from majority classes. Therefore, we propose to generate oracle bone character images based on CycleGAN.

# 3 Method

To our knowledge, OBC306 is the existing largest dataset for oracle bone characters recognition. However, it suffers a severe problem of class imbalance due to the scarcity of the unearthed inscriptions. Our aim is to generate samples for minority classes in OBC306 using GAN-based methods. The vanilla GAN model [9] generates samples without designated class labels. The generated sample could belong to any class. However, our target is to generate samples belonging to designated classes. One possible way is to train a GAN model for each class. Unfortunately, training one model for each class requires a large amount of samples which is in conflict with minority classes. This contradiction makes the task based on the vanilla GAN model be challenging. The propose of cGAN [13] makes it possible to generate samples conditioning on specific class labels. cGAN is trained to learn the mapping from latent code to the corresponding image which is especially complex for a domain like oracle characters which have complicated glyphs. Furthermore, it also requires considerable number of samples from each class. *pix2pix* [14] is one successful application of cGAN in the image-to-image translation task. It uses aligned image pairs from two different domains (such as gray scale and color) to train the model to learn the mapping from one domain image to the other. The major limitation of *pix2pix* is its requirement for aligned image pairs. Even though this limitation seems not to be a hinder in the context of our task and a paired image-to-image translation network can be trained to generate higher quality images, our task benefits a lot from the unpaired image-to-image training. Because being different from other image generation tasks which pursue only the high quality of images, our main goal is to improve the recognition accuracy, and it's mostly influenced by the diversity of the data. CycleGAN [12] tries to learn to the mapping between two domain collections without aligned image pairs. This working mechanism is beneficial to the oracle character images generation task especially for minority classes which could utilize rich stroke variety information from majority classes. Based on the analysis above, A CycleGAN-based data augmentation method for improving oracle bone characters recognition is proposed in this work. Two data domains are assumed as glyph images and real character images respectively.

## 3.1 Statistics of OBC306

OBC306 [8] consists of 309551 instances belonging to 306 categories. There are 70 categories with more than 1000 samples, including 5 categories with more than 10000 images. 236 categories with less than 1000 samples, including 85 categories with less than 100 samples and 29 categories with only one sample. Figure 2 illustrates the distribution of data in OBC306. As can be seen, the distribution is quite imbalanced. We set 1000/category as the target of augmentation.

## 3.2 Extraction of the Glyphs of Oracle Characters

We extract the glyph of oracle from the real samples. The process for extraction of the glyph is quite simple, being binarization, noise erasing and image inverse

**Fig. 2.** Illustration of data distribution in OBC306.

if required. Figure 3 presents several cases of the extracted glyph. The glyph images will be one of the image collections of CycleGAN.

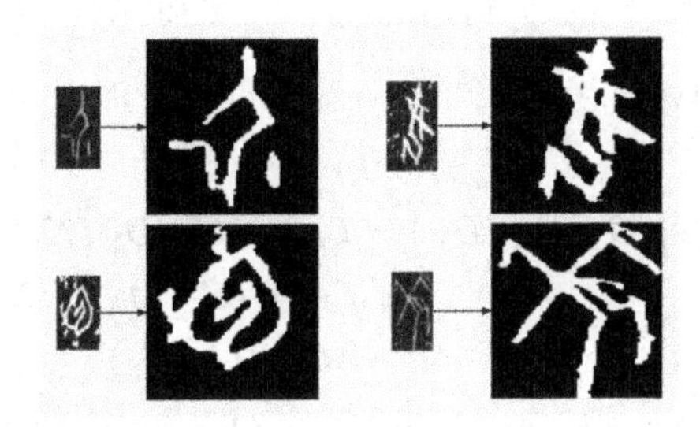

**Fig. 3.** Some examples of the extracted glyph.

### 3.3 CycleGAN

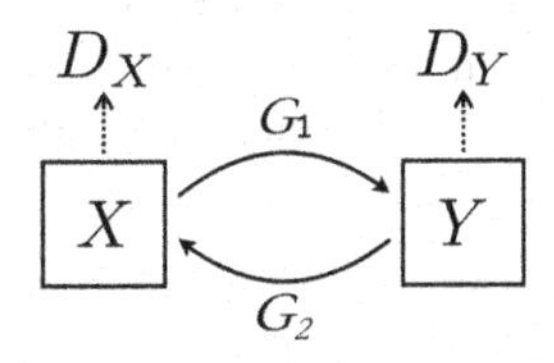


**Fig. 4.** The framework of CycleGAN.

Given two image collections $X$ and $Y$, CycleGAN aims to learn the mapping between $X$ and $Y$. As illustrated in Fig. 4, CycleGAN consists of two generators $G_1 : X \rightarrow Y$ and $G_2 : Y \rightarrow X$, and two discriminators $D_X$ and $D_Y$. $G_1$ tries to generate images $G_1(x)$ which are indistinguishable from images $\in Y$ and $D_Y$ tries to distinguish between generated samples $G_1(x)$ and real samples $\in Y$. The same case is for $G_2$ and $D_X$. In order to constrain the learned mapping further, the consistency loss (Eq. 1) is introduced in addition to the adversarial

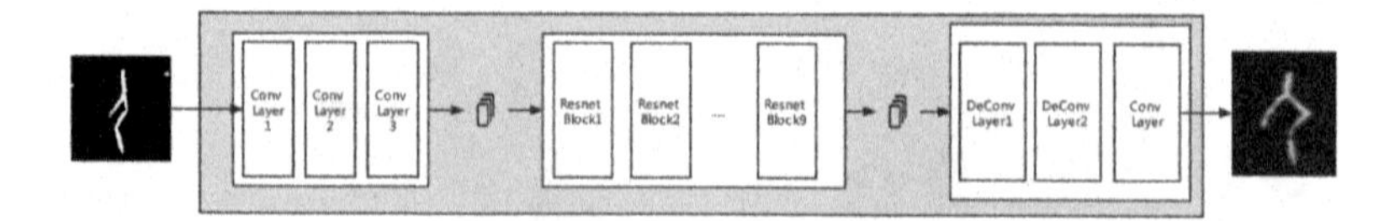


**Fig. 5.** The network architecture of the generator used in this work.

loss (Eq. 2). The consistency loss term tries to make the model work as: $x \rightarrow G_1(x) \rightarrow G_2(G_1(x)) \approx x$ and $y \rightarrow G_2(y) \rightarrow G_1(G_2(y)) \approx y$. The full objective is shown as Eq. 3.

$$\begin{aligned} \mathcal{L}_{\text{cyc}}(G_1, G_2) = {} & \mathbb{E}_{x \sim p_{\text{data}}(x)} \left[ \|G_2(G_1(x)) - x\|_1 \right] \\ & + \mathbb{E}_{y \sim p_{\text{data}}(y)} \left[ \|G_1(G_2(y)) - y\|_1 \right] \end{aligned} \tag{1}$$

$$\begin{aligned} \mathcal{L}_{\text{GAN}}\left(G_1, D_Y, X, Y\right) = {} & \mathbb{E}_{y \sim p_{\text{data}}(y)} \left[\log D_Y(y)\right] \\ & + \mathbb{E}_{x \sim p_{\text{data}}(x)} \left[\log\left(1 - D_Y(G_1(x))\right]\right. \end{aligned} \tag{2}$$

$$\begin{aligned} \mathcal{L}\left(G_1, G_2, D_X, D_Y\right) = {} & \mathcal{L}_{\text{GAN}}\left(G_1, D_Y, X, Y\right) \\ & + \mathcal{L}_{\text{GAN}}\left(G_2, D_X, Y, X\right) \\ & + \lambda \mathcal{L}_{\text{cyc}}(G_1, G_2) \end{aligned} \tag{3}$$

As CycleGAN does not require paired samples, we randomly select 1000 real images and 1000 glyph images to train a CycleGAN model. Both generators and discriminators are implemented with neural networks. We use the same network architecture with literature [12]. The generator consists of three convolution layers for extracting features from input images, nine residual blocks which could translate feature vectors from the source domain to the target domain, two deconvolution layers and one convolution layer for restoring low-level features from feature vectors and generating images following the distribution of the target domain. The architecture is presented in Fig. 5. For the discriminator (Fig. 6), we use five convolution layers among which the last one is used to produce a 1-dimensional output telling input images are real or fake.

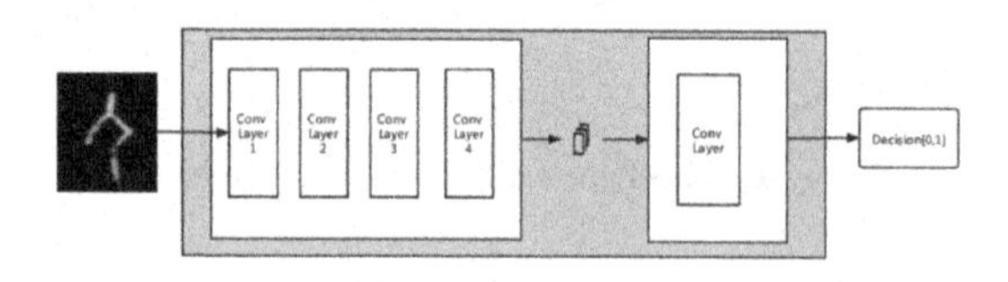


**Fig. 6.** The network architecture of the discriminator used in this work.

### 3.4 Samples Generation for Categories with Few Instances

Our object is to generate instances for oracle characters with few real instances to solve the problem of class imbalance. 1000/category is the target of augmentation. However for some categories with few real instances, the corresponding glyph images are quite few also. To achieve the target, we enrich the glyph images set first via general augmentation methods such as rotation, noise generation, occlusion, etc. Then feed these glyph images to multiple trained CycleGAN models to generate oracle character images. For example, assume that we need to generate 900 instances for some category. First step is to get 30 glyph images via general augmentation methods if required and next is to feed these 30 glyph images to 30 trained CycleGAN models to yield 900 instances.

## 4 Experiments

### 4.1 Experiments on Oracle Character Images Generation

**Settings.** As stated previously, we evaluate the proposed method on the OBC306 dataset. We randomly select 1000 real images and 1000 glyph images (unpaired) to train a CycleGAN model. The model is trained for 200 epochs with Adam optimizer [15]. For the first 100 epochs, the learning rate is set to 0.0002; it linearly decays to zero over the next 100 epochs.

**Metrics.** GAN-train and GAN-test as defined in [17] are used as quantitative metrics to evaluate the quality of generated samples.

- **GAN-train** represents the accuracy of a classification model trained with the synthetic samples and evaluated on real samples.
- **GAN-test** is the accuracy of a classification model trained with real data, and evaluated on synthetic data.
- **GAN-base** denotes the accuracy of a baseline network trained and tested on real data [18].

Via comparing GAN-train accuracy and GAN-base accuracy, GAN-test accuracy and GAN-base accuracy, we can check the quality of the generated samples. Ideally, both GAN-train and GAN-test should be close to GAN-base accuracy.

**Results.** We carry out samples augmentation for 236 categories with less than 1000 instances. Totally, there are 185362 samples generated. Figure 7 compares the generated samples with the real samples of the same category. From the perceptual point of view, there is still a discrepancy between the real ones and the generated ones.

We also take an automatic quantitative measure on the generated samples using GAN-train and GAN-test. As CycleGAN models are trained with samples randomly selected, in other words, with unbalanced data samples, we question that if the quality of generated samples is different or not among categories with

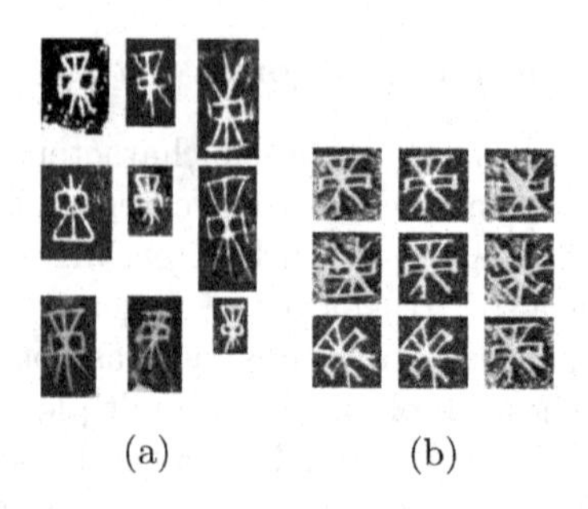

Fig. 7. Comparison of the real samples (a) and the generated samples (b).

variable numbers of real instances. Thus, experiments are carried out respectively on categories with different numbers of real instances. Specifically, we select 5 categories which have around 500 real instances to conduct Experiment $A$ and 5 categories which have 10–100 real instances to conduct Experiment $B$. From Table 1, we can see in Experiment $A$, GAN-train is relatively close to GAN-base (87.4% VS 83.2%) which tells that CycleGAN models generate for the selected 5 categories high-quality images with similar diversity to the real data. In Experiment $B$, GAN-train is 9.5% less than GAN-base (81.6% VS 72.1%) which means the generated samples for the selected categories lost some patterns compared to the real data. For both cases, GAN-test is not far away from GAN-base and is not larger than GAN-base telling that the trained CycleGAN models do not have the problem of over-fitting. In conclusion, CycleGAN tends to generate samples of higher quality for categories with a lot of real instances.

**Table 1.** Results of GANtrain and GANtest experiments. $A$: experiment on categories with around 500 real instances; $B$: experiment on categories with 10–100 real instances

| Experiment | GAN-base | GAN-train | GAN-test |
|---|---|---|---|
| $A$ | 87.4% | 83.2% | 84.3% |
| $B$ | 81.6% | 72.1% | 76.8% |

### 4.2 Experiments on Oracle Characters Recognition

**Settings.** We adopt AlexNet [5], ResNet50 [16] and Inception-v4 [20] which have shown good performance on OBC306 as the classification models. For each category, 80% of the samples are used for training, and the remaining 20% are used for test. Classifiers are trained with Adam optimizer also.

To evaluate the contribution of generated samples to recognition task, we set three experiments. The three experiments use the same test data. In terms of training data, Experiment I uses samples from the original OBC306 dataset only.

Experiment II uses samples from the original OBC306 dataset + samples generated with traditional methods. Experiment III uses samples from the original OBC306 dataset + samples generated via CycleGAN. The numbers of training data for Experiment II & III are the same, larger than the number of training data in Experiment I .

**Metrics.** To account for long-tail effects, we measure the performance of the model with regard to multiple metrics as same as [19].

- **Mean accuracy** denotes the accuracy that is calculated by averaging the accuracy of per class.
- **Total accuracy** refers to the overall accuracy of all the test samples on all classes.

**Table 2.** Results of oracle characters recognition on the OBC306 dataset.

| Experiment | Model | Mean accuracy | | | Total accuracy |
|---|---|---|---|---|---|
| | | Top 1 | Top 5 | Top 10 | |
| I: Original OBC306 | AlexNet [8] | 65.24% | 74.22% | 79.15% | — |
| | ResNet50 [8] | 68.77% | 78.67% | 81.32% | — |
| | Inception-v4 [8] | 69.51% | 78.11% | 81.16% | 86.41% |
| | Inception-v4 (277 classes)* | 75.62% | — | — | 89.81% |
| II: Traditional augmentation | AlexNet | 66.41% | 75.24% | 80.10% | — |
| | ResNet50 | 71.89% | 80.27% | 81.74% | — |
| | Inception-v4 | 74.61% | 82.92% | 84.23% | 89.92% |
| III: CycleGAN-based augmentation | AlexNet | 70.17% | 77.22% | 82.71% | — |
| | ResNet50 | 77.63% | 87.15% | 91.83% | — |
| | Inception-v4 | **83.44%** | **91.39%** | **93.13%** | **92.02%** |
| IV: Mix-up augmentation (277 classes)* | Inception-v4 [19] | 79.02% | — | — | 91.56% |

* The results of these two rows are extracted directly from the published literature [19]. They remove 29 classes which have only one instance and carry out the experiments on the remaining 277 classes. The rest of the results in this table are produced by us.

**Results.** Table 2 presents recognition accuracies in different experiments. From the statistics, we can see the samples generated by CycleGAN could improve the recognition accuracy dramatically: mean accuracy by approximately 14% (69.51% → 83.44%), total accuracy by approximately 6% (86.41% → 92.02%). However, the samples generated by traditional methods only improve the mean accuracy by around 5% (69.51% → 74.61%) and total accuracy by around 3% (86.41% → 89.92%). [19] proposed a mix-up augmentation strategy to synthesize samples by fusing features linearly from two examples. This method synthesizes samples in the feature space which makes it difficult to evaluate the quality of the synthesized samples explicitly. Furthermore, our method outperforms the mix-up strategy in terms of both mean and total accuracy. While their total accuracy is improved by around 2% (89.81% → 91.56%). Our total accuracy is improved by around 6% (86.41% → 92.02%).

Here is one point we need to clarify. For 29 classes with only one sample, we put them in the test set. Thus, there is no data from these classes to train CycleGAN models. But, we feed glyph images of these classes into trained CycleGANs to generate samples (1000 per class). These operations maybe influence the mean accuracy a lot. However, the total accuracy will not be affected much as there are around 60000 samples in the test set. Furthermore, the total accuracy is improved by around 6% which verifies the effectiveness of the proposed method.

To further analyse the effects of generated samples on the recognition accuracy of classes with variable numbers of real instances, we design a new recognition experiment. (1) Firstly randomly select 5 categories which have 1000–1100 real instances. (2) Secondly remove a part of the real instances to make these 5 categories be minority classes. (3) Use the traditional augmentation method and the CycleGAN-based method respectively to recover the original quantity. (4) ResNet50 is adopted as the classifier. Figure 8 shows the recognition results under different settings. As can be seen, when 90% real data removed, CycleGAN-based method exhibits significantly better performance than the traditional method. When 70% real data removed, CycleGAN-based method slightly outperforms the traditional method. When 50% real data removed, there is no obvious discrepancy between the two methods. Form this experiment, we could tell that the CycleGAN-based augmentation method has a good performance for minority classes. For classes with around 500 real instances, only a slight improvement is observed. The result of the case using 50% real data is close to the case using 100% real data which means the classifier learns the patterns of oracle characters with around 500 samples each category.

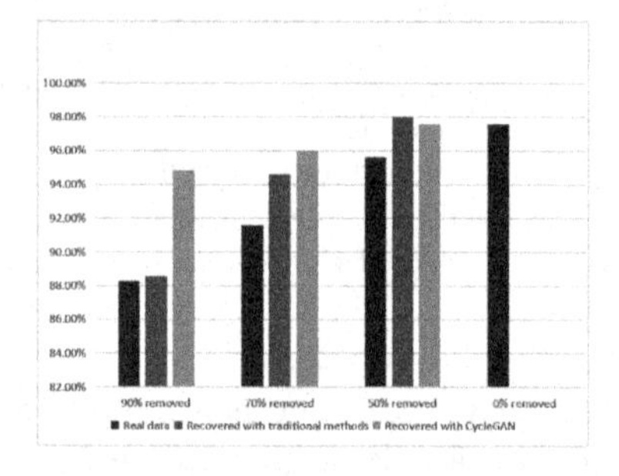


**Fig. 8.** Results of oracle characters recognition on specific classes.

## 5 Conclusion

In this work, we propose a CycleGAN-based data augmentation method to solve the problem of class imbalance in the existing oracle characters dataset. Totally, 185362 samples are generated which could be a supplementary to the existing dataset. With these generated samples, the SOTA results of recognition task on OBC306 are improved greatly in terms of mean accuracy and total accuracy.

However, some limitations exist in our work. Currently, we extract the glyph from real samples only and take some simple transformations if required. In fact, during the training process of CycleGAN, the features learned from real samples also guide the structure of glyph. Thus, extracting the glyph from generated samples could be an interesting attempt in the future. Another point is there is no attempt to solve the problem of large intra-class variance which will be left to the future.

## Appendix

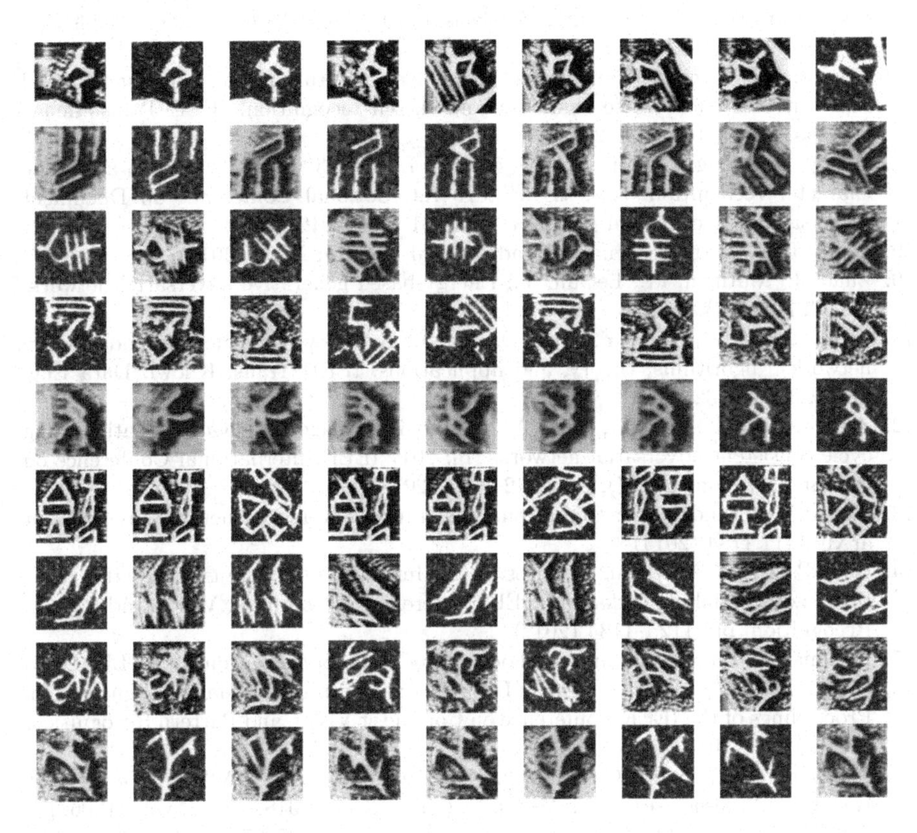

**Fig. 9.** Illustration of samples generated via CycleGAN. The samples in a row belong to the same category.

## References


1. Flad, R.-K.: Divination and power: a multi-regional view of the development of oracle bone divination in early China. Curr. Anthropol. **49**(3), 403–437 (2008)
2. Li, F., Woo, P.-Y.: The coding principle and method for automatic recognition of Jia Gu wen characters. Int. J. Hum Comput Stud. **53**(2), 289–299 (2000)

3. Li, Q.-S., Yang, Y.-X., Wang, A.-M.: Recognition of inscriptions on bones or tortoise shells based on graph isomorphism. Comput. Appl. Eng. Educ. **47**(8), 112–114 (2011)
4. Meng, L.: Two-stage recognition for oracle bone inscriptions. In: International Conference on Image Analysis and Processing, pp. 672–682 (2017)
5. Krizhevsky, A., Sutskever, I., Hinton, G.-E.: ImageNet classification with deep convolutional neural networks. In: International Conference on Neural Information Processing Systems (2012)
6. Zhang, Y.-K., Zhang, H., Liu, Y.-G., Yang, Q., Liu, C.-L.: Oracle character recognition by nearest neighbor classification with deep metric learning. In: 2019 International Conference on Document Analysis and Recognition (ICDAR), pp. 309–314 (2019)
7. Guo, J., Wang, C., Roman-Rangel, E., Chao, H., Rui, Y.: Building hierarchical representations for oracle character and sketch recognition. IEEE Trans. Image Process. **25**(1), 104–118 (2015)
8. Huang, S., Wang, H., Liu, Y., Shi, X., Jin, L.: OBC306: a large-scale oracle bone character recognition dataset. In: 2019 International Conference on Document Analysis and Recognition (ICDAR), pp. 681–688 (2019)
9. Goodfellow, I., et al.: Generative adversarial nets. In: NIPS (2014)
10. Zhao, J., Mathieu, M., LeCun, Y.: Energy-based generative adversarial network. In: ICLR (2017)
11. Gui, J., Sun, Z., Wen, Y., Tao, D., Ye, J.: A review on generative adversarial networks: algorithms, theory, and applications. IEEE Trans. Knowl. Data Eng. **35**(4), 3313–3332 (2021)
12. Zhu, J., Park, T., Isola, P., Efros, A.-A.: Unpaired image-to-image translation using cycle-consistent adversarial networks. In: 2017 IEEE International Conference on Computer Vision (ICCV), pp. 2242–2251 (2017)
13. Mirza, M., Osindero, S.: Conditional generative adversarial nets. arXiv preprint arXiv:1411.1784 (2014)
14. Isola, P., Zhu, J.-Y., Zhou, T., Efros, A.-A.: Image-to-image translation with conditional adversarial networks. In: IEEE Conference on Computer Vision and Pattern Recognition, pp. 1125–1134 (2017)
15. Kingma, D., Ba, J.: Adam: a method for stochastic optimization. In: ICLR (2015)
16. He, K., Zhang, X., Ren, S., Sun, J.: Deep residual learning for image recognition. In: Proceedings of the IEEE Conference on Computer Vision and Pattern Recognition, pp. 770–778 (2016)
17. Shmelkov, K., Schmid, C., Alahari, K.: How good is my GAN? In: Ferrari, V., Hebert, M., Sminchisescu, C., Weiss, Y. (eds.) ECCV 2018. LNCS, vol. 11206, pp. 218–234. Springer, Cham (2018). https://doi.org/10.1007/978-3-030-01216-8_14
18. Bissoto, A., Valle, E., Avila, S.: The six fronts of the generative adversarial networks. arXiv preprint arXiv:1910.13076 (2019)
19. Li, J., Wang, Q.-F., Zhang, R., Huang, K.: Mix-up augmentation for oracle character recognition with imbalanced data distribution. In: Proceedings of International Conference on Document Analysis and Recognition (2021)
20. Szegedy, C., Ioffe, S., Vanhoucke, V., Alemi, A.-A.: Inception-v4, inception-ResNet and the impact of residual connections on learning. In: Proceedings of the Thirty-First AAAI Conference on Artificial Intelligence (2017)

# Local-Global Interaction and Progressive Aggregation for Video Salient Object Detection

Dingyao Min[1], Chao Zhang[2,3](✉), Yukang Lu[1], Keren Fu[1](✉), and Qijun Zhao[1]

[1] College of Computer Science, Sichuan University, Chengdu 610065, China
{fkrsuper,qjzhao}@scu.edu.cn
[2] Intelligent Policing Key Laboratory of Sichuan Province, Luzhou 646000, China
[3] Sichuan Police College, Luzhou 646000, China
galoiszhang@gmail.com

**Abstract.** Video salient object detection (VSOD) aims at locating and segmenting visually distinctive objects in a video sequence. There still exist two problems that are not well handled in VSOD. First, facing unequal and unreliable spatio-temporal information in complex scenes, existing methods only exploit local information from different hierarchies for interaction and neglect the role of global saliency information. Second, they pay little attention to the refinement of the modality-specific features by ignoring fused high-level features. To alleviate the above issues, in this paper, we propose a novel framework named IANet, which contains local-global interaction (LGI) modules and progressive aggregation (PA) modules. LGI locally captures complementary representation to enhance RGB and OF (optical flow) features mutually, and meanwhile globally learns confidence weights of the corresponding saliency branch for elaborate interaction. In addition, PA evolves and aggregates RGB features, OF features and up-sampled features from the higher level, and can refine saliency-related features progressively. The sophisticated designs of interaction and aggregation phases effectively boost the performance. Experimental results on six benchmark datasets demonstrate the superiority of our IANet over nine cutting-edge VSOD models.

**Keywords:** Video salient object detection · saliency detection · local-global interaction · progressive aggregation

## 1 Introduction

Video salient object detection (VSOD) aims to model the mechanism of human visual attention and locate visually distinctive objects in a video sequence, which often serves as an important pre-processing step for many downstream vision tasks, such as video compression [10], video captioning [17], and person re-identification [31].

Compared with image salient object detection (ISOD) which only exploits spatial information in a static image, there exists temporal information in a


M. Tanveer et al. (Eds.): ICONIP 2022, CCIS 1793, pp. 101–113, 2023.
https://doi.org/10.1007/978-981-99-1645-0_9

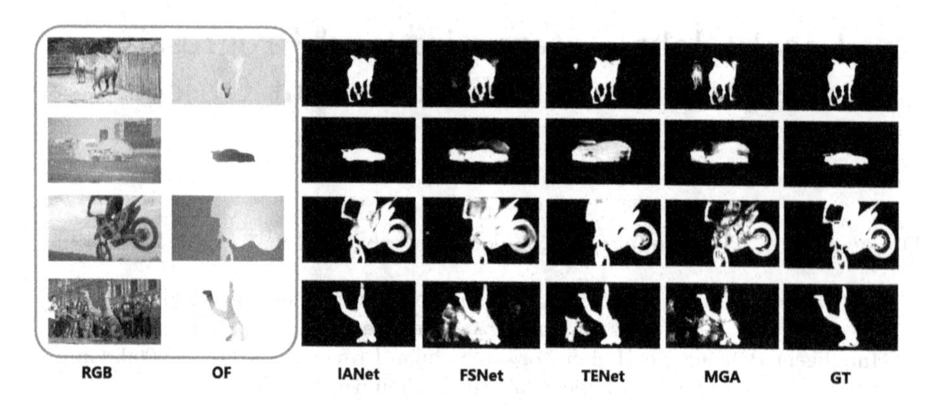


**Fig. 1.** Examples of challenging cases in video salient object detection. Compared to other state-of-the-art (SOTA) methods, *e.g.*, FSNet [11], TENet [21], and MGA [14], our model shows better detection performance under, *e.g.*, foreground-background visual similarity, environmental interference, and fast object motion. GT means ground truth.

video sequence. Therefore, VSOD needs to capture motion cues. However, more available information also brings more challenges. First, as shown in Fig. 1, there are many complex scenarios in real-world scenarios. In the $1^{st}$ and $2^{nd}$ rows, existing methods [11,14,21] have difficulty in identifying salient objects from RGB images due to foreground-background visual similarity and environmental interference. The $3^{rd}$ row shows the salient object is clear in the RGB image but motion cues are misleading in the OF image due to fast motion. The $4^{th}$ row shows both RGB and OF images hold complex cluttered background. In contrast, our proposed method can accurately segment salient objects by exploiting both local and global information in various scenarios.

Second, the early deep model [29] first applied the fully convolutional network (FCN) to extract spatial and temporal features. After that, optical flow and long-short term memory (LSTM) are used to extract more robust temporal features. Nevertheless, they still adopted some simple strategies to fuse features without any refinement. [3,11] firstly concatenated RGB and OF features, and then adopted a traditional UNet-like decoder to get the final prediction. [6,22] extracted temporal features by LSTM and used addition or concatenation operation to integrate features. These operations directly fuse multi-modal and multi-level features without considering level-specific and modality-specific characteristic, and therefore are insufficient to mine discriminative saliency-related information from multi-modal and multi-level features.

In this work, we propose a novel framework, *i.e.*, IANet, which mainly considers addressing the aforementioned problem through more adequate investigation on interaction of cross-modal complementary information as well as aggregation of multi-modal and multi-level features. First, we propose a local-global interaction (LGI) module, which contains a local mutual enhancement (LME) module and a global weight generator (GWG). LGI can fully exploit complementary information between RGB and OF features in both local and global scopes. Specially, in the local scope, LME is designed to cross-enhance RGB and OF features hierarchically. Meanwhile, GWG can learn dynamic channel-wisely weight vectors in the global scope, which represent confidence weights of the corresponding saliency branch. Liu *et al.* [23] also proposed a way to generate the corresponding weights by using single-modal and multi-level features, which ignored the role of multi-modal and multi-level global information. Different from previous methods [3,11,14,23], we fully exploit the correlation of global features.

Second, a progressive aggregation (PA) module is employed to aggregate multi-modal and multi-level features in a progressive refinement way. The fact that is neglected by previous methods [3,11,23] is that features from the higher level provide discriminative semantic and saliency-related information, which can refine modality-specific features to focus on saliency-related regions and suppress background distractors. Therefore, we first evolve and aggregate RGB/OF features and up-sampled features from the higher level by element-wise multiplication and maximization, respectively. To fuse the features obtained above, we utilize the channel attention mechanism [9] to adaptively select informative channel features, aiming to control information flows from multi-modal and multi-level features.

Overall, our main contributions are two-fold:

- We propose a *local-global* interaction (LGI) module to excavate complementary RGB and OF features to enhance each other *locally* and automatically learn the confidence weights of the corresponding saliency branch *globally*.
- We design a progressive aggregation (PA) module, which can first evolve and aggregate modality-specific (RGB/OF) features and up-sampled features from the higher level, and then fuse the above features by emphasizing meaningful features along channel dimensions with the attention mechanism. As such, our model can make full use of the higher level features to refine the modality-specific features to further boost performance.

## 2 Related Work

Traditional video salient object detection methods [1,26,30] mainly rely on handcrafted features and heuristic models. Due to the limitation of hand-crafted features and the low efficiency of heuristic models, these methods cannot handle complex scenarios and are hard to apply in practice.

With the development of deep learning, convolutional neural networks (CNNs) were employed into VSOD. Wang *et al.* [29] concatenated the current frame and output of the previous frame as input of a dynamic FCN to explore intra-frame temporal information. Song *et al.* [22] and Fan *et al.* [6] modeled spatio-temporal information with ConvLSTM. Gu *et al.* [7] proposed a pyramid constrained self-attention module to capture temporal information in video segments directly. Optical flow-based methods [3,11,14,21] used a two-stream fashion to extract RGB and OF features, respectively. Li *et al.* [14] adopted a one-way motion-guided mode, using motion information to enhance spatial information. Ren *et al.* [21] designed a novel Triple Excitation Network, which reinforces the training from three aspects, namely RGB, OF, and online excitations. Chen *et al.* [3] proposed a framework which adaptively captures available information from spatial and temporal cues. Recently, Ji *et al.* [11] achieved bidirectional message transfer by a full-duplex strategy.

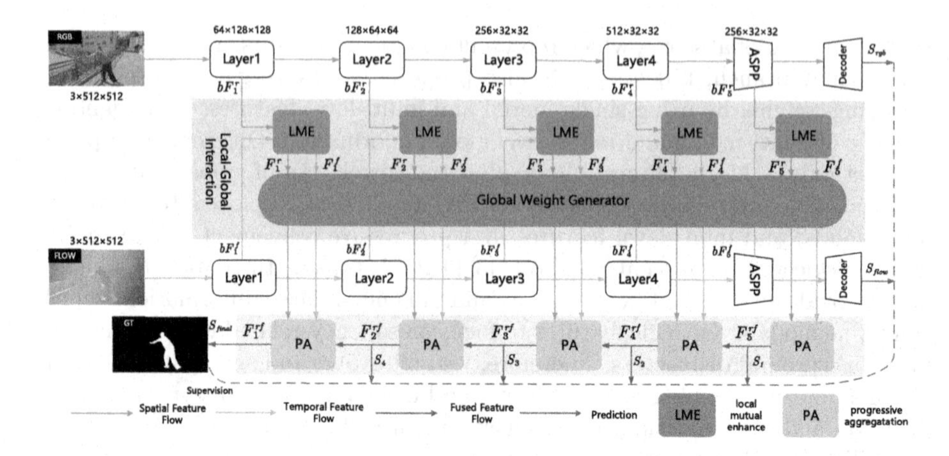


**Fig. 2.** Overall architecture of the proposed IANet.

## 3 The Proposed Method

### 3.1 Overview of Network Architecture

The overall framework of the proposed method is shown in Fig. 2. We propose a local-global interaction and progressive aggregation network (IANet) for video salient object detection. The network consists of symmetric ResNet-34 [8] backbones, local-global interaction (LGI) modules, and progressive aggregation (PA) modules. Given a pair of RGB and optical flow images, the symmetric backbones extract RGB and OF features at five different layers. After the top-down features extraction from the backbones, the multi-level RGB and OF features are fed to LGI that is composed of local mutual enhancement (LME) module and global weight generator (GWG). Finally, RGB features, OF features and up-sampled features from the higher level are independently forwarded to PA module. Details are described below.

### 3.2 Symmetric RGB and Flow Streams

Initially, we employ two symmetric ResNet-34 backbones as encoders to extract features from RGB and OF images. Let the feature outputs of the last four RGB/OF ResNet hierarchies be denoted as $bF_i^m$ $(m \in \{r, f\}, i = 1, \ldots, 4)$. The symbols 'r' and 'f' mean RGB-related features and OF-related features, respectively. Following [14], we apply an atrous spatial pyramid pooling (ASPP [2]) module after the fourth layer. The feature after ASPP module is input into the decoder to generate $S_{rgb}/S_{flow}$. The decoder used in this paper contains three convolutional layers to realize channel reduction. Finally, we obtain five-level RGB and OF features, namely $bF_i^m$ $(m \in \{r, f\}, i = 1, \ldots, 5)$.

### 3.3 Local-Global Interaction (LGI)

LGI is structured as two parts: LME module and GWG module. In realistic scenarios, both RGB and OF information are sometimes unreliable and unequal. Existing bi-modal methods [3,11,14] only focus on local interaction strategies and overlook the role of global information. To exploit complementary information between RGB and OF features and simultaneously balance the importance of RGB and OF features to the final saliency map, the obtained RGB and OF features are first locally enriched by LME. Different from previous methods [3,11], after the local interaction, we further leverage the encoded global saliency information to dynamically scale all five-level features with adaptive weights.

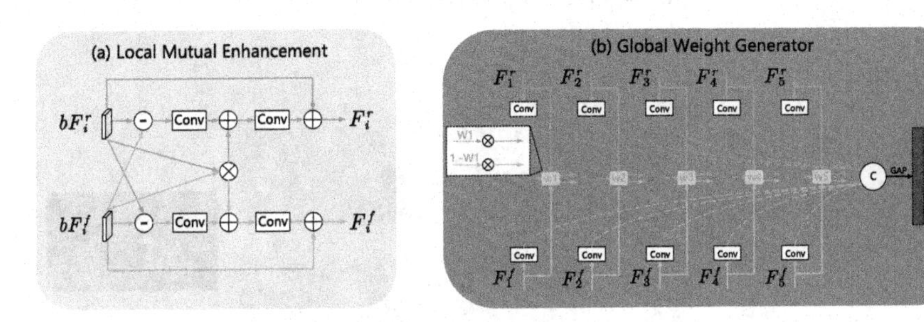


**Fig. 3.** Two parts of local-global interaction (LGI) module, namely LME and GWG. Here, "C" denotes feature concatenation, and $\ominus$, $\oplus$ and $\otimes$ denote element-wise subtraction, element-wise addition, and multiplication, respectively.

**Local Mutual Enhancement (LME).** The structure of LME is shown in Fig. 3(a). For each saliency branch, we firstly extract complementary parts by the mutual subtraction operation, as follows:

$$\begin{aligned} C_i^r &= \mathcal{F}_{conv}(bF_i^f \ominus bF_i^r), \\ C_i^f &= \mathcal{F}_{conv}(bF_i^r \ominus bF_i^f), \end{aligned} \tag{1}$$

where $\mathcal{F}_{conv}$ denotes convolution operation. In addition, the common parts are usually able to locate salient objects accurately. So we extracts the common parts between $bF_i^r$ and $bF_i^f$ by element-wise multiplication, then combines them with $C_i^r$ and $C_i^f$ respectively by element-wise addition. The re-calibrated RGB and OF features can be written as:

$$\begin{aligned} F_i^r &= \mathcal{F}_{conv}(C_i^r + bF_i^r \otimes bF_i^f) + bF_i^r, \\ F_i^f &= \mathcal{F}_{conv}(C_i^f + bF_i^r \otimes bF_i^f) + bF_i^f, \end{aligned} \tag{2}$$

where $\otimes$ is element-wise multiplication with broadcasting.

LME takes a selective enhancement strategy, where redundant information will be suppressed to avoid contamination between features by element-wise multiplication and important features will complement each other by element-wise subtraction. The visual examples are presented in Fig. 4. As shown, OF features help RGB features focus on saliency-related regions, and in turn RGB features help OF features eliminate part of background noise to some extent.

**Global Weight Generator (GWG).** The structure of GWG is shown in Fig. 3(b). We first squeeze all feature maps into 64 channels. The compressed features are uniformly upsampled to the same scale. We concatenate RGB and OF features to get $F_r$ and $F_f$ with 320 channels, respectively. Specifically, global average pooling (GAP) is employed to $F_r$ and $F_f$. Next, we concatenate the two tensors into one tensor, where $[\cdot, \cdot]$ means the concatenation operation:

$$w = [GAP(F_r), GAP(F_f)]. \tag{3}$$

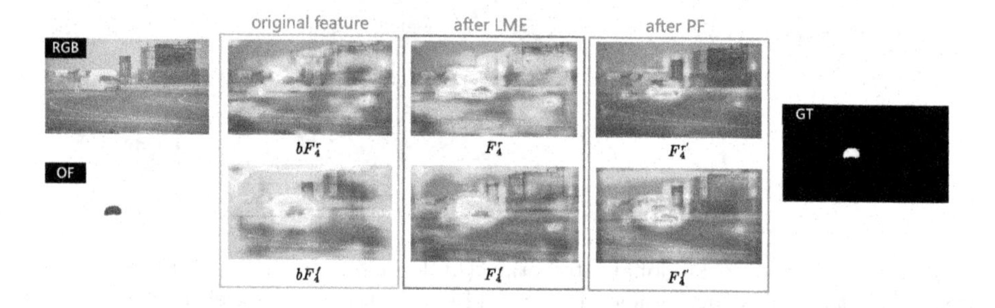


**Fig. 4.** Visualization of the features before and after LME and pre-fusion.

The obtained tensor $w \in \mathbb{R}^{640 \times 1 \times 1}$ contains global information, which reflects the contribution of $F_i^r / F_i^f$ $(i = 1, \ldots, 5)$ to final performance of the model. We further use $w$ to learn adaptive weights for RGB and OF branches respectively. For each layer $i \in \{1, \ldots, 5\}$ of the RGB branch, we can obtain its corresponding adaptive weights by feeding $w$ to two consecutive fully connected layers:

$$w_i = \sigma(FC_i(ReLU(FC(w))), \tag{4}$$

where $FC$ and $FC_i$ represent two fully connected layers (the latter is $i$-th layer specific), $ReLU$ is non-linear activation, and $\sigma$ is a sigmoid function that scales weights to interval $(0, 1)$. The derived $w_i \in \mathbb{R}^{c_i \times 1 \times 1}$ has the same number of channels as the $i$-th layer feature, which can channel-wisely scale $F_i^r$ / $F_i^f$ as:

$$F_i^r \leftarrow w_i \otimes F_i^r, F_i^f \leftarrow (1 - w_i) \otimes F_i^f. \tag{5}$$

The obtained $F_i^r$ and $F_i^f$ are the feature maps dynamically scaled with adaptive weights $w_i$ and $1 - w_i$, respectively. Benefiting from GWG, we can obtain features that are more valuable for final saliency prediction. Specifically, by learning adaptively, the model will leverage more RGB features if the RGB images contribute more to final prediction, and vice versa.

### 3.4 Progressive Aggregation (PA)

In this section, we describe the structure of PA (see Fig. 5). Given the features $\{F_i^r, F_i^f, F_{i+1}^{rf}\}$ $(i = 1, \ldots, 4)$ with the same size, we propose a PA module to progressively evolve and aggregate the above features. $F_i^r$ and $F_i^f$ denote RGB and OF features of the current layer, which contain unclear semantic information and redundant details. $F_{i+1}^{rf}$ denotes fused features from the higher level by using up-sampling operation and two convolutional layers, which usually provide rich semantic information. Previous methods [3,11,23] first fuse RGB and OF features and then combine them with the higher-level features in a UNet-like way. This manner neglects the fact that the higher-level features can refine the modality-specific features. Therefore, before the final fusion phase, we perform pre-fusion (PF) on $F_i^r$ and $F_{i+1}^{rf}$, $F_i^f$ and $F_{i+1}^{rf}$ by element-wise multiplication and maximization, respectively:

$$\begin{aligned} F_i^{r'} &= [(F_i^r \otimes F_{i+1}^{rf}), Max(F_i^r, F_{i+1}^{rf})], \\ F_i^{f'} &= [(F_i^f \otimes F_{i+1}^{rf}), Max(F_i^f, F_{i+1}^{rf})]. \end{aligned} \tag{6}$$

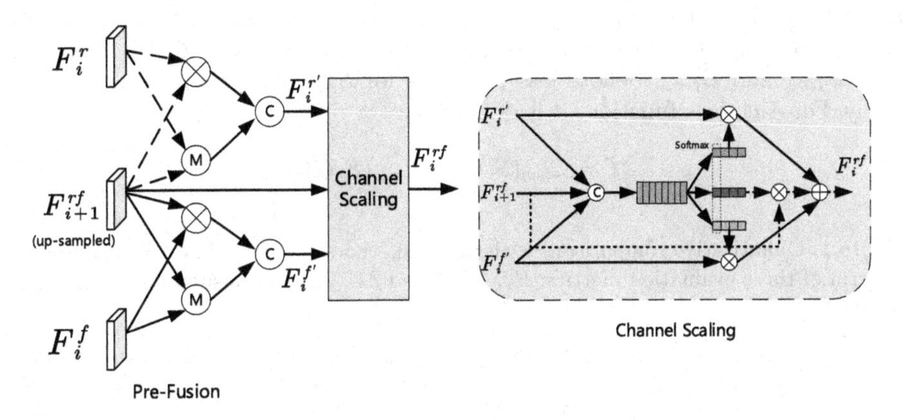


**Fig. 5.** The scheme of progressive aggregation (PA) module, where "C" denotes feature concatenation, and ⊕, ⊗ and "M" denote element-wise addition, multiplication, and maximization, respectively.

As shown in Fig. 4, the features after pre-fusion have clearer salient regions, while the cluttered backgrounds are concurrently reduced, simultaneously.

After that, to fuse the features obtained above, we utilize the channel attention mechanism [9] to adaptively select informative channel features, namely channel scaling (CS). We first use channel attention to control the information flows from $\{F_i^{r'}, F_i^{f'}, F_{i+1}^{rf}\}$:

$$w_c = FC(GAP([F_i^{r'}, F_i^{f'}, F_{i+1}^{rf}])). \tag{7}$$

Then, we employ another three FC layers to generate the weight vectors with the same number of channels as $\{F_i^{r'}, F_i^{f'}, F_{i+1}^{rf}\}$, respectively. Further, we utilize channel-wise softmax operation to generate final adaptive weights $\{w_1^c, w_2^c, w_3^c\}$ $\in \mathbb{R}^{c\times 1\times 1}$. At last, we use $\{w_1^c, w_2^c, w_3^c\}$ to adjust $\{F_i^{r'}, F_i^{f'}, F_{i+1}^{rf}\}$ as below:

$$F_i^{rf} = w_1^c \otimes F_i^{r\prime} + w_2^c \otimes F_i^{f\prime} + w_3^c \otimes F_{i+1}^{rf}, \tag{8}$$

where $F_i^{rf}$ is the fused features that will be passed to the next PA module. Notably, since $F_5^r$ and $F_5^f$ are already the highest level features, we directly fuse them by the above channel attention mechanism to generate $F_5^{rf}$.

### 3.5 Loss Function

Given four side outputs in the network, the overall loss can be written as:

$$\mathcal{L}_{total} = \mathcal{L}_{final} + \mathcal{L}_{rgb} + \mathcal{L}_{flow} + \sum_{i=1}^{4} \lambda_i \mathcal{L}_i, \tag{9}$$

where $\lambda_1, \lambda_2, \lambda_3$ and $\lambda_4$ balance the effect of each loss function on the training process. In our experiments, $\lambda_1, \lambda_2, \lambda_3$ and $\lambda_4$ are set to 0.4,0.4, 0.8 and 0.8, respectively. The side output of each stage corresponds to $\mathcal{L}_i(i = 1, \ldots, 4)$. $\mathcal{L}_{rgb}, \mathcal{L}_{flow}$ and $\mathcal{L}_{final}$ denote loss functions for $S_{rgb}, S_{flow}$ and $S_{final}$, respectively. For each loss function, it is defined as:

$$\mathcal{L} = \mathcal{L}_{bce}(S, G) + \mathcal{L}_{iou}(S, G), \tag{10}$$

**Table 1.** Quantitative comparisons with SOTA methods on five public VSOD datasets in term of three evaluation metrics: $S_\alpha$, $F_\beta^{\max}$ and $M$. The best results are highlighted in **bold**.

| | Metric | SCNN [24] | FGR [13] | PDBM [22] | SSAV [6] | MGA [14] | PCSA [7] | TENet [21] | CAG [3] | FSNet [11] | IANet |
|---|---|---|---|---|---|---|---|---|---|---|---|
| | Year | 2018 | 2018 | 2018 | 2019 | 2019 | 2020 | 2020 | 2021 | 2021 | 2022 |
| ***DAVIS*** | $S_\alpha \uparrow$ | 0.761 | 0.838 | 0.882 | 0.893 | 0.913 | 0.868 | 0.916 | 0.906 | 0.920 | **0.927** |
| | $F_\beta^{\max} \uparrow$ | 0.679 | 0.783 | 0.855 | 0.861 | 0.893 | 0.880 | 0.904 | 0.898 | 0.907 | **0.918** |
| | $M \downarrow$ | 0.077 | 0.043 | 0.028 | 0.028 | 0.022 | 0.022 | 0.019 | 0.018 | 0.020 | **0.016** |
| ***FBMS*** | $S_\alpha \uparrow$ | 0.794 | 0.809 | 0.851 | 0.879 | 0.912 | 0.868 | **0.915** | 0.870 | 0.890 | 0.892 |
| | $F_\beta^{\max} \uparrow$ | 0.762 | 0.767 | 0.821 | 0.865 | **0.909** | 0.837 | 0.897 | 0.858 | 0.888 | 0.888 |
| | $M \downarrow$ | 0.095 | 0.088 | 0.064 | 0.040 | **0.026** | 0.040 | 0.026 | 0.039 | 0.041 | 0.033 |
| ***SegV2*** | $S_\alpha \uparrow$ | - | 0.664 | 0.864 | 0.851 | **0.895** | 0.866 | 0.868 | 0.865 | 0.870 | 0.891 |
| | $F_\beta^{\max} \uparrow$ | - | 0.748 | 0.808 | 0.798 | 0.840 | 0.811 | 0.810 | 0.826 | 0.806 | **0.857** |
| | $M \downarrow$ | - | 0.169 | 0.024 | 0.023 | 0.024 | 0.024 | 0.027 | 0.027 | 0.025 | **0.013** |
| ***ViSal*** | $S_\alpha \uparrow$ | 0.847 | 0.861 | 0.907 | 0.942 | 0.945 | 0.946 | 0.946 | - | 0.928 | **0.947** |
| | $F_\beta^{\max} \uparrow$ | 0.831 | 0.848 | 0.888 | 0.938 | 0.942 | 0.941 | 0.948 | - | 0.913 | **0.951** |
| | $M \downarrow$ | 0.071 | 0.045 | 0.032 | 0.021 | 0.016 | 0.017 | 0.014 | - | 0.022 | **0.011** |
| ***DAVSOD*** | $S_\alpha \uparrow$ | 0.680 | 0.701 | - | 0.755 | 0.757 | 0.741 | 0.780 | 0.762 | 0.773 | **0.784** |
| | $F_\beta^{\max} \uparrow$ | 0.494 | 0.589 | - | 0.659 | 0.662 | 0.656 | 0.664 | 0.670 | 0.685 | **0.710** |
| | $M \downarrow$ | 0.127 | 0.095 | - | 0.084 | 0.079 | 0.086 | 0.074 | 0.072 | 0.072 | **0.067** |
| ***VOS*** | $S_\alpha \uparrow$ | 0.704 | 0.715 | 0.817 | 0.819 | 0.807 | 0.828 | - | - | 0.703 | **0.829** |
| | $F_\beta^{\max} \uparrow$ | 0.609 | 0.669 | 0.742 | 0.742 | 0.743 | 0.747 | - | - | 0.659 | **0.762** |
| | $M \downarrow$ | 0.109 | 0.097 | 0.078 | 0.074 | 0.069 | 0.065 | - | - | 0.108 | **0.063** |

where $\mathcal{L}_{bce}$ and $\mathcal{L}_{iou}$ are binary cross-entropy loss and intersection over union loss [20], respectively. $G$ and $S$ denote the ground truth and predicted saliency map, respectively.

## 4 Experiments

### 4.1 Datasets and Metrics

To demonstrate the effectiveness of our method, we conduct experiments on six public VSOD benchmatk datasets, including DAVIS [19], FBMS [16], SegV2 [12], ViSal [28], DAVSOD [6], VOS [15]. For quantitative evaluation, we adopt three metrics, *i.e.* S-measure [4] ($S_\alpha, \alpha = 0.5$), max F-measure [5] ($F_\beta^{\max}, \beta^2 = 0.3$), MAE [18] ($M$). See the related papers for specific definitions.

### 4.2 Implementation Details

In all experiments, the input RGB and optical flow images are uniformly resized to $512 \times 512$. Following [3], optical flow is rendered by the state-of-the-art optical flow prediction method [25]. The optical flow corresponding to the current frame is generated from the current frame and the previous frame. Different from the previous methods [7,14,21], we train our model with a simple one-step end-to-end strategy on the training set of DUTS [27], DAVIS, DAVSOD, instead of a multi-stage strategy. Since DUTS is a static image dataset, we fill the input optical flow with zeros. We use Adam optimizer and set batch size 8 in all experiments. The initial learning rate of the backbone and other parts is set to 5e-4 and 5e-3, respectively. The learning rate decays by 0.1 in the 15th epoch. The entire training process takes 20 epochs. The inference time of IANet is 34 FPS on Titan X (Pascal) GPU.

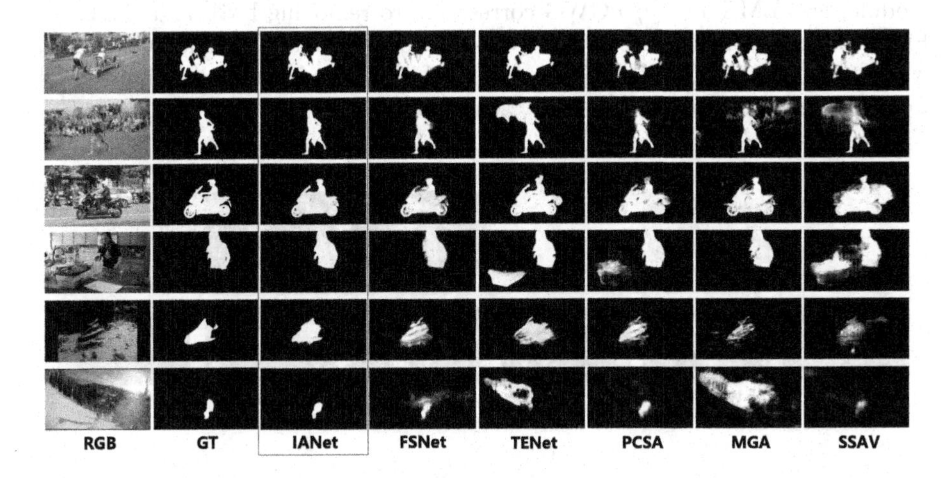


**Fig. 6.** Visual comparisons of the proposed method with SOTA models.

### 4.3 Comparisons to SOTAs

We compare our proposed IANet with 9 SOTA deep learning-based VSOD methods: SCNN [24], FGR [13], PDBM [22], SSAV [6], MGA [14], PCSA [7], TENet [21], CAG [3] and FSNet [11]. For fair comparison, we use the evaluation code from [6] to get results in our experimets.

**Quantitative Evaluation.** As shown in Table 1, our proposed IANet achieves the best result overall. Specifically, our method exceeds other SOTA methods on DAVIS, SegV2, ViSal, DAVSOD, and VOS. The performance on FBMS is relatively worse since we did not use the train set of FBMS for training. Notably, on the most challenging dataset DAVSOD, our method achieves a significant improvement compared with the second-best TENet (0.710 *vs.* 0.664, 0.067 *vs.* 0.074 in terms of $F_\beta^{\max}$ and $M$ respectively). Note we do not employ any post-processing compared with TENet. This shows that our method performs well in various scenarios.

**Visual Evaluation.** Visual comparisons are shown in Fig. 6. Overall, for several complex scenarios, such as complex boundary ($1^{st}$ row), cluttered fore-background ($2^{nd}$, $4^{th}$ and $6^{th}$ rows), multiple moving objects ($3^{rd}$ row) and low contrast ($5^{th}$ row), our method can better locate and segment salient objects with fine-grained details.

### 4.4 Ablation Study

To verify the impact of our key modules, we conduct experiments on DAVIS and DAVSOD datasets, which include the largest numbers of video sequences and the most complex/diverse scenarios. Quantitative results are shown in Table 2.

**Local-Global Interaction (LGI) Validation.** We compare three variants: *w/o* LGI, *w/o* LME, *w/o* GWG. *w/o* LGI represents our full model without LGI module. *w/o* LME and *w/o* GWG correspond to removing LME and GWG. In Fig. 7, the prediction results of three variants contain noise and holes. Compared with the full model, the results indicate that the proposed LGI module is essential for improving the performance as presented in Table 2.

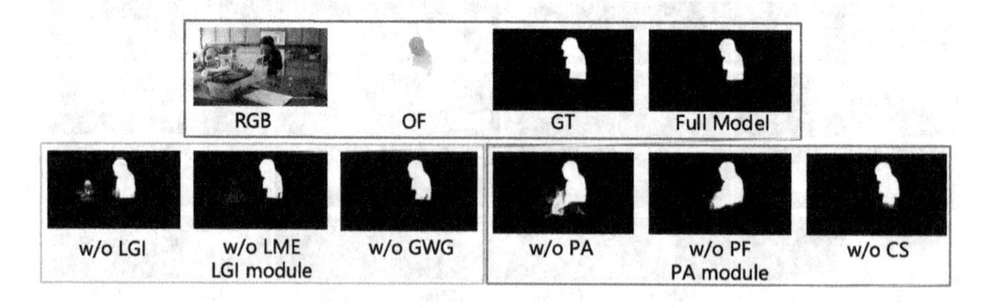


**Fig. 7.** Visual comparisons of different model variants.

**Table 2.** Ablation studies of different model variants from our IANet. The best results are highlighted in **bold**.

| Modules | Models | DAVIS | | | DAVSOD | | |
|---|---|---|---|---|---|---|---|
| | | $S_\alpha \uparrow$ | $F_\beta^{\max} \uparrow$ | $M \downarrow$ | $S_\alpha \uparrow$ | $F_\beta^{\max} \uparrow$ | $M \downarrow$ |
| | **Full model** | **0.927** | **0.918** | **0.015** | **0.784** | **0.710** | **0.067** |
| LGI | w/o LGI | 0.919 | 0.908 | 0.016 | 0.767 | 0.690 | 0.073 |
| | w/o LME | 0.923 | 0.912 | 0.016 | 0.778 | 0.703 | 0.068 |
| | w/o GWG | 0.923 | 0.912 | 0.015 | 0.775 | 0.702 | 0.068 |
| PA | w/o PA | 0.917 | 0.906 | 0.017 | 0.765 | 0.691 | 0.074 |
| | w/o PF | 0.921 | 0.911 | 0.016 | 0.776 | 0.705 | 0.069 |
| | w/o CS | 0.921 | 0.914 | 0.017 | 0.778 | 0.701 | 0.071 |
| GLI | global-local | 0.925 | 0.916 | 0.015 | 0.775 | 0.694 | 0.069 |

**Progressive Aggregation (PA) Validation.** We compare three variants: *w/o* PA, *w/o* PF, *w/o* CS. *w/o* PA denotes the fusion of features using the U-Net like approach. *w/o* PF and *w/o* CS denote removal of two sub-modules respectively. Observing Fig. 7 and Table 2, our proposed progressive aggregation strategy improves the model performance. The visual results reflect that three baselines generate foreground noise and incomplete salient object. The results show that our proposed progressive aggregation strategy improves the model performance.

**About Interaction Order.** Besides, we also compare a reverse interaction pattern for the LGI module, namely global-local interaction (GLI), which conducts GWG before LME. From Table 2, the results show a certain degree of degradation. This proves that it is more reasonable to perform the local interaction before the global interaction. GLI may introduce too much global noise by performing the global interaction first.

In summary, the ablation studies demonstrate the effectiveness and advantages of the proposed modules. In addition, the results of the ablation experiments also demonstrate that careful feature interaction and well-designed feature aggregation could effectively improve the performance of VSOD.

## 5 Conclusion

In this paper, we propose a novel video SOD network equipped with local-global interaction and progressive aggregation. The former adopts a new interaction strategy to achieve full interaction of multi-modal features, whereas the latter progressively refines saliency-related features. The experimental results demonstrate that both modules boost the performance of the final model. The elaborate designs of the interaction and aggregation phases enable our approach to achieve overall state-of-the-art performance on six benchmark datasets.

**Acknowledgements.** This work was supported by the NSFC (62176169, 621761 70, 61971005), SCU-Luzhou Municipal Peoples Government Strategic Cooperation Project (2020CDLZ-10), and Intelligent Policing Key Laboratory of Sichuan Province (ZNJW2022KFMS001, ZNJW2022ZZMS001).

## References

1. Chen, C., Li, S., Wang, Y., Qin, H., Hao, A.: Video saliency detection via spatial-temporal fusion and low-rank coherency diffusion. IEEE Trans. Image Process. **26**(7), 3156–3170 (2017)
2. Chen, L.C., Papandreou, G., Kokkinos, I., Murphy, K., Yuille, A.L.: Deeplab: Semantic image segmentation with deep convolutional nets, atrous convolution, and fully connected crfs 40(4), 834–848 (2017)
3. Chen, P., Lai, J., Wang, G., Zhou, H.: Confidence-guided adaptive gate and dual differential enhancement for video salient object detection. In: 2021 IEEE International Conference on Multimedia and Expo (ICME). pp. 1–6. IEEE (2021)
4. Fan, D.P., Cheng, M.M., Liu, Y., Li, T., Borji, A.: Structure-measure: A new way to evaluate foreground maps. In: Proceedings of the IEEE international conference on computer vision. pp. 4548–4557 (2017)
5. Fan, D.P., Gong, C., Cao, Y., Ren, B., Cheng, M.M., Borji, A.: Enhanced-alignment measure for binary foreground map evaluation. arXiv preprint arXiv:1805.10421 (2018)
6. Fan, D.P., Wang, W., Cheng, M.M., Shen, J.: Shifting more attention to video salient object detection. In: Proceedings of the IEEE/CVF Conference on Computer Vision and Pattern Recognition, pp. 8554–8564 (2019)
7. Gu, Y., Wang, L., Wang, Z., Liu, Y., Cheng, M.M., Lu, S.P.: Pyramid constrained self-attention network for fast video salient object detection. In: Proceedings of the AAAI Conference on Artificial Intelligence (2020)
8. He, K., Zhang, X., Ren, S., Sun, J.: Deep residual learning for image recognition. In: Proceedings of the IEEE Conference on Computer Vision and Pattern Recognition, pp. 770–778 (2016)
9. Hu, J., Shen, L., Sun, G.: Squeeze-and-excitation networks. In: Proceedings of the IEEE Conference on Computer Vision and Pattern Recognition, pp. 7132–7141 (2018)
10. Itti, L.: Automatic foveation for video compression using a neurobiological model of visual attention. IEEE TIP **13**(10), 1304–1318 (2004)
11. Ji, G.P., Fu, K., Wu, Z., Fan, D.P., Shen, J., Shao, L.: Full-duplex strategy for video object segmentation. In: ICCV (2021)
12. Li, F., Kim, T., Humayun, A., Tsai, D., Rehg, J.M.: Video segmentation by tracking many figure-ground segments. In: Proceedings of the IEEE International Conference on Computer Vision, pp. 2192–2199 (2013)
13. Li, G., Xie, Y., Wei, T., Wang, K., Lin, L.: Flow guided recurrent neural encoder for video salient object detection. In: Proceedings of the IEEE Conference on Computer vision and Pattern Recognition, pp. 3243–3252 (2018)
14. Li, H., Chen, G., Li, G., Yizhou, Y.: Motion guided attention for video salient object detection. In: Proceedings of International Conference on Computer Vision (2019)
15. Li, J., Xia, C., Chen, X.: A benchmark dataset and saliency-guided stacked autoencoders for video-based salient object detection. IEEE Trans. Image Process. **27**(1), 349–364 (2017)

16. Ochs, P., Malik, J., Brox, T.: Segmentation of moving objects by long term video analysis. IEEE Trans. Pattern Anal. Mach. Intell. **36**(6), 1187–1200 (2013)
17. Pan, Y., Yao, T., Li, H., Mei, T.: Video captioning with transferred semantic attributes. In: CVPR, pp. 6504–6512 (2017)
18. Perazzi, F., Krähenbühl, P., Pritch, Y., Hornung, A.: Saliency filters: Contrast based filtering for salient region detection. In: 2012 IEEE Conference on Computer Vision and Pattern Recognition, pp. 733–740. IEEE (2012)
19. Perazzi, F., Pont-Tuset, J., McWilliams, B., Van Gool, L., Gross, M., Sorkine-Hornung, A.: A benchmark dataset and evaluation methodology for video object segmentation. In: Proceedings of the IEEE Conference on Computer Vision and Pattern Recognition, pp. 724–732 (2016)
20. Rahman, M.A., Wang, Y.: Optimizing intersection-over-union in deep neural networks for image segmentation. In: ISVC, pp. 234–244 (2016)
21. Ren, S., Han, C., Yang, X., Han, G., He, S.: TENet: triple excitation network for video salient object detection. In: Vedaldi, A., Bischof, H., Brox, T., Frahm, J.-M. (eds.) ECCV 2020. LNCS, vol. 12350, pp. 212–228. Springer, Cham (2020). https://doi.org/10.1007/978-3-030-58558-7_13
22. Song, H., Wang, W., Zhao, S., Shen, J., Lam, K.-M.: Pyramid dilated deeper ConvLSTM for video salient object detection. In: Ferrari, V., Hebert, M., Sminchisescu, C., Weiss, Y. (eds.) ECCV 2018. LNCS, vol. 11215, pp. 744–760. Springer, Cham (2018). https://doi.org/10.1007/978-3-030-01252-6_44
23. Su, Y., Wang, W., Liu, J., Jing, P., Yang, X.: Ds-net: dynamic spatiotemporal network for video salient object detection. arXiv preprint arXiv:2012.04886 (2020)
24. Tang, Y., Zou, W., Jin, Z., Chen, Y., Hua, Y., Li, X.: Weakly supervised salient object detection with spatiotemporal cascade neural networks. IEEE Trans. Circuits Syst. Video Technol. **29**(7), 1973–1984 (2018)
25. Teed, Z., Deng, J.: RAFT: Recurrent All-Pairs Field Transforms for Optical Flow. In: Vedaldi, A., Bischof, H., Brox, T., Frahm, J.-M. (eds.) ECCV 2020. LNCS, vol. 12347, pp. 402–419. Springer, Cham (2020). https://doi.org/10.1007/978-3-030-58536-5_24
26. Tu, W.C., He, S., Yang, Q., Chien, S.Y.: Real-time salient object detection with a minimum spanning tree. In: Proceedings of the IEEE conference on computer vision and pattern recognition. pp. 2334–2342 (2016)
27. Wang, L., Lu, H., Wang, Y., Feng, M., Wang, D., Yin, B., Ruan, X.: Learning to detect salient objects with image-level supervision. In: Proceedings of the IEEE Conference on Computer Vision and Pattern Recognition. pp. 136–145 (2017)
28. Wang, W., Shen, J., Shao, L.: Consistent video saliency using local gradient flow optimization and global refinement. IEEE Trans. Image Process. **24**(11), 4185–4196 (2015)
29. Wang, W., Shen, J., Shao, L.: Video salient object detection via fully convolutional networks. IEEE Trans. Image Process. **27**(1), 38–49 (2017)
30. Xi, T., Zhao, W., Wang, H., Lin, W.: Salient object detection with spatiotemporal background priors for video. IEEE Trans. Image Process. **26**(7), 3425–3436 (2016)
31. Zhao, R., Ouyang, W., Wang, X.: Unsupervised salience learning for person re-identification. In: CVPR. pp. 3586–3593 (2013)

# A Fast Stain Normalization Network for Cervical Papanicolaou Images

Jiawei Cao[1], Changsheng Lu[2], Kaijie Wu[1](✉), and Chaochen Gu[1]

[1] Shanghai Jiao Tong University, Shanghai, China
{cjw333,kaijiewu,jacygu}@sjtu.edu.cn
[2] The Australian National University, Canberra, Australia
Changsheng.Lu@anu.edu.au

**Abstract.** The domain shift between different styles of stain images greatly challenges the generalization of computer-aided diagnosis (CAD) algorithms. To bridge the gap, color normalization is a prerequisite for most CAD algorithms. The existing algorithms with better normalization effect often require more computational consumption, resisting the fast application in large-size medical stain slide images. This paper designs a fast normalization network (FTNC-Net) for cervical Papanicolaou stain images based on learnable bilateral filtering. In our FTNC-Net, explicit three-attribute estimation and spatially adaptive instance normalization are introduced to guide the model to transfer stain color styles in space accurately, and dynamic blocks are adopted to adapt multiple stain color styles. Our method achieves at least 80 fps over 1024×1024 images on our experimental platform, thus it can synchronize with the scanner for image acquisition and processing, and has advantages in visual and quantitative evaluation compared with other methods. Moreover, experiments on our cervical staining image dataset demonstrate that the FTNC-Net improves the precision of abnormal cell detection.

**Keywords:** cervical stain images · stain style transfer · stain color normalization · bilateral filtering

## 1 Introduction

Color normalization reduces the domain gap of different stain colors and improves the accuracy of computer-aided diagnosis (CAD) algorithms for pathology. Compared with hematoxylin and eosin stain (H&E stain) in histopathology, Papanicolaou stain (Pap stain) is generally applied in cytology [20], including cervical cancer screening and detection [1]. The basic principle of Pap stain is to distinguish between acidophilic and basophilic components and obtain detailed chromatin patterns [18]. Thus, Papanicolaou stain's cell foreground color attributes are different from H&E stain. Meanwhile, the analysis efficiency will be degraded for processing whole slide images, usually gigabytes, if the normalization algorithm is slow. Therefore, designing a fast stain normalization method is beneficial for cytology applications.


M. Tanveer et al. (Eds.): ICONIP 2022, CCIS 1793, pp. 114–126, 2023.
https://doi.org/10.1007/978-981-99-1645-0_10

Since the pathological CAD algorithms mainly focus on histology, the stain normalization algorithms are mainly applied to H&E stain histological images. Reinhard *et al.* [19] converted input images from RGB space to LAB space and then transformed the distribution of inputs into the targets through linear conversion. Macenko *et al.* [16] implemented color normalization by finding the two most significant singular values of the singular value decomposition (SVD) as stain basis vectors. Vahadane *et al.* [25] used sparse non-negative matrix factorization stain separation method to find better stain basis vectors. Nadeem *et al.* [17] proposed mathematically robust stain style transfer by multimarginal Wasserstein barycenter. Shaban *et al.* [21] proposed StainGAN based on Cycle-GAN [30] for color normalization. Tellez *et al.* [23] performed various color data augmentation on input images and used a U-net to achieve a self-supervised normalization. The above methods are for H&E stain images and do not consider the characteristics of cervical Pap stain images.

Recently, there are also works on cervical cytology images. Chen [3] achieved normalization that keeps both hue and structure consistencies by LAB spatial foreground masks and GAN method. Kang *et al.* [9] realized a fast color normalization model by multi-layer convolution and StainGAN [21]. It is difficult for the method in [3] to process images in real-time, and the method in [9] cannot exceed the performance of StainGAN due to the network architecture design.

Transfer learning [12,13] is one effective approach to address the color domain gap in stain images. We find that neural style transfer [4,26] can process images more quickly and be applied to multi-domain transformation. Earlier style transfer works were mainly applied between photorealistic and artistic images, which is unsuitable for our medical task. Recently, more and more style transfer works between photorealistic images have been proposed [10,11,29]. Xia *et al.* [27] proposed a real-time style transfer network based on HDR-Net[5], which reserves the main structure of the network in HDR-Net[5] and adds a series of style transfer blocks to achieve photorealistic style transfer.

Therefore, we propose a real-time normalization network for cervical stain images (FTNC-Net). In this paper, the main contribution:

- A bilateral learning three-branch network, including the low-resolution splatting branch, the autoencoder branch, and the full-resolution guided branch, is designed to realize real-time style transfer on 1024×1024 images.
- Explicit three-attribute masks and pseudo-probability maps, including background, eosinophil, and basophil, are estimated and guide spatially adaptive instance normalization to achieve a better style transfer effect.
- Dynamic blocks are introduced to handle multiple style transfer problems.

## 2 Methodology

### 2.1 Network Architecture

We propose a bilateral learning network to achieve color normalization for cervical Pap stain images. As shown in Fig. 1, the proposed network architecture

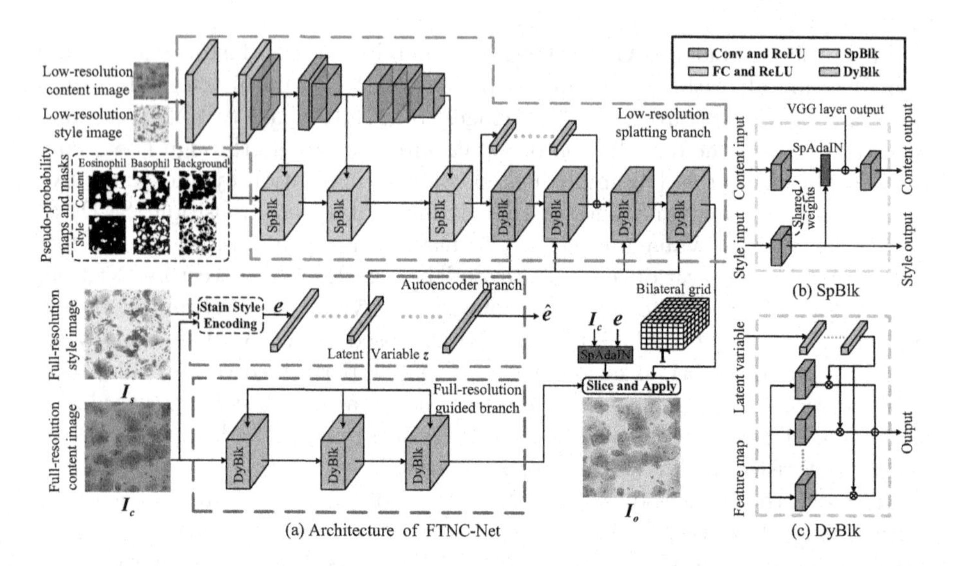


**Fig. 1.** Proposed architecture of our fast normalization network (FTNC-Net). (a) The overall pipeline; (b) Splatting block (SpBlk) in the low-resolution branch; (c) Dynamic block (DyBlk) in both low-resolution and full-resolution branches. (Color figure online)

contains three branches: low-resolution splatting branch (blue dotted frame), autoencoder (AE) branch (green dotted frame), and full-resolution guided branch (purple dotted frame). The splatting branch receives low-resolution content images and style images with their pseudo-probability maps and masks to generate bilateral grids $\Gamma$. The AE branch takes the feature vectors $e$ extracted from the content and the style images through stain style encoding as the input to generate a latent variable $z$ and the reconstructed $\hat{e}$. The full-resolution branch takes full-resolution content images as input, and the guided maps are produced after a series of nonlinear transformations by multiple dynamic blocks. The output images are generated by applying the per-pixel linear transformation, whose weights are sliced out from bilateral grids and guided maps.

**Pseudo-probability Map and Spatially Adaptive Instance Normalization.** Adaptive instance normalization (AdaIN) [7,8] is one of the critical operations for style transfer, which transfers the mean and standard deviation of target features to source features globally. The cytological image of Pap stain has a more apparent color distribution in the space, but the AdaIN does not leverage color distribution information. Since resulting color space distribution deterioration by AdaIN, we propose to use spatial masks and probability maps to guide AdaIN called spatially adaptive instance normalization (SpAdaIN):

$$\mathrm{SpAdaIN}(x_{ij}, y_{ij}) = \sigma_{ij}^{y}\left(\frac{x_{ij} - \mu_{ij}^{x}}{\sigma_{ij}^{x}}\right) + \mu_{ij}^{y} \tag{1}$$

where $x_{ij}$ and $y_{ij}$ are the pixels at $(i, j)$ of the content and style features, respectively, and $\mu_{ij}^x$ and $\sigma_{ij}^x$ are the mean and standard deviation in $x_{ij}$. Taking $\mu_{ij}^x$ and $\mu_{ij}^y$ as an example, their formulations are:

$$\mu_{ij}^x = P_{\text{eos}}(x_{ij})\mu_{\text{eos}}^x + P_{\text{bas}}(x_{ij})\mu_{\text{bas}}^x + P_{\text{bg}}(x_{ij})\mu_{\text{bg}}^x \quad (2)$$

$$\mu_{ij}^y = P_{\text{eos}}(x_{ij})\mu_{\text{eos}}^y + P_{\text{bas}}(x_{ij})\mu_{\text{bas}}^y + P_{\text{bg}}(x_{ij})\mu_{\text{bg}}^y \quad (3)$$

where $P_{\text{eos}}$, $P_{\text{bas}}$ and $P_{\text{bg}}$ are the pseudo-probability of eosinophil, basophil and background categories, respectively. $\mu_{\text{eos}}^x$, $\mu_{\text{bas}}^x$ and $\mu_{\text{bg}}^x$ are the mean of eosinophil, basophil, and background categories of the content feature, which is calculated by the guidance of the mask. Since the HSV color space distinguishes eosinophil, basophil, and background better than the LAB color space, we consider the HSV to estimate the mask and pseudo-probability map. Pseudo-probability map of three categories is generated from the hue channel $h$ and saturation channel $s$, where threshold values of $h$ and $s$ are $\eta$ and $\theta$, respectively. We normalize all channels of HSV into [0, 1] and set $\theta = 0.09$, $\eta = 0.7$, $\epsilon = 10$. The estimation mask is a particular case of the pseudo-probability map. When $\epsilon$ approaches infinity, the pseudo-probability map is the three-attribute mask:

$$Prob(c, \psi) = \begin{cases} f(\frac{\psi - c}{max(\psi - c)}) & c < \psi \\ 0 & c = \psi \\ f(\frac{\psi - c}{max(c - \psi)}) & c > \psi \end{cases} \quad (4)$$

$$f(x) = \frac{1}{1 + e^{-\epsilon x}} \quad (5)$$

$$P_{\text{bg}} = Prob(s, \theta) \quad (6)$$

$$P_{\text{eos}} = (1 - P_{\text{bg}})(1 - Prob(h, \eta)) \quad (7)$$

$$P_{\text{bas}} = (1 - P_{\text{bg}})Prob(h, \eta) \quad (8)$$

**Low-resolution Splatting Branch and Full-resolution Guided Branch.** Similar to the structure described in [27], content and style features are exacted from the top VGG [22] path, specifically parts of $vgg19$. Then feature of $Conv1\_1$, $Conv2\_1$, $Conv3\_1$, and $Conv4\_1$ from VGG is fed into three splatting blocks with SpAdaIN to generate the intermediate feature. Each splatting block is shown in Fig. 1(b). The input of the splatting block is a content feature and style feature. After the convolution layer of shared weights, SpAdaIN is carried out. The transformed feature is summed up with top VGG results and then sent to the next convolution layer. A bilateral grid is obtained by local and global nonlinear transformation for the content output of the last splatting block. In the full-resolution branch, the full-resolution content image is input, and a guided map is generated after a series of nonlinear transformations. We replace the convolutional layers in local transformation and full-res branch with dynamic blocks compared to the original network structure.

**Stain Style Encoding and Dynamic Block.** As for our stain style encoding, pixels are divided into three categories: eosinophil, basophil, and background.

Then their mean and standard deviation on the RGB three channels are calculated to obtain 18-dimensional stain style encoding vectors through estimated 3-attribute masks. We set the stain style encoding vectors as the input of the AE branch to generate latent variable $z$. Since the effect of the original network [27] for multiple stain style transfer is unsatisfactory. We replace the convolutional layers of the last half in the low-resolution branch and all the convolutional layers in the high-resolution branch with dynamic blocks, as Fig. 1(a) is shown. The dynamic block is a learnable weighted parallel convolution structure with multilayer perception (MLP), as shown in Fig. 1(c). The MLP transforms latent variables $z$ to get weights for parallel layers, and the number of parallel convolutional layers in every dynamic block is set as four. The different content and style image pairs possess different stain style encodings, resulting in different latent variables $z$, which guide the network to process images dynamically.

### 2.2 Loss Function

Our approach includes style loss, content loss, bilateral grid regularization loss, and reconstruction loss. The style loss is the $L_2$ distance of stain style encoding vectors between style image $I_s$ and output image $I_o$, namely $\mathcal{L}_s = \|E(I_s) - E(I_o)\|_2$ ($E$ is stain style encoding in Fig. 1(a)). The content loss is $\mathcal{L}_c = \|F(I_c) - F(I_o)\|_2$, where $F(I_c)$ and $F(I_o)$ are the feature maps of content image $I_c$ and output image $I_o$ from $Conv4\_1$ of pre-trained VGG19. The bilateral grid regularization loss is $\mathcal{L}_{reg} = \sum_{i=1}^{3} \|\Gamma - \Gamma_i\|_2$ which limits the degree of bilateral grid variation, where $\Gamma$ is the bilateral grid and $\Gamma_i$ is the neighborhood of $\Gamma$ in depth, height and width directions. The reconstruction loss is $\mathcal{L}_{rec} = \|e - \hat{e}\|_2$ ($e$ is the encoding from AE). Overall, our total loss function is

$$\mathcal{L} = \mathcal{L}_s + \lambda_1 \mathcal{L}_c + \lambda_2 \mathcal{L}_{reg} + \lambda_3 \mathcal{L}_{rec} \tag{9}$$

where we set $\lambda_1 = \lambda_2 = 0.01$ and $\lambda_3 = 0.1$ in our paper.

## 3 Experiments

Our dataset comes from two hospitals and we select 6 different color styles of images as the training dataset and testing dataset for experiments, in which each category includes 200 images for training and 20 images for testing, and the pixel size is 1024 × 1024. The visualization results of the training dataset by T-sne are shown in Fig. 2. These images are all cropped from whole slide images scanned at 20×Na0.75. The CPU of our experimental platform is Ryzen 3600, GPU is RTX3090, and the operating system is Ubuntu 18.04.

### 3.1 Normalization Evaluation

We use the Adam optimizer with an initial learning rate of $10^{-3}$. The learning rate steps down to $10^{-5}$, and the total epoch is 50 to train the model. Meanwhile, we apply HSV jitters in saturation and value channels for input images

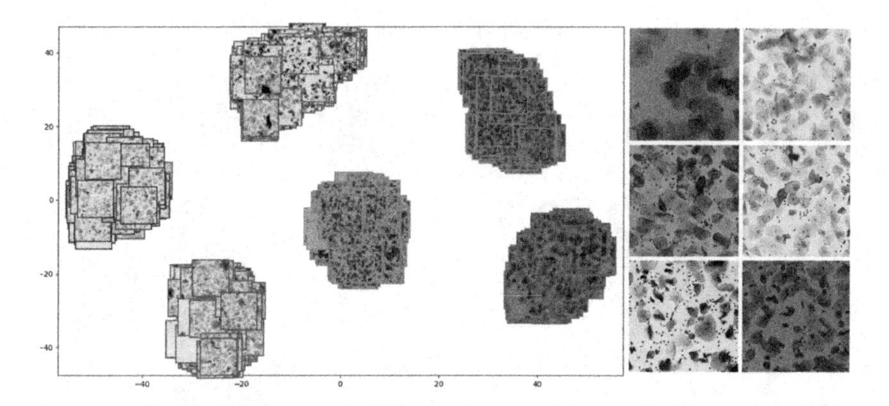

**Fig. 2.** The data distribution visualization by t-SNE and 6 sampled images.

**Table 1.** Stain transfer result comparison. Each score is mean ± standard deviation.

| Methods | SSIM (↑) | Gram Loss (↓) | Time (sec.) (↓) |
|---|---|---|---|
| Reinhard [19] | $0.8991 \pm 0.0562$ | $0.0098 \pm 0.0047$ | $0.0481 \pm 0.0030$ |
| Macenko [16] | $0.9022 \pm 0.0565$ | $0.0091 \pm 0.0048$ | $1.1117 \pm 0.0375$ |
| Vahdane [25] | $\mathbf{0.9112 \pm 0.0484}$ | $0.0090 \pm 0.0046$ | $3.0196 \pm 0.0248$ |
| Ours | $0.9101 \pm 0.0080$ | $\mathbf{0.0080 \pm 0.0040}$ | $\mathbf{0.0111 \pm 0.0003}$ |

to augment the data stain style. The ranges are $[-0.25, 0.25]$ and $[-0.2, 0.2]$ for saturation and value channels, respectively.

For non-aligned data, the quantitative evaluation of style transfer results has always been a problem. Referring to [6,29], we use the pre-trained VGG to calculate the distance of gram matrix in *conv*4_1 between the output results and the target style images to evaluate the stain color style similarity and calculate structural similarity (SSIM) index to evaluate the degree of structural retention of color transferred images and content images. The processing time per image for different methods is also recorded. We compare ours with Reinhard [19], Macenko [16], and Vahdane [25]. The index results are shown in Table 1 and visualization results are shown in Fig. 3.

The StainGAN [21] is also evaluated on our dataset, where the densities of squamous cells or neutrophils for different stain style categories are inconsistent. These inconsistencies cause the discriminator to guide the generator to generate the distributions of squamous cells and neutrophils similar to the target category. However, this changes the structure of the cervical stain images. More experimental images are described in the **Appendix**.

Since the FTNC-Net specifies the attributes of pixels in the stain images, our method achieves a better visual result than others, as shown in Fig. 3. Moreover, Gram loss in quantitative evaluation is also the smallest. The Macenko [6] and Vahdane [25] methods are initially designed for H&E stain, and they are less

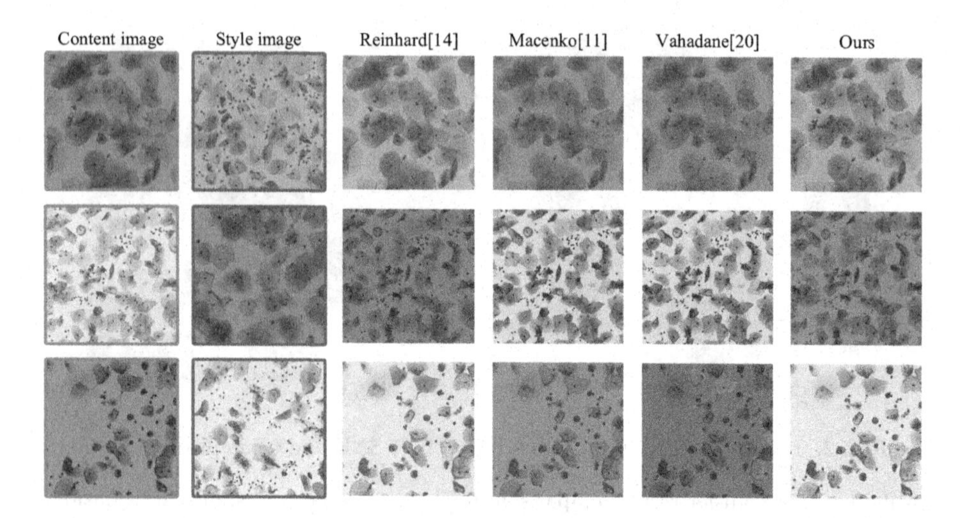


**Fig. 3.** Stain normalization results of source images by different normalization methods.

**Table 2.** Stain transfer time with different resolution. The running time is mean ± standard deviation (sec.).

| Image size | Reinhard [19] | Macenko [16] | Vahdane [25] | Ours |
|---|---|---|---|---|
| $1024 \times 1024$ | $0.0481 \pm 0.0030$ | $1.1117 \pm 0.0375$ | $3.0196 \pm 0.0248$ | $\mathbf{0.0111 \pm 0.0003}$ |
| $2048 \times 2048$ | $0.2211 \pm 0.0146$ | $4.1454 \pm 0.1404$ | $5.6676 \pm 0.0778$ | $\mathbf{0.0362 \pm 0.0010}$ |
| $3072 \times 3072$ | $0.5715 \pm 0.0333$ | $9.0254 \pm 0.2756$ | $9.8801 \pm 0.1792$ | $\mathbf{0.0779 \pm 0.0021}$ |

expressive on content and style images with significant style differences in Pap stain. The models may select the wrong color vector basis, as shown in the third row of Fig. 3.

The last column of Table 1 represents the processing time per 1024*1024 size image, excluding the image reading and writing time. The FTNC-Net is the fastest processing method among the comparisons. In order to explore the relationship between the processing speed of the model and the pixel size of the input image, we test on two additional image sizes for different methods. The results are in Table 2.

### 3.2 Abnormal Cell Detection

We take pathologist-annotated 1958 atypical squamous cells of undetermined significance (ASCUS) and 1641 low-grade squamous intraepithelial lesions (LSIL) cells as training dataset for abnormal cell detection experiments. These cervical stain images for training are similar in color style. 100 ASCUS and 100 LSIL with different styles are used as the test dataset. Compared to traditional cell detection that mainly uses image processing techniques like contour extraction, circle or ellipse detection [14,15], the deep learning methods such as FCOS and

**Table 3.** Average precision (AP) for abnormal cell detection with or without color normalization (*norm.*). After applying our color *norm.* in ASCUS and LSIL, AP boosts.

| Method | Label | w/o *norm.* | Reinhard [25] | Macenko [16] | Vahdane [25] | Ours |
|---|---|---|---|---|---|---|
| FCOS [24] | ASCUS | 26.0 | 28.4 | 29.3 | 28.5 | **33.8** |
| | LSIL | 37.9 | 41.8 | 38.0 | 37.7 | **42.1** |
| Reppoints [28] | ASCUS | 29.5 | 26.6 | 31.1 | 27.8 | **34.9** |
| | LSIL | 46.8 | 48.7 | 49.1 | 42.1 | **54.1** |

Reppoints, are more robust. The effectiveness of the FTNC-Net is demonstrated by comparing the average precision of detection models without and with color normalization on the testing dataset, are shown in Table 3. The abnormal cell detection modules are implemented by MMDetection [2].

### 3.3 Ablation Studies

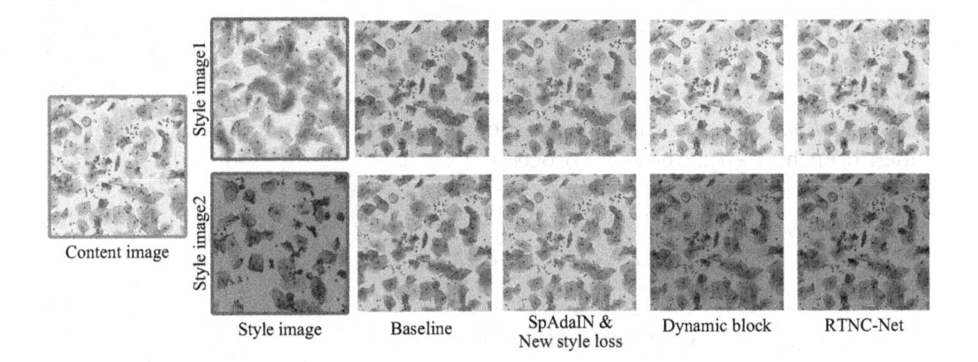


**Fig. 4.** Stain normalization results with different blocks in FTNC-Net.

To illustrate the roles of different designs in the FTNC-Net, we conduct a series of experiments, and results are shown in Fig. 4. We take the model in [27] as the baseline. The experiments from the baseline show that the training results appear cross-color without constraints of spatial color attributes. The baseline model performs poorly for multiple color style cervical images. For medical tasks, the pre-trained VGG weights for loss function are difficult to obtain. Adding the SpAdaIN blocks and changing the style loss solve the problem of cross-color, but the performance of various color styles is still unsatisfactory. Thus, dynamic blocks are added to guide the network to transfer between different styles through stain style encoding to improve the performance of multi-style transfer. However, the original style still influences the transformed image. The previously illustrated blocks in the model are all designed to generate better linear transformation weights. The input of the linear transform is changed by directly transforming the full-resolution content image and style image through SpAdaIN, which reduces the differences caused by the color style of the content images.

## 4 Conclusion

This paper proposes a fast normalization method for Pap stain cervical cell slides based on learnable bilateral filtering. Due to the use of an explicit three-attribute representation and SpAdaIN, our method achieves non-aligned style transfer more precisely. Compared with the previous encoder-decoder structure, our low-resolution and high-resolution branch parallel operations reduce the computational complexity of deep learning networks and realize fast stain style transfer. The FTNC-Net makes it possible to process style transfer online for scanners and has the potential ability to generate various stain styles by replacing the AE with the variational autoencoder (VAE). Further, our method improves the precision of abnormal cells detection algorithms.

## Appendix

Our network (FTNC-Net) is splitted into three branches: the low-resolution splatting ($Lsp$) branch, the autoencoder ($Ae$) branch, and the full-resolution guided branch ($Fgd$). The total process:

$$z = AE(concat(Encoding(I_c), Encoding(I_s))) \tag{10}$$

$$\Gamma = Lsp(I_{lc}, I_{ls}, attr_c, attr_s, z) \tag{11}$$

$$m = Fgd(I_c, z) \tag{12}$$

$$k, b = GridSample(m, \Gamma) \tag{13}$$

$$I_o = k * SpAdaIn(I_c, I_s, attr_c, attr_s) + b \tag{14}$$

The input of $Ae$ is a vector concatenated by two stain style encodings from the full-resolution content image $I_c$ and style image $I_s$. $Lsp$ generates the bilateral grid $\Gamma$ by the low-resolution content image $I_{lc}$, the low-resolution style image $I_{ls}$, 3-attribute estimation ($attr_c$, $attr_s$) of $I_c$ and $I_s$, and the latent variable $z$. $Fgd$ produces the guided map $m$ by $I_c$ and $z$. The $GrideSample$ computes the linear transformation weights $k$ and $b$ using $m$ values and pixel locations from $\Gamma$, and this Eq. (13) called *slice* connects $Lsp$ and $Fgd$. The output image $I_o$ is obtained by applying the linear weights calculated in Eq. (13) to the output of $SpAdaIn$, and this process is called *apply*, where $SpAdaIn$ reduces the influence of the stain color of $I_c$.

The detailed architecture of FTNC-Net is shown in Table 4 and 5:

**Table 4.** Architecture of $Lsp$ and $Fgd$

| **Architecture of** $Lsp$ | | | | |
|---|---|---|---|---|
| layer | | Input | Filter/Stride | Output Size |
| $SpBlk1$ | $conv1\_1$ | $conv1\_1(vgg19)$ | $3 \times 3/2$ | $128 \times 128 \times 8$ |
| | $conv\_s1$ | $conv2\_1(vgg19)$ | $1 \times 1/1$ | $128 \times 128 \times 8$ |
| | $conv1\_2$ | $conv1\_1(SpBlk1)$ | $3 \times 3/1$ | $128 \times 128 \times 8$ |
| $SpBlk2$ | $conv2\_1$ | $conv1\_2(SpBlk1)$ | $3 \times 3/2$ | $64 \times 64 \times 16$ |
| | $conv\_s2$ | $conv3\_1(vgg19)$ | $1 \times 1/1$ | $64 \times 64 \times 16$ |
| | $conv2\_2$ | $conv2\_1(SpBlk2)$ | $3 \times 3/1$ | $64 \times 64 \times 16$ |
| $SpBlk3$ | $conv3\_1$ | $conv2\_2(SpBlk2)$ | $3 \times 3/2$ | $32 \times 32 \times 32$ |
| | $conv\_s3$ | $conv4\_1(vgg19)$ | $1 \times 1/1$ | $32 \times 32 \times 32$ |
| | $conv3\_2$ | $conv3\_1(SpBlk3)$ | $3 \times 3/1$ | $32 \times 32 \times 32$ |
| $DyBlk\_h1$ | $conv1\_1$ | $conv3\_2(SpBlk3)$ | $3 \times 3/2$ | $16 \times 16 \times 64$ |
| | $conv1\_2$ | $conv3\_2(SpBlk3)$ | $3 \times 3/2$ | $16 \times 16 \times 64$ |
| | $conv1\_3$ | $conv3\_2(SpBlk3)$ | $3 \times 3/2$ | $16 \times 16 \times 64$ |
| | $conv1\_4$ | $conv3\_2(SpBlk3)$ | $3 \times 3/2$ | $16 \times 16 \times 64$ |
| | $fc\_d$ | $fc\_e(Ae)$ | – | 4 |
| $DyBlk\_h2$ | $conv2\_1$ | $sum(DyBlk\_h1)$ | $3 \times 3/1$ | $16 \times 16 \times 64$ |
| | $conv2\_2$ | $sum(DyBlk\_h1)$ | $3 \times 3/1$ | $16 \times 16 \times 64$ |
| | $conv2\_3$ | $sum(DyBlk\_h1)$ | $3 \times 3/1$ | $16 \times 16 \times 64$ |
| | $conv2\_4$ | $sum(DyBlk\_h1)$ | $3 \times 3/1$ | $16 \times 16 \times 64$ |
| | $fc\_d$ | $fc\_e(Ae)$ | – | 4 |
| $global$ | $conv1$ | $sum(DyBlk\_h2)$ | $3 \times 3/2$ | $8 \times 8 \times 64$ |
| | $conv2$ | $conv1(global)$ | $3 \times 3/2$ | $4 \times 4 \times 64$ |
| | $fc1$ | $conv2(global)$ | – | 256 |
| | $fc2$ | $fc1(global)$ | – | 128 |
| | $fc3$ | $fc2(global)$ | – | 64 |
| | $fc4$ | $fc3(global)$ | – | 64 |
| $DyBlk\_l1$ | $conv1\_1$ | $sum(DyBlk\_h2)$ | $3 \times 3/2$ | $16 \times 16 \times 64$ |
| | $conv1\_2$ | $sum(DyBlk\_h2)$ | $3 \times 3/2$ | $16 \times 16 \times 64$ |
| | $conv1\_3$ | $sum(DyBlk\_h2)$ | $3 \times 3/2$ | $16 \times 16 \times 64$ |
| | $conv1\_4$ | $sum(DyBlk\_h2)$ | $3 \times 3/2$ | $16 \times 16 \times 64$ |
| | $fc\_d$ | $fc\_e(Ae)$ | – | 4 |
| $DyBlk\_l2$ | $conv2\_1$ | $sum(DyBlk\_l1)$ | $3 \times 3/1$ | $16 \times 16 \times 64$ |
| | $conv2\_2$ | $sum(DyBlk\_l1)$ | $3 \times 3/1$ | $16 \times 16 \times 64$ |
| | $conv2\_3$ | $sum(DyBlk\_l1)$ | $3 \times 3/1$ | $16 \times 16 \times 64$ |
| | $conv2\_4$ | $sum(DyBlk\_l1)$ | $3 \times 3/1$ | $16 \times 16 \times 64$ |
| | $fc\_d$ | $fc\_e(Ae)$ | – | 4 |
| $DyBlk\_l3$ | $conv3\_1$ | $sum(DyBlk\_l2, global)$ | $3 \times 3/1$ | $16 \times 16 \times 64$ |
| | $conv3\_2$ | $sum(DyBlk\_l2, global)$ | $3 \times 3/1$ | $16 \times 16 \times 64$ |
| | $conv3\_3$ | $sum(DyBlk\_l2, global)$ | $3 \times 3/1$ | $16 \times 16 \times 64$ |
| | $conv3\_4$ | $sum(DyBlk\_l2, global)$ | $3 \times 3/1$ | $16 \times 16 \times 64$ |
| | $fc\_d$ | $fc\_e(Ae)$ | – | 4 |
| $DyBlk\_l4$ | $conv4\_1$ | $sum(DyBlk\_l3)$ | $3 \times 3/1$ | $16 \times 16 \times 96$ |
| | $conv4\_2$ | $sum(DyBlk\_l3)$ | $3 \times 3/1$ | $16 \times 16 \times 96$ |
| | $conv4\_3$ | $sum(DyBlk\_l3)$ | $3 \times 3/1$ | $16 \times 16 \times 96$ |
| | $conv4\_4$ | $sum(DyBlk\_l3)$ | $3 \times 3/1$ | $16 \times 16 \times 96$ |
| | $fc\_d$ | $fc\_e(Ae)$ | – | 4 |
| $fc_d$ | $fc1$ | $fc\_e(Ae)$ | – | 16 |
| | $fc2$ | $fc1$ | – | 16 |
| | $fc3$ | $fc2$ | – | 8 |
| | $fc4$ | $fc3$ | – | 8 |
| | $fc5$ | $fc4$ | – | 8 |
| | $fc6$ | $fc5$ | – | 4 |
| | $fc7$ | $fc6$ | – | 4 |

*(continued)*

**Table 4.** (*continued*)

| **Architecture of $Fgd$** | | | | |
|---|---|---|---|---|
| $DyBlk\_f1$ | $conv1\_1$ | $I_c$ | $3 \times 3/1$ | $1024 \times 1024 \times 16$ |
| | $conv1\_2$ | $I_c$ | $3 \times 3/1$ | $1024 \times 1024 \times 16$ |
| | $conv1\_3$ | $I_c$ | $3 \times 3/1$ | $1024 \times 1024 \times 16$ |
| | $conv1\_4$ | $I_c$ | $3 \times 3/1$ | $1024 \times 1024 \times 16$ |
| | $fc\_d$ | $fc\_e(Ae)$ | – | 4 |
| $DyBlk\_f2$ | $conv2\_1$ | $sum(DyBlk\_f1)$ | $3 \times 3/1$ | $1024 \times 1024 \times 16$ |
| | $conv2\_2$ | $sum(DyBlk\_f1)$ | $3 \times 3/1$ | $1024 \times 1024 \times 16$ |
| | $conv2\_3$ | $sum(DyBlk\_f1)$ | $3 \times 3/1$ | $1024 \times 1024 \times 16$ |
| | $conv2\_4$ | $sum(DyBlk\_f1)$ | $3 \times 3/1$ | $1024 \times 1024 \times 16$ |
| | $fc\_d$ | $fc\_e(Ae)$ | – | 4 |
| $DyBlk\_f3$ | $conv3\_1$ | $sum(DyBlk\_f2)$ | $3 \times 3/1$ | $1024 \times 1024 \times 16$ |
| | $conv3\_2$ | $sum(DyBlk\_f2)$ | $3 \times 3/1$ | $1024 \times 1024 \times 16$ |
| | $conv3\_3$ | $sum(DyBlk\_f2)$ | $3 \times 3/1$ | $1024 \times 1024 \times 16$ |
| | $conv3\_4$ | $sum(DyBlk\_f2)$ | $3 \times 3/1$ | $1024 \times 1024 \times 16$ |
| | $fc\_d$ | $fc\_e(Ae)$ | – | 4 |
| conv4 | | $sum(DyBlk\_f33)$ | $1 \times 1/1$ | $1024 \times 1024 \times 1$ |

**Table 5.** Architecture of $Ae$

| layer | | Input | Output Size | layer | | Input | Output Size |
|---|---|---|---|---|---|---|---|
| $fc\_e$ | $fc1$ | $I_c$ | 16 | $fc\_d$ | $fc1$ | $fc4(fc\_e)$ | 16 |
| | $fc2$ | $fc1(fc\_e)$ | 16 | | $fc2$ | $fc1(fc\_d)$ | 16 |
| | $fc3$ | $fc2(fc\_e)$ | 16 | | $fc3$ | $fc2(fc\_d)$ | 16 |
| | $fc4$ | $fc3(fc\_e)$ | 16 | | $fc4$ | $fc3(fc\_d)$ | 18 |

Due to the constrain of GPU memory, StainGAN[1] is implemented on images cropped into 320×320 from 1024×1024. As the Fig. 5 shown, the discriminator in GAN guides the generator to generate the nonexistent neutrophils to match the style image, which changes the structure of the content image. And the results of StainGAN appear cross-color since there is no color consistency limits.

[1] https://github.com/xtarx/StainGAN.

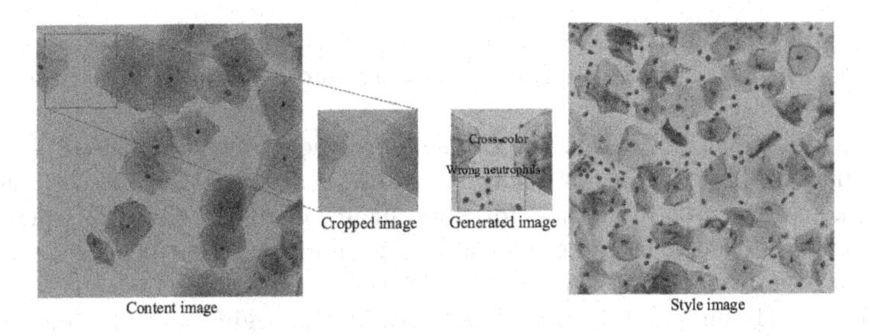


**Fig. 5.** A sample from StainGAN.

## References

1. Abbas, A., Aster, J., Kumar, V.: Robbins Basic Pathology, 9th edn. Elsevier, Amsterdam (2012)
2. Chen, K., et al.: MMDetection: open mmlab detection toolbox and benchmark. arXiv preprint arXiv:1906.07155 (2019)
3. Chen, X., et al.: An unsupervised style normalization method for cytopathology images. Comput. Struct. Biotechnol. J. **19**, 3852–3863 (2021)
4. Gatys, L.A., Ecker, A.S., Bethge, M.: Image style transfer using convolutional neural networks. In: Proceedings of the IEEE Conference on Computer Vision and Pattern Recognition, pp. 2414–2423 (2016)
5. Gharbi, M., Chen, J., Barron, J.T., Hasinoff, S.W., Durand, F.: Deep bilateral learning for real-time image enhancement. ACM Trans. Graph. (TOG) **36**(4), 1–12 (2017)
6. Hong, K., Jeon, S., Yang, H., Fu, J., Byun, H.: Domain-aware universal style transfer. In: Proceedings of the IEEE/CVF International Conference on Computer Vision, pp. 14609–14617 (2021)
7. Huang, X., Belongie, S.: Arbitrary style transfer in real-time with adaptive instance normalization. In: Proceedings of the IEEE International Conference on Computer Vision, pp. 1501–1510 (2017)
8. Huang, X., Liu, M.Y., Belongie, S., Kautz, J.: Multimodal unsupervised image-to-image translation. In: Proceedings of the European Conference on Computer Vision (ECCV), pp. 172–189 (2018)
9. Kang, H., et al.: Stainnet: a fast and robust stain normalization network. Front. Med. **8** (2021). https://doi.org/10.3389/fmed.2021.746307
10. Li, Y., Fang, C., Yang, J., Wang, Z., Lu, X., Yang, M.H.: Universal style transfer via feature transforms. Adv. Neural Inf. Process. Syst. **30**, 1–11 (2017)
11. Li, Y., Liu, M.Y., Li, X., Yang, M.H., Kautz, J.: A closed-form solution to photorealistic image stylization. In: Proceedings of the European Conference on Computer Vision (ECCV), pp. 453–468 (2018)
12. Lu, C., Gu, C., Wu, K., Xia, S., Wang, H., Guan, X.: Deep transfer neural network using hybrid representations of domain discrepancy. Neurocomputing **409**, 60–73 (2020)
13. Lu, C., Wang, H., Gu, C., Wu, K., Guan, X.: Viewpoint estimation for workpieces with deep transfer learning from cold to hot. In: Cheng, L., Leung, A.C.S., Ozawa, S. (eds.) ICONIP 2018. LNCS, vol. 11301, pp. 21–32. Springer, Cham (2018). https://doi.org/10.1007/978-3-030-04167-0_3

14. Lu, C., Xia, S., Huang, W., Shao, M., Fu, Y.: Circle detection by arc-support line segments. In: 2017 IEEE International Conference on Image Processing (ICIP), pp. 76–80. IEEE (2017)
15. Lu, C., Xia, S., Shao, M., Fu, Y.: Arc-support line segments revisited: an efficient high-quality ellipse detection. IEEE Trans. Image Process. **29**, 768–781 (2019)
16. Macenko, M., et al.: A method for normalizing histology slides for quantitative analysis. In: 2009 IEEE International Symposium on Biomedical Imaging: From Nano to Macro, pp. 1107–1110. IEEE (2009)
17. Nadeem, S., Hollmann, T., Tannenbaum, A.: Multimarginal wasserstein barycenter for stain normalization and augmentation. In: Martel, A.L., et al. (eds.) MICCAI 2020. LNCS, vol. 12265, pp. 362–371. Springer, Cham (2020). https://doi.org/10.1007/978-3-030-59722-1_35
18. Raju, K.: Evolution of pap stain. Biomed. Res. Therapy **3**(2), 1–11 (2016)
19. Reinhard, E., Adhikhmin, M., Gooch, B., Shirley, P.: Color transfer between images. IEEE Comput. Graph. Appl. **21**(5), 34–41 (2001)
20. Schulte, E.: Standardization of biological dyes and stains: pitfalls and possibilities. Histochemistry **95**(4), 319 (1991)
21. Shaban, M.T., Baur, C., Navab, N., Albarqouni, S.: Staingan: stain style transfer for digital histological images. In: 2019 IEEE 16th International Symposium on Biomedical Imaging (ISBI 2019), pp. 953–956. IEEE (2019)
22. Simonyan, K., Zisserman, A.: Very deep convolutional networks for large-scale image recognition. arXiv preprint arXiv:1409.1556 (2014)
23. Tellez, D., et al.: Quantifying the effects of data augmentation and stain color normalization in convolutional neural networks for computational pathology. Med. Image Anal. **58**, 101544 (2019)
24. Tian, Z., Shen, C., Chen, H., He, T.: FCOS: fully convolutional one-stage object detection. In: Proceedings of the IEEE/CVF International Conference on Computer Vision, pp. 9627–9636 (2019)
25. Vahadane, A., et al.: Structure-preserving color normalization and sparse stain separation for histological images. IEEE Trans. Med. Imaging **35**(8), 1962–1971 (2016)
26. Wu, X., Lu, C., Gu, C., Wu, K., Zhu, S.: Domain adaptation for viewpoint estimation with image generation. In: 2021 International Conference on Control, Automation and Information Sciences (ICCAIS), pp. 341–346. IEEE (2021)
27. Xia, X., et al.: Joint bilateral learning for real-time universal photorealistic style transfer. In: Vedaldi, A., Bischof, H., Brox, T., Frahm, J.-M. (eds.) ECCV 2020. LNCS, vol. 12353, pp. 327–342. Springer, Cham (2020). https://doi.org/10.1007/978-3-030-58598-3_20
28. Yang, Z., Liu, S., Hu, H., Wang, L., Lin, S.: Reppoints: point set representation for object detection. In: Proceedings of the IEEE/CVF International Conference on Computer Vision, pp. 9657–9666 (2019)
29. Yoo, J., Uh, Y., Chun, S., Kang, B., Ha, J.W.: Photorealistic style transfer via wavelet transforms. In: Proceedings of the IEEE/CVF International Conference on Computer Vision, pp. 9036–9045 (2019)
30. Zhu, J.Y., Park, T., Isola, P., Efros, A.A.: Unpaired image-to-image translation using cycle-consistent adversarial networks. In: Proceedings of the IEEE International Conference on Computer Vision, pp. 2223–2232 (2017)

# MEW: Evading Ownership Detection Against Deep Learning Models

Wenxuan Yin and Haifeng Qian(✉)

East China Normal University, Shanghai, China
hfqian@admin.ecnu.edu.cn

**Abstract.** Training deep neural network (DNNs) requires massive computing resources and data, hence the trained models belong to the model owners' Intellectual Property (IP), and it is very important to defend against the model stealing attack. Recently, a well-known approach named Dataset Inference (DI) claimed that by measuring the distance from the sample to the decision boundary, it can be determined whether the theft has occurred.

In this paper, we show that DI is not enough for IP protection. To demonstrate this, we propose a new system called MEW, which combines the Model Inversion (MI) attack and Elastic Weight Consolidation (EWC) to evade the detection of DI. We first use the pre-trained adversary model to generate a data pool and adaptively select samples to approximate the Fisher Information Matrix of the adversary model. Then we use an adaptation of EWC to slightly fine-tune the adversary model which moves it decision boundary slightly. Our empirical results demonstrate that the adversary model evaded the DI detection with 40 samples. We also lay out the limitations of MEW and discuss them at last.

**Keywords:** Deep learning · Model Stealing attack · Dataset inference

## 1 Introduction

Deep Learning models are dedicated to solving many challenging problems in areas like computer vision and natural language processing in various products and services. It is well known that training a well-performing deep learning model requires numerous computing resources and time [1]. For the academic or financial purpose, model owners can upload their models to the cloud server providing services to the public. In most cases, model parameters and structures are invisible to the user, the user can only get the output (label or probabilities) corresponding to his or her input, which allows malicious users to steal the intellectual property (IP) of the victim model through the exposed interface. This threat is named model stealing attack [16]. Moreover, recent attacks [7,15,17] have shown that it is possible for an adversary to obtain a duplication that is highly close to (up to 99%) even outperforms the victim model in original classification tasks. Since the adversary can obtain a function-similar duplication of the victim, it incurs highly urgency to protect the victim's IP.


M. Tanveer et al. (Eds.): ICONIP 2022, CCIS 1793, pp. 127–136, 2023.
https://doi.org/10.1007/978-981-99-1645-0_11

To prevent the model stealing attack, the victim could select differential privacy [12] or watermark [3,8] as defense strategies. But they have some shortcomings, such as decreasing accuracy or requiring retraining [10]. Recently, Maini [13] proposed a method named Dataset Inference which exploits private training data's unique features to make model ownership verification to safeguard the intellectual property of models. Experiments have shown that the Dataset Inference can verify the model was stolen with 95% confidence by only exposing 10 private training samples of the victim model.

In this paper, we present a countermeasure MEW, which weakens the impact of Dataset Inference on the adversary. With the help of MEW, DI cannot appease security requirements (expose private samples as little as possible) and efficiency requirements (claim the model is stolen with high probability) simultaneously. We find the key of DI depends on the model similarity (like decision boundary) brought by training set knowledge. Once the adversary fine-tunes its own model with some processed data, the detection confidence of DI will drop significantly. However, the test accuracy would also drop. This raises an important question: How to find the most suitable sample which altering adversary's decision boundary rather than decreasing the model accuracy? We propose two techniques to solve it. Firstly, we bring Model Inversion [6] into mind and make some modifications. Depending on the adversary model, we could generate a data pool similar to the original private training set. Then, we divide the data pool into two parts based on the similarity with the private training set. The other technique is adapted from Elastic Weight Consolidation (EWC) [11], to slow down the training speed of weights related to original training tasks, making the decision boundary move slightly. With the help of the proposed system, we can successfully evade the detection by DI. In general, our contributions in this paper are two-fold:

- ***The Proposal of MEW.*** We propose MEW, a useful system to evade DI's detection. We observe the success of DI is mainly attributed to model similarity (like decision boundary) brought by training set knowledge. Once the adversary fine-tunes its model with highly informative data, the detection confidence of DI will drop significantly.
- ***Empirical Results.*** We provide experimental results to show the effectiveness of our system. The results show that our method makes the adversary model hide from DI with as little as 40 random private samples, even making it completely invalid.

## 2 Related Work

### 2.1 Model Stealing

Model stealing attack aims to steal part of the intellectual property of the victim model through obtaining a copy with similar functions. Generally, according to the different permission levels of the adversary, it can be roughly divided into three main methods:

**Query-Only Attack ($\mathcal{A}_Q$):** In this setting, the victim model can only be queried in the form of APIs. The adversary obtains the knowledge of the victim model by querying the API. There are two types of attack depending on the forms returned by the API: Label-only attack [4] and Logit-only attack [15]. (e.g. Label or Probabilities).

**Data-Accessible Attack ($\mathcal{A}_D$):** In this setting, the adversary can obtain the dataset of the victim. They could use knowledge distillation [9] to train a substitute model or train a new model from scratch.

**Model-Accessible Attack ($\mathcal{A}_M$):** In this setting, the adversary has authority over the entire model, including the internal structure, hyper-parameters, and gradients of the victim model. The adversary can train a duplication through zero-knowledge distillation [14], or use a local independent dataset to fine-tune the victim model.

### 2.2 Dataset Inference

Dataset Inference makes ownership verification by identifying whether the adversary model has involved knowledge of the victim model learned from its private training dataset. Firstly, considering a $N$-class task, for every sample $(x, y)$ in the training dataset, DI first generates its distance $\delta$ to each class. Based on the access to the internal gradients of the victim model, DI performs two kinds of generation methods: MinGD and Blind Walk, which represent the White-box and Black-box, respectively. In MinGD, DI obtains the minimum distance to target class m $\delta_m$ by $min_\delta d(x, x+\delta)\ s.t. f(x+\delta) = m$. In Blind Walk, DI first chooses an initial direction $\delta$, and takes $k$ steps in this direction until $f(x+k\delta) = m$. The $\delta_m = k\delta$ is a proxy representing the distance from $y$ to $m$. In general, the metric $(\delta_1, \delta_2, ..., \delta_N)$ is the feature embedding of the model.

Moreover, the victim randomly chooses some number of samples from its private training dataset and public dataset and calculates the distance vectors of its own model. The vectors will be labeled as -1 or 1 (1 means in the private training dataset otherwise -1) and be used to train a binary classifier. In the verification phase, the victim selects equal number of samples from private and public dataset, conduct the hypothesis test based on the binary classifier. If the confidence scores of private samples are significantly greater than those of public samples, the adversary model is marked as stolen from the victim.

Dataset Inference is capable of determining whether the intellectual property of the victim model learned by private dataset has been theft. However, the capability critically depends on the memory and inheritance of decision boundary. Once the decision boundary of adversary model moves slightly, DI would feature the same sample differently. Therefore, the binary classifier would make misjudge.

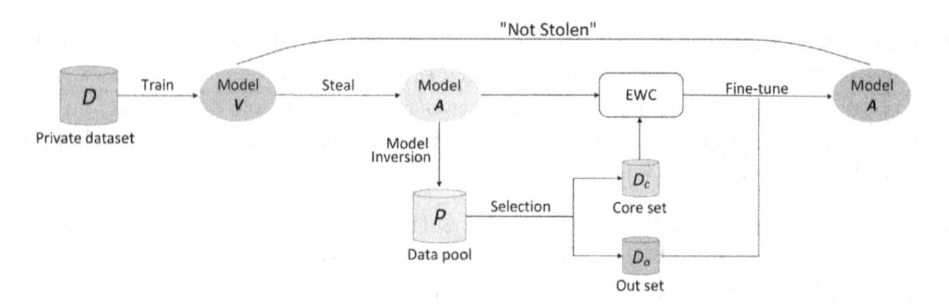


**Fig. 1.** The proposed MEW system firstly inverse the adversary model and get a data pool. Then it divide the data pool into two parts, one for EWC to calculate the penalty term of the loss function, the other for fine-tuning the adversary model $A$. After that, $A$ can bypass the detection of victim.

# 3 MEW

## 3.1 Overview

Based on the understanding in Sect. 2, we proposed a method called MEW, to help an adversary model evade the detection of DI without any knowledge about victim model, Fig. 1 illustrates the overview of MEW.

Firstly, we inverse the pre-trained adversary model $f_A$ and obtain a data pool $P$ containing synthetic images similar to the private training data. However, due to the catastrophic forgetting, the test accuracy may drastically reduce if we fine-tune the adversary model with the data pool directly. Thus, we adopt Elastic Weight Consolidation(EWC) [11], a method that slows down the learning speed of specific model parameters which overcome the sense of catastrophic forgetting. It requires part of original training data to calculate the Fisher Information matrix. In our setting, the adversary has no extra information about the victim except the copy model. Fortunately, we have a data pool inversed by adversary model similar to the private training data. So we could select the samples with highest probability to approximate the Fisher Information matrix. Combining the proposed two methods, we could evade the detection of DI, more details would be shown below.

## 3.2 Inversion and Selection

The adversary has no extra knowledge about the victim including data distribution or model parameters, except its copy of the victim model. Inspired by the model inversion attack, we can recover the images from the adversary model. The model inversion scheme we used is based on [18], but different from it. We replace the well-trained model with our adversary model. By inputting an initial random noise $x$ with equal size of training data and a target label $y$, the recovered image from the adversary model $f_A$ is synthesized by optimizing these three terms:

Class prior is usually used for class-conditional image generation. Given the target label $y$, it minimize the cross entropy loss:

$$\mathcal{L}_{cp} = CE(f_A(x), y) \tag{1}$$

BN Regularization is firstly introduced in [18] to regularize the distribution of synthetic image. Formally, the regularization is represented as the divergence between the mean and variance of synthetic images $\mathcal{N}(\mu_l(x), \sigma_l^2(x))$ and batch normalization statistics $\mathcal{N}(\mu_l, \sigma_l^2)$ of the adversary model as follows:

$$\mathcal{L}_{bn} = \sum_l D(\mathcal{N}(\mu_l(x), \sigma_l^2(x)), \mathcal{N}(\mu_l, \sigma_l^2)) \tag{2}$$

Adversarial Distillation [5] is devoted to the diversity of synthetic images, where the $x$ is forced to maximize the Kullback-Leibler divergence between teacher model $f_t(x; \theta_t)$ and student model $f_s(x; \theta_s)$.

$$\mathcal{L}_{adv} = -KL(f_t(x)/\tau || f_s(x)/\tau) \tag{3}$$

Combined the above techniques could lead to the inversion framework:

$$\mathcal{L} = \alpha \cdot \mathcal{L}_{cp} + \beta \cdot \mathcal{L}_{bn} + \gamma \cdot \mathcal{L}_{adv} \tag{4}$$

where $\lambda$, $\beta$ and $\gamma$ represent the penalty coefficient. Intuitively, the adversary model has stolen the internal parameters of the victim model so the synthetic images obtained by restoring the adversary model are similar to the victim private dataset.

Consequently, to obtain some samples approximating the Fisher Information matrix and fine-tune the adversary model, we split the data pool into two parts by the classification loss. The synthetic images are sorted according to classification loss in ascending order, and the first K samples are selected as the most similar samples. We denoted them as a core set $D_C$, another part is denoted as $D_O$.

### 3.3 Fine-tune with EWC

To maintain the accuracy of the previous task, EWC aims to retard the learning speed of parameters which is important to the previous task. We first calculate the Fisher information matrix $F = [F_{ij}]_{n \times n}$ of the previous task. The element of $F$ is calculated as:

$$F_{ij} = \mathbb{E}_{x \sim D}[\frac{\partial log f(x|\theta)}{\partial \theta_i} \cdot \frac{\partial log f(x|\theta)}{\partial \theta_j}] \tag{5}$$

where $f(x|\theta)$ is the output of the model with parameters $\theta$ for input $x$, and $D$ represents the whole training dataset. To measure the contribution each parameter made, EWC selects the diagonal of the Fisher information matrix of the prior task:

$$F_i = F_{ii} = \mathbb{E}_{x \sim D}[\frac{\partial log f(x|\theta)}{\partial \theta_i}^2 |_{\theta=\theta^*}] \tag{6}$$

where $\theta^*$ means the model parameters of prior tasks. When the pre-trained model faces a new task, to preserve the memory of the prior task, the model parameters $\theta$ should be close to the previous parameters $\theta^*$. Therefore, we set a penalty term proportional to the gap between model parameters $\theta$ and $\theta^*$ which is related to $F_i$. So, we add a regularization term into loss function for training a new task as:

$$\mathcal{L}_{EWC} = \mathcal{L}_0 + \sum_{i}^{n} \frac{\lambda}{2} F_i (\theta_i - \theta_i^*)^2 \tag{7}$$

where $\mathcal{L}_{EWC}$ is the loss function , and $\mathcal{L}_0$ is the loss optimizing the new task (i.e. cross-entropy loss). $\lambda$ is a parameter that controls the importance of prior tasks. $F_i$ is the diagonal of the Fisher information matrix of private tasks. $i$ represents each parameter.

However, we do not have access to the prior task training data so we cannot calculate the $F_i$. Fortunately, we gained two datasets in the previous step which are drawn from a similar distribution of private training datasets. Compared with $D_O$, $D_C$ and victim training set are more similar. To solve the problem, we select the $D_C$ to approximate the Fisher information matrix.

In our setting, the adversary first obtains the intellectual property of the victim model, which means the adversary model can handle the original task. Given that the $D_O$ contains different knowledge from the private dataset, the EWC enables the model to update parameters that control the test accuracy slightly, while the decision boundary has been modified sufficiently. We will show the result in Sect. 4, the adversary can evade the detection of Dataset Inference with only 40 samples.

# 4 Experiments

## 4.1 Dataset

**CIFAR.** CIFAR10 is composed of 32×32 colored images in 10 classes with 50000 training images and 10000 testing images. CIFAR100 has the same distribution as CIFAR10, but it has some differences with 100 classes. Each class contains 500 training images and 100 testing images.

## 4.2 Experimental Setup

**Adversary.** Following the setting in [13], we consider 6 different adversary illustrated in Sect. 2.1.

**Models Architecture.** The victim model is a Wide residual network [19] with 28 depth and 10 widen factors (WRN-28-10). Considering fine-tune attacker, we directly fine-tune in the victim model, so its model architecture is WRN-28-10. For the other attackers, we choose WRN-16-1 on CIFAR10 and WRN-16-2 on CIFAR100.

**Table 1.** Accuracy(%) of victim model and six adversary models on training dataset $D$, testing set $T$ from CIFAR10 and CIFAR100. The subscript $b$ and $a$ means before and after EWC.

| | CIFAR10 | | | CIFAR100 | | |
|---|---|---|---|---|---|---|
| | $D$ | $T_b$ | $T_a$ | $D$ | $T_b$ | $T_a$ |
| Victim | 100.0 | 95.5 | – | 100.0 | 81.5 | – |
| Distillation | 97.1 | 90.7 | 89.0 | 97.9 | 74.5 | 72.0 |
| Diff. Architecture | 97.8 | 92.0 | 90.6 | 100.0 | 78.5 | 76.5 |
| Zero-shot | 92.7 | 88.2 | 87.1 | 64.8 | 57.0 | 56.5 |
| Fine-tune | 99.9 | 95.6 | 94.7 | 99.1 | 80.5 | 78.9 |
| Label-query | 87.2 | 85.2 | 83.6 | 60.9 | 56.5 | 55.5 |
| Logit-query | 93.6 | 91.9 | 90.8 | 71.7 | 61.5 | 58.0 |

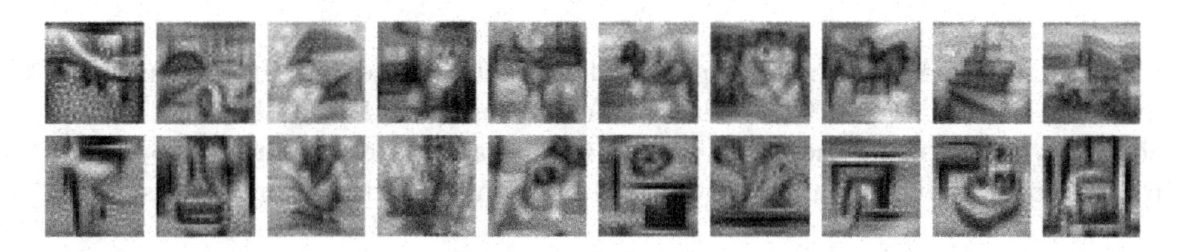

**Fig. 2.** 32×32 images generated by inverting the fine-tune adversary model trained on CIFAR10. 1st row: inversed samples on core set. 2st row: inversed samples on out set. All images are correctly classified, namely: plane, car, bird, cat, deer, dog, frog, horse, ship and truck.

**Model Training.** The victim model are trained for 100 epochs with SGD optimizer and an initial learning rate of 0.1, momentum of 0.9, weight decay of $10^{-4}$. Depending on the type of attack, we adopt different training methods. For the query-accessible attacker, we use unlabeled TinyImages [2] for 20 epochs. Fine-tune attack is similar to the query-accessible attack but training epochs are 5. For the data-accessible attack, we directly train the two models in the original training dataset for 100 epochs. In the case of zero-shot learning, we use the data-free adversarial distillation [5] method and train the model for 500 epochs.

### 4.3 Evaluation Results

**Inversion Image.** Based on [18], after a simple grid search, we found $\lambda = 1$, $\beta = 2.5 \cdot 10^{-3}$, $\gamma = 10.0$ work best for the inversion framework. For all six pre-trained adversary model, we inverse it and get 50k synthetic images. Depending on the classification loss, we select 10k samples with the lowest loss into the core set $D_C$, the others compose the out set $D_O$. As shown in Fig. 2, the pre-trained adversary model produce samples that are similar to the original datasets. The samples on $D_C$ are more nature and realistic.

**Table 2.** Results of Ownership Tester p-value (lower is better) using 40 samples on six threat models.

| Model Stealing Attack | | CIFAR10 | | | | CIFAR100 | | | |
|---|---|---|---|---|---|---|---|---|---|
| | | Before | | After | | Before | | After | |
| | | MinGD | Rand | MinGD | Rand | MinGD | Rand | MinGD | Rand |
| $\mathcal{V}$ | Victim | $10^{-34}$ | $10^{-19}$ | – | – | $10^{-21}$ | $10^{-34}$ | – | – |
| $\mathcal{A}_D$ | Distillation | $10^{-8}$ | $10^{-8}$ | 0.20 | 0.42 | $10^{-4}$ | $10^{-6}$ | 0.61 | 0.37 |
| | Diff. Architecture | $10^{-10}$ | $10^{-11}$ | 0.34 | 0.45 | $10^{-9}$ | $10^{-24}$ | 0.83 | 0.92 |
| $\mathcal{A}_M$ | Zero-Shot | $10^{-4}$ | $10^{-5}$ | 0.18 | 0.11 | $10^{-3}$ | $10^{-2}$ | 0.28 | 0.61 |
| | Fine-tuning | $10^{-9}$ | $10^{-8}$ | 0.54 | 0.50 | $10^{-14}$ | $10^{-24}$ | 0.95 | 0.99 |
| $\mathcal{A}_Q$ | Label-query | $10^{-5}$ | $10^{-3}$ | 0.11 | 0.16 | $10^{-2}$ | $10^{-2}$ | 0.47 | 0.46 |
| | Logit-query | $10^{-7}$ | $10^{-7}$ | 0.21 | 0.13 | $10^{-8}$ | $10^{-2}$ | 0.69 | 0.47 |
| $\mathcal{I}$ | Independent | 0.96 | 0.94 | – | – | 0.78 | 0.96 | – | – |

**Classification Accuracy.** We present the results of accuracy before and after fine-tune with EWC on CIFAR10, CIFAR100 in Table 1. We observe that the adversary model using the EWC would decrease a little test accuracy around 1% to 2%. Fortunately, the adversary gets a safer and covert copy model instead.

**Ownership Detection.** We make dataset inference in different 6 attack methods before and after fine-tune. Table 2 shows p-value, which illustrates the confidence of the hypothesis test of whether a model was stolen. If the p-value is below a significant level $\alpha$, the adversary model would be labeled stolen. Besides, we also make DI in victim model and independent model trained by separate dataset which is shown on the table as Victim and Independent.

From Table 2, DI points out theft of all six adversary. But after MEW processes, the adversary evade the detection successfully. Among the six attack methods, fine-tune adversary model which is consistently captured by DI has the most conspicuous decline from $10^{-24}$ to 0.99 at most. It means the adversary model was regarded as 'independent' by DI. This is expected because fine-tune model has the decision boundary that is most similar to the victim model. Once it is modified, it will most likely deviate away from the victim model decision boundary. The model trained with a different architecture also successfully evades detection with a relatively high probability of 0.34 in MinGD and 0.45 in Blind Walk. Partly because different model architectures own distinct storage methods of intellectual property which indicate unique decision boundaries. The discrepancy would be amplified after processing by MEW. The other adversary model that are trained by the feedback from the victim model were marked as 'stolen' with at least 99.9% probability in CIFAR10 and 99% in CIFAR100. However, after processing with MEW, all of them evaded DI detection successfully with 89% in CIFAR10 and 72% in CIFAR100 at most.

Figure 3 show that DI claimed the model was stolen with more than 99% probability when only 40 samples were leaked on CIFAR10 and CIFAR100. After

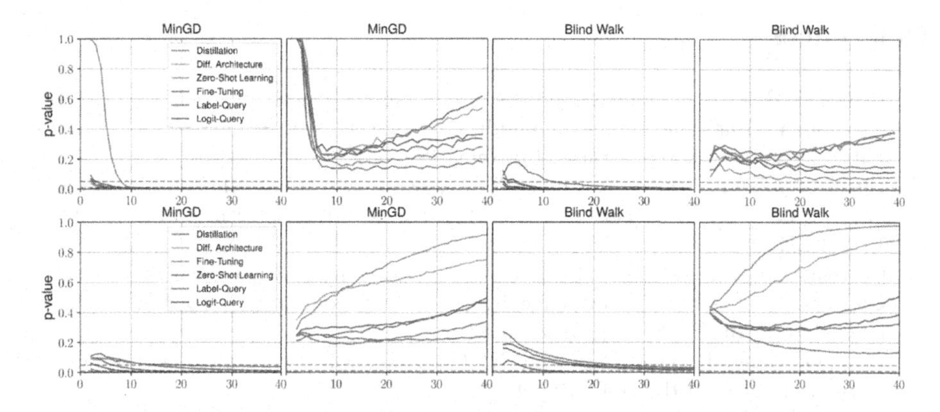


**Fig. 3.** p-values against the number of revealed samples in CIFAR10 and CIFAR100. Significance level $\alpha = 0.05$ and 0.01 have been indicated by a dashed line in the figure. The first row and the second row represent the results on CIFAR10 and CIFAR100 respectively. The first and third figure represent the original adversary model under victim detection in MinGD and Blind Walk respectively. The second and fourth figure show the effect of the adversary model after our MEW processing.

fine-tuning the adversary model on the out set, the p-value makes a substantial increase. DI cannot claim model ownership with a probability of more than 90% with exposing 40 samples which means the two models exist significant difference. To achieve the same verification effect, the victim should expose more samples even can not make a judgment at all.

## 5 Discussion

We acknowledge that the effectiveness of MEW depends on the quality and diversity of the synthetic images. The high similarity of images between the core set and the private training set helps to inhibiting the catastrophic forgetting which brings a relatively small decrease of accuracy. However, the suitable gap of images between the out set and the private training set makes the decision boundary move sufficiently. So it is hard to generate proper images and quantitatively analyze their quality. We consider it as interesting future work.

## 6 Conclusion

In this work, we conduct a new system MEW to evade the detection of DI. To achieve our goals, we first use the pre-trained adversary model to generate a data pool and adaptive select highly similar synthetic images to approximate the previous task Fisher Information Matrix. Then we use an adaption of EWC to slightly fine-tune the adversary model decision boundary. Our empirical results demonstrate that the adversary model evades the DI detection.

## References

1. Brown, T.B., et al.: Language models are few-shot learners. In: NeurIPS (2020)
2. Carmon, Y., Raghunathan, A., Schmidt, L., Liang, P., Duchi, J.C.: Unlabeled data improves adversarial robustness. In: NeurIPS (2019)
3. Chen, G., Chen, S., Xiao, Y., Zhang, Y., Lin, Z., Lai, T.H.: SGXPECTRE: stealing intel secrets from SGX enclaves via speculative execution. In: Euro S&P, pp. 142–157. IEEE (2019)
4. Correia-Silva, J.R., Berriel, R.F., Badue, C., de Souza, A.F., Oliveira-Santos, T.: Copycat CNN: stealing knowledge by persuading confession with random non-labeled data. In: IJCNN, pp. 1–8 (2018)
5. Fang, G., Song, J., Shen, C., Wang, X., Chen, D., Song, M.: Data-free adversarial distillation. arXiv:1912.11006 (2019)
6. Fredrikson, M., Jha, S., Ristenpart, T.: Model inversion attacks that exploit confidence information and basic countermeasures. In: CCS, pp. 1322–1333 (2015)
7. He, X., Lyu, L., Sun, L., Xu, Q.: Model extraction and adversarial transferability, your BERT is vulnerable!. In: Proceedings of the 2021 Conference of the North American Chapter of the Association for Computational Linguistics: Human Language Technologies, pp. 2006–2012 (2021)
8. He, X., Xu, Q., Lyu, L., Wu, F., Wang, C.: Protecting intellectual property of language generation APIS with lexical watermark. In: AAAI (2022)
9. Hinton, G., Vinyals, O., Dean, J.: Distilling the knowledge in a neural network. arXiv:1503.02531 (2015)
10. Jia, H., Choquette-Choo, C.A., Chandrasekaran, V., Papernot, N.: Entangled watermarks as a defense against model extraction. In: USENIX (2021)
11. Kirkpatrick, J., et al.: Overcoming catastrophic forgetting in neural networks. PNAS **114**(13), 3521–3535 (2016)
12. Lee, T., Edwards, B., Molloy, I., Su, D.: Defending against neural network model stealing attacks using deceptive perturbations. In: S & P Workshop, pp. 43–49 (2019)
13. Maini, P., Yaghini, M., Papernot, N.: Dataset inference: ownership resolution in machine learning. In: ICLR (2021)
14. Micaelli, P., Storkey, A.: Zero-shot knowledge transfer via adversarial belief matching. In: NeurIPS (2019)
15. Orekondy, T., Schiele, B., Fritz, M.: Knockoff nets: stealing functionality of black-box models. In: CVPR (2019)
16. Tramèr, F., Zhang, F., Juels, A., Reiter, M.K., Ristenpart, T.: Stealing machine learning models via prediction APIs. In: USENIX (2016)
17. Xu, Q., He, X., Lyu, L., Qu, L., Haffari, G.: Beyond model extraction: imitation attack for black-box NLP APIs. arXiv:2108.13873 (2021)
18. Yin, H., et al.: Dreaming to distill: Data-free knowledge transfer via deepinversion. In: CVPR (2020)
19. Zagoruyko, S., Komodakis, N.: Wide residual networks. In: BMVC (2016)

# Spatial-Temporal Graph Transformer for Skeleton-Based Sign Language Recognition

Zhengye Xiao[1], Shiquan Lin[1], Xiuan Wan[1], Yuchun Fang[1(✉)], and Lan Ni[2(✉)]

[1] School of Computer Engineering and Science, Shanghai University, Shanghai, China
{xiaozy,funterlin,hideinsoul,ycfang}@shu.edu.cn
[2] College of Liberal Arts, Shanghai University, Shanghai, China
yclannimail@shu.edu.cn

**Abstract.** For continuous sign language recognition (CSLR), the skeleton sequence is insusceptible to environmental variances and achieves much attention. Previous studies mainly employ hand-craft features or the spatial-temporal graph convolution networks for skeleton modality and neglect the importance of capturing the information between distant nodes and the long-term context in CSLR. To learn more robust spatial-temporal features for CSLR, we propose a Spatial-Temporal Graph Transformer (STGT) model for skeleton-based CSLR. With the self-attention mechanism, the human skeleton graph is treated as a fully connected graph, and the relationship between distant nodes can be established directly in the spatial dimension. In the temporal dimension, the long-term context can be learned easily due to the characteristic of the transformer. Moreover, we propose graph positional embedding and graph multi-head self-attention to help the STGT distinguish the meanings of different nodes. We conduct the ablation study on the action recognition dataset to validate the effectiveness and analyze the advantages of our method. The experimental results on two CSLR datasets demonstrate the superiority of the STGT on skeleton-based CSLR.

**Keywords:** Continuous sign language recognition · Transformer · Graph neural network

## 1 Introduction

The participation of deaf people in social activities has been an essential measure for deaf people's social integration and benefits their physical and mental health. Therefore, Continuous Sign Language Recognition (CSLR) has received more and more attention as an intelligent communication method between deaf and ordinary people.

Previous studies mainly focus on the RGB modality to solve the CSLR problem. However, the RGB-based methods are easily interfered with by the environment and human clothes. Skeleton sequence, which is not well explored in CSLR, is a promising modality to solve this problem. The skeleton sequence provides coordinates of key joints in the human body and hands, which ensure the model


M. Tanveer et al. (Eds.): ICONIP 2022, CCIS 1793, pp. 137–149, 2023.
https://doi.org/10.1007/978-981-99-1645-0_12

focus on the trajectory and eliminate the negative influence of the environment, such as background and illumination. More importantly, the skeleton data can be easily collected with the development of camera arrays, Kinect, and pose estimation methods.

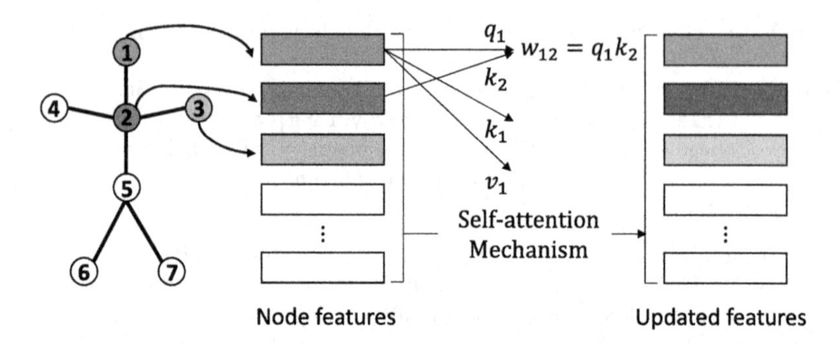


**Fig. 1.** An example of updating node features. Node features are converted into query, key, and value vectors. The weights of the edges are computed by the query and key vectors of different nodes.

The existing skeleton modality methods for CSLR mainly use the hand-craft features [21,24] or Spatial-Temporal Graph Convolutional Networks (ST-GCNs) [1], which are one of the most popular methods in skeleton-based action recognition. However, we think they neglect two important aspects of CSLR. The first one is that the path lengths between the key points of the left and right hands are very long, so it is essential to focus on the model's ability to capture the information between distant nodes. ST-GCNs usually employ the high-order adjacency matrices to establish associations between the key points with a long path, but in the extreme case, the over-smoothing problem [11] will lead to a decline in generalization performance. The second one is that sign language is one of the natural languages, so context is essential to word recognition. Previous methods usually employ temporal convolution to integrate the information of consecutive frames but neglect the context information.

To learn more robust spatial-temporal features for CSLR, we propose the Spatial-Temporal Graph Transformer (STGT) for skeleton-based CSLR. The STGT is composed of spatial and temporal transformers. The spatial dimension situation is shown in Fig. 1. With the self-attention mechanism [18] in spatial transformer, we can treat the human skeleton data as a fully connected graph, and the weights of each edge are learned adaptively. As a result, each node in the graph can receive the message from other nodes directly without the limitation of the adjacency matrix. Moreover, we propose the graph positional embedding and graph multi-head self-attention mechanism to prevent the network from confusing the meaning of different nodes. In the temporal dimension, the original transformer is employed and aggregates the long-term sequence information to learn the context information in CSLR. The spatial and temporal transformers are stacked alternately to learn the spatial-temporal features. Finally, sequence learning module is used to recognize the gloss (*i.e.*, the words in CSLR) sequence.

To verify our method, we conduct experiments on both action recognition and CSLR datasets. The experimental results demonstrate the superiority of the proposed STGT method.

The contribution of this work are summarized as follows:

1. We propose a novel Spatial-Temporal Graph Transformer (STGT) model. The relationship between distant nodes and long-term context can be captured with the spatial and temporal self-attention mechanism.
2. We propose graph positional embedding and graph multi-head self-attention for the spatial transformer to solve the problem of confusing the meanings of different nodes in the fully connected graph, and the generalization of STGT improves.

## 2 Proposed Method

### 2.1 Preliminaries

**Notations.** The human skeleton data can be donated as a graph $\mathcal{G} = (\mathcal{V}, \mathcal{E})$, where $\mathcal{V} = \{v_i\}_{i=1}^{N}$ is the set of $N$ nodes containing joints information, and $\mathcal{E} = \{e_i\}_{i=1}^{M}$ is the set of edges. The adjacency matrix of $\mathcal{G}$ is represented as $A \in \mathbb{R}^{N \times N}$, where $A_{i,j} = 1$ if there is an edge between $v_i$ and $v_j$, and 0 otherwise. In our method, $\mathcal{G}$ is an undirect graph and $A$ is symmetric. Each action is represented as a $T$-frame skeleton node sequence which contains a node features set $\mathcal{X} = \{x_{t,n} \in \mathbb{R}^{C \times 1} | 1 \leq t \leq T, 1 \leq n \leq N\}$, where $x_{t,n}$ is the $C$ dimension feature of node $v_n$ at timestep $t$. $x_{t,n}$ is usually composed of 2D or 3D coordinates of joints and confidence.

**Self-Attention.** The transformer replaces the recurrent units with the multi-head self-attention mechanism and achieves excellent performance in long sequence. For an embedding sequence matrix $Z = [z_1, z_2, \ldots, z_n]$, self-attention mechanism first learns to project $Z$ into the query matrix $Q = f_Q(Z)$, key matrix $K = f_K(Z)$, and value matrix $V = f_V(Z)$, where $f_Q$, $f_K$ and $f_V$ are projection functions. The self-attention mechanism is computed in Eq. (1).

$$\text{Attention}(Z) = \text{softmax}(\frac{QK^T}{\sqrt{d_k}})V, \tag{1}$$

where $d_k$ is the dimension of each query, and $1/\sqrt{d_k}$ improves the numerical stability for the scaled-dot product term.

The multi-head self-attention mechanism runs $H$ self-attention operations and projects the concatenated outputs, which are computed in Eq. (2).

$$\text{MutiHead}(Z) = \text{Concat}(\text{head}_1, \ldots, \text{head}_\text{H})W, \tag{2}$$

where $\text{head}_\text{h} = \text{Attention}_\text{h}(Z)$, and $W$ is a parameter matrix which merges the concatenated outputs. Multi-head attention fuses the information from different representations at different positions and improves the generalization.

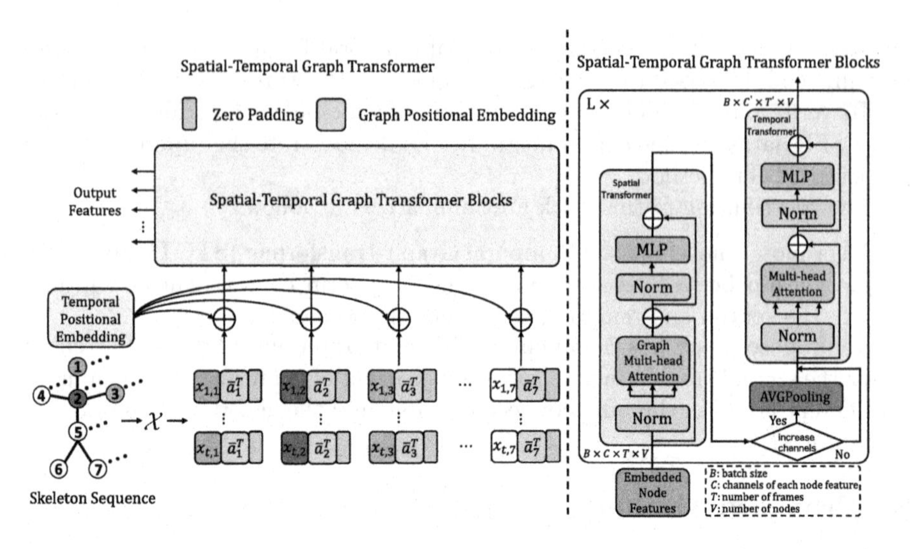


**Fig. 2.** The framework of the proposed model and illustrations of the spatial transformer and the temporal transformer.

## 2.2 Model Overview

The overview of the STGT is shown in Fig. 2. For a skeleton sequence input, we first concatenate the graph positional embeddings with node features and then add temporal positional embeddings. After that, these embedded node features are sent into spatial-temporal graph transformer blocks. Each block composes of a spatial transformer and a temporal transformer. Next, the spatial transformers and temporal transformers are stacked alternately. Finally, the spatial-temporal features are extracted. We elaborate on the spatial transformer and the temporal transformer in the following sections.

## 2.3 Spatial Transformer

The spatial transformer module treats the skeleton data as a fully connected graph and extracts the spatial interaction among nodes at each timestep. However, since each node is connected to all other nodes, the network may treat different nodes equally, such as the head node and hand nodes. So directly running the original transformer architecture on spatial nodes is not feasible. The spatial information of nodes will be lost. To solve this problem, we further propose the graph positional embedding and graph multi-head self-attention mechanism.

**Graph Positional Embedding.** To make the network distinguish the nodes that represent different parts, we propose concatenating graph positional embedding with the node features. Human skeleton joints are naturally in the form of a graph, and the adjacency matrix can reflect the connection relationship between nodes so that we can take the $n$-th row of the normalized adjacency matrix $\bar{A}$ in Eq. (3) as the graph positional embedding of $v_n$.

$$\bar{A} = \tilde{D}^{-\frac{1}{2}} \tilde{A} \tilde{D}^{-\frac{1}{2}}, \tag{3}$$

where $\tilde{A} = A + I$ is the skeleton graph with self-loops, $\tilde{D}$ is the diagonal degree matrix of $\tilde{A}$. If $\bar{A} = [\bar{a}_1, \bar{a}_2, \ldots, \bar{a}_N]$, the node features with graph positional embedding is represented in Eq. (4).

$$x^e_{t,n} = x_{t,n} \circ \bar{a}^T_n, \tag{4}$$

where $\circ$ is the concatenation operation and $\bar{a}_n$ is the $n$-th column of $\bar{A}$.

**Graph Self-attention Mechanism.** Although the human skeleton graph is not reasonable for some particular actions, it can also be considered good prior information and helps the network distinguish the different nodes. The proposed graph self-attention mechanism is computed in Eq. (5).

$$\text{GAttention}(Z) = \text{softmax}(\frac{QK^T}{\sqrt{d_k}} + \bar{A})V. \tag{5}$$

and the graph multi-head self-attention mechanism is computed in Eq. (6).

$$\text{GMutiHead}(Z) = \text{Concat}(\text{head}_1, \ldots, \text{head}_\text{H})W, \tag{6}$$

where $\text{head}_\text{h} = \text{GAttention}_\text{h}(Z)$. The values of elements in $\bar{A}$ are large enough to influence the attention in the early epochs and improve the generalization in the late epochs.

The inputs of the spatial transformer are the nodes at the same timestep. We pad the embedded features with zeros so that the dimension of the features can be divided by the number of heads in multi-head self-attention. Similar to the transformer encoder, a spatial transformer consists of a graph multi-head self-attention block and a Multilayer Perceptron (MLP) block. To get rid of the warmup step and reduce the complexity of setting hyperparameters, we put layernorm (LN) before each block, and the residual connections are set after every block [2,19]. The computation progress of the spatial transformer is represented in Eq. (7).

$$\begin{aligned} Z^{(t)}_0 &= [x^e_{t,1} \circ m; x^e_{t,2} \circ m; \cdots; x^e_{t,N} \circ m], \\ Z'^{(t)}_l &= \text{GMultihead}(\text{LN}(Z^{(t)}_{l-1})) + Z^{(t)}_{l-1}, \\ Z^{(t)}_l &= \text{MLP}(\text{LN}(Z'^{(t)}_l)) + Z'^{(t)}_l, \end{aligned} \tag{7}$$

where $Z^{(t)}_l$ represents the node features at timestep $t$ in the $l$-th layer, $l = 2i-1, i = 1, \ldots, L$, and $m$ is the zero paddings. The odd layers are spatial transformer modules.

### 2.4 Temporal Transformer

The temporal transformer module extracts the long-term sequence information. The inputs of the temporal transformer are the same node at different timesteps. These nodes form a sequence, and this sequence is similar to the sequence of word embeddings in NLP, so we directly employ the original transformer encoder as our temporal transformer, which consists of a multi-head self-attention block and an MLP block. Layernorm is set before each block, and the residual connection is put after each block. The computation progress of the temporal transformer is represented in Eq. (8).

$$\begin{aligned} Z_0^{(n)} &= [x_{1,n}^e \circ m; x_{2,n}^e \circ m; \cdots ; x_{T,n}^e \circ m] + E, \\ Z_l^{'(n)} &= \text{Multihead}(\text{LN}(Z_{l-1}^{(n)})) + Z_{l-1}^{(n)}, \\ Z_l^{(n)} &= \text{MLP}(\text{LN}(Z_l^{'(n)})) + Z_l^{'(n)}, \end{aligned} \tag{8}$$

where $Z_l^{(n)}$ represents the node features for node $n$ at different timesteps in the $l$-th layer, $l = 2i, i = 1, \ldots, L$. The even layers are temporal transformer modules. $E$ in Eq. (9) is the temporal positional embedding proposed in [18], which can reflect the distance between different nodes in the sequence.

$$\begin{aligned} E_{t,2i} &= \sin(t/10000^{2i/d}), \\ E_{t,2i+1} &= \cos(t/10000^{2i/d}), \end{aligned} \tag{9}$$

where $i$ is the index of dimension, and $d$ is the dimension of $x_{t,n} \circ \bar{a}_n^T$.

### 2.5 Action and Continuous Sign Language Recognition

The output features of the spatial-temporal graph transformer blocks is $Z_l \in \mathbb{R}^{B \times C' \times T' \times N}$, where $B$ is the batch size, $C'$ is the output channels and $T'$ is the output frames. For action recognition, we employ a global pooling for the channel dimension to get one-dimension features $Z_l' \in \mathbb{R}^{B \times C'}$. Then the features are fed into an MLP block to get the probabilities for each class. For CSLR, an average pooling is performed on the vertex dimension to achieve the sequence features $Z_l' \in \mathbb{R}^{B \times T' \times C'}$. Then, a Bidirectional Long Short Term Memory (BiLSTM) network is applied in sequence recognition.

## 3 Experiments

### 3.1 Dataset

We employ NTU RGB+D 60 [15] as the action recognition evaluation dataset, which is composed of 60 different action classes. The skeleton sequences in the NTU RGB+D 60 are captured by Kinect with 25 key joints of bodies. Forty different subjects perform the actions in the dataset. We conduct our experiments on the Cross-Subject (X-sub) protocol, where the 40 subjects are split into training and testing groups.

We evaluate our method on the Phoenix-2014 [10] and Chinese Sign Language (CSL) [8] datasets. PHOENIX-2014 is one of the biggest and most popular benchmarks for CSLR. The dataset is collected from the real German weather forecast, leading to poor image quality and extreme imbalance in the frequency of occurrence of glosses. There are 5,672 videos in Train, 540 videos in Dev, and 629 videos in Test, respectively. The vocabulary volume is 1,295, of which 1,232 glosses exist in Train. It contains videos of 9 different signers, and we evaluate our method on the multi-signer dataset.

CSL is one of the biggest Chinese Sign Language datasets collected in the laboratory with high-resolution images. It contains 100 sentences. Each sentence is performed by 50 signers and repeated by five times. To compare with other methods fairly, we use 5,000 videos of CSL. The 50 signers perform each sentence only once. We conduct the signer-independent experiments. The former 40 signers are used for training, and the latter ten ones are used for testing. So there are 4,000 videos for training and 1,000 videos for testing.

We use OpenPose [3] to estimate the two-dimension coordinates of the total 54 body and hand skeletons. The coordinates with confidence are taken as the input feature. Unfortunately, due to the poor image quality, some frames in the PHOENIX-2014 dataset cannot be correctly estimated. So we use linear interpolation to estimate the failed frames from successful neighbor frames.

We employ classification accuracy as the evaluation metric for NTU-RGB+D 60 dataset. In Phoenix-2014 and CSL datasets, Word Error Rate (WER), which measures the similarity between two sentences, is used as the evaluation metric. Moreover, since the ground truth of the CSL dataset is natural language, we employ the metrics in Neural Machine Translation (NMT), including BLEU-1, ROUGE-L, METEOR, and CIDEr, to evaluate STGT on the CSL dataset.

### 3.2 Implementation Details

For the NTU RGB+D 60 dataset, the STGT model comprises 9 spatial-temporal graph transformer blocks. We increase the dimension of features at the 1st, 4-th, and 7-th spatial transformers. The first, second, and last three blocks have 64, 128, and 256 channels for output, respectively. We perform an average pooling among the adjacent three frames with two steps after the 4-th and 7-th spatial transformers. The loss function is cross-entropy with label smoothing of value $\epsilon = 0.1$. We employ the Adam optimizer with an initial training rate of 0.0003. The learning rate is decayed with 0.1 at epochs 30 and 40. The number of epochs is set as 50. In the experiment, we find that weight decay can improve performance while dropout is harmful, so we set weight decay as 0.0001 and dropout as 0. All skeleton sequences are padded to $T = 300$ frames by replaying the actions. The inputs are preprocessed with normalization following [16].

For both CSLR datasets, The first three blocks have 128 channels for output. The following three blocks have 256 channels for output. And the last three blocks have 512 channels for output. Other settings are the same as the settings in NTU RGB+D 60 dataset. The loss function is connectionist temporal classification (CTC) [6] loss, which is widely used in speech recognition. We employ the Adam optimizer with an initial training rate of 0.0003. The learning rate is decayed with 0.1 at epochs 10 and 20. The number of epochs is set as 40, and the batch size is set as 1.

**Table 1.** Ablation study on graph positional embedding and graph multi-head attention.

| Graph Positional Embedding | Graph Multi-head Attention | NTU RGB+D 60 Accuracy(%) | CSL WER(%) |
|---|---|---|---|
| | ✓ | 81.9 | 1.9 |
| ✓ | | 82.6 | 1.8 |
| ✓ | ✓ | **82.9** | **1.4** |

**Table 2.** Classification accuracy of some action classes.

| **Action Classes** | **Accuracy(%)** | |
|---|---|---|
| | STGT | AGC-LSTM [17] |
| Hugging other person | 93.8 | – |
| Cheer up | 89.8 | – |
| Kicking something | 95.3 | – |
| Pointing to something with finger | 78.6 | 71.0 |
| Pat on back of other person | 88.8 | 68.0 |
| Reading | 42.3 | 63.0 |
| Writing | 40.1 | 63.0 |
| Playing with phone/tablet | 57.1 | 63.0 |
| Typing on a keyboard | 61.5 | 69.0 |

### 3.3 Ablation Study

We analyze the effectiveness of the proposed graph positional embedding and graph multi-head attention in the STGT on the X-sub config of NTU RGB+D 60 and CSL datasets. Moreover, we report some action class accuracy recognized by the STGT to find out the advantages and shortages of STGT.

**Graph Positional Embedding.** In Table 1, the accuracy improves by 1.0% and WER decreases by 0.5% compared with the method without graph positional embedding. If the meaning of key points is not embedded in the features, STGT is difficult to determine the meaning of each key point and will lead to confusion. Hence, the graph positional embedding retains the information of graph structure and is unique, it can effectively help STGT distinguish the meaning of different key points and improve the generalization of the model.

**Graph Multi-head Attention.** Using graph multi-head attention brings 0.3% improvements on accuracy and 0.4% decrease on WER, which is shown in Table 1. The improvement on NTU RGB + D 60 dataset is small. We think it is due to the small number of key points extracted by Kinect, and the self-attention mechanism is enough to learn the human skeleton map with graph positional embedding. However, for the CSLR problem, there are many key points, and the graph multi-head attention can play a better initialization.

**Classification Accuracy of Some Action Classes.** We report the classification accuracy of some action classes in Table 2. Benefiting from the spatial transformer, the actions including 'hugging other person', 'cheer up' and 'kick something', which need to consider the relationship between distant joints, can be classified well. AGC-LSTM [17] provided a confusion matrix for action classes that are difficult to classify. Compared with AGC-LSTM, our model can classify 'pointing to something with finger' and 'patting on back of other person' very well. A spatial and temporal transformer union can better capture the long-term trajectory information. While for static actions such as 'reading' and 'writing', AGC-LSTM performs better due to feature augmentation.

**Table 3.** Classification accuracy comparison against mainstream methods on the NTU RGB+D 60 skeleton dataset.

| **Methods** | **NTU-RGB+D 60** | |
|---|---|---|
| | X-Sub(%) | Inference GPU Memory |
| VA-LSTM [22] | 79.4 | – |
| IndRNN [12] | 81.8 | – |
| ARRN-LSTM [23] | 81.8 | – |
| ST-GCN [20] | 81.5 | 11 GB |
| STGT (ours) | **82.9** | 3.5 GB |

**Table 4.** The experimental results on the Phoenix-2014 dataset.

| Methods | Modality | Dev. WER(%) (before relaxation) | Dev (after relaxation) | | Test (after relaxation) | |
|---|---|---|---|---|---|---|
| | | | WER(%) | sub/del/ins | WER(%) | sub/del/ins |
| DecoupleGCN [5] | Skeleton | 54.0 | 50.4 | 29.9/17.0/3.5 | 49.4 | 29.7/15.8/3.9 |
| MS-G3D [13] | Skeleton | 50.6 | 46.5 | 26.4/**16.5**/3.6 | **44.2** | 25.8/**15.5**/**2.9** |
| ST-GCN-RES [16] | Skeleton | 50.3 | 46.7 | 25.7/18.2/**2.8** | 45.7 | **25.1**/17.7/**2.9** |
| STGT | Skeleton | **49.7** | **46.2** | **23.5**/18.2/4.5 | 45.0 | 23.8/16.7/4.4 |
| I3D [4] | RGB | – | 35.2 | –/13.0/4.2 | 35.3 | –/13.1/4.1 |

## 3.4 Results on Action Recognition

We compare our model with other mainstream methods in Table 3. Compared with traditional LSTM architecture methods, the proposed transformer architecture method STGT can outperform them in X-sub. ST-GCN is a baseline in GCN architecture methods, and STGT can perform better. Benefiting from the transformer architecture, the other advantage of STGT is that it takes up less GPU memory at inference. Under the batch size of 16, the needed GPU memory of STGT is four times less than ST-GCN.

### 3.5 Results on Continuous Sign Language Recognition

We compare the other three outstanding methods in the field of the skeleton-based action recognition, including MS-G3D [13], Decouple-GCN [5], and ST-GCN-RES. ST-GCN-RES is ST-GCN with a residual adjacency matrix [16].

The experimental results on the Phoenix-2014 dataset are shown in Table 4. The authors of the Phonix-2014 dataset provide a relaxation method to reduce the difficulty of recognition by treating the words with similar meanings but challenging to recognize as synonyms of simple words. These words are treated as the same class. So 'before relaxation' column does not use the relaxation method, and 'after relaxation' column uses the relaxation method. STGT outperforms other methods on the skeleton modality before relaxation on Dev. Moreover, we find that the WER of STGT is 0.9% lower than MS-G3D before relaxation, but STGT can only outperform MS-G3D 0.3% after relaxation on the Dev and performs worse on the Test. This result demonstrates that STGT can recognize hard words better, but the relaxation method weakens this ability. Although STGT does not achieve the best performance on Test, it still outperforms the other two methods. The experimental results demonstrate the effectiveness of the STGT.

**Table 5.** The experimental results on the CSL dataset.

| Methods | Modality | CIDEr | BLEU-1 | ROUGE-L | METEOR | WER(%) |
|---|---|---|---|---|---|---|
| MS-G3D [13] | Skeleton | 8.886 | 0.967 | 0.964 | 0.656 | 4.9 |
| DecoupleGCN [5] | Skeleton | 9.052 | 0.973 | 0.972 | 0.675 | 4.0 |
| ST-GCN-RES [16] | Skeleton | 9.096 | 0.977 | 0.977 | 0.680 | 3.1 |
| HLSTM+attn [7] | RGB | 9.084 | 0.948 | 0.951 | 0.703 | 10.2 |
| BAE+attn [9] | RGB | 9.037 | 0.933 | 0.934 | 0.706 | 7.4 |
| IAN [14] | RGB | 9.342 | 0.980 | 0.981 | 0.713 | – |
| STMC [24] | RGB | – | – | – | – | 2.1 |
| STGT | Skeleton | **9.411** | **0.991** | **0.991** | **0.734** | **1.4** |

The experimental results on the CSL dataset is shown in Table 5. The STGT outperforms other methods on the skeleton modality. The WER of STGT is 3.5% lower than MS-G3D, proving that the STGT can learn better spatial-temporal features. STGT also has a 1.7% lower WER than the second-lowest model ST-GCN-RES. In the metrics of NMT, the proposed method also demonstrates large advantages compared with other methods. These experiments prove the effectiveness of the STGT. Compared with the RGB modalities, we find that the skeleton modality methods are superior to most RGB modalities. This is because the data in CSL dataset is collected in the lab and has high resolution so that OpenPose can estimate the coordinates correctly. Moreover, the experimental protocol in the CSL dataset is signer-independent (*i.e.*, the training signers do not exist in the test set). The skeleton modality is not easily affected by the environment, so it has some advantages compared with the RGB modality. Notably, the STGT has better performance than the STMC [24], which is the state-of-the-art method in the CSL dataset. The experiments on both CSLR datasets prove that STGT is powerful to extract long-term context and information between distant nodes in high-resolution situations.

## 4 Conclusion

This paper proposes a novel Spatial-Temporal Graph Transformer model for skeleton-based CSLR. STGT employs spatial transformers to treat the skeleton data as a fully connected graph, and the weights of edges are learned adaptively. The temporal transformer can aggregate temporal information. The union of spatial and temporal transformers can effectively strengthen the relationship between distant joints and capture trajectory information. We propose graph positional embedding and graph multi-head attention to help the model distinguish the joints with different meanings. The ablation study on NTU RGB+D 60 proves the effectiveness of the proposed schemes of graph positional embedding and graph multi-head attention. The proposed STGT demonstrates superior performance on the Phoenix-2014 and CSL datasets, which proves its advantages on skeleton-based CSLR.

**Acknowledgment.** The work is supported by the National Natural Science Foundation of China under Grant No.: 61976132, 61991411 and U1811461, and the Natural Science Foundation of Shanghai under Grant No.: 19ZR1419200.

We appreciate the High Performance Computing Center of Shanghai University and Shanghai Engineering Research Center of Intelligent Computing System No.: 19DZ2252600 for providing computing resources.

## References

1. de Amorim, C.C., Macêdo, D., Zanchettin, C.: Spatial-temporal graph convolutional networks for sign language recognition. In: Tetko, I.V., Kůrková, V., Karpov, P., Theis, F. (eds.) ICANN 2019. LNCS, vol. 11731, pp. 646–657. Springer, Cham (2019). https://doi.org/10.1007/978-3-030-30493-5_59
2. Baevski, A., Auli, M.: Adaptive input representations for neural language modeling. In: International Conference on Learning Representations (2018)
3. Cao, Z., Hidalgo, G., Simon, T., Wei, S.E., Sheikh, Y.: Openpose: realtime multi-person 2D pose estimation using part affinity fields. IEEE Trans. Pattern Anal. Mach. Intell. **43**(1), 172–186 (2019)
4. Carreira, J., Zisserman, A.: Quo vadis, action recognition? a new model and the kinetics dataset. In: proceedings of the IEEE Conference on Computer Vision and Pattern Recognition, pp. 6299–6308 (2017)
5. Cheng, K., Zhang, Y., Cao, C., Shi, L., Cheng, J., Lu, H.: Decoupling GCN with DropGraph module for skeleton-based action recognition. In: Vedaldi, A., Bischof, H., Brox, T., Frahm, J.-M. (eds.) ECCV 2020. LNCS, vol. 12369, pp. 536–553. Springer, Cham (2020). https://doi.org/10.1007/978-3-030-58586-0_32

6. Graves, A., Fernández, S., Gomez, F., Schmidhuber, J.: Connectionist temporal classification: labelling unsegmented sequence data with recurrent neural networks. In: Proceedings of the 23rd International Conference on Machine Learning, pp. 369–376 (2006)
7. Guo, D., Zhou, W., Li, H., Wang, M.: Hierarchical LSTM for sign language translation. In: Proceedings of the AAAI Conference on Artificial Intelligence, vol. 32 (2018)
8. Huang, J., Zhou, W., Zhang, Q., Li, H., Li, W.: Video-based sign language recognition without temporal segmentation. In: Proceedings of the AAAI Conference on Artificial Intelligence, vol. 32 (2018)
9. Huang, S., Ye, Z.: Boundary-adaptive encoder with attention method for Chinese sign language recognition. IEEE Access **9**, 70948–70960 (2021)
10. Koller, O., Forster, J., Ney, H.: Continuous sign language recognition: towards large vocabulary statistical recognition systems handling multiple signers. Comput. Vis. Image Underst. **141**, 108–125 (2015)
11. Li, Q., Han, Z., Wu, X.M.: Deeper insights into graph convolutional networks for semi-supervised learning. In: Thirty-Second AAAI Conference on Artificial Intelligence (2018)
12. Li, S., Li, W., Cook, C., Zhu, C., Gao, Y.: Independently recurrent neural network (indrnn): Building a longer and deeper RNN. In: Proceedings of the IEEE Conference on Computer Vision and Pattern Recognition, pp. 5457–5466 (2018)
13. Liu, Z., Zhang, H., Chen, Z., Wang, Z., Ouyang, W.: Disentangling and unifying graph convolutions for skeleton-based action recognition. In: Proceedings of the IEEE/CVF Conference on Computer Vision and Pattern Recognition, pp. 143–152 (2020)
14. Pu, J., Zhou, W., Li, H.: Iterative alignment network for continuous sign language recognition. In: Proceedings of the IEEE/CVF Conference on Computer Vision and Pattern Recognition, pp. 4165–4174 (2019)
15. Shahroudy, A., Liu, J., Ng, T.T., Wang, G.: NTU RGB+D: a large scale dataset for 3D human activity analysis. In: Proceedings of the IEEE Conference on Computer Vision and Pattern Recognition, pp. 1010–1019 (2016)
16. Shi, L., Zhang, Y., Cheng, J., Lu, H.: Two-stream adaptive graph convolutional networks for skeleton-based action recognition. In: Proceedings of the IEEE Conference on Computer Vision and Pattern Recognition, pp. 12026–12035 (2019)
17. Si, C., Chen, W., Wang, W., Wang, L., Tan, T.: An attention enhanced graph convolutional LSTM network for skeleton-based action recognition. In: Proceedings of the IEEE Conference on Computer Vision and Pattern Recognition, pp. 1227–1236 (2019)
18. Vaswani, A., et al.: Attention is all you need. In: Advances in Neural Information Processing Systems, pp. 5998–6008 (2017)
19. Xiong, R., et al.: On layer normalization in the transformer architecture. arXiv preprint arXiv:2002.04745 (2020)
20. Yan, S., Xiong, Y., Lin, D.: Spatial temporal graph convolutional networks for skeleton-based action recognition. In: AAAI (2018)
21. Zhang, J., Zhou, W., Xie, C., Pu, J., Li, H.: Chinese sign language recognition with adaptive hmm. In: 2016 IEEE International Conference on Multimedia and Expo (ICME), pp. 1–6. IEEE (2016)
22. Zhang, P., Lan, C., Xing, J., Zeng, W., Xue, J., Zheng, N.: View adaptive recurrent neural networks for high performance human action recognition from skeleton data. In: Proceedings of the IEEE International Conference on Computer Vision, pp. 2117–2126 (2017)

23. Zheng, W., Li, L., Zhang, Z., Huang, Y., Wang, L.: Relational network for skeleton-based action recognition. In: 2019 IEEE International Conference on Multimedia and Expo (ICME), pp. 826–831. IEEE (2019)
24. Zhou, H., Zhou, W., Zhou, Y., Li, H.: Spatial-temporal multi-cue network for continuous sign language recognition. In: Proceedings of the AAAI Conference on Artificial Intelligence, vol. 34, pp. 13009–13016 (2020)

# Combining Traffic Assignment and Traffic Signal Control for Online Traffic Flow Optimization

Xiao-Cheng Liao, Wen-Jin Qiu, Feng-Feng Wei, and Wei-Neng Chen(✉)

South China University of Technology, Guangzhou, China
cwnraul634@aliyun.com

**Abstract.** With the continuous development of urbanization, traffic congestion has become a key problem that plagues many large cities around the world. As new information technologies like the Internet of Things and the mobile Internet develop, the interconnection between vehicles and road facilities provides a new mechanism to improve transportation efficiency. In this paper, we adopt the mechanism of vehicle-road coordination, and propose a new dynamic traffic flow optimization approach that combines the traffic assignment method and traffic signal control method together. For traffic assignment, a gene expression programming (GEP) based online navigation algorithm is proposed to generate a generalized navigation rule for the vehicles on the road network. Each vehicle can dynamically select an appropriate route for itself through the navigation rule based on its own states and information about the nearby road network. For traffic signal control, the Maximum Throughput Control (MTC) method is adopted. MTC checks the states of the intersections periodically and greedily takes the action that maximum the throughput of the intersections. By combining these two methods, the vehicle-road coordination mechanism can significantly improve the efficiency of city traffic flow optimization. The experimental results yielded based on the CityFlow simulator verify the effectiveness of the proposed approach.

**Keywords:** Traffic assignment · Traffic signal control · Gene expression programming (GEP)

## 1 Introduction

With the increasing number of vehicles in large cities worldwide, severe traffic congestion has become quite normal in metropolitans all over the world. It is traffic congestion that cause more serious vehicle emission and air pollution.

This work was supported in part by the National Key Research and Development Project, Ministry of Science and Technology, China (Grant No. 2018AAA0101300), and in part by the National Natural Science Foundation of China under Grants 61976093. The research team was supported by the Guangdong Natural Science Foundation Research Team No. 2018B030312003.


M. Tanveer et al. (Eds.): ICONIP 2022, CCIS 1793, pp. 150–163, 2023.
https://doi.org/10.1007/978-981-99-1645-0_13

Meanwhile, citizens have to suffer from long commute time due to traffic congestion. To alleviate the contradiction between urban traffic demand and supply, an important and straight forward way is making full use of the existing urban traffic resources. It may also become the only method when the city is too crowded to get enough space to build new roads or widen the existing roads.

Presently, drivers prefer choosing routes with shortest distance or lowest cost according to the recommendation from some online map tools like Amap or Google Maps. It may lead to a situation that a large number of vehicles concentrate in several specific routes. At the same time, an unreasonable traffic signal control strategy may worsen the urban traffic conditions as it will dramatically slow down the speed of vehicles through the intersections. So, to make full use of the existing urban traffic resources, it is necessary to guide vehicles to select appropriate routes and design a reasonable traffic signal control strategy.

Guiding vehicles to choose appropriate routes that minimize the average travel time, or in other words, traffic assignment, has been studied by researchers for a long time [1–3]. It can be classified into two categories according to the methods used to solve it, namely: 1) analytical-based methods [4–6] and 2) simulation-based methods [7–9]. Analytical-based methods assign vehicles to different routes according to analytical formulations [10]. Simple as it is, analytical-based methods are popular in the last few decades due to their scalability. However, enormous attention has been attached to simulation-based methods recently, as they are able to capture the dynamics and uncertainties of the real traffic flow [11]. As for the field of traffic signal control, there are also numbers of researches done on it [12]. Conventional methods [13–16] formulate traffic signal control problem as an optimization problem and use some optimization techniques to get better results. For examples, TUC [16] adopts some numerical optimization techniques to minimize vehicle travel time. Max-pressure [17] is a greedy method that always takes the action that maximizes the throughput of the network. Reinforcement learning techniques [18–20] have become another popular way to control the traffic signal recently. Most RL methods treat intersections as RL agents and minimize the average travel time of vehicles through the coordination among agents [21].

Although lots of work has been done in the field of traffic assignment and traffic signal control, there is little work combine both of them together. Meanwhile, the development of new information technologies like Internet of Things and the mobile Internet makes the interconnection between vehicles and road facilities becomes closer. Vehicle-road coordination is more convenient and requires further developments. In this paper, we propose a method that takes the advantages of traffic assignment and traffic signal control. Firstly, we obtain a navigation rule for simulation-based traffic assignment through gene expression programming (GEP) [22]. With the help of this rule, we can achieve dynamic traffic assignment in a decentralized way with local information only. Then, we employ a greedy method to control the traffic signal in the road network. This method checks the state of each intersection periodically. Based on the feedback, an action that maximizes the throughput of the intersections will be adopted and keeps till next check. With these two methods, the urban traffic resources are used in a more effective way as the experimental results show that using both methods is significantly better than using either of them. The contributions of this work are listed as follows:

1) We adopt GEP to train an online navigation rule. With the help of this rule, vehicles can select appropriate routes dynamically according to the local information only. 2) A method that combines the traffic assignment and traffic signal control together is proposed in this paper. The result of the experiments show that the combination is meaningful. 3) All of our experiments are conducted on CityFlow [23], an open-source traffic simulator that supports traffic flow simulation and traffic signal control, making our method more realistic.

The rest of the paper is organized as follows. Section 2 introduces the online traffic flow optimization problem and gives the major notations of it. Section 3 proposes an effective algorithm named GEP-based vehicle navigation algorithm with heuristic strategy to solve this problem. Experiments and the results are discussed in Sect. 4. Finally, Sect. 5 concludes this paper.

## 2 Problem Definition

In this section, some important definitions related to the model and problem to be solved in this paper are introduced. Based on these definitions, we formulate the online traffic flow optimization problem in a mathematical way.

### 2.1 Road Network

**Definition 1.** *(Directed graph G): In vehicle routing problem, road networks are usually defined as a directed graph $G = (V, E)$. Different from the traditional graph where intersections are defined as nodes and roads are defined as edges, we define roads as nodes $V$ in our work. A directed edge $(i, j)$ represents the connection between node $i$ and node $j$, meaning that road $i$ can go directly to road $j$.*

**Definition 2.** *(Node v): Each node $v$ ($v \in V$) contains multiple properties. It can be represented as a quad, $v = (L_{left}, L_{straight}, L_{right}, len)$, where $L_{left}$, $L_{straight}$ and $L_{right}$ are the set of left-turn lanes, straight lanes and right-turn lanes for road $v$, respectively. $len$ is the length of road $v$. For simplicity, in our work, we call the lanes in the same set of a road homogeneous lanes, otherwise, heterogeneous lanes.*

**Definition 3.** *(Origin node set O and destination node set D of a road network): The origin node $o_i (o_i \in O)$ refers to the node with no in-degree in the node set $V$, which means that there is no way to reach this road in the road network. An destination node $d_j (d_j \in D)$ is a node that has no out-degree in the node set $V$. It is said this road could not reach any other roads. They can be represented as:*

$$O = \{o_i \mid d_{in}(o_i) = 0 \wedge o_i \in V\} \tag{1}$$

$$D = \{d_j \mid d_{out}(d_j) = 0 \wedge d_j \in V\} \tag{2}$$

An example of the road network is shown in the Fig. 1. There are fourteen roads in the double-intersection road network, represented as $V = \{v_1, v_2, \cdots, v_{14}\}$. Every road has three types of lanes, representing different driving directions: left turn, straight and right turn. In Fig. 1a, there is only one lane for each type. Vehicles need to drive on the right side of the road, and U-turns are not allowed. As we can see, vehicles in road 1 can reach road 14 by turning right, go straight to road 12 and reach road 10 by turning left. Also, vehicles in road 10 can reach road 5, road 7 and road 9. Therefore, the road network has the following edges: (1,14), (1,12), (1,10), (10,5), (10,7), (10,9) and so on. The directed graph constructed from the road network is shown in Fig. 1b

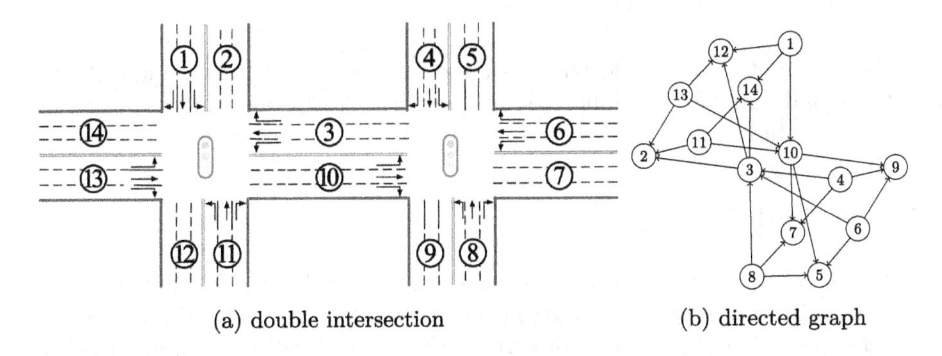


(a) double intersection (b) directed graph

**Fig. 1.** The road network of double intersection and its corresponding directed graph.

## 2.2 Online Traffic Flow Optimization Problem

**Definition 4.** *(Vehicle $\Psi$ ): The vehicle $\Psi_k$ can be represented as:*

$$\Psi_k = \{v_o^k, v_c^k, L^k, v_d^k, t_o^k, t_d^k\} \tag{3}$$

where $v_o^k$ is the origin node, $v_c^k$ is the current node, $L^k$ is the lane where vehicle $\Psi_K$ is located, $v_d^k$ is the destination node, $t_o^k$ is the time when the vehicle $\Psi_K$ enters the road network and $t_d^k$ is the time when it leaves the road network.

**Definition 5.** *(Vehicle travel time): The traveling time of a vehicle is defined as the time it takes from entering the road network to leaving the road network. The time consumption of a vehicle $\Psi_k$ is majorly related to the length of route traveled by the vehicle and the traffic flow on the route.*

$$T_{travel}(\Psi_k) \propto \left( \sum_{(i',j') \in E'} flow_{(i',j')}, \sum_{i' \in V'} len_{i'} \right) \tag{4}$$

where $V'$ represents the set of nodes that the vehicle $\Psi_k$ reached and $E'$ represents the set of edges that $\Psi_k$ travels from the origin node $v_o^k$ to the destination node $v_d^k$. $flow_{(i',j')}$ represents the traffic flow from node $i'$ to node $j'$, and $len_{i'}$ is the length of the road $i'$.

**Definition 6.** *(OTFO: Online traffic flow optimization problem): In the OTFO problem, we suppose that there is a steady stream of vehicles entering the road network from nodes in the origin node set to nodes in destination node set. Our goal is to minimize the average travel time of all vehicles.*

$$\min \frac{1}{n} \sum_{k=1}^{n} T_{travel}(\Psi_k) \tag{5}$$

To achieve the goal in formula (5), we combine the traffic assignment and traffic signal control techniques together, which is described in the following section in detail.

# 3 Method

In this section, we introduce the method used to solve the OTFO problem. Firstly, we introduce the traffic signal control strategy. Then, a GEP-based online vehicle navigation algorithm is presented. Finally, the whole dynamic decision-making process that combines the above two methods is demonstrated.

## 3.1 Max-Throughput Control

To solve the OTFO problem, traffic signal control is an important factor to be considered. A reasonable traffic signal control strategy enables vehicles to move through the road network in a more quick way.

In real road network, the behaviors of vehicles are complicated. For simplicity, we assume that vehicles are only allowed to change to the other homogeneous lanes of the current road. It is forbidden that vehicles change to heterogeneous lanes of the current road. Under such assumption, whenever a vehicle passes through a intersection, the process of selecting a lane in next road is exactly the process of choosing the next-next road. For example, in Fig. 1, when there is a vehicle in the straight lane of road 13, it can only reach to road 10 in next step as the vehicle can't turn into the left-turn lane or the right-turn lane. When it comes to the moment that the vehicle passes through the intersection, it will select the left-turn of road 10 as the next lane when the destination node of the vehicle is road 5. As for the behaviours of the traffic signal, we assume that turn

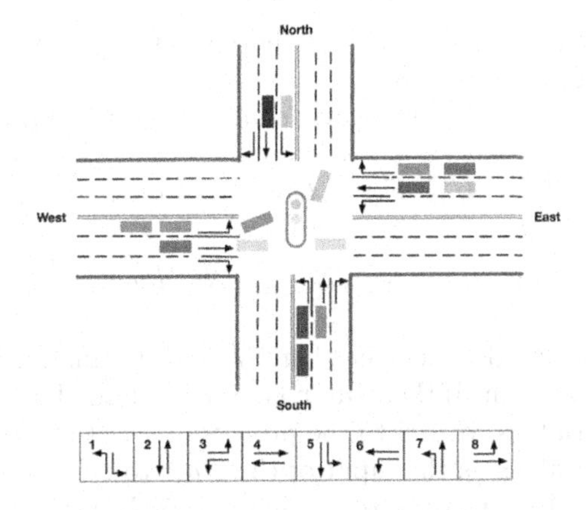


**Fig. 2.** Traffic signal phase

right is always allowed in every intersection. In addition, we consider 8 types of traffic signal phase as shown in Fig. 2.

In the OTFO problem, there is a steady stream of vehicles entering the road network. As we will plan the route of each vehicle in a dynamic way, it is difficult to get a reasonable traffic signals strategy by pre-calculating the offset among different intersections. So, we adopt a greedy method, Max-Throughput Control (MTC), to control the traffic signal of all intersections with the purpose of maximizing the throughput. For every intersection, MTC counts the number of waiting cars in each lane periodically. Then, for each intersection, MTC calculates the number of vehicles that can pass the intersection under each traffic signal phase. At last, MTC chooses the phase with maximum vehicles influenced for every intersection, which will be kept in the next period. The cycle of adjustment is set as a small value (10 s) to adapt to the dynamic traffic flow in our experiment.

### 3.2 GEP-based Online Vehicle Navigation Algorithm

Vehicle routing is another important factor to be considered in the OTFO problem. We need to determine appropriate routes for vehicles to balance the traffic flow on each road. To achieve this goal, we propose a GEP-based online vehicle navigation algorithm to get a navigation rule for vehicles.

With the help of the navigation rule, every vehicle can select an appropriate lane of the next road according to some information during the whole journey. The related variables can be described as follows:

$$[N_{L^k}, N_{v_p(L_i)}, N_{v_q}, len_{v_q}, S(v_q, v_d^k)] \tag{6}$$

where $N_{L^k}$ indicates the number of vehicles in current lane, $N_{v_p(L_i)}$ contains the number of vehicles in each lane $L_i$ of the next road $v_p$ where vehicle $\Psi_K$ is going to, $N_{v_q}$ is the number of vehicles on the next-next road $v_q$, $len_{v_q}$ is the length of node $v_q$, $S(v_q, v_d^k)$ donates the shortest distance between node $v_q$ and the destination node $v_d^k$ of vehicle $\Psi_K$.

The decision process of the navigation rule is shown below:

$$L' = F(v_c^k,\ L^k,\ v_d^k, Rule) \tag{7}$$

where

$$F = \underset{L_i \in v_p.L}{\arg\min}\ Rule(N_{L_k}, N_{v_p(L_i)}, N_{v_q}, len_{v_q}, S(v_q, v_d^k)) \tag{8}$$

where the current road $v_c^k$, current lane $L^k$ and destination $v_d^k$ of the vehicle $\Psi_K$ and *Rule* is set as input. If there are $n$ lanes in each road, there are $n$ different five-dimensional variables related to the process of selection. We use each set of five-dimensional variables as the input of the *Rule* to estimate the cost of each lane. The function $F$ returns the cost-minimum lane $L'$ as the next lane to enter.

Designing an effective navigation rule is challenging, since the traffic model in CityFlow is considered as a black box. We are unable to infer any specific information about *Rule*. In fact, it is almost impossible to find an exact match for this rule manually. So, we turn to GEP for help as GEP is suitable for solving symbolic optimization problems [27] we encountered.

In our method, navigation rules are encoded as chromosomes in GEP. Every chromosome contains exactly one gene. Each gene represents a navigation rule. Gene is composed of a head and a tail. The element of head can be picked from function set and terminal set. The element of tail can be picked from the terminal set only. The function set and terminal set are defined as formula (9) and formula (10).

$$FS = \{+, -, \times, \div\} \tag{9}$$

$$TS = \{N_{L_k}, N_{v_p(L_i)}, N_{v_q}, len_{v_q}, S(v_q, v_d^k), other\ constants\} \tag{10}$$

The fitness value is set as the average travel time of all vehicles in a period of time (3600 s for example). NP chromosomes are constructed randomly to form the initial population of GEP. In each iteration, we adopt selection, crossover and mutation operations to update the population. Specifically, we use three-element tournament selection strategy to select the best chromosome for the offspring operation. For crossover operation, we apply both one-point crossover and two-point crossover to generate new chromosomes. To ensure each new chromosome produced by mutation is valid, mutation operator should follow the principle that symbols in head can be replaced by symbols from either the function set or the terminal set, but symbols in tail can only be replaced by symbols from the terminal set. In our work, we apply four mutation operators : 1) point mutation; 2)inversion mutation; 3) insertion sequence (IS) mutation and 4) root insertion sequence (RIS) mutation. More detail about GEP can be found in [22].

### 3.3 Decision-Making Process

The whole decision process of our method combines the two techniques introduced above. When there is a vehicle entering the road network, it will drive on a randomly selected lane of its origin road. When the vehicle comes to an intersection, it will find the best lane of the next road to go to through *Rule*, according to the information of itself and the information of the road network nearby. But the action is not always be taken at once as the vehicle has to obey the signal of the traffic light. As mentioned above, we adopt the MTC to control the traffic signal of all intersections. For every intersection, it takes the phase with maximum vehicles influenced for a period of time. When a vehicle reach the destination, it will be removed from the road network and the travel time of the vehicle will be recorded.

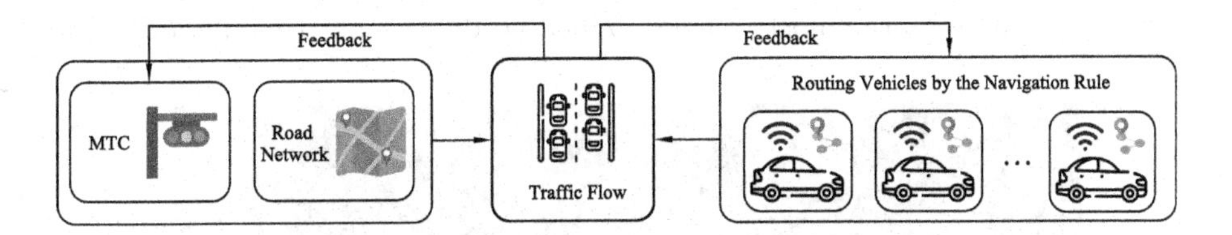


**Fig. 3.** Cooperation between MTC and the navigation rule.

The cooperation between the MTC and the navigation rule is shown in Fig. 3. The traffic flow in the road network is directly influenced by the MTC and the navigation rule. Meanwhile, the traffic flow also feedback to the MTC and the navigation rule simultaneously. In fact, the navigation rule takes the traffic signal control strategy into consideration. Traffic signal has great impact on the number of vehicles on the eight roads through it. As the navigation rule takes the number of vehicles in the nearby road network into consideration, the result of the MTC also has impact on it. When the vehicles reach to the next intersection, the traffic signal may need to change to another phase to adapt to the incoming vehicles. From the perspective of vehicle, the time consumption of a vehicle is majorly related to the length of route traveled by the vehicle and the waiting time of all intersections in its route. They are the result of the navigation rule and MTC respectively. Through the cooperation of these two modules, we can get a lower average travel time of all vehicles as shown in next section.

## 4 Experiment

In this section, we conduct several groups of experiments on CityFlow [23], an open-source traffic simulator that supports traffic flow simulation and traffic signal control, to evaluate the performance of the method we proposed. Before that, we will introduce the experimental setting, compared methods and the navigation rule extracted by the GEP-based vehicle navigation algorithm.

### 4.1 Settings

**Parameters Setting:** 1) Hyper-parameters of GEP: We set the population size NP = 32 and the maximum number of iterations as 100. The probabilities of four different mutation operators are listed as follows. Point-mutation rate $p_m = 0.05$, Inversion-mutation rate $p_{im} = 0.2$, IS-mutation rate $p_{is} = 0.1$, RIS-mutation rate $p_{ris} = 0.1$. The probabilities of one-point crossover rate is 0.4 and 0.6 for two-point crossover. In the *otherconstants* in terminal set is set as { 2, 3, 5, 7, 11, 13, 17, 19, 23, 29}.
2) Traffic light signal control cycle: In the Max-Throughput Control strategy, the traffic light at the intersection is adjusted every 10 s [17], and the traffic light is kept as the one with the most vehicles influenced during that time period.

**Fig. 4.** Gudang Sub-district, 16 intersections with 80 nodes (i.e., roads) and 192 directed edges, Hangzhou, China

**Road Network:** We used the real road network data of Gudang Sub-distict in Hangzhou [18,24]. As shown in Fig. 4. We model this road network as a directed graph $G = (V, E)$ where a node $v$ represents a road, and the directed edge $(i, j)$ represents the connection between node $i$ and node $j$. There are 16 intersections, 80 nodes and 192 directed edges in this road network. There are 16 entering roads and 16 exiting roads in this network, we set many of them (see next paragraph for more details) as the origin nodes and destination nodes respectively. To mention again, the nodes in our work are not intersections but roads. The intersections only provide the topological relationship between the nodes that a road leads to which other road in one intersection.

**Scenarios:** We randomly select 12 pairs of ODs from the origin set and destination set as the OD pairs of vehicles (without putting back sampling). An origin node corresponds to exact one destination node. A sampling result is used as a scenario. There are a total number of $(C_{16}^{12})^2 * A_{12}^{12}$ scenarios. So, the probability of scenarios repetition is very small.

### 4.2 Compared Methods

As there are few methods proposed to solve OTFO problem, we adapt the following two heuristic algorithms, 1)Dijkstra's algorithm [25] and 2) A* algorithm [26] to solve OTFO problem to compare with our method.

In Dijkstra's algorithm, all vehicles always follow the shortest path. The rule can be described as:

$$Dij_{rule} = f\left(len_{v_q}, S(v_q, v_d^k)\right) = \underset{v_q \in V}{\arg\min}\ \left(len_{v_q} + S(v_q, v_d^k)\right) \tag{11}$$

$len_{v_q}$ is the road length of the next-next road node $v_q$ the vehicle wolud like to go, $S(v_q, v_d^k)$ donates the shortest distance between node $v_q$ and the destination node $v_d^k$. Because $v_p(Li)$ decides to $v_q$, finding the best $v_q$ available backwards to find the best lane $L_i(L_i \in v_p.L)$ to be entered next.

A* algorithm is an algorithm that finds the lowest cost of passage for paths with multiple nodes in a graph. In this paper, we define the estimation function of the A* algorithm as:

$$A^*_{rule} = g\left(N_{v_p(L_i)}, N_{v_q}, len_{v_q}, S(v_q, v_d^k)\right) \tag{12}$$

$$g = \underset{L_i \in v_p.L}{\arg\min}\left(k\left(len_{v_q} + S(v_q, v_d^k)\right)\right) \tag{13}$$

where

$$k = \log(N_{v_p(L_i)} + N_{v_q} + 1) + 1 \tag{14}$$

$k$ is the coefficient associated with the vehicles' numbers on lane $v_p(L_i)$ and road $v_q$. By considering these factors, we want to get such a situation that the traffic flow on the road network is more balanced.

### 4.3 The Extracted Rule

We randomly select 12 pairs of ODs from Hangzhou road network as the vehicles' navigation origin-destination pairs (without putting back sampling). We train our navigation rules by GEP with a time interval of 5, i.e., one vehicle enters to the road network at each origin node every 5 s, simulated on Cityflow for 1000 s. The simplified form of the exacted rule is as follows.

$$Rule = h\left(N_{L_k}, N_{v_p(L_i)}, len_{v_q}, S(v_q, v_d^k)\right) \tag{15}$$

where

$$h = \underset{L_i \in v_p.L}{\arg\min}\left\{\frac{17N_{L_k}}{6} + N_{v_p(L_i)} + len_{v_q} + S(v_q, v_d^k) + \frac{283}{6}\right\} \tag{16}$$

It can be seen in the extracted rule, the smaller $N_{L_k}$, $N_{v_p(L_i)}$, $len_{v_q}$ and $S(v_q, v_d^k)$ of a lane, the higher priority it has. Moreover, compared with formula (8), the trained rule without the participation of parameter $N_{v_q}$. It is reasonable for the following two reasons. First, $N_{v_q}$ represents the number of vehicles on the next-next road $v_q$ at the current time, when $\Psi_K$ reaches node $v_q$, the number of vehicles on the road has changed. Second, $N_{v_q}$ is the number of vehicles in all lanes of road $v_q$. It does not reflect the congestion of the lane that vehicle $\Psi_K$ will choose in the next decision.

### 4.4 Experimental Results

To show the effectiveness of our method, we conduct comparative experiments under two different intervals. The simulation time is set as 3600 s. The experimental results show that our algorithm is very effective in solving the OTFO problem. Details are introduced as follows.

**1) Results of experiments with $\boldsymbol{Interval} = 5\,s$**
We compared Dijkstra's algorithm, A* algorithm and our algorithm without MTC and with MTC strategy in this subsection. In Table 1, each row represents the average result of 30 independent replications of a scenario. Without MTC, our method gives significantly better results than the other two methods.

After adding MTC strategy, all three algorithms improve significantly. Our algorithm is better than A* algorithm in all scenarios. However, our algorithm perform better than Dijkstra's algorithm on only 6 scenarios. We deduce that the reason is that when the interval is large, the total number of vehicles is not very large. In such a situation, the congestion occurs rarely when MTC strategy is added. So, the total length of the road traversed is more important than traffic flow. As Dijkstra's algorithm only considers the shortest path, it can get a good result. For further comparison, we shorten the interval to generate more complicated scenarios in the following experiments.

**Table 1.** Results of experiments with Interval = 5 s.

| Scenarios | Dij | Dij-MTC | A* | A*-MTC | Ours | Ours-MTC |
|---|---|---|---|---|---|---|
| 1 | 886.45 | **431.11** | 960.83 | 474.03 | 717 | 447.49 |
| 2 | 796.71 | **380.82** | 896.76 | 439.64 | 711.58 | 390.57 |
| 3 | 1668.29 | 612.18 | 1182.18 | 564.82 | 843 | **516.16** |
| 4 | 924.11 | 499.64 | 909.52 | 465.55 | 653.37 | **410.32** |
| 5 | 1380.26 | 638.24 | 1015.46 | 525.47 | 710.83 | **456.63** |
| 6 | 1058.28 | **457.38** | 1043.6 | 476.19 | 757.6 | 460.71 |
| 7 | 1282.82 | 500.02 | 1081.23 | 523.37 | 808.95 | **466.68** |
| 8 | 1180.39 | 778.5 | 1012.9 | 628.95 | 817.52 | **474.47** |
| 9 | 980.4 | 542.84 | 987.9 | 488.8 | 627.76 | **418.59** |
| 10 | 1116.38 | **369.75** | 903.03 | 395.1 | 726.66 | 401.26 |

**Table 2.** Results of experiments with Interval = 3 s.

| Scenarios | Dij | Dij-MTC | A* | A*-MTC | Ours | Ours-MTC |
|---|---|---|---|---|---|---|
| 1 | 1529.8 | 1451.98 | 1621.53 | 1387.79 | 1252.12 | **862.26** |
| 2 | 1309.58 | 1069.38 | 1393.54 | 814.51 | 1102.43 | **587.79** |
| 3 | 2178.84 | 1185.43 | 1796.57 | 976.54 | 1390.36 | **835.42** |
| 4 | 1496.13 | 1309.23 | 1418.62 | 795 | 993.7 | **597.71** |
| 5 | 1945.16 | 1405.64 | 1608.67 | 854.64 | 1131.47 | **696.53** |
| 6 | 1718.6 | 1717.89 | 1612.99 | 948.12 | 1230.81 | **884.9** |
| 7 | 1733.27 | 906.38 | 1574.68 | 859.4 | 1324.52 | **708.11** |
| 8 | 1750.82 | 1885.76 | 1520.97 | 1019.73 | 1362.08 | **784.01** |
| 9 | 1483.5 | 1010.96 | 1513.1 | 803.11 | 959.16 | **605.19** |
| 10 | 1479.45 | 647.87 | 1384.43 | **604.13** | 1134.74 | 635.26 |

**2) Results of experiments with $Interval = 3\,s$**
In this subsection, we shorten the interval to 3 s. It means that for every origin node, there is a vehicle enters the road network every 3 s. The total number of vehicles are 1.7 times that of the previous experiment, resulting in a set of exceedingly challenging scenarios. From Table 2, we can see that A* algorithm performs better than Dijkstra's algorithm in most scenarios, with or without MTC strategy. The manually designed A* algorithm in this work takes the real-time traffic information into consideration, which makes itself more suitable for challenging scenarios. Moreover, it can be seen in the Table 2 that our algorithm with MTC strategy performs significantly better than the other two algorithms on all the scenarios. It proves that our algorithm is more effective when the traffic flow is more complex and the optimization space is larger.

## 5 Conclusion

In this paper, we use a method combining traffic assignment and traffic signal control together to solve the OTFO problem. The proposed approach is able to route vehicles dynamically and can fully integrate traffic light strategies to achieve better performance. Numerical experimental results show that, compared with other algorithms, in our approach, the cooperation between the traffic assignment method and traffic signal control method is better. Thus, the average travel time is smaller, especially in more congested scenarios.

In the future, we would like to consider more practical models like vehicles can turn into heterogeneous lanes, vehicles can turn around. We also intent to apply reinforcement learning techniques to solve the OTFO problem.

## References


1. Wardrop, J.G.: Road paper: some theoretical aspects of road traffic research. Proc. Inst. Civil Eng. **1**(3), 325–362 (1952)
2. Beckmann, M., Mcguire, C.B., Winsten, C.B.: Studies in the Economics of Transportation. Yale University Press, New Haven (1956)
3. Smith, M.J.: The stability of a dynamic model of traffic assignment-an application of a method of Lyapunov. Transp. Sci. **18**(3), 245–252 (1984)
4. Roughgarden, T., Tardos, É.: How bad is selfish routing? J. ACM (JACM) **49**(2), 236–259 (2002)
5. Akamatsu, T., Wada, K., Iryo, T., Hayashi, S.: A new look at departure time choice equilibrium models with heterogeneous users. Transp. Res. Part B: Methodol. **148**, 152–182 (2021)
6. Osawa, M., Fu, H., Akamatsu, T.: First-best dynamic assignment of commuters with endogenous heterogeneities in a corridor network. Transp. Res. Part B: Methodol. **117**, 811–831 (2018)
7. Shou, Z., Chen, X., Fu, Y., Di, X.: Multi-agent reinforcement learning for Markov routing games: a new modeling paradigm for dynamic traffic assignment. Transp. Res. Part C: Emerg. Technol. **137**, 103560 (2022)
8. Han, S., Fang, S., Wang, X., Chen, X., Cai, Y.: A simulation-based dynamic traffic assignment model for emergency management on the hangzhou bay bridge. In ICCTP 2010: Integrated Transportation Systems: Green, Intelligent, Reliable, pp. 883–895 (2010)
9. Tian, Y., Chiu, Y.C., Gao, Y.: Variable time discretization for a time-dependent shortest path algorithm. In 2011 14th International IEEE Conference on Intelligent Transportation Systems (ITSC), pp. 588–593. IEEE (2011)
10. Peeta, S., Ziliaskopoulos, A.K.: Foundations of dynamic traffic assignment: the past, the present and the future. Netw. Spat. Econ. **1**(3), 233–265 (2001)
11. Larsson, T., Patriksson, M.: An augmented Lagrangean dual algorithm for link capacity side constrained traffic assignment problems. Transp. Res. Part B: Methodol. **29**(6), 433–455 (1995)
12. Zhao, D., Dai, Y., Zhang, Z.: Computational intelligence in urban traffic signal control: a survey. IEEE Trans. Syst. Man Cybern. Part C (Appl. Rev.) **42**(4), 485–494 (2011)
13. Hunt, P.B., Robertson, D.I., Bretherton, R.D., Royle, M.C.: The SCOOT on-line traffic signal optimisation technique. Traff. Eng. Control **23**(4) (1982)
14. Koonce, P., Rodegerdts, L.: Traffic signal timing manual (No. FHWA-HOP-08-024). United States. Federal Highway Administration (2008)
15. Lowrie, P.R.: SCATS: a traffic responsive method of controlling urban traffic control. Roads Traff. Author. (1992)
16. Diakaki, C., Papageorgiou, M., Aboudolas, K.: A multivariable regulator approach to traffic-responsive network-wide signal control. Control Eng. Pract. **10**(2), 183–195 (2002)
17. Varaiya, P.: The max-pressure controller for arbitrary networks of signalized intersections. In: Advances in Dynamic Network Modeling in Complex Transportation Systems, pp. 27–66. Springer, New York (2013). https://doi.org/10.1007/978-1-4614-6243-9_2
18. Wei, H., et al.: Colight: learning network-level cooperation for traffic signal control. In Proceedings of the 28th ACM International Conference on Information and Knowledge Management, pp. 1913–1922 (2019)

19. Chu, K.F., Lam, A.Y., Li, V.O.: Traffic signal control using end-to-end off-policy deep reinforcement learning. IEEE Trans. Intell. Transp. Syst. **23**, 7184–7195 (2021)
20. Ying, Z., Cao, S., Liu, X., Ma, Z., Ma, J., Deng, R.H.: PrivacySignal: privacy-preserving traffic signal control for intelligent transportation system. IEEE Trans. Intell. Transp. Syst. **23**, 1629–16303 (2022)
21. Noaeen, M., et al.: Reinforcement learning in urban network traffic signal control: a systematic literature review. Expert Syst. Appl. **199**, 116830 (2022)
22. Ferreira, C.: Gene expression programming: a new adaptive algorithm for solving problems (2001). arXiv preprint cs/0102027
23. Tang, Z., et al.: Cityflow: a city-scale benchmark for multi-target multi-camera vehicle tracking and re-identification. In: Proceedings of the IEEE/CVF Conference on Computer Vision and Pattern Recognition, pp. 8797–8806 (2019)
24. Gudang Sub-distict. https://www.openstreetmap.org/#map=14/30.2813/120.1034. Accessed 1 June 2022
25. Dijkstra, E.W.: A note on two problems in connexion with graphs. Numerische mathematik **1**(1), 269–271 (1959)
26. Hart, P.E., Nilsson, N.J., Raphael, B.: A formal basis for the heuristic determination of minimum cost paths. IEEE Trans. Syst. Sci. Cybern. **4**(2), 100–107 (1968)
27. Branke, J., Hildebrandt, T., Scholz-Reiter, B.: Hyper-heuristic evolution of dispatching rules: a comparison of rule representations. Evol. Comput. **23**(2), 249–277 (2015)

# Convolve with Wind: Parallelized Line Integral Convolutional Network for Ultra Short-term Wind Power Prediction of Multi-wind Turbines

Ruiguo Yu[1,2,3], Shaoqi Xu[4], Jian Yu[1,2,3], Zhiqiang Liu[1,2,3](✉), and Mei Yu[1,2,3]

[1] College of Intelligence and Computing, Tianjin University, Tianjin, China
{rgyu,yujian,tjubeisong,yumei}@tju.edu.cn
[2] Tianjin Key Laboratory of Cognitive Computing and Application, Tianjin, China
[3] Tianjin Key Laboratory of Advanced Networking, Tianjin, China
[4] Tianjin International Engineering Institute, Tianjin University, Tianjin, China
xushaoqi@tju.edu.cn

**Abstract.** Wind power prediction(WPP) is an efficient way to facilitate reliable wind power integration. Recent studies have shown the effectiveness of spatio-temporal features in WPP tasks. And convolutional neural network(CNN) is a classical spatial feature extraction method. However, in the WPP problem, CNN adopts distance as an indicator of correlation strength in wind power feature extraction, which ignores the influence of wind direction. In this paper, parallelized line integral convolutional network(PLICN) module, is proposed to replace CNN for feature extraction and prediction in WPP tasks. Based on physical principle, we assume that wind turbines in one prevailing wind direction have higher spatio-temporal correlation than others. So, those neighbor turbines in one dominant wind direction are selected as a line feature for line convolution afterwards in PLICN module. Experiments show that, the MSE of PLICN module is 2.00% $\sim$ 17.00% lower than that of CNN in ultra-short-term WPP tasks.

**Keywords:** Wind power prediction · Spatio-temporal feature · Artificial neural networks · Time series analysis

## 1 Introduction

According to global wind report 2021 released by Global Wind Energy Council(GWEC), 93GW of new wind power capacity was installed in 2020 with a 53% year-on-year increase [4]. As global wind power capacity installed growing rapidly, reliable wind power integration is more and more important to the stability and security of power grid. Accurate wind power prediction(WPP) is an effective way to avoid harm from the stochastic nature of wind sources.

---

This work is supported by National Natural Science Foundation of China (Grant No. 61976155).


M. Tanveer et al. (Eds.): ICONIP 2022, CCIS 1793, pp. 164–175, 2023.
https://doi.org/10.1007/978-981-99-1645-0_14

Deep learning methods are the most popular data-driven methods for WPP in recent years. It is suitable for modeling complex nonlinear relationships and performs well on WPP problems. According to the workflow, there are two key steps in data-driven methods: feature extraction and relationship learning.

**The feature extraction step** extracts features that are highly correlated with the prediction task from the data sequence, and the features are used in the subsequent relationship learning process. Feature extraction methods can be divided into two categories: classical methods and deep learning methods.

There are two strategies for classical feature extraction methods. The first is to select effective features from candidates. A typical method is phase space reconstruction (PSR) [9,14]. The second is to construct new features from original features, A typical method is principal component analysis(PCA) [7].

Deep learning methods are widely used in feature extraction tasks. CNN [19], graph convolutional network (GCN) [20] and graph attention network (GAT) [1] are widely used in spatial feature extraction, while RNN [16], LSTM [17], GRU [13] and their derivatives are often used to extract temporal features.

**The relationship learning step** captures the change pattern of feature. The relationship in WPP can be divided into spatial, temporal and spatio-temporal relationship. Typical spatial relationship learning methods include CNN [19], GCN [10], GAT [1], etc. These methods are also suitable for spatial feature extraction. Typical methods for temporal relationship learning include RNN [16], LSTM [17], GRU [13], etc. Spatio-temporal relationship learning is the focus of research in recent years, and typical methods include FC-CNN [18], ConvLSTM [15], spatial temporal GCN [11] and so on.

CNN is an effective method for spatial feature extraction which is applied in various methods. However, in the WPP problem, CNN adopts euclidean distance as a measure of the strength of spatial correlation(As shown in Fig. 1(a)) in feature extraction. It fails to combine the physical characteristics of the wind. Wind turbines located in the same dominant wind direction have stronger spatio-temporal correlation(As shown in Fig. 1(b)).

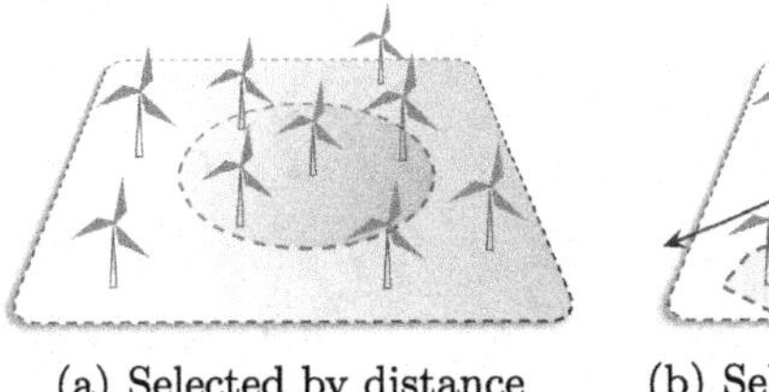

(a) Selected by distance

(b) Selected by wind direction

**Fig. 1.** Feature selection strategy

In this paper, parallelized line integral convolutional network(PLICN) is proposed for WPP tasks. The novelty and contribution are detailed as follows:

**Feature Extraction:** Combined with wind direction, adjacent data points in a dominant wind direction are selected as a line feature for prediction task afterwards. The line feature has a better performance than distance based feature in ultra-short-term wind power prediction.

**Prediction Model:** A novel architecture, PLICN, is proposed to replace CNN for feature extraction and relationship learning in WPP tasks.

The rest of this paper is arranged as follows. Section 2 introduces a spatio-temporal correlation assumption and some analysis in order to explain the motivation of the proposed method. Section 3 introduces the dataset and reveals the principles of the method we proposed. Section 4 exhibits the result of comparison experiment with other methods. Section 5 draws the conclusion and puts forward future work prospects.

## 2 Correlation Assumption and Spatio-temporal Analysis

Spatio-temporal correlation widely exists among different wind turbines in the region. Both wind direction and distance are the main factors for the strength of the correlation. The correlation attenuates with the increase of distance. The two wind turbines located in the same dominant wind direction have stronger spatio-temporal correlation than that of other wind turbines at the same distance. As illustrated in Fig. 2, inspired by [3], the correlation assumption can be summarized as: the turbines in the same prevailing wind direction(A and B) have stronger spatio-temporal correlation than other turbines which are not(B and C).

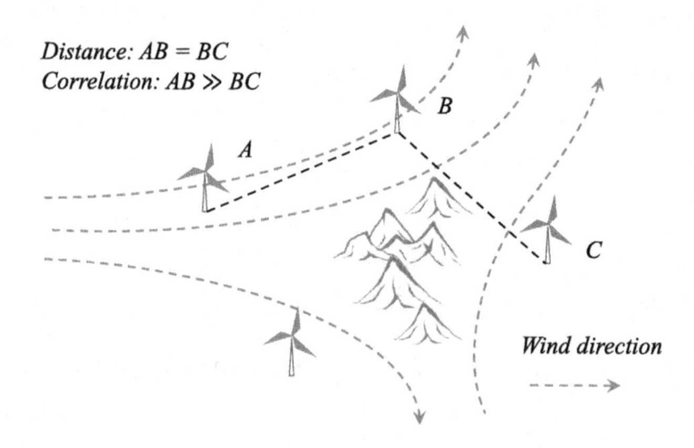


**Fig. 2.** Correlation assumption [3]

Although the details of the airflow above wind farm are very complicated, on the whole it still complies with some principles such as time delay [8] and wake effect [12] as shown in Fig. 3. In order to simplify the description of wind motion, it is typically assumed that wind propagates between two sites without much changes in its current amplitude and profile within a short time period [5].

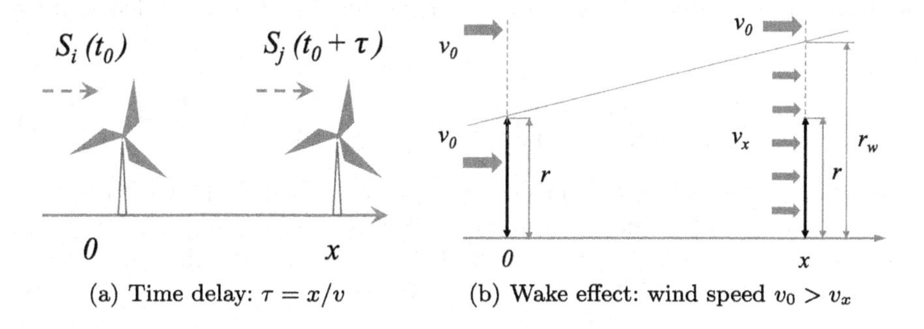


(a) Time delay: $\tau = x/v$

(b) Wake effect: wind speed $v_0 > v_x$

**Fig. 3.** Time delay and wake effect of wind

## 2.1 Time Delay

The wind farm covers a large area in general. Therefore, it may take several minutes for wind to travel from one wind turbine to another. The future status of the downstream turbine is strongly correlated with the current status of the upstream turbine. Ideally, assuming the wind speed $v$ is constant, the time delay $\tau$ of wind travelling between two sites with distance $x$ in the prevailing wind direction can be simply represented as: $\tau = x/v$.

Wind turbines located in the dominant wind direction have stronger spatio-temporal correlations. As shown in Fig. 3(a), The state of upstream wind turbine at $t_0$ is recorded as $S_i(t_0)$. The state of the downstream wind turbine after the delay $\tau$ is recorded as $S_j(t_0 + \tau)$. The smaller the delay $\tau$, the smaller the difference between $S_i(t_0)$ and $S_j(t_0 + \tau)$. However, when the distance between two wind turbines is the same, the delay $\tau$ may be different. The time delay $\tau$ can be affected by many factors such as distance, wind speed, wind direction, etc. The time delay reflects the spatio-temporal correlation between wind turbines to a certain extent. The smaller the delay $\tau$, the stronger the spatio-temporal correlation.

Therefore, considering the wind direction factor in the correlation assumption, we select the power values of adjacent wind turbines located in the same dominant wind direction and construct them as line features for ultra-short-term wind power prediction tasks. Line features remove noisy data points that are weakly correlated with the WPP task at the same distance.

## 2.2 Wake Effect

In a wind farm, wind drives the front wind turbine to rotate and generate electricity. After the wind leaves the turbine, a wake is generated and the wind speed decreases, making the efficiency of the rear wind turbines in the same wind direction reduced. Wake effect is a factor that cannot be ignored in spatio-temporal analysis of wind power generation. As illustrated in Fig. 3(b), the deficit of wind velocity caused by a single wake on a flat area can described by Jenson model [8]. The wind speed $v_x$ behind wind turbine at position 0 can be calculated by Eq. (1) [8]:

$$\begin{cases} k = 0.5/ln(h/z) \\ v_x = v_0(1 - (1 - \sqrt{1-c})(r/(r+kx))^2) \end{cases} \tag{1}$$

where $r$ is the length of a turbine blade, $h$ is the height of wind turbine, $z$ is the surface roughness and $c$ is the drag and lift ratio coefficient. Equation (1) shows that $v_x$ is not constant if wake effects are considered. So time delay $\tau$ can be re-represent as Eq. (2) [8]:

$$\tau = \int_0^x \frac{dx}{v_x} = \int_0^x \frac{dx}{v_0[1 - (1 - \sqrt{1-c})(r/(r+kx))^2]} \tag{2}$$

Equation (2) shows that the delay $\tau$ under the influence of the wake effect is affected by the distance $x$, and the relationship is very complicated. In actual situations, it will be more complicated by the interference of various factors, and it is difficult to quantify. Therefore, we propose the PLICN module, which learns the complex spatio-temporal relationship between wind turbines in a data-driven manner. The line convolution layer in the PLICN module takes the line features as input, and learns the weight representation of the distance factor in the correlation assumption during model training process.

## 3 Parallelized Line Integral Convolutional Network

PLICN module combines the ideas of LIC [2] with deep learning. It convolves line features of all sites in parallel. And its parameters can be updated in training process. A PLICN module includes two key steps: line feature extraction and line convolution. The principle of PLICN is detailed as follows:

### 3.1 Line Feature Extraction

Before performing line feature extraction, the raw data needs to be preprocessed. The preprocessing steps mainly include:

1) According to the location of the wind turbine, the output power sequence of the wind turbine group is reorganized into a power matrix, so that the data can be quickly obtained through a two-dimensional index.
2) Power matrix missing data completion.
3) For each station in the area, according to the wind field, start from the station and move forward or backward at the corresponding wind speed, and record the new position reached after a specific time interval. These new positions

will be used as indices to sample the power matrix to obtain power values. The line feature is obtained by combining the power value of each station itself and the power value of the front and rear positions.

In order to speed up the subsequent model training process and facilitate the gradient update of parameters, it is necessary to normalize the original power data according to Eq. (3):

$$p = \frac{power}{capacity} \tag{3}$$

where *power* is the original power data, *capacity* is the rated power of each wind turbine itself, and $p$ is the normalized power data ranging from 0 to 1. In addition, the output of the model also ranges from 0 to 1, and the predicted power values can be obtained by multiplying them by the *capacity* of the corresponding site.

The wind and power data comes from Wind Integration National Dataset (wind) [6], provided by National Renewable Energy Laboratory (NERL), USA. The dataset contains simulated wind meteorological and power data under $2km$ resolution between 2007 and 2012 for the entire continental United States. In this paper, we select a subset that contains wind and power data in the area ranging from 42.5°N to 42.9°N in latitudes, and from 95.1°W to 95.4°W in longitudes. The rectangular area we selected contains 205 sites in total. These sites form a matrix with missing elements. To meet the requirements of subsequent indexing and convolution operations, it is necessary to perform missing data completion operation.

Each site(include missing points) in the matrix is given a two-dimensional coordinate $(y, x)$ in bottom-to-top, left-to-right order. Missing values are completed by linear interpolation based on distance according to Eq. (4)–(5).

$$data_t = \sum_{i \in D} weight_i \times data_i \tag{4}$$

$$weight_i = \frac{\sum_{j \in D, j \neq i} d_{jt}}{(|D| - 1) \sum_{j \in D} d_{jt}} \tag{5}$$

where $d_{jt}$ is the distance between site $j$ and $t$, and $t$ is the target site to be interpolated. $D$ is the set of all sites within a certain distance from site $t$, and $|D|$ is the number of sites in the collection. The $data_t$ to be interpolated include wind speed, power and wind direction. After interpolation, according to the wind speed and direction of each site, find the previous position and next position on the wind direction path within a certain time interval, which can be calculated by Eq. (6). Time interval can be 5 min or 10 min.

$$\overrightarrow{displacement} = \overrightarrow{speed} \times interval \tag{6}$$

Then the previous position index and next position index of each site can be determined by Eq. (7):

$$\begin{cases} index_{prev} = index_{curr} - \overrightarrow{displacement}/grid\_size \\ index_{next} = index_{curr} + \overrightarrow{displacement}/grid\_size \end{cases} \tag{7}$$

where *index* is a two-dimensional coordinate, which can be used to access power values of these new locations for line convolution afterwards. The value of *grid_size* is 2 km.

Next step is to obtain the power values $p_{prev}, p_{next}$ corresponding to the coordinate $index_{prev}, index_{next}$ through sampling operation. There are two methods in sampling operation. One is linear interpolation: find a number of points that are closest to the new point, and then interpolate according to the distance to get its power value. The other is nearest sampling, which uses the power value of the point closest to the new point. We adopt the second method in our experiment for its low time complexity.

Mark the corresponding power value of a station itself as $p_{curr}$, then the line feature of the station is $f_k = (p_{prev}, p_{curr}, p_{next})$. Line feature extraction algorithm is shown in Algorithm 1.

**Algorithm 1:** Line feature extraction algorithm

**Input**: Power matrix: *matrix*, Wind speed filed: $V$, Coordinate collection of each site: $\mathcal{K}$

**Output**: Collection of line features for each site: $\mathcal{F}$

1 Initialization: $interval = 5min$, $\mathcal{F} = \varnothing$ ;
2 Normalize the raw power data according to equation (3);
3 Complete the missing power data according to equation (4) ;
4 **for** *each coordinate* $k(y_k, x_k)$ *in* $\mathcal{K}$ **do**
5     Get the wind speed at site $k$: $v_k = V[k]$;
6     Get the power value at site $k$: $p_k = matrix[k]$;
7     Calculate *displacement* according to equation (6);
8     Calculate coordinates: $i(y_i, x_i)$, $j(y_j, x_j)$ according to equation (7);
9     Sampling: $p_i = matrix[i], p_j = matrix[j]$;
10     Construct line feature of site $k$: $f_k = (p_i, p_k, p_j)$;
11     Add $f_k$ to the set $\mathcal{F}$;
12 **end**
13 Output set $\mathcal{F}$;
14 return

### 3.2 Line Convolution

If the line feature of site $k$ includes n-order adjacent points, denote the line feature as $f_k(p_0, ..., p_n, ..., p_{2n})$, where $p_n$ is the power value of site $k$ itself. In the line convolution layer of the PLICN module, the convolution kernel is denoted as $C(w_0, ..., w_n, .., w_{2n})$, its size is the same as the line feature and the kernel $C$ is weight-sharing in one PLICN module. Denote the output of $f_k$ after line convolution as $f_{LC}(f_k)$, then the line convolution is shown in Eq. (8), where $\sigma$ is activation function.

$$f_{LC}(f_k) = \sigma(\sum_{i=0}^{2n} w_i p_i + bias) \tag{8}$$

PLICN, which is similar to CNN, is able to get a wider receptive field by stacking PLICN layers to achieve more accurate results. A 2-layers PLICN structure is shown in Fig. 4.

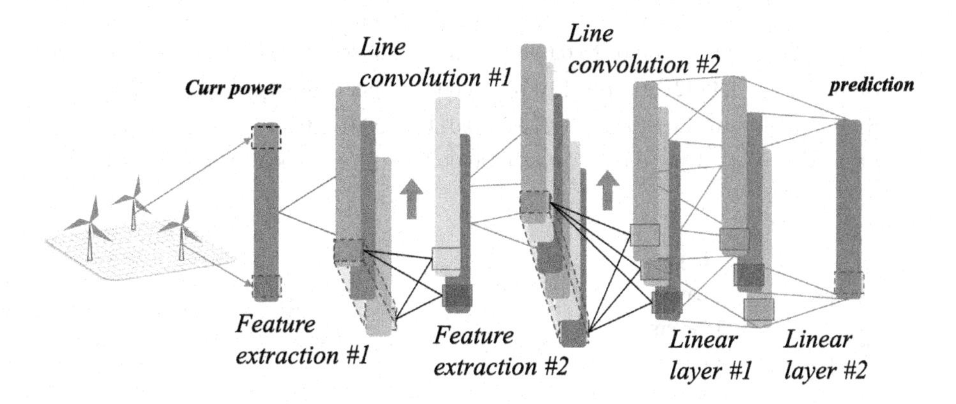


**Fig. 4.** The architecture of 2-layers PLICN

## 4 Experiment Results and Discussions

In our experiments, we select the power and wind data for the whole year of 2012 as the training set. And we select the data from January 2013 as the test set. The data interval is 5 min, focusing on ultra-short-term wind power prediction. Under these circumstances, there are 105,121 training samples and 8,641 test samples in total. We implemented the proposed method with Python 3.7.4 and pytorch 1.7.1.

The prediction accuracy is evaluated by mean squared error (MSE):

$$MSE = \frac{1}{m}\sum_{i=1}^{m}(y_i - \hat{y}_i)^2 \tag{9}$$

### 4.1 PLICN with Different Layers

We set the layer number of PLICN to 1, 2, 3, and 4 respectively to predict the power value after 1–6 time steps. The number of channels of the 1st, 2nd, 3rd and 4th layers of PLICN modules are 32, 64, 128 and 128 respectively. Time step scale is set to 5 min. We conducted 24 sets of experiments in total, and each set of experiments was repeated 5 times. The MSE of the predicted results is shown in Table 1 and Fig. 5.

Figure 5 illustrates the tendency of MSE loss with the prediction time scale and the number of model layers. As the forecast time step increases, the uncertainty of airflow will increase, and it is logical that the forecast error will gradually increase. In addition, as the number of model layers increases, the accuracy

**Table 1.** Parameters and MSE values of different layers and time steps

| MSE_Power | Layer_num = 1 | Layer_num = 2 | Layer_num = 3 | Layer_num=4 |
|---|---|---|---|---|
| Step = 1 | $0.2428 \pm 0.013$ | $0.1205 \pm 0.003$ | $0.1160 \pm 0.005$ | $0.1106 \pm 0.002$ |
| Step = 2 | $0.5105 \pm 0.036$ | $0.3747 \pm 0.003$ | $0.3582 \pm 0.002$ | $0.3555 \pm 0.004$ |
| Step = 3 | $0.8473 \pm 0.042$ | $0.6927 \pm 0.010$ | $0.6586 \pm 0.002$ | $0.6510 \pm 0.009$ |
| Step = 4 | $1.1742 \pm 0.019$ | $1.0176 \pm 0.003$ | $0.9839 \pm 0.008$ | $0.9584 \pm 0.006$ |
| Step = 5 | $1.5253 \pm 0.025$ | $1.3504 \pm 0.011$ | $1.2891 \pm 0.007$ | $1.2728 \pm 0.007$ |
| Step = 6 | $1.8370 \pm 0.057$ | $1.6684 \pm 0.003$ | $1.6172 \pm 0.005$ | $1.5871 \pm 0.019$ |

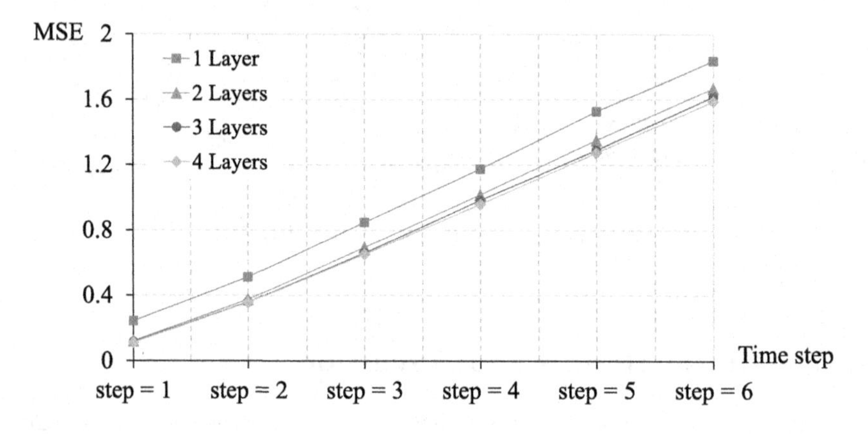


**Fig. 5.** MSE curves of different layers and time steps

of the prediction results will gradually improve, but the improvement will become smaller and smaller. However, increasing the number of model layers will lead to a rapid increase in the amount of parameters, resulting in an increment in training time and a rapid expansion of model size. When the layer number of model reaches 4, the improvement only increases by 1.86% compared to the 3-layer PLICN, but the parameter amount increases by 118%. Therefore, we choose a 3-layer PLICN for comparison experiment.

### 4.2 Contrast Experiment

Similar to CNN, PLICN is a basic network for spatial feature extraction and spatial relationship learning. So we choose other three basic networks, MLP, CNN and LSTM, that are commonly used for wind power prediction as benchmarks to verify the validity of PLICN. The layer number of the three benchmarks and PLICN are 3, and the number of nodes or channels in layers 1, 2, and 3 are 32, 64, and 128, respectively. The input of MLP, CNN and PLICN is a vector composed of the power values of all wind turbines at one moment. The time window length of LSTM is set to 6, so its input is a matrix composed of the power value vectors of all stations at 6 consecutive moments. The experiment results are shown in Table 2 and Fig. 6, each data in the table is the average of the corresponding 5 experiment results.

**Table 2.** Parameters and MSE values of different layers and time steps

| MSE_Power | 3-layers MLP | 3-layers CNN | 3-layers LSTM | 3-layers PLICN |
|---|---|---|---|---|
| Parameter | 55,176 | 103,489 | 63,752 | **41,857** |
| Step = 1 | 0.8131 | 0.1397 | 0.9035 | **0.1160** |
| Step = 2 | 0.9674 | 0.3996 | 0.9785 | **0.3582** |
| Step = 3 | 1.0519 | 0.6971 | 1.0264 | **0.6586** |
| Step = 4 | 1.2502 | 1.0039 | 1.1392 | **0.9839** |
| Step = 5 | 1.4522 | 1.2890 | **1.2642** | 1.2891 |
| Step = 6 | 1.7242 | 1.5834 | **1.3254** | 1.6172 |

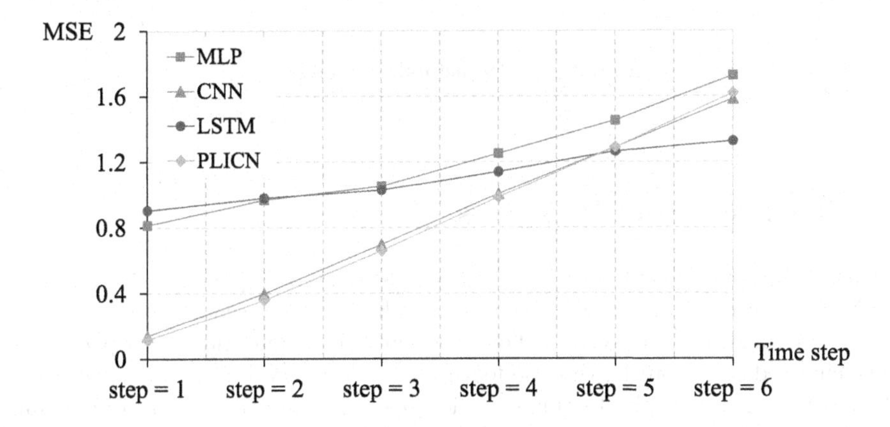


**Fig. 6.** MSE curves of 4 models at different time steps

As shown in Fig. 6, the MSE values of the three all increase as the number of prediction steps becomes longer. Among them, the prediction accuracy of MLP is the worst. When the prediction time steps are 1, 2, 3 and 4, PLICN performs better than CNN. The MSE value of the former is 17.00%, 10.35%, 5.52% and 2.00% lower than the latter respectively. When the prediction step is greater than 4, the performance of PLICN is not as good as CNN. According to the experiment results, PLICN is more suitable for short-term power prediction. In terms of model parameters, the parameters number of 3-layers CNN is 103,489, while that of 3-layers PLICN is 41,857, which is 59.55% less than that of CNN.

Wind is unstable. Airflow has inertia in a short period of time. It will maintain its original state of motion for a period of time, but this inertia will gradually disappear over time. Therefore, when the prediction time step becomes long enough, the direction of the wind may have deviated greatly from the original direction. However, PLICN relies on the strong correlation feature selected by the wind speed field at the previous moment and it is no longer valid after the wind direction changes significantly(as shown in Fig. 7). Therefore, in long time step prediction, the wind direction changes in a wide range, distance based feature selection strategy(CNN for example) may have a better effect. However, the line feature selected by PLICN may lose important and highly relevant feature points. So PLICN is more suitable for short-term forecasting situations.

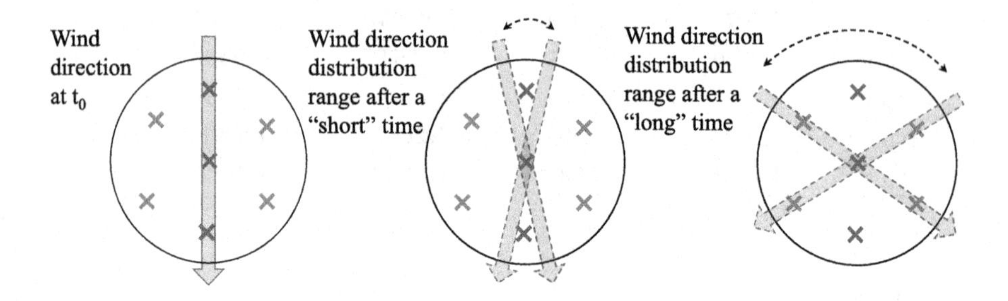


**Fig. 7.** PLICN instability analysis

## 5 Conclusion and Prospects

In ultra-short-term wind power prediction tasks, PLICN performs better than CNN with fewer parameters. When the prediction time scale exceeds 20 min, due to the unstable nature of airflow, the wind direction and wind speed may have changed significantly, the line feature is no longer valid, so the accuracy of PLICN will drop sharply. In this situation, time correlation is more important. LSTM has a better performance at this time.

Next step, we plan to combine the advantages of PLICN and LSTM, enable PLICN to handle multi-time data series, enhance its prediction accuracy on a longer time scale.

## References

1. Aykas, D., Mehrkanoon, S.: Multistream graph attention networks for wind speed forecasting. In: 2021 IEEE Symposium Series on Computational Intelligence (SSCI), pp. 1–8. IEEE (2021)
2. Cabral, B., Leedom, L.C.: Imaging vector fields using line integral convolution. In: Proceedings of the 20th Annual Conference on Computer Graphics and Interactive Techniques, pp. 263–270 (1993)
3. Cheng, L., Zang, H., Xu, Y., et al.: Augmented convolutional network for wind power prediction: a new recurrent architecture design with spatial-temporal image inputs. IEEE Trans. Ind. Inf. **17**(10), 6981–6993 (2021)

4. Council, G.W.E.: Gwec—global wind report 2021. Global Wind Energy Council, Brussels, Belgium (2021)
5. Dowell, J., Weiss, S., Hill, D., et al.: Short-term spatio-temporal prediction of wind speed and direction. Wind Energy **17**(12), 1945–1955 (2014)
6. Draxl, C., Clifton, A., Hodge, B.M., et al.: The wind integration national dataset (wind) toolkit. Appl. Energy **151**, 355–366 (2015)
7. Feng, C., Cui, M., Hodge, B.M., et al.: A data-driven multi-model methodology with deep feature selection for short-term wind forecasting. Appl. Energy **190**, 1245–1257 (2017)
8. Han, X., Qu, Y., Wang, P., et al.: Four-dimensional wind speed model for adequacy assessment of power systems with wind farms. IEEE Trans. Power Syst. **28**(3), 2978–2985 (2012)
9. Hu, R., Hu, W., Gökmen, N., et al.: High resolution wind speed forecasting based on wavelet decomposed phase space reconstruction and self-organizing map. Renew. Energy **140**, 17–31 (2019)
10. Khodayar, M., Wang, J.: Spatio-temporal graph deep neural network for short-term wind speed forecasting. IEEE Trans. Sustain. Energy **10**(2), 670–681 (2019). https://doi.org/10.1109/TSTE.2018.2844102
11. Li, Y., Lang, J., Ji, L., et al.: Weather forecasting using ensemble of spatial-temporal attention network and multi-layer perceptron. Asia-Pac. J. Atmos. Sci. **57**(3), 533–546 (2021)
12. Lundquist, J., DuVivier, K., Kaffine, D., et al.: Costs and consequences of wind turbine wake effects arising from uncoordinated wind energy development. Nat. Energy **4**(1), 26–34 (2019)
13. Niu, Z., Yu, Z., Tang, W., et al.: Wind power forecasting using attention-based gated recurrent unit network. Energy **196**, 117081 (2020)
14. Song, Z., Tang, Y., Ji, J., et al.: Evaluating a dendritic neuron model for wind speed forecasting. Knowl.-Based Syst. **201**, 106052 (2020)
15. Sun, Z., Zhao, M.: Short-term wind power forecasting based on VMD decomposition, ConvLSTM networks and error analysis. IEEE Access **8**, 134422–134434 (2020)
16. Wilms, H., Cupelli, M., Monti, A., et al.: Exploiting spatio-temporal dependencies for rnn-based wind power forecasts. In: 2019 IEEE PES GTD Grand International Conference and Exposition Asia (GTD Asia), pp. 921–926. IEEE (2019)
17. Wu, Y.X., Wu, Q.B., Zhu, J.Q.: Data-driven wind speed forecasting using deep feature extraction and LSTM. IET Renew. Power Gener. **13**(12), 2062–2069 (2019)
18. Yu, R., Liu, Z., Li, X., et al.: Scene learning: deep convolutional networks for wind power prediction by embedding turbines into grid space. Appl. Energy **238**, 249–257 (2019)
19. Yu, Y., Han, X., Yang, M., et al.: Probabilistic prediction of regional wind power based on spatiotemporal quantile regression. IEEE Trans. Ind. Appl. **56**(6), 6117–6127 (2020)
20. Zhang, H., Yan, J., Liu, Y., et al.: Multi-source and temporal attention network for probabilistic wind power prediction. IEEE Trans. Sustain. Energy **12**(4), 2205–2218 (2021)

# Bottom-Up Transformer Reasoning Network for Text-Image Retrieval

Zonghao Yang, Yue Zhou(✉), and Ao Chen

Shanghai JiaoTong University, Shanghai, China
zhouyue@sjtu.edu.cn

**Abstract.** Image-text retrieval is a complicated and challenging task in the cross-modality area, and lots of experiments have made great progress. Most existing researches process images and text in one pipeline or are highly entangled, which is not practical and human-friendly in the real world. Moreover, the image regions extracted by Faster-RCNN are highly over-sampled in the image pipeline, which causes ambiguities for the extracted visual embeddings. From this point of view, we introduce the Bottom-up Transformer Reasoning Network (BTRN). Our method is built upon the transformer encoders to process the image and text separately. We also embed the tag information generated by Faster-RCNN to strengthen the connection between the two modalities. Recall at K and normalized discounted cumulative gain metric (NDCG) metrics are used to evaluate our model. Through various experiments, we prove our model can reach state-of-the-art results.

**Keywords:** Text-image retrieval · Transformer encoder · Tag embedding · Double-Stream architecture

## 1 Introduction

Text-image retrieval aims to give a query text and search for the image which has the most similar semantic information (text to image retrieval) or vice versa (image to text retrieval) [5,8,9,12,17,25]. The core challenge of this task is to find the proper embedding space which can reserve and represent the semantic meaning of the image and the text. Many works have been proposed to solve this challenge. Based on the difference in the network architecture, text-image retrieval can be further divided into Double-Stream Architecture and Singe-Stream Architecture. Double-Stream Architecture means images and text are regarded as two different data streams, which are respectively transmitted into two different subnetworks. Then they are fused through an interactive module. For example, Anderson et al. [12] firstly propose bottom-up attention by using Faster-RCNN on the image to make the proposal regions represent an image and get outstanding performance. Wang et al. [27] more focus on exploring the interactions between images and text before calculating similarities in a joint space. Recently, the state-of-arts (SOTA) are mainly based on transformers [24]. ViLBERT [17] and LXMERT [23] adopt a more conservative double-stream architecture. Image and text are processed in two pipelines, while the embedding


M. Tanveer et al. (Eds.): ICONIP 2022, CCIS 1793, pp. 176–187, 2023.
https://doi.org/10.1007/978-981-99-1645-0_15

features of image and text only interact at the end of the model. Li et al. [13] add the Graph Convolutions into the network and improve the performance. Diao et al. [7] further this work by adding the filter network to neglect the less meaningful word in the text.

Despite the great performance on the text-image retrieval task, the performance of these works remains very inefficient when facing the large-scale text-image database. The reason is that the architectures in these works are highly entangled, so if we want to retrieve images by a specific text (or vice versa), we need to calculate all the similarity scores of the whole dataset and descent them. Besides, these methods also overlook the overlap of the proposal regions problem when embed the images.

For this reason, we proposed our Bottom-up transformer reasoning network (BTRN) to solve this problem. Firstly, we utilize transformer encoders to separately embed the images and text to avoid this problem. Then we continue to use transformer encoders to reason them and get high semantics information of the two pipelines. Finally, we create a similarity matrix by the output and calculate the score in many different criteria.

Besides, the regions are extracted via Faster-RCNN object detectors and are highly over-sampled [14]. When the regions highly overlap, the region features are not easily distinguishable only by the proposal features and spatial information(see Sect. 3.2). To solve this problem, we utilize and embed the tag information into the image pipeline at the first beginning, the way of which is not the same as any work, to our best knowledge.

Above all, our work can be summarized as follows:

- We propose a Bottom-up Transformer Reasoning Network (BTRN), in which we embed images and text in two separate pipelines and then project them together in the latter network. Then we establish a similarity matrix and calculate the score by different criteria.
- In the image pipeline, we utilize and embed the tag information extracted by Faster-RCNN to acquire more specific semantic information of the highly overlapped proposal regions.
- Besides Recall at K, We design a proper similarity function for the NDCG metric to evaluate our model. We show that our model can reach state-of-the-art results on the text-image retrieval task. Our model outperforms about 2.1% on the image retrieval and about 1.4% on the MS-COCO datasets (Recall at 1).

## 2 Related Work

### 2.1 Text-Image Retrieval

Embedding the image and text into one joint space without message loss is key to text-image retrieval. One universal method is to preprocess images and text respectively and project the two modalities into a joint space. Lee et al. [2] embed image regions rather than the whole image and get great performance.

Wang et al. [27] improve the performance by proposing a gate module focusing on cross-modal messages. Li et al. [13] propose a visual reasoning pipeline with Graph Convolution Networks on the image regions. Diao et al. [7] further this work and introduce the filtration module to reduce the influence of unimportant text words. LXMERT [23] uses transformers to embed the image and text and interact at the end. However, these models are highly entangled in architecture. They are inefficient in large datasets since they cannot embed the text and images into separate pipelines.

Our work is related to the previous works [19,20,29,31]. We introduce a Double-Stream Architecture with two relatively independent pipelines in which the text and images are embedded separately to get representative features. Besides, we add tag information of the image regions to solve the overlap problem. To our best knowledge, it is first seen in the Double-Stream Architecture. It indeed boosts the capacity of our model through experiment (see Sect. 6).

### 2.2 Text-Image Retrieval Evaluation Metrics

The universal way to evaluate the text-image retrieval model is Recall at K, which measures the proportion of queries whose ground-truth match ranks within the top K retrieval items. We adopt R@1, R@5, and R@10 in our work.

Besides, inspired by [3,19,20], we give a descending weight on the retrieval results considering the high relevance retrieval even if they are not ground-truth. We think this metric is more human-friendly in the real word text-image retrieval (see Sect. 3.4).

## 3 Bottom-Up Transformer Reasoning Network

In this section, we introduce the detailed architecture of our Bottom-up Transformers Reasoning Network, as shown in Fig. 1. Generally, our model preprocesses images and text into two separate pipelines and reasons them after a linear projection layer. Finally, we use global features or a similarity matrix to calculate the similarity scores.

### 3.1 Image and Text Features Initialization

Specifically, the input of our model is $I = r_0, r_1, ...r_n$ for $n$ salient image regions and $C = w_0, w_1, ...w_m$ for $m$ text words. Both the $I$ and $C$ are generated by the SOTA pre-trained method. More specifically, we use Faster-RCNN and BERT, respectively.

We use bottom-up features, training by Faster-RCNN [22] with ResNet-101 on the Visual Genome dataset [11]. As for the text, we compare and decide to use BERT [6] to extract word embedding features. We think the text features already have the context information thanks to the transformer architecture in BERT.

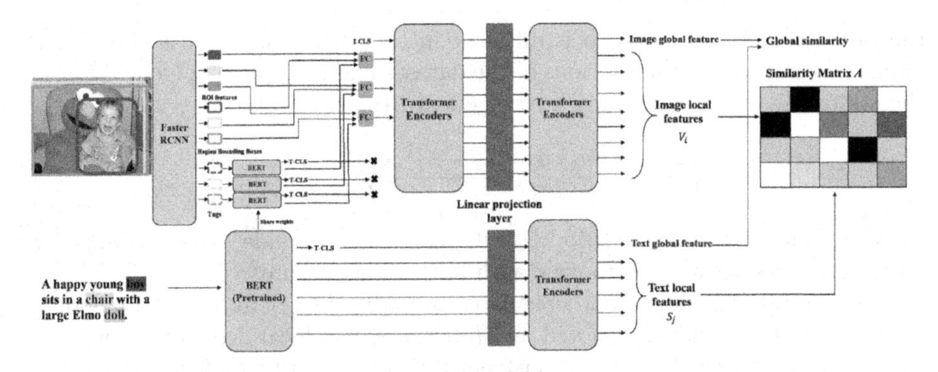


**Fig. 1.** Overview of our Bottom-up Transformer Reasoning Network (BTRN). The number of the Transformer Encoders is illustrated in Sect. 5.2. The image and text features are extracted by Faster-RCNN and BERT, respectively. We also utilize tag information to strengthen the semantic meaning in the image pipeline, as illustrated in Sect. 3.2. The I-CLS (Image-Classification) and T-CLS (Text-Classification), the output of which serves as global features, are the first position in the transformer encoder. The final text-image similarity score is calculated via the global features or the similarity matrix.

## 3.2 Bottom-Up Transformer Reasoning Network (BTRN)

As described in Fig. 1, our model is based on transformer encoders. The reasoning network uses a stack of transformer encoders to embed both image and text pipelines. Thanks to its self-attention, transformer encoders can reason the input feature sets disregarding their intrinsic nature. In detail, we take the salient image regions and caption words as input.

Inspired by the BERT, we utilize the special token in the first place of the transformer encoders (I-CLS and T-CLS) both on the image and text pipelines, which we think carries the global description(features). In this way, we expand the number of image regions to $m+1$ and the number of text words to $n+1$. We set the I-CLS token and the T-CLS token to $r_0$ and $w_0$, respectively. We initialize the I-CLS token and the T-CLS token zero vector. After the reasoning module, the self-attention mechanism can update the token with context information. Then we use one linear projection layer to embed them into the same dimension. After experiments (See Sect. 6.2), we use another transformer encoder to reason them without weight sharing. We regard the output of the $r_0$ and $w_0$ as the global description(features), while the output of the $\{r_1, ...r_{n+1}\}$ and $\{w_1, ...w_{m+1}\}$ are treated as local description(features).

More specifically, we use the bottom-up features and their corresponding spatial information like the universal method [12,14]. Besides, only the features above cannot tell the difference when two image regions highly overlap since the spatial and bottom-up features are almost identical. For example, in Fig. 1 the region features for doll and chair are not distinguishable, as their regions highly overlap. The semantic information (tags of the proposal region extracted by Faster-RCNN) can fix the problem. Although the rest information is almost the same, the tag information is different in this situation. We use the BERT model, training by all the captions of the dataset and max-pooling (see Sect. 6.1) them into the image features. The image features are as follows:

$$c = \left\{ \frac{x_1}{W}, \frac{x_2}{W}, \frac{y_1}{H}, \frac{y_2}{H}, \frac{(x_2 - x_1) \times (y_2 - y_1)}{WH}, tag \right\} \tag{1}$$

We use a fully-connected, ReLU, fully-connected module (share weights for all regions) to concatenate above c with bottom-up features.

As illustrated in Fig. 1, we calculate similarity $S(i, c)$ by cosine for global description. The result is denoted as *Match loss*. For local description, we calculate the similarity $S(i, c)$ by constructing the similarity matrix $A$ with the output of the last transformer encoder. $A$ is constructed as:

$$A_{ij} = \frac{v_i^T s_j}{\left|v_i^T\right| \left|s_j\right|} \qquad i \in (1, n+1), j \in (1, m+1) \tag{2}$$

where $v_i$ is output of the $r_i$ and $s_j$ is output of the $w_i$ in the last layer. Following [19], we use max-sum pooling to get the final similarity score. For maxing over rows and summing over columns, we denote it as *MrSc*, while *McSr* on the vice versa. We also introduce the *Align loss*, which calculates the sum of the *MrSc* and *McSr*.

$$S^{MrSc} = \sum_j \max_i A_{ij} \tag{3}$$

$$S^{McSr} = \sum_i \max_j A_{ij} \tag{4}$$

$$S^{Align} = S^{MrSc} + S^{McSr} \tag{5}$$

### 3.3 Learning

Following [8,13], this work uses a hinge-based triplet ranking loss with hard negative samples to train the model. The loss function is:

$$L_m(i, c) = \max_{c'} \left[\alpha + S\left(i, c'\right) - S(i, c)\right]_+ + \max_{i'} \left[\alpha + S\left(i', c\right) - S(i, c)\right]_+ \tag{6}$$

The $[x]_+$ means $max(x, 0)$. For efficiency [8], the hard negatives are found in the mini-batch rather than the whole dataset. The hard negative images $i'$ and text $c'$ are as follows:

$$i' = \arg\max_{j \neq i} S(i, c) \tag{7}$$

$$c' = \arg\max_{d \neq c} S(i, d) \tag{8}$$

### 3.4 NDCG Metric for Text-Image Retrieval

In the text-image retrieval task, most works utilize the Recall@K metric to measure the performance. The Recall@K metric measures the proportion of queries whose ground-truth match ranks within the top K retrieval items. This metric is suitable when the query is very specific. It is expected to find an item that matches perfectly with the query. However, it seems too strict in real-world search engines. Since most users often give a general description other than a detailed one, and they are not often searching for the exact result. They often choose one among several high-relevance results as different use.

So, we use another metric that is often used in information retrieval, the Normalized Discounted Cumulative Gain (NDCG). The NDCG metric re-evaluates the retrieval ranking list. The core of NDCG is ensuring the high-relevance item takes a higher position in the retrieval ranking list. In other words, decreasing weight is used in the retrieval results to guarantee that the relevance of the items decrease.

The NDCG metric is defined as follows:

$$\mathrm{NDCG}_p = \frac{\mathrm{DCG}_p}{\mathrm{IDCG}_p}, \quad \text{where} \quad DCG_p = \sum_{i=1}^{p} \frac{\mathrm{rel}_i}{\log_2(i+1)} \tag{9}$$

where p is the number of the items in the list, $rel_i$ is a positive number representing the relevance between the $i$-th item of the retrieved results and the query. And the $IDCG_p$ is the best possible ranking.

As for the $rel_i$, we can achieve it by calculating the similarity scores between the query caption and the captions associated with a certain image. Following [19,20], $rel_i$ is defined as $\mathrm{rel}_i = \tau\left(\bar{C}_i, C_j\right)$, where $C_j$ is the query text and $\bar{C}_i$ is the captions associated with the image $I_i$, and $\tau : \mathbb{S} \times \mathbb{S} \rightarrow [0,1]$ is a similarity function with normalization. In this way, we avoid to generate a large relevance matrices between the images and captions and are able to calculate the similarity function efficiently.

Above all, in image retrieval, $\mathrm{rel}_i = \tau\left(\bar{C}_i, C_j\right)$, where $\bar{C}_i$ is the set of captions associated with one retrieval image, $C_j$ is the query caption. In text retrieval $\mathrm{rel}_i = \tau\left(\bar{C}_j, C_i\right)$, where $\bar{C}_j$ is the set of captions associated with the query image $I_j$ and $C_i$ is the one retrieval caption.

Following [3,19], we use SPICE and ROUGE-L as function $\tau$ to calculate the similarity scores. SPICE focus on high-level semantic features between words and concepts. ROUGE-L is often used in the longest common sub-sequence and focuses on syntactic constructions.

## 4 Efficiency

One main objective of our work is to extract the image and text features efficiently. So it is necessary to build a network that can extract and embed the image and text features in several. Two separate pipelines can disentangle the visual and text modalities. Moreover, it is also relatively efficient to calculate the similarity scores.

In detail, if $K$ is the number of images and $L$ is the number of sentences in the database, our work, as well as other Double-Stream Architecture, has a feature space complexity of $O(K)+O(L)$. And the time complexity of the feature extraction stage is also $O(K) + O(L)$.

However, the Singe-Stream Architecture [4,26,28] and other works that entangle the two modalities take $O(KL)$ as a feature space, as well as the retrieval time complexity. As illustrated in Sect. 1, these works are inefficient in facing the real-world datasets or being deployed to real-world search engines. Some works [18] try to fix this issue by embedding the image and text on two pipelines and interacting on the very last end. In this scenario, the time complexity of a new query needs another $O(K)$ or $O(L)$. We need to re-evaluate the last attention layers in the best case.

Regarding the similarity score, our work uses straightforward ways in order not to cause additional time costs or memories burden. Our *Match Loss* uses simple dot products which can be calculated and ranked efficiently when deployed in real-world search engines. Other losses also use simple dot products and sums, which avoid complex layers of memories.

The designs of the two disentangled pipelines and simple similarity function score enable us to acquire the retrieval results efficiently, given a new query.

## 5 Experiments

**Table 1.** Experiment results on MS-COCO 1k test set

| Model | Image To Text Retrieval | | | | | Text To Image Retrieval | | | | |
|---|---|---|---|---|---|---|---|---|---|---|
| | R@1 | R@5 | R@10 | SPICE | R-L | R@1 | R@5 | R@10 | SPICE | R-L |
| SCAN [12] | 72.7 | 94.8 | 98.4 | - | - | 58.8 | 88.4 | 94.8 | - | - |
| CAMP [27] | 72.3 | 94.8 | 98.3 | - | - | 58.5 | 87.9 | 95.0 | - | - |
| VSRN [13] | 76.2 | 94.8 | 98.2 | 0.748 | 0.704 | 62.8 | 89.7 | 95.1 | 0.732 | 0.637 |
| SGRAF [7] | 79.6 | 96.2 | 98.5 | - | - | 63.2 | 90.7 | 96.1 | - | - |
| PFAN [26] | 76.5 | 96.3 | 99.0 | - | - | 61.6 | 89.6 | 95.2 | - | - |
| TERAN [20] | 80.2 | 96.6 | 99.0 | 0.756 | 0.720 | 67.0 | 92.2 | 96.9 | 0.747 | 0.680 |
| BTRN | | | | | | | | | | |
| *Match* | 71.1 | 93.9 | 97.8 | 0.737 | 0.700 | 58.2 | 88.8 | 95.3 | 0.730 | 0.656 |
| *McSr* | 77.4 | 95.8 | 98.6 | 0.749 | 0.713 | 64.2 | 91.9 | 96.6 | 0.737 | 0.667 |
| *Align* | 76.9 | 96.2 | 98.9 | 0.751 | 0.715 | 64.4 | 91.5 | 96.5 | 0.735 | 0.662 |
| *MrSc* | **81.6** | **98.2** | **99.5** | **0.757** | **0.726** | **69.1** | **93.7** | **97.4** | **0.751** | **0.694** |

To prove the effectiveness of our model on the existing vision-language datasets, we train our Bottom-up Transformer Reasoning Network on Flickr30k

**Table 2.** Experiment results on Flickr30k 1k test set

| Model | Image To Text Retrieval | | | | | Text To Image Retrieval | | | | |
|---|---|---|---|---|---|---|---|---|---|---|
| | R@1 | R@5 | R@10 | SPICE | R-L | R@1 | R@5 | R@10 | SPICE | R-L |
| SCAN [12] | 67.4 | 90.3 | 95.8 | - | - | 48.6 | 77.7 | 85.2 | - | - |
| CAMP [27] | 68.1 | 89.7 | 95.2 | - | - | 51.5 | 77.1 | 85.3 | - | - |
| VSRN [13] | 71.3 | 90.6 | 96.0 | - | - | 54.7 | 81.8 | 88.2 | - | - |
| SGRAF [7] | 77.8 | 94.1 | 97.4 | - | - | 58.5 | 83.0 | 88.8 | - | - |
| PFAN [26] | 70.0 | 91.8 | 95.0 | - | - | 50.4 | 78.7 | 86.1 | - | - |
| TERAN [20] | **79.2** | 94.4 | 96.8 | 0.707 | 0.636 | 63.1 | 87.3 | 92.6 | 0.695 | 0.577 |
| BTRN | | | | | | | | | | |
| *Match* | 63.4 | 88.4 | 94.0 | 0.674 | 0.554 | 48.9 | 79.7 | 88.2 | 0.673 | 0.554 |
| *McSr* | 75.4 | 94.4 | 97.8 | 0.709 | 0.636 | 60.6 | 87.5 | 92.8 | 0.690 | 0.578 |
| *Align* | 73.8 | **94.7** | 97.2 | 0.706 | 0.632 | 60.5 | 87.6 | 92.6 | 0.689 | 0.570 |
| *MrSc* | 75.1 | 94.4 | **98.5** | **0.712** | **0.671** | **64.0** | **89.0** | **93.8** | **0.695** | **0.583** |

datasets [30] and MS-COCO datasets [16]. We compare our results on these datasets with the state-of-the-art approaches' results on the same metric, as shown in Table 1 and 2.

### 5.1 Datasets and Protocols

Flickr30k dataset contains 31,783 images. Each image has five corresponding unique captions. Following the split setting in [10], we choose 1014 different images as the validation set and another 1000 as the test set, which the rest used as the training set. MS-COCO dataset contains 123,287 images, annotated by five different captions. We split 5000 images for validation, another 5000 for test and the remains for training. To save space, we do not show the result in MS-COCO 5k test set.

For text-image retrieval, we evaluate our performance by the wildly-used metrics Recall at K(R@K). Besides, we also evaluate our performance by the NDCG metric. We use ROUGE-L [15] and SPICE [1] as text-image pair relevance. We set the NDCG parameter $p = 25$ as in [3].

### 5.2 Implementation Details

For text pipeline, our work uses the pre-trained BERT model trained on mask language task on the English sentences and continues training the BERT model on the captions of the two datasets. The size of these text embeddings is 768. For the image pipeline, our work uses Anderson's model to extract the image features. We reserve all the proposal regions as long as their IOU relevance is above 0.2. The other settings are the same as Anderson's [2]. The size of image embeddings is 2048. The linear projection layer is 1024 dimensional.

We use four transformer encoders on the image pipeline to get the image description and another two after a linear projection layer through our experiment. The drop-out is set to 0.1. We trained our model for 30 epochs, following [8,13,19,20,29,31]. Our work uses Adam optimizer and the learning rate is set to $1e-5$ for the first 15 epochs and $1e-6$ for the rest. The margin $\alpha$ of hinge-based loss is set to 0.2 in our work.

### 5.3 Experimental Result

We prove the effectiveness of our model by comparing the Recall at K metric and the NDCG metric with other state-of-the-art methods in Table 1. We can see that our model can achieve and surpass the SOTA in some cases. We finally choose the model trained in *MrSc* loss to represent our BTRN model. More specially, the R@1, R@5, and R@10 metric improve 1.4%, 2.0%, 0.6% when text retrieval while 2.1%, 1.5%, 0.5% improve when image retrieval, on the MS-COCO 1k test set. Similarly, the R@K and NDCG improve both retrieval situations in the Flickr30k 1k test set.

### 5.4 Qualitative Result and Analysis

Query : A dog running through a grassy field.

Query : A man playing musical instruments.

**Fig. 2.** Examples of Image Retrieval by our model. The second example shows that even though the ground truth match is not in the first position, the retrieval results still have high relevance.

Figure 2 shows some cases that use text as queries to retrieve images. A red box frames the ground truth image. We can see that some retrieval results provide strong relevance even if the ground truth box is not in the first position.

# 6 Ablation Study

In our work, we re-evaluate our model through a series of ablation studies. We try to analyze the impact of different ways of tag embedding. Then we explore the impact of weight-sharing in the transformer reasoning phase. Furthermore, we also explore the model trained in *MrSc loss* and test it in *match loss* to verify whether the local similarly can improve the model capacity. The relevant results are shown in Table 3.

**Table 3.** Ablation studies on Flickr30k 1k test set.

| Model | Text Retrieval | | | Image Retrieval | | |
|---|---|---|---|---|---|---|
| | R@1 | R@5 | R@10 | R@1 | R@5 | R@10 |
| BTRN *MrSc loss* | | | | | | |
| Max encoding | 75.1 | 94.4 | **98.5** | **64.0** | **89.0** | **93.8** |
| Mean encoding | 75.6 | 93.8 | 97.1 | 61.9 | 88.2 | 92.9 |
| Without tags | **75.7** | **94.6** | 97.1 | 61.2 | 88.1 | 92.8 |
| *MrSc share-w* | 75.3 | 93.9 | 96.9 | 62.1 | 87.8 | 92.8 |
| *Match loss* | 71.1 | 93.9 | 97.8 | 58.2 | 88.8 | 95.3 |
| *Match+MrSc* | 74.4 | 93 | 96.7 | 59.4 | 85.7 | 91.4 |

## 6.1 Averaging Versus Maxing

When facing the tag embedding in the visual pipeline, we also try to use average pooling for the tag features rather than max pooling. The result is shown in the first two rows of Table 3. It suggests that the max-pooling layer can better reserve and distinguish the different semantics information of a phrase consisting of multiple words than the average-pooling layer. However, there is a little loss in R@1 and R@5 in the text retrieval task, which may explain that the semantic information of tags does not perfectly match the captions. However, in general, the max-pooling layer is the most suitable for embedding the tag information.

## 6.2 Sharing Weights in Reasoning Phase

We also consider sharing the weights of the last two transformer encoders. Sharing weights can not only strengthen the connection of two pipelines but also reduce the model size. The result is shown in the first and fourth row in Table 3. We can see that there is no need to share the weights in the reasoning phase since the features from the projection layer have had enough semantic information.

## 6.3 Others

We also try to combine different training strategies to get better performance or improve one with another. In other words, we want to verify whether the local features can better understand the fine-grained semantic information. For this purpose, we train the model with *MrSc loss* and test it with *match loss*. The result is shown in the last two rows in Table 3. It confirms our idea that the local features can help understand the fine-grained semantic information.

## 7 Conclusion

Based on transformer encoders, we present our novel Bottom-up Transformer Reasoning Network (BTRN), which processes the image and text in two pipelines. We utilize tag information of the highly overlapped proposal regions to solve the different semantics problem, which can better extract local features of the images. Then we create a similarity matrix and calculate the score in many different criteria. Besides R@K, We also design a suitable similarity function for the NDCG metrics to re-evaluate our model.

We demonstrate the effectiveness of our model by testing it on the Flickr30k and MS-COCO datasets. Results on the benchmark datasets show that our model can reach and outperform SOTA methods on some tasks.

## References

1. Anderson, P., Fernando, B., Johnson, M., Gould, S.: SPICE: semantic propositional image caption evaluation. In: Leibe, B., Matas, J., Sebe, N., Welling, M. (eds.) ECCV 2016. LNCS, vol. 9909, pp. 382–398. Springer, Cham (2016). https://doi.org/10.1007/978-3-319-46454-1_24
2. Anderson, P., et al.: Bottom-up and top-down attention for image captioning and vqa (2017)
3. Carrara, F., Esuli, A., Fagni, T., Falchi, F., Moreo Fernández, A.: Picture it in your mind: generating high level visual representations from textual descriptions. Inf. Retrieval J, 208–229 (2017). https://doi.org/10.1007/s10791-017-9318-6
4. Chen, H., Ding, G., Liu, X., Lin, Z., Liu, J., Han, J.: Imram: Iterative matching with recurrent attention memory for cross-modal image-text retrieval. In: Proceedings of the IEEE/CVF Conference on Computer Vision and Pattern Recognition, pp. 12655–12663 (2020)
5. Chen, Y.C., et al.: Uniter: learning universal image-text representations (2019)
6. Devlin, J., Chang, M.W., Lee, K., Toutanova, K.: Bert: pre-training of deep bidirectional transformers for language understanding (2018)
7. Diao, H., Zhang, Y., Ma, L., Lu, H.: Similarity reasoning and filtration for image-text matching (2021)
8. Faghri, F., Fleet, D.J., Kiros, J.R., Fidler, S.: Vse++: improving visual-semantic embeddings with hard negatives. In: British Machine Vision Conference (2018)
9. Ge, X., Chen, F., Jose, J.M., Ji, Z., Wu, Z., Liu, X.: Structured multi-modal feature embedding and alignment for image-sentence retrieval. In: Proceedings of the 29th ACM International Conference on Multimedia, pp. 5185–5193 (2021)
10. Karpathy, A., Fei-Fei, L.: Deep visual-semantic alignments for generating image descriptions. IEEE Trans. Pattern Anal. Mach. Intell., 664–676 (2016)
11. Krishna, R., Zhu, Y., Groth, O., Johnson, J., Li, F.F.: Visual genome: Connecting language and vision using crowdsourced dense image annotations. Int. J. Comput. Vis. **123**(1) (2017)

12. Lee, K.H., Xi, C., Gang, H., Hu, H., He, X.: Stacked cross attention for image-text matching (2018)
13. Li, K., Zhang, Y., Li, K., Li, Y., Fu, Y.: Visual semantic reasoning for image-text matching. In: 2019 IEEE/CVF International Conference on Computer Vision (ICCV) (2019)
14. Li, X., et al.: Oscar: object-semantics aligned pre-training for vision-language tasks (2020)
15. Lin, C.Y.: Rouge: a package for automatic evaluation of summaries. In: Proceedings of the Workshop on Text Summarization Branches Out (WAS 2004) (2004)
16. Lin, T.Y., Maire, M., Belongie, S., Hays, J., Zitnick, C.L.: Microsoft coco: Common objects in context. In: European Conference on Computer Vision (2014)
17. Lu, J., Batra, D., Parikh, D., Lee, S.: Vilbert: pretraining task-agnostic visiolinguistic representations for vision-and-language tasks (2019)
18. Macavaney, S., Nardini, F.M., Perego, R., Tonellotto, N., Goharian, N., Frieder, O.: Efficient document re-ranking for transformers by precomputing term representations. In: arXiv (2020)
19. Messina, N., Falchi, F., Esuli, A., Amato, G.: Transformer reasoning network for image-text matching and retrieval (2020)
20. MessinaNicola, AmatoGiuseppe, EsuliAndrea, FalchiFabrizio, GennaroClaudio, Marchand-MailletStéphane: fine-grained visual textual alignment for cross-modal retrieval using transformer encoders. ACM Transactions on Multimedia Computing, Communications, and Applications (TOMM) (2021)
21. Nguyen, M.D., Nguyen, B.T., Gurrin, C.: A deep local and global scene-graph matching for image-text retrieval. arXiv preprint arXiv:2106.02400 (2021)
22. Ren, S., He, K., Girshick, R., Sun, J.: Faster r-cnn: towards real-time object detection with region proposal networks. IEEE Trans. Pattern Anal. Mach. Intell. **39**(6), 1137–1149 (2017)
23. Tan, H., Bansal, M.: Lxmert: Learning cross-modality encoder representations from transformers (2019)
24. Vaswani, A., et al.: Attention is all you need. In: arXiv (2017)
25. Wang, H., Zhang, Y., Ji, Z., Pang, Y., Ma, L.: Consensus-aware visual-semantic embedding for image-text matching. In: Vedaldi, A., Bischof, H., Brox, T., Frahm, J.-M. (eds.) ECCV 2020. LNCS, vol. 12369, pp. 18–34. Springer, Cham (2020). https://doi.org/10.1007/978-3-030-58586-0_2
26. Wang, Y., Yang, H., Qian, X., Ma, L., Fan, X.: Position focused attention network for image-text matching (2019)
27. Wang, Z., Liu, X., Li, H., Sheng, L., Yan, J., Wang, X., Shao, J.: Camp: cross-modal adaptive message passing for text-image retrieval. In: 2019 IEEE/CVF International Conference on Computer Vision (ICCV) (2020)
28. Xu, X., Wang, T., Yang, Y., Zuo, L., Shen, F., Shen, H.T.: Cross-modal attention with semantic consistence for image-text matching. IEEE Trans. Neural Networks Learn. Syst. **31**(12), 5412–5425 (2020)
29. Yao, T., Pan, Y., Li, Y., Mei, T.: Exploring visual relationship for image captioning. In: European Conference on Computer Vision (2018)
30. Young, P., Lai, A., Hodosh, M., Hockenmaier, J.: From image descriptions to visual denotations: New similarity metrics for semantic inference over event descriptions. Nlp.cs.illinois.edu (2014)
31. Zheng, Z., Zheng, L., Garrett, M., Yang, Y., Shen, Y.D.: Dual-path convolutional image-text embedding (2017)

# Graph Attention Mixup Transformer for Graph Classification

Jiaxing Li, Ke Zhang, Xinyan Pu, and Youyong Kong(✉)

Jiangsu Provincial Joint International Research Laboratory of Medical Information Processing, School of Computer Science and Engineering, Southeast University, Nanjing, China
{jiaxing_li,kylenz,puxinyan,kongyouyong}@seu.edu.cn

**Abstract.** Transformers have been successfully applied to graph representation learning due to the powerful expressive ability. Yet, existing Transformer-based graph learning models have the challenge of overfitting because of the huge number of parameters compared to graph neural networks (GNNs). To address this issue, we propose an end-to-end regularized training scheme based on Mixup for graph Transformer models called Graph Attention Mixup Transformer (GAMT). We first apply a GNN-based learnable graph coarsening method to align the number of nodes and preserve local neighborhood information. Then, we mix the pairwise graph attention weights and node embeddings as additional synthetic training samples to train a graph Transformer model with smoother decision boundaries. Finally, we introduce contrastive learning to further regularize the graph Transformer model. Extensive experiments on real-world graph classification benchmark datasets show that GAMT improves the performance and generalization ability of graph Transformer models without additional computing resource consumption.

**Keywords:** Graph Transformer · Overfitting · Mixup · Graph Classification

## 1 Introduction

Transformer-based methods [14] have shown the most powerful performance in the fields of natural language processing [2] and computer vision [3]. Recently, due to the powerful expressive ability of Transformer, researchers have begun to introduce Transformer architecture into graph representation learning, and have demonstrated performance beyond graph neural networks (GNNs) in tasks such as node classification [4], graph classification [8], and graph prediction [21]. However, while it shows strong performance, a problem that cannot be ignored is that the model based on the Transformer architecture will bring a huge number of parameters compared to GNNs. In the case of limited training data, it is prone to overfitting.


M. Tanveer et al. (Eds.): ICONIP 2022, CCIS 1793, pp. 188–199, 2023.
https://doi.org/10.1007/978-981-99-1645-0_16

When training deep neural networks, applying regularization techniques such as data augmentation is essential to prevent overfitting and improve the generalization ability of models. Recently, Mixup and its variants [16,25] as a data-independent and effective regularization method based on interpolation, have been theoretically and experimentally shown to improve the generalization ability of deep neural networks [23,24]. However, due to the irregularity of graph structure data and different numbers of nodes between different graphs, it is challenging to directly apply Mixup to graph structure data. To avoid the problem of different numbers of nodes when mixing graphs, a simple approach is to mix fixed-size graph representation produced by the READOUT function [19]. However, this approach ignores the topological relationship between node pairs. We find that the attention weights naturally generated in Transformer can be well suited for capturing relationships between nodes, so we introduce Mixup into the graph Transformer models to improve the generalization ability. As far as we know, there is no regularization method specifically designed to improve the generalization ability of graph Transformer models.

In this paper, we propose a novel regularization method specifically designed for training graph Transformer models, called Graph Attention Mixup Transformer (GAMT). Specifically, we utilize a GNN-based graph coarsening method to assign all nodes from each graph to a fixed number of clusters to form new nodes, while preserving local neighborhood information as much as possible. Then, feed the embeddings of coarsened nodes into Transformer encoders to calculate graph attention, where we mix the attention weight matrices and embedding value matrices of different graphs (see Fig. 1). The attention weight captures the interaction between different nodes and node embeddings represent the characteristics of each node itself. Therefore, mixing graph attention considers both of the above information at the same time. Moreover, we introduce contrastive learning to further regularize the graph Transformer model. The synthetic sample generated by Mixup can naturally be used as the positive sample of the given sample, and others different from the given sample are negative samples. Then we construct an additional contrastive loss to maximize the similarity between the positive pair. In summary, the main contributions of this paper are as follows:

- We propose GAMT, an end-to-end regularized training scheme based on Mixup to improve the generalization ability of graph Transformer models. To the best of our knowledge, this is the first work specially designed to solve the challenge of the insufficient generalization ability of graph Transformer models.
- We innovatively apply the Mixup operation to graph attention and introduce contrastive self-supervised learning as a regularization term to further regularize graph Transformer models.
- Extensive experiments on real-world graph classification benchmark datasets show that GAMT improves the performance and generalization ability of graph Transformer models without additional computing resource consumption.

## 2 Related Work

### 2.1 Graph Transformers

The existing graph neural networks update node representations by aggregating features from the neighbors, which have achieved great success in node classification and graph classification [5,7,15]. However, with Transformer's excellent performance in natural language processing [2] and computer vision [3], researchers have begun to apply Transformer to graph representation learning tasks to further consider the long-range information. Transformer cannot directly use the structural information of a graph like GNNs, so the current work of graph Transformer is mainly focused on encoding the structural information of graphs [4,9,21]. Different from the existing graph Transformer efforts, we focus more on improving the generalization ability of graph Transformer models.

### 2.2 Mixup and Its Variants

Mixup [16,25] is an efficient interpolation-based data augmentation method to regularize deep neural networks, which generates additional virtual samples from adjacent training sample distributions to expand the support for training distribution. Mixup-based training methods improve the performance in image classification [24], text classification [23] and other tasks. Recently, Mixup has been used by researchers in graph representation learning to solve overfitting problems. GraphMix [18] uses Mixup for semi-supervised node classification by sharing parameters with GNN through an additional fully-connected network. [19] propose to augment the training data of graph classification by mixing fixed-size graph representation after READOUT function.

## 3 Methodology

In this section, we describe the proposed Graph Attention Mixup Transformer (GAMT) in detail, which can effectively regularize the graph Transformer models and improve the generalization performance. The overall architecture of GAMT is shown in Fig. 1.

### 3.1 Graph Coarsening

A graph $\mathcal{G} = (\mathcal{V}, \mathcal{E})$ can be represented by an adjacency matrix $A \in \{0,1\}^{N \times N}$ and a node attributes matrix $X \in \mathbb{R}^{N \times d}$ with d-dimensional features. Given a set of labeled graphs $G = \{(\mathcal{G}_1, y_1), ..., (\mathcal{G}_n, y_n)\}$, the goal of graph classification is to learn a mapping $f(\cdot)$ that maps graphs to the set of labels. However, due to the irregularity of graph data, given a set of graphs, the number of nodes and edges might be quite different, so it is challenging to directly apply Mixup to graphs. To solve the above problems, we utilize a GNN-based graph coarsening method to align nodes from each graph, that is, to assign nodes to a fixed number of clusters.

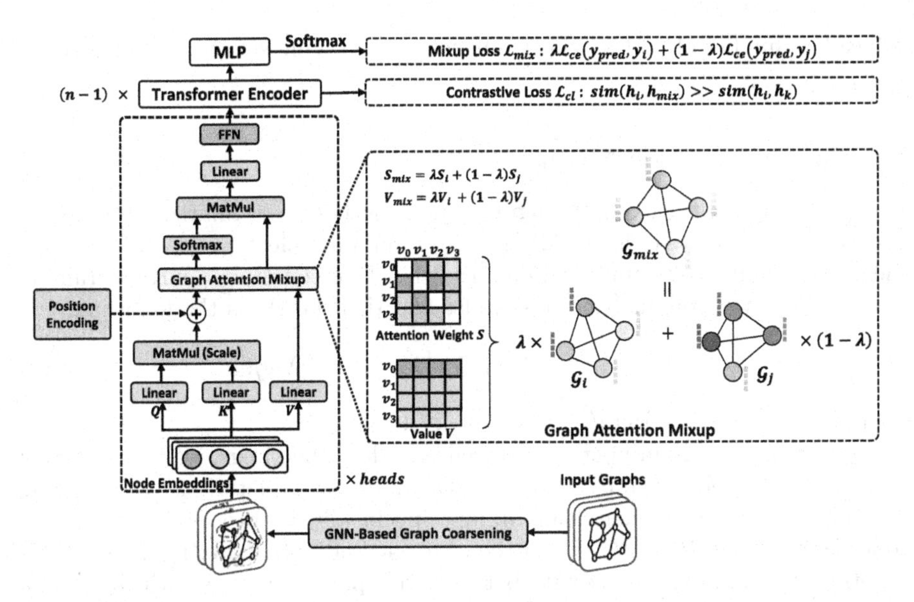


**Fig. 1.** Architecture of the proposed Graph Attention Mixup Transformer (GAMT). The input original graphs are coarsened to obtain the coarsened graphs and their embeddings, and then input them to the first layer of Transformer Encoder. The right part of the figure shows the process of graph attention Mixup, mixing the attention weights $S$ and node representations $V$ between graph $\mathcal{G}_i$ and $\mathcal{G}_j$. Finally, the contrastive loss and classification loss are calculated respectively after the remaining $n-1$ layers of Transformer encoders.

Specifically, given a graph $\mathcal{G} = (X, A)$ with $n$ nodes, the graph coarsening operation divides the current nodes into $m$ clusters, and each cluster $C_i$ represents a new node. Inspired by DiffPool [22], we use the output of a $l$-layer GNN to learn the cluster assignment matrix $C$ over the nodes:

$$C = \text{softmax}\left(\text{GNN}_1^{(l)}\left(A, X, W_1^{(l)})\right)\right), \tag{1}$$

where $C \in \mathbb{R}^{n \times m}$, and it represents the weight that the current node is assigned to the new node after coarsening. If we directly coarsen the graph at this time, it will cause the current node to lose part of the local neighborhood information. Therefore, we need to add another GNN to capture the local neighborhood information of the current node:

$$H = \text{GNN}_2^{(l)}(A, X, W_2^{(l)})). \tag{2}$$

Then we use the learned assignment matrix $C$ to perform coarsening on the nodes of the graph $\mathcal{G}$ to obtain a new graph $\mathcal{G}'$:

$$H' = C^T H \in \mathbb{R}^{m \times d}, \quad A' = C^T A C \in \mathbb{R}^{m \times m}. \tag{3}$$

In this way, we perform the above-mentioned coarsening process on all input graphs to ensure the consistency of the number of nodes, and at the same time reduce the number of nodes, which is beneficial to saving subsequent calculation costs.

### 3.2 Graph Attention Mixup

Mixup [25] is an effective data augmentation method to regularize deep neural network. The proposed idea of Mixup is to randomly select two samples $(x_i, y_i)$ and $(x_j, y_j)$ from the training data distribution $\mathcal{D}$ to construct a virtual training sample by convex combination of samples and their labels as the following:

$$\tilde{x} = \lambda x_i + (1 - \lambda)x_j, \tilde{y} = \lambda y_i + (1 - \lambda)y_j, \tag{4}$$

where $\lambda \in [0, 1]$ or $\lambda \sim \text{Beta}(\alpha, \alpha)$ with $\alpha \in (0, \infty)$.

Instead of directly mixing the coarsened node embeddings, we focus on how to consider the relationship between nodes and the characteristics of the node itself at the same time when using Mixup. Specifically, we process the coarsened nodes embeddings $H' \in \mathbb{R}^{m \times d}$ and the coarsened adjacency matrix $A' \in \mathbb{R}^{m \times m}$ denoting the connectivity strength between each pair of clusters with the vanilla Transformer encoder architecture to perform the graph attention Mixup proposed above. Specifically, we first perform a linear projection and a layer Normalization for $H'$ to normalize the node embeddings:

$$H_{in} = \text{LayerNorm}(H'W_h), \tag{5}$$

where $W_h \in \mathbb{R}^{d \times d_{in}}$ is a learnable weight matrix and $d_{in}$ is the dimension of Transformer encoders. Then we parametrize the self-attention with three projection matrices $W_q \in \mathbb{R}^{d \times d_k}, W_k \in \mathbb{R}^{d \times d_k}$ and $W_v \in \mathbb{R}^{d \times d_v}$. To better encode topological structure information into self-attention, we add the coarsened adjacency matrix $A'$ by a linear projection as a bias term for the attention weight:

$$S = \frac{(H_{in}W_q)(H_{in}W_k)^{\mathrm{T}}}{\sqrt{d_k}} + A'W_a, \tag{6}$$

where $S$ is the attention weight capturing the similarity between different nodes with their topological information, and $W_a \in \mathbb{R}^{m \times m}$ denotes a learnable weight for mapping the topological information. Next, we perform the Mixup operation on the attention weight $S$ and the value matrix $V = H_{in}W_v$ for graph $\mathcal{G}_i$ and $\mathcal{G}_j$:

$$\tilde{S} = \lambda S_i + (1 - \lambda)S_j, \tilde{V} = \lambda V_i + (1 - \lambda)V_j. \tag{7}$$

Here $\lambda \sim \text{Beta}(\alpha, \alpha)$ is the mixing weight sampled from the Beta distribution with $\alpha \in (0, \infty)$ being a hyperparameter. Therefore, in each multi-head attention (MHA) layer $head_i$, we use the mixed $\tilde{S}$ and $\tilde{V}$ to calculate self-attention for subsequent calculations:

$$head_i = \text{softmax}(\tilde{S})\tilde{V}. \tag{8}$$

Then, we perform the above operations on each corresponding head, and then concatenate the outputs to feed into the feed-forward network (FFN) with a layer normalization:

$$\begin{aligned} \mathrm{MHA}(H_{in}) &= \mathrm{Concat}(head_1, ..., head_h), \\ H^{(1)} &= \mathrm{LayerNorm}(\mathrm{FFN}(\mathrm{MHA}(H_{in}))), \end{aligned} \tag{9}$$

where $H^{(1)}$ is the output of the first Transformer encoder layer with Graph Attention Mixup operation, and the next $n-1$ layers of Transformer encoders are standard without Mixup.

Finally, after passing through $n$ layers of Transformer encoders, we obtain the mixed node representations $H^{(n)}$ with local information and global information. To generate a fixed-size graph-level representation, we use a READOUT function that aggregates the node representations into the entire graph representation $h_{mix}$:

$$\boldsymbol{h}_{mix} = \mathrm{READOUT}\left(\left\{h_v^{(n)} \mid v \in \mathcal{V}'\right\}\right). \tag{10}$$

with a READOUT function, we can simply use the average or sum over all node representations $h_v^{(n)}, \forall v \in \mathcal{V}'$ from the coarsened graph $\mathcal{G}'$.

### 3.3 Graph Contrastive Learning with Mixup

As a data augmentation technology, Mixup naturally brings additional self-supervised information while extending the training distribution [17]. To make full use of the self-supervised information between original samples and mixed samples, we use the idea of contrastive learning to further regularize the graph Transformer model.

Specifically, we define two tasks on the graph Transformer model. The first task is the classification task of augmented samples, where we use the mixed samples obtained in the previous section to perform the graph classification task. We feed the mixed graph-level representation $h_{mix}$ into the MLP layer with softmax classifier to predict the label:

$$y_{pred} = \mathrm{softmax}(\mathrm{MLP}(h_{mix})). \tag{11}$$

Then, the Mixup loss $\mathcal{L}_{mix}$ is defined as a mixture of the cross-entropy loss function $\mathcal{L}_{ce}$ over the $y_i$ and $y_j$ with the mixing ratio $\lambda$:

$$\mathcal{L}_{mix} = \lambda\mathcal{L}_{ce}(y_{pred}, y_i) + (1-\lambda)\mathcal{L}_{ce}(y_{pred}, y_j). \tag{12}$$

The other is to build a contrastive self-supervised learning (CSSL) task based on the augmented samples brought by Mixup, which is shared with the Transformer Encoders weights of the graph classification task. The mixed sample $\mathcal{G}_{mix}$ is obtained by mixing the graphs $\mathcal{G}_i$ and $\mathcal{G}_j$, so the distance between the graph representations $h_i$ and $h_{mix}$ embedded by Transformer Encoders is closer than that of other samples like $h_k(k \neq i)$, namely:

$$\mathrm{sim}(h, h_{mix}) > \mathrm{sim}(h, h_k). \tag{13}$$

Thus the CSSL task can be used as a regularizer that relies on Mixup to further regularize the model. We use the loss function of InfoNCE [12] to maximize the similarity between positive sample pairs ($h_i$ and $h_{mix}$), and minimize the similarity between negative sample pairs (others). As a regularization term, the contrastive loss $L_{cl}$ can be defined as:

$$\mathcal{L}_{cl} = -\log \frac{\exp\left(\operatorname{sim}\left(h_i, h_{mix}\right)/\tau\right)}{\sum_{k=1[k \neq i]}^{N} \exp\left(\operatorname{sim}\left(h_i, h_k\right)/\tau\right)}, \tag{14}$$

where $\operatorname{sim}(a, b) = ab^{\top}$ denotes the inner product between two vectors and $\tau$ is a temperature parameter. Finally, we combine these objectives to train the whole process in an end-to-end way as:

$$\mathcal{L} = \mathcal{L}_{mix} + \beta \mathcal{L}_{cl}, \tag{15}$$

where $\beta$ is a hyperparameter tocontrol the contribution of the contrastive learning loss. In this way, we not only use Mixup for data augmentation to regularize the graph Transformer, but also realize its additional value for contrastive self-supervised learning.

# 4 Experiments

## 4.1 Experimental Setup

**Datasets.** We select 6 commonly used benchmark datasets from TU datasets [11] including 4 biochemical datasets (PROTEINS, NCI1, NCI109 and Mutagenicity) and 2 social network datasets (IMDB-BINARY and IMDB-MULTI) to evaluate the proposed method, which are summarized in Table 1. These datasets have been widely used for benchmarking of graph classification tasks.

**Baselines.** We first select the five most commonly used GNN-based methods as the baselines. These methods all show powerful performance in graph classification tasks, which are GCN [7], GAT [15], GraphSAGE [5], GIN [20] and ASAP [13]. In addition, since the proposed GAMT is designed for regularizing graph Transformers, we further benchmark GAMT against recent Transformer-based graph representation learning approaches, namely Transformer [14], GT [4], GraphiT [10], GMT [1]. The experimental results of the baselines in Table 1 are partly taken from the original papers, and the remaining experimental results of other datasets are calculated by us.

**Parameters Settings.** For all experiments, we refer to the previous work [22] by performing a 10-fold cross-validation to evaluate GAMT following the 80% training, 10% validation and 10% test dataset splits, and report the mean accuracy and standard deviation over 10 folds. The proposed GAMT is implemented with PyTorch, and we use the Adam optimizer [6] with $\lambda = 1$ to optimize the models. We select the best model on the validation set by performing

**Table 1.** Test accuracy (%) of different methods on graph classification. The reported results are mean and standard deviations over 10 folds. The last three rows are our proposed method with different position encodings. The baselines values with * means that the results are calculated by us. For each dataset, we highlight the best performing method.

| | **Method** | PROTEINS | NCI1 | NCI109 | Mutagenicity | IMDB-B | IMDB-M |
|---|---|---|---|---|---|---|---|
| | Graphs | 1,113 | 4,110 | 4,127 | 4,337 | 1,000 | 1,500 |
| | Avg.Nodes | 39.06 | 29.87 | 29.68 | 30.32 | 19.77 | 13.00 |
| | Avg.Edges | 72.82 | 32.30 | 32.13 | 30.77 | 96.53 | 65.94 |
| GNNs | GCN* | $75.4 \pm 2.8$ | $74.9 \pm 3.3$ | $74.0 \pm 2.9$ | $80.3 \pm 2.4$ | $73.4 \pm 2.9$ | $49.7 \pm 5.6$ |
| | GAT* | $74.9 \pm 3.4$ | $74.4 \pm 2.7$ | $73.3 \pm 2.6$ | $80.0 \pm 2.1$ | $73.0 \pm 3.8$ | $50.8 \pm 3.8$ |
| | GraphSAGE* | $75.0 \pm 4.1$ | $76.2 \pm 2.7$ | $73.7 \pm 4.8$ | $79.9 \pm 2.8$ | $71.8 \pm 4.0$ | $50.8 \pm 3.0$ |
| | GIN* | $74.4 \pm 3.7$ | $78.0 \pm 1.5$ | $76.3 \pm 2.3$ | $80.8 \pm 1.5$ | $72.7 \pm 4.9$ | $50.5 \pm 5.5$ |
| | ASAP* | $75.4 \pm 3.7$ | $74.7 \pm 3.6$ | $72.8 \pm 3.2$ | $79.1 \pm 2.8$ | $71.9 \pm 3.9$ | $50.1 \pm 3.5$ |
| Transformer | Transformer | $75.6 \pm 4.9$ | $70.0 \pm 4.5$ | $72.6 \pm 2.2$* | $74.3 \pm 1.6$* | $72.0 \pm 3.9$* | $47.8 \pm 3.3$* |
| | GT-LapPE | $74.6 \pm 2.7$ | $78.9 \pm 1.1$ | $74.9 \pm 2.3$* | $79.5 \pm 2.1$* | $71.2 \pm 4.3$* | $49.6 \pm 2.9$* |
| | GraphiT-Adj | $72.4 \pm 4.9$ | $79.7 \pm 2.0$ | $76.1 \pm 2.2$* | $80.4 \pm 1.7$* | $74.2 \pm 4.2$* | $49.8 \pm 3.1$* |
| | GraphiT-3RW | $76.2 \pm 4.4$ | $77.6 \pm 3.6$ | $76.3 \pm 2.2$* | $80.7 \pm 2.2$* | $73.4 \pm 4.7$* | $50.7 \pm 2.8$* |
| | GMT | $75.1 \pm 0.6$ | $76.5 \pm 2.0$* | $74.6 \pm 2.8$* | $80.8 \pm 1.6$* | $73.5 \pm 0.8$ | $50.7 \pm 0.8$ |
| Ours | GAMT | $76.0 \pm 2.9$ | $81.5 \pm 1.4$ | $80.2 \pm 2.0$ | $83.1 \pm 1.7$ | $73.9 \pm 3.6$ | $51.6 \pm 2.9$ |
| | GAMT-Adj | $76.4 \pm 5.3$ | $\mathbf{82.6 \pm 1.4}$ | $80.3 \pm 1.6$ | $83.5 \pm 1.5$ | $\mathbf{74.3 \pm 3.8}$ | $\mathbf{52.0 \pm 3.6}$ |
| | GAMT-3RW | $\mathbf{76.8 \pm 4.6}$ | $82.4 \pm 1.6$ | $\mathbf{80.6 \pm 2.0}$ | $\mathbf{84.0 \pm 2.2}$ | $74.1 \pm 2.7$ | $51.8 \pm 2.4$ |

a random search on the following hyperparameter settings: (1) initial learning rate $\in \{0.01, 0.001, 0.0001\}$; (2) batch size $\in \{64, 128, 256, 512\}$; (3) hidden layer dimensions $\in \{64, 128, 256\}$; (4) number of Transformer encoder layers $\in \{3, 6, 9, 12\}$; (5) number of attention heads $\in \{1, 2, 4, 8\}$. The hyperparameter $\alpha$ of the Beta distribution in Mixup is set to 1.0, and the regularization ratio parameter $\beta$ of the contrastive learning loss is set to 0.1.

**Model Configurations.** To further fully demonstrate the power of our proposed approach with a more fair comparison, we implement three GAMT variants, namely *GAMT*, *GAMT-Adj*, and *GAMT-3RW*. The main difference between the three GAMT variants is the use of different position encoding to eliminate the additional influence of position encoding on experimental results. The position encoding we use is the adjacency matrix and 3-step random walk kernel [10].

### 4.2 Results and Analysis for Graph Classification

The results from Table 1 show the average prediction accuracy and standard deviations of our method and baseline methods on six datasets, which demonstrates that the proposed GAMT outperforms all GNN-based and Transformer-based baselines across all datasets. For biochemical datasets, such as PROTEINS and Mutagenicity, our method performs more than 1.5% better than the best base-

**Table 2.** Test accuracy (%) of ablation studies. GC denotes GNN-based graph coarsening, and CL denotes contrastive learning.

| Ablation | Modules | | | Datasets | | | |
|---|---|---|---|---|---|---|---|
| | GC | Mixup | CL | PROTEINS | NCI1 | NCI109 | Mutagenicity |
| $A_1$ | ✗ | ✗ | ✗ | 74.8 ± 3.5 | 78.7 ± 1.7 | 77.2 ± 3.1 | 76.4 ± 3.1 |
| $A_2$ | ✓ | ✗ | ✗ | 74.9 ± 2.3 | 79.6 ± 2.1 | 77.5 ± 2.2 | 80.1 ± 1.8 |
| $A_3$ | ✓ | ✓ | ✗ | 76.1 ± 3.4 | 80.5 ± 2.2 | 78.9 ± 2.0 | 82.7 ± 1.8 |
| GAMT | ✓ | ✓ | ✓ | **76.4 ± 5.3** | **82.6 ± 1.4** | 80.3 ± 1.6 | **83.5 ± 1.5** |

line, which shows that graph attention Mixup is valuable for improving the performance of graph Transformer models in graph classification tasks. Especially in NCI1 and NCI109, our method has made a huge improvement, which further shows that our method is powerful. We can see that although the Transformer-based methods achieve strong performance in graph classification tasks, they are still inferior to GNNs in some small-scale datasets. This is the meaning of our work. From the perspective of regularization models, we combine the advantages of GNNs and Transformers, applying Mixup to achieve the best performance of the graph Transformer models.

### 4.3 Ablation Studies

**Ablation Experiments.** To verify that the modules we proposed are helpful for performance improvement, we conduct ablation studies on GAMT by removing graph coarsening, Mixup, and contrastive learning. The experimental results are shown in Table 2, which show that these three modules all play an important role in improving the performance. The GNN-based graph coarsening aggregates local neighborhood information, so Transformer can focus more on capturing long-range information. Graph Attention Mixup expands the training distribution by mixing both node attributes and topological relationships. Finally, contrastive learning can capture the differences between graphs as extra information to further regularize the model.

**GAMT with Different Mixup Variants.** In addition, we also compare different Mixup variants to show that our proposed Graph Attention Mixup outperforms existing Mixup methods. The experimental results are shown in Fig. 2, we compare the experimental results of Input Mixup [25], Manifold Mixup [19] and our Graph Attention Mixup respectively. The results show that Mixup is helpful for improving the model classification accuracy, but our Graph Attention Mixup has the most significant improvement. This is because Graph Attention Mixup not only leverages node attributes, but also considers the attention relationship with topological information between different nodes when mixing the pairwise graphs, which helps to generate mixed samples that are closer to the true distribution.

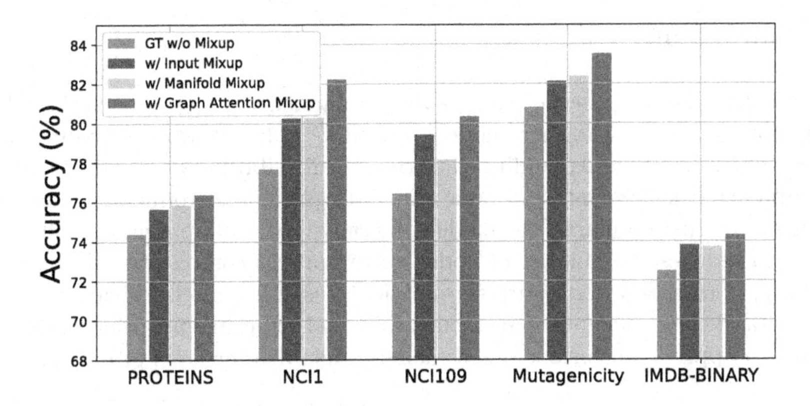


**Fig. 2.** Test Accuracy (%) on four datasets with different Mixup variants, where the red part represents our proposed Graph Attention Mixup. (Color figure online)

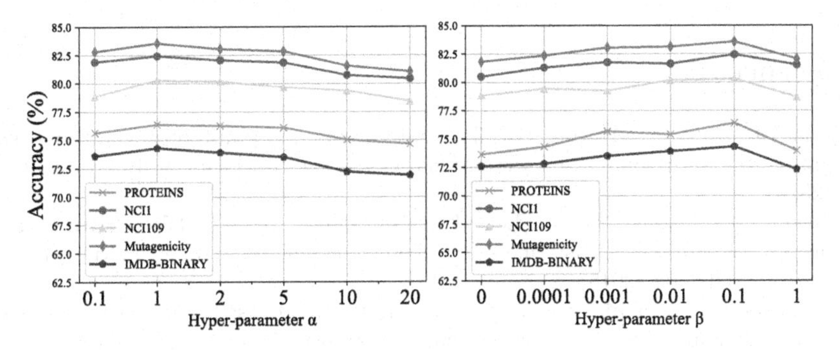


**Fig. 3.** Sensitivity analysis results of hyperparameters $\alpha$ and $\beta$ on different datasets.

### 4.4 Sensitivity Analysis

We conduct a series of experiments to investigate the sensitivity of parameters in GAMT. First, we analyze the hyperparameter $\alpha$ in Mixup, which determines the Mixup ratio $\lambda$. We select 6 values $\in \{0.5, 1, 2, 5, 10, 20\}$ for experiments. The experimental results are shown in Fig. 3, which shows that when $\alpha$ is increased, it will bring a worse effect. This is because a larger $\alpha$ will make $\lambda$ have a larger probability close to around 0.5, while in practical experiments the mixing term will perform better as a small perturbation.

We further analyze the regularization proportion $\beta$ of the contrastive loss, and the results are shown in Fig. 3. It can be seen from the results that when $\beta$ gradually increases from 0, the classification accuracy increases. This is because the contrastive learning provides additional information for better training of the model. However, when it increases to a certain extent, it will distract the attention of the main classification loss, so the classification accuracy will decrease when $\beta = 1$.

# 5 Conclusion

In this paper, we propose an end-to-end regularized training scheme based on Mixup for graph Transformer models, called Graph Attention Mixup Transformer. The purpose of expanding the training distribution is achieved by mixing graph attention weights and node embeddings while introducing contrastive learning to further regularize the model. Not only that, but applying graph coarsening also reduces the number of nodes and improves computational efficiency. GAMT performs very competitively on widely used graph classification benchmark datasets and occupies less memory in real experiments, which makes it possible to apply Transformer-based models when dealing with graphs at scale.

**Acknowledgement.** This work is supported by grant BE2019748 Natural Science Foundation of Jiangsu Province and grant 31800825, 31640028 National Natural Science Foundation of China.

# References

1. Baek, J., Kang, M., Hwang, S.J.: Accurate learning of graph representations with graph multiset pooling. In: International Conference on Learning Representations (2020)
2. Devlin, J., Chang, M.W., Lee, K., Toutanova, K.: BERT: pre-training of deep bidirectional transformers for language understanding. arXiv preprint arXiv:1810.04805 (2018)
3. Dosovitskiy, A., et al.: An image is worth 16x16 words: transformers for image recognition at scale. In: International Conference on Learning Representations (2020)
4. Dwivedi, V.P., Bresson, X.: A generalization of transformer networks to graphs. CoRR abs/2012.09699 (2020)
5. Hamilton, W.L., Ying, R., Leskovec, J.: Inductive representation learning on large graphs. In: Proceedings of the 31st International Conference on Neural Information Processing Systems, pp. 1025–1035 (2017)
6. Kingma, D.P., Ba, J.: Adam: a method for stochastic optimization. arXiv preprint arXiv:1412.6980 (2014)
7. Kipf, T.N., Welling, M.: Semi-supervised classification with graph convolutional networks. In: International Conference on Learning Representations (2017)
8. Kong, Y., et al.: Spatio-temporal graph convolutional network for diagnosis and treatment response prediction of major depressive disorder from functional connectivity. Hum. Brain Mapp. **42**(12), 3922–3933 (2021)
9. Kreuzer, D., Beaini, D., Hamilton, W.L., Létourneau, V., Tossou, P.: Rethinking graph transformers with spectral attention. In: Beygelzimer, A., Dauphin, Y., Liang, P., Vaughan, J.W. (eds.) Advances in Neural Information Processing Systems (2021). https://openreview.net/forum?id=huAdB-Tj4yG
10. Mialon, G., Chen, D., Selosse, M., Mairal, J.: Graphit: encoding graph structure in transformers (2021)
11. Morris, C., Kriege, N.M., Bause, F., Kersting, K., Mutzel, P., Neumann, M.: Tudataset: a collection of benchmark datasets for learning with graphs. arXiv preprint arXiv:2007.08663 (2020)

12. Oord, A.v.d., Li, Y., Vinyals, O.: Representation learning with contrastive predictive coding. arXiv preprint arXiv:1807.03748 (2018)
13. Ranjan, E., Sanyal, S., Talukdar, P.: Asap: adaptive structure aware pooling for learning hierarchical graph representations. In: Proceedings of the AAAI Conference on Artificial Intelligence, vol. 34, pp. 5470–5477 (2020)
14. Vaswani, A., et al.: Attention is all you need. In: Advances in Neural Information Processing Systems, pp. 5998–6008 (2017)
15. Veličković, P., Cucurull, G., Casanova, A., Romero, A., Liò, P., Bengio, Y.: Graph attention networks. In: International Conference on Learning Representations (2018)
16. Verma, V., et al.: Manifold mixup: better representations by interpolating hidden states. In: International Conference on Machine Learning, pp. 6438–6447. PMLR (2019)
17. Verma, V., Luong, T., Kawaguchi, K., Pham, H., Le, Q.: Towards domain-agnostic contrastive learning. In: International Conference on Machine Learning, pp. 10530–10541. PMLR (2021)
18. Verma, V., e tal.: Graphmix: improved training of GNNs for semi-supervised learning. In: Proceedings of the AAAI Conference on Artificial Intelligence, vol. 35, pp. 10024–10032 (2021)
19. Wang, Y., Wang, W., Liang, Y., Cai, Y., Hooi, B.: Mixup for node and graph classification. In: Proceedings of the Web Conference 2021, pp. 3663–3674 (2021)
20. Xu, K., Hu, W., Leskovec, J., Jegelka, S.: How powerful are graph neural networks? In: International Conference on Learning Representations (2018)
21. Ying, C., e tal.: Do transformers really perform badly for graph representation? In: Beygelzimer, A., Dauphin, Y., Liang, P., Vaughan, J.W. (eds.) Advances in Neural Information Processing Systems (2021). https://openreview.net/forum?id=OeWooOxFwDa
22. Ying, Z., You, J., Morris, C., Ren, X., Hamilton, W., Leskovec, J.: Hierarchical graph representation learning with differentiable pooling. In: Advances in Neural Information Processing Systems (2018). https://proceedings.neurips.cc/paper/2018/file/e77dbaf6759253c7c6d0efc5690369c7-Paper.pdf
23. Yoon, S., Kim, G., Park, K.: Ssmix: saliency-based span mixup for text classification. In: Findings of the Association for Computational Linguistics: ACL/IJCNLP 2021, Online Event, 1–6 August 2021, pp. 3225–3234. Association for Computational Linguistics (2021)
24. Yun, S., Han, D., Oh, S.J., Chun, S., Choe, J., Yoo, Y.: Cutmix: regularization strategy to train strong classifiers with localizable features. In: Proceedings of the IEEE/CVF International Conference on Computer Vision (ICCV), October 2019
25. Zhang, H., Cisse, M., Dauphin, Y.N., Lopez-Paz, D.: Mixup: beyond empirical risk minimization. In: International Conference on Learning Representations (2018)

# Frequency Spectrum with Multi-head Attention for Face Forgery Detection

Parva Singhal, Surbhi Raj(✉), Jimson Mathew, and Arijit Mondal

Indian Institute of Technology Patna, Bihta, India
{parva_2011cs10,surbhi_2021cs36,jimson,arijit}@iitp.ac.in

**Abstract.** Incredible realistic fake faces can be easily created using various Generative Adversarial Networks (GANs). For detecting GAN-generated fake images, various popular heavy-weight convolutional neural networks (CNNs) and vision transformers (ViTs) are developed with high accuracy. These methods have high computational costs and require large datasets for training. Their accuracy drops when tested on unseen datasets. To overcome these limitations, we have proposed a light-weight robust model with a significantly lesser number of parameters (approx 0.2 million). We combine the multi-head attention of the transformer with features extracted through frequency and laplacian spectrum of an image. It processes both global and local information of the image for forgery detection. It requires a very small dataset for training and shows better generalization results on unseen datasets. We have performed our experiments on smaller sets of DFFD dataset and tested on larger sets. The proposed model achieves an accuracy of 99%, having significantly fewer parameters in in-domain settings and thus is computationally less expensive. It is also tested on unseen datasets in cross GANs setting with an accuracy that is at par with the existing state-of-the-art, albeit heavy model ResNet-50 and other light-weight models such as MobileNetV3, SqueezeNet, and MobileViT.

**Keywords:** Fake Face Detection · Generative Adversarial Networks (GAN) · Fast Fourier Transform (FFT) · Laplacian of Gaussian (LoG) · Multi-head Attention · Vision Transformer (ViT)

## 1 Introduction

The creation, manipulation and alteration of digital faces can be done effortlessly using handy and easily accessible image-editing tools and commercial applications. These tools and applications are being used to create fake pornography and morphed videos or images which are circulated over social media to defame someone. This has led to increase in crimes over the internet and has impacted our society negatively. As a result, it is critical to develop a model which can detect fake face images irrespective of the manipulation techniques used.

Over the past few years, four types of face manipulation techniques have received significant research attention [28]. The first is entire face synthesis. In this technique, GANs like StyleGAN [13], StyleGANv2 [14] are used to generate


M. Tanveer et al. (Eds.): ICONIP 2022, CCIS 1793, pp. 200–211, 2023.
https://doi.org/10.1007/978-981-99-1645-0_17

new faces. The second technique is identity swap in which the face of one individual is replaced with the face of another individual. The third one is attribute manipulation which includes editing basic facial features like adding makeup, changing eye color, adding glasses, etc. GANs like StarGAN [1], StarGANv2 [2] are used to perform such manipulation. The last one is expression swap which transforms the facial expression of one individual with the facial expression of another individual.

In the literature, several experiments are performed by researchers using vision transformer [5], efficientnet [27] etc. for fake face detection and diverse techniques are used for fake face generation from different GANs. The available state-of-the-art models lack the generalization ability and their accuracy drops when tested on unknown datasets. These are big and complex models which claim very high accuracy. The high performances of these models come at the expense of extremely large number of weight parameters. However, for many real-world applications, we need a lightweight and fast model that uses diversified datasets to run on resource-constrained mobile devices [20].

The objective of the present paper is to develop a lightweight and memory efficient model with an accuracy which is similar to the state-of-the-art models but at a significantly lower parameter cost. We have developed a model with about 0.2 million parameters by combining the knowledge of frequency spectrum domain with multi-headed attention of transformer for forgery detection. Figure 1 presents the sample of images of the dataset that we have used in our experiments. The proposed model exhibits higher generalization ability and is independent of the specific GANs for the generation of fake face images.

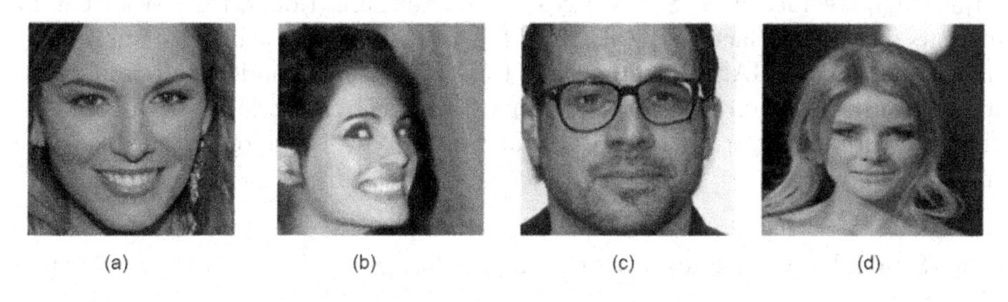


**Fig. 1.** DFFD dataset images- (a) PGGAN (Fake), (b) StarGAN (Fake), (c) StyleGAN (Fake) and (d) CelebA (Real).

This paper has three major contributions. First, the proposed model is a very lightweight, robust, and efficient model which can be used for general-purpose image classification problems on mobile devices. Second, the model has a good generalization ability as we have tested the model on unknown datasets on which the model is not trained. Third, the model requires very fewer data for training unlike ViT and other CNN models which require larger datasets for training. Thus, the proposed model works well even when data is scarce.

This paper is structured as follows. In Sect. 2, we present a comprehensive overview of several methods of fake face generation and detection. Section 3 describes details of the proposed methodology and its components. The performance evaluation of our framework as well as comparisons with other CNN models on the DFFD [4] dataset are presented in Sect. 4. Finally, we provide concluding observations in Sect. 5 and highlight the potential developments.

## 2 Related Work

In recent years, fake face detection has become an emerging and interesting area of research in the field of image processing. Researchers have contributed to the field of face forgery detection by proposing various models to identify the real or fake faces and curtail the spread of fake face images.

### 2.1 Datasets Generation Using GANs

Manipulated face images can be easily generated by using GANs. In the beginning, only low-resolution images could be generated using generative adversarial networks. The creation of high-resolution images triggered mode collapse, and later GANs gradually remedied the problem. PGGAN [12] was developed as a new training strategy that involves starting with low resolution and gradually improving it by layering the networks. It results in the creation of higher-resolution image. StyleGAN [13] is a PGGAN extension that suggests a new generator structure. It can boost the resolution of created photographs and adjust high-level features. StyleGAN's AdaIN normalisation, on the other hand, exhibits droplet artefacts. StyleGAN2 [14] corrects this flaw while also improving image quality. StarGAN [1] is a method that uses a single model to do image-to-image translations across various domains. Various GAN models are proposed in the literature for generating fake faces and has a wide scope of further research in this field.

### 2.2 Spatial-Based Face Forgery Detection

Nguyen et al. presented an in-depth study of techniques for creation and detection of deepfakes, its challenges and research trends [22]. In recent years, researchers utilized the texture description globally by exploiting the gram matrix [6] for fake face detection. Liu et al. focused on improving the robustness and generalization ability of the model by proposing a novel framework architecture which also facilitates further result interpretations and analysis [19]. The improvement in the generalization ability of the model was propounded by Xuan et al. [30]. They preprocessed the image before the deep networks for pixel level similarity between fake and real images [30]. The experiment was performed on fake datasets generated by DCGAN [24], WGAN-GP [7] and PGGAN [12]. The two-stage training process was utilized for pairwise learning using the contrastive loss to discriminate the real and fake features of images [10]. This two stream network was trained on different GANs(DCGAN, WGAN, PGGAN) generated images. Researchers also focused their work on improving the generalizability of the model. Wang et al. focused on this problem by applying pre and post processing and data augmentation with Resnet50 as classifiers and trained their model on PGGAN dataset and evaluated on dataset of images from other GAN models [29]. Face X-ray [18] focuses primarily on the blending stage, which is included in most face forgeries, and hence achieves state-of-the-art performance in raw video transferability. Its performance significantly declines when encountering low-resolution photos, and fails in generalizability. Thus, the performances of all these spatial domain approaches are highly dependent on the quality of images of datasets and its data distribution.

### 2.3 Frequency-Based Face Forgery Detection

Researchers also started exploring the analysis of the frequency spectrum for finding the artifacts of forged images. It is significantly applied in a wide variety of applications in computer vision. Stuchi et al. proposed slicing of images into small blocks by using distinct frequency filtering bands for extracting both global and local information for image classification tasks [26]. Also, Durall et al. were the first to employ the analysis of the frequency domain by averaging the amplitude of each frequency band with DFT, and then classifier was added to identify the artifacts in fake images [15]. This approach has an advantage that it does not require huge data for training but its accuracy drops for low-resolution images. Qian et al. showed the implication of DCT to extract the frequency domain features for identifying fake faces which performed well on low quality images and in-domains setting but failed in generalization [23]. Also, Jeong et al. focused on the generalizability by leveraging the artifacts in the frequency spectrum of GAN-generated images for fake detection [11].

### 2.4 Vision Transformer(ViT)

Recently, researchers detected the Deepfake videos with vision transformer [5] and achieved excellent accuracy. Cocomini et al. combined the variants of vision transformers with the convolutional networks, efficientnet B0 [27] as patch extractor which further feeds to ViT and utilized the fake video datasets for the experiments [3]. Miao et al. used a similar approach by encoding the inter-patch relationship by transformers for learning the local forgery features and improving the generalizability of the model [21]. Similarly, Heo et al. added the distillation mechanism with the vision transformer model for detecting deepfakes [9]. These models achieved high accuracy but requires extremely large-scale datasets for training.

## 3 Proposed Methodology

### 3.1 Architecture

Figure 2 elucidates the Frequency Spectrum Multi-Head Attention Network for fake face detection. The architecture is divided in two phases. First, we extract

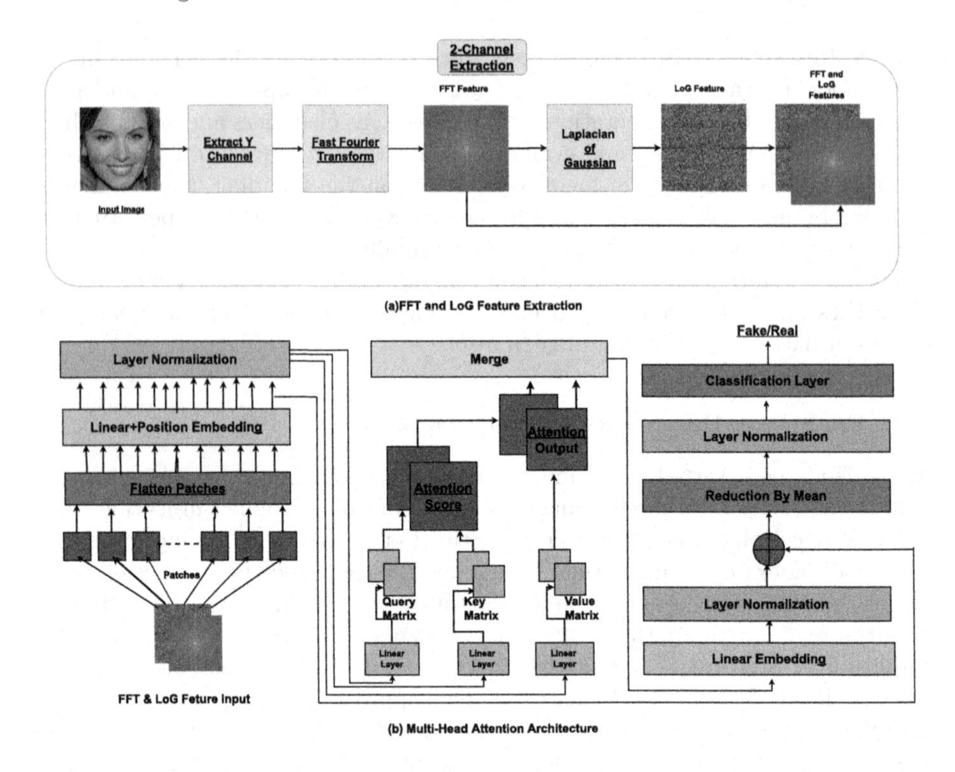


**Fig. 2.** The pipeline of the proposed architecture.

features in frequency domain from RGB image using Fast Fourier Transform (FFT) and change across values in frequency domain using Laplacian of Gaussian (LoG). Both these feature maps are concatenated and provided as an input to our deep-neural architecture. We try to explore global information from these feature maps and thereby used Multi-head Attention. We experimented with ViT but on exploration identified that using only a single Multi-head Attention Layer along with residual network and normalization, we can detect GAN-generated face images with high accuracy.

### 3.2 Two-Channel Extraction

Figure 3 shows the different characteristics of frequency features for authentic and GAN generated face images. We convert the RGB image in YCbCr channel format and for our convenience, consider the Y channel only. Fast Fourier Transform is applied to convert the spatial domain features to the frequency domain. The Y channel is subjected to Fourier transformation as in the following equation.

$$F(k,l) = \frac{1}{N^2} \sum_{i=0}^{N-1} \sum_{j=0}^{N-1} f(i,j) e^{-i2\pi(ki/N + lj/N)} \tag{1}$$

Here, $N \times N$ is the image size, $f(i,j)$ is the value in spatial domain and $F(k,l)$ is its corresponding value in frequency domain. To adjust the FFT output's frequency distribution, we shift the coordinates to relocate the origin at the center.

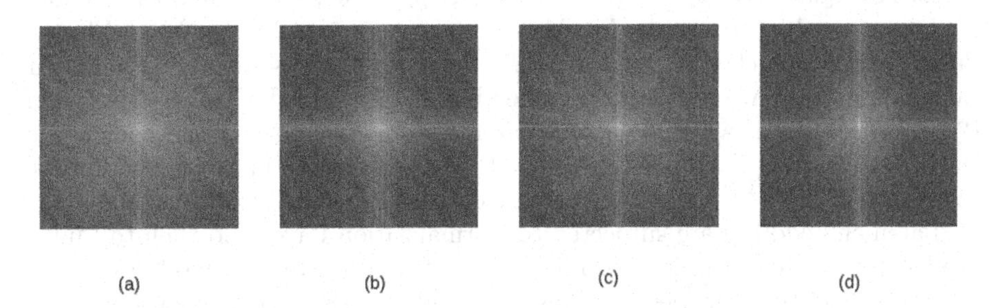


**Fig. 3.** FFT output for (a) PGGAN (Fake), (b) StarGAN (Fake), (c) StyleGAN (Fake) and (d) CelebA (Real).

The second channel feature is obtained using the Laplacian operation. Equation 2 shows the Laplacian operation.

$$L(x,y) = \frac{\mathrm{d}^2 I}{\mathrm{d}x^2} + \frac{\mathrm{d}^2 I}{\mathrm{d}y^2} \tag{2}$$

Here, $L(x,y)$ is the Laplacian of pixel value of image $I$ at $(x,y)$. It is a spatial second derivative operation which highlights the rapid intensity change. When used as a kernel for an RGB image, it detects edges, which means change in values across pixels. To get the information regarding how frequency spectrum changes, we apply this operation on the FFT channel feature which extracts changes in frequency measurement. We apply the noise removal technique and use the Gaussian function. This complete second channel feature extraction is referred to as LoG, i.e., Laplacian of Gaussian Operation.

Figure 2(a), shows the process for two-channel frequency features. Before the first phase, input image $X \in R^{C \times h \times w}$ is resized to $X_n \in R^{C \times h' \times w'}$, where C is the number of channels, $(h,w)$ is the original resolution and $(h',w')$ is the new resolution of image i.e. $224 \times 224$. Based on proposed model, the resized image is converted to YCbCr channel and Y channel, $X_y \in R^{1 \times h' \times w'}$ is extracted. $X_y$ is input to FFT and successively to LOG, which also result in the same dimension. Both features are concatenated where the first feature contains frequency domain values while the second contains values concerning its changes, resulting $X_f \in R^{2 \times h' \times w'}$. It is input to Multi-head Attention, discussed in the next sub-section. The dimension of the final output of first phase is $2 \times 224 \times 224$.

### 3.3 Multi-head Attention

To capture the forgery pattern from the features obtained in subsection-3.2, we apply Multi-head Attention. Figure 2(b), shows the proposed architecture. Before applying to the attention, the input feature $X_f \in R^{2 \times h' \times w'}$ is split into patches of equal sizes $X_f p_i \in R^{2 \times P \times P}$, where $p_i$ is the patch 'i' and the number of patches are $N_p = (h' \times w')/P^2$. These patches are passed to the block Flatten Patches, where patches are flattened. The flattened patches are passed to Linear Embedding layer, where position embedding is also added. Each patch size of P*P changes to E, where P is the patch size of 8, and E is the embedding feature size of 224. Finally, combining all the patches with linear embedding we get $X_f p \in R^{(N_p+1) \times E}$, where the dimension of this matrix is 785 × 224.

Patch embeddings are subjected to Normalization Layer and then to Multi-head Attention, consisting of two heads, i.e., two self-attention modules. Three learnable features, Queries Q, Keys K, and Values V of dimension $Q \in R^{2 \times (N_p+1) \times E/2}$, $K \in R^{2 \times (N_p+1) \times E/2}$, $V \in R^{2 \times (N_p+1) \times E/2}$ are used. Patch embeddings are input to three fully connected linear layer each for Query, Key, and Value. Each Query, Key and Value is of dimension 2 × 785 × 112. Over each epoch, these linear layers get updated. Using Query and Key matrix, we generate attention score which in-turn used with Value matrix to obtain attention output having dimension 2 × 785 × 112. These attention matrices are merged using merge layer. Equation 3 refers to the Attention matrix calculation.

$$A = softmax(\frac{QK^T}{\sqrt{E}})V \tag{3}$$

The obtained matrices are passed to Linear and Normalization Layer. We add the output generated from Normalization layer and the patch embeddings generated at beginning. This act as a residual network which leads to better performance in training. The output dimension is same as patch embeddings.

The residual connection output is input to the block Reduction by Mean. Here, the embeddings are averaged over each patch and generate $X_m \in R^{1 \times E}$ of dimension 1 × 224. The generated result is input to Normalization layer and then to Classification layer. The output of the classification layer contains LogSoftmax values for fake and real. The max of these values indicate whether the image is fake or real and update the model accordingly based on loss.

## 4 Experiment and Results

### 4.1 Settings

**Dataset.** To train the model and assess its performance, we use DFFD i.e., Diverse Fake Face Dataset [4]. The dataset is publicly available and contains a large number of GAN generated images. Among the various GAN generated images available, we focus on PgGAN v1, PgGAN v2, StyleGAN-FFHQ, StyleGAN-CelebA, StarGAN and FaceApp for fake face image dataset and CelebA for real image dataset. We use 12000 fake and 12000 real face images for training and evaluate the model on 15000 fake and 15000 real face images.

**Dataset Preprocessing.** The images are resized to 224 * 224 and transformed into frequency and Laplacian features as a pre-processing step. Frequency and Laplacian transformation of an image takes 0.051888 s. The obtained values are input to the Two-Channel Extraction phase.

**Hyper Parameter Used.** We set the kernel size of Gaussian operation to 3 * 3. The batch size for training the images is set to 16. We use SGD optimizer and set learning rate to 0.001. We train the model for 300 epochs and propagate LogSoftmax values backward with loss function as cross-entropy.

### 4.2 Results

We describe the results of the testing phase. Apart from combined testing, we perform the experiments in cross-GAN settings for generalization and show TSNE plots. The model's overall performance was measured using accuracy matrix. For each class performance, their recall value is observed. Table 1 shows instances of dataset used to train and test the model for generalization.

**Table 1.** Dataset instances for cross-GAN

| Data Type | Train | Test |
|---|---|---|
| PgGAN | 5000 | 2000 |
| StyleGAN | 5000 | 2000 |
| StarGAN | 5000 | 2000 |
| FaceApp | 0 | 2000 |
| CelebA | 5000 | 2000 |

**Model Accuracy, Cross-GAN Result and TSNE-Plots.** The trained accuracy obtained is 99% and the testing accuracy is 99%. The complexity of the model in terms of trainable parameters is around 0.2 million. Figure 4 shows the TSNE plots which our model produced based on its last layer features. The plots clearly show two clusters which differentiate fake and real images. Table 2 presents the recall values in Cross-GAN testing of the proposed model where it is trained on one type of GAN and tested on other types of GAN.

We visualize the proposed model by using Grad-CAM which highlights the region responsible for classification [25]. Figure 5 shows the different frequency distribution for real and fake face images. In fake face images, the distribution captured is denser while in the real face, it is sparse.

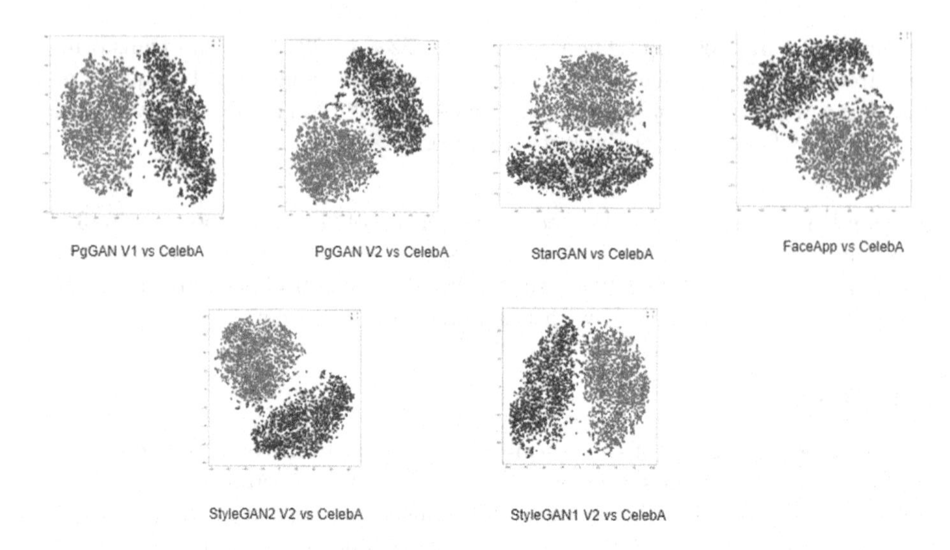


**Fig. 4.** TSNE plots between various GAN's and CelebA images

**Table 2.** Cross-GAN recall between PgGAN, StarGAN, StyleGAN & FaceApp

| Train⇓ Test⇒ | PgGAN | StyleGAN | StarGAN | FaceApp |
|---|---|---|---|---|
| PgGAN | | 0.97 | 0.98 | 0.98 |
| StyleGAN | 0.99 | | 0.99 | 0.99 |
| StarGAN | 0.96 | 0.96 | | 0.96 |

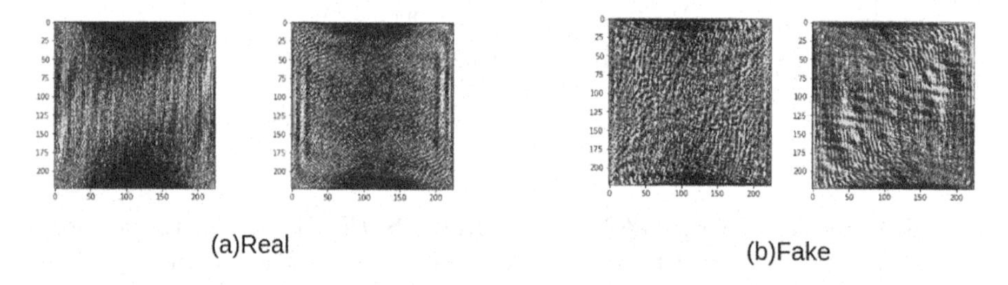


**Fig. 5.** GradCam output for (a) Real (b) Fake

Figure 6 shows the comparison of the proposed model with other CNN models like ResNet-50 [8], MobileViT [20], MobileNetV3 [16] and SqueezeNet [17] with respect to number of parameters and accuracy. We can observe that though the accuracy of the proposed model is similar to that of the other state-of-the-art models, the number of parameters is significantly less. Also, the training time is significantly less compared to other models. Thus, the proposed model is computationally effective and memory efficient.

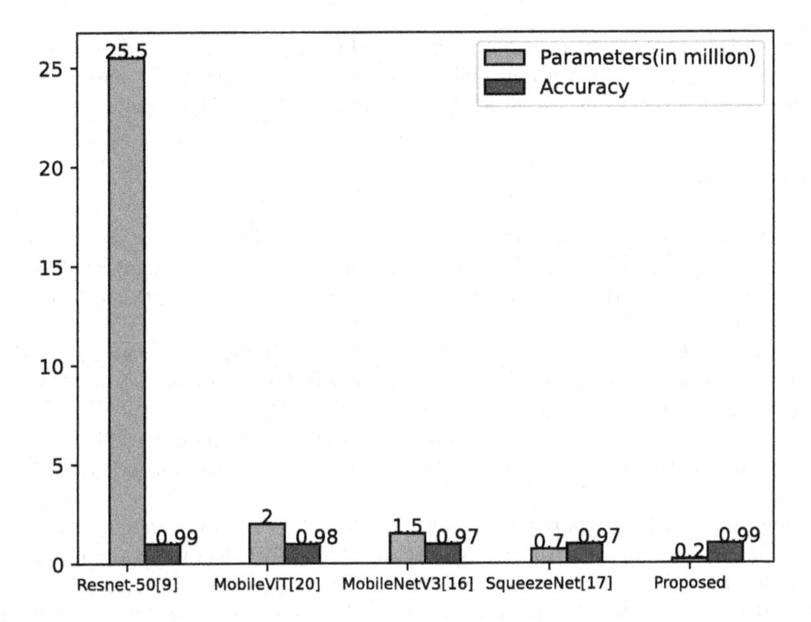


**Fig. 6.** Comparison of the proposed model with the SOTA models

### 4.3 Ablation Study

We also perform ablation experiments on the same dataset to assess the impact of frequency and Laplacian features with the multi-head attention network. When we use only Laplacian features, the accuracy of the model drops significantly. This suggests that the frequency channel is advantageous for our model. When we use only frequency features with the multi-head attention network, the accuracy is 96%. Further, we apply the Laplacian operation on these frequency features to capture the change in the frequency spectrum and then use it with the multi-head attention network. Its accuracy is improved to 99% and it is highly proficient in detecting fake face images.

## 5 Conclusion and Future Work

In this paper, we have proposed a novel lightweight model for the detection of fake face images. In our experiment, we have used frequency domain instead of standard RGB values which reduces the feature maps and shows significant performance improvement. It requires lesser data for training and achieves similar accuracy to state-of-the-art models with lower parameters. Though we have obtained good results with a low parameter model but the dataset used for training the model is limited to GAN-generated facial images only. In the future, it may be further optimized for low-quality images and other deepfake datasets. The proposed model also achieves high accuracy in cross-GAN settings. It can be used for many real-world applications of vision tasks and can be easily deployed on mobile devices for further use.

**Acknowledgement.** This research is supported through the project, Centre of Excellence in Cyber Crime Prevention against Women and Children- AI based Tools for Women and Children Safety, from Ministry of Home Affairs, Government of India.

## References

1. Choi, Y., Choi, M., Kim, M., Ha, J.W., Kim, S., Choo, J.: StarGAN: unified generative adversarial networks for multi-domain image-to-image translation. In: Proceedings of the IEEE Conference on Computer Vision and Pattern Recognition, pp. 8789–8797 (2018)
2. Choi, Y., Uh, Y., Yoo, J., Ha, J.W.: StarGAN v2: diverse image synthesis for multiple domains. In: Proceedings of the IEEE/CVF Conference on Computer Vision and Pattern Recognition, pp. 8188–8197 (2020)
3. Coccomini, D., Messina, N., Gennaro, C., Falchi, F.: Combining efficientnet and vision transformers for video deepfake detection. arXiv preprint arXiv:2107.02612 (2021)
4. Dang, H., Liu, F., Stehouwer, J., Liu, X., Jain, A.K.: On the detection of digital face manipulation. In: Proceedings of the IEEE/CVF Conference on Computer Vision and Pattern Recognition, pp. 5781–5790 (2020)
5. Dosovitskiy, A., et al.: An image is worth 16x16 words: transformers for image recognition at scale. arXiv preprint arXiv:2010.11929 (2020)
6. Gatys, L., Ecker, A.S., Bethge, M.: Texture synthesis using convolutional neural networks. In: Advanced Neural Information Processing System, vol. 28, pp. 262–270 (2015)
7. Gulrajani, I., Ahmed, F., Arjovsky, M., Dumoulin, V., Courville, A.: Improved training of Wasserstein GANs. arXiv preprint arXiv:1704.00028 (2017)
8. He, K., Zhang, X., Ren, S., Sun, J.: Deep residual learning for image recognition. In: Proceedings IEEE Conference on Computer Vision and Pattern Recognition, pp. 770–778 (2016)
9. Heo, Y.J., Choi, Y.J., Lee, Y.W., Kim, B.G.: Deepfake detection scheme based on vision transformer and distillation. arXiv preprint arXiv:2104.01353 (2021)
10. Hsu, C.C., Zhuang, Y.X., Lee, C.Y.: Deep fake image detection based on pairwise learning. Appl. Sci. **10**(1), 370 (2020)
11. Jeong, Y., Kim, D., Min, S., Joe, S., Gwon, Y., Choi, J.: BIHPF: bilateral high-pass filters for robust deepfake detection. In: Proceedings of the IEEE/CVF Winter Conference on Applications of Computer Vision, pp. 48–57 (2022)
12. Karras, T., Aila, T., Laine, S., Lehtinen, J.: Progressive growing of GANs for improved quality, stability, and variation. arXiv preprint arXiv:1710.10196 (2017)
13. Karras, T., Laine, S., Aila, T.: A style-based generator architecture for generative adversarial networks. In: Proceedings of the IEEE/CVF Conference on Computer Vision and Pattern Recognition, pp. 4401–4410 (2019)
14. Karras, T., Laine, S., Aittala, M., Hellsten, J., Lehtinen, J., Aila, T.: Analyzing and improving the image quality of styleGAN. In: Proceedings of the IEEE/CVF Conference on Computer Vision and Pattern Recognition, pp. 8110–8119 (2020)
15. Kaushik, R., Bajaj, R.K., Mathew, J.: On image forgery detection using two dimensional discrete cosine transform and statistical moments. Procedia Comput. Sci. **70**, 130–136 (2015)
16. Koonce, B.: Mobilenetv3. In: Koonce, B. (ed.) Convolutional Neural Networks with Swift for Tensorflow, pp. 125–144. Springer, Berkeley (2021). https://doi.org/10.1007/978-1-4842-6168-2_11

17. Koonce, B.: Squeezenet. In: Koonce, B. (ed.) Convolutional Neural Networks with Swift for Tensorflow, pp. 73–85. Springer, Berkeley (2021). https://doi.org/10.1007/978-1-4842-6168-2_7
18. Li, L., et al.: Face x-ray for more general face forgery detection. In: Proceedings of the IEEE/CVF Conference on Computer Vision and Pattern Recognition, pp. 5001–5010 (2020)
19. Liu, Z., Qi, X., Torr, P.H.: Global texture enhancement for fake face detection in the wild. In: Proceedings of the IEEE/CVF Conference on Computer Vision and Pattern Recognition, pp. 8060–8069 (2020)
20. Mehta, S., Rastegari, M.: Mobilevit: light-weight, general-purpose, and mobile-friendly vision transformer. arXiv preprint arXiv:2110.02178 (2021)
21. Miao, C., Chu, Q., Li, W., Gong, T., Zhuang, W., Yu, N.: Towards generalizable and robust face manipulation detection via bag-of-local-feature. arXiv preprint arXiv:2103.07915 (2021)
22. Nguyen, T.T., Nguyen, C.M., Nguyen, D.T., Nguyen, D.T., Nahavandi, S.: Deep learning for deepfakes creation and detection: a survey. arXiv preprint arXiv:1909.11573 (2019)
23. Qian, Y., Yin, G., Sheng, L., Chen, Z., Shao, J.: Thinking in frequency: face forgery detection by mining frequency-aware clues. In: Vedaldi, A., Bischof, H., Brox, T., Frahm, J.-M. (eds.) ECCV 2020. LNCS, vol. 12357, pp. 86–103. Springer, Cham (2020). https://doi.org/10.1007/978-3-030-58610-2_6
24. Radford, A., Metz, L., Chintala, S.: Unsupervised representation learning with deep convolutional generative adversarial networks. arXiv preprint arXiv:1511.06434 (2015)
25. Selvaraju, R.R., Cogswell, M., Das, A., Vedantam, R., Parikh, D., Batra, D.: Grad-cam: visual explanations from deep networks via gradient-based localization. In: Proceedings of the IEEE International Conference on Computer Vision, pp. 618–626 (2017)
26. Stuchi, J.A., et al.: Improving image classification with frequency domain layers for feature extraction. In: 2017 IEEE 27th International Workshop on Machine Learning for Signal Processing (MLSP), pp. 1–6. IEEE (2017)
27. Tan, M., Le, Q.: Efficientnet: rethinking model scaling for convolutional neural networks. In: International Conference on Machine Learning, pp. 6105–6114. PMLR (2019)
28. Tolosana, R., Vera-Rodriguez, R., Fierrez, J., Morales, A., Ortega-Garcia, J.: Deepfakes and beyond: a survey of face manipulation and fake detection. Inf. Fusion **64**, 131–148 (2020)
29. Wang, S.Y., Wang, O., Zhang, R., Owens, A., Efros, A.A.: CNN-generated images are surprisingly easy to spot... for now. In: Proceedings of the IEEE/CVF Conference on Computer Vision and Pattern Recognition (CVPR), June 2020
30. Xuan, X., Peng, B., Wang, W., Dong, J.: On the generalization of gan image forensics. In: Sun, Z., He, R., Feng, J., Shan, S., Guo, Z. (eds.) CCBR 2019. LNCS, vol. 11818, pp. 134–141. Springer, Cham (2019). https://doi.org/10.1007/978-3-030-31456-9_15

# Autoencoder-Based Attribute Noise Handling Method for Medical Data

Thomas Ranvier(✉), Haytham Elgazel, Emmanuel Coquery, and Khalid Benabdeslem

Univ Lyon, UCBL, CNRS, INSA Lyon, LIRIS, UMR 5205, 43 bd du 11 Novembre 1918, 69622 Villeurbanne, France
{thomas.ranvier,haytham.elghazel,khalid.benabdeslem}@univ-lyon1.fr, emmanuel.coquery@liris.cnrs.fr

**Abstract.** Medical datasets are particularly subject to attribute noise, that is, missing and erroneous values. Attribute noise is known to be largely detrimental to learning performances. To maximize future learning performances, it is primordial to deal with attribute noise before performing any inference. We propose a simple autoencoder-based preprocessing method that can correct mixed-type tabular data corrupted by attribute noise. No other method currently exists to entirely handle attribute noise in tabular data. We experimentally demonstrate that our method outperforms both state-of-the-art imputation methods and noise correction methods on several real-world medical datasets.

**Keywords:** Data Denoising · Data Imputation · Attribute Noise · Machine Learning · Deep Learning

## 1 Introduction

Medical studies are particularly subject to outliers, erroneous, meaningless, or missing values. In most real-life studies, not solely limited to the medical field, the problem of incomplete data and erroneous data is unavoidable. Those corruptions can occur at any data collection step. They can be a natural part of the data (patient noncompliance, irrelevant measurement, etc.) or appear from corruption during a later data manipulation phase [15]. Regardless of their origin, those corruptions are referred to as "noise" in the following work. Noise negatively impacts the interpretation of the data, be it for a manual data analysis or training an inference model on the data. The goal of a machine learning model is to learn inferences and generalizations from training data and use the acquired knowledge to perform predictions on unseen test data later on. Thus, the quality of training data on which a model is based is of critical importance, the less noisy the data is, the better results we can expect from the model.

Noise can be divided into two categories, namely class noise and attribute noise [17]. Class noise corresponds to noise in the labels, e.g. when data points are labeled with the wrong class, etc. Attribute noise, on the other hand, corresponds


M. Tanveer et al. (Eds.): ICONIP 2022, CCIS 1793, pp. 212–223, 2023.
https://doi.org/10.1007/978-981-99-1645-0_18

to erroneous and missing values in the attribute data, that is, the features of the instances. Attribute noise tends to occur more often than class noise in real-world data [13,15,17]. Despite this fact, compared to class noise, very limited attention has been given to attribute noise [17]. In real-world medical data, the probability of mislabeled data in a survival outcome context is quite low. We focused our work on attribute noise to maximize prediction performance while trying to compensate for a lack of appropriate methods within the literature.

The problem of imputing missing values has been vastly addressed in the literature, one can choose from many imputation methods to complete its data depending on its specific needs [9]. Imputation methods only address part of the attribute noise problem, they can handle missing values but cannot handle erroneous values, which can be highly detrimental to imputation results. Those methods have been widely researched, but methods able to deal with erroneous values have been less researched and can be considered incomplete at the moment [15].

Handling erroneous values can be done in three main ways: using robust learners that can learn directly from noisy data and naturally compensate or partially ignore the noise, filtering methods that remove data points that are classified as noisy, and polishing methods that aim to correct noisy instances. Robust learners are models that are less sensitive to noise in the data than classic models but they present several disadvantages [13]. They usually have limited learning potential compared to other learners. Using robust learners is not useful if we aim to perform anything else than the task the learner will solve. Filtering methods aim to detect which instances are noisy in order to delete them from the training set [13,17]. By training a learner on this cleaned set it can learn inferences without being disturbed by erroneous values and outliers, which eventually leads to better prediction performances on test data. The third way to deal with erroneous values is to use a polishing method [11], which corrects instances detected as noisy. Such a method can correct erroneous values on small datasets but lacks scalability for larger datasets containing more features [13]. Those three methods are able to deal with erroneous values and outliers, but are not able to deal with incomplete data, they only address part of the attribute noise problem.

At the moment the only way to handle attribute noise in its entirety is to use a combination of an imputation method followed by a noise correction method, to the best of our knowledge the literature lacks a method that would be able to perform both those tasks at once. Real-world data and especially medical data are subject to attribute noise in its entirety, it is important to conceive an approach able to handle the totality of attribute noise.

In this paper, we propose a preprocessing method based on autoencoders that deals with attribute noise in its entirety in real-world tabular medical data containing both quantitative and qualitative features (mixed-type data). Our method is able to learn from incomplete and noisy data to produce a corrected version of the dataset. It does not require any complete instance in the dataset and can truly handle attribute noise by performing both missing values completion and correction of erroneous values at the same time. We conduct extensive experiments on an imputation task on real-world medical data to compare our method to other state-of-the-art methods and obtain competitive and even significantly better results on classification tasks performed on the corrected data. We extend our experiments to show that our method can both complete missing data while correcting erroneous values, which further improves the obtained results.

The complete source code used to conduct the experiments is available at the following github repository[1].

The rest of the paper is organized as follows: we first present related work of data imputation and noise correction in both tabular and image data, especially in the medical field, in Sect. 2. Then, we present and explain our proposed approach in Sect. 3. Section 4 shows our experimental results compared to both data imputation and noise correction state-of-the-art methods. Finally, we conclude with a summary of our contributions.

## 2 Related Work

Denoising is vastly researched in the image field, in the image medical domain it is easy to find recent reviews and methods to correct noisy images [6]. Correction of tabular data on the other hand is less researched, only the imputation part seems to attract lots of attention. In this paper, we are especially focused on methods that can be applied to mixed-type tabular data.

Recently, lots of autoencoder-based imputation methods have been researched [8]. An autoencoder is a machine learning algorithm that takes an input $x \in \mathbb{R}^d$, with $d$ the number of features, and learns an intermediate representation of the data noted $z \in \mathbb{R}^h$, with $h$ the size of the newly constructed latent space. Then, from the intermediate representation $z$, the model reconstructs the original data $x$, we note the model output $\hat{x}$. During its training, the reconstruction error between $x$ and $\hat{x}$ is minimized.

One of those autoencoder-based imputation methods is MIDA: Multiple Imputation using Denoising Autoencoders, introduced in 2018 [4]. Unlike most autoencoder-based methods, which are usually applied to images, MIDA has been successfully applied to tabular data. This imputation method learns from a complete training dataset and can then be applied to unseen incomplete test data to impute the missing values. The authors assume that in order to learn how to impute missing values MIDA must learn from complete data. However, in this paper our experimental protocol does not provide a clean dataset to train on, therefore we show that MIDA obtains satisfactory results when properly parameterized, even when learning on incomplete data.

We want to show that autoencoders can not only be used to impute missing values, but also to correct erroneous values that are part of the observed values.

[1] https://github.com/ThomasRanvier/Autoencoder-based_Attribute_Noise_Handling_Method_for_Medical_Data.

It is easier to correct erroneous values in an image than it is in tabular data. In images, pixels from a close neighborhood are related to each other, which might not be true for arbitrarily ordered features in tabular data. As stated earlier correction of images is a very active research domain. Recently, Ulyanov et al. introduced a new approach called "Deep Image Prior [12]." This innovative approach uses autoencoders to restore images, but does not use the original data $X$ as model input, instead, the autoencoder is given pure noise as input and is trained to reconstruct the original corrupted data $X$ from the noise. In that way, the model is no longer considered an autoencoder but a generative model, however, in practice the model keeps the same architecture. Therefore, the only information required to correct the input image is already contained in the image itself. By stopping the training before complete convergence it is possible to obtain a cleaner image than the original corrupted image. Ulyanov et al. showed that their approach outperforms other state-of-the-art methods on a large span of different tasks.

In this paper, we aim to conceive a method that would be able to correct mixed-type tabular data, we aim to use the lessons from [4] and [12] to conceive a method able to handle attribute noise as a whole in a preprocessing step.

## 3 A Method to Truly Handle Attribute Noise

Our method is based on a deep neural architecture that is trained to reconstruct the original data from a random noise input. We note the original data with its attribute noise $X \in \mathbb{R}^{n \times d}$, with $n$ the number of instances in the dataset and $d$ the number of features. We note the deep generative model $\hat{X} = f_{\Theta}(\cdot)$, with $\Theta$ the model parameters that are learned during training and $\hat{X}$ the model output, in our case the model output is a reconstruction of $X$. The input of the model is noted $Z \in \mathbb{R}^{n \times d}$ and has the same dimension as $X$, which keeps our model a kind of autoencoder. The model is trained to reconstruct $X$ using the following loss term: $L(X, \hat{X}) = ||(\hat{X} - X) \odot M||^2$, where $\odot$ is the Hadamard product (element-wise product) and $M \in \mathbb{R}^{n \times d}$ is a binary mask that stores the locations of missing values in $X$, $M_{ij} = 1$ if $X_{ij}$ is observed and $M_{ij} = 0$ if $X_{ij}$ is missing. Applying the mask $M$ to the loss ensures that the loss is only computed on observed values, in this way the reconstruction $\hat{X}$ will fit the observed values in $X$, while missing values will naturally converge to values that are statistically consistent given the learned data distribution. Figure 1 shows how the model is fitted to the original data $X$ with the application of the binary mask $M$ during training.

To determine the right step at which to stop training, we define a stopping condition based on the evolution of a given metric. In a supervised setting, for example, we regularly compute the AUC on a prediction task performed on the reconstructed data $\hat{X}$, which gives us the evolution of the quality of the reconstruction. We stop the training when the AUC degrades for a set number of iterations, then we obtain a reconstruction with consistent imputations and noise correction, which provides better data quality than the original data. If no

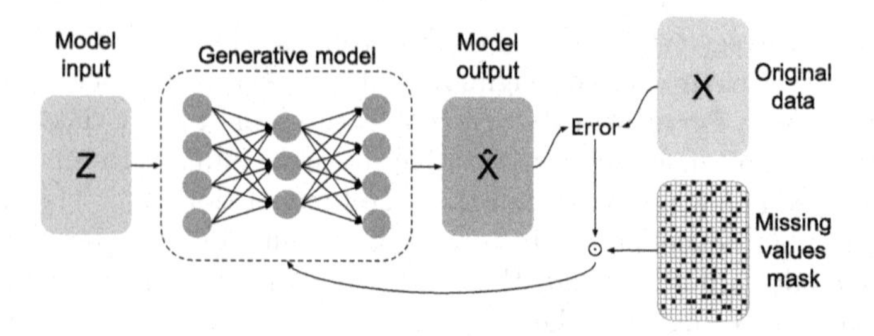


**Fig. 1.** The model parameters are trained so that the model learns to reconstruct the original data. Stopping the training phase after enough iterations without overfitting outputs a reconstruction $\hat{X}$ cleaner than the original data $X$.

supervision is possible, the only metric that can be used to determine when to stop training is the loss value, which gives correct results but is quite limited since it is harder to determine when overfitting starts.

In practice, we set the input of the model either as random noise or as the original data depending on the empirically obtained results. We note that on datasets containing large amounts of features, using 1D convolutions instead of classic fully-connected layers tends to give better results, it helps our method to scale on datasets with large amounts of data and features.

In a supervised context our method is able to brilliantly handle attribute noise as a whole thanks to the use of an early-stop conditioned on a supervised metric chosen for its ability to represent the quality of imputations at each step. Stopping training after enough learning, but right before overfitting, allows our method to reach a point where the reconstructed data $\hat{X}$ contains less noise than the original data $X$ while containing all important information from $X$.

## 4 Experimental Results

### 4.1 Used Datasets

We ran our experiments on three real-life medical mixed-type tabular datasets naturally containing missing values. We evaluated our method and compared our results to other state-of-the-art methods on those medical datasets.

- NHANES, US National Health and Nutrition Examination Surveys: Those are surveys conducted periodically by the US NHCS to assess the health and nutritional status of the US population [1]. We used data from studies spanning from years 2000 to 2008, with 95 features and about 33% missing values. We selected the "diabetes" feature as a class and randomly selected 1000 samples from both outcomes to evaluate the quality of the data correction on a classification task on this class.
- COVID19: This dataset was publicly released with the paper [14], it contains medical information collected in early 2020 on pregnant and breastfeeding women. We based our data preprocessing on the one realized in the original

paper, we selected only the measurements from the last medical appointment for each patient. After preprocessing, we obtain a dataset composed of 361 patients with 76 features, with about 20% missing data. We evaluate the quality of the data correction on a classification task on the survival outcome, 195 patients have survived and 166 are deceased.
- Myocardial infarction complications: This medical dataset is available on the UCI machine learning repository, it was publicly released with the paper [3]. It is composed of 1700 patients with 107 features, with about 5% missing values. We evaluate the quality of the data correction on a classification task on the survival outcome, 1429 patients have survived and 271 are deceased.

### 4.2 Used Metrics

We evaluate the quality of the obtained correction on classification tasks. As can be seen from the previous section, the medical datasets we used are not all balanced, the Myocardial dataset is especially imbalanced. In such a context it is important to choose metrics that are not sensitive to imbalance.

In a medical context where we aim to predict the outcome between sane and sick, it is extremely important not to classify sick patients as sane since it would be very detrimental for them not to get an appropriate medical response. In machine learning terms we are in cases where false positives on the negative class would be less detrimental than false negatives, therefore we should aim to minimize false negatives.

Appropriate metrics, in this case, are the AUC: Area Under the Receiver Operating Characteristic (ROC) Curve, and the balanced accuracy. The AUC corresponds to the area under the ROC curve obtained by plotting the true positive rate (recall) against the false positive rate (1-specificity). An AUC score of 1 would mean that the classifier gives true positives 100% of the time, whereas a value of 0.5 means that the classifier is no better than a random prediction. The balanced accuracy is defined as the average of recall obtained on each class, which is simply the average of the true positive rate between all the classes.

### 4.3 Experimental Protocol

Our experiments aim to compare our method to other state-of-the-art methods for both imputation and noise correction tasks. All our experiments are evaluated using the balanced accuracy and the AUC metrics. We repeated each experiment 10 times with 10 different stochastic seeds to set up the random state of non-deterministic methods. For each experiment we compare the performances of each method to ours using t-tests. We use the results from those statistical tests to determine if our method is significantly better, even, or significantly worse than each other method, based on a $p$-value set at 0.05.

We first evaluate our method on an imputation task on the three medical datasets previously described. Each of those datasets is missing part of its data, we compare the quality of the data imputation by training a decision tree on a classification task on each dataset after imputation.

The capacity of our method to impute missing values on incomplete and noisy data is assessed by introducing artificial noise in the datasets. Noise is artificially added to the data by randomly replacing a certain rate of attribute values with a random number drawn from a uniform distribution, as described in Zhu et al. [17]. We compare the results at the following noise rates: 0/5/10/15/20/40/60%.

Finally, to assess the effectiveness of our method to both complete missing values while correcting erroneous values, we introduce artificial noise in the naturally incomplete datasets. We apply our method and compare its results to those obtained by a sequential execution (*i.e.* pipeline) of an imputation method followed by a noise correction method.

The results from our method are compared to those of other state-of-the-art imputation methods:

- MEAN, MEDIAN and KNN: We used the "SimpleImputer" and "KNNImputer" classes from the python library "scikit-learn"[2].
- MICE: Multivariate Imputation by Chained Equations has been introduced in 2011 in [2]. This is a very popular method of imputation because it provides fast, robust, and good results in most cases. We used the implementation from the experimental "IterativeImputer" class from "scikit-learn".
- GAIN: Generative Adversarial Imputation Nets, introduced recently in [16], two models are trained in an adversarial manner to achieve good imputation. We used the implementation from the original authors[3].
- SINKHORN: An optimal transport based method for data imputation introduced in [7] We used the implementation from the original authors[4].
- SOFTIMPUTE: The SOFTIMPUTE algorithm has been proposed in 2010 [5], it iteratively imputes missing values using an SVD. We used the public re-implementation by Travis Brady of the Mazumder and Hastie's package[5].
- MISSFOREST: An iterative imputation method based on random forests introduced in 2012 in [10]. We used the "MissForest" class from the python library "missingpy"[6].
- MIDA: Multiple Imputation Using Denoising Autoencoders has been recently proposed in [4]. We implemented MIDA using the author description from the original paper and the code template supplied in this public gist[7]

The following noise correction methods are used as comparison:

- SFIL: Standard Filtering, which we implemented such as described in [11].
- SPOL: Standard Polishing, which we also implemented such as described in [11].
- PFIL, PPOL: Improved versions of SFIL and SPOL where the noisy instances to filter or polish are identified using the Panda noise detection method [13].

[2] https://scikit-learn.org.
[3] https://github.com/jsyoon0823/GAIN.
[4] https://github.com/BorisMuzellec/MissingDataOT.
[5] https://github.com/travisbrady/py-soft-impute.
[6] https://pypi.org/project/missingpy/.
[7] https://gist.github.com/lgondara/18387c5f4d745673e9ca8e23f3d7ebd3.

### 4.4 Results

In this subsection, we present and analyze the most important comparative results between our proposed method and state-of-the-art methods. The entire experimental results with statistical significance can be found in our code.

**Imputation on Incomplete Medical Data.** With our first experiment, we show that our method can impute missing values in real-world medical datasets.

Table 1 shows the results on the three real-world medical datasets. We can see that our method obtains very competitive results on all datasets. We obtain significantly better results than other state-of-the-art methods in most cases for both metrics. The only cases in which our method performs significantly worse are against KNN and MICE on COVID data on the balanced accuracy metric. This shows that our method is able to impute missing values on incomplete real-world medical mixed-type tabular data with results as good as other state-of-the-art imputation methods and even better in most cases.

**Imputation on Incomplete and Noisy Medical Data.** Our second experiment shows that our method can impute missing values in real-world medical datasets in a noisy context. We artificially add noise to the data at various rates: 0/5/10/15/20/40/60%, and evaluate each imputation method at each noise level.

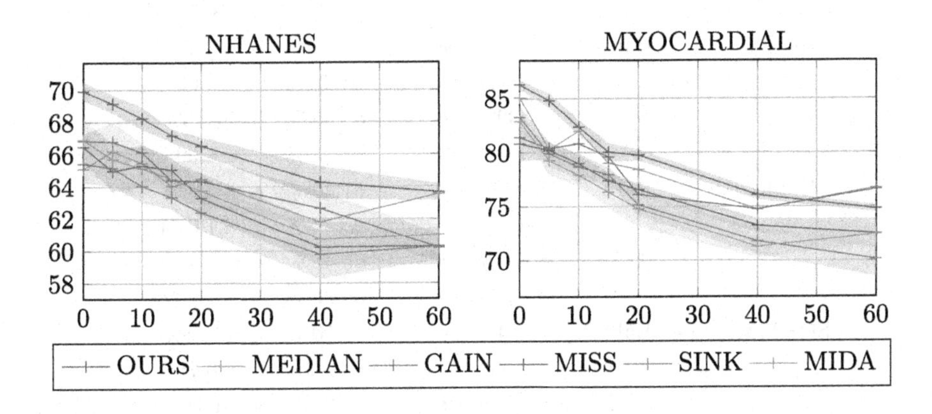


**Fig. 2.** AUC results on imputation on incomplete and noisy medical data

Figure 2 shows AUC results obtained on NHANES and MYOCARDIAL data at each noise rate against several imputation methods. In both cases, we note that our method globally obtains significantly better results than other methods. The performance of all methods drops when the noise level increases, which is expected. On NHANES data, our method performs largely better than others until a noise rate of 60%, where the MEDIAN imputation gets similar results to ours. This can probably be explained by the fact that, with a noise level that high,

**Table 1.** Comparative study on an imputation task. BalACC corresponds to the balanced accuracy, AUC is the area under the ROC curve. Our method is compared to each other using t-tests with a $p$-value of 0.05, when our method is significantly better it is indicated by $\bullet$, even by $\equiv$, and significantly worse by $\circ$.

| Model | Metric | MYOCARDIAL | | NHANES | | COVID | |
|---|---|---|---|---|---|---|---|
| OURS | BalACC | **77.91%**<br>**±1.12%** | | **64.17%**<br>**±0.36%** | | 86.84%<br>±1.23% | |
| | AUC | **86.28%**<br>**±0.42%** | | **69.92%**<br>**±0.56%** | | **92.95%**<br>**±0.93%** | |
| MEAN | BalACC | 77.30%<br>±0.00% | ≡ | 60.35%<br>±0.00% | • | 85.91%<br>±0.00% | • |
| | AUC | 85.09%<br>±0.00% | • | 66.10%<br>±0.00% | • | 91.20%<br>±0.00% | • |
| KNN | BalACC | 68.83%<br>±0.00% | • | 63.00%<br>±0.00% | • | **88.08%**<br>**±0.00%** | ∘ |
| | AUC | 78.94%<br>±0.00% | • | 67.78%<br>±0.00% | • | 91.53%<br>±0.00% | • |
| GAIN | BalACC | 63.89%<br>±2.21% | • | 61.36%<br>±0.53% | • | 85.14%<br>±0.91% | • |
| | AUC | 74.22%<br>±1.11% | • | 66.85%<br>±0.40% | • | 91.36%<br>±0.73% | • |
| MICE | BalACC | 76.55%<br>±0.00% | • | 61.70%<br>±0.00% | • | 87.98%<br>±0.00% | ∘ |
| | AUC | 81.39%<br>±0.00% | • | 67.30%<br>±0.00% | • | 92.43%<br>±0.00% | ≡ |
| MISSFOREST | BalACC | 73.00%<br>±0.87% | • | 61.40%<br>±1.03% | • | 85.15%<br>±1.67% | • |
| | AUC | 80.82%<br>±1.60% | • | 66.48%<br>±0.90% | • | 91.30%<br>±1.20% | • |
| SOFTIMPUTE | BalACC | 77.24%<br>±0.99% | ≡ | 61.70%<br>±0.93% | • | 84.48%<br>±0.78% | • |
| | AUC | 84.88%<br>±0.77% | • | 66.93%<br>±1.08% | • | 91.12%<br>±0.85% | • |
| SINKHORN | BalACC | 75.66%<br>±1.22% | • | 60.77%<br>±0.98% | • | 86.82%<br>±1.49% | ≡ |
| | AUC | 83.26%<br>±1.01% | • | 65.42%<br>±1.18% | • | 91.48%<br>±1.13% | • |
| MIDA | BalACC | 75.09%<br>±0.70% | • | 62.15%<br>±1.26% | • | 85.55%<br>±1.12% | • |
| | AUC | 82.87%<br>±0.78% | • | 66.91%<br>±1.30% | • | 91.67%<br>±0.62% | • |

it is nearly impossible to impute coherent values other than the median or mean value for each feature. We can observe the same pattern on MYOCARDIAL data, with the difference that GAIN seems to have learned how to adapt to such an amount of noise in this case. Those results show that on low to high noise rates, our method can impute missing values while correcting erroneous values. It provides better data correction than most other methods. At extreme noise rates naive methods might provide better results.

**Comparison with the Combination of Imputation and Noise Correction Methods.** The last experiment compares our method results to those obtained from the combination of an imputation method followed by a noise correction method. We chose MICE as the state-of-the-art imputation method since it obtains competitive results against ours in a not noisy context. We then apply the four noise correction methods SFIL, PFIL, SPOL, and PPOL.

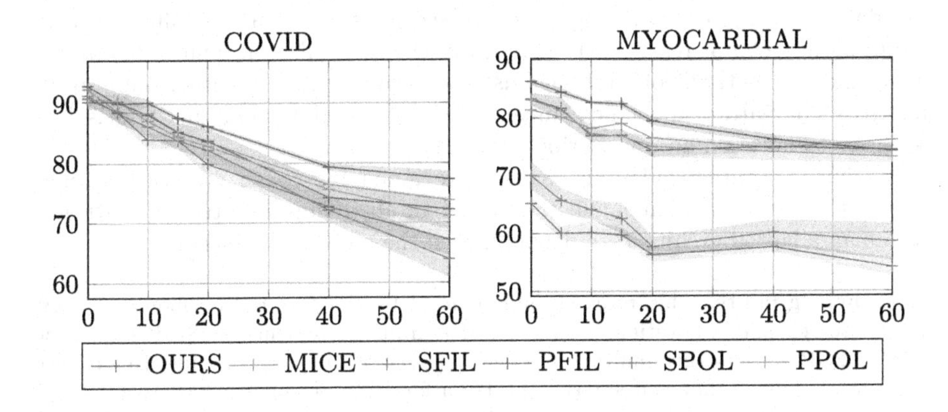


**Fig. 3.** AUC results on combination of imputation and noise correction

Figure 3 show AUC results obtained on COVID and MYOCARDIAL data at each noise rate. We note that SFIL and SPOL perform worse than the Panda alternative of both those methods at all noise rates. We also note that for both datasets the other state-of-the-art noise correction methods give very poor results as soon as the noise level reaches more than 5%, at higher noise rates the data quality is better before noise correction than after. For COVID data, all methods yield similar results at low noise levels, with our method on top with a very slight advantage. At high rates, however, our method gives significantly better results than all other methods. For MYOCARDIAL data, the opposite pattern can be observed, our method gives significantly better results up until a noise rate of 40%, after which MICE imputation is slightly better. This experiment completes the conclusions drawn from the second experiment, our method provides very good data correction, up until the noise rate becomes too extreme, at that point, simpler methods achieve slightly better results. The fact that the opposite is observed on COVID data is probably due to a remarkable original data quality, which would explain why our method becomes significantly better only at higher noise levels.

## 5 Conclusion

Handling attribute noise means imputing missing values while correcting erroneous values and outliers. This phenomenon is of critical importance in medical data, where attribute noise is especially present and detrimental to analysis and learning tasks on the data. No method in the literature is capable of handling attribute noise in its entirety in mixed-type tabular data. Many methods exist to impute missing values while some other methods can correct erroneous values, but none are able to do both.

In this paper, we propose an autoencoder-based preprocessing approach to truly handle attribute noise. Our method produces a corrected version of the original dataset by imputing missing values while correcting erroneous values without requiring any complete or clean instance in the original data. Our experiments show that our method competes against and even outperforms other imputation methods on real-world medical mixed-type tabular data. Our method is less sensitive to noise on an imputation task.

Finally, as autoencoder approaches are amenable to an empirical tuning phase, we plan to implement in the future an algorithm able to automatically define an adapted architecture depending on the dataset dimensions.

**Acknowledgments.** This research is supported by the European Union's Horizon 2020 research and innovation program under grant agreement No 875171, project QUALITOP (Monitoring multidimensional aspects of QUAlity of Life after cancer ImmunoTherapy - an Open smart digital Platform for personalized prevention and patient management).

## References

1. Barnard, J., Meng, X.-L.: Applications of multiple imputation in medical studies: from AIDS to NHANES. Stat. Meth. Med. Res. **8**(1). ISSN 0962–2802. https://doi.org/10.1177/096228029900800103
2. van Buuren, S., Groothuis-Oudshoorn, K.: mice: multivariate Imputation by Chained Equations in R. Journal of Statistical Software **45**(3), 1–67 (2011). ISSN 1548–7660. https://doi.org/10.18637/jss.v045.i03
3. Golovenkin, S.E., et al.: Trajectories, bifurcations, and pseudo-time in large clinical datasets: applications to myocardial infarction and diabetes data. GigaScience **9**(11), giaa128, November 2020. ISSN 2047–217X. https://doi.org/10.1093/gigascience/giaa128
4. Gondara, L., Wang, K.: MIDA: multiple imputation using denoising autoencoders. In: Phung, D., Tseng, V.S., Webb, G.I., Ho, B., Ganji, M., Rashidi, L. (eds.) PAKDD 2018. LNCS (LNAI), vol. 10939, pp. 260–272. Springer, Cham (2018). https://doi.org/10.1007/978-3-319-93040-4_21

5. Mazumder, R., Hastie, T., Tibshirani, R.: Spectral regularization algorithms for learning large incomplete matrices. J. Mach. Learn. Res. JMLR **11**, 2287–2322 (2010)
6. Sagheer, S.V.M., George, S.N.: A review on medical image denoising algorithms. Biomed. Sig. Process. Control **61** (2020). ISSN 1746–8094. https://doi.org/10.1016/j.bspc.2020.102036
7. Muzellec, B., Josse, J., Boyer, C., Cuturi, M.: Missing data imputation using optimal transport. In: Proceedings of the 37th International Conference on Machine Learning, pp. 7130–7140. PMLR, November 2020. ISSN: 2640–3498 (2020)
8. Pereira, R.C., Santos, M., Rodrigues, P., Abreu, P.H.: Reviewing autoencoders for missing data imputation: technical trends, applications and outcomes. J. Artif. Intell. Res. **69**, December 2020. https://doi.org/10.1613/jair.1.12312
9. Stef, V.B.: Flexible Imputation of Missing Data, 2nd edn.. Chapman & Hall (2018)
10. Stekhoven, D.J., Bühlmann, P.: MissForest-non-parametric missing value imputation for mixed-type data. Bioinformatics **28**(1) (2012). ISSN 1367–4803. https://doi.org/10.1093/bioinformatics/btr597
11. Teng, C.M.: Polishing Blemishes: issues in data correction. IEEE Intell. Syst. **19**(2) (2004). ISSN 1941–1294. https://doi.org/10.1109/MIS.2004.1274909. Conference Name: IEEE Intelligent Systems
12. Ulyanov, D., Vedaldi, A., Lempitsky, V.: Deep image prior. Int. J. Comput. Vis. **128**(7), 1867–1888 (2020). https://doi.org/10.1007/s11263-020-01303-4
13. Van Hulse, J.D., Khoshgoftaar, T.M., Huang, H.: The pairwise attribute noise detection algorithm. Knowl. Inf. Syst. **11**(2), 171–190 (2007). ISSN 0219–1377, 0219–3116. https://doi.org/10.1007/s10115-006-0022-x
14. Yan, l., et al.: An interpretable mortality prediction model for COVID-19 patients. Nat. Mach. Intell. **2**(5), 283–288 (2020). ISSN 2522–5839. https://doi.org/10.1038/s42256-020-0180-7
15. Yang, Y., Wu, X., Zhu, X.: Dealing with predictive-but-unpredictable attributes in noisy data sources. In: Boulicaut, J.-F., Esposito, F., Giannotti, F., Pedreschi, D. (eds.) PKDD 2004. LNCS (LNAI), vol. 3202, pp. 471–483. Springer, Heidelberg (2004). https://doi.org/10.1007/978-3-540-30116-5_43
16. Yoon, J., Jordon, J., Schaar, M.: GAIN: missing data imputation using generative adversarial nets. In: Proceedings of the 35th International Conference on Machine Learning, pp. 5689–5698. PMLR, July 2018. ISSN: 2640–3498
17. Zhu, X., Wu, X.: Class noise vs. attribute noise: a quantitative study. Artif. Intell. Rev. **22**(3), 177–210 (2004). ISSN 1573–7462. https://doi.org/10.1007/s10462-004-0751-8

# A Machine-Reading-Comprehension Method for Named Entity Recognition in Legal Documents

Xinrui Zhang and Xudong Luo(✉)

Guangxi Key Lab of Multi-Source Information Mining and Security, School of Computer Science and Engineering, Guangxi Normal University, Guilin 541004, China
luoxd@mailbox.gxnu.edu.cn

**Abstract.** Named Entity Recognition (NER) is essential for helping people quickly grasp legal documents. To recognise nested and non-nested entities in legal documents, in this paper, we propose a Machine-Reading-Comprehension (MRC) method, which is integrated with biaffine attention and graph-based dependency parsing. Specifically, we regard an NER task as a $\langle Query, Context, Answer\rangle$ triple and construct a query statement for each entity according to the annotation guideline notes. Then, we use the BERT pre-trained language model to encode *Query* and *Context*, fused with the rotary position embedding after feature mapping. Next, we use the biaffine attention to score each subsequence of a text. Finally, we use a balanced softmax to decide whether or not a subsequence is an entity. We do many experiments to show that our model has achieved good results in recognising nested and non-nested entities. We also do some experiments to demonstrate the effectiveness of some components in the entire model we propose.

**Keywords:** Machine reading comprehension · Named entity recognition · Biaffine attention · Legal document · Artificial intelligence and law

## 1 Introduction

The number of legal documents is enormous and information-intensive, and various countries are vigorously promoting the construction of intelligent justice [3,13,22]. However, the studies related to various tasks of natural language processing in the legal field are still in their early stage. Among these tasks, named entity recognition (NER) is one of the most critical ones. In fact, NER is the basis of downstream tasks such as information extraction, judicial summarisation, and judicial Question and Answer (Q&A). The task of NER in the legal domain is locating and classifying nouns and phrases characteristic of the legal domain in unstructured legal texts into pre-specified entity categories [11]. The higher the accuracy rate of NER is, the more effective and reliable an intelligent legal system with NER is. However, besides non-nested (flat) named entities, a


M. Tanveer et al. (Eds.): ICONIP 2022, CCIS 1793, pp. 224–236, 2023.
https://doi.org/10.1007/978-981-99-1645-0_19

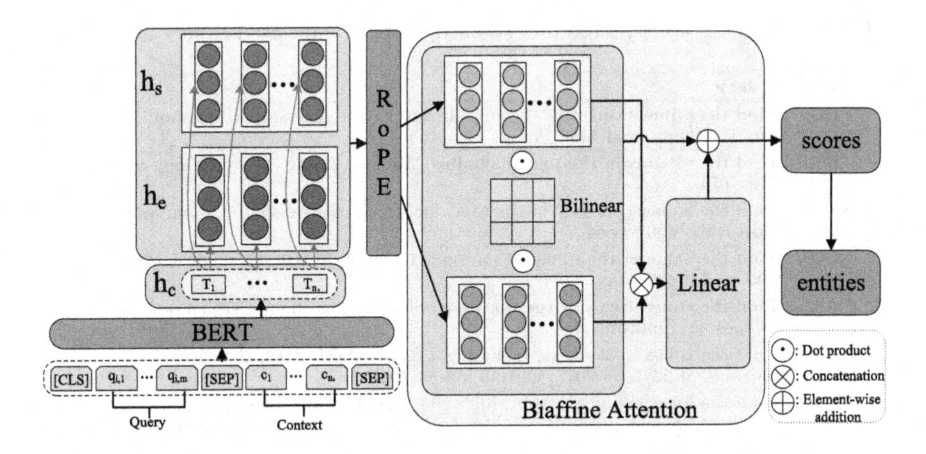


**Fig. 1.** Overall architecture of our model

legal document typically contains many nested entities. Therefore, to increase the accuracy of NER in a legal document, we need to significantly guarantee a high accuracy of recognising nested entities.

There exist some methods for recognising nested entities. For example, the sequence labelling methods assign multiple categories to a token using a hierarchical approach. Wang *et al.* [19] extracted entities iteratively from inner ones to outermost ones in an inside-to-outside way. On the contrary, Shibuya *et al.* [15] used an outside-to-inside way. However, in these methods, multiple layers may lead to the transfer of errors from one layer to the next. Thus, many researchers [6,10] give up the sequence labelling methods and turn to the span-based methods. They usually use two-stage methods for NER. Nevertheless, these methods still could have the problem of incorrect transmission between stages. More specifically, a wrong choice of candidate span in the first stage may lead to an erroneous determination of entity type in the second stage.

To address the issue, in this paper, we propose a Machine Reading Comprehension (MRC) method for NER. The method integrates the idea of reading comprehension tasks with graph-based dependency parsing. Specifically, we construct query statements for each entity type according to annotation guideline notes and reconstructed the dataset according to the triad format $\langle Query, Context, Answer \rangle$, where the lookup of each type of entity span is considered as the *Answer* to the *Query*. Also, we establish the dependencies of the start and end of each span for the specific answer finding and calculate the score using the biaffine attention.

Figure 1 shows the overall architecture of our model. Our start and end sequences are different feature mappings of the BERT [4] output. Then, both start and end sequences incorporate a Rotary Position Embedding (RoPE) [17] that encodes the position information of a character relative to others, allowing the model to recognise the boundary positions of the span of an entity better. Next, the start and end sequences with RoPE added are put to biaffine attention to calculate the scores of spans. Finally, we use a balanced softmax [16] in the classification to pick out entity spans.

**Table 1.** Queries constructed for ten entity types (translated from Chinese)

| Type | Query |
|---|---|
| NHCS | Find the crime suspect in the text, including the defendants Moumou Li, Moujia Zhang, and Mou Wang. |
| NHVI | Find the victims in the text, including Moumou Wang, Moujia Yang, and Mou Liu. |
| NCSM | Find the stolen currency in the text, including cash, cash 100 yuan, and cash RMB 400 yuan. |
| NCGV | Find the value of the items in the text, including RMB 200 yuan, RMB 600, and 5,890 yuan. |
| NASI | Find the stolen items in the text, including mobile phones, stolen mobile phones, and motorcycles. |
| NT | Find the times in the text, including a day in April 2018, the early morning of 22 June 2018, and 23 March 2018. |
| NS | Find the locations in the text, including Bengbu City, Bengshan District** Street, Room **Unit**, **Village**, and Gulou District, the home of Mouceng Yang. |
| NO | Find the organisation in the text, including Public Security Agency, **County Price Certification Center, and Project Department. |
| NATS | Find the crime tools in the text, including screwdrivers, motorcycles, and scissors. |
| NCSP | Find the theft profits in the text, including 100 yuan, RMB 300 yuan, and 50,000 yuan |

The main contributions of this paper are as follows. (1) We propose a new NER model by integrating the MRC idea with biaffine attention so that it can recognise nested entities well. (2) To the best of our knowledge, we are the first to integrate MRC, biaffine attention and RoPE for NER. (3) We have done extensive experiments to show that the model proposed in this paper outperforms the state-of-art baseline methods.

The rest of the paper is organised as follows. Section 2 presents how we convert an NER task to an MRC task. Section 3 discusses BERT fine-tuning for our purpose. Section 4 explains how we recognise entities with biaffine attention and RoPE. Section 5 experimentally evaluates our model against several state-of-art baseline methods. Section 6 compares our work with related work to show how our work advances the state-of-art on this topic. Finally, Sect. 7 concludes this paper with future work.

## 2 NER as MRC

This section will explain how to convert a NER task to an MRC task.

### 2.1 Query Statement

To convert a NER task to an MRC task, we first need to construct a query statement for each entity type. For a NER task, the entities are extracted by answering this query in the given text. Inspired by Li *et al.* [12], we construct query statements following annotation guideline notes. Each query statement contains a description of this entity type and the three most common examples of entities belonging to this entity type. These three most common entity examples are the most frequent three according to our statistics on the training data. For example, for a given text: "Beijing is China's capital city." If an entity of type "location" needs to be extracted, the query "Please find the locations in this text, including Beijing, Chengdu, and Xi'an." can be constructed. Suppose there are $k$ entity types $Y = \{y_1, \cdots, y_k\}$, then the query statement corresponding to each entity type is $\vec{q}_y$. We use LegalCorpus (a publicly available dataset) in this paper, which is from the track of information extraction of the Challenge of AI in Law Competition in 2021 (CAIL 2021).[1] The LegalCorpus dataset contains ten entity types. Table 1 shows the queries we constructed for these ten entity types.

### 2.2 Dataset of Query, Context, and Answer

Then, we construct the dataset in the MRC format of $\langle Query, Context, Answer\rangle$. The original data format in LeaglCorpus is $\langle ID, Context, Entities\rangle$, where *ID* is the unique number of each text data, *Context* is the content of the text, and *Entities* are the entities contained in the text. *Entities* is a list of entity-type and entity-value pairs, *i.e.*,

$$Entities = \{\langle y_i, (s_j, e_j)\rangle \mid i = 1, \cdots, n_t, j = 1, \cdots, n_e\}, \tag{1}$$

where $y_i$ is the type of entity $j$ in the data with *ID*; $s_j$ and $e_j$ are the start and end of entity $j$, respectively; $n_t$ is the total number of all the types in the data with *ID*; and $n_e$ is the total number of all the entities in the data with *ID*. Thus, for $\langle ID, Context, Entities\rangle$ where *Entities* is given by formula (1), we can define a piece of QCA (Query, Context, and Answer) data as follows:

$$QCA = \{\langle \vec{q_i}, \vec{c}, (s_j, e_j)\rangle \mid i = 1, \cdots, n_t, j = 1, \cdots, n_e\}, \tag{2}$$

where $\vec{q_i}$ is the query of type $i$; $\vec{c}$ is the *Context* of $\langle Query, Context, Answer\rangle$; and $s_j$ and $e_j$ are the start and end of entity $j$ of type $i$, respectively.

## 3 BERT Fine-Tuning

This section will discuss how to fine-tune the BERT pre-trained language model for obtaining the token embeddings of a legal document.

First, we convert the input format of our task to that of BERT. Specifically, given a *Query-Context* pair $\langle \vec{q_i}, \vec{c}\rangle$, where *Query* $\vec{q_i} = (q_{i,1}, \cdots, q_{i,m})$ and *Context* $\vec{c} = (c_1, \cdots, c_{n_c})$, we convert it into the following tokens:

$$[CLS], q_{i,1}, \cdots, q_{i,m}, [SEP], c_1, \cdots, c_{n_c}, [SEP],$$

where $m$ and $n_c$ are the numbers of tokens in *Query* and *Context*, respectively.

Then BERT encodes the tokens to get a representation with rich semantic information, *i.e.*,

$$(H_{q_i}, H_c) = \text{BERT}([CLS], q_{i,1}, \cdots, q_{i,m}, [SEP], c_1, \cdots, c_{n_c}, [SEP]), \tag{3}$$

[1] http://cail.cipsc.org.cn/task9.html?raceID=7.

where $H_{q_i}$ is an $m \times D$ real-number matrix and $H_c$ is an $n_c \times D$ real number matrix, and $D$ is the dimension of the BERT output.

In the BERT output, we discard $H_{q_i}$ and only take the feature representation of *Context* $H_c$ to locate the *Answer*. Since $H_{q_i}$ contains the features of the entity, which is equivalent to adding prior knowledge to the *Context*, $H_c$ is richer than the case that Query $\vec{q_i}$ is not input into BERT.

# 4 Extracting Answer Spans

After obtaining the feature representation $H_c$ of Context $\vec{c}$, our model is going to recognise the entity (*i.e.*, the answer span) from $H_c$. This section explains how to use biaffine attention and RoPE to recognise the entities.

## 4.1 Biaffine Attention

Now we discuss how to use biaffine attention to score the start-to-end of entities. Inspired by Yu *et al.* [21], we use the dependency parsing model [5] for NER. The model calculates scores for the start-to-end dependencies of a text subsequence. In the NER task, we score the start-to-end of a text subsequence and determine whether or not this subsequence is an entity based on this scoring.

First, the feature encoding $H_c$ of Context $\vec{c}$ is passed through a fully connected layer (*i.e.*, Dense) to obtain tensor $H_0$. Then it is mapped into two different tensor spaces to obtain two feature representations $h_s$ and $h_e$ of Context $\vec{c}$, *i.e.*,

$$H_0 = Dense(H_c), \tag{4}$$

where $H_0$ is an $n_c \times 2d$ real-number matrix and $d$ is a hyper-parameter;

$$(h_s, h_e) = Maps(H_0), \tag{5}$$

where $h_s$ and $h_e$ are two $n_c \times d$ real-number matrixies. Finally, we calculate the score $S(j)$ of each subsequence $j$ of *Context* as follows:

$$S(j) = h_s(j)^T U h_e(j) + W(h_s(j) \oplus h_e(j)) + b, \tag{6}$$

where $U$ is a $d \times k \times d$ real-number matrix of the possibility of being an entity that start from $s(j)$ to $e(j)$, $W$ is the weight, and $b$ is the bias, and $\oplus$ is a concatenation operation.

## 4.2 Add Rotary Position Embedding

Now we integrate Rotary Position Embedding (RoPE) [17] with formula (6). $H_c$ carries the absolute position information of Context $\vec{c}$, but the model does not know the relative position information between each token. Therefore, inspired by Su *et al.* [17], we integrate the RoPE with the biaffine attention, which can further mark the relative position relationship between tokens. Firstly, in RoPE, the rotation based on absolute position is applied to $h_s$ and $h_e$ sequences (see formula (5)), *i.e.*,

$$h'_s = RoPE(h_s), \tag{7}$$
$$h'_e = RoPE(h_e), \tag{8}$$

Then, in the biaffine attention's calculation, the multiplication calculation between $h_s$ and $h_e$ implicitly infers the relative position information. Thus, we modify formula (6) to:

$$S(j) = h'_s(j)^T U h'_e(j) + W(h'_s(j) \oplus h'_e(j)) + b. \tag{9}$$

### 4.3 Train and Test

According to Su [16], we use a balanced softmax as the loss function, *i.e.*,

$$L = \ln\left(e^{-S_0} + \sum_{j\in\Omega_{neg}} e^{-S(j)}\right) + \ln\left(e^{S_0} + \sum_{j\in\Omega_{pos}} e^{S(j)}\right), \tag{10}$$

where $\Omega_{neg}$ is the set of non-entities and $\Omega_{pos}$ is the set of entities. At training time, the model is trained in the way of end-to-end by minimising the loss calculated by formula (10).

The number of all subsequences of Context $\vec{c} = (c_1, \cdots, c_{n_c})$ is $n_c(n_c+1)/2$, which can cause a serious imbalance between the number of entities and non-entities. Su extends softmax plus cross-entropy to multi-label classification by introducing an additional 0 type. The score of this 0 type is $S_0$. All scores of the target types are bigger than $S_0$, and all scores of the non-target types are less than $S_0$, turning multi-label classification into a two-by-two comparison of target and non-target types scores.

At test time, we only consider span $j$ that satisfies $S(j) > S_0$ as an entity for output.

## 5 Experiments

This section will experimentally evaluate and analyse our model.

### 5.1 Dataset

In our experiments, we use the QCA dataset that we reconstructed from the LegalCorpus[2] dataset in Subsect. 2.2. We split the QCA dataset into the training set, the validation set, and the testing set at the ratio of 8:1:1. Table 2 shows the dataset statistics of QCA. Table 3 shows the statistics of named entities in the QCA dataset, and Table 4 shows an example of data containing entities.

[2] http://cail.cipsc.org.cn/task9.html?raceID=7.

**Table 2.** Dataset statistics of QCA

| Training data | Validation data | Testing data | Total |
|---|---|---|---|
| 41,970 | 5,250 | 5,250 | 52,470 |

**Table 3.** The statistic of named entities in the QCA dataset

| Entity type | Label | Total number | Nested number |
|---|---|---|---|
| Crime Suspect | NHCS | 6, 463 | 67 |
| Victim | NHVI | 3, 108 | 1, 017 |
| Stolen Money | NCSM | 915 | 49 |
| Goods Value | NCGV | 2, 090 | 49 |
| Stolen Items | NASI | 5, 781 | 462 |
| Time | NT | 2, 765 | 20 |
| Site | NS | 3, 517 | 545 |
| Organisation | NO | 806 | 37 |
| Crime tool | NATS | 735 | 20 |
| Theft of profit | NCSP | 481 | 15 |

**Table 4.** Example of data (translated from Chinese)

| | |
|---|---|
| **Query:** | Find the value of the items in the text, including RMB 200 yuan, RMB 600, and 5,890 yuan |
| **Context:** | Dai Guobang's complete collection of paintings and drawings of Chinese stories in white collector's book worth RMB 449 |
| **Type:** | NCGV |
| **Span:** | ["16;17"] |

### 5.2 Baselines

We compare our model with the following three classical baselines.

- BERT-Biaffine: Yu *et al.* [21] give a method to treat NER as dependency parsing. They use biaffine to calculate span scores. For a fair comparison, we replace the softmax cross-entropy they use with the balanced softmax used in our paper.
- BERT-BiLSTM-CRF: Dai *et al.* [2] use this model to efficiently recognise named entities from medical documents. BiLSTM consists of a forward LSTM [7] and a backward LSTM, which are used to establish contextual information. CRF [8] is used for sequence label order constraints.
- BERT-Softmax: This model uses BERT for feature encoding and a softmax layer to predict the labels.

### 5.3 Experimental Setting

We conducted our experiments on a single GPU (Tesla V100). We use Pytorch to build our model with a pre-trained model with the Chinese BERT-wwm-ext version [1]. In all experiments, we use BertAdam as an optimiser during training and set the learning rate as 1e-5, batch size as 32, and epochs as 50. We use the Fast Gradient Method (FGM) for adversarial training. We use accuracy on the validation set to achieve early stopping for all the experiments.

**Table 5.** The precision, recall, and F1-score comparison of our model with three baseline methods

| Models | **Our model** | | | BERT-Biaffine | | | BERT-BiLSTM-CRF | | | BERT-Softmax | | |
|---|---|---|---|---|---|---|---|---|---|---|---|---|
| NE | (%) | | | | | | | | | | | |
| | $P$ | $R$ | $F_1$ | $P$ | $R$ | $F_1$ | $P$ | $R$ | $F_1$ | $P$ | $R$ | $F_1$ |
| NHCS | 94.00 | 96.91 | 95.43 | 91.68 | 95.11 | 93.37 | 94.90 | 96.91 | 95.89 | 94.48 | 97.56 | **95.99** |
| NHVI | 94.57 | 95.18 | **94.87** | 91.67 | 95.50 | 93.54 | 91.98 | 95.82 | 93.86 | 93.33 | 94.53 | 93.93 |
| NCSM | 77.78 | 84.85 | 81.16 | 78.64 | 81.82 | 80.20 | 74.31 | 81.82 | 77.88 | 76.32 | 87.88 | **81.69** |
| NCGV | 98.60 | 97.25 | **97.92** | 96.79 | 96.79 | 96.79 | 97.67 | 96.33 | 97.00 | 96.77 | 96.33 | 96.55 |
| NASI | 84.68 | 86.28 | **85.47** | 77.18 | 83.53 | 80.23 | 75.16 | 78.90 | 76.99 | 74.88 | 79.76 | 77.24 |
| NT | 93.59 | 91.64 | **92.61** | 89.19 | 91.99 | 90.57 | 93.86 | 90.59 | 92.20 | 91.67 | 91.99 | 91.83 |
| NS | 90.75 | 87.96 | **89.33** | 85.29 | 87.68 | 86.46 | 65.78 | 75.91 | 70.48 | 66.26 | 75.91 | 70.76 |
| NO | 92.59 | 96.15 | **94.34** | 89.87 | 91.03 | 90.45 | 89.74 | 89.74 | 89.74 | 79.78 | 91.03 | 85.03 |
| NATS | 79.45 | 96.67 | **87.22** | 73.02 | 76.67 | 74.80 | 78.57 | 91.67 | 84.61 | 77.03 | 95.00 | 85.07 |
| NCSP | 90.91 | 86.21 | 88.50 | 92.31 | 82.76 | 87.27 | 86.89 | 91.38 | **89.08** | 87.93 | 87.93 | 87.93 |
| Overall | **90.77** | **91.93** | **91.35** | 86.71 | 90.13 | 88.39 | 84.49 | 88.29 | 86.35 | 84.04 | 88.89 | 86.40 |

**Table 6.** Model ablation tests on the test set

| Model | $P$ | $R$ | $F_1$ |
|---|---|---|---|
| **Full model** | **90.77** | **91.93** | **91.35** |
| - MRC | $89.45_{(-1.32)}$ | $90.96_{(-0.97)}$ | $90.19_{(-1.16)}$ |
| - RoPE | $90.72_{(-0.05)}$ | $91.74_{(-0.19)}$ | $91.23_{(-0.12)}$ |

### 5.4 Results

The evaluation metrics used in this paper are Precision, Recall, and F1-score. In addition, we use the exact-match standard: an entity is correct only if its location range and category both are correct.

Table 5 shows the benchmark experimental results and Fig. 2 visualises the comparison. Our model's overall $PRF_1$ value is higher than all the baselines. It is also largely higher than the baselines in identifying ten types of entities. Achieving the highest $F_1$ scores on NHVI, NCGV, NASI, NT, NS, NO, and NATS entity-types, which are +0.94%, +0.46%, +5.24%, +0.41%, +2.87%, +3.89%, and +2.15% over the best baseline model, respectively.

We also performed a series of ablation experiments to analyse the effects of different components of our model to investigate the advantages of our entire model for the NER task. Table 6 shows the result of our ablation experiments. Specifically, We conducted ablation experiments with MRC and RoPE removed and analysed their effects separately.

(1) *Not Treating NER as MRC*: After we do not consider the NER task as an MRC task, the $F_1$ score decreases by 1.16%, demonstrating the MRC formulation's validity. Furthermore, we analyse that the query encodes prior information about the entity for Context $\vec{c}$, enhancing the features.
(2) *Not Using RoPE*: If not using RoPE, the $F_1$ score decreases by 0.12%, which indicates that adding RoPE further enhances the location of the start and end of the entity, which improves recognition accuracy.

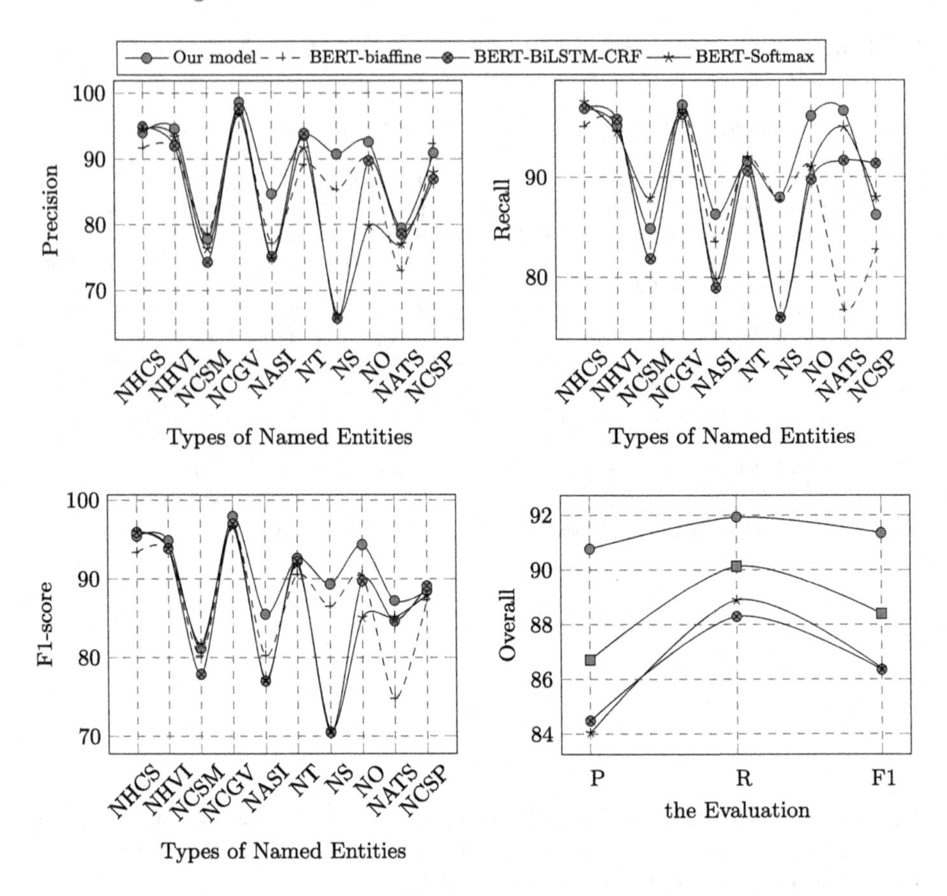


**Fig. 2.** The Precision, recall, and F1-score intuitive comparison of different methods

### 5.5 Case Study

We have conducted a case study to demonstrate that our model can recognise both nested and non-nested entities. As shown in Fig. 3, it is an example from the test set, where entities marked in red indicate incorrectly recognised entities.

- For the non-nested entities in the example, our model and the baseline models can accurately recognise entity boundaries of types NHCS, NCGV, and NHVI. That is because these entity types are composed of shorter words and are not nested by other entities, so all models recognise them well.
- For nested entities, the NASI type entity "a rose gold iPhone 7 from Moumou Yao" is nested with the NHVI type entity "Moumou Yao", making it longer and more fine-grained than other types of entities. Thus, the corresponding model for NER must understand the semantic information better and more accurately detect the entity's boundaries. As shown in Fig. 3, both our model and the BERT-Biaffine model (which can detect entity boundaries) accurately

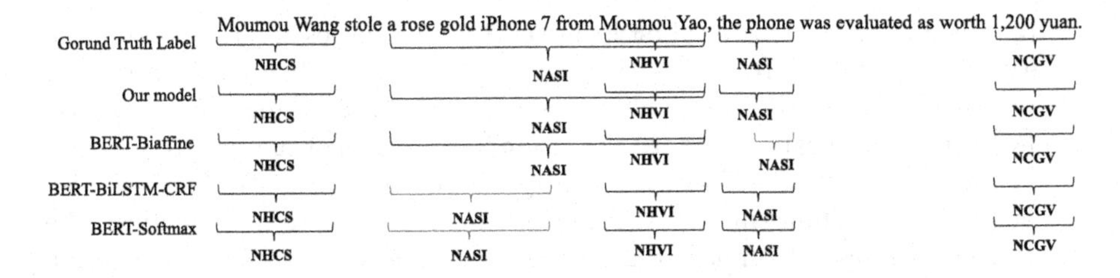


**Fig. 3.** Case Study

recognise the boundaries of this nested entity. In contrast, the other two labelling-based models cannot.

- Our model outperforms the BERT-Biaffine model for fine-grained entities with referents such as the entity "the phone". The BERT-BiLSTM-CRF can capture the information of such referred fine-grained entities due to the ability of BiLSTM to capture long-range contextual semantic information.

# 6 Related Work

This section will compare our work with related work to show how our work advances the state-of-art of the research area of NER.

## 6.1 Nested Named Entity Recognition

(1) *Sequence Labelling-Based Methods*: In 2021, Wang *et al.* [19] used a hierarchical approach to recognise nested entities, explicitly excluding the effect of optimal paths. Their experiments show that recognising the innermost entities first outperforms the traditional scheme of recognising the outermost entities first. However, this approach still suffers from error transmission between layers. In 2021, Zhou *et al.* [24] proposed a progressive multi-task learning framework to reduce the layer error transfer problem. However, the sequence labelling-based methods have natural disadvantages in the recognition of nested entities. Because sequence labelling can only set one tag per character, all of the above hierarchical methods suffer from some error transfer problems. To address these issues, many span-based works have developed in recent years. The span-based methods recognise in terms of sub-sequences, with little need to consider the cascading of nested entities. Our work is also span-based.

(2) *Span-Based Methods*: In 2021, Li *et al.* [9] proposed a novel span-based joint model that traverses all the possible text spans to recognise entities. The model uses a MLP to classify whether the span is an entity span and what the entity type is. Unlike them, we use biaffine attention to score the possible text spans and then determine whether the span is an entity based on the scoring. In 2021, Shen *et al.* [14] proposed a two-stage method for

recognising nested entities: first locate entity boundaries and then determine entity types for NER. In locating entity boundaries, they used filters and regressors to improve the quality of span candidates. Unlike them, we use RoPE to incorporate the relative position information of the span to improve further the accuracy of locating entity boundaries. In 2021, Li *et al.* [10] proposed a modular interaction network for recognise nested entities. The method first uses biaffine attention during boundary detection to generate a feature representation of each possible boundary location. Then it uses softmax to obtain the probability of whether the span is an entity. However, we use a balanced softmax to determine whether or not the span is an entity, which removes the class imbalance between entities and non-entities to some extent.

### 6.2 Machine Reading Comprehension

MRC is to extract an answer to a question from a text relevant to the question, while NER is to extract entities from a document. As a result, some researchers treat a NER problem as an MRC problem. In particular, since large-scale pre-trained language models such as BERT provide a more efficient and convenient way to encode these natural language processing tasks, many researchers have based BERT to convert NER tasks into MRC tasks. For example, in 2020, Li *et al.* [12] constructed three probability matrices of entities' starts, ends, and spans. This method requires significantly more time and memory consumption than our approach because ours only requires the construction of a single matrix for entity recognition. Zhao *et al.* [23] focused on relationship extraction rather than the NER task. In 2021, Sun *et al.* [18] formulated the NER task as an MRC task, but they directly used linear layers to classify the head and tail of entities. However, unlike them, we used a dependency resolution approach to calculate the dependency strength of the start and end of entities. In 2021, Xiong *et al.* [20] used the same annotation guide to construct the query as we did. Moreover, they used softmax to calculate the start and end probabilities of the spans and then took the subsequence with the highest probability as an entity. However, this approach is not applicable to the case where the text contains multiple entities of the same type. In this case, we do better than them because our method can identify multiple entities of the same type by calculating the span score and thus obtain the entity with a score higher than the corresponding threshold.

## 7 Conclusion

The NER of legal documents is crucial to the construction of intelligent justice. In this paper, we proposed an MRC method for NER of legal documents. The method integrates biaffine attention with rotated position embedding to encode relative position information to enhance further the recognition of the start and end of an entity. We also used a balanced softmax to eliminate the category imbalance problem of entities and non-entities in a legal document. The extensive experiments show that our model achieves competitive performance compared to some state-of-art baseline methods. However, in the future, it is worth investigating how to reduce the model's time and space complexity and apply this model to documents in other domains such as medical science.

**Acknowledgements.** This work was supported by Research Fund of Guangxi Key Lab of Multi-source Information Mining & Security (22-A-01-02).

## References

1. Cui, Y., Che, W., Liu, T., Qin, B., Wang, S., Hu, G.: Revisiting pre-trained models for Chinese natural language processing. In: Findings of the Association for Computational Linguistics: EMNLP 2020, pp. 657–668 (2020)
2. Dai, Z., Wang, X., Ni, P., Li, Y., Li, G., Bai, X.: Named entity recognition using BERT BiLSTM CRF for Chinese electronic health records. In: 2019 12th International Congress on Image and Signal Processing, Biomedical Engineering and Informatics, pp. 1–5 (2019)
3. De Sanctis, F.M.: Artificial intelligence and innovation in Brazilian justice. Int. Ann. Criminol. **59**(1), 1–10 (2021)
4. Devlin, J., Chang, M.W., Lee, K., Toutanova, K.: BERT: pre-training of deep bidirectional transformers for language understanding. In: Proceedings of the 17th Annual Conference of the North American Chapter of the Association for Computational Linguistics: Human Language Technologies, vol. 1, pp. 4171–4186 (2019)
5. Dozat, T., Manning, C.D.: Deep biaffine attention for neural dependency parsing. In: Proceedings of the 5th International Conference on Learning Representations, pp. 1–8 (2017)
6. Eberts, M., Ulges, A.: Span-based joint entity and relation extraction with transformer pre-training. In: Proceedings of the 24th European Conference on Artificial Intelligence. Frontiers in Artificial Intelligence and Applications, vol. 325, pp. 2006–2013 (2019)
7. Hochreiter, S., Schmidhuber, J.: Long short-term memory. Neural Comput. **9**(8), 1735–1780 (1997)
8. Lafferty, J., McCallum, A., Pereira, F.C.: Conditional random fields: probabilistic models for segmenting and labeling sequence data. In: Proceedings of the 18th International Conference on Machine Learning, pp. 282–289 (2001)
9. Li, F., Lin, Z., Zhang, M., Ji, D.: A span-based model for joint overlapped and discontinuous named entity recognition. In: Proceedings of the 59th Annual Meeting of the Association for Computational Linguistics and the 11th International Joint Conference on Natural Language Processing, vol. 1, pp. 4814–4828 (2021)
10. Li, F., et al.: Modularized interaction network for named entity recognition. In: Proceedings of the 59th Annual Meeting of the Association for Computational Linguistics and the 11th International Joint Conference on Natural Language Processing, vol. 1, pp. 200–209 (2021)
11. Li, J., Sun, A., Han, J., Li, C.: A survey on deep learning for named entity recognition. IEEE Trans. Knowl. Data Eng. **34**(1), 50–70 (2020)
12. Li, X., Feng, J., Meng, Y., Han, Q., Wu, F., Li, J.: A unified MRC framework for named entity recognition. In: Proceedings of the 58th Annual Meeting of the Association for Computational Linguistics, pp. 5849–5859 (2020)

13. Re, R.M., Solow-Niederman, A.: Developing artificially intelligent justice. Stanford Technol. Law Rev. **22**, 242 (2019)
14. Shen, Y., Ma, X., Tan, Z., Zhang, S., Wang, W., Lu, W.: Locate and label: A two-stage identifier for nested named entity recognition. In: Proceedings of the 59th Annual Meeting of the Association for Computational Linguistics and the 11th International Joint Conference on Natural Language Processing, vol. 1, pp. 2782–2794 (2021)
15. Shibuya, T., Hovy, E.: Nested named entity recognition via second-best sequence learning and decoding. Trans. Assoc. Comput. Linguist. **8**, 605–620 (2020)
16. Su, J.: Extend "softmax+cross entropy" to multi-label classification problem. https://kexue.fm/archives/7359 (2020)
17. Su, J., Lu, Y., Pan, S., Wen, B., Liu, Y.: RoFormer: enhanced transformer with rotary position embedding. arXiv preprint arXiv:2104.09864 (2021)
18. Sun, C., Yang, Z., Wang, L., Zhang, Y., Lin, H., Wang, J.: Biomedical named entity recognition using BERT in the machine reading comprehension framework. J. Biomed. Inform. **118**, 103799 (2021)
19. Wang, Y., Shindo, H., Matsumoto, Y., Watanabe, T.: Nested named entity recognition via explicitly excluding the influence of the best path. In: Proceedings of the 59th Annual Meeting of the Association for Computational Linguistics and the 11th International Joint Conference on Natural Language Processing, vol. 1, pp. 3547–3557 (2021)
20. Xiong, Y., et al.: Improving deep learning method for biomedical named entity recognition by using entity definition information. BMC Bioinform. **22**(1), 1–13 (2021)
21. Yu, J., Bohnet, B., Poesio, M.: Named entity recognition as dependency parsing. In: Proceedings of the 58th Annual Meeting of the Association for Computational Linguistics, pp. 6470–6476 (2020)
22. Zekos, G.I.: Advanced Artificial Intelligence and Robo-Justice. Springer, Cham (2022). https://doi.org/10.1007/978-3-030-98206-5
23. Zhao, T., Yan, Z., Cao, Y., Li, Z.: Asking effective and diverse questions: a machine reading comprehension based framework for joint entity-relation extraction. In: Proceedings of the Twenty-Ninth International Conference on International Joint Conferences on Artificial Intelligence, pp. 3948–3954 (2021)
24. Zhou, B., Cai, X., Zhang, Y., Yuan, X.: An end-to-end progressive multi-task learning framework for medical named entity recognition and normalization. In: Proceedings of the 59th Annual Meeting of the Association for Computational Linguistics and the 11th International Joint Conference on Natural Language Processing, vol. 1, pp. 6214–6224 (2021)

# Cross-Modality Visible-Infrared Person Re-Identification with Multi-scale Attention and Part Aggregation

Li Fan[1], Shengrong Gong[1,2](✉), and Shan Zhong[2]

[1] Northeast Petroleum University, Daqing, China
[2] Changshu Institute of Technology, Changshu, China
shrgong@cslg.edu.cn

**Abstract.** In the cross-modality visible-infrared person re-identification (VI-ReID) task, the cross-modality matching degree of visible-infrared images is low due to the large difference in cross-modality image features. Existing methods often impose constraints on the original pixels or extracted features to extract discriminative features, which are prone to introduce irrelevant background clutter and have a weak ability to extract cross-modality invariant features. This paper proposes an end-to-end neural network called multi-scale attention part aggregation network (MSAPANet). The framework consists of an intra-modality multi-scale attention (IMSA) module and a fine-grained part aggregation learning (FPAL). IMSA module is used to mine intra-modality attention-enhanced discriminative part features and suppress background feature extraction. FPAL fuses fine-grained local features and global semantic information through channel-spatial joint soft attention (CSA) to efficiently extract cross-modality shared features. Experiments were carried out on SYSU-MM01 and RegDB, two common datasets for VI-ReID, and the results show that under various settings, our method outperforms the reference current state-of-the-art methods. In this paper, by designing the network structure, the network mine the intra-modality and inter-modality salient information at the same time, improving the discriminative performance of fine-grained features in the channel and spatial dimensions, promoting modality-invariant and discriminative feature representation learning for VI-ReID tasks.

**Keywords:** Cross-modality visible-infrared person re-identification · Person re-identification · Channel-spatial joint soft attention · Multi-scale · Shareable features

## 1 Introduction

VI-ReID [3,24] task is to match the persons appearing under different cameras with the given visible (or infrared) person image. Compared with the traditional single modality (RGB-RGB) [9,27,33,34] person Re-ID, VI-ReID is more suitable for real-world environments because it can reduce the limitations of night or low-light conditions. VI-ReID is affected by background noise, large variation


M. Tanveer et al. (Eds.): ICONIP 2022, CCIS 1793, pp. 237–248, 2023.
https://doi.org/10.1007/978-981-99-1645-0_20

(person pose and visual angle, etc.) of intra-modality person images, and large modal gap differences. Therefore, reducing the modal difference and learning discriminative features from heterogeneous images are the key points for studying VI-ReID tasks.

A large number of researchers have paid attention to VI-ReID, and achieved substantial performance improvements with effective methods. In early explorations of the VI-ReID task, to achieve the goal of focusing on as much intra-modality salient information as possible and reducing cross-modality image differences, the most common strategy was to investigate the feature extraction [13,22,24,29,30]. However, global feature learning of heterogeneous images is not robust to additional background noise, and the ability to mine modality invariant information is weak. As a result, the researchers considered the metric learning (ML) method that projects heterogeneous person images into the common spatial to solve the VI-ReID task. With the development of deep learning, researchers have applied generative adversarial network (GAN)-based image generation methods [15,20,22] to the VI-ReID task, to reduce the cross-domain gap between visible and infrared images. However, the network of GAN designed by the image generation method is complicated, which will bring a lot of irrelevant noise to the sample, and there are more redundant parameters in the network, which requires a lot of training time and computational cost.

Since the part feature learning for single-modality person Re-ID [10,11] is not suitable for capturing local features in VI-ReID across the channel gap, this paper designs the MSAPANet to learn cross-modality discriminative shared features. MSAPANet consists of two main modules: the IMSA module and the FPAL. By using the dependency association of part features [21], the IMSA module can mine intra-modality important local feature cues. The FPAL decomposes global features and then fuses the CSA module to enhance the semantic information, improving the network's performance in recognizing fine-grained part features in the channel and spatial dimensions, and maximizing focus on cross-modality shared features.

## 2 Related Work

**Cross-modality VI-ReID.** In the visible-infrared modality, feature learning is a necessary step for similarity measurement, early models of feature learning [12] were done by training contours or local descriptors, and most research in recent years has focused on designing convolutional neural networks (CNN) to enhance visual representation and learns deep features. Ye et al. [29,30] designed two-stream networks for VI-ReID, for extracting specific modality features. Based on the two-stream network, Liu et al. [13] proposed a mid-level feature merging method. Different from the processing methods at the feature extraction level, many researchers apply the image conversion method based on the GAN in VI-ReID. Wang et al. [20] proposed an end-to-end AlignGAN, which uses the feature and pixel alignment works together to bridge the visible-infrared modal gap. Choi et al. [2] proposed a Hi-CMD method to automatically separate the ID discriminant and ID exclusion factors from two images.

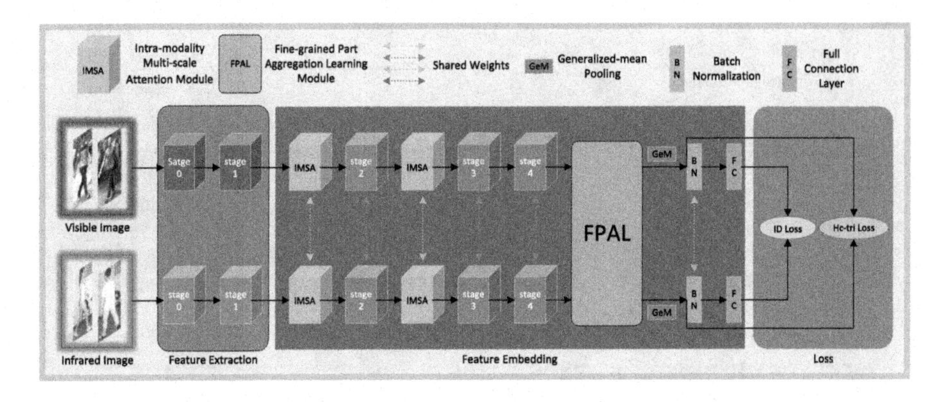


**Fig. 1.** Multi-scale attention part aggregation network (MSAPANet).

**Attention Mechanisms.** In 2017, Hu Jie et al. proposed SENet [6], which simulates cross-channel relationships in feature maps by learning the weights of each channel in the network. Woo et al. [23] integrated the spatial attention mechanism again on top of the previous channel attention and proposed CBAM, which utilizes global average pooling and global max pooling in both channel and spatial dimensions. Li et al. [10] proposed a harmonious attention model that retags the misaligned pedestrian images while improving the matching accuracy, maximizing the effective information of different levels of visual attention for person Re-ID.

# 3 Method

## 3.1 Overview

The VI-ReID task widely uses the ResNet50 two-stream network [5,24,27] as the baseline of implementation. To deal with the problem of low matching of different modality images and difficult extraction of cross-modal shareable features, this paper designs the MSAPANet. MSAPANet includes the IMSA module and the FPAL module, with the overall architecture shown in Fig. 1. MSAPANet uses stage 0 and stage 1 of ResNet50 [5] as feature extractors, their weights are independent. stage 2, stage 3, and stage 4 are used as feature embedding parts, and the two modalities share the network parameters of the feature embedding part. Inspired by [27], we choose generalized-mean (GeM) pooling to obtain fine-grained features, adding a batch normalization (BN) layer to learn shareable feature representations.

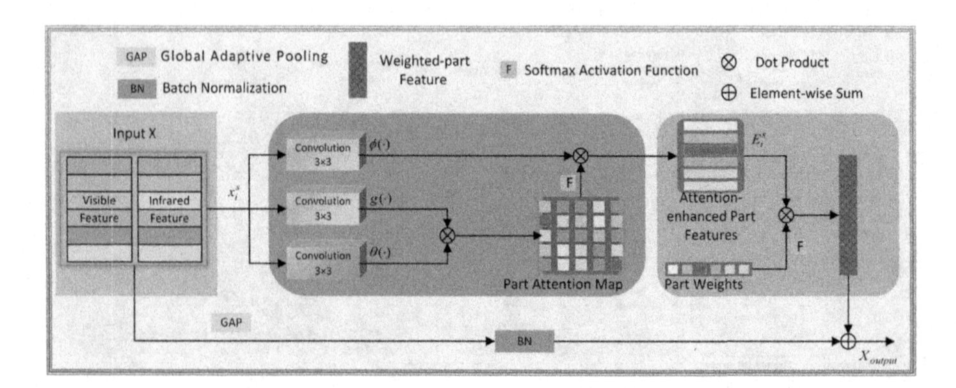


**Fig. 2.** Intra-modality multi-scale attention (IMSA) module.

To maximize separation class spacing and learn salient discriminative features, we combine hetero-center triplet loss $L_{hc\_tri}$ [35] and identity loss $L_{id}$ [8] as the baseline learning objective. The total loss is expressed as:

$$L = L_{hc\_tri} + L_{id} \tag{1}$$

### 3.2 Intra-modality Multi-scale Attention Module

Inspired by no-local attention [21], this paper designs the IMSA module that focuses on intra-modality local attention. Compared with methods that only focus on global features [22], the IMSA module can mine the contextual cues of intra-modality part features. As shown in Fig. 2, the input of the IMSA module is the feature map extracted from stage1 and stage2, denoted as $X \in R^{h \times w \times c}$ ($h \times w$ denotes the input feature map size, $c$ denotes the channel dimension). First, the feature map is horizontally divided into $s$ scales by region pooling, and the obtained part features are represented as $x_i^s$, and each part of the same feature map will be input into the $3 \times 3$ convolution g$(\cdot)$, $\theta(\cdot)$ and $\phi(\cdot)$. To amplify the part attention difference and enhance the feature discriminability, the softmax function is used to normalize the part attention. The attention-enhanced part features $E_i^s$ are represented by the inner product of the part features embedded in the convolution $\phi(\cdot)$ and the computed part attention $A_{i,j}^s \in [0, 1]^{s \times s}$, defined as:

$$E_i^s = a_i^s \times \phi(x_i^s) \tag{2}$$

where, $\phi(x_i^s) = w^\phi \times x_i^s$, $w^\phi$ are the weights of the convolution, and $a_i^s \in \{A_{i,j}^s\}^{s \times s}$ is the part attention. Simple average pooling or concatenation of part features may cause the superposition of noisy parts, so we feed the original features into the global adaptive pooling (GAP) [31] and BN [19,28] layers through another branch, and then weighted aggregation with learnable attention-enhanced part features.

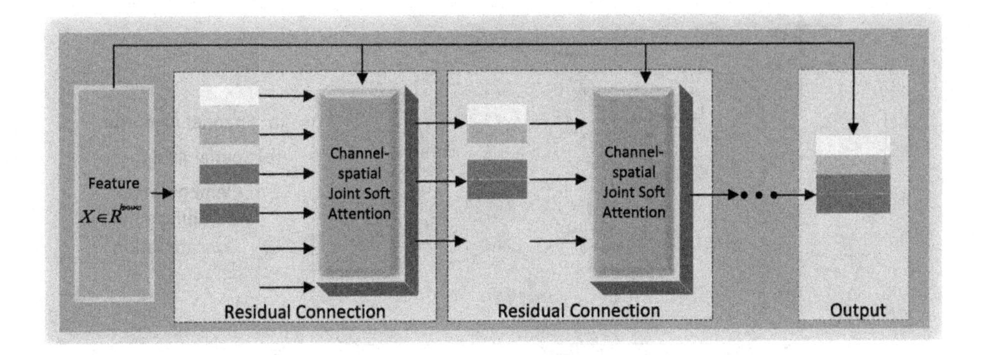


**Fig. 3.** Fine-grained part aggregation (FPA) module.

### 3.3 Fine-Grained Part Aggregation Learning

To obtain fine-grained image information, and learn inter-modality shared features, this paper designs the FPAL method, FPAL including the fine-grained part aggregation (FPA) Module and CSA module, to fuse the fine-grained local information into the global feature representation to enhance cross-modal features.

**Fine-Grained Part Aggregation Module.** As shown in Fig. 3, the output features from stage 4 in MSAPANet are divided horizontally by scale $r$ using the ceiling operation to obtain fine-grained small-scale part features [32], denoted as $\frac{h}{r} \times w \times c$. At different scale stages, part features at each scale are combined in pairs with adjacent features to form a larger representation. In a coarser-grained stage, the CSA module further exploits the local semantic information. In the last stage, all enhanced shareable part features are integrated and a final global representation with a scale of 1 is output. The part feature aggregation process is expressed as:

$$x_{t,r}^{k} = F\left\{x_{t,r-1}^{l-1 \to k} || x_{t,r-1}^{l \to k}; x_{0,r}^{k}\right\} \tag{3}$$

where $x_{0,r}^{k}$ denotes the input part feature of the $k$th part with scale $r$, $x_{t,r}^{k}$ denotes the output enhanced aggregated feature with scale $r$, $||$ denotes the cascade operation in height dimension, $\to$ represents the correspondence, $k = (1, 2, 3, \ldots, n)$, and $k$ is 1 when the output is the final global feature representation. For example, the $k$th part feature of scale $r$ is obtained by combining the $l$th part feature of its predecessor (scale $r - 1$) and its adjacent $l - 1$th part feature.

**Channel-Spatial Joint Soft Attention Module.** To obtain the spatially structured information and form a unified representation with the fine-grained feature encoding of the image, inspired by [10,17,23,26], this paper learns two kinds of attention in a joint but independent way. As shown in Fig. 4, one branch of the CSA module is soft attention which combines channel attention and spatial

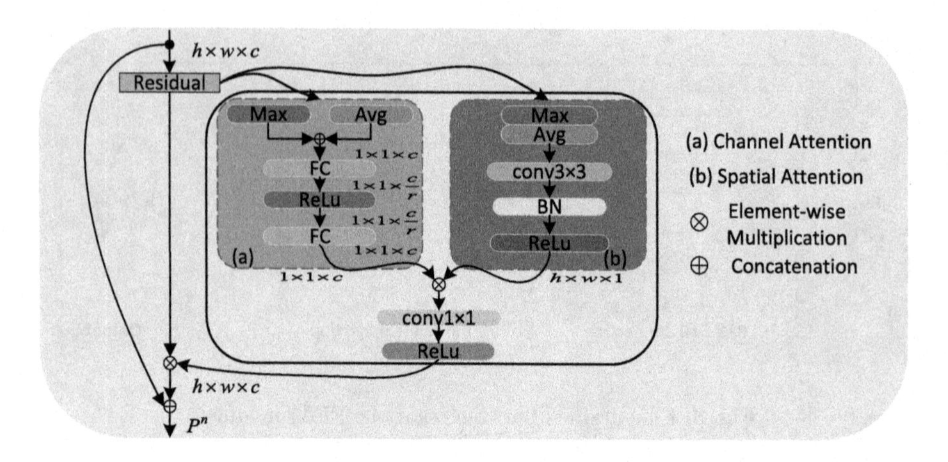


**Fig. 4.** Channel-spatial joint soft attention (CSA) module.

attention, and the other branch input is the part feature from the FPA module. The CSA module not only alleviates the large differences in cross-modal features but is also computationally efficient, it can be used as a plug-and-play module. Deploying a $1 \times 1$ convolution to combine the two types of attention, after tensor multiplication of the matrix, the soft attention is normalized to between 0.5 and 1 using a sigmoid function, and the resulting joint soft attention feature $P^n$ is represented as:

$$P^n = \sigma(w^t(S^n \otimes C^n)) \tag{4}$$

where $\sigma$ represents the sigmoid activation function, $w^t$ represents the $1 \times 1$ convolution, $S^n$ represents the 2D spatial attention map, $C^n$ represents the 1D channel attention map, $S^n \in R^{h\times w\times 1}$, $C^n \in R^{1\times 1\times c}$, $n$ represent the scale level. $\otimes$ represents the corresponding multiplication of matrix elements.

*Channel Attention.* Channel attention is shared in the height and width dimensions, which focus on discrimination and salient features. As shown in Fig. 4 (a), To address the problem of underutilization of relevant information in features, two different channel feature descriptors, $C_{\text{avg}}$ and $C_{\max}$, are generated using the average set and the maximum set. To capture the channel dependencies using the generated aggregated information, the channels are activated using the nonlinear ReLu function [10] after dimensionality reduction in the full connection layer. The channel dimension is again converted back to the original dimension using another full connection layer, and the channel attention $C$ is obtained after activation, shown as:

$$C = \delta(w^2_{\text{FC}}(\delta(w^1_{\text{FC}}(C_{avg} + C_{\max})))) \tag{5}$$

where, $w^1_{\text{FC}}$, $w^2_{\text{FC}}$ respectively represent the corresponding weights of the two fully connection layers [6], $\delta$ represents the ReLu activation function.

*Spatial Attention.* Spatial attention focuses on the position of features, uses the spatial relationship between features to generate an attention map, and shares them across the channel dimension, effectively highlighting the information domain position. As shown in Fig. 4 (b), the two-dimensional vectors, $S_{\text{avg}}$ and $S_{\text{max}}$ are obtained by average pooling and max pooling along the channel axis and connect them along the channel axis, go through 3×3 convolution and BN layer, highlight the position of emphasis or suppression, and generate spatial attention $S$, and the calculation process is as:

$$S = \delta(w^{s}_{\text{conv}}(S_{\text{avg}}||S_{\text{max}})) \tag{6}$$

where $w^{s}_{conv}$ represents the learnable parameters in the convolutional layer. $\delta$ represents the RuLe activation function, $||$ represents the concatenation operation.

# 4 Experiments

## 4.1 Experimental Settings

**Datasets and Evaluation Protocol.** The method proposed in this paper will have experimented on two public datasets (SYSU-MM01 [24] and RegDB [16]). The SYSU-MM01 [24] dataset collects images of 491 pedestrians through 4 visible light monitors and 2 infrared monitors. In the testing phase, two modes of all search and indoor search are set. The RegDB [16] dataset collects the images of 412 target persons by a monitor with dual modes of visible and infrared, each person collects 10 infrared and 10 visible images respectively. The test mode sets visible-infrared and infrared-visible modes. Cumulative matching characteristics (CMC) [15,24] and mean average precision (mAP) [7,24] are used as two criteria in this experiment, The detailed CMC rank-k (k = 1,10) accuracy is listed.

**Implementation Details.** The model is tested on the PyTorch framework, using ResNet50 [5,24] as the baseline network for feature extraction [4,8], and the stride of the last convolutional block is set to 1 to obtain a finer-grained feature map for the feature embedding stage. The size of the input image is set to $288 \times 144$. In this experiment, the SGD [4,14,25] optimizer is used to update the network parameters, the weight decay is $5 \times 10^{-4}$, the momentum is 0.9, the initial learning rate is set to 0.01, and the training total 80 epochs, linearly increasing the learning rate to 0.1 for the first 20 epochs, set to 0.01 for the next 40 epochs, and 0.001 for the last 40 epochs.

## 4.2 Comparison with State-of-Art Methods

Compared with some popular VI-ReID methods, the experimental results on the SYSU-MM01 and RegDB datasets are shown in Table 1. In all search mode of the SYSU-MM01 dataset, our model has a rank-1 accuracy and mAP score of 1.13%

**Table 1.** Performance comparison results with state-of-the-art methods on dataset SYSU-MM01 and RegDB.

| Setting | All Search | | | Indoor Search | | | Visible to Infrared | | | Infrared to Visible | | |
|---|---|---|---|---|---|---|---|---|---|---|---|---|
| Method | rank-1 | rank-10 | mAP | rank-1 | rank-10 | mAP | rank-1 | rank-10 | mAP | rank-1 | rank-10 | mAP |
| Zero-Pad [24] | 14.80 | 54.12 | 15.95 | 20.58 | 68.38 | 26.92 | 17.75 | 34.21 | 18.90 | 16.63 | 34.68 | 17.82 |
| HCML [29] | 14.32 | 53.16 | 16.16 | 24.52 | 73.25 | 30.08 | 24.44 | 47.53 | 20.08 | 21.70 | 45.02 | 22.24 |
| BDTR [30] | 27.32 | 66.96 | 27.32 | 31.92 | 77.18 | 41.86 | 33.56 | 58.61 | 32.76 | 32.92 | 58.46 | 31.96 |
| D2RL [22] | 28.90 | 70.60 | 29.20 | – | – | – | 43.40 | 66.10 | 44.10 | – | – | – |
| AlignGAN [20] | 42.40 | 85.00 | 40.70 | 45.90 | 87.60 | 54.30 | 57.90 | – | 53.60 | 56.30 | | 53.40 |
| Hi-CMD [2] | 34.94 | 77.58 | 35.94 | – | – | – | 70.93 | 86.39 | 66.04 | – | – | – |
| DDAG [31] | 54.75 | 90.39 | 53.02 | 61.02 | 94.06 | 67.98 | 69.34 | 86.19 | 63.46 | 68.06 | 85.15 | 61.80 |
| G2DA [18] | 57.07 | 90.99 | 55.05 | 63.70 | 94.06 | 69.83 | 71.72 | 87.13 | 65.90 | 69.50 | 84.87 | 63.88 |
| NFS [1] | 56.91 | 91.34 | 55.45 | 62.79 | 96.53 | 69.79 | 80.54 | 91.96 | 72.10 | 77.95 | 90.45 | 69.79 |
| **MSAPANet(Ours)** | **58.04** | **92.31** | **58.91** | **64.40** | **95.82** | **71.61** | **82.91** | **92.75** | **74.33** | **79.93** | **91.95** | **72.71** |

and 3.46%, respectively, showing the best performance; in the indoor search mode, we outperform representative methods. In two experimental settings using the RegDB dataset infrared and visible images as query sets, The performance is improved over the reference method. The experimental results show that our method effectively reduces the intra-modal and inter-modal differences in VI-ReID, proving that the MSAPANet can effectively extract cross-modal shared features and learn modality-invariant discriminative semantic information.

### 4.3 Ablation Study

We have explored the effectiveness of the proposed module on the datasets RegDB and SYSU-MM01. The experiments are set as an infrared-visible mode on RegDB and all search mode on SYSU-MM01.

**The Effectiveness of Each Model.** BS represents the improved ResNet50 baseline two-stream network, M represents the IMSA module, and F represents the FPAL module. The experimental results are shown in Table 2. The experimental results show that the performance of our improved network is better than the baseline network [9,24,33], the IMSA module and the FPAL model also improve the network performance. The method of hierarchical feature fusion is conducive to mining modal shared semantic information, which can reduce the modal difference of VI-ReID.

**Table 2.** Evaluate the validity of each model.

| Method | | | RegDB | | SYSU-MM01 | |
|---|---|---|---|---|---|---|
| BS | M | F | rank-1 | mAP | rank-1 | mAP |
| √ | × | × | 60.80 | 57.12 | 50.77 | 49.18 |
| √ | √ | × | 66.54 | 61.18 | 52.70 | 50.11 |
| √ | × | √ | 72.19 | 71.15 | 54.18 | 53.40 |
| √ | √ | √ | 76.19 | 70.36 | 58.04 | 55.91 |

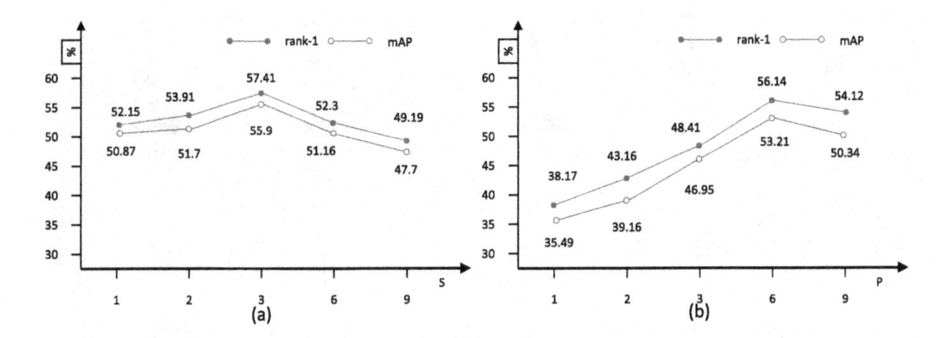


**Fig. 5.** Influence of the number of feature partitions on IMSA model and FPAL model.

**Parameter Effectiveness of Hierarchical Feature Aggregation.** To explore the effectiveness of feature partitioning in the IMSA and FPAL modules, the impact of different parameter configurations intra-model and cross-modal shared feature extraction performance was evaluated. We conduct experiments on the dataset SYSU-MM01, Fig. 5 (a) shows the effect of the part divisions on the performance of the IMSA module, and Fig. 5 (b) shows the performance of the FPAL module in hierarchical partitioning at different scales. The experimental results show that a larger division scale can capture more fine-grained salient features, but if the granularity is too fine, it will reduce the performance, and feature hierarchical re-fusion has better performance for learning fine-grained features.

**Cross-Modality Fine-Grained Shared Feature Visualization.** One of the keys to improving the VI-ReID task is to train a network that can extract more discriminative cross-modality sharable features. Figure 6 shows the pixel pattern map of the IMSA module. The experiment visualizes the prominent feature regions on the image, Fig. 6 (a) (c) shows the pedestrian salient features captured in the baseline network, and Fig. 6 (b) (d) is the visual result of the attention of 4 different identity images under the IMSA module. Another key to VI-ReID task is to extract cross-modal shared features, Fig. 7 (a) (e) respectively represent two modal images of two different identities, and Fig. 7 (b) (d) represents the visualization of the attention overlaid on the original image under the action of the baseline network, Fig. 7 (c) shows the visualization of the co-attention of two modal pictures under the action of the FPAL module. The results show that the IMSA module can capture more discriminative features for pedestrian images, and can achieve good results in both modes. Under the action of the FPAL module, the common correlation information in the two modalities is more prominent. The FPAL module can extract inter-modality shared features, and suppress other regions through part aggregation operation.

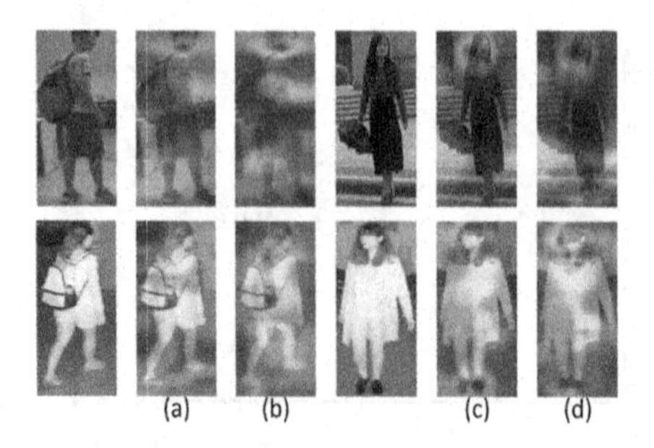


Fig. 6. Visualization of IMSA module on dataset SYSU-MM01.

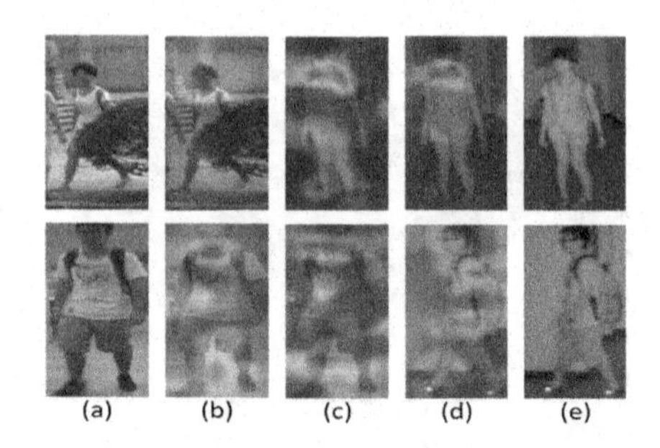


Fig. 7. Visualization of FPAL module on dataset SYSU-MM01.

## 5 Conclusion

In this paper, we propose a Multi-Scale Attention Part Aggregation Network (MSAPANet), that efficiently explores cross-modal correlations. MSAPANet's innovation is reflected in its two components: One is the IMSA module, which focuses on intra-modality local features and performs attention enhancement on them, mining saliency-enhanced part features. Compared with various methods of ML, the IMSA module is more robust to the interference caused by background clutter, and reduces the interference of low-discriminative irrelevant features; Second is the FPAL module, which focuses on cross-modal shared features in a cascade framework, performs deep extraction of shareable fine-grained features between heterogeneous images. Compared with image generation methods such as AliGAN and Hi-CMD, the FPAL module does not rely on complex image generation methods, effectively reducing the complexity of network design. Related experiments were conducted on two public datasets of VI-ReID (RegDB and SYSU-MM01), and multiple sets of experiments demonstrated, MSAPANet can effectively extract intra-modal saliency information and cross-modal shared features, and the network exhibited excellent performance benefited from the combined effect of IMSA module and FPAL module. In further work, we plan to investigate the distance loss of different modes and further expand the dataset.

**Acknowledgements.** This work was supported by the National Natural Science Foundation of China (61972059, 61773272, 62102347), China Postdoctoral Science Foundation(2021M69236), Key Laboratory of Symbolic Computation and Knowledge Engineering of Ministry of Education, Jilin University (93K172017K18), Natural Science Foundation of Jiangsu Province under Grant (BK20191474, BK20191475, BK20161268), Qinglan Project of Jiangsu Province (No.2020).

## References

1. Chen, Y., Wan, L., Li, Z., Jing, Q., Sun, Z.: Neural feature search for RGB-infrared person re-identification. In: Proceedings of the IEEE/CVF Conference on Computer Vision and Pattern Recognition, pp. 587–597 (2021)

2. Choi, S., Lee, S., Kim, Y., Kim, T., Kim, C.: Hi-CMD: hierarchical cross-modality disentanglement for visible-infrared person re-identification. In: Proceedings of the IEEE/CVF Conference on Computer Vision and Pattern Recognition, pp. 10257–10266 (2020)
3. Feng, Z., Lai, J., Xie, X.: Learning modality-specific representations for visible-infrared person re-identification. IEEE Trans. Image Process. (2020)
4. Gao, G., Shao, H., Wu, F., Yang, M., Yu, Y.: Leaning compact and representative features for cross-modality person re-identification. In: World Wide Web, pp. 1–18 (2022)
5. He, K., Zhang, X., Ren, S., Sun, J.: Deep residual learning for image recognition. IN: 2016 IEEE Conference on Computer Vision and Pattern Recognition (CVPR) (2016)
6. Hu, J., Shen, L., Sun, G.: Squeeze-and-excitation networks. In: Proceedings of the IEEE Conference on Computer Vision and Pattern Recognition, pp. 7132–7141 (2018)
7. Huang, Z., Liu, J., Li, L., Zheng, K., Zha, Z.J.: Modality-adaptive mixup and invariant decomposition for RGB-infrared person re-identification. arXiv preprint arXiv:2203.01735 (2022)
8. Jambigi, C., Rawal, R., Chakraborty, A.: MMD-ReiD: a simple but effective solution for visible-thermal person ReID. arXiv preprint arXiv:2111.05059 (2021)
9. Leng, Q., Ye, M., Tian, Q.: A survey of open-world person re-identification. IEEE Trans. Circuits Syst. Video Technol. 1092–1108 (2019)
10. Li, W., Zhu, X., Gong, S.: Harmonious attention network for person re-identification. In: 2018 IEEE/CVF Conference on Computer Vision and Pattern Recognition (2018)
11. Li, Y., He, J., Zhang, T., Liu, X., Zhang, Y., Wu, F.: Diverse part discovery: occluded person re-identification with part-aware transformer (2021)
12. Liao, S., Hu, Y., Zhu, X., Li, S.Z.: Person re-identification by local maximal occurrence representation and metric learning. In: Proceedings of the IEEE Conference on Computer Vision and Pattern Recognition, pp. 2197–2206 (2015)
13. Liu, H., Cheng, J., Wang, W., Su, Y., Bai, H.: Enhancing the discriminative feature learning for visible-thermal cross-modality person re-identification. Neurocomputing **398**, 11–19 (2020)
14. Liu, H., Ma, S., Xia, D., Li, S.: Sfanet: a spectrum-aware feature augmentation network for visible-infrared person reidentification. IEEE Trans. Neural Netw. Learn. Syst. (2021)
15. Moon, H., Phillips, P.J.: Computational and performance aspects of PCA-based face-recognition algorithms. Perception **30**(3), 303–321 (2001)
16. Nguyen, D.T., Hong, H.G., Kim, K.W., Park, K.R.: Person recognition system based on a combination of body images from visible light and thermal cameras. Sensors **17**(3), 605 (2017)
17. Rao, Y., Chen, G., Lu, J., Zhou, J.: Counterfactual attention learning for fine-grained visual categorization and re-identification. In: Proceedings of the IEEE/CVF International Conference on Computer Vision, pp. 1025–1034 (2021)
18. Wan, L., Sun, Z., Jing, Q., Chen, Y., Lu, L., Li, Z.: G2DA: geometry-guided dual-alignment learning for RGB-infrared person re-identification. arXiv preprint arXiv:2106.07853 (2021)
19. Wang, F., et al.: Residual attention network for image classification. In: Proceedings of the IEEE Conference on Computer Vision and Pattern Recognition, pp. 3156–3164 (2017)

20. Wang, G., Zhang, T., Cheng, J., Liu, S., Yang, Y., Hou, Z.: RGB-infrared cross-modality person re-identification via joint pixel and feature alignment (2019)
21. Wang, X., Girshick, R., Gupta, A., He, K.: Non-local neural networks. In: Proceedings of the IEEE Conference on Computer Vision and Pattern Recognition, pp. 7794–7803 (2018)
22. Wang, Z., Wang, Z., Zheng, Y., Chuang, Y.Y., Satoh, S.: Learning to reduce dual-level discrepancy for infrared-visible person re-identification. In: 2019 IEEE/CVF Conference on Computer Vision and Pattern Recognition (CVPR) (2019)
23. Woo, S., Park, J., Lee, J.Y., Kweon, I.S.: CBAM: convolutional block attention module. In: Proceedings of the European Conference on Computer Vision (ECCV), pp. 3–19 (2018)
24. Wu, A., Zheng, W.S., Yu, H.X., Gong, S., Lai, J.: RGB-infrared cross-modality person re-identification. In: 2017 IEEE International Conference on Computer Vision (ICCV) (2017)
25. Wu, Q., et al.: Discover cross-modality nuances for visible-infrared person re-identification. In: Proceedings of the IEEE/CVF Conference on Computer Vision and Pattern Recognition, pp. 4330–4339 (2021)
26. Yang, J., et al.: Learning to know where to see: a visibility-aware approach for occluded person re-identification. In: Proceedings of the IEEE/CVF International Conference on Computer Vision, pp. 11885–11894 (2021)
27. Ye, M., Shen, J., Lin, G., Xiang, T., Hoi, S.: Deep learning for person re-identification: a survey and outlook. IEEE Trans. Pattern Anal. Mach. Intell. 1 (2021)
28. Ye, M., Lan, X., Leng, Q., Shen, J.: Cross-modality person re-identification via modality-aware collaborative ensemble learning. IEEE Trans. Image Process. **29**, 9387–9399 (2020)
29. Ye, M., Lan, X., Li, J., Yuen, P.: Hierarchical discriminative learning for visible thermal person re-identification. In: Proceedings of the AAAI Conference on Artificial Intelligence, vol. 32 (2018)
30. Ye, M., Lan, X., Wang, Z., Yuen, P.C.: Bi-directional center-constrained top-ranking for visible thermal person re-identification. IEEE Trans. Inf. Forensics Secur. **15**, 407–419 (2019)
31. Ye, M., Shen, J., J. Crandall, D., Shao, L., Luo, J.: Dynamic dual-attentive aggregation learning for visible-infrared person re-identification. In: Vedaldi, A., Bischof, H., Brox, T., Frahm, J.-M. (eds.) ECCV 2020. LNCS, vol. 12362, pp. 229–247. Springer, Cham (2020). https://doi.org/10.1007/978-3-030-58520-4_14
32. Zhang, C., Liu, H., Guo, W., Ye, M.: Multi-scale cascading network with compact feature learning for RGB-infrared person re-identification. In: 2020 25th International Conference on Pattern Recognition (ICPR), pp. 8679–8686. IEEE (2021)
33. Zhang, Z., Lan, C., Zeng, W., Jin, X., Chen, Z.: Relation-aware global attention for person re-identification. In: 2020 IEEE/CVF Conference on Computer Vision and Pattern Recognition (CVPR) (2020)
34. Zheng, Z., Yang, X., Yu, Z., Zheng, L., Yang, Y., Kautz, J.: Joint discriminative and generative learning for person re-identification. 2019 IEEE/CVF Conference on Computer Vision and Pattern Recognition (CVPR) (2020)
35. Zhu, Y., Yang, Z., Wang, L., Zhao, S., Hu, X., Tao, D.: Hetero-center loss for cross-modality person re-identification. Neurocomputing **386**, 97–109 (2020)

# Bearing Fault Diagnosis Based on Dynamic Convolution and Multi-scale Gradient Information Aggregation Under Variable Working Conditions

Yimeng Long[1], Zhaowei Shang[1], and Lingzhi Zhao[2](✉)

[1] College of Computer Science, Chongqing University, Chongqing 400044, China
{longyimeng,szw}@cqu.edu.cn

[2] Zhuhai Huafa Group Science and Technology Research Institute Co., Ltd., Zhuhai 519000, China
zlz1229@126.com

**Abstract.** In modern industrial production, the working conditions of rotating machinery tend to change with different speeds and loads, and the raw signals are easily polluted by noise interference. To face these challenges, in this paper, we propose a dynamic perception network with gradient information aggregation (DGNet) for machinery fault diagnosis, which contributes to the improvement of generalization ability and noise resistance. Specifically, we use a two-branch structure for fault diagnosis. First, dynamic perception is regarded as the main branch, which contributes to increasing the representation capability of our model. With an attention mechanism and kernel aggregation, this approach can adaptively learn the fault distribution between different signals and achieve good generalization ability. Then, we use a gradient aggregation branch to fuse multi-scale structural information into dynamic perception, filtering low-frequency noise via global structural constraints. Experimental results on the Case Western Reserve University (CWRU) dataset show that our proposed DGNet outperforms the state-of-the-art approaches in terms of generalization ability and noise resistance.

**Keywords:** Fault diagnosis · Variable working conditions · Residual convolutional neural networks · Dynamic convolution · Gradient aggregation

## 1 Introduction

Rotating machinery is the driving force of industrial production, as it has been widely used in various mechanical equipment and currently plays an increasingly important role. As one of the important parts of rotating machinery, rolling

This work was supported by Smart Community Project Based on Machine Vision and Internet of Things Platform (Grant No. ZH22017002200003PWC).


M. Tanveer et al. (Eds.): ICONIP 2022, CCIS 1793, pp. 249–263, 2023.
https://doi.org/10.1007/978-981-99-1645-0_21

bearings have a tremendous impact on the performance, stability, and life cycle of the entire machine. Applying fault diagnostic techniques can not only reduce the effect of damage or failure of rotating machinery on the reliability and safety of the entire mechanical system but also reduce any economic lossess [1]. Therefore, it is of great significance to research the fault diagnosis of mechanical equipment.

In 2006, Hinton et al. [2] proposed the concept of deep learning for the first time and constructed a network structure that contained multiple hidden layers, which enhanced the learning ability of sample features and improved the network performance. The deep learning [3] model can automatically learn fault features from data and has more powerful feature extraction and nonlinear feature characterization capabilities. In recent years, it has been gradually applied to the field of mechanical fault diagnosis [4–6]. In view of the above advantages, in recent years, Shao et al. [7] used a wavelet function as the nonlinear activation function to design a wavelet autoencoder (WAE), built a deep WAE with multiple WAEs, and used an extreme learning machine (ELM) as a classifier to diagnose bearing faults. He et al. [8] proposed the envelope spectra of the resampled data used as the feature vectors and a classifier model based on Gaussian RBM for bearing fault diagnosis. Gan et al. [9] employed a characterization based on the wavelet packet energy of a raw signal, and a hierarchical diagnosis network based on a deep confidence network was used for bearing fault diagnosis.

Currently, the focus of fault diagnosis research is on deep CNNs. According to the dimension of the processing object, CNN can be divided into one-dimensional (1D) and two-dimensional (2D) modes. In 2D models, Wen et al. [10] converted the fault signals to RGB image format as the input datatype based on ResNet-50 and proposed a network with a depth of 51 convolutional layers for fault diagnosis. Sun et al. [11] used multi-scale signal feature information extracted by DTCWT to form a matrix and combined it with a CNN for gear fault diagnosis. However, both the dimension-raising transformation and the 2D convolution calculation increase the computational burden, which has an impact on the fault diagnosis efficiency.

In 1D models, Ince et al. [12] presented a fast and accurate motor condition monitoring and fault-detection system using a 1D-CNN that has an inherent adaptive design. Levent et al. [13] proposed a system based on a 1D-CNN that takes raw time-series sensor data as input for real-time fault detection and monitoring. Compared with the 2D processing method, the 1D structure is relatively simple, and it requires less time and computational resources in the training and testing processes and has the advantage of having less algorithmic complexity. According to the above references, although it is proved that the 1D-CNN-based model has advantages in the number of parameters and the floating-point computations, as the fault diagnosis demands of improving adaptability and noise resistance under dynamic working conditions are becoming more complicated, the current fault diagnosis methods based on 1D-CNNs have the following problems:

In actual industrial production, the working conditions of rotating machinery is complex changeable, the vibration signal collected by the sensor will also be affected by the production environment, and it is easily mixed with all types of noise. Under these complex working conditions, the distribution of training and test data is often quite different. At this time, it is difficult for 1D-CNN method to obtain better reasoning performance in the latter based on the feature distribution learned in the former. Therefore, due to the distribution similarity between training and test data, this method has poor fault diagnosis performance under complex working conditions.

To address the aforementioned problems, this paper proposes a residual CNN model based on dynamic perception convolutional and gradient information aggregation for rolling bearing fault diagnosis. The contributions made by this work are as follows:

We use the dynamic convolution to adaptively learn the fault distribution between different signals, which can improve the accuracy of fault classification and generalization ability in different working conditions. For noise signals, we use multi-scale gradient aggregation branches to fuse gradient information to effectively filter noise and improve the noise resistance of our model. We apply our proposed model to fault diagnosis with working condition changes and noise data. The final comparison results of fault diagnosis show that our model has achieved better performance than the existing methods based on deep learning.

# 2 Method

## 2.1 Framework

As shown in Fig. 1, first, our machinery fault diagnosis method takes a raw vibration signal fragment as input; then, we perform data augmentation according to the overlapping sampling method. Finally, our proposed DGNet can classify the fault after end-to-end training, which consists of two parts: dynamic perception and gradient aggregation.

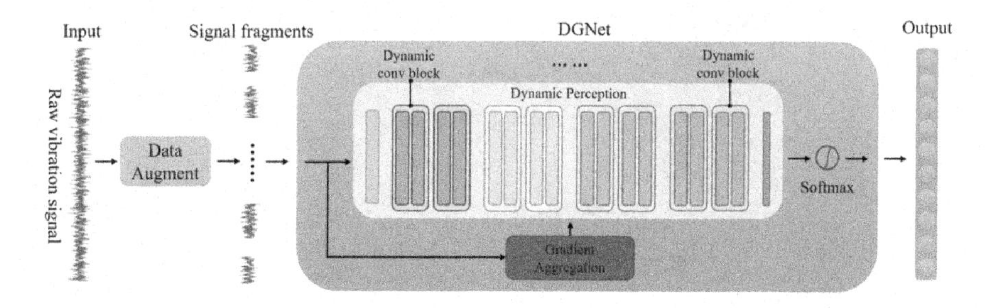


**Fig. 1.** Architecture of our proposed method.

## 2.2 Dynamic Perception

Dynamic convolution is proposed to achieve a better balance between the network performance and computational burden within the scope of efficient neural networks. With the attention mechanism and kernel aggregation, it can increase the representation capability of a network with limited depth and width, which does not add extra layers to the whole network [14]. Thus, in our dynamic perception branch, we adopt dynamic convolution for each residual block to adaptively learn the different fault distributions with multiple convolution kernels and increase the generalization performance of our model between different fault signals.

As shown in Fig. 1, the whole branch regards ResNet18 as the backbone, and the processed fault signal sequence is taken as the input. First, a basic convolution is performed with a kernel size of $11 \times 1$. Then, we employ the residual blocks in four stages, and each stage is composed of a specific number of blocks. Different from ResNet, we replace each residual block with a dynamic convolution, which can provide an adaptive representation for different fault signals. Finally, a fully connected layer is applied to obtain a tensor of $10 \times 1$, and after the softmax operation, we obtain the fault classification of each signal.

**Dynamic Convolution.** To increase the performance of fault classification and the generalization ability to different speeds and loads, we use dynamic convolution in each of the convolution blocks, which enhances the representation capability of our network. In Fig. 2, the dynamic convolution consists of two parts: attention and multiple kernel aggregation.

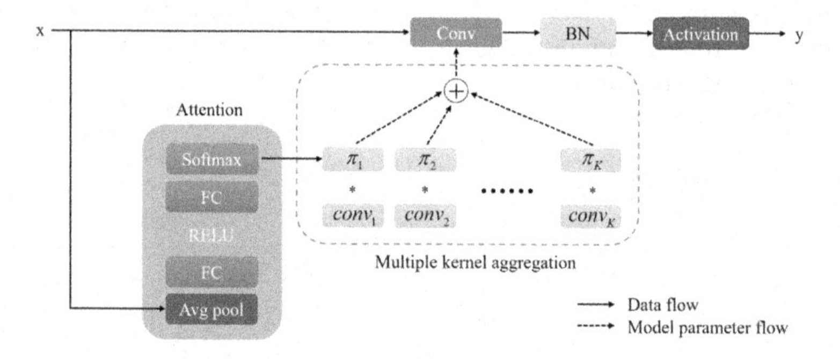


**Fig. 2.** A dynamic convolution block.

For the attention module, we first squeeze the global spatial information by global average pooling. Then, two fully connected layers (with a RELU operation between them) and softmax are adopted to generate normalized attention weights for $K$ convolution kernels. Unlike SENet [15], the attention of dynamic convolution is generated over convolution kernels, which costs little for each attention. Assuming that the input feature map is of size $L \times C_i$, we obtain Multi-Adds $O(\tau(x))$ as follows:

$$O(\tau(x)) = LC_i + C_i^2/4 + C_iK/4 \tag{1}$$

Because of the small kernel size, we can aggregate convolution kernels efficiently for kernel aggregation. For $K$ convolution kernel aggregation (kernel size of $D_K \times D_K$), extra Multi-Adds of $KC_iC_0D_K^2 + KC_0$ are needed with $C_i$ input channels and $C_0$ output channels. Finally, after multiple kernel aggregation, we employ batch normalization and a RELU to generate the output feature map $y$.

**Softmax Temperature.** In the conventional task of classification, softmax can enlarge the gap between output values through the index and then normalize the output of a vector close to one hot. The difference between the values of the vector is distinctive, and the hard output makes it difficult for the model to learn the approximate distribution. Therefore, the softmax temperature $\tau$ is applied to the attention layer for dynamic convolution, which can contribute to softening the distance between output features. Based on the softmax, the softening form can be expressed as follows:

$$p_i = \frac{\exp(s_i/\tau)}{\sum_j \exp(s_j/\tau)} \tag{2}$$

where $\tau$ is the softmax temperature. Obviously, when the temperature approaches 0, the output distribution will converge to a one-hot vector with distinct values. When approaching infinity, the output distribution of the function is close. Thus, in dynamic convolution, we can set an appropriate $\tau$ to soften the output of the attention layer, which can help avoid falling into local optimization and is conducive to training.

### 2.3 Gradient Aggregation

For 2D images, gradient information is conducive to the description and representation of their edge boundary, which is beneficial for the extraction of structural information. In 1D fault signal classification, we can use gradient information to capture structural features that are helpful for the representation of high-frequency information. Similar to edge boundaries, this approach can easily reduce noise interference and improve the robustness of the model in our fault signal classification task. Thus, we employ a gradient aggregation branch to extract structural features of 1D signals, which increases the noise immunity of our DGNet. Specifically, the structure of the network is shown in Fig. 3.

Our gradient aggregation branch consists of three multi-scale branches and a trunk composition branch, and each multi-scale branch combines a middle layer into the composition branch [16]. By fusing different scales of information through multiple branches, the system can effectively capture the multi-scale structural representation of fault signals, filtering interference of terrible noise.

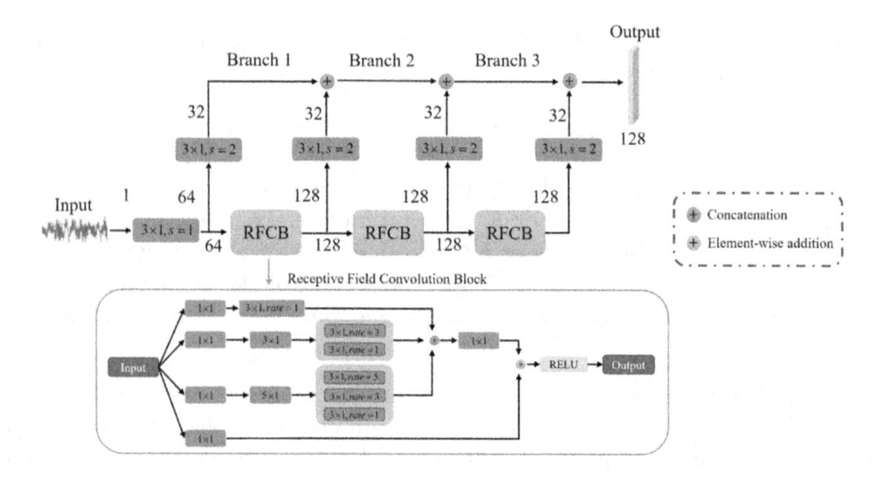


**Fig. 3.** The branch of gradient aggregation.

To maintain global structural integrity, we use multiple receptive field mechanisms such as RBF [17] in the composition branch to form our receptive field convolution block (RFCB), which contributes to enhancing the perception field of vision and covering a wider range of learning areas on both sides of the signal sequence.

Unlike RBF, where a convolution with a single dilation rate is used after each kernel convolution, we employ multiple dilated convolutions for RFCB in a more flexible way to compensate for the problem of missing information caused by the single dilation rate. Specifically, each multi-scale branch is composed of a standard convolution with a kernel size of $3 \times 1$, and corresponding to each branch, we apply RFCB to the composition branch. First, the feature map of $H \times W \times C$ is taken as input for four groups, and the kernels of different sizes are applied to Groups 1 to 3. After applying basic convolutions of $1 \times 1$, $3 \times 1$, and $5 \times 1$, we employ group dilated convolution to capture multiple receptive fields for each branch. Then, Group 4 is regarded as the shortcut that can be concatenated with Groups 1 to 3. In this way, we can ultimately generate global structural features with multiple receptive fields of aggregation.

# 3 Experiments and Analysis

## 3.1 Introduction to the Dataset

**Dataset Description.** To validate the effectiveness of the proposed DGNet model, the bearing fault dataset provided by CWRU [18] was used. The main body of the experimental rig is a Reliance Electric motor [19]. The testing bearings consist of the drive end bearing (SKF 6205) and the fan end bearing (SKF 6203), manually implanting single point failures of different sizes, such as 0.007–0.040 in. in diameter, by using the electro-discharge machining technique. After the speed was stabilized, the vibration acceleration signal data of the motor load subjected to 0–3 hp were recorded. In this paper, the 12 kHz drive-end bearing fault data and the normal baseline data are used for fault classification. There are three bearing fault types, including inner race faults (IRFs), ball faults (BFs), and outer race faults (ORFs). For each fault type, there are three fault diameters, namely, 0.007 in., 0.014 in., and 0.021 in. (1 in. = 25.4 mm). As a result, including the normal data, there are 10 classifications in this dataset.

**Data Augment.** The best way to enhance the generalization ability of machine learning models is to use more training samples [20]. Due to the limited experimental data, this paper performs data augmentation according to the overlapping sampling method in references [21] and [22]. In other words, when samples are acquired from the raw signal, there is an overlap between each segment of the signal and its subsequent segment, as shown in Fig. 4.

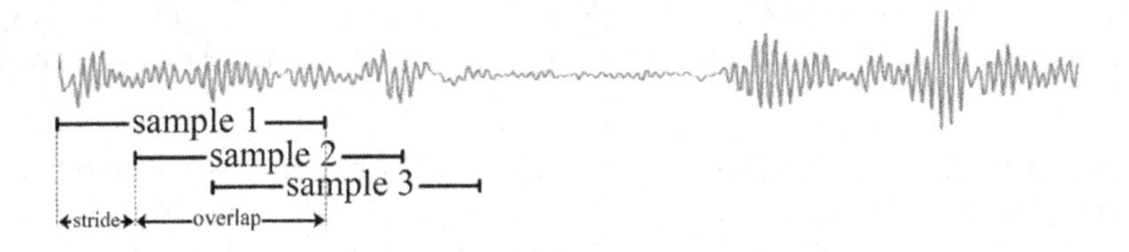


**Fig. 4.** Schematic diagram of overlapping sampling.

Here, stride denotes the interval step between two adjacent samples, which is determined according to the length of the raw vibration signal. The number of samples obtained after overlapping sampling is calculated as follows:

$$Samples = \frac{total\_point - window\_size}{stride} + 1 \quad (3)$$

where *total_point* and *window_size* denote the length of the original signal and the number of individual sample sampling points, respectively. If a raw signal has a length of 121,265, then the stride is set to 130, and the signal can be segmented into $(121{,}265 - 1{,}024)/130 + 1 = 926$ samples of length $1024 \times 1$. In the experiment, a total of 900 samples are selected for each classification of data, and the dataset is divided into the training set and the test set, with 800 training samples and 100 test samples randomly selected.

**Multiload Datasets.** To test the accuracy and generalization ability of the model in this paper under compound operating conditions with variable loads, the dataset is divided into A, B, C, and D according to the hp value of the motor load, which correspond to the data collected under loads of 0, 1, 2, and 3 hp, respectively. In datasets A, B, C, and D, each category contains 800 training

**Table 1.** Failure categories for the CWRU dataset.

| Datasets | Motor Load (HP) | Fault Diameter (Inches) | Training samples | Testing samples | Fault types | Label |
|---|---|---|---|---|---|---|
| A/B/C/D | 0/1/2/3 | 0 | 800/800/800/800 | 100/100/100/100 | Normal | 0 |
| | | 0.007 | 800/800/800/800 | 100/100/100/100 | IRF | 1 |
| | | 0.014 | 800/800/800/800 | 100/100/100/100 | IRF | 2 |
| | | 0.021 | 800/800/800/800 | 100/100/100/100 | IRF | 3 |
| | | 0.007 | 800/800/800/800 | 100/100/100/100 | BF | 4 |
| | | 0.014 | 800/800/800/800 | 100/100/100/100 | BF | 5 |
| | | 0.021 | 800/800/800/800 | 100/100/100/100 | BF | 6 |
| | | 0.007 | 800/800/800/800 | 100/100/100/100 | ORF | 7 |
| | | 0.014 | 800/800/800/800 | 100/100/100/100 | ORF | 8 |
| | | 0.021 | 800/800/800/800 | 100/100/100/100 | ORF | 9 |

samples and 100 test samples. Specific information on all of the data required to generate the experiments on the different datasets is shown in Table 1.

In the experiments, we train the model using the dataset A training set and test it using datasets A, B, C, and D, denoted ACC, A-B, A-C, and A-D, respectively. Similarly, we train and test it using datasets B, C, and D, respectively.

**Noise Dataset.** The CWRU dataset used in this paper was collected in the indoor laboratory, which has a low level of noise. However, in practical industrial applications, the vibration signal measured by the accelerometer is easily polluted. Thus, using the raw CWRU dataset might not accurately reflect the classification performance of our DGNet in real working conditions. To verify the fault diagnosis robustness of DGNet in noisy data, we add white Gaussian noise to the raw signal. The signal-to-noise ratio (SNR, unit dB) is used to measure the severity of the noise, which is defined as follows:

$$SNR = 10\lg(\frac{P_{signal}}{P_{noise}}) \tag{4}$$

where $P_{signal}$ and $P_{noise}$ represent the strength of the raw and noise signals, respectively.

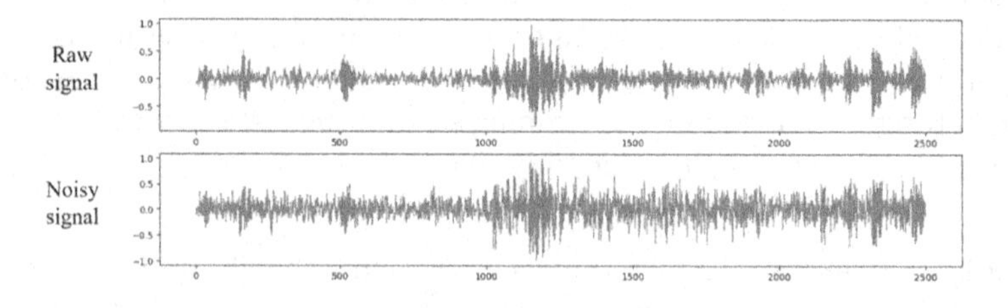


**Fig. 5.** The effect of noise on the signal.

Furthermore, Fig. 5 shows the vibration signal of the ball fault 0.014 in. and the noise signal after adding 0 dB white Gaussian noise. Compared with the raw signal, it is obvious that the peak area of the noise signal has become difficult to distinguish. In the experiment, in terms of the noise resistance, we use noise-free data for model training and test the data with three different decibels of white Gaussian noise: 0 dB, 5 dB, and 10 dB, to simulate fault diagnosis under real working conditions.

### 3.2 Result Analysis

In the convolutional network, the first layer basically embeds the data into a new larger vector space, and increasing the convolutional kernel size of the first convolutional layer can preserve the information of the raw signal samples as much as possible, thus improving the generalization ability of the model. To maintain the detailed feature extraction capability of the model, all of the convolution kernel sizes except for the first layer are kept at 3. The experimental results are shown in Fig. 6.

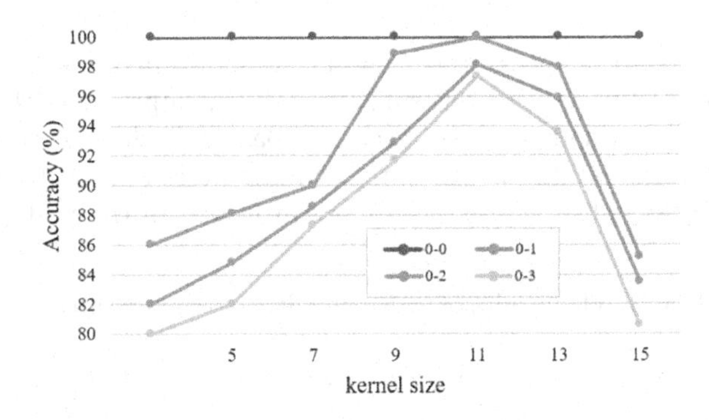


**Fig. 6.** The results of different first layer convolutional kernel sizes.

As shown in Fig. 6, we assume that 0–0 to 0–3 correspond to A-A to A-D, the generalization ability of the model gradually increases when the size of the convolution kernel increases from 3 to 11 in the first layer and then decreases when it exceeds 11. The reason is that a convolution kernel with a large scale could lead to an excessively strong smoothing ability, which will reduce the extraction performance of local features. To improve the generalization ability and retain the extraction performance of the details, we set the scale of the first layer convolution kernel to 11 in subsequent experiments.

Table 2 shows the classification accuracy and generalization ability when using dynamic convolutions with different $K$. First, the dynamic convolution approach outperforms the standard convolution, even with a small $K = 2$. This finding demonstrates the strength of our method. In addition, the accuracy stops increasing once $K$ is larger than 12. The reason is that as $K$ increases, even though the model has more expression power, it becomes more difficult to optimize all of the convolution kernels and attention simultaneously, and the network is more prone to over-fitting.

**Table 2.** The results of different numbers of convolution kernels ($K$).

| K | ACC (%) | Generalization ability (%) | | |
|---|---|---|---|---|
| | | A-B | A-C | A-D |
| / | 99.90 | 65.62 | 62.59 | 59.21 |
| 2 | 100.00 | 79.06 | 79.19 | 72.06 |
| 4 | 100.00 | 84.64 | 81.72 | 79.15 |
| 6 | 100.00 | 90.72 | 89.11 | 87.43 |
| 8 | 100.00 | 95.97 | 95.43 | 94.94 |
| 10 | 100.00 | 99.16 | 97.93 | 97.90 |
| 12 | 100.00 | 100.00 | 98.97 | 98.94 |
| 14 | 100.00 | 99.90 | 98.12 | 98.45 |
| 16 | 100.00 | 99.89 | 98.28 | 98.13 |

General softmax is a special case ($\tau = 1$). As shown in Table 3, as $\tau$ increases from 1 to 10, the accuracy and generalization ability improve. With the increase in $\tau$, the output of the softmax layer becomes sparser. Such nearly even attention can promote kernel learning in the early training period to improve the performance. Specifically, when $\tau = 10$, the model will obtain the best performance.

**Table 3.** The results of different softmax temperatures ($\tau$).

| Temperature | ACC (%) | Generalization ability (%) | | |
|---|---|---|---|---|
| | | A-B | A-C | A-D |
| $\tau = 1$ | 100.00 | 94.63 | 92.00 | 90.14 |
| $\tau = 5$ | 100.00 | 97.91 | 96.20 | 95.94 |
| $\tau = 10$ | 100.00 | 99.99 | 98.48 | 98.21 |
| $\tau = 15$ | 100.00 | 98.99 | 97.53 | 96.36 |
| $\tau = 20$ | 100.00 | 97.89 | 96.23 | 95.03 |
| $\tau = 30$ | 100.00 | 95.28 | 93.23 | 92.77 |

We test the effect of the gradient aggregation branch on the algorithm's performance with the whole branch. As illustrated in Table 4, compared with no gradient aggregation, the method with the branch performs better when there are different proportions of noise fault signals.

Specifically, in the noise group of 10 dB, the fault classification accuracy of the method with gradient aggregation is improved by 2.61% compared with the method without it. As the proportion of noise in the signal increases, the gradient

**Table 4.** The results of the gradient aggregation branch on the noise resistance performance.

| Method | ACC (%) | 10 dB (%) | 5 dB (%) | 0 dB (%) |
|---|---|---|---|---|
| ours | 100.00 | 99.92 | 99.21 | 95.46 |
| / | 100.00 | 97.38 | 92.53 | 80.62 |

aggregation branch has more obvious advantages, and it brings an 18.41% accuracy improvement. Because of the structural features captured by the branch, our DGNet can better filter more noise, which indicates that our gradient aggregation branch helps to work better under different levels of noise.

**Table 5.** The effect of multiscale branches and RFCB on our model.

| Method | ACC (%) | 10 dB (%) | 5 dB (%) | 0 dB (%) |
|---|---|---|---|---|
| No branch + RFCB | 100.00 | 96.02 | 93.21 | 85.46 |
| 1 branch + RFCB | 100.00 | 97.38 | 95.73 | 88.31 |
| 2 branches + RFCB | 100.00 | 98.83 | 97.37 | 90.27 |
| 3 branches + RFCB (ours) | 100.00 | 99.91 | 99.20 | 95.44 |
| 3 branches | 100.00 | 96.78 | 94.03 | 87.47 |

Then, to further verify the effectiveness of the multi-scale structure in our gradient aggregation, we use branches of 0 to 3 for the four groups shown in Table 5. It is obvious that more branches in multi-scale structures can achieve better performance. Specifically, 3 branches in gradient aggregation have improved by 11.70% compared with no branch. Multi-scale structures can fuse both low-level features that represent details and high-level semantic information, which is helpful for the characterization of the feature distribution of fault signals. In addition, we removed the RFCB module in the composition branch to test the effectiveness of multiple receptive fields. The results in Line 5 of the table indicate that without the RFCB module, the performance decreases relatively strongly, which proves that multiple receptive fields contribute to fault classification. With a greater receptive field, our DGNet can better capture global features to generate structural information, and the multiple receptive fields can make up for the feature loss by a single dilation rate.

### 3.3 Comparison with Other Models

According to the above experimental results and a series of training experiments, we set the relevant hyperparameters of training are illustrated. The initial learning rate is 0.1, the weight decay is 1e−4, and the dropout rate is 0.1, we train 400 epochs and preserve the model with the highest accuracy. We compare DGNet with recent fault diagnosis methods [10,23–29] based on deep learning in terms of noise resistance and generalization ability, which are shown in Tables 6 and 7, respectively. All of the experiments are conducted on two NVIDIA GTX 1080 GPUs with PyTorch.

**Table 6.** Comparison of the accuracy and noise immunity of different methods.

| Method | ACC (%) | 10 dB (%) | 5 dB (%) | 0 dB (%) |
|---|---|---|---|---|
| DGNet | 100.00 | 99.93 | 98.98 | 95.19 |
| Reference [23] | 100.00 | 99.87 | 96.53 | 92.68 |
| Reference [26] | 100.00 | 98.23 | 96.67 | 92.15 |
| Reference [25] | 100.00 | 97.90 | 91.25 | 82.18 |
| Reference [28] | 100.00 | 97.24 | 89.16 | 77.20 |
| Reference [10] | 99.99 | 82.32 | 75.80 | 62.58 |
| Reference [29] | 99.64 | 97.23 | 93.40 | 90.78 |
| Reference [27] | 99.33 | 89.37 | 72.08 | 55.67 |
| Reference [24] | 98.89 | 86.07 | 80.22 | 72.47 |

As shown in Table 6, DGNet has great performance on noisy data. Especially in the case of a small SNR, it can still achieve high accuracy. In terms of noise resistance, compared with references [23] and [26], which maintain good performance, our method still obtains an advantage of 0.05–3.31%. From the above results, it is obvious that our proposed DGNet performs state-of-the-art approaches in terms of accuracy and noise resistance.

**Table 7.** Comparison of the accuracy and generalization ability of different methods.

| Generalizationability (%) | DGNet | Ref [23] | Ref [26] | Ref [25] | Ref [28] | Ref [10] | Ref [29] | Ref [27] | Ref [24] |
|---|---|---|---|---|---|---|---|---|---|
| ACC | 100.00 | 100.00 | 100.00 | 100.00 | 100.00 | 99.99 | 99.64 | 99.33 | 98.89 |
| A-B | 100.00 | 98.12 | 97.78 | 97.26 | 98.32 | 94.59 | 94.14 | 95.09 | 91.33 |
| A-C | 98.97 | 96.77 | 96.38 | 95.97 | 97.25 | 92.90 | 92.35 | 88.97 | 90.38 |
| A-D | 98.84 | 94.71 | 92.89 | 92.56 | 93.78 | 90.66 | 87.18 | 79.29 | 85.20 |
| B-A | 99.20 | 97.25 | 98.62 | 96.69 | 94.17 | 96.87 | 93.48 | 94.31 | 90.76 |
| B-C | 99.62 | 96.47 | 96.74 | 96.11 | 97.50 | 93.77 | 92.99 | 92.47 | 89.80 |
| B-D | 98.79 | 94.78 | 93.84 | 90.99 | 94.48 | 91.10 | 89.62 | 91.47 | 89.52 |
| C-A | 98.83 | 94.57 | 91.81 | 91.48 | 93.12 | 92.14 | 89.17 | 89.30 | 88.68 |
| C-B | 99.76 | 96.97 | 96.79 | 94.86 | 96.84 | 93.69 | 90.75 | 95.62 | 92.68 |
| C-D | 98.92 | 95.94 | 98.45 | 95.73 | 97.67 | 92.91 | 91.46 | 93.89 | 92.08 |
| D-A | 98.13 | 93.99 | 91.89 | 91.02 | 93.65 | 85.60 | 81.18 | 85.37 | 86.85 |
| D-B | 98.95 | 96.71 | 92.51 | 93.94 | 95.79 | 89.67 | 90.14 | 89.23 | 89.64 |
| D-C | 99.96 | 97.97 | 94.71 | 96.20 | 98.07 | 88.63 | 93.53 | 91.25 | 90.00 |
| AVG | 99.49 | 98.12 | 97.78 | 97.26 | 95.89 | 91.88 | 94.14 | 90.52 | 91.33 |

As shown in Table 7, DGNet shows the best generalization ability under all types of variable load conditions. Specifically, the accuracy of our method is improved by 4.13–19.55% compared with others under a load change of 0 hp-3 hp, and the average generalization ability is as great as 99.49%. In other words, in the fault diagnosis of practical industrial production machinery with load changes, our DGNet has the best generalization ability.

## 4 Conclusions

In this paper, we propose a dynamic perception network with gradient information aggregation, a model-based 1D-CNN for machinery fault diagnosis. Compared with 1D-CNN, the inference of 2D-CNN processing methods has low efficiency, and this aspect will damage the expression of local features without good time-frequency correlation. Due to the complex working conditions, the previous 1D-CNN methods have poor generalization ability for different speeds and loads and cannot solve the recognition of noise fault signals well. Thus, we use a two-branch structure for fault diagnosis. First, we adopt dynamic perception as the main branch. With an attention mechanism and kernel aggregation, it can be helpful to increase the representation capability of our model. Then, we fuse the multi-scale structural information into the dynamic perception through a gradient aggregation branch, which helps to impose structural constraints to filter low frequency noise. Experiments on the CWRU dataset, and ablation studies testify to the effectiveness of our method, which indicates that under complex working conditions of different loads, speeds, and noise interference, our method outperforms the state-of-the-art methods in terms of generalization ability and noise resistance.

## References


1. Ma, S., Cai, W., Shang, Z., Liu, G.: Lightweight deep residual CNN for fault diagnosis of rotating machinery based on depthwise separable convolutions. IEEE Access **7**, 57023–57036 (2019)
2. Hinton, G.E., Salakhutdinov, R.R.: Reducing the dimensionality of data with neural networks. Science **313**(5789), 504–507 (2006)
3. Lecun, Y., Bengio, Y., Hinton, G.: Deep learning. Nature **521**(7553), 436–444 (2015)
4. Che, C., Wang, H., Ni, X., Lin, R.: Hybrid multimodal fusion with deep learning for rolling bearing fault diagnosis. Measurement **173**, 108655 (2021)
5. Zou, Y., Liu, Y., Deng, J., Zhang, W.: A novel transfer learning method for bearing fault diagnosis under different working conditions. Measurement **171**, 108767 (2021)
6. Zhao, Z., Qiao, B., Wang, S., Shen, Z., Chen, X.: A weighted multi-scale dictionary learning model and its applications on bearing fault diagnosis. J. Sound Vib. **446**(28), 429–452 (2019)
7. Shao, H., Jiang, H., Li, X., Wu, S.: Intelligent fault diagnosis of rolling bearing using deep wavelet auto-encoder with extreme learning machine. Knowl. Based Syst. **140**(15), 1–14 (2018)
8. He, X., Wang, D., Li, Y., Zhou, C.: A novel bearing fault diagnosis method based on gaussian restricted Boltzmann machine. Math. Probl. Eng. **2016**(2957083), 1–8 (2016)
9. Gan, M., Wang, C., Zhu, C.: Construction of hierarchical diagnosis network based on deep learning and its application in the fault pattern recognition of rolling element bearings. Mech. Syst. Sig. Process. **72**, 92–104 (2016)

10. Wen, L., Li, X., Gao, L.: A transfer convolutional neural network for fault diagnosis based on ResNet-50. Neural Comput. Appl. **32**(10), 6111–6124 (2019). https://doi.org/10.1007/s00521-019-04097-w
11. Sun, W.F., et al.: An intelligent gear fault diagnosis methodology using a complex wavelet enhanced convolutional neural network. Materials. **10**(7), 790 (2017)
12. Ince, T., Kiranyaz, S., Eren, L., Askar, M., Gabbouj, M.: Real-time motor fault detection by 1-D convolutional neural networks. IEEE Trans. Ind. Electron. **63**(11), 7067–7075 (2016)
13. Levent, E., Turker, I., Serkan, K.: A generic intelligent bearing fault diagnosis system using compact adaptive 1D CNN classifier. J. Sig. Proc. Syst. **91**(2), 179–189 (2019)
14. Chen, Y., Dai, X., Liu, M., Chen, D., Yuan, L., Liu, Z.: Dynamic convolution: attention over convolution kernels. In: IEEE Conference on Computer Vision and Pattern Recognition, pp. 11027–11036 (2020). https://doi.org/10.1109/CVPR42600.2020.01104
15. Hu, J., Shen, L., Albanie, H., Sun, G., Wu, E.: Squeeze-and-excitation networks. IEEE Trans. Pattern Anal. Mach. Intell. **42**(8), 2011–2023 (2020)
16. Dai, Q., Fang, F., Li, J., Zhang, G., Zhou, A.: Edge-guided composition network for image stitching. Pattern Recogn. **118**, 108019 (2021)
17. Liu, S., Huang, D., Wang, Y.: Receptive field block net for accurate and fast object detection. In: Ferrari, V., Hebert, M., Sminchisescu, C., Weiss, Y. (eds.) ECCV 2018. LNCS, vol. 11215, pp. 404–419. Springer, Cham (2018). https://doi.org/10.1007/978-3-030-01252-6_24
18. Smith, W.A., Randall, R.B.: Rolling element bearing diagnostics using the Case Western Reserve University data: a benchmark study. Mech. Syst. Sig. Process. **64**, 100–131 (2015)
19. Case Western Reserve University (CWRU) Bearing Data Center. http://csegroups.case.edu/bearingdatacenter/pages/download-data-file. Accessed 27 Apr 2021
20. Goodfellow, I., Bengio, Y., Courville, A.: Deep Learning. MIT Press, Cambridge (2016)
21. Jia, F., Lei, Y., Lin, J., Zhou, X., Lu, N.: Deep neural networks: a promising tool for fault characteristic mining and intelligent diagnosis of rotating machinery with massive data. Mech. Syst. Sig. Process. **72**, 303–315 (2016)
22. Chen, K., Zhou, X., Fang, J., Zheng, P., Wang, J.: Fault feature extraction and diagnosis of gearbox based on EEMD and deep briefs network. Int. J. Rotating Mach. **5**, 1–10 (2017)
23. Li, Z., Wang, Y., Ma, J.: Fault diagnosis of motor bearings based on a convolutional long short-term memory network of Bayesian optimization. IEEE Access. **9**, 97546–97556 (2021)
24. Long, Y., Zhou, W., Luo, Y.: A fault diagnosis method based on one-dimensional data enhancement and convolutional neural network. Measurement **180**, 109532 (2021)
25. Zhang, D., Zhou, T.: Deep convolutional neural network using transfer learning for fault diagnosis. IEEE Access **9**, 43889–43897 (2021)
26. Luo, J., Huang, J., Li, H.: A case study of conditional deep convolutional generative adversarial networks in machine fault diagnosis. J. Intell. Manuf. **32**, 407–425 (2021)
27. Gao, S., Pei, Z., Zhang, Y., Li, T.: Bearing fault diagnosis based on adaptive convolutional neural network with Nesterov momentum. IEEE Sens. J. **21**(7), 9268–9276 (2021)

28. Zhang, J., Sun, Y., Guo, L., Gao, H., Hong, X.: A new bearing fault diagnosis method based on modified convolutional neural networks. Chin. J. Aeronaut. **33**(2), 439–447 (2020)
29. Han, T., Zhang, L., Yin, Z., Tan, A.C.C.: Rolling bearing fault diagnosis with combined convolutional neural networks and support vector machine. Measurement **177**, 109022 (2021)

# Automatic Language Identification for Celtic Texts

Olha Dovbnia[1], Witold Sosnowski[2], and Anna Wróblewska[1(✉)]

[1] Faculty of Mathematics and Information Science,
Warsaw University of Technology, Warsaw, Poland
anna.wroblewska1@pw.edu.pl

[2] Faculty of Information Technology,
Polish-Japanese Academy of Information Technology, Warsaw, Poland
witold.sosnowski@hotmail.com

**Abstract.** Language identification is an important Natural Language Processing task. Though, it has been thoroughly researched in the literature, some issues are still open. This work addresses the identification of the related low-resource languages on the example of the Celtic language family. In this work, we collected a new dataset for the combination of the selected languages: Irish, Scottish, Welsh and also English.

Language identification is a classification task, so we applied supervised models such as SVM and neural networks. Traditional statistical features were tested alongside the output of clustering, autoencoder, and topic modelling methods. The analysis showed that the unsupervised features could serve as a valuable extension to the n-gram feature vectors. It led to an improvement in performance for more entangled classes. The best model achieved 98% F1 score and 97% MCC. The dense neural network consistently outperformed the SVM model.

**Keywords:** Natural language identification · Natural language processing · Low-resource languages · Machine learning · Classification

## 1 Introduction

Language Identification (LI) approaches the problem of automatic recognition of specific natural languages [6]. LI has multiple applications in the modern world. The original use case is routing source documents to language-appropriate Natural Language Processing (NLP) components, such as machine translation and dialogue systems [18]. Moreover, LI is a valuable component of the corpus creation pipelines, especially for the low-resource languages. Text crawled from the web resources usually needs to be separated by language to create a meaningful corpus. Most of the other NLP methods assume input text is monolingual; in that case, routing of the input documents is crucial.

LI can be a relatively easy task for the most popular languages with the abundance of available resources. Nevertheless, there are languages with fewer native speakers and less research focus; such languages are considered low-resource. Low-resource languages and related languages continuously pose a problem for


M. Tanveer et al. (Eds.): ICONIP 2022, CCIS 1793, pp. 264–275, 2023.
https://doi.org/10.1007/978-981-99-1645-0_22

language identification [5,6]. This paper focuses on applying language identification methods to a family of low-resource languages on the example of the Celtic language group. The main problem with the low-resource languages is the unavailability of high-quality corpora. In our research, we created a corpus of three Celtic languages. It contains Irish, Scottish, and Welsh texts. Moreover, the corpus is extended with a small proportion of English samples. This work also focuses on the preparation of a language identification model and the evaluation of the different feature extraction methods. In particular, we explored the possibility of the application of the output of the unsupervised learning models as a feature representation for the classification task. To solve the lack of large corpora for low-resource languages, we proved that the features extracted from unsupervised methods ensure high performance on the reduced labelled set size. Consequently, this study validates that it is possible to save resources on data annotation, determine the language class of low-resource languages more efficiently, and ensure that the data for further processing is of high quality. Thus, in this study, our main contributions are:

- Collecting the dataset of three Celtic languages (Irish, Scottish, Welsh) and English (Sect. 3),
- Preparing a method to identify these languages from the Celtic family and differentiate them from English (Sect. 4),
- Exploring and evaluating the applicability of unsupervised models as a feature extraction technique and experimenting with the reduced labelled dataset to save human labour work (Sect. 5).[1]

In the following sections, we discussed the related research (Sect. 2), the dataset creation process (Sect. 3), the proposed approach (Sect. 4), the results of the experiments (Sect. 5), and conclusions (Sect. 6).

## 2 Related Work

Language Identification faces unique challenges compared to other NLP tasks. Most NLP problems assume that the language of the source data is known. Therefore, it is possible to apply language-specific approaches to tokenization and grammar representation. Moreover, the text belonging to the same language can be written with varying orthography and encodings.

The most popular features in LI literature are n-grams, both on character and word level [4]. While most of the articles address them in some way, more unusual features appear as well. [9] enriched n-gram classification with text statistics and POS-tagging to differentiate varieties of the Dutch language. Table 1 presents a summary of articles working on related low-resource languages.

[13] dealt with Irish and Welsh languages. However, they had a bit different task compared to us, resulting in the different dataset and the application

[1] Our dataset and source code is available at https://github.com/grant-TraDA/celtic_languages_detection.

**Table 1.** Results from research papers on similar tasks – related low-resource languages identification. Note: char LM – character language model

| Article | Languages | # of Classes | Classifier | Accuracy | F1 |
|---|---|---|---|---|---|
| [13] | Irish, Welsh, Breton | 3 pairs | Char LM | 74–89% | 32–63% |
| [15] | Philippines local languages, English | 8 | SVM | 92% | 97% |
| [5] | Uyghur, Kazakh | 2 | MaxEnt | 96% | 95% |
| [14] | Indian local languages | 9 | Weak Labels | 98% | 98% |

of other models. They attempted to perform language identification on code-switching samples, i.e., to identify segments of different languages in one sample. Their classifiers performed the binary classification on language pairs of a Celtic language and English. The dataset collected and used in their work was scarce (40 to 50 samples per language pair), so authors had to use features based on word lists. [15] performed classification of the Philippines languages without code-switching. Their dataset had similar class sizes to the ones chosen in our work, but the number of classes was higher. Additionally, the level of closeness between the classified languages is not known, and several language groups can be included. Another paper addresses the separation of the Uyghur and Kazakh short texts [5]. These languages are also low-resource and closely-connected. The similarities between these languages at the word level are about 90%. A large corpus was created and professionally annotated in this work, ensuring high quality of the data.

In our research, the task of LI was performed on languages from the Celtic family. The whole language family has a deficient number of native speakers, making this language group low-resource. Additionally, we applied unsupervised models as feature extraction methods to explore the possibility of reducing the need in an extensive labelled dataset (using a large amount of unlabelled data with a smaller proportion of the labelled entries). Therefore, we addressed the issues of the related low-resource languages and unsupervised LI.

## 3 Our Collected Dataset

We considered the most popular Celtic languages: Irish, Scottish Gaelic, and Welsh. Irish and Scottish belong to the same language subgroup. Consequently, they are likely to be more challenging to distinguish. English, on the other hand, comes from a completely different language family. The corpus for a combination of these Celtic languages had not existed. In this work, it was aggregated from resources available online for each of the languages. Irish data comes from the collection of historical texts (up to 1926) [16] which was the only open corpus of Irish that we had found. The data was not prepared for immediate processing and required scraping from the web page. We divided the text into sentences and filtered by language using a FastText model [3,7,8] as some English sentences appeared in the data. English sentences were filtered out according to a threshold set experimentally to 0.5. The source corpus for Scottish Gaelic [17] was annotated at code-switching points of Scottish and English. It required scraping and sentencing. During sentencing, an additional category of code-mixed text emerged and was filtered out. Publicly available language resources for Welsh were the most scarce. Three sources were combined: raw data from research by [2]; dataset of Welsh phrases [12]; historical texts [11]. After sentencing, the same FastText annotation as for Irish was applied with a new experimental threshold set to 0.8. English represents a language class that is commonly mixed with these Celtic languages. Thus, English sentences were taken from a reduced version of British National Corpus [1]. Additional quality checks were not necessary in this case.

Class sizes of the final dataset were chosen similar to those used by [15], in which the authors performed LI on the related low-resource languages similar to our task. Table 2 shows our corpus statistics in detail.

**Table 2.** Final dataset statistics

| | Total | Irish | Scottish | Welsh | English |
|---|---|---|---|---|---|
| Sentences | 9,969 | 2,689 | 2,582 | 3,098 | 1,600 |
| Unique sentences | 9,942 | 2,688 | 2,568 | 3,086 | 1,600 |
| Words | 172,655 | 46,495 | 52,745 | 48,688 | 24,727 |
| Average words/sentence | 17.31 | 17.29 | 20.42 | 15.71 | 15.45 |

## 4 Our Approach

We applied supervised models for language classification and explored the possibility to utilize the advantages of the unsupervised learning in the LI pipeline. An unsupervised model identified the hidden patterns in the whole dataset and transformed the data into a latent representation. Later, the created structural information was employed as the features for the classification models. The following kinds of features were prepared and then tested in different combinations:

- Individual characters encoded as integers in a dictionary (bag-of-characters),
- n-grams (bi-grams, tri-grams) – the most popular features mentioned in the LI literature [4],
- Text statistics – the average word length in the input and the average number of consonants per word,
- Clustering outputs – k-means (initial number of clusters: 4), Birch (Balanced Iterative Reducing and Clustering using Hierarchies), agglomerative clustering, and Gaussian Mixture models based on the n-grams and text statistics (then the classifiers take as an input results of the all four clustering outputs),

- 2-dimensional variational autoencoder (VAE) hidden representation – the trained VAE architecture took as an input text encoded as a sequence of characters and forwarded it into 4 fully-connected encoder and decoder layers (see Fig. 1),
- Vector with probabilities of the input sample belonging to each of the Latent Dirichlet Analysis (LDA) topics (the number of topics was set to 4).

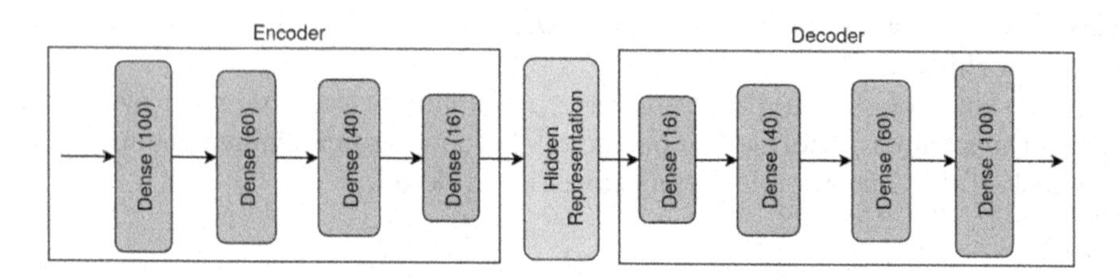


**Fig. 1.** VAE architecture used in our experiments. Numbers in parentheses stand for the number of neurons in a layer

We tested two settings – LI pipelines: (1) classification with a full training set, (2) preparing unsupervised output on the full training dataset, and then utilizing supervised classification with the reduced training dataset. Both settings were tested on the same test set. We experimented with a support vector machine (SVM), a dense neural network (NN), a convolutional neural network (CNN) on the features described above. Then, we compared the achieved results with a large pre-trained English model: RoBERTa [10] encoder provided by the *huggingface* library as the pretrained language model *roberta-large*. We tested case-sensitive RoBERTa encoder with the dropout and classification head[2]. The input consisted of at most 512 tokens. The training was performed with the AdamW optimiser with the scheduler and the weight decay of 0.01. We set the batch size of 32 and the dropout rate in all the layers was 0.1. For testing the performance of the encoder, we used F1-weighted. Each fine-tuning experiment was repeated for the prepossessed and plain input text.

The same preprocessing was applied in all our experiments. It involved the removal of punctuation, special symbols, digits, and the application of lowercasing but the words were not stemmed nor lemmatised. During the experiments, 80% of the dataset was taken as the training set and 20% as the testing set. We used a random constant train-test split of our dataset in all experiments. In the reduced set scenario, we get 30% of training dataset for classification models (in each test it was the same part of training dataset).

[2] https://huggingface.co/docs/transformers/model_doc/roberta#transformers.RobertaForSequenceClassification.

# 5 Results

## 5.1 Data Insights from the Unsupervised Models

To get a better understanding of the relationships in the multilingual data, we applied the unsupervised learning. Figure 2 shows the statistical feature space – n-grams and text statistics – transformed into a 2-dimensional representation using the Principal Component Analysis (PCA). With this initial test, we discovered that in our corpus Welsh has the closest connection to English, but these languages remain mostly separable from each other. Moreover, they are independent of the Irish-Scottish pair. Distinguishing between Irish and Scottish appears to be a more demanding problem as they are not linearly separable in this space; however, they are distinct from Welsh and English.

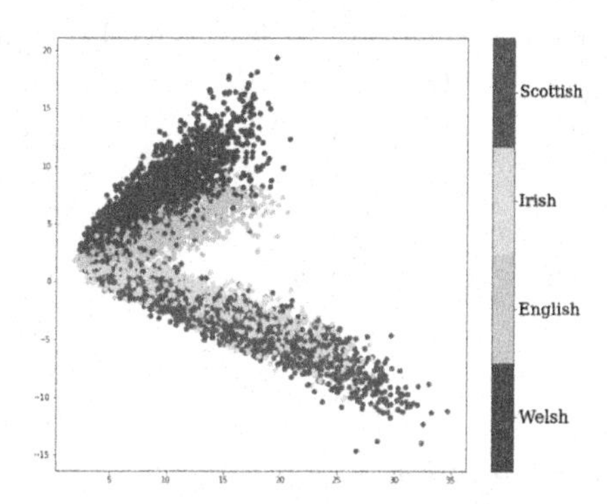


**Fig. 2.** 2D statistical features representation: n-grams and text statistics. The features form natural clusters for the four considered languages, however the distinction between Scottish and Irish is more challenging

Figure 3 shows the groups (clusters) found by the chosen clustering models, which were trained based on the statistical features: n-grams and text statistics. The hierarchical clustering models (see Fig. 3a and 3b) identified the clusters vaguely corresponding to Welsh and English, but the two other languages proved to be challenging. The Gaussian mixture provides the best-fitted representation (Fig. 3d) among all clustering methods. This model makes errors only with a cluster mapping to the Irish language wrongly classifying most of its samples as Scottish. K-means algorithm (Fig. 3c) has identified clusters with similar distribution as the agglomerative approach but with more sharp cluster borders.

The LDA model we created operated on the word n-grams (bi-grams and tri-grams). LDA topics 1 and 2 are related closely, while topics 3 and 4 are separated from the other ones. We discovered that for topics 1 and 3, most of the terms belong to Irish and English, respectively (see Table 3). The topic 2 contains n-grams that simultaneously appear in all considered Celtic languages. The topic 4 includes a combination of words belonging mostly to Welsh and English.

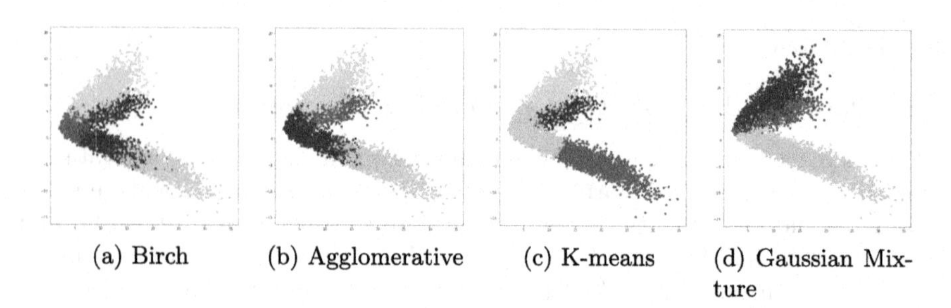

(a) Birch (b) Agglomerative (c) K-means (d) Gaussian Mixture

**Fig. 3.** Clusters found by the clustering algorithms transformed with PCA. The colours represent the identified groups without any language mapping

**Table 3.** LDA top-5 terms per topic

| Topic 1 (Irish) | Topic 2 (common Celtic words) | Topic 3 (English) | Topic 4 (Welsh and English) |
|---|---|---|---|
| agus | an | and | na |
| an | is | the | yn |
| na | air | of | ar |
| do | ach | to | fy |
| air | ar | for | la |

## 5.2 Supervised Experiments

We tested character encodings as the input vectors for the neural models: dense network (NN) and convolutional neural network (CNN). Table 4 shows that CNN is a definite leader. Therefore, language identification is best performed on small text fragments (CNN filter sizes were set to 2, 3, 4 characters) as opposed to individual character features. We measured our results mainly with mean F1 score, and Matthew's correlation coefficient (MCC), because the dataset is slightly imbalanced – see Table 2.

**Table 4.** Neural networks' results on characters encodings.

| Model | Accuracy | MCC | Mean F1 |
|---|---|---|---|
| NN | 92% | 89% | 92% |
| CNN | **94%** | **92%** | **94%** |

In the subsequent experiments, we used the character n-grams as the input representation for classification models: a support vector machine (SVM) and a dense network (NN). They are the most popular feature extraction and classification approaches for LI task. Two text statistics measures extended them: the average word length and the average number of consonants per word for each sample. However, these text statistics had a deteriorating effect on the classification performed by SVM and NN – see Table 5. The NN model using only n-gram features looks especially promising, even better than previous models reported in Table 4. The NN models have higher results than SVM models on MCC and F1 score for the Irish and Scottish classes, which was identified before as the most challenging.

**Table 5.** Results for models trained on 100% of the labelled training set. Note: the results are measured on an independent test set.

| Feature | Accuracy | MCC | Mean F1 | F1 (Irish) | F1 (Scottish) |
|---|---|---|---|---|---|
| **SVM** | | | | | |
| n-grams | 97% | 96% | 97% | 96% | 96% |
| n-grams+text stat | 96% | 95% | 97% | 95% | 96% |
| **NN** | | | | | |
| n-grams | 98% | 98% | 99% | 98% | 98% |
| n-grams+text stat | 98% | 97% | 98% | 97% | 98% |
| clusters | 76% | 68% | 77% | 64% | 51% |
| VAE | 46% | 25% | 36% | 30% | 59% |
| LDA | 64% | 52% | 65% | 56% | 51% |
| clusters+n-grams | 97% | 96% | 97% | 95% | 96% |
| VAE+n-grams | 98% | 97% | 98% | 97% | 98% |
| LDA+n-grams | 98% | 97% | 98% | 97% | 98% |
| RoBERTa-text preprocessed | 97% | 97% | 97% | 97% | 97% |
| RoBERTa-original text | **99%** | **99%** | **99%** | **99%** | **99%** |

### 5.3 Semi-supervised Experiments

In this experiment, we evaluated the impact of the unsupervised feature extraction approaches and compared them with fine-tuned pre-trained large RoBERTa model. Classification models using clustering output achieved the best performance among the models based on the exclusive unsupervised features. When considering combined features, VAE output and n-grams are the best; however, they are slightly worse than the sole n-gram-based models (Table 5). These results are marginally better than English RoBERTa pre-trained model when the input text is preprocessed and slightly worse than English RoBERTa pre-trained model with the original text as an input without any preprocessing. Figure 4 also shows additional analysis of the results after grouping results for different lengths of input samples from the test set. For this purpose, test observations were first sorted by length and then grouped into 10 bins, each containing the same number of sequences of similar length. The x-axis indicates the rounded average number of tokens from the sentences in a given bin. The models in the groups preserve similar characteristics as the results for the whole test dataset. Noteworthy, for the longer input texts, the RoBERTa models are a bit worse. Nevertheless, it can be beneficial to have a huge pre-trained model of a similar language, English, compared with the other languages from the Celtic family. Yet, it is not always possible in the case of low-resource languages.

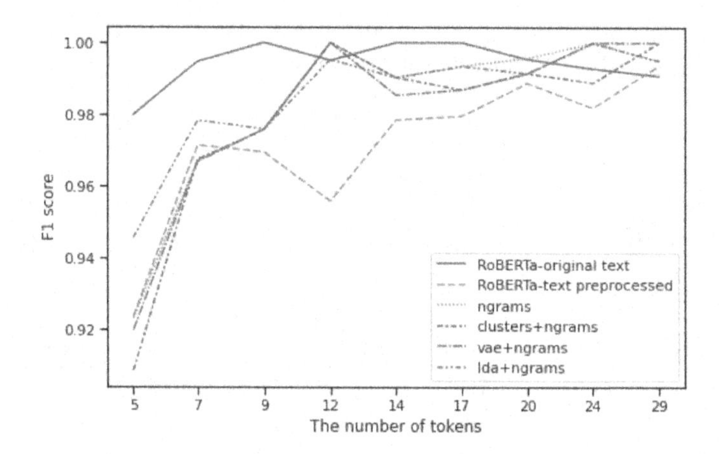


**Fig. 4.** Results for neural networks and RoBERTa models trained on the original and preprocessed text given for input samples with different number of tokens in the test set

### 5.4 Experiment with a Reduced Labelled Set

Finally, we evaluated the possibility of the application of the unsupervised features trained on a large amount of unlabelled data to create a classifier trained on a smaller labelled dataset. The previous experiments showed that acceptable – comparable but slightly worse – performance level is achieved when the unsupervised features are combined with the n-grams. So these models will be compared to the best supervised models that were using only the n-grams on the reduced labelled dataset. In this experiment, 30% of the training set was used for classifiers, while unsupervised features were prepared on the whole training set without labels; the testing set remained unchanged.

When training the classifiers only on the 30% of the labelled training set, the drop in MCC for the best-supervised models was minor. For the neural network, VAE latent representation combined with n-grams achieved the best results among the unsupervised features (Table 6). However, it could not beat the only

**Table 6.** Results for neural networks trained on 30% of the labelled training set. Note: unsupervised features were prepared on the full training dataset; the results are measured on an independent test set.

| Feature | Accuracy | MCC | Mean F1 | F1 (Irish) | F1 (Scottish) |
|---|---|---|---|---|---|
| clusters+n-grams | 95% | 94% | 96% | 94% | 95% |
| VAE+n-grams | 98% | 97% | 98% | 97% | 98% |
| LDA+n-grams | 96% | 95% | 96% | 95% | 97% |
| only n-grams | 98% | 97% | 98% | 97% | 98% |
| RoBERTa-text preprocessed | 96% | 96% | 96% | 96% | 96% |
| RoBERTa-original text | **99%** | **99%** | **99%** | **99%** | **99%** |

n-gram supervised model, reaching the same results level. It was unclear whether the classifier uses the unsupervised features or its predictions are based solely on the n-grams, given the similarity of the performance results to the supervised models. We trained both supervised and semi-supervised (VAE) models using 30% of the labelled training set. The confusion matrices in Table 7 show their performance. While the classification of Welsh and English is unchanged with the addition of the features based on the VAE representation, a slight improvement in the separation of the Irish-Scottish pair is observable. It means that the VAE features are considered and have a beneficial impact on the language classes that were the hardest to separate in all our experiments.

**Table 7.** Confusion matrices of the best NN models on the 30% of the labelled training set (W – Welsh, E – English, I – Irish, S – Scottish)

| | | Predicted labels | | | | | | | |
|---|---|---|---|---|---|---|---|---|---|
| | | VAE+n-grams (30%) | | | | n-grams (30%) | | | |
| | | W | E | I | S | W | E | I | S |
| True labels | Welsh (W) | 611 | 2 | 0 | 0 | 611 | 2 | 0 | 0 |
| | English (E) | 0 | 300 | 0 | 1 | 0 | 300 | 0 | 1 |
| | Irish (I) | 2 | 9 | 512 | 11 | 2 | 9 | 510 | 13 |
| | Scottish (S) | 0 | 1 | 9 | 528 | 0 | 1 | 11 | 526 |

## 6 Conclusions

This paper explored how to identify the Celtic languages and differentiate them from English. The dataset collected for this work contained three languages: Irish, Scottish, Welsh and a smaller English portion. Moreover, the paper compared LI results using various settings (e.g. with n-grams, preprocessed and word statistics). We showed that a neural network operating on the unsupervised and n-gram features could separate even closely-related languages, such as those in the Celtic language family. The dense neural network with n-gram input showed a better performance than SVM, and other networks with unsupervised features combined with n-grams. Noteworthy, for all classifiers, some language classes posed a continuous challenge. The predictions for the Irish and Scottish classes had the lowest performance. However, the VAE features with n-gram input presented a slight improvement in the separation of the Irish-Scottish pair.

The proposed approach to use the unsupervised models' output as the feature vectors for classification can be beneficial when the labelled data is scarce, like in this scenario of low-resource languages. These results confirm that the acceptable performance can be preserved even with a relatively small dataset, making the corpus annotation process less labour-intensive and expensive.

**Acknowledgments.** The research was funded by the Centre for Priority Research Area Artificial Intelligence and Robotics of Warsaw University of Technology within the Excellence Initiative: Research University (IDUB) programme (grant no 1820/27/Z01/POB2/2021).

## References

1. BNC Consortium: British national corpus, baby edition. Oxford Text Archive (2007). http://hdl.handle.net/20.500.12024/2553
2. Ellis, N.C., O'Dochartaigh, C., Hicks, W., Morgan, M., Laporte, N.: Raw files from Cronfa electroneg o Gymraeg (2001). https://www.bangor.ac.uk/canolfanbedwyr/ceg.php.en
3. FastText: Fasttext language identification model (2016). https://fasttext.cc/docs/en/language-identification.html
4. Hanani, A., Qaroush, A., Taylor, S.: Identifying dialects with textual and acoustic cues. In: Proceedings of the Fourth Workshop on NLP for Similar Languages, Varieties and Dialects (VarDial), pp. 93–101. Association for Computational Linguistics, Valencia, Spain (2017). https://doi.org/10.18653/v1/W17-1211. https://www.aclweb.org/anthology/W17-1211
5. He, J., Huang, X., Zhao, X., Zhang, Y., Yan, Y.: Discriminating between similar languages on imbalanced conversational texts. In: Proceedings of the Eleventh International Conference on Language Resources and Evaluation (LREC 2018). European Language Resources Association (ELRA), Miyazaki, Japan (2018). https://www.aclweb.org/anthology/L18-1497
6. Jauhiainen, T., Lui, M., Zampieri, M., Baldwin, T., Lindén, K.: Automatic language identification in texts: a survey. J. Artif. Intell. Res. **65**, 11675 (2018). https://doi.org/10.1613/jair.1.11675
7. Joulin, A., Grave, E., Bojanowski, P., Douze, M., Jégou, H., Mikolov, T.: Fasttext.zip: compressing text classification models. arXiv preprint arXiv:1612.03651 (2016)
8. Joulin, A., Grave, E., Bojanowski, P., Mikolov, T.: Bag of tricks for efficient text classification. arXiv preprint arXiv:1607.01759 (2016)

9. van der Lee, C., van den Bosch, A.: Exploring lexical and syntactic features for language variety identification. In: Proceedings of the Fourth Workshop on NLP for Similar Languages, Varieties and Dialects (VarDial), pp. 190–199. Association for Computational Linguistics, Valencia, Spain (2017). https://doi.org/10.18653/v1/W17-1224. https://www.aclweb.org/anthology/W17-1224
10. Liu, Y., et al.: Roberta: a robustly optimized BERT pretraining approach. arxiv 2019. arXiv preprint arXiv:1907.11692 (1907)
11. Testunau, L.O.: Welsh historical texts (2020). http://testunau.org/testunau/outlineeng.htm
12. Testunau, L.O.: Welsh phrases downloaded from BBC (2020). http://testunau.org/testunau/phrasebank/phrase.htm
13. Minocha, A., Tyers, F.: Subsegmental language detection in celtic language text. In: Proceedings of the First Celtic Language Technology Workshop, pp. 76–80. Association for Computational Linguistics and Dublin City University, Dublin, Ireland (2014). https://doi.org/10.3115/v1/W14-4612. https://www.aclweb.org/anthology/W14-4612
14. Palakodety, S., KhudaBukhsh, A.: Annotation efficient language identification from weak labels. In: Proceedings of the Sixth Workshop on Noisy User-generated Text (W-NUT 2020), pp. 181–192. Association for Computational Linguistics, Online (2020). https://doi.org/10.18653/v1/2020.wnut-1.24. https://www.aclweb.org/anthology/2020.wnut-1.24
15. Regalado, R.V.J., Agarap, A.F., Baliber, R.I., Yambao, A., Cheng, C.K.: Use of word and character n-grams for low-resourced local languages. 2018 International Conference on Asian Language Processing (IALP), pp. 250–254 (2018). https://doi.org/10.1109/IALP.2018.8629235
16. Royal Irish Academy: Historical Irish corpus 1600–1926 (2020). http://corpas.ria.ie/index.php?fsg_function=1
17. University of Glasgow: Corpas na gàidhlig, digital archive of Scottish Gaelic (2020). https://dasg.ac.uk/corpus/
18. Zhang, Y., Riesa, J., Gillick, D., Bakalov, A., Baldridge, J., Weiss, D.: A fast, compact, accurate model for language identification of codemixed text. In: Proceedings of the 2018 Conference on Empirical Methods in Natural Language Processing, pp. 328–337. Association for Computational Linguistics, Brussels, Belgium (2018). https://doi.org/10.18653/v1/D18-1030. https://www.aclweb.org/anthology/D18-1030

# Span Detection for Kinematics Word Problems

Savitha Sam Abraham[1(✉)], Deepak P[2,3], and Sowmya S. Sundaram[4]

[1] Örebro University, Örebro, Sweden
savitha.sam-abraham@oru.se
[2] Queen's University Belfast, Belfast, UK
deepaksp@acm.org
[3] IIT Madras, Chennai, India
[4] L3S Research Center, Hannover, Germany
sundaram@l3s.de

**Abstract.** Solving kinematics word problems is a specialized task which is best addressed through bespoke logical reasoners. Reasoners, however, require structured input in the form of kinematics parameter values, and translating textual word problems to such structured inputs is a key step in enabling end-to-end automated word problem solving. Span detection for a kinematics parameter is the process of identifying the smallest span of text from a kinematics word problem that has the information to estimate the value of that parameter. A key aspect differentiating kinematics span detection from other span detection tasks is the presence of multiple inter-related parameters for which separate spans need to be identified. State-of-the-art span detection methods are not capable of leveraging the existence of a plurality of inter-dependent span identification tasks. We propose a novel neural architecture that is designed to exploit the inter-relatedness between the separate span detection tasks using a single joint model. This allows us to train the same network for span detection over multiple kinematics parameters, implicitly and automatically transferring knowledge across the kinematics parameters. We show that such a joint training delivers an improvement of accuracies over real-world datasets against state-of-the-art methods for span detection.

## 1 Introduction

Solving word problems such as those from kinematics is quite a hard task involving text data. It requires exploitation of intricate domain knowledge such as those of kinematics equations, and are best addressed using bespoke reasoning-based solvers. For example, answering questions such as:

> *'How long would it take for a ball dropped from a 300 m tall tower to fall to the ground?'*

requires usage of knowledge from physics domain. State-of-the-art reasoners [1] take such a problem as input and generate a solution using the reasoning prowess embedded within them. A word problem from kinematics domain would describe,


M. Tanveer et al. (Eds.): ICONIP 2022, CCIS 1793, pp. 276–288, 2023.
https://doi.org/10.1007/978-981-99-1645-0_23

in English, a scenario involving a single object or multiple objects in motion. Reasoning-based solvers rely on the availability of an equivalent structured specification of the problem statement as shown below:

> **Initial state (S0)**: [($vertical\ velocity = 0\ m/s$), ($vertical\ position = 300\ m$), ($vertical\ acceleration = -9.8\ m/s^2$)]
> **Final State (S1)**: [($vertical\ position = 0\ m$), ($time = ?$)]

Viewing it from the perspective of identifying pertinent text spans to enable the text-to-structured conversion, observe that: (i) the span, *'ball dropped'* in the problem statement, implies that the vertical component of velocity is 0 $m/s$ initially, and (ii) the span, *'300 m'*, indicates the ball's vertical position initially and (iii) the span *'ground'* implies that the final vertical position is 0 $m$.

**Task Setting:** Given a word problem and a kinematics parameter (e.g., vertical position), we are interested in identifying the smallest span[1] of text from the word problem that contains all the information in order to estimate the value of that kinematics parameter. Two aspects of the task make it infeasible to use pattern-matching based span extraction.

*First*, values of the specific parameters are not necessarily explicitly specified in the word problem. For example, the spans of text representing the facts $vertical\ velocity = 0\ m/s$ and $vertical\ position = 0\ m$ in the above problem do not have any numerical quantities in them. Even when numerical quantities are specified explicitly, the domains of values overlap across different parameters, and the context determines the span-to-parameter mapping. Consider:

> *'A dart is thrown horizontally with a speed of 10 m/s towards the bull's eye. It hits a point on the rim vertically below the bull's eye, 0.2 s later. How far below the bull's eye does the dart hit?'*

Here, the span *'How far below'* is associated with vertical position, the span *'10 m/s'* is associated with horizontal velocity, while *'thrown horizontally'* is associated with vertical velocity (implies $vertical\ velocity = 0\ m/s$).

*Second*, word problems are often written in plain text and the number of different ways of expressing the same information are as numerous as the language allows. In view of these challenges that make it infeasible to construct a simple pattern-based extractor, we consider span identification for kinematics parameters using data driven approach as a natural alternative.

**Research Gap:** If these span detection tasks were to be treated as a sequence labelling problem, state-of-the-art architectures such as [3] for sequence labelling may be employed. A straightforward way of using such techniques would be to learn a separate model for each kinematics parameter. The kinematics word problem domain presents a very useful opportunity to improve over such a scenario, that of the existence of multiple but related kinematics parameters for which the

[1] *Smallest* avoids degenerate solutions that classify the entire text as span.

span detection task needs to be performed. Arguably, the span detector for the *vertical velocity* parameter can leverage some of the knowledge embedded within the detector for *horizontal velocity*. Such inter-task learning is hardly straightforward within current state-of-the-art sequence labelling frameworks that leverage deep learning methodologies. This is so since the learner is usually designed as one that is fed with *(whole text, relevant span)* pairs to learn a mapping $f$ : *input text* $\rightarrow$ *relevant span*; the learner is thus not cognizant of the kinematics parameter (and the nuances of the parameter) for which it is learning to identify spans for. Although there are certain multi-task learning architectures proposed for jointly learning multiple sequence labelling tasks as in [2], they represent a task by an identifier (that does not represent any task specific information), thus learning a mapping $f$ : *input text, task-id* $\rightarrow$ *relevant span.* To our best knowledge, there is no existing technique that can exploit semantic relationships between such inter-related tasks in span detection.

**Our Contributions:** In this paper, we propose the first method to exploit semantic inter-task relationships in multi-parameter span detection for kinematics word problems. In concrete terms, we design a framework that takes in *(whole text, kinematics parameter information, relevant span)* triplets in order to learn a mapping $f$ : *(input text, kinematics parameter information)* $\rightarrow$ *relevant span.* This allows for cross-parameter (across different kinematics parameters) learning, which would enable us to exploit the relationship between the different kinematics parameters to improve the accuracy of span detection for all the kinematics parameters. While our framework is inspired by and evaluated on span detection tasks over kinematics word problems, our architecture is potentially usable for other domains with inter-related span detection tasks.

## 2 Problem Definition

The vast majority of kinematics word problems in e-learning scenarios involve a body in motion within a two-dimensional space. The information about the body and its movement is usually fully specified using six parameters, its horizontal and vertical position (*hp* and *vp*), its horizontal and vertical velocity (*hv* and *vv*) and its horizontal and vertical acceleration (*ha* and *va*).

We now define the task of supervised span identification for kinematics word problems, for a particular chosen kinematics parameter, say $p$. First, we start with a training set, $\mathcal{T}_p$, comprising a set of kinematics word problems $\mathcal{W}$ along with their spans for the parameter $p$, $\mathcal{S}_p$.

$$\mathcal{T}_p = [\mathcal{W} = \{\dots, W, \dots\}, \mathcal{S}_p = \{\dots, \{.., [W_p^s, W_p^e], ..\}, \dots\}] \tag{1}$$

The elements of $\mathcal{S}_p$ have one-to-one correspondence with those in $\mathcal{W}$, with each element of $\mathcal{S}_p$ being a set of pairs, $\{..,[W_p^s, W_p^e], ..\}$, indicating the start and end positions of the various spans[2] for $p$, within the the word problem $W$.

[2] $p$ could have multiple spans within the same word problem, for example, one relating to the initial state, and another relating to the final state.

The current state-of-the-art in span detection can be easily leveraged in order to learn a span detection model, $\mathcal{M}_p$, using $\mathcal{T}_p$ as the training data; separate models would be learnt for each parameter $p$. In this paper, we consider learning a single model over the training sets across parameters $\mathcal{T} = \{\ldots, \mathcal{T}_p, \ldots\}$ in order to learn a cross-parameter joint model $\mathcal{M}$, which can then perform span prediction for every specified kinematics parameter it has been trained on. Informally, $\mathcal{M}$, when fed with $(W, p)$, would be able to predict the spans corresponding to $p$ from within the word problem $W$, i.e., it would predict $\{.., [W_p^s, W_p^e], ..\}$.

## 3 Related Work

We describe related works from two directions: recent NLP works on word problem solving and recent works on span detection.

**NLP in Automated Word Problem Solvers:** Most of the recent works in automated word problem solving focus on solving simple math word problems involving single linear equations [14], unlike our domain. Many existing kinematics word problem solvers [5,9] completely bypass the natural language understanding step by taking the structured specification of the word problem as input. Some other solvers like [4] relax this strict assumption somewhat by expecting users to manually represent the original word problem in a simplified English format. This is a tedious task for the user as it requires her to try different formulations of the word problem until it is accepted by the system as a valid reformulation. In short, NLP-based transformation of a kinematics word problem to a formal representation has not attracted significant research interest (to our best knowledge). It is in this context that we consider span identification for kinematics parameters using data-driven approaches in this paper.

**Span Detection by Sequence Labelling:** In recent times, much advancements in sequence labelling has been due to neural networks using layers of LSTMs (Long Short Term Memory) and BI(directional)-LSTMs. One of the pioneering works in this space [8] proposes a variety of neural architectures for the sequence labelling task, passing it through a recurrent layer (LSTM or Bi-LSTM) and then forwarding the output on to a Conditional Random Field (CRF) to produce a sequence labelling output. The work in [10], besides using BI-LSTM, leverages a convolutional neural network (CNN) to learn a character-based representation of the word, leading to an architecture codenamed BI-LSTM-CNN-CRF. From our initial experiments, we observed that the simpler BI-LSTM-CRF yields better results for kinematics span detection, arguably due to character-level representations being not very useful for detecting long text spans in our case. More recently, Changpinyo et al. [2] proposed multi-task learning architectures to jointly learn multiple sequence tagging tasks by augmenting a sequence labelling model (like BI-LSTM CRF) with "task embeddings", essentially a task identifier. This approach, by construction, puts all tasks as equally (un)related. Instead, we represent each task by some task specific information that is incorporated into the model in such a way that it can be used in detecting the relevant spans. They also capture relatedness between the tasks. Other span detection architectures like Bidirectional Attention Flow Network (BIDAF) [11] are designed to return a single span for each query. Since a word problem may have multiple spans associated with a kinematics parameter, BIDAF is not suitable for our task. Span detection may be considered as analogous to *slot-filling* as explored in generic text scenarios [7,13]; however, the nuances underlying the (kinematics) word problem domain make our task specification significantly different.

# 4 JKS: Our Approach

Our approach, **J***oint Learning for* **K***inematics* **S***pan Detection*, builds upon the state-of-the-art through two novel mechanisms:

- **Embedding Parameter Information:** We devise a method to inform sequence labelling frameworks of the kinematics parameter for which it is performing span detection by injecting parameter-relatedness information into the word problem representation. While we build upon a particular sequence neural network based labeller in our description, the spirit of the approach is generalizable across any sequence labelling method.
- **Joint Training:** The extended framework, fed with (kinematics) parameter-relatedness information, is then trained using cross-parameter training data. We outline a cross-parameter training method that trains the neural network so that it can then predict parameter-specific spans for the same kinematics word problem.

## 4.1 Span Detection Through Parameter-Enriched Representations

BI-LSTM-CRF [8] is a span detection model that we build upon in this paper; this also serves as a baseline method in our empirical evaluation. BI-LSTM-CRF is a span detection framework that is trained using training data of the form [*text sequence* $\rightarrow$ *expected span*]. Given our task of kinematics parameter span detection, we can learn a *separate BI-LSTM-CRF for each kinematics parameter* $p$ by feeding it with the training data $\mathcal{T}_p$. Each BI-LSTM-CRF, subjected to such (kinematics) parameter-specific training, is then useful only for span prediction for the particular parameter.

Here, we describe a generalization of the BI-LSTM-CRF model, named JKS, to render it applicable for generalized cross-parameter span detection. Figure 1 shows how BI-LSTM-CRF architecture is extended to incorporate task specific information into it (newly added layers are in blue). JKS enriches the input word problem representation with parameter-specific information so that the downstream processing within the BI-LSTM and CRF layers can make use of it in span detection. The JKS architecture expects training data objects of the form $(W, \mathcal{E}(p)) \rightarrow \{.., [W_p^s, W_p^e].., \})$, where $\mathcal{E}(p)$ is some information of the kinematics parameter $p$. This allows us to design a cross-parameter training method that we shall describe in the next section. The additional information $\mathcal{E}(p)$ is a vector from the same space as the word-embeddings. In particular, it may be a vector denoting a word or a phrase that is closely related to the kinematics parameter $p$. A simple construction of $\mathcal{E}(p)$ would be to choose a vector corresponding to words/phrases that are semantically related to the kinematics parameter. Section 5 (see Table 2) gives some examples of words/phrases that are related to some kinematics parameters of relevance. When the phrase comprises multiple words, we will pick the pre-trained embeddings of each word in the phrase, and model $\mathcal{E}(p)$ as the mean of the embeddings corresponding to the separate words that form the phrase.

The word problem $W$ is first passed on to an embedding layer, which involves simply looking up a pre-trained word embedding vector for every word in $W$. These may be 100-dimensional GloVe embeddings[3] from a word embedding model trained over a large cross-domain corpus. This would yield a $L \times 100$ matrix where $L$ is the maximum length of a kinematics word problem in the dataset. In the original BI-LSTM-CRF architecture, this $L \times 100$ matrix is directly passed onto the BI-LSTM layer. The key differentiator of JKS is that the input embedding $\mathcal{E}(p)$ is compared with each of the $L$ word embeddings to derive a cosine similarity quantification; the similarity to $\mathcal{E}(p)$ is then appended to each of the $L$ word embeddings of the input word problem resulting in $L$ 101-dimensional word embeddings (100d from the pre-trained embeddings, and an additional number denoting the cosine similarity with $\mathcal{E}(p)$). This $L \times 101$ matrix is then passed on to a BI-LSTM layer. A BI-LSTM layer takes this matrix and outputs a sequence of contextual representation of the words, which would be an $L \times 400$ matrix. A dense feed forward network projects the output of BI-LSTM onto a layer of size equal to the size of labels; four labels, *begin*, *within*, *end* and *outside*, are appropriate for our task. The CRF layer takes this projection as input, and predicts a probability distribution over the set of all labels. During the training phase, this prediction is compared against the expected labels and errors are propagated all the way back through the CRF, the dense layer, and then the BI-LSTM.

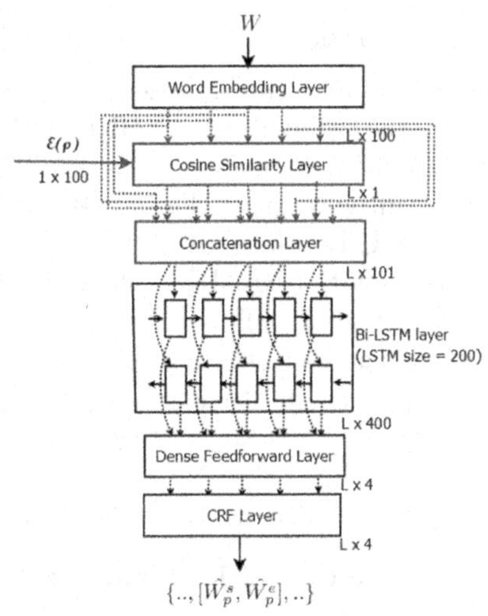


**Fig. 1.** JKS Architecture - BI-LSTM-CRF with the representation extension indicated in blue (Color figure online).

[3] http://nlp.stanford.edu/projects/glove.

### 4.2 Cross-Parameter Training of JKS

With the JKS architecture, described above (Sect. 4.1), designed to process training inputs of the form $(W, \mathcal{E}(p)) \rightarrow \{.., [W_p^s, W_p^e], ..\})$, cross-parameter training in JKS is straightforward. Consider a word problem $W$ that has labelling for two parameters $p1$ and $p2$; this will be used to create two training data elements:

$$[(W, \mathcal{E}(p1)) \rightarrow \{.., [W_{p1}^s, W_{p1}^e], ..\}] \qquad [(W, \mathcal{E}(p2)) \rightarrow \{.., [W_{p2}^s, W_{p2}^e], ..\}] \quad (2)$$

The training is simply performed over the training set formed from across different kinematics parameters. While we have illustrated the case for two parameters, this model can be extended straightforwardly to any number of parameters.

### 4.3 Discussion

The JKS architecture interleaves a cosine similarity and concatenation layer in between the embedding and BI-LSTM layers, providing the BI-LSTM layer with $L \times 101$ matrix as against the $L \times 100$ matrix used in the BI-LSTM-CRF framework. Intuitively, given that spans for a parameter $p$ would likely comprise words that are highly related to $p$, they are likely to have a high value in their cosine similarities with $\mathcal{E}(p)$. JKS, when trained with cross-parameter training sets, has an abundance of such parameter relevance information which makes it capable of exploiting such higher-level abstract cues implicitly within the learning process without having to do any explicit design to facilitate transfer of information across the different kinematics parameters.

## 5 Experiments and Results

### 5.1 Datasets and Experimental Setup

**Datasets:** We got a number of parameter spans labelled for the kinematics word problems, collected from internet sources and mechanics textbooks, by human labellers. Our focus was on velocity and position parameters since they are present for most word problems we had in our dataset. Based on the availability of labellings, we chose three kinematics parameters for our empirical evaluation. For the horizontal velocity (hv) and vertical position (vp) parameters, we had the spans marked for 210 word problems and for vertical velocity (vv), we had spans marked for 63 word problems. The datasets will be made public.

**Experimental Setup:** With our datasets being reasonably modest in size, we chose to use a round-robin evaluation strategy instead of a fixed test/train split. Towards that, we partition the dataset into $k$ splits (we set $k = 5$), and conducted $k$ different training/test evaluations, with each of the $k$ splits used for testing once. within each evaluation, one of the $k$ splits is chosen as the test set and another as the development set, with the remaining forming the training set. The test accuracies over each of these $k$ experiments would be averaged to obtain a single value across the $k$ settings. The hyper-parameters we used are listed in Table 1. We used the AdaDelta algorithm [12] for training.

**Table 1.** Hyper-parameters

| Hyper-parameter | Value |
|---|---|
| LSTM state size | 200 |
| Batch size | 10 |
| Gradient clipping | 5.0 |
| Initial learning rate | 0.01 |
| Decay rate | 0.05 |
| Number of epochs | 75 |

**Table 2.** Symbol-Phrase

| Task | Symbol | Phrase |
|---|---|---|
| $hv$ | hp | horizontal pace |
| | lv | lateral velocity |
| | hv | horizontal velocity |
| $vp$ | pp | plumb position |
| | vp | vertical position |
| | vd | vertical distance |
| $vv$ | vs | vertical speed |
| | ps | plumb speed |
| | vv | vertical velocity |

## 5.2 Evaluation Measure and Baselines

**Evaluation Measure:** The prediction made by a model on a test set problem is evaluated by measuring the extent of overlap between the actual span from the manual labelling and the span predicted by the model. For a particular kinematics parameter $p$ and word problem $W$, let $[W_p^s, W_p^e]$ denote the actual labelled span, and $[\hat{W_p^s}, \hat{W_p^e}]$ be the predicted span. We will use $|[X,Y]|$ to denote the number of words contained in the span $[X,Y]$. Now, a simple overlap metric between the predicted and labelled spans may be defined as follows:

$$O([W_p^s, W_p^e], [\hat{W_p^s}, \hat{W_p^e}]) = \frac{|[W_p^s, W_p^e] \cap [\hat{W_p^s}, \hat{W_p^e}]|}{max(|[W_p^s, W_p^e]|, |[\hat{W_p^s}, \hat{W_p^e}]|)} \tag{3}$$

where the intersection between spans is itself a span that comprises the sequence of common words across them. Given the construction, this overlap measure is in $[0,1]$, where 0.0 is when there is no overlap between the predicted and labelled spans, and 1.0 when the spans are identical. This measure is averaged across all predicted spans in the test dataset for each kinematics parameter.

**Baselines:** We compare our approach against three baselines. The first is the BI-LSTM-CRF technique [8] (abbreviated BLC), instantiated separately for each kinematics parameter. A similar parameter-specific instantiation can also be performed using JKS, by simply training separately with each parameter-specific dataset. Such an instantiation would yield separate models for each kinematics parameter (much like BLC); we will abbreviate this to SKS, as a shorthand for *Separate Kinematics Span Detection* (as against *joint* in JKS). SKS is the second baseline. The comparative evaluation between JKS and SKS enables studying the improvement achieved by means of the joint training as against that by the parameter-enriched representation alone. The third baseline is the multi-task learning architecture in [2]. Changpinyo et al. [2] proposed two architectures for Multi-Task Learning, one where task embedding is incorporated into the encoder (a BI-LSTM) by concatenating it with the input to the encoder (referred to as TE+ENC), and one where it is incorporated into the decoder (Dense-CRF layers) by concatenating it with the output of the encoder (referred to as

**Table 3.** Comparing JKS trained over the combination of *hv* and *vp* datasets vs. Baseline for *hv* span prediction. Table 2 maps symbols (pp,vd,..) to phrases.

| $\mathcal{E}(vp)$ | $\mathcal{E}(hv)$ | *hv* overlap accuracy | | | | | | | |
|---|---|---|---|---|---|---|---|---|---|
| | | JKS | TE+ENC-Sem | TE+DEC-Sem | TE+ENC-ID | TE+DEC-ID | SKS | BLC | Improvement (%) |
| pp | hp | **0.4684** | 0.4165 | 0.2199 | 0.3963 | 0.1595 | 0.4144 | 0.4304 | 8.8289 |
| pp | hv | **0.4938** | 0.3897 | 0.2172 | 0.3963 | 0.1595 | 0.4390 | 0.4304 | 12.4829 |
| pp | lv | **0.4945** | 0.3897 | 0.2172 | 0.3963 | 0.1595 | 0.4551 | 0.4304 | 8.6574 |
| vp | hp | **0.4854** | 0.4165 | 0.2199 | 0.3963 | 0.1595 | 0.4144 | 0.4304 | 12.7788 |
| vp | hv | **0.4830** | 0.3897 | 0.2172 | 0.3963 | 0.1595 | 0.4390 | 0.4304 | 10.0227 |
| vp | lv | **0.4791** | 0.3897 | 0.2172 | 0.3963 | 0.1595 | 0.4551 | 0.4304 | 5.2735 |
| vd | hp | **0.4776** | 0.3856 | 0.2088 | 0.3963 | 0.1595 | 0.4144 | 0.4304 | 10.9665 |
| vd | hv | **0.4554** | 0.4012 | 0.2072 | 0.3963 | 0.1595 | 0.4390 | 0.4304 | 3.7357 |
| vd | lv | **0.4609** | 0.4012 | 0.2072 | 0.3963 | 0.1595 | 0.4551 | 0.4304 | 1.2744 |
| Average | | **0.4775** | 0.3977 | 0.2146 | 0.3963 | 0.1595 | 0.4361 | 0.4304 | 8.2245 |

TE+DEC). Since both JKS and TE+ENC, TE+DEC do multi task learning, but only differ in the way task specific information is incorporated into the BI-LSTM-CRF model, comparing them allow us to evaluate the effectiveness of the technique used by JKS in incorporating semantic task specific information. The task embedding originally used in TE+ENC and TE+DEC is learned from a task-id. We name these architectures TE+ENC-ID and TE+DEC-ID. To make the comparison more fair, we additionally adapt TE+ENC and TE+DEC so that it learns task embeddings from semantically related phrases ($\mathcal{E}(p)$), instead of a task identifier, yielding techniques TE+ENC-Sem and TE+DEC-Sem.

**Table 4.** Comparing JKS trained over the combination of *hv* and *vp* datasets vs. Baseline for *vp* span prediction. Table 2 maps symbols (pp,vd,..) to phrases.

| $\mathcal{E}(vp)$ | $\mathcal{E}(hv)$ | *vp* overlap accuracy | | | | | | | |
|---|---|---|---|---|---|---|---|---|---|
| | | JKS | TE+ENC-Sem | TE+DEC-Sem | TE+ENC-ID | TE+DEC-ID | SKS | BLC | Improvement (%) |
| pp | hp | **0.5601** | 0.3889 | 0.1595 | 0.4187 | 0.2304 | 0.5282 | 0.5492 | 1.9847 |
| pp | hv | **0.5603** | 0.4105 | 0.1587 | 0.4187 | 0.2304 | 0.5282 | 0.5492 | 2.0211 |
| pp | lv | **0.5746** | 0.4105 | 0.1587 | 0.4187 | 0.2304 | 0.5282 | 0.5492 | 4.6249 |
| vp | hp | **0.5879** | 0.3889 | 0.1595 | 0.4187 | 0.2304 | 0.5296 | 0.5492 | 7.0466 |
| vp | hv | **0.5507** | 0.4105 | 0.1587 | 0.4187 | 0.2304 | 0.5296 | 0.5492 | 0.2731 |
| vp | lv | **0.5910** | 0.4105 | 0.1587 | 0.4187 | 0.2304 | 0.5296 | 0.5492 | 7.6110 |
| vd | hp | **0.5792** | 0.4583 | 0.1634 | 0.4187 | 0.2304 | 0.5135 | 0.5492 | 5.4624 |
| vd | hv | **0.5766** | 0.4234 | 0.1613 | 0.4187 | 0.2304 | 0.5135 | 0.5492 | 4.9890 |
| vd | lv | **0.5817** | 0.4234 | 0.1613 | 0.4187 | 0.2304 | 0.5135 | 0.5492 | 5.9176 |
| Average | | **0.5735** | 0.4138 | 0.1599 | 0.4187 | 0.2304 | 0.5237 | 0.5492 | 4.4367 |

### 5.3 Empirical Evaluation over *hv and vp*

We now evaluate the performance of $JKS$, trained over a combination of two kinematics parameters: *horizontal velocity* ($hv$) and *vertical position* ($vp$), against baselines. The BLC and SKS methods process the *hv* and *vp* training datasets separately to create separate models, which are then separately used for testing over the corresponding test datasets. The remaining baselines are trained over data from both *hv* and *vp*. SKS, TE+ENC-Sem, TE+DEC-Sem and JKS need

**Table 5.** Comparing *vv* accuracies of JKS trained over the combination of *hv*, *vp* and *vv* datasets vs. Baseline. Table 2 maps symbols (pp,vd,..) to phrases.

| $\mathcal{E}(vp)$ | $\mathcal{E}(hv)$ | $\mathcal{E}(vv)$ | *vv* overlap accuracy | | | | | | | |
|---|---|---|---|---|---|---|---|---|---|---|
| | | | JKS | TE+ENC-Sem | TE+DEC-Sem | TE+ENC-ID | TE+DEC-ID | SKS | BLC | Improvement (%) |
| pp | hp | ps | **0.3462** | 0.2649 | 0.0219 | 0.2890 | 0.0349 | 0.1739 | 0.2198 | 19.7923 |
| pp | hp | vv | **0.3262** | 0.3232 | 0.0219 | 0.2890 | 0.0349 | 0.1828 | 0.2198 | 0.9282 |
| pp | hp | vs | 0.2346 | 0.2649 | 0.0219 | **0.2890** | 0.0349 | 0.2425 | 0.2198 | −18.8235 |
| pp | lv | ps | **0.3317** | 0.2785 | 0.0219 | 0.2890 | 0.0349 | 0.1739 | 0.2198 | 14.7750 |
| pp | lv | vv | **0.3363** | 0.0919 | 0.0219 | 0.2890 | 0.0349 | 0.1828 | 0.2198 | 16.3667 |
| pp | lv | vs | 0.2403 | 0.2785 | 0.0219 | **0.2890** | 0.0349 | 0.2425 | 0.2198 | −16.8512 |
| pp | hv | ps | **0.3091** | 0.2785 | 0.0219 | 0.2890 | 0.0349 | 0.1739 | 0.2198 | 6.9550 |
| pp | hv | vv | **0.3438** | 0.0919 | 0.0219 | 0.2890 | 0.0349 | 0.1828 | 0.2198 | 18.9619 |
| pp | hv | vs | **0.3504** | 0.2785 | 0.0219 | 0.2890 | 0.0349 | 0.2425 | 0.2198 | 21.2456 |
| vp | hp | ps | **0.3572** | 0.2649 | 0.0219 | .2890 | 0.0349 | 0.1739 | 0.2198 | 23.5986 |
| vp | hp | vv | 0.2688 | 0.3232 | 0.0219 | **0.2890** | 0.0349 | 0.1828 | 0.2198 | −16.8316 |
| vp | hp | vs | **0.3413** | 0.2649 | 0.0219 | 0.2890 | 0.0349 | 0.2425 | 0.2198 | 18.0968 |
| vp | hv | ps | 0.2837 | 0.2785 | 0.0219 | **0.2890** | 0.0349 | 0.1739 | 0.2198 | −1.8339 |
| vp | hv | vv | **0.3394** | 0.0919 | 0.0219 | 0.2890 | 0.0349 | 0.1828 | 0.2198 | 17.4394 |
| vp | hv | vs | **0.3696** | 0.2785 | 0.0219 | 0.2890 | 0.0349 | 0.2425 | 0.2198 | 27.8892 |
| vp | lv | ps | **0.3291** | 0.2785 | 0.0219 | 0.2890 | 0.0349 | 0.1739 | 0.2198 | 13.8754 |
| vp | lv | vv | **0.2940** | 0.0919 | 0.0219 | 0.2890 | 0.0349 | 0.1828 | 0.2198 | 1.7301 |
| vp | lv | vs | **0.3224** | 0.2785 | 0.0219 | 0.2890 | 0.0349 | 0.2425 | 0.2198 | 11.5570 |
| vd | hp | ps | **0.3541** | 0.2782 | 0.0219 | 0.2890 | 0.0349 | 0.1739 | 0.2198 | 22.5259 |
| vd | hp | vv | **0.2966** | 0.28 | 0.0219 | 0.2890 | 0.0349 | 0.1828 | 0.2198 | 2.6297 |
| vd | hp | vs | **0.3189** | 0.2782 | 0.0219 | 0.2890 | 0.0349 | 0.2425 | 0.2198 | 10.3460 |
| vd | hv | ps | **0.3219** | 0.2245 | 0.0219 | 0.2890 | 0.0349 | 0.1739 | 0.2198 | 11.3840 |
| vd | hv | vv | **0.2929** | 0.0699 | 0.0219 | 0.2890 | 0.0349 | 0.1828 | 0.2198 | 1.3494 |
| vd | hv | vs | 0.2585 | 0.2245 | 0.0219 | **0.2890** | 0.0349 | 0.2425 | 0.2198 | −10.5536 |
| vd | lv | ps | **0.3207** | 0.2245 | 0.0219 | 0.2890 | 0.0349 | 0.1739 | 0.2198 | 10.9688 |
| vd | lv | vv | **0.2947** | 0.0699 | 0.0219 | 0.2890 | 0.0349 | 0.1828 | 0.2198 | 1.9723 |
| vd | lv | vs | **0.3227** | 0.2245 | 0.0219 | 0.2890 | 0.0349 | 0.2425 | 0.2198 | 11.6608 |
| Average | | | **0.3150** | 0.2287 | 0.0219 | 0.289 | 0.0349 | 0.1997 | 0.2198 | 8.1909 |

semantic information corresponding to the kinematics parameter $p$ (denoted as $\mathcal{E}(p)$) to be part of the training process; our choices of $\mathcal{E}(hv)$ and $\mathcal{E}(vp)$ are listed in Table 2. The results over all these combinations for *hv* and *vp* is shown in Tables 3 and 4 respectively, with the best overlap value for each parameter highlighted in each row. The improvement achieved by JKS over the next best method is indicated in percentage in the last column. BLC, TE+ENC-ID and TE+DEC-ID do not depend on $\mathcal{E}(.)$s and thus have the same value across the rows. The results unmistakably establish the pre-eminence of the JKS method over the baselines, with the trends across various methods being consistent across various $\mathcal{E}(.)$s. The performance of JKS surpasses that of TE+ENC-Sem and TE+DEC-Sem, that also use task specific information in the form of phrases in $\mathcal{E}(hv)$ and $\mathcal{E}(vp)$. This implies that the technique used by JKS to incorporate task specific information into the model (using cosine similarity) is better than the technique used by the latter architectures. It is interesting to note that there is a decline in the SKS span detection accuracy of *vp* when compared to BLC, indicating that the parameter information (i.e., $\mathcal{E}(.)$) do not help to improve the span detection much. This could also be partly due to the size of the datasets; larger datasets that contain enough redundancy to enable learning meaningful correlations between parameter information and span labellings could potentially widen the gap between SKS and BLC, though that is subject to verification.

### 5.4 Empirical Evaluation over *hv, vp and vv*

We now consider the full dataset that we have, that comprising all three parameters: *horizontal velocity* ($hv$), *vertical position* ($vp$) and *vertical velocity* ($vv$). Of these, $vv$ has significantly sparser labelled data. This makes it interesting to observe and quantify how much the joint learning using JKS comprising a relatively large dataset of spans from $hv$ and $vp$ and a small dataset of spans for $vv$ is able to improve upon the baselines trained over just the $vv$ data. We use three phrases *plumb speed*, *vertical speed* and *vertical velocity* to form $\mathcal{E}(vv)$. We experiment with various combinations of the phrases in $\mathcal{E}(hv)$, $\mathcal{E}(vp)$ and $\mathcal{E}(vv)$. The $vv$ overlap accuracies for all combinations is shown in Table 5. Along the lines of our expectation, JKS significantly outperforms the baselines BLC and SKS as JKS has an abundance of spans of related parameters $hv$ and $vp$ to make use of, whereas the latter architectures are left to work with the few scores of $vv$ spans supplied to them. Similarly the better performance of JKS as against TE+ENC-Sem, TE+DEC-Sem, TE+ENC-ID and TE+DEC-ID can be explained as before as in Sect. 5.3. The impact of adding a small number of $vv$ spans in improving upon the JKS $hv$ and $vp$ span predictions (observed from Sect. 5.3) may be expected to be small; this is so since $hv$ and $vp$ that already have an abundance of data are unlikely to benefit significantly from a much smaller set of problems from $vv$. Our empirical observation of JKS trained over all three datasets was found to be in line with this observation, and the corresponding trends of $hv$ and $vp$ overlap accuracies were very much like those observed in Tables 3 and 4; thus, we omit those results for space constraints.

### 5.5 Discussion

Based on the analysis, we may conclude the following:

- **Joint training improves span detection:** Through modelling of semantic information, JKS is able to exploit the task inter-relationships to deliver gains in span detection accuracies.
- **Sparse parameters also benefit from joint training:** Our second experiment indicates that parameters with very limited labelled data benefit significantly from joint training.
- **Effective representation of task:** The better performance of JKS as compared to other multi-task learning architectures like TE+ENC and TE+DEC, shows that the learning framework in JKS is more effective.

**Applicability of BERT Embeddings:** BERT [6], a state-of-the-art contextual language model, has become quite popular. We analyzed the performance of the proposed JKS model with BERT embeddings and observed that JKS with

GloVe embeddings performed better, upto 40%, than JKS with BERT embeddings. We believe that this is so since the contextual information from general corpora (Wikipedia and BooksCorpus [15]) encoded by the pre-trained BERT model is unsuited to our specific task domain of kinematics. Procuring a large dataset of kinematics text and learning BERT over them, would be a promising way forward to exploit contextual information within the JKS framework.

## 6 Conclusions and Future Work

We considered the novel task of exploiting inter-task semantic dependencies for span detection, inspired by the scenario of kinematics word problem solving. We developed a neural architecture, JKS, which uses a joint training framework to incorporate task-relevant semantic information during the training process. This enables JKS to learn a single model for multiple parameters, implicitly and automatically transferring knowledge across parameters. Through an empirical evaluation over real-world datasets over several existing and adapted baselines, we illustrate that JKS is able to achieve significant gains in span detection accuracy. These observations illustrate the superiority of JKS for the task.

**Future Work:** We are currently exploring the suitability of JKS to other reasoning-heavy problem domains such as kinetics and those from mathematical disciplines. Another potential research direction would be towards predicting the temporal relations between the spans within a word problem, which would further ease the operation of the downstream reasoner.

## References

1. Abraham, S.S., Khemani, D.: Hybrid of qualitative and quantitative knowledge models for solving physics word problems. In: FLAIRS (2016)
2. Changpinyo, S., Hu, H., Sha, F.: Multi-task learning for sequence tagging: an empirical study. arXiv:1808.04151 (2018)
3. Chiu, J., Nichols, E.: Named entity recognition with bidirectional LSTM-CNNs. Trans. Assoc. Comput. Linguist. **4**(1), 357–370 (2016)
4. Clark, et al.: Capturing and answering questions posed to a knowledge-based system. In: K-CAP (2007)
5. De Kleer, J.: Multiple representations of knowledge in a mechanics problem-solver. In: Readings in Qualitative Reasoning About Physical Systems, pp. 40–45 (1990)
6. Devlin, J., Chang, M.W., Lee, K., Toutanova, K.: BERT: pre-training of deep bidirectional transformers for language understanding. arXiv:1810.04805 (2018)
7. Huang, L., Sil, A., Ji, H., Florian, R.: Improving slot filling performance with attentive neural networks on dependency structures. In: EMNLP (2017)
8. Huang, Z., Xu, W., Yu, K.: Bidirectional LSTM-CRF models for sequence tagging. arXiv:1508.01991 (2015)
9. Kook, H.J., Novak, G.S.: Representation of models for expert problem solving in physics. IEEE Trans. Knowl. Data Engg. **3**(1), 48–54 (1991)

10. Ma, X., Hovy, E.: End-to-end sequence labeling via bi-directional LSTM-CNNs-CRF. In: ACL, vol. 1 (2016)
11. Seo, M., Kembhavi, A., Farhadi, A., Hajishirzi, H.: Bidirectional attention flow for machine comprehension. arXiv:1611.01603 (2016)
12. Zeiler, M.D.: ADADELTA: an adaptive learning rate method. arXiv:1212.5701 (2012)
13. Zhang, C., Li, Y., Du, N., Fan, W., Yu, P.S.: Joint slot filling and intent detection via capsule neural networks. In: ACL (2019)
14. Zhang, D., Wang, L., Xu, N., Dai, B.T., Shen, H.T.: The gap of semantic parsing: a survey on automatic math word problem solvers. arXiv:1808.07290 (2018)
15. Zhu, Y., et al.: Aligning books and movies: towards story-like visual explanations by watching movies and reading books. In: ICCV (2015)

# Emotion-Aided Multi-modal Personality Prediction System

Chanchal Suman[1(✉)], Sriparna Saha[1], and Pushpak Bhattacharyya[2]

[1] Department of Computer Science and Engineering, Indian Institute of Technology Patna, Bihta, India
{1821cs11,sriparna}@iitp.ac.in

[2] Department of Computer Science and Engineering, Indian Institute of Technology Bombay, Mumbai, India
pb@cse.iitb.ac.in

**Abstract.** Cyber forensics, personalized services, and recommender systems require the development of automatic personality prediction systems. Current paper works on developing a multi-modal personality prediction system from videos considering three different modalities, text, audio and video. The emotional state of a user helps in revealing the personality. Based on this cue, we have developed an emotion-aided personality prediction system in a multi-modal setting. Using the IBM tone analyzer, the existing ChaLearn-2017 dataset is enriched with emotion labels and those are used as an additional feature set in the proposed neural architecture for automatic personality prediction. Different features from video, audio, and text are extracted using CNN architectures and finally, the emotion labels are concatenated with the extracted feature set before feeding them to the sigmoid layer. For experimentation purposes, our enriched dataset is used. From the obtained results, it can be concluded that the concatenation of emotion labels as an additional feature set yields comparative results.

**Keywords:** Personality · Emotion · Convolutional Neural Network · Auxiliary

## 1 Introduction

The combination of an individual's emotions, behavior, thought patterns, and motivation makes up their personality. Our personalities impact our lives, influencing our mental health, life choices, desires, and happiness. Thus, enhanced personal assistants, job screening, recommender systems, political forecasting, forensics, specialized counseling, psychological studies, and other applications of automatic personality prediction exist [5]. It comes under the applications of authorship analysis task [11–14]. The most commonly used measures in the literature are the big-five personality traits, namely Neuroticism, Extraversion, Agreeableness, Openness, and Conscientiousness. Extraversion characterizes a person's assertiveness, sociability, talkativeness, excitability, and high emotional expressiveness. Neuroticism reveals the sadness, moodiness, and emotional instability of the person in question. Whether a person is trustworthy, straightforward,


M. Tanveer et al. (Eds.): ICONIP 2022, CCIS 1793, pp. 289–301, 2023.
https://doi.org/10.1007/978-981-99-1645-0_24

modest, generous, unreliable, complicated, or boastful is determined by their agreeableness. Conscientiousness indicates whether a person is well-organized or not. Their openness describes the nature of a person's imagination and creativity. These traits are described in yes/no form [5].

The handiness of cameras, cheap storage, and high-performance computing have led to an increase in human-computer interaction. Automatically computed personality is an essential clue for assessing the person's internal qualities from a given video. Our perceptions of other people's emotions and personalities are heavily influenced by their facial appearances [6,7,21]. In [21], researchers have shown the joint impact of apparent personality and facial emotion.

Motivated by these findings, we have tried to analyze the effect of utilizing emotion features in CNN-based model. Feature extraction from images (primarily facial features) is mainly performed using Convolutional neural network (CNN) based architectures [3]. Pre-trained CNN architectures like VGG-16 and VGG-face have also been used for extracting features from images. Different audio features such as the Logfbank, Mel-Frequency Cepstral Coefficients (MFCC), pitch, Zero Crossing Rate (ZCR), and loudness have also been extracted from OpenSMILE [8]. Late fusion and early fusion-based techniques have been used for combining different modalities [1–4,19]. Motivated by these findings, we have tried to analyze the effect of emotion features in our developed model using different CNN architectures. The developed emotion-aware personality prediction system takes help from the emotion information, learns the emotion and personality features, and then fuses them for the classification.

To perform the experiments, ChaLearn-2017 dataset has been considered. There are personality scores for each user. But, this dataset does not have the emotion labels for any user [2]. To the best of our knowledge, there is no existing dataset annotated with both personality scores and emotion labels. This has motivated us to annotate the existing ChaLearn-2017 dataset with emotion labels for each sample. We have performed all the experiments on our enriched ChaLearn-2017 dataset. An accuracy of 91.35% is achieved using our proposed approach, which is comparable to the systems having no information about emotion. Below, we have discussed the main contributions of this work:

- Due to the unavailability of a dataset with personality and emotion labels, we have annotated the bench-mark ChaLearn-2017 data (popularly used for personality prediction task) with emotion labels using a semi-supervised approach. Thus, we create a new dataset having both personality and emotion labels.
- Proposal of an emotion-aided multi-modal personality prediction system where we have shown the effect of utilizing emotion information in predicting the personality of different users considering audio, video and text information.

## 2 Related Works

Apparent personality prediction competition was held by CVPR in 2017, and 2016 [1]. Graphical approaches have also been employed in developing such systems.

In [20], TrigNet, a psycholinguistic knowledge-based tripartite graph network, has been created for predicting personality traits. This network consists of a tripartite graph network and a BERT-based graph initializer. Graphical encoding of CNN has also been used for recognizing the personality [9].

In [1,2], an audiovisual network with the analysis of different features like visual, acoustic, and audiovisual is presented. Similarly, a pre-trained CNN is used for extracting facial and ambient features for the final prediction [19]. The main reason behind the usage of CNN architectures for the prediction is the usefulness of facial information in the video [17]. Mainly late fusion-based techniques have been employed for fusing different modalities [1,16,19]. Using early fusion, the tri-modal fusion is implemented in [4]. Authors in [15], used different pre-trained CNN architectures like ResNet, VGGish CNN, MTCNN, and CNN over bi-grams for extracting different features from visual, audio, and textual modalities.

The way we perceive other people's emotions and personalities is heavily influenced by how they look (facial appearances). Researchers in [6,7,21] analyzed the behavior of emotion features over personality pre diction task. The role of physical appearance over first impression is examined in [6,7]. The joint impact of apparent personality and facial emotion is shown in [21], using an end-to-end neural architecture. A deep siamese network is developed which is optimized using multi-task learning framework. Motivated by these findings, we have tried to analyze the effect of emotional features in this work.

To analyze the effect of emotion features in the personality prediction technique, we have chosen the concat-based multimodal system proposed in [15], as our base model. Using Multi-task Cascaded Convolutional Networks (MTCNN) and ResNet, facial and ambient features are extracted from the videos. VGG-ish CNN is used for extracting features from the audio. Text features are extracted from CNN using bigrams. Finally, concatenation is used for fusing visual, audio, and text features before feeding them to the sigmoid layer. We have taken this system as our base model for fusing emotional features. This model is referred to as concat-based model in next sections.

## 3 Emotion-Enriched Personality Prediction Corpus

A single-track competition was held by ECCV in 2017 to quantitatively assess the apparent big five personality traits. There are 10,000 15-second YouTube videos (having tri-modal information) in the dataset. Text, audio, and visual are the three modalities of the data samples. There are 10, 000 human-centered short video samples in the dataset. For each video, the ground truth values for five traits are fractional scores in the range of 0 to 1. A transcript is also available for each video. There are 6000 videos in the training data and 2000 in both the testing and validation data [19].

Table 1 contains some of the data samples (transcripts). For the first sample, the personality score for neuroticism is higher (0.82), than 0.7 for agreeableness. For all the other three traits, the scores are higher than 0.5. For the second sample, the score for neuroticism is 0.54, and for openness is 0.5. For the other three traits, the scores are lower than 0.5. In this way, for each of the samples, the corresponding scores for different personality traits are present in the dataset.

**Table 1.** Some samples (Transcripts) of the personality prediction dataset, AcronymE:Extraversion, A:Agreeableness, C:Conscientiousness, O:Openness, N:Neuroticism

| Transcript | E | A | C | O | N |
|---|---|---|---|---|---|
| was great in the '50s, but over time really doesn't hold up. Astronauts are in outer space. They lose ... One of the ... There's like a two man space rocket. The one guy goes out to fix something, and | 0.6 | 0.7 | 0.53 | 0.61 | 0.82 |
| I could just ... and be in Los Angeles, and didn't have to take a 9 h flight there. So super power | 0.44 | 0.41 | 0.44 | 0.5 | 0.54 |
| Food, you can't really complain. So, yeah that's what I like most about Christmas. My favorite Christmas song from Sarah Rose. Is it Sarah Rose or Sarai Rose? I presume it's Sarah Rose | 0.36 | 0.42 | 0.30 | 0.23 | 0.31 |
| Use herbs in ... Maybe 80% of my spells and rituals and stuff. I actually started out really- | 0 | 0.05 | 0.20 | 0.03 | 0.44 |

Emotion plays a vital role in predicting the apparent personality traits of a user [21]. To the best of our knowledge, there does not exist any standard dataset having both personality and emotion labels. IBM Watson Tone Analyzer[1] is used for generating tone labels for each transcript in the ChaLearn-17 dataset. This tone analyzer categorized the transcript in a multi-labeled format having seven classes, namely:- fear, anger, sadness, joy, analytical, tentative, and confident. Cognitive linguistic analysis is used for classifying the input samples into possible tone classes. The tone analysis is done using both the sentence and the document level. The tone classification is done only if there is 50 percent confidence in the decision; otherwise, the transcript is considered neutral. We aim to annotate the complete dataset. Thus, the dataset samples are fed to the tone analyzer for generating the class labels. Only 656 samples out of all the 10000 samples are labeled as neutral.

The manual correction has been incorporated after finding annotations from the tone analyzer. We double-checked the input samples and each of the assigned labels. The majority of the samples were labeled correctly. We had to manually correct the labels of 73 samples that were sarcastic and difficult to identify. Because 73 out of 10,000 samples is a minimal number, the chosen tone analyzer

[1] https://www.ibm.com/watson/services/tone-analyzer/.

**Table 2.** Some samples (Transcript) of the enriched personality prediction dataset, Acronym F:Fear, A: Anger, S:Sadness, J:Joy, C:Confident, An:Analytical, T:Tentative

| Transcript | A | F | J | S | An | C | T |
|---|---|---|---|---|---|---|---|
| I love that it's available on Amazon now. What has been going on? We just finished up our homecoming tour for 2015. It was absolutely phenomenal. We got some incredible footage, we upgraded | 0 | 0 | 1 | 0 | 0 | 0 | 0 |
| Communicate with you guys, because sometimes I feel like routine videos and stuff like that can come off as fake, and like I don't want to come off as fake | 0 | 0 | 1 | 0 | 0 | 1 | 0 |
| Communicate with you guys, because sometimes I feel like routine videos and stuff like that can come off as fake, and like I don't want to come off as fake | 0 | 0 | 0 | 0 | 0 | 1 | 0 |
| This is how raw this is, is I've actually one of the coolest parts of an Apple product, in my opinion, is the unboxing because the attention to detail and I have a blog post talking about what I learned while camping out for | 0 | 0 | 1 | 0 | 1 | 0 | 0 |

can be used to label the tones correctly. The existing personality dataset is enriched with emotion labels in this way. This dataset, which includes personality and emotion labels, is now being used to conduct experiments.

We have shown the annotated data samples in Table 2. For each sample, the labels for fear, sadness, anger, joy, analytical, confident, and tentative are tabulated. For the first sample, all the emotion labels are zero except the sadness label. Similarly, the labels for sadness and the confident are one, and for others are zero. The label for the second last sample is one for analytical, and for others are zero.

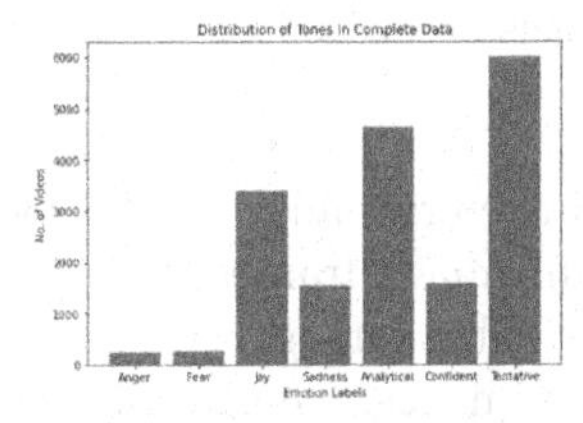


**Fig. 1.** Frequency Distribution of Emotions in the Personality Data

Figure 1 represents the statistics for the number of instances of different tones displayed by different personality owners in the complete dataset. The distributions of these emotion labels are uniform for all the training, testing, and validation dataset. It can be seen that most of the labels fall in the category of tentative, analytical, and joy. After these categories, the descending order of emotion labels are confident, sadness, anger, and fear.

# 4 The Proposed Approach

Our developed prediction system works in three stages, i) extraction of personality features, ii) fusion of personality and emotion features, and iii) personality prediction.

## 4.1 Extraction of Personality Features

Firstly, we have extracted the features for predicting personality using the concat-based model as discussed in Sect. 2. The visual features are extracted from ResNet, audio features are extracted from the VGG-ish CNN, and the textual features are extracted from the CNN-based architecture. The emotion labels of the sample are fused in the second last layer of the developed system. Below, we have discussed the fusion of these features in detail.

## 4.2 Fusion of Personality and Emotion Features

In order to develop an emotion-aware personality prediction system, we need to fuse the emotion and personality features. The fusion of these two feature sets has been performed in each input mode separately. We have developed four different personality prediction models, of which three are based on single-modal and the fourth one is multi-modal.

For each data sample, a tone vector is generated using the emotion labels. There are seven emotion labels for a single sample. Thus, a tone vector of size seven is generated for each of the samples. In this way, the tone vector generated for the first data sample shown in Table 2 is [0,0,1,0,0,0,0]. Similarly, the tone vector for each of the samples is created. Below, we have discussed the developed models in detail.

- Video-based emotion-aware personality prediction model: Firstly, the video samples are converted into image frames. Equally spaced, ten image frames are chosen as visual input. These images are fed to the ResNet for extracting ambient features from the images. Finally, all ten features are concatenated together to represent the final visual feature vector. After that, the 7-category-based tone vector of that video is concatenated with the final visual feature vector. Finally, the concatenated feature vector is fed to the sigmoid layer for the final prediction of personality scores. This model is shown in Fig. 2(a).
- Audio-based emotion-aware personality prediction model: 15 equally spaced audio signals are extracted from the given video samples. The audio features extracted from the mel spectogram are considered as the input for the audio modality. The features are then passed through the VGG-ish CNN

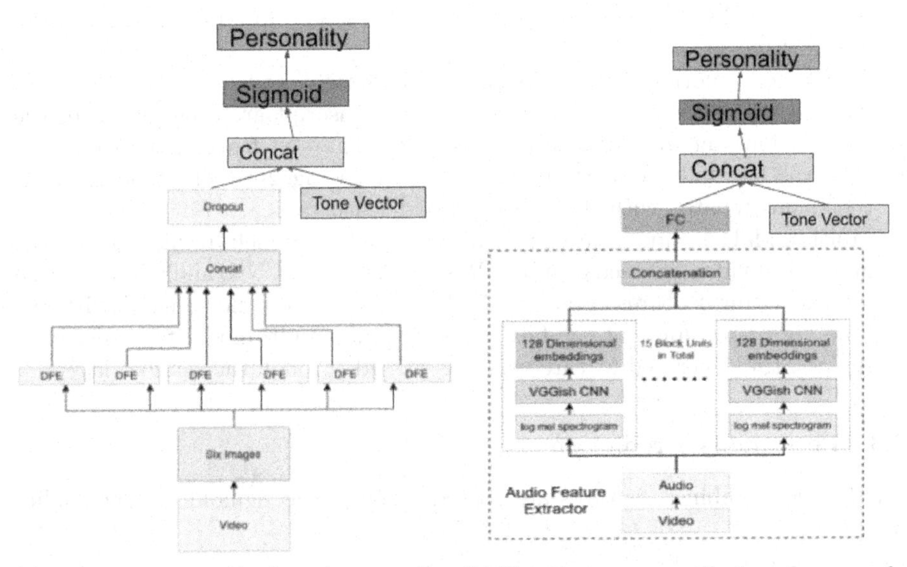


(a) Emotion-aware video based personality prediction model

(b) Emotion-aware audio based personality prediction model

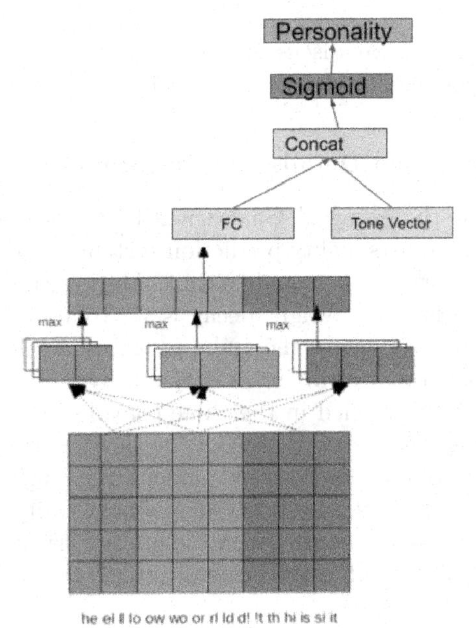


(c) Emotion-aware text based personality prediction model

**Fig. 2.** Developed Emotion-aware Personality Prediction Models

architecture. Similar to the visual-based model, the tone vectors are concatenated with the audio features for the final feature representation, as shown in Fig. 2(b).

- Text-based emotion-aware personality prediction model: The character n-grams (n=2) based representations of the transcript are directly fed to the convolutional neural network-based system. The text features are extracted from the last fully connected layer as shown in Fig. 2(c). The tone vector is then concatenated with the extracted text features.
- Multi-modal emotion-aware personality prediction model: In this model, text, audio, and video are used as input. ResNet, VGG-ish CNN, and CNN are used for extracting features. Finally, all the features are concatenated with the tone vector to represent the final emotion-aided multi-modal feature vector. At last, the generated feature vector is passed through the sigmoid layer.

### 4.3 Personality Prediction

Finally, the combined features are passed through the sigmoid for personality prediction.

## 5 Results and Discussion

In this section, we have discussed the results obtained from our developed emotion-aided personality prediction models.

### 5.1 Implementation Details and Performance Evaluation

We have utilized the enriched Chalearn-2017 dataset for assessing the developed emotion aware personality prediction systems. There are continuous values between 0 and 1 for each personality trait. Thus, mean squared error loss with L2 regularization is used. Weight decay is set to 1e-5. Adam optimizer with a learning rate of 1e-4 is set for the audio modality, while it is set to 1e-5 for the visual. To prevent from overfitting, a dropout layer of 0.5 is applied. Filter sizes of 52, 28, and 14 are applied in the three convolution layers with a dropout of 0.5. 20 is used as batch size in each of the developed systems. Early stopping with the patience of 8 is used to save model from overfitting.

We have calculated the accuracy of a single trait value of a video as one minus the absolute difference between the ground truth value and the predicted value of that trait [19]. Thus for mean accuracy, it is calculated for all the five traits for each of the user.

### 5.2 Results

We have shown the performance of our developed models in Table 3. The accuracy values obtained for the emotion-aided visual, audio, text-char, and multi-modal systems are 91.16%, 90.64%, 88.96%, and 91.35%, respectively. In Table 3, we have also shown the accuracy values attained by these models without using any emotion information. The performance of the emotion-aware audio-based model is same as the audio models without the use of any emotion information. An improvement of 0.24% is reported in the emotion-aware text based model. On the other hand, there are decreases of 0.09%, and 0.08% in the case of emotion-aware visual, and emotion-aware multimodal systems, respectively, over the models without using any emotion information. Thus it can be concluded that the performances of emotion-aware systems are comparative.

**Table 3.** Performance comparison of developed emotion-aware and without emotion-aware Systems

| Model | Accuracy with emotion | Accuracy without emotion |
|---|---|---|
| Visual | 91.16 | 91.27 |
| Audio | 90.64 | 90.64 |
| Text-char | 88.96 | 88.72 |
| Visual+Text+Audio (Concat) | 91.35 | 91.43 |

**Table 4.** Class-wise accuracies (in %) for the proposed model and the previous works on test data; Acronym: E: Extraversion, Ag: Agreeableness, A: Average, N: Neuroticism, C: Conscientiousness, O: Openness

| | Average | E | Ag | C | N | O |
|---|---|---|---|---|---|---|
| DCC [2] | 91.1 | 91.1 | 91.0 | 91.4 | 90.9 | 91.1 |
| evolgen [10] | 91.2 | 91.5 | 91.2 | 91.2 | 91.0 | 91.2 |
| NJU-LAMDA [19] | 91.3 | 91.3 | 91.3 | 91.7 | 91.0 | 91.2 |
| Concat-based approach [15] | 91.43 | 91.73 | 91.32 | 91.92 | 91.03 | 91.11 |
| PersEmoN [21] | 91.7 | 92.0 | 91.4 | 92.1 | 91.4 | 91.5 |
| Ours (Emotion-aware multi-modal system) | 91.35 | 91.55 | 91.39 | 91.86 | 90.98 | 91.05 |

### Comparison with Other Works

Table 4, shows the comparative performance of our proposed approach with other existing works. Those works are listed below:

- NJU-LAMDA: VGG-16 is used for obtaining the visual features. Finally, averaging technique is applied for the calculation of the final results [19].
- Evolgen: In this work, CNN and LSTM are used for extracting the audio, and visual features for the personality trait prediction [10], respectively.
- DCC: In this work, ResNet is used for extracting the audio-visual features, followed by a fully connected layer [2].

- Concat-base approach: In this approach, a multi-modal prediction system is developed which uses ResNet, MTCNN, and CNN-based techniques [15].
- PersEmoN: In this approach, an end-to-end multi-tasking-based personality prediction system is developed. It shows the joint relationship of personality and emotion [21].

The current state-of-the-art (SOTA) system over ChaLearn-17 dataset is PersEmon, which uses the emotion feature while predicting the personality scores. The second-best work is based on the concatenation of audio, visual, and text features [15]. Different approaches have been used for combining the three modalities. Average accuracy of 91.43% is achieved using the concatenation method.

The results mentioned in Table 4 illustrate that the developed emotion-aware concatenation-based model achieves a mean accuracy of 91.35%. On the other hand, the current SOTA yields an accuracy of 91.7%, and the second-best work achieved 91.43%. In this way, it can be concluded that the developed emotion-aware system achieves comparable performance with respect to other existing systems. The accuracy difference between PersEmoN and our developed system is approx. 0.35%. The improved performance of PersEmoN can be attributed to the usage of implicit feature learning capacity of the end-to-end neural architecture of PersEmoN. Another possible reason for the reduced performance by the proposed model is the generation of emotion labels of the videos in semi-supervised way.

From the obtained results, it can also be concluded that emotion features are helping in the personality prediction, as compared to the individual audio and text modalities. In the case of visual and multi-modal systems, addition of emotion labels leads to decrease in performance. Overall, we can say that the results obtained from the emotion-aware systems are not degrading the performance much. This shows the efficacy of emotion labels while predicting the personality of a user.

## 6 Qualitative Analysis

The comparative results shown in Table 3 show negligible differences in the performance of the emotion-aware system with respect to the simple multi-modal system. This motivated us to perform a qualitative analysis of the Chalearn-2017 dataset. The manual analyses of the data samples are carried out.

According to correlational analyses, Extraversion is positively related to the intensity, frequency, and duration of positive emotions, while neuroticism is positively related to the frequency and duration of negative emotions. In this way, the duration of positive emotions is the strongest predictor of extraversion, while the frequency of negative emotions is the strongest predictor of neuroticism [18]. Thus, it can be said that these two traits are opposite in nature. The manual analysis shows that there are 3903 users having higher values of extraversion as well as neuroticism. Also, 1206 users show joyous behavior along with a higher score for neuroticism. Inspired by these findings, we removed those 3903 users having higher extraversion and neuroticism scores. The developed four models are again trained using the reduced dataset. The results achieved using the reduced dataset are shown in Table 5. From the results, it is seen that the accuracy value for video-based model increases in comparison to the accuracy achieved over original data as shown in Table 3. On the other hand, the accuracy values for all the other models have decreased. But still, the performance of emotion-aware models are lesser than the model having no information about the emotion labels. This makes us conclude that we need to develop more stronger models for recognising the emotion as well as personality prediction system (like multi-tasking based approaches).

**Table 5.** Performance comparison of developed emotion-aware and without emotion-aware systems using reduced dataset

| Model | Accuracy with emotion | Accuracy without emotion |
|---|---|---|
| Visual | 91.8 | 91.92 |
| Audio | 89.65 | 89.62 |
| Text | 88.35 | 88.7 |
| Visual+Text+Audio (Concat) | 91.3 | 91.37 |

## 7 Conclusion and Future Works

The behavior, mental health, emotion, life choices, social nature, and thought patterns of an individual are revealed by personality. Current paper reports about the development of an emotion-aided multi-modal personality prediction system, which uses different pre-trained CNNs like ResNet, and VGGish CNN for extracting features from the textual, audio, and visual modalities. In addition to these features, emotion labels are used as auxiliary features in the network. All the features extracted from the three modalities and emotion features are concatenated together. Finally, the concatenated feature set is then used in the neural network setting for the final prediction. The experimental results illustrate that, addition of these emotion features yields similar performance as comparative to the models having no information about emotion.

In future, we will use depression, tone, and other features for analysing the behavior of the personality prediction system. Multi-tasking based models can also be developed for learning emotion-aided features.

**Acknowledgement.** Dr. Sriparna Saha gratefully acknowledges the Young Faculty Research Fellowship (YFRF) Award, supported by Visvesvaraya Ph.D. Scheme for Electronics and IT, Ministry of Electronics and Information Technology (MeitY), Government of India, being implemented by Digital India Corporation (formerly Media Lab Asia) for carrying out this research.

## References

1. Güçlütürk, Y., et al.: Multimodal first impression analysis with deep residual networks. IEEE Trans. Affect. Comput. **9**(3), 316–329 (2017)
2. Güçlütürk, Y., Güçlü, U., van Gerven, M.A.J., van Lier, R.: Deep impression: audiovisual deep residual networks for multimodal apparent personality trait recognition. In: Hua, G., Jégou, H. (eds.) ECCV 2016. LNCS, vol. 9915, pp. 349–358. Springer, Cham (2016). https://doi.org/10.1007/978-3-319-49409-8_28
3. Gürpınar, F., Kaya, H., Salah, A.A.: Combining deep facial and ambient features for first impression estimation. In: Hua, G., Jégou, H. (eds.) ECCV 2016. LNCS, vol. 9915, pp. 372–385. Springer, Cham (2016). https://doi.org/10.1007/978-3-319-49409-8_30
4. Kampman, O., Barezi, E.J., Bertero, D., Fung, P.: Investigating audio, video, and text fusion methods for end-to-end automatic personality prediction. In: Proceedings of the 56th Annual Meeting of the Association for Computational Linguistics (Volume 2: Short Papers), pp. 606–611 (2018)
5. Mehta, Y., Majumder, N., Gelbukh, A., Cambria, E.: Recent trends in deep learning based personality detection. Artif. Intell. Rev. **53**(4), 2313–2339 (2019). https://doi.org/10.1007/s10462-019-09770-z
6. Naumann, L.P., Vazire, S., Rentfrow, P.J., Gosling, S.D.: Personality judgments based on physical appearance. Pers. Soc. Psychol. Bull. **35**(12), 1661–1671 (2009)
7. Olivola, C.Y., Todorov, A.: Fooled by first impressions? reexamining the diagnostic value of appearance-based inferences. J. Exp. Soc. Psychol. **46**(2), 315–324 (2010)
8. Polzehl, T., Moller, S., Metze, F.: Automatically assessing personality from speech. In: 2010 IEEE Fourth International Conference on Semantic Computing, pp. 134–140. IEEE (2010)
9. Shao, Z., Song, S., Jaiswal, S., Shen, L., Valstar, M., Gunes, H.: Personality recognition by modelling person-specific cognitive processes using graph representation. In: Proceedings of the 29th ACM International Conference on Multimedia, pp. 357–366 (2021)
10. Subramaniam, A., Patel, V., Mishra, A., Balasubramanian, P., Mittal, A.: Bimodal first impressions recognition using temporally ordered deep audio and stochastic visual features. In: Hua, G., Jégou, H. (eds.) ECCV 2016. LNCS, vol. 9915, pp. 337–348. Springer, Cham (2016). https://doi.org/10.1007/978-3-319-49409-8_27
11. Suman, C., Chaudhari, R., Saha, S., Kumar, S., Bhattacharyya, P.: Investigations in emotion aware multimodal gender prediction systems from social media data. IEEE Trans. Comput. Soc. Syst. **PP**(99), 1–10 (2022). https://doi.org/10.1109/TCSS.2022.3158605
12. Suman, C., Naman, A., Saha, S., Bhattacharyya, P.: A multimodal author profiling system for tweets. IEEE Trans. Comput. Soc. Syst. **8**(6), 1407–1416 (2021). https://doi.org/10.1109/TCSS.2021.3082942
13. Suman, C., Raj, A., Saha, S., Bhattacharyya, P.: Authorship attribution of microtext using capsule networks. IEEE Trans. Comput. Soc. Syst. **9**(4), 1038–1047 (2022). https://doi.org/10.1109/TCSS.2021.3067736
14. Suman, C., Saha, S., Bhattacharyya, P.: An attention-based multimodal Siamese architecture for tweet-user verification. IEEE Transactions on Computational Social Systems, pp. 1–9 (2022). https://doi.org/10.1109/TCSS.2022.3192909
15. Suman, C., Saha, S., Gupta, A., Pandey, S.K., Bhattacharyya, P.: A multi-modal personality prediction system. Knowl.-Based Syst. **236**, 107715 (2022)

16. Tellamekala, M.K., Giesbrecht, T., Valstar, M.: Apparent personality recognition from uncertainty-aware facial emotion predictions using conditional latent variable models. In: 2021 16th IEEE International Conference on Automatic Face and Gesture Recognition (FG 2021), pp. 1–8. IEEE (2021)
17. Ventura, C., Masip, D., Lapedriza, A.: Interpreting CNN models for apparent personality trait regression. In: Proceedings of the IEEE Conference on Computer Vision and Pattern Recognition Workshops, pp. 55–63 (2017)
18. Verduyn, P., Brans, K.: The relationship between extraversion, neuroticism and aspects of trait affect. Personality Individ. Differ. **52**(6), 664–669 (2012)
19. Wei, X.S., Zhang, C.L., Zhang, H., Wu, J.: Deep bimodal regression of apparent personality traits from short video sequences. IEEE Trans. Affect. Comput. **9**(3), 303–315 (2017)
20. Yang, T., Yang, F., Ouyang, H., Quan, X.: Psycholinguistic tripartite graph network for personality detection. arXiv preprint arXiv:2106.04963 (2021)
21. Zhang, L., Peng, S., Winkler, S.: PersEmoN: a deep network for joint analysis of apparent personality, emotion and their relationship. IEEE Trans. Affect. Comput. **13**, 298–305 (2019)

# Kernel Inversed Pyramidal Resizing Network for Efficient Pavement Distress Recognition

Rong Qin[1], Luwen Huangfu[2,3], Devon Hood[4], James Ma[5], and Sheng Huang[1,6](✉)

[1] School of Big Data and Software Engineering, Chongqing University, Chongqing 400044, China
{qinrongzxxlxy,huangsheng}@cqu.edu.cn

[2] Fowler College of Business, San Diego State University, San Diego, CA 92182, USA
lhuangfu@sdsu.edu

[3] Center for Human Dynamics in the Mobile Age, San Diego State University, San Diego, CA 92182, USA

[4] College of Science, San Diego State University, San Diego, CA 92182, USA
dhood1037@sdsu.edu

[5] Department of Business Analysis, University of Colorado at Colorado Springs, Colorado 80918, USA
jma@uccs.edu

[6] Ministry of Education Key Laboratory of Dependable Service Computing in Cyber Physical Society, Chongqing University, Chongqing 400044, China

**Abstract.** Pavement Distress Recognition (PDR) is an important step in pavement inspection and can be powered by image-based automation to expedite the process and reduce labor costs. Pavement images are often in high-resolution with a low ratio of distressed to non-distressed areas. Advanced approaches leverage these properties via dividing images into patches and explore discriminative features in the scale space. However, these approaches usually suffer from information loss during image resizing and low efficiency due to complex learning frameworks. In this paper, we propose a novel and efficient method for PDR. A light network named the Kernel Inversed Pyramidal Resizing Network (KIPRN) is introduced for image resizing, and can be flexibly plugged into the image classification network as a pre-network to exploit resolution and scale information. In KIPRN, pyramidal convolution and kernel inversed convolution are specifically designed to mine discriminative information across different feature granularities and scales. The mined information is passed along to the resized images to yield an informative image pyramid to assist the image classification network for PDR. We applied our method to three well-known Convolutional Neural Networks (CNNs), and conducted an evaluation on a large-scale pavement image dataset named CQU-BPDD. Extensive results demonstrate that KIPRN can generally improve the pavement distress recognition of these CNN models and show that the simple combination of KIPRN and EfficientNet-B3 significantly outperforms the state-of-the-art patch-based method in both performance and efficiency.


M. Tanveer et al. (Eds.): ICONIP 2022, CCIS 1793, pp. 302–312, 2023.
https://doi.org/10.1007/978-981-99-1645-0_25

**Keywords:** Pavement Distress Recognition · Image Classification · Resizing Network

# 1 Introduction

Pavement distress is one of the largest threats to modern road networks and, as such, Pavement Distress Recognition (PDR) is an important aspect in maintaining logistics infrastructure. Traditionally, pavement distress recognition is done manually by professionals, and this requires a large overhead of labor and extensive domain knowledge [1]. Given the complex and vast network of roadways, it is almost impossible to accomplish the pavement distress inspection manually. Therefore, automating the pavement distress recognition task is essential.

In recent decades, many methods have been proposed to address this issue through the use of computer vision. Conventional approaches often utilize rudimentary image analysis, hand-crafted features, and traditional classifiers [5,15,18,20,27]. For example, Salman et al. [20] proposed a crack detection method based on the Gabor filter, and Li et al. [15] developed a neighboring difference histogram method to detect conventional, human visual, pavement disease. The main problem of these approaches is that the optimizations of feature extraction and image classification steps are separated or even omitted from any learning process. Inspired by the remarkable success of deep learning, numerous researchers have employed such models to solve this problem [2,10,11,14,17,26]. For example, Gopalakrishnan et al. [11] employed a Deep Convolutional Neural Network (DCNN) trained on the ImageNet database, and transferred that learning to automatically detect cracks in Hot-Mix Asphalt (HMA) and Portland Cement Concrete (PCC) surfaced pavement images. Fan et al. [10] proposed a novel road crack detection algorithm that is based on deep learning and adaptive image segmentation. However, these approaches often overlook many key characteristics of pavement images, such as the high image resolution and the low ratio of distressed to non-distressed areas.

Huang et al. [13] presented a Weakly Supervised Patch Label Inference Network with Image Pyramid (WSPLIN-IP) to solve those problems. The WSPLIN-IP exploited high resolution information and scale space of images by dividing an image pyramid into patches for label inference via weakly supervised learning and achieved promising performances in comparison with other state-of-the-art approaches. However, its patch collection strategy and the complex patch label inference processes can lead to low efficiency in practical applications. Moreover, as the mainstream PDR approach, the CNN-based methods often need to resize images to a uniform size for the CNNs, where traditional resizing algorithms, such as bilinear interpolation, are employed. As a result, image resizing is completely independent of model optimization, which inevitably leads to the loss of some discriminative information. A few existing related studies, such as [19], are often difficult to apply to pavement images due to the high sensitivity of pavement diseases to deformation and the need for multi-scale input.

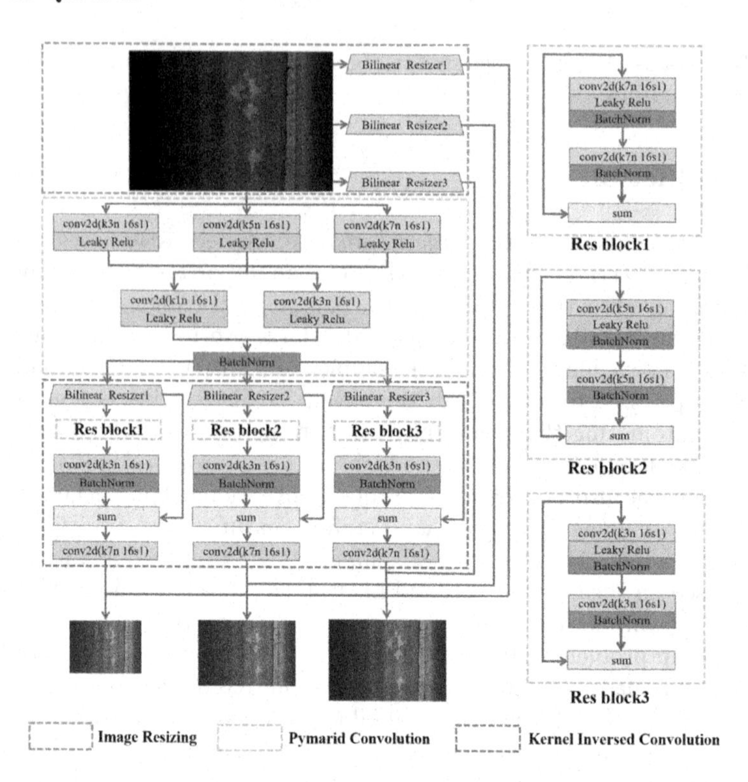


**Fig. 1.** The network architecture of Kernel Inversed Pyramidal Resizing Network (KIPRN).

Inspired by the idea of Resizing Network [24], we elaborate a light image resizing network named the Kernel Inversed Pyramidal Resizing Network (KIPRN) to address these issues. The KIPRN can be integrated into any deep learning-based model as a self-contained supplemental module and be optimized with the model as one whole integration, and it learns to retain the information, and compensates for the information loss caused by the image resizing based on the bilinear interpolation. As shown in Fig. 1, KIPRN employs pyramidal convolutions [9] to extract information from the original pavement images with different granularity, and then produces a three-layer image pyramid for each input image with our designed kernel inversed ResBlock. Pyramidal convolutions enable the mining of more resolution information with different sizes of convolutions, while the kernel inversed ResBlock can better mine the scale information by enlarging the differences between relative receptive fields in different resizing branches. Finally, the produced image pyramid will be input into the subsequent deep learning-based PDR model to exploit the scale space again without greatly increasing computational burdens.

We evaluated the KIPRN on a large-scale pavement image dataset named CQU-BPDD [26]. Extensive results show that our method generally boosts many deep learning-based PDR models. Moreover, our enhanced EfficientNet-B3 not only achieved state-of-the-art performances, but also obtained prominent advantages in efficiency compared with the WSPLIN-IP, which is a recent state-of-the-art PDR approach that also considers the resolution and scale information and utilizes the EfficientNet-B3 as its backbone network. The main contributions of our work can be summarized as follows:

- We propose a novel resizing network named KIPRN that can boost any deep learning-based PDR approach by exploiting the resolution and scale information of images. Moreover, the KIPRN will not require significant computational cost. To the best of our knowledge, our work is the first attempt to use a deep learning model to study image resizing in pavement distress analysis.
- We propose kernel inversed ResBlocks, which applies the smaller convolutional kernels to the larger feature maps, while applying the larger convolutional kernels to the smaller feature maps in a size-inversion way. It implicitly enlarges the scale space and, thereby, enhances the scale information exploitation.
- Extensive results demonstrate that the KIPRN can generally improve deep learning-based PDR approaches and achieve a state-of-the-art performance without greatly increasing computational burdens.

## 2 Methodology

### 2.1 Problem Formulation and Overview

Pavement distress recognition is an image classification task to classify the images of damaged pavements into different distress categories. Let $X = \{x_i\}_{i=1}^{n}$ and $Y = \{y_i\}_{i=1}^{n}$ be the pavement images and their corresponding labels respectively. $y_i$ is a $C$-dimensional one-hot vector, where $C$ is the number of pavement distress categories. $y_{ij}$ represents the $j$-th element of $y_i$. If the $j$-th element is the only non-zero element, it indicates that the corresponding pavement image has the $j$-th type of pavement distress. The goal of pavement distress recognition is to train a PDR model to recognize pavement distress in a given pavement image.

In deep learning-based PDR, the high-resolution images are often resized into a fixed size to meet the input or efficiency requirement of these models. Moreover, some studies also show that exploiting the scale information of pavement images can benefit pavement distress recognition. Thus, image resizing is an inevitable process in pavement distress recognition based on deep learning. However, the conventional linear interpolation-based image resizing process is independent of the optimization of the pavement distress model, and, thereby, often causes the loss of discriminative information.

To address this issue, we propose an end-to-end network named KIPRN for training to resize the pavement images and, thereby, aiding the PDR model. KIPRN can be plugged into any deep learning-based image classification models as a pre-network and optimized with these models. The pavement distress recognition process based on the KIPRN can be represented as follows,

$$\hat{y} \leftarrow \mathrm{softmax}(P_\phi(\Gamma_\theta(x))), \tag{1}$$

where $\Gamma(\cdot)$ and $P(\cdot)$ are the mapping functions of the KIPRN and the PDR model respectively, while $\phi$ and $\theta$ are their corresponding parameters. As shown in Fig. 1, the KIPRN consists of three modules, namely Image Resizing (IR), Pyramidal Convolution (PC) and Kernel Inversed Convolution (KIC). On the one hand, the original pavement images will be resized into different sizes through IR. On the other hand, PC extracts the features of original pavement images from different granularities, and then KIC mines the scale information of the extracted features with different branches via inversed kernels. The mined information will be compensated into the resized images to yield a three-layer image pyramid for each pavement image. Finally, the produced image pyramid will be input into the subsequent deep learning-based PDR model to accomplish the label inference.

### 2.2 Image Resizing

The Image Resizing (IR) module is used to resize the original pavement image into $m$ different sizes. Here, we set $m = 3$. This process can be represented as follows,

$$\mathcal{I} = H(x), \tag{2}$$

where $H(\cdot)$ is any chosen traditional image resizing algorithm, and $\mathcal{I} = \{I_j\}_{j=1}^{m}$ is the collection of resized images. $I_j$ is the $j$-th resized pavement image. We followed [24] and adopted bilinear interpolation as the image resizing algorithm. The KIPRN will compensate the subsequently learned discriminative and scale information into $\mathcal{I}$ to generate a pavement image pyramid that is more conducive for a deep learning-based PDR model to correctly recognize pavement distress.

### 2.3 Pyramidal Convolution

The next step of the KIPRN is to leverage a Pyramidal Convolution (PC) module to mine and preserve the relevant information of the original images from different feature granularities. The golden dash-line rectangle in Fig. 1 shows the details of the PC module, which is a two-layers of pyramidal convolution [9]. The first layer adopts three convolution kernels, whose sizes are $3 \times 3$, $5 \times 5$, and $7 \times 7$ respectively, while two convolution kernels of the second layer are $1 \times 1$ and $3 \times 3$ respectively. Let $Q(\cdot)$ be the mapping function of the PC module where $\eta$ is its corresponding parameter. The pyramidal feature map can be generated as follows,

$$f = Q_{\eta}(x), \tag{3}$$

which sufficiently encodes the detailed features of pavement images under different granularities.

### 2.4 Kernel Inversed Convolution

Once we obtain the feature map $f$, a Kernel Inversed Convolutional (KIC) module is designed to better exploit the scale information of pavement images by enlarging the differences in the receptive fields in different convolution branches, as shown in the red dash-line rectangle in Fig. 1. In KIC, the feature map $f$ is resized into $m$ different sizes, which are identical to the ones of those resized images in the image resizing module,

$$\hat{F} = H(f) = \{\hat{f}_j\}_{j=1}^{m}, \tag{4}$$

where $\hat{F} = \{\hat{f}_j\}_{j=1}^{m}$ is the collection of the resized feature maps, and $\hat{f}_j$ is the resized feature map corresponding to the resized image $I_j$. Thereafter, these resized feature maps $\hat{F}$ are fed into different convolution branches to produce the information compensations for different resized images, and yield the final image pyramid. The whole image pyramid generation process can be represented as follows,

$$S = R_{\zeta}(\hat{F}) + \mathcal{I} = \{s_j\}_{j=1}^{m}, \tag{5}$$

where $S = \{s_j\}_{j=1}^{m}$ is the generated image pyramid, $R(\cdot)$ is the mapping function of the Kernel Inversed Convolution (KIC) module, and $\zeta$ is its learned parameters.

To better retain the discriminative information of pavement images across different scales, we adopted an idea from [4], and elaborate a series of kernel inversed ResBlocks for feature learning. In these ResBlocks, as shown in the orange dash-line rectangles in Fig. 1, the smaller convolution kernel is applied to the larger feature map, which enables the mining of the local detailed information of images, while the larger convolution kernel is applied to the smaller feature map, which enables the capturing of the global structural features of images. In other words, the kernel inversed convolution enlarges the perception range in the scale space, and thereby preserves more conducive information for the solution of the subsequent task. In this manner, the whole KIPRN process is a composition of these modules, $\Gamma = H \circ Q \circ R$, where $\theta = \{\eta, \zeta\}$.

### 2.5 Multi-scale Pavement Distress Recognition

In the final step, the generated image pyramid was input into an image classification network $P_{\phi}(\cdot)$ to predict the distress label of a given pavement image as follows,

$$\hat{y} \leftarrow \text{softmax}(\sum_{j=1}^{m} P(s_j)) = \text{softmax}(P(R(H(Q(x))) + H(x))). \tag{6}$$

The KIPRN was optimized with the image classification network together in an end-to-end manner. Let $\mathcal{L}(\cdot, \cdot)$ be the loss function of the image classification network. The optimal parameters of the KIPRN and the image classification network can be obtained by solving the following programming problem,

$$\{\hat{\theta}, \hat{\phi}\} \leftarrow \arg\min_{\theta,\phi} \sum_{i=1}^{n} \mathcal{L}(y_i, \hat{y}_i). \tag{7}$$

In this study, we chose three well known Convolutional Neural Networks (CNNs), namely EfficientNet-B3, ResNet-50 and Inception-v1, as the image classification networks to validate the effectiveness of the KIPRN.

## 3 Experiment

### 3.1 Dataset and Setup

A large-scale pavement dataset named CQU-BPDD [26] was employed for evaluation. It consists of 43,861 typical pavement images and 16,795 diseased pavement images and includes seven different types of distresses, namely alligator crack, crack pouring, longitudinal crack, massive crack, transverse crack, raveling, and repair. Since pavement distress recognition is a follow-up task of pavement distress detection, we only used the diseased pavement images for training and testing. For a fair comparison, we followed the data split strategy in [13], so that 5,140 images were selected for training while the remaining ones were used for testing.

As shown in Table 1, we chose three traditional shallow learning methods, five well-known CNNs, two classical vision transformers and WSPLIN-IP as baselines. For all deep learning models, we adopted AdamW [16] as the optimizer and the learning rate was set to 0.0001. The input images were resized by the KIPRN into three resolutions, $300 \times 300$, $400 \times 400$, and, $500 \times 500$, to yield the image pyramid. Recognition accuracy is used as the performance metric following [13].

**Table 1.** The recognition accuracies of different methods on CQU-BPDD.

| Method | Type | Accuracy |
|---|---|---|
| RGB + RF [3] | Single-scale | 0.305 |
| HOG [7] + SVM [6] | Single-scale | 0.318 |
| VGG-16 [21] | Single-scale | 0.562 |
| Inception-v3 [23] | Single-scale | 0.716 |
| ViT-S/16 [8] | Single-scale | 0.750 |
| ViT-B/16 [8] | Single-scale | 0.753 |
| WSPLIN-IP[6] [13] | Multi-scale | 0.837 |
| WSPLIN-IP [13] | Multi-scale | 0.850 |
| ResNet-50 [12] | Single-scale | 0.712 |
| ResNet-50 | Multi-scale | 0.786 |
| **Ours** + ResNet-50 | Multi-scale | 0.827 |
| Inception-v1 [22] | Single-scale | 0.726 |
| Inception-v1 | Multi-scale | 0.781 |
| **Ours** + Inception-v1 | Multi-scale | 0.803 |
| EfficientNet-B3 [25] | Single-scale | 0.786 |
| EfficientNet-B3 | Multi-scale | 0.830 |
| **Ours** + EfficientNet-B3 | Multi-scale | **0.861** |

The reproduced results with the codes provided by the original authors under our experimental settings.

### 3.2 Pavement Disease Recognition

Table 1 tabulates the pavement distress recognition accuracies of different methods on the CQU-BPDD dataset. The results show that the KIRPN significantly boosts

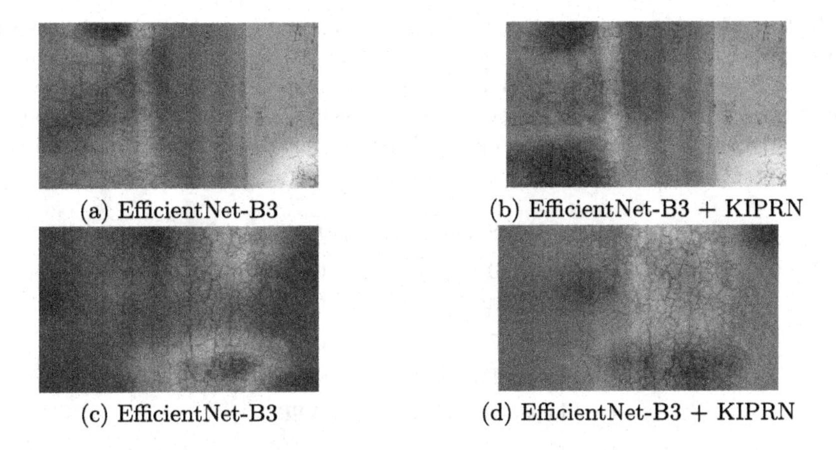


**Fig. 2.** Comparison of Class Activation Mapping (CAM) between EfficientNet-B3 and EfficientNet-B3 + KIPRN.

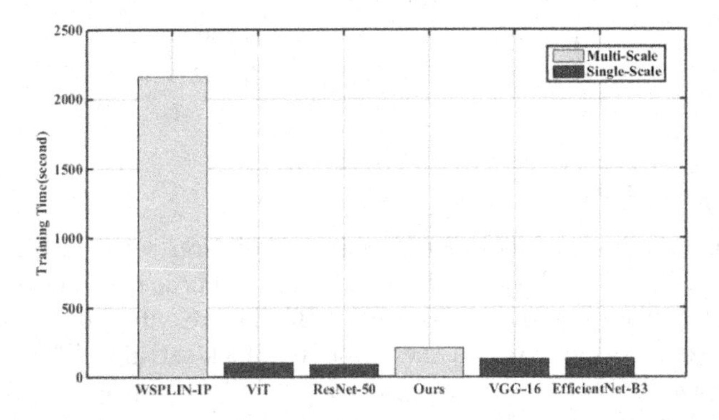


**Fig. 3.** The training time of models per epoch on the CQU-BPDD dataset.

all three chosen CNN-based image classification models. The KIPRN improves the accuracy of EfficientNet-B3, ResNet-50 and Inception-v1 by 7.5%, 11.5% and 7.7% respectively. Moreover, EfficientNet-B3 enhanced by the KIPRN achieved the best performance among all the methods. As shown in Fig. 2, the Class Activation Mapping (CAM) of the EfficientNet-B3 enhanced by the KIPRN is better than that of the original EfficientNet-B3. Another interesting observation regarding the multi-scale approaches is they often perform better than their single-scale versions, even though they simply use bilinear interpolation to resize images. The second best performed method, WSPLIN-IP, another multi-scale approach, adopted EfficientNet-B3 as its backbone networks. However, EfficientNet-B3 enhanced by KIPRN achieved 2.4% accuracy gains over WSIPLIN-IP under the same experimental settings. Overall, the multi-scale approaches based on the KIPRN consistently outperformed the versions based on the bilinear interpolation. These results clearly reveal two facts, namely 1) exploiting scale space of images can improve pavement distress recognition, and 2) the KIPRN enables learning conductive information for distress recognition during image resizing.

**Table 2.** Different settings of ResBlocks and pyramidal convolution in the KPRIN deployed with EfficientNet-B3. The i × i indicates the size of convolution kernel in Resblock. Forward is the strategy that applies a small kernel to the small feature map. Inversed is the inversed strategy of Forward, which is the strategy adopted in our method. Pyconv is pyramidal convolution. First and Last puts Pyconv on the first or the last two convolution layers.

| Moudule | Setting/Design | Accuracy |
|---|---|---|
| Resblocks | 3×3 | 0.844 |
| Resblocks | 5×5 | 0.852 |
| Resblocks | 7×7 | 0.847 |
| Resblocks | Forward | 0.857 |
| Resblocks | Inversed | **0.861** |
| Pyconv | All | 0.828 |
| Pyconv | None | 0.855 |
| Pyconv | Resblock | 0.845 |
| Pyconv | Last | 0.830 |
| Pyconv | First | **0.861** |

### 3.3 Ablation Study

Table 2 tabulates the impacts of different convolution settings on the performance of the KIPRN deployed on EfficientNet-B3. These experiments were conducted on KIPRN deployed on EfficientNet-B3. The results show that the kernel inversed strategy outperforms all the other ResBlock settings. This verifies that our strategy can capture more information in the scale space by enlarging the differences in relative receptive fields. Another interesting observation is that putting pyramidal convolution on the first two layers led to the best performances. We attribute this to the fact that pyramidal convolutions are more capable of capturing low-level visual details than learning the abstract semantic features that the last layers are designed for.

Figure 3 reports the training times of different models per epoch. The observations show that EfficientNet-B3 enhanced by the KIPRN (ours) has similar training efficiency compared to the single-scale EfficientNet-B3, while enjoying the 10X training speeds over WISPLIN-IP, which is also a multi-scale approach.

## 4 Conclusion

In this study, we proposed an end-to-end resizing network named the KIPRN, which can boost any deep learning-based PDR approach by assisting it to better exploit the resolution and scale information of images. The KIPRN consists of IR, PC and KIC. The KIPRN can be integrated into any deep learning-based PDR model in a plug-and-play way and optimizes together with a deep learning-based PDR model in an end-to-end manner. Extensive results show that our method generally boosts many deep learning-based PDR models. In our future work, we will attempt to compensate the features to the original pavement image in a better way than simple addition.

**Acknowledgments.** This study was supported by the San Diego State University 2021 Emergency Spring Research, Scholarship, and Creative Activities (RSCA) Funding distributed by the Division of Research and Innovation, as well as XSEDE EMPOWER Program under National Science Foundation grant number ACI-1548562.

## References

1. Benedetto, A., Tosti, F., Pajewski, L., D'Amico, F., Kusayanagi, W.: FDTD simulation of the GPR signal for effective inspection of pavement damages. In: Proceedings of the 15th International Conference on Ground Penetrating Radar, pp. 513–518. IEEE (2014)
2. Bhagvati, C., Skolnick, M.M., Grivas, D.A.: Gaussian normalisation of morphological size distributions for increasing sensitivity to texture variations and its applications to pavement distress classification. In: CVPR (1994)
3. Breiman, L.: Random forests. Mach. learn. **45**(1), 5–32 (2001)
4. Chen, C.F., Fan, Q., Mallinar, N., Sercu, T., Feris, R.: Big-little net: an efficient multi-scale feature representation for visual and speech recognition. arXiv preprint arXiv:1807.03848 (2018)
5. Chou, J., O'Neill, W.A., Cheng, H.: Pavement distress classification using neural networks. In: Proceedings of IEEE International Conference on Systems, Man and Cybernetics, vol. 1, pp. 397–401. IEEE (1994)
6. Cortes, C., Vapnik, V.: Support-vector networks. Mach. Learn. **20**(3), 273–297 (1995)
7. Dalal, N., Triggs, B.: Histograms of oriented gradients for human detection. In: 2005 IEEE Computer Society Conference on Computer Vision and Pattern Recognition (CVPR 2005), vol. 1, pp. 886–893. IEEE (2005)
8. Dosovitskiy, A., et al.: An image is worth 16x16 words: transformers for image recognition at scale. arXiv preprint arXiv:2010.11929 (2020)
9. Duta, I.C., Liu, L., Zhu, F., Shao, L.: Pyramidal convolution: rethinking convolutional neural networks for visual recognition. arXiv preprint arXiv:2006.11538 (2020)
10. Fan, R., et al.: Road crack detection using deep convolutional neural network and adaptive thresholding. In: 2019 IEEE Intelligent Vehicles Symposium (IV), pp. 474–479. IEEE (2019)
11. Gopalakrishnan, K., Khaitan, S.K., Choudhary, A., Agrawal, A.: Deep convolutional neural networks with transfer learning for computer vision-based data-driven pavement distress detection. Constr. Build. Mater. **157**, 322–330 (2017)
12. He, K., Zhang, X., Ren, S., Sun, J.: Deep residual learning for image recognition. In: Proceedings of the IEEE Conference on Computer Vision and Pattern Recognition, pp. 770–778 (2016)
13. Huang, G., Huang, S., Huangfu, L., Yang, D.: Weakly supervised patch label inference network with image pyramid for pavement diseases recognition in the wild. In: ICASSP 2021–2021 IEEE International Conference on Acoustics, Speech and Signal Processing (ICASSP), pp. 7978–7982. IEEE (2021)

14. Li, B., Wang, K.C., Zhang, A., Yang, E., Wang, G.: Automatic classification of pavement crack using deep convolutional neural network. Int. J. Pavement Eng. **21**(4), 457–463 (2020)
15. Li, Q., Liu, X.: Novel approach to pavement image segmentation based on neighboring difference histogram method. In: 2008 Congress on Image and Signal Processing, vol. 2, pp. 792–796. IEEE (2008)
16. Loshchilov, I., Hutter, F.: Fixing weight decay regularization in Adam (2018)
17. Naddaf-Sh, S., Naddaf-Sh, M.M., Kashani, A.R., Zargarzadeh, H.: An efficient and scalable deep learning approach for road damage detection. In: 2020 IEEE International Conference on Big Data (Big Data), pp. 5602–5608. IEEE (2020)
18. Nejad, F.M., Zakeri, H.: An expert system based on wavelet transform and radon neural network for pavement distress classification. Expert Syst. Appl. **38**(6), 7088–7101 (2011)
19. Recasens, A., Kellnhofer, P., Stent, S., Matusik, W., Torralba, A.: Learning to zoom: a saliency-based sampling layer for neural networks. In: Ferrari, V., Hebert, M., Sminchisescu, C., Weiss, Y. (eds.) ECCV 2018. LNCS, vol. 11213, pp. 52–67. Springer, Cham (2018). https://doi.org/10.1007/978-3-030-01240-3_4
20. Salman, M., Mathavan, S., Kamal, K., Rahman, M.: Pavement crack detection using the Gabor filter. In: 16th international IEEE Conference on Intelligent Transportation Systems (ITSC 2013), pp. 2039–2044. IEEE (2013)
21. Simonyan, K., Zisserman, A.: Very deep convolutional networks for large-scale image recognition. arXiv preprint arXiv:1409.1556 (2014)
22. Szegedy, C., et al.: Going deeper with convolutions. In: Proceedings of the IEEE Conference on Computer Vision and Pattern Recognition, pp. 1–9 (2015)
23. Szegedy, C., Vanhoucke, V., Ioffe, S., Shlens, J., Wojna, Z.: Rethinking the inception architecture for computer vision. In: Proceedings of the IEEE Conference on Computer Vision and Pattern Recognition, pp. 2818–2826 (2016)
24. Talebi, H., Milanfar, P.: Learning to resize images for computer vision tasks. In: Proceedings of the IEEE/CVF International Conference on Computer Vision, pp. 497–506 (2021)
25. Tan, M., Le, Q.: Efficientnet: rethinking model scaling for convolutional neural networks. In: International Conference on Machine Learning, pp. 6105–6114. PMLR (2019)
26. Tang, W., Huang, S., Zhao, Q., Li, R., Huangfu, L.: Iteratively optimized patch label inference network for automatic pavement disease detection. arXiv preprint arXiv:2005.13298 (2020)
27. Wang, C., Sha, A., Sun, Z.: Pavement crack classification based on chain code. In: 2010 Seventh International Conference on Fuzzy Systems and Knowledge Discovery, vol. 2, pp. 593–597. IEEE (2010)

# Deep Global and Local Matching Network for Implicit Recommendation

Wei Yang[1,2](✉), Yiqun Chen[1,2], Jun Sun[3], and Yan Jin[3]

[1] University of Chinese Academy of Sciences, Beijing, China
weiyangvia@gmail.com, chenyiqun2020@ia.ac.cn
[2] Institute of Automation, Chinese Academy of Sciences, Beijing, China
[3] Bytedance Technology, Beijing, China

**Abstract.** In real recommendation scenarios, users often have implicit behaviors including clicks, rather than explicit behaviors. In order to solve the matching problem under implicit data, many researchers have proposed methods based on neural networks, mainly including representation learning and matching function learning methods. However, these methods do not take into account the diverse preferences of users, and there is no fine-grained modeling matching relationship. In this paper, we consider the matching problem from a new perspective and propose a novel deep global and local matching network (DeepGLM) model. In detail, DeepGLM introduces multi-aspect representations to express the user's various preferences, and calculates the global matching degree between user and item through the hierarchical interactive matching module. Then, the attention mechanism is adopted to calculate the local matching relationship based on feature interactions. In addition, the gating mechanism is used to control the effective transmission of global and local matching information. Extensive experiments on four real-world datasets show significant improvements of our proposed model over the state-of-the-art methods.

## 1 Introduction

With the development of the Internet, recommender systems have been widely used in scenarios such as e-commerce, music, advertising, and news [1]. In practical applications, users usually do not give explicit preferences to items (e.g. rating, review) [10], so it is often difficult to obtain user explicit behaviors, while implicit user behavior (e.g. click) occurs more often [4,9]. The implicit behavior of users does not reveal preference information, which is often difficult to model, but it plays an important role in helping to study user preferences. Therefore, it is of great value and significance to be able to make effective recommendations based on implicit data [12,19].

With the rise of deep learning [18,25], deep neural networks (DNNs) have been widely used in various tasks due to their powerful fitting capabilities [13,24]. There mainly includes methods based on representation learning and matching function learning [4,7]. However, the methods based on representation learning


M. Tanveer et al. (Eds.): ICONIP 2022, CCIS 1793, pp. 313–324, 2023.
https://doi.org/10.1007/978-981-99-1645-0_26

limit the representation of the matching relationship due to the use of inner products, and the methods based on matching function learning are lacking in capturing low-order relationship. DeepCF [4] adopts both representation learning-based methods and matching function learning-based methods, combining their advantages and overcoming their disadvantages. However, many previous work simply used DNN to learn nonlinear matching functions, without considering more fine-grained matching relationships.

Compared with the previous work, we propose a new way of learning matching relationship. First, we introduce implicit multi-aspect representations to express the user's various preferences, and further calculate the global matching degree between user and item through the hierarchical interactive matching module. Second, since the user ID can be seen as a reflection of the user's identity information, we learn the local matching relationship based on the constructed implicit multi-aspect representations. Specifically, we use the attention mechanism to perceive the importance of different aspect through user ID information, and further calculate the local matching relationship between user and item in various aspect spaces. Finally, we use a gating mechanism to fuse the learned global matching and local matching information.

The main contributions of our work can be summarized as follows:

- We propose a global matching learning module, which is based on multiple aspect representation. By using hierarchical interactive matching module to learn the multiple interactions between aspects, the model can effectively learn the global matching information.
- We propose a novel model named Deep Global and Local Matching network (DeepGLM), which can effectively learn the matching relationship between user and item. In addition, we design an auxiliary loss to help the model learn more effectively.
- We conducted comprehensive experiments on four real-world datasets. The experimental results fully demonstrated the effectiveness of our proposed models which outperform other state-of-the-art methods.

## 2 Related Work

Collaborative filtering algorithms are widely used in recommendation systems [22]. The neighborhood-based collaborative filtering method [20] recommends items liked by users with the same interest to target users based on historical click data, without using explicit scoring. Because the neighborhood-based collaborative filtering methods are relatively simple and cannot handle the sparsity problem well, many methods based on matrix factorization (MF) [15] have been proposed. MF solves the problem of data sparsity, and introduces an optimized learning process to make the model more scalable. The Weighted Regularized Matrix Factorization (WR-MF) model [12] introduces the user's preference and confidence to model implicit feedback. BPR [19] introduces the maximum a posteriori estimation method to optimize pairwise loss based on implicit feedback. eALS [10] weights missing data according to item popularity, which solve the deficiencies of WR-MF for different negative feedback preferences.

AutoRec [21] propose the use of an autoencoder to learn the compression vector representation of the rating matrix, including user-based and item-based AutoRec. DMF [23] is based on MF and uses neural networks to learn implicit representations of users and items. Although methods based on representation learning can perform more complex nonlinear representations of users and items, it is not very accurate to use inner products to calculate matching scores linearly. Due to the limitation of the inner product to calculate the matching score, NeuMF [9] combines generalized matrix factorization for modeling low-order relationships and multilayer perceptron (MLP) for modeling high-order relationships to learn matching functions. ONCF [7] adopts outer product operation to model pairwise correlations between embedding dimensions on the basis of NCF. Under the ONCF framework, ConvNCF [7] leverages convolutional neural network (CNN) [17] to learn high-order correlations among embedding dimensions from locally to globally in a hierarchical way. DeepCF [4] combines the method based on representation learning and matching function learning. LightGCN [6] learns by linearly propagating user embedding and item embedding on the user-item interaction graph, which is easy to implement and train.

## 3 DeepGLM Model

This section describes the proposed DeepGLM model in detail. First, we give an overview of the overall structure of DeepGLM, which is shown in Fig. 1. Second, we propose a global matching learning module, which can learn the overall matching information between user and item. Then, we describe the local specific matching learning module for learning user-specific and item-specific matching information. Next, we propose a gating mechanism to fuse global and local matching information. Finally, the training optimization of is presented.

### 3.1 Global Matching Learning

The user's historically clicked items implicitly reflect the user's global preferences. We calculate the global matching information between the user and item based on the user and item rating.

**Multi-Aspect Representation Module.** The historical interactions between a user and items reflect the user's personal preference, which should be diverse. For example, users may not only care about whether the price of the product is cheap, but also whether the product is convenient to use. Price and practicality are different aspects that users care about. User ratings consist of items the user has interacted with in history, and should belong to multiple identical aspects that the user likes. Therefore, we use multiple networks to model multiple implicit aspect, which are defined as follows:

$$z_j^u = f_j(W_j^u e_r^u + b_j) \tag{1}$$

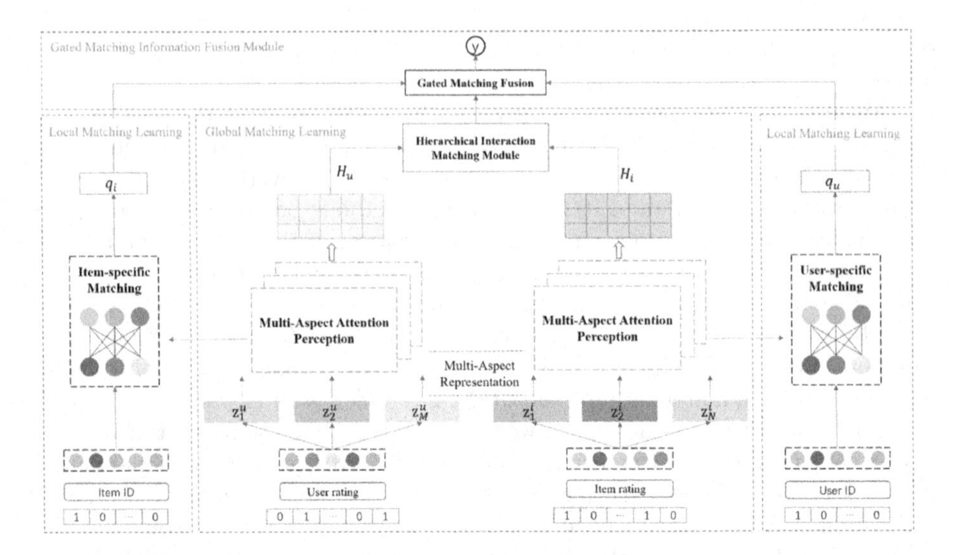


**Fig. 1.** The architecture of the Deep Global and Local Matching Network.

where $z_j^u$ represents the $j$-th aspect of the item that the user likes, $f_j$represents the neural network that learns the $j$-th aspect. $e_r^u$ denotes the embedding representation of user rating.

In order to fully represent the user's multi-aspect information, we construct low and high order implicit representations through multi-layer fully connected neural networks. For the $j$-th aspect $z_j^u$, the multi-layer information representation is constructed as follows:

$$h_{j,l}^u = g_l(W_l^u z_j^u + b_l) \tag{2}$$

where $h_{j,l}^u$ denotes the implicit representation of the $j$-th aspect at the $l$-th layer. $f_l$ represents the neural network that learns the $l$-th representation. Further, we stack the implicit representations of $L$ layers as the hierarchical representation matrix of the $j$-th aspect:

$$H_j^u = stack([h_{j,1}^u, h_{j,2}^u, \ldots, h_{j,L}^u]) \tag{3}$$

where $H_j^u$ denotes the hierarchical representation matrix of the $j$-th aspect. *stack* represents the function that stack representations by column.

**Item-Aspect Perception Module.** According to user rating, multiple aspects of user concern can be constructed based on items that the user has interacted with. Considering that users pay different attention to different aspects when selecting items, aspects of different items are of different importance. Therefore, we fuse the representation matrix based on the ID information of the current item as follows:

$$H_u = \sum_{l=1}^{H_u} \gamma_j H_j^u$$
$$\gamma_j = \frac{e^{a_j^u e_2^u}}{\sum_{l=1}^{L} e^{a_j^u e_2^u}} \tag{4}$$

where $H_u$ represents the global aspect representation matrix based on item multi-aspect perception. $e_2^u$ denotes the embedding representation of user ID. $a_j^u$ represents the average representation information of the $j$-th aspect, which is calculated as follows:

$$a_j^u = average\_pooling(H_j^u) \tag{5}$$

where *average_pooling* represents a function that averages the matrix by column. Using the same method as above, we can get the global aspect representation $H_i$ based on user multi-aspect perception.

**Hierarchical Interaction Matching Module.** Considering that there are multiple item aspects favored by users when selecting an item, there are also selected user aspects when an item is selected by users. In order to fully explore the matching relationship between user and item at different aspects, we use attention mechanism to perceive matching information. We define query, key and value in attention mechanism as follows:

$$Q = W_q^i H_i, K = W_k^u H_u, V = W_v^u H_u \tag{6}$$

where $Q$ represents the query in the attention mechanism, and $K$ and $V$ represent key and value. $W_q^i$, $W_k^u$ and $W_v^u$ are parameters. Taking into account the diversity in the construction of relationships between multilevel representations, we adopt the attention mechanism as follows:

$$A = Attention(Q, K, V) = softmax(\frac{QK}{\sqrt{d_k}})V \tag{7}$$

where $A$ denotes the representation of multi-aspcet perception. $d_k$ refers to the matrix dimension. Further, we perform the *averagepooling* operation on $A$ to obtain the fusion representation as:

$$p = average\_pooling(A) \tag{8}$$

where $p$ represents matching information under hierarchical attention perception.

### 3.2 Local Matching Learning

The ID of the user is used as the symbol of the user, which indicates the identity information. We can get item-specific matching information by establishing the relationship between the ID information of the candidate item and the user rating.

The implicit multi-aspect representation $z_j^u$ reflects the characteristics of users who have clicked on the item. In order to judge the connection between the current user and the users who have clicked in the past, we use the attention mechanism [26] to perceive the importance as follows:

$$\alpha_j^u = ReLU(W_j^u[z_j^i; e_2^u; z_j^i \odot e_2^u; z_j^i - e_2^u] + b_j^u) \tag{9}$$

where $\alpha_j^u$ represents the importance of the $j$-th implicit aspect to the current user. $e_2^u$ denotes the embedding representation of user ID.

Further, considering that multiple aspects actually represent multiple common characteristics of users, we establish a matching relationship between the current user and multiple aspects that contain different user characteristics. Based on the learned aspect score reflecting the importance of the current user, we construct user-perceived aspect matching as follows:

$$\begin{aligned} q_u &= \sum_{l=1}^{N} \beta_j^u (z_j^i \odot e_2^u) \\ \beta_j^u &= \frac{e^{\alpha_j^u}}{\sum_{l=1}^{N} e^{\alpha_j^u}} \end{aligned} \tag{10}$$

where $q_u$ represents user-specific local matching representation. $\beta_j^u$ is the attention score. In the same way, we can get the item-specific local matching representation $q_i$.

### 3.3 Gated Information Fusion

Inspired by neural attentive item similarity (NAIS) [8], we propose a gated information fusion module, which control the delivery of information. Modeled on gated recurrent neural network [3,11], we take the global matching information $p$ as the historical hidden state and the local specific matching information $x = [q_u, q_i]$ as the input to obtain the fusion information as follows:

$$\begin{aligned} \widetilde{p} &= tanh(W_p x + U_p(r_g \odot p) + b_h) \\ p_g &= z_g \odot p + (1 - z_g) \odot \widetilde{p} \end{aligned} \tag{11}$$

where $r_g$ represents the reset gate and $z_g$ represents the update gate, which are defined as follows:

$$\begin{aligned} z_g &= \sigma(W_z x + U_z p + b_z) \\ r_g &= \sigma(W_r x + U_r p + b_r) \end{aligned} \tag{12}$$

Finally, we get the prediction result of the model:

$$r = \sigma(W_g p_g + b_g) \tag{13}$$

where $r$ denotes the output probability.

### 3.4 Training Optimization

In this paper, we mainly predict the probability of interaction between a user and an item from the perspective of matching probability. Following NCF [9], We define the first cross-entropy loss function as follows:

$$Loss_1 = -y_{u,i}log(r_{u,i}) - (1 - y_{u,i})log(1 - r_{u,i}) \tag{14}$$

where $r_{u,i} \in [0, 1]$ represents the predicted match probability. In addition, we constructed implicit multiple aspect representations based on user ratings and item ratings. In order to ensure the difference between different aspect representations, we introduce regularized auxiliary loss to assist in the generation of multiple aspects, which is defined as follows:

$$Loss_2 = -\lambda_1 \sum_{a=1}^{M} \sum_{b=a+1}^{M} \frac{z_a^u \cdot z_b^u}{|z_a^u||z_b^u|} - \lambda_2 \sum_{a=1}^{N} \sum_{b=a+1}^{N} \frac{z_a^i \cdot z_b^i}{|z_a^i||z_b^i|} \tag{15}$$

where $z_a^u$ and $z_b^u$ represent the $a$-th and $b$-th implicit aspects learned from user ratings. $\lambda_1$ and $\lambda_2$ represent hyperparameters, which are used to control the degree of regularization. Finally, we combine the cross-entropy loss and the regularized auxiliary loss to get the total loss as follows:

$$Loss = \frac{1}{N} \sum_{(u,i) \in I \cup I^-}^{N} (Loss_1 + Loss_2) \tag{16}$$

where $Loss$ represents the total loss of the model, consisting of cross-entropy loss and regularized auxiliary loss.

## 4 Experiments

In this section, we conduct extensive experiments to answer the following questions:

- **RQ1** How does our DeepGLM perform as compared with state-of-the art recommendation methods?
- **RQ2** Are the proposed global matching learning and local matching learning module helpful for improving the recommendation performance?
- **RQ3** How do the key hyperparameter affect DeepGLM's performance?

### 4.1 Experimental Settings

**Datasets.** We adopt four real-world benchmark datasets for evaluation. The statistics of the four datasets are shown in Table 1, and their descriptions are as follows:

**AMusic**[1] **Dataset** [16] is a classic dataset composed of music product data collected from Amazon. It records information about users, items and ratings.

**AToy**[2] **Dataset** [16] is a dataset composed of toy product data collected from Amazon. It records information about users, items and ratings.

**MovieLens**[3] **Dataset** [5] is a widely adopted benchmark dataset in movie recommendation, which records users' ratings of different movies.

**Last.FM**[4] **Dataset** [2] contains social networking, tagging, and music artist listening information from a set of 2000 users from Last.fm online music system.

**Table 1.** Statistics of the evaluation datasets.

| Datasets | #instances | #users | #items |
|---|---|---|---|
| AMusic | 46,087 | 1,776 | 12,929 |
| AToy | 84,642 | 3,137 | 33,953 |
| MovieLens | 1,000,209 | 6,040 | 3,706 |
| Last.FM | 69,149 | 1,741 | 2,665 |

**Baselines.** To evaluate the performance, we compared the proposed DeepGLM with the following baselines:

- **Itempop** [19] ranks items based on popularity by counting the number of interactions. This recommendation algorithm is often used as a benchmark.
- **BPR** [19] based on maximum posterior estimator optimizes the MF model with a pairwise ranking loss, which is tailored to learn from implicit feedback.
- **eALS** [10] designs a learning algorithm based on alternating least squares, which is one of the optimal models of matrix factorization.
- **DMF** [23] is a classic method based on representation learning, which learns the implicit representation of features through neural networks.
- **NeuMF** [9] is a benchmark based on matching function learning. NeuMF learns high-order nonlinear matching functions through MLP, and learns low-order matching relationships with GMF.
- **ConvNCF** [7] is a model based on matching function learning, which uses outer product to generate a two-digit interaction map, and employs a convolutional neural network to learn high-order correlations.
- **DeepCF** [4] is a state-of-the-art matching model. DeepCF combines the method based on representation learning and matching function learning.
- **LightGCN** [6] is a state-of-the-art matching model based on graph learning, LightGCN linearly propagates user embedding and item embedding on the user-item interaction graph, which is easy to implement and train.

[1] http://jmcauley.ucsd.edu/data/amazon/.
[2] http://jmcauley.ucsd.edu/data/amazon/.
[3] https://grouplens.org/datasets/movielens/.
[4] http://www.lastfm.com.

**Parameter Settings.** To determine the model hyperparameters, we randomly sample an interaction for each user as a validation set to fine-tune the parameters. All methods were optimized with Adam [14], where the batch size was set to 256 with considering both training time and convergence rate. The learning rate was searched in [0.0001, 0.0005, 0.001]. We tested the embedding size from 8 to 512. To prevent overfitting, we used dropout for neural network models. The dropout ratio was searched in [0, 0.1, 0.2, 0.3, 0.4, 0.5], respectively. We mainly use Hit Ratio (HR@k) and Normalized Discounted Cumulative Gain (NDCG@k) to measure the recommendation effect. In this work, we truncated the ranked list at 10 for both metrics.

**Table 2.** Experimental results of different methods in term of HR@10 and NDCG@10.

| Models | AMusic | | AToy | | Ml-1m | | LastFM | |
|---|---|---|---|---|---|---|---|---|
| | HR | NDCG | HR | NDCG | HR | NDCG | HR | NDCG |
| Itempop | 0.2483 | 0.1304 | 0.2840 | 0.1518 | 0.4535 | 0.2542 | 0.6628 | 0.3862 |
| BPR | 0.3653 | 0.2300 | 0.3675 | 0.2389 | 0.6900 | 0.4194 | 0.8187 | 0.5098 |
| eALS | 0.3711 | 0.2352 | 0.3717 | 0.2434 | 0.7018 | 0.4280 | 0.8265 | 0.5162 |
| DMF | 0.3744 | 0.2149 | 0.3535 | 0.2016 | 0.6565 | 0.3761 | 0.8840 | 0.5804 |
| NeuMF | 0.3891 | 0.2391 | 0.3650 | 0.2155 | 0.7210 | 0.4387 | 0.8868 | 0.6007 |
| ConvNCF | 0.3946 | 0.2572 | 0.3860 | 0.2464 | 0.7238 | 0.4402 | 0.8951 | 0.6161 |
| DeepCF | 0.4116 | 0.2601 | 0.4150 | 0.2513 | 0.7253 | 0.4416 | 0.8975 | 0.6186 |
| LightGCN | 0.4203 | 0.2685 | 0.4274 | 0.2513 | 0.7280 | 0.4437 | 0.8991 | 0.6215 |
| DeepGLM | **0.4341** | **0.2773** | **0.4530** | **0.2751** | **0.7306** | **0.4464** | **0.9107** | **0.6249** |

### 4.2 Overall Performance (RQ1)

In order to verify the effectiveness of our proposed DeepGLM, we conducted comprehensive experiments on four datasets. Table 2 reports the performance comparison results. Figure 2 shows the performance of Top-K recommended lists where the ranking position K ranges from 1 to 10. We have the following observations: First, our proposed DeepGLM performs best and achieves state-of-the-art on four datasets. Although DeepCF is one of the best matching models as a unified matching learning framework, it is still limited to simple superposition of neural networks. The experimental results strongly prove the effectiveness of our model. Second, with the increase of k, our model not only maintains the optimal effect all the time, but the experimental results prove that our model has good consistency and effectiveness. In addition, the improvement of the model effect increases with the increase of k.

### 4.3 Effectiveness of Each Module (RQ2)

To verify the effectiveness of each module of DeepGLM, we conduct ablation experiments to study. The experimental results are shown in Table 3. We can

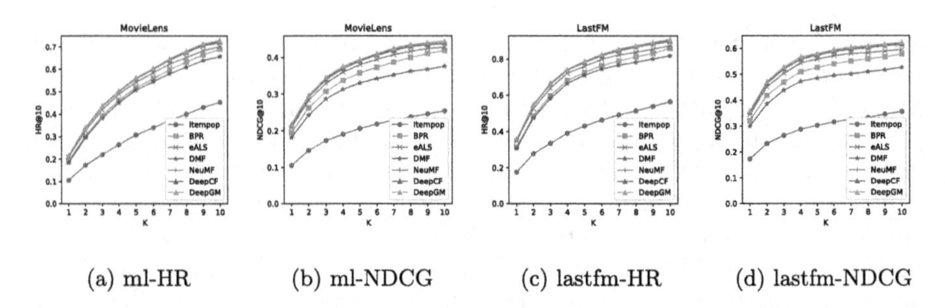


(a) ml-HR (b) ml-NDCG (c) lastfm-HR (d) lastfm-NDCG

**Fig. 2.** Evaluation of Top-K item recommendation where K ranges from 1 to 10 on the MovieLens and LastFM datasets.

have the following observation conclusions: First, the effect of the global matching learning module is equivalent to that of DeepCF, indicating that our implicit multi-interest modeling is effective. Second, the effect of replacing the gating mechanism with simple concatenate and vector addition is also better than that of a single GML, which shows that the fusion of global matching learning and local matching learning is effective and necessary. Third, the concatenate and vector addition are not as effective as our model using the gating mechanism.

**Table 3.** Performance of DeepGLM with different module. LML: Only with the local matching learning module. GML: Only with the global matching learning module. With addition: DeepGLM replaces the gated network with vector addition. With concat: DeepGLM replaces the gated network with vector connection. Without aux_loss: DeepGLM removes the auxiliary $loss_2$.

| Model | AMusic | | LastFM | |
|---|---|---|---|---|
| | HR | NDCG | HR | NDCG |
| LML | 0.3927 | 0.2567 | 0.8879 | 0.6028 |
| GML | 0.4125 | 0.2651 | 0.9015 | 0.6143 |
| with addition | 0.4248 | 0.2688 | 0.9049 | 0.6198 |
| with concat | 0.4267 | 0.2705 | 0.9078 | 0.6205 |
| without aux_loss | 0.4281 | 0.2714 | 0.9085 | 0.6213 |
| DeepGLM | **0.4341** | **0.2773** | **0.9107** | **0.6249** |

## 4.4 Hyper-parameter Study (RQ3)

**Negative Sampling Rate.** We sampled different numbers of negative samples for each positive sample, and the experimental results on the AMusic and LastFM datasets are shown in Table 4. We can see that the optimal number of negative samples for both datasets is between 4 and 8.

**Table 4.** Performance of DeepGLM with different negative sampling rate.

| Negatives | AMusic | | LastFM | |
|---|---|---|---|---|
| | HR | NDCG | HR | NDCG |
| 1 | 0.3173 | 0.1448 | 0.7556 | 0.4836 |
| 2 | 0.4025 | 0.2342 | 0.8762 | 0.5915 |
| 3 | 0.4286 | 0.2701 | 0.9061 | 0.6095 |
| 4 | 0.4341 | 0.2773 | 0.9107 | 0.6249 |
| 5 | **0.4357** | **0.2803** | 0.9055 | 0.6126 |
| 6 | 0.4328 | 0.2748 | **0.9118** | **0.6312** |
| 7 | 0.4339 | 0.2767 | 0.9072 | 0.6190 |
| 8 | 0.4317 | 0.2745 | 0.9107 | 0.6257 |

## 5 Conclusion

In this paper, we propose a novel model named deep global and local matching network (DeepGLM) based on global matching learning and local matching learning. Comprehensive experiments on four real-world datasets have fully proved the effectiveness of our model. In future work, we consider introducing auxiliary information such as reviews and images to further enhance matching relationship modeling.

## References

1. Bobadilla, J., Ortega, F., Hernando, A., Gutiérrez, A.: Recommender systems survey. Knowl.-Based Syst. **46**, 109–132 (2013)
2. Cantador, I., Brusilovsky, P., Kuflik, T.: Second workshop on information heterogeneity and fusion in recommender systems (hetrec2011). In: Proceedings of the fifth ACM Conference on Recommender Systems, pp. 387–388 (2011)
3. Chung, J., Gulcehre, C., Cho, K., Bengio, Y.: Empirical evaluation of gated recurrent neural networks on sequence modeling. arXiv preprint arXiv:1412.3555 (2014)
4. Deng, Z.H., Huang, L., Wang, C.D., Lai, J.H., Philip, S.Y.: DeepCF: a unified framework of representation learning and matching function learning in recommender system. In: Proceedings of the AAAI Conference on Artificial Intelligence, vol. 33, pp. 61–68 (2019)
5. Harper, F.M., Konstan, J.A.: The MovieLens datasets: history and context. ACM Trans. Interact. Intell. Syst. (TIIS) **5**(4), 1–19 (2015)
6. He, X., Deng, K., Wang, X., Li, Y., Zhang, Y., Wang, M.: LightGCN: simplifying and powering graph convolution network for recommendation. In: Proceedings of the 43rd International ACM SIGIR Conference on Research and Development in Information Retrieval, pp. 639–648 (2020)
7. He, X., Du, X., Wang, X., Tian, F., Tang, J., Chua, T.S.: Outer product-based neural collaborative filtering. arXiv preprint arXiv:1808.03912 (2018)
8. He, X., He, Z., Song, J., Liu, Z., Jiang, Y.G., Chua, T.S.: NAIS: neural attentive item similarity model for recommendation. IEEE Trans. Knowl. Data Eng. **30**(12), 2354–2366 (2018)

9. He, X., Liao, L., Zhang, H., Nie, L., Hu, X., Chua, T.S.: Neural collaborative filtering. In: Proceedings of the 26th International Conference on World Wide Web, pp. 173–182 (2017)
10. He, X., Zhang, H., Kan, M.Y., Chua, T.S.: Fast matrix factorization for online recommendation with implicit feedback. In: Proceedings of the 39th International ACM SIGIR Conference on Research and Development in Information Retrieval, pp. 549–558 (2016)
11. Hochreiter, S., Schmidhuber, J.: Long short-term memory. Neural Comput. **9**(8), 1735–1780 (1997)
12. Hu, Y., Koren, Y., Volinsky, C.: Collaborative filtering for implicit feedback datasets. In: 2008 Eighth IEEE International Conference on Data Mining, pp. 263–272. IEEE (2008)
13. Huang, T., Zhang, Z., Zhang, J.: FiBiNET: combining feature importance and bilinear feature interaction for click-through rate prediction. In: Proceedings of the 13th ACM Conference on Recommender Systems, pp. 169–177 (2019)
14. Kingma, D.P., Ba, J.: Adam: a method for stochastic optimization. arXiv preprint arXiv:1412.6980 (2014)
15. Koren, Y., Bell, R., Volinsky, C.: Matrix factorization techniques for recommender systems. Computer **42**(8), 30–37 (2009)
16. Lakkaraju, H., McAuley, J., Leskovec, J.: What's in a name? Understanding the interplay between titles, content, and communities in social media. In: Proceedings of the International AAAI Conference on Web and Social Media, vol. 7 (2013)
17. Lawrence, S., Giles, C.L., Tsoi, A.C., Back, A.D.: Face recognition: a convolutional neural-network approach. IEEE Trans. Neural Networks **8**(1), 98–113 (1997)
18. LeCun, Y., Bengio, Y., Hinton, G.: Deep learning. Nature **521**(7553), 436–444 (2015)
19. Rendle, S., Freudenthaler, C., Gantner, Z., Schmidt-Thieme, L.: BPR: Bayesian personalized ranking from implicit feedback. arXiv preprint arXiv:1205.2618 (2012)
20. Sarwar, B., Karypis, G., Konstan, J., Riedl, J.: Item-based collaborative filtering recommendation algorithms. In: Proceedings of the 10th international conference on World Wide Web, pp. 285–295 (2001)
21. Sedhain, S., Menon, A.K., Sanner, S., Xie, L.: AutoRec: autoencoders meet collaborative filtering. In: Proceedings of the 24th international conference on World Wide Web, pp. 111–112 (2015)
22. Su, X., Khoshgoftaar, T.M.: A survey of collaborative filtering techniques. In: Advances in Artificial Intelligence 2009 (2009)
23. Xue, H.J., Dai, X., Zhang, J., Huang, S., Chen, J.: Deep matrix factorization models for recommender systems. In: IJCAI, vol. 17, pp. 3203–3209. Melbourne, Australia (2017)
24. Yang, W., Hu, T.: DFCN: an effective feature interactions learning model for recommender systems. In: Jensen, C.S., et al. (eds.) DASFAA 2021. LNCS, vol. 12683, pp. 195–210. Springer, Cham (2021). https://doi.org/10.1007/978-3-030-73200-4_13
25. Zhang, S., Yao, L., Sun, A., Tay, Y.: Deep learning based recommender system: a survey and new perspectives. ACM Comput. Surv. (CSUR) **52**(1), 1–38 (2019)
26. Zhou, G., et al.: Deep interest evolution network for click-through rate prediction. In: Proceedings of the AAAI conference on artificial intelligence, vol. 33, pp. 5941–5948 (2019)

# A Bi-hemisphere Capsule Network Model for Cross-Subject EEG Emotion Recognition

Xueying Luan, Gaoyan Zhang(✉), and Kai Yang

Tianjin Key Laboratory of Cognitive Computing and Application, College of Intelligence and Computing, Tianjin University, Tianjin, China
{zhanggaoyan,kai_y}@tju.edu.cn

**Abstract.** Cognitive neuroscience research has revealed that electroencephalography (EEG) has a strong correlation with human emotions. However, due to the individual differences in EEG signals, the traditional models have the shortcoming of poor generalization ability. Based on the discovery that responses to emotional stimuli in the cerebral hemispheres is asymmetric, in this paper, we propose a Bi-hemispheric Capsule Network (Bi-CapsNet) Model for cross-subject EEG emotion recognition. Specifically, we firstly use a long short term memory (LSTM) layer to learn the asymmetry of emotion expression between the left and right hemispheres of the human brain and the deep representations of all the EEG electrodes'signals in different frequency bands. In order to capture the relationship between EEG channels more detailedly, a special mechanism called routing-by-agreement mechanism has been implemented between LSTM and EmotionCaps. We also use a domain discriminator working corporately with the EmotionCaps to reduce the domain shift between the source domain and the target domain. In addition, we propose a method to reduce the uncertainty of predictions on the target domain data by minimizing the entropy of the prediction posterior. Finally, the cross-subject EEG emotion recognition experiments conducted on two public datasets, SEED and SEED-IV, were that when the length of the EEG data samples is 1 s, the proposed model can obtain better results than most methods on the SEED-IV dataset and also achieves state-of-the-art performance on the SEED dataset.

**Keywords:** EEG emotion recognition · Cerebral hemisphere asymmetry · Long short term memory (LSTM) · Routing-by-agreement mechanism · Uncertainty

## 1 Introduction

Emotion is an advanced activity of the brain. It is a complex psychological and physiological mechanism gradually formed by human beings in the process of adapting to the social environment [1]. Affective computing is an important research direction in modern human-computer interaction. It aims to improve the ability of the computer to recognize, understand and correctly respond to


M. Tanveer et al. (Eds.): ICONIP 2022, CCIS 1793, pp. 325–336, 2023.
https://doi.org/10.1007/978-981-99-1645-0_27

human emotional states. Emotion recognition is the basic problem in affective computing. Traditional emotion recognition methods rely on non-physiological signals, such as facial expressions [2], speech [3], and body postures [4], which are not applicable for severe patients or the physically disabled. A new field of emotion recognition is called affective brain-computer interfaces (aBCIs) [5]. In aBCIs, electroencephalography (EEG) [6] is the most widely used physiological signal. Because EEG is the spontaneous physiological signal of the human brain, EEG is not affected by subjective will and provides an objective and reliable way for emotion recognition.

The traditional EEG-based emotion recognition methods need to collect and annotate enough EEG signals to establish a model based on a specific subject. However, the acquisition of EEG signals is a time-consuming and high-cost work, and the individual differences of EEG signals limit the generalization of the model among subjects. To address the problems of limited labeled data and individual differences in EEG emotion recognition, the following issues are worth to be studied in depth. One is how to extract significant emotional features from EEG signals, and the other one is how to develop a more generalized emotion recognition model. In recent years, embedding adaptive modules into deep networks to learn transferable representations has become a research highlight in cross-subject EEG emotion recognition. For example, Li et al. [7] applied Deep Adaptation Network (DAN) for dealing with the cross-subject EEG emotion recognition on SEED [8] and SEED-IV [9], and achieved better performance than traditional machine learning algorithms. In the past several years, many researchers have proposed to take cerebral hemispheric asymmetry as the prior knowledge to extract features or develop models, which effectively enhance the performance of EEG emotion recognition. In recent years, based on this prior knowledge, bi-hemispheres domain adversarial neural network (BiDANN) [10] and bi-hemispheric discrepancy model (BiHDM) [11] have been proposed to verify the discrimination and effectiveness of asymmetric information in EEG emotion recognition. Furthermore, There is obvious correlativity between the EEG frequency band and the emotional mechanism of the brain. Zheng et al. [8] explored critical frequency bands that can dramatically affect the effect of emotion recognition by examining the weight distribution learned by Deep Belief Network (DBN).

Considering the hemispheric laterality responses of the human brain in perceiving different emotional stimuli, in this paper, we propose a Bi-hemispheric Capsule Network (Bi-CapsNet) Model. Specifically, we capture the differences between hemispheres and the approximate relationship between EEG channels in each frequency band through the LSTM layer. The routing-by-agreement mechanism captures more detailed connections between EEG channels through the transformation matrix, and can distinguish the significant difference information reflected by the human brain under different emotional states. Since the source domain EEG data training network may be unreliable in predicting the target domain EEG data, we use two methods to reduce the uncertainty of predictions on the target domain data. On the one hand, considering the data distribution shift of EEG signals, we introduce a global domain discriminator to mitigate the domain shift between source domain data and target domain data, which will enable the feature learning process to generate emotion related but domain invariant data representation. On the other hand, we propose a regularization method that reduces the uncertainty of predictions on the target domain, so as to better generalize in cross-subject classification tasks. Different from [12], our method explicitly regularizes the prediction results by minimizing the prediction information entropy of the target domain data during the transfer learning process. Finally, we carry out experiments to verify the effectiveness of our Bi-CapsNet model.

## 2 The Proposed Method

The proposed Bi-CapsNet model has three modules, namely, LSTM, Emotion-Caps and global domain discriminator. The details are illustrated in Fig. 1.

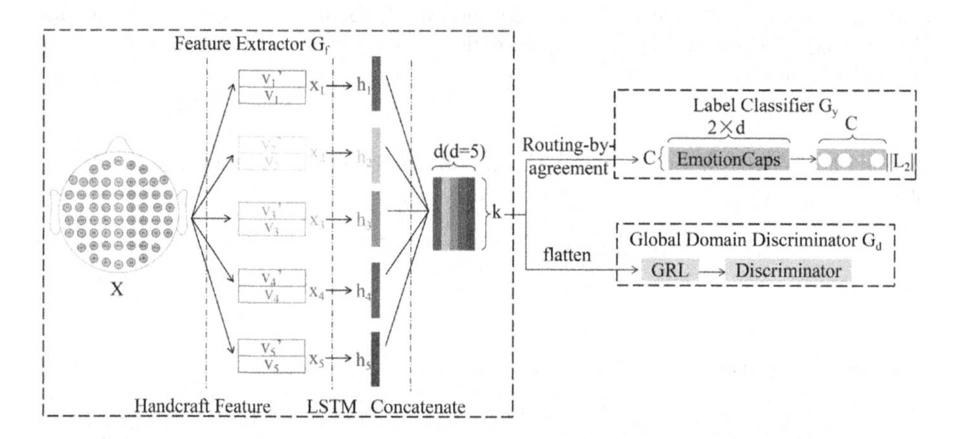


**Fig. 1.** The framework of Bi-CapsNet. Bi-CapsNet consists of a deep feature extractor $G_f$, a label classifier $G_y$, a global domain discriminator $G_d$. GRL stands for Gradient Reversal Layer.

### 2.1 The Bi-hemisphere Matrix

Let $X = [x_1, x_2, \cdots, x_d] \in R^{2\times d\times n}$ denotes an EEG sample, where $x_i \in R^{2\times n}$ represents the Bi-hemisphere Matrix (BiM) of the i-th frequency band of EEG signal, and d and n represent the number of EEG frequency bands and electrodes respectively. Figure 2 shows the detailed construction process of BiM. We divide the 62 electrodes into three parts based on the spatial locations of the electrodes, namely, the left hemisphere region, the right hemisphere region, and the shared region. The corresponding EEG electrodes in the left and right hemispheres are exchanged to construct the auxiliary vector $v_i^{'} \in R^n$ of the eigenvector $v_i \in R^n$

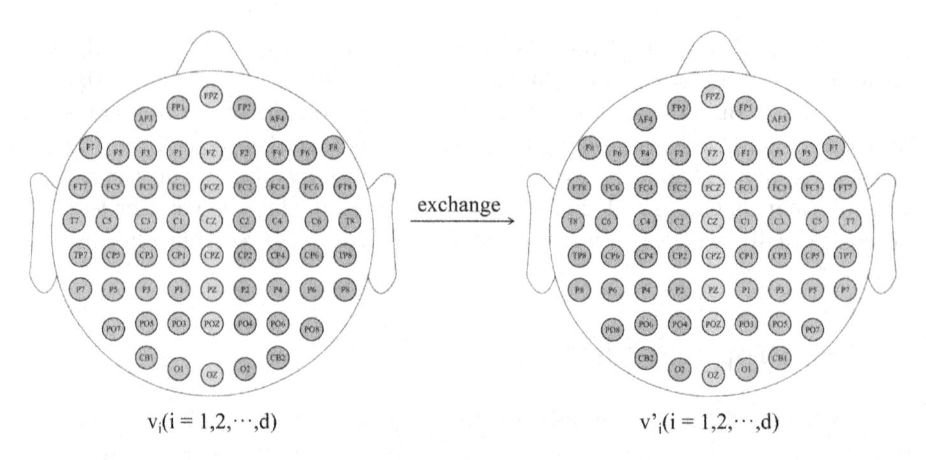


**Fig. 2.** An illustration of the divisions of the 62 electrodes into 3 clusters, where the same color denotes the electrodes are grouped into the same brain region. Exchange the electrodes in the left and right brain regions, that is, the auxiliary vector $v_i^{'}$ of the original vector $v_i$.

of the i-th frequency band. The BiM is composed of the manual eigenvector $v_i$ of the i-th frequency band and its auxiliary vector $v_i^{'}$, i.e.,

$$x_i = [v_i^{'}, v_i], i = 1, 2, \cdots, d \tag{1}$$

## 2.2 Asymmetric Feature Learning

The long short term memory (LSTM) network [13] is a special type of recurrent neural network (RNN) [14], which can make use of the previous context. In this paper, we use this feature of LSTM to learn the differential expression of left and right hemispheres by means of auxiliary vectors. By applying the LSTM to the BiM of each frequency band, the asymmetric difference between the two hemispheres is learned and the spatial relationship feature of the electrode is preliminarily extracted, which can be expressed as:

$$f(x_i) = [h_i^{'}, h_i] \in R^{2\times k} \tag{2}$$

where $f(\cdot)$ denotes the LSTM feature learning operation and k is the hidden units dimension of regional spatial LSTM. After the above differential feature learning process, we take the last vector $h_i$ of LSTM as the feature representation. In this case, the final eigenvector can be expressed as:

$$U = [\sigma(h_1), \sigma(h_2), \cdots, \sigma(h_d)]^T \in R^{k\times d} \tag{3}$$

where $\sigma(\cdot)$ indicates nonlinear operation, i.e. tanh function.

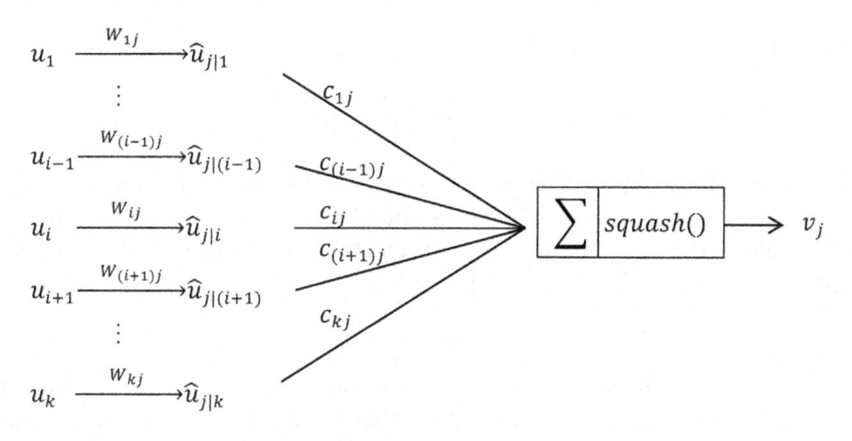


**Fig. 3.** Routing-by-agreement mechanism.

## 2.3 EmotionCaps

The final module is EmotionCaps [15]. After the feature extraction phase, for each data sample, we obtained k capsule outputs, each of which is a d-dimensional vector, which can be expressed as:

$$U = [u_1, u_2, \cdots, u_k] \in R^{k \times d} \tag{4}$$

The feature extraction module and Emotioncaps are connected through the routing-by-agreement mechanism [16]. This process not only captures the spatial relationship by transformation matrices, but also routes the information between capsules by reinforcing connections of those capsules to obtain a high grade of agreement. In the following, we provide the details of this process and show it in Fig. 3.

First of all, the input vector $u_i(i = 1, 2, \cdots, k)$ is multiplied by the weight matrix $W_{ij}(j = 1, 2, \cdots, C)$ to obtain a higher-level vector, which can be regarded as that each capsule neuron outputs to a neuron in the next layer with different strength connections. This step can be formulated as:

$$\hat{u}_{j|i} = W_{ij} u_i \tag{5}$$

After that, all prediction vectors $\hat{u}_{j|i}$ from the lower capsule are weighted and added to obtain $s_j$. This step is to measure the importance of low-level emotional features to high-level emotional features. This step can be expressed as:

$$s_j = \sum_i c_{ij} \hat{u}_{j|i} \tag{6}$$

where $c_{ij}$ is the coupling coefficient determined by the iterative dynamic routing process. The sum of the coupling coefficients between the i-th capsule of the previous layer and all emotional capsules in the next layer is 1. The coupling coefficient $c_{ij}$ can be computed as:

$$c_{ij} = \frac{exp(b_{ij})}{\sum_k exp(b_{ik})} \tag{7}$$

The core of routing-by-agreement mechanism lies in the update method of parameter $b_{ij}$. By measuring the consistency between the current output $v_j$ of j-th capsule in the upper layer and the predicted $\hat{u}_{j|i}$ of i-th capsule, the initial coupling coefficient is iteratively improved.

$$b_{ij} \leftarrow b_{ij} + v_j \cdot \hat{u}_{j|i} \tag{8}$$

In the last step, the non-linear function known as "squash" is used to limit the length of the vector $v_j$ between 0 and 1, so that the length of the eigenvector can represent the probability of the existence of the entity. The "squash" function is defined as follows:

$$v_j = \frac{||s_j||^2}{1 + ||s_j||^2} \frac{s_j}{||s_j||} \tag{9}$$

The EmotionCaps has C $(2 \times b)$D emotional capsules , corresponding to C emotional states. The loss function of the classifier is the sum of all emotional capsules losses:

$$L_c(X; \theta_f, \theta_c) = \sum_{c=1}^{C} T_c max(0, m^+ - ||v_c||)^2 + \lambda(1 - T_c max(0, m||v_c|| - m^-)^2 \tag{10}$$

where $T_c$ indicates whether class c exists. We set $m^+$, $m^-$, and $\lambda$ to 0.9, 0.1, and 0.5 respectively. We use the length of the final eigenvector to represent the probability of the existence of the corresponding emotion category.

### 2.4 Discriminator

In the cross-subject EEG emotion recognition, the emotion recognition model trained based on the source subjects'EEG data may not be suitable for the EEG data of the target subject'EEG data. The use of global discriminator aims to align the feature distributions between the source domain and the target domain. To be specific we predefine the source domain data set $X^S = \{X_1^S, X_2^S, \cdots, X_{M1}^S\}$ and target domain data set $X^T = \{X_1^T, X_2^T, \cdots, X_{M2}^T\}$ where $M_1$ and $M_2$ denote the number of source samples and target samples, respectively. The domain difference is eliminated by maximizing the above loss function of the discriminator:

$$L_d(X_i^S, X_j^T; \theta_f, \theta_d) = -\sum_{i=1}^{M_1} logP(0|X_i^S) - \sum_{j=1}^{M_2} logP(1|X_j^T) \tag{11}$$

here $\theta_d$ represents the learnable parameter of the discriminator, while $P(0|X_i^S)$ and $P(1|X_j^T)$ are the probability that $X_i^S$ belongs to the source domain and $X_j^T$ belongs to the target domain, respectively.

### 2.5 The Entropy Loss

In order to make the prediction of EEG data in the target domain more reliable, we use an entropy regularizer [17] to minimize the entropy of the network's predictions, which will bring better empirical performance.

$$L_{ent}(X_j^T;\theta_f,\theta_c)=\sum_{j=1}^{M_2}H(y) \tag{12}$$

where, $H(\cdot)$ is the entropy of the probability distribution $y$ over all C classes for an input $X_j^T$. This loss will make the probability of one class much larger than that of other classes, thus reducing the prediction uncertainty. In the transfer learning process, the decision boundary given by the optimizer itself is modified, which is conducive to explicitly regularize the predictions.

### 2.6 The Optimization of Bi-CapsNet

The overall loss function of Bi-CapsNet can be expressed as follows:

$$\begin{aligned} minL(X^S,X^T;\theta_f,\theta_c,\theta_d) = minL_c(X^S;\theta_f,\theta_c) + maxL_d(X^S,X^T;\theta_f,\theta_d) \\ +minL_{ent}(X^T;\theta_f,\theta_c) \end{aligned} \tag{13}$$

By minimizing $L(X^S,X^T;\theta_f,\theta_c,\theta_d)$, we can achieve better emotion category prediction for training data samples. By maximizing $L_d(X^S,X^T;\theta_f,\theta_d)$, we can achieve domain invariant features to mitigate domain differences in EEG emotion recognition. By minimizing $L_{ent}(X^T;\theta_f,\theta_c)$, we can increase the classification stability of the target domain data. Our goal is to find the optimal parameter that minimizes the loss function $L(X^S,X^T;\theta_f,\theta_c,\theta_d)$. We introduce the gradient reversal layer (GRL) for the discriminator to transform the maximization $L_d(X^S,X^T;\theta_f,\theta_d)$ problem into the minimization problem, so that the SGD method can be used to optimize the parameters.

## 3 Experiments

### 3.1 Datasets

We conduct experiments on two publicly available emotion recognition datasets, SEED and SEED-IV, to evaluate the proposed Bi-CapsNet method.

For both datasets, the raw EEG signals are recorded using an ESI NeuroScan System with 62 electrode channels placed according to the international 10-20 system. The EEG data are collected when the participants are watching kinds of emotional film clips. The raw EEG data is down-sampled 200 Hz and then processed with a band-passfilter between 1 75 Hz. We directly use the pre-computed differential entropy (DE) features smoothed by linear dynamic systems (LDS) in SEED and SEED-IV as the input to feed our model. In contrast to other features, DE features are more discriminative in EEG emotion recognition.

SEED dataset [8] comprises the EEG data of 15 subjects (7 males), and each subject contains the EEG data recorded from three sessions. Every session contains 15 trials of EEG samples covering three emotion classes, i.e., positive, negative and neutral emotions (five trials for each emotion). For SEED, there are three duplicate sessions for each subject. In our experiment, we only use the first session for each subject since the first one reflects more reliable emotions than the later two sessions. As a result, there are totally 3394 data samples in the SEED dataset.

SEED-IV dataset [9] also contains 15 subjects (7 males), and each subject has three sessions. Every session contains 24 trials of EEG samples covering four emotion classes, i.e., neutral, sad, fear, and happy emotions (six trials for each emotion). Since SEED-IV dataset has three different sessions for each subject, we evaluate our model using data from all three sessions. Thus, there are totally 2505 data samples in the SEED-IV dataset.

### 3.2 Implementation Details

In this experiment, we investigate the cross-subject EEG emotion recognition problem, in which the training EEG data samples and the testing ones come from different subjects. To this end, we adopt the leave-one-subject-out (LOSO) cross-validation strategy [18] to evaluate the performance of the proposed Bi-CapsNet method, in which we use the EEG signals of one subject as testing data and the EEG signals of the remaining subjects as training data. After each subject has been used once as testing subject, the average accuracy of all the subjects is then calculated as the final result.

The feature dimension of the input samples is $5 \times 62$ (62 EEG channels by 5 frequency bands) for both datasets. For our method, we adopt SGD optimizer to optimize the parameters. The batch size is set to 32 empirically. For the LSTM layer, there is one hidden layer with 512 units. We train the model for 50 epochs and set maximum number of iteration to 3. For SEED, we set the learning rate to 0.003, and for SEED-IV, we set the learning rate to 0.006. Moreover, for SEED dataset, we adopt the source domain data selection strategy, which is not adopted for SEED-IV dataset. This strategy means that in the process of model training, except that the first epoch uses all the source domain data, the source domain data used by each subsequent epoch is the source domain data fitting the model after the previous epoch training. We implement our method via the Pytorch framework of version 3.9.2.

### 3.3 Results and Discussions

To validate the superiority of Bi-CapsNet, we compare other methods in the same experimental setup, as shown in Table 1. All methods compared in our study are representative methods in previous studies. When the length of the training EEG data is 1 s, the performance of our model is worse than RGNN [12] on SEED-IV but better than BiHDM [11] on SEED.

**Table 1.** The classification performance for cross-subject EEG emotion recognition on SEED and SEED-IV datasets.

| Model | SEED | SEED-IV |
|---|---|---|
| SVM [19] | 56.73/16.29 | 37.99/12.52 |
| TCA [20] | 63.64/14.88 | 56.56/13.77 |
| A-LSTM [21] | 72.18/10.85 | 55.03/09.28 |
| DGCNN [22] | 79.95/09.02 | 52.82/09.23 |
| DAN [7] | 83.81/08.56 | 58.87/08.13 |
| BiDANN-S [10] | 84.14/06.87 | 65.59/10.39 |
| BiHDM [11] | 85.40/07.53 | 69.03/08.66 |
| RGNN [12] | 85.30/06.72 | **73.84/08.02** |
| Bi-CapsNet(Our model) | **85.92(07.44)*** | 67.74(08.78) |

* Denotes the experiment results obtained are based on the source domain data selection strategy.

To see the confusions of Bi-CapsNet in recognizing different emotions, we depict the confusion matrices corresponding to the experimental results of Bi-CapsNet. From Fig. 4, we can observe that neutral emotions are largely identified on both SEED and SEED-IV datasets. This suggests that compared with positive emotion and negative emotion (including sad emotion and fear emotion), neutral emotion has more similar brain responses in different people. Besides, From Fig. 4(b) corresponding to the SEED-IV dataset, we can see that sadness and fear are both negative emotions, so they are more likely to be confused, which is largely attributed to the more similar emotional mechanisms of sad emotion and fear emotion.

Since our entropy regularizer regularizes the prediction results by minimizing the prediction information entropy of the target domain data, in our ablation experiment, the entropy regularizer is also removed while removing the domain discriminator. From Table 2, we can see that the performance of our model has a greater variance when the domain discriminator is deleted, which validates the importance of the domain discriminator in improving the robustness of the Bi-CapsNet model against cross-subject variations. Moreover, we conduct ablation experiments to investigate the contribution of the entropy regularizer in our model, as illustrated in Table 2. Supposing that the entropy regularizer is removed, the performance of our model deteriorates, verifying the positive impact of the entropy regularizer on the performance of our model. This performance gain verifies our hypothesis about uncertainty that the source domain EEG data training network may not be reliable in predicting the target domain EEG data.

The source domain data selection strategy has significantly improved the performance of the model on the SEED dataset, which indicates the effectiveness of the strategy in capturing domain commonalities. However, the premise of using this strategy is that there are enough data in each category, so that the model can better obtain more trustworthy and common data in the training process. In the experiment using the SEED-IV dataset, we do not use this strategy because compared with the SEED dataset, the SEED-IV dataset includes more trials and

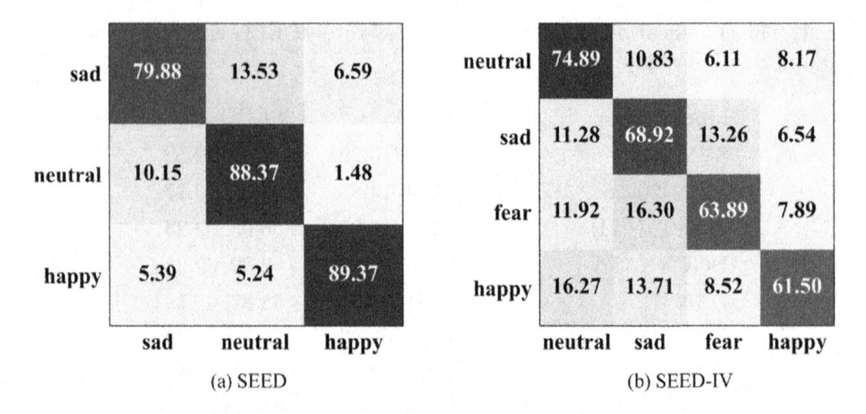


**Fig. 4.** The confusion matrices of the Bi-CapsNet method in the cross-subject EEG emotion recognition experiment.

contains more kinds of emotional states. Therefore, the SEED-IV dataset needs more data to apply this strategy. However, the reality is that in SEED-IV, each subject has only 2505 data samples, while each subject in the SEED dataset has 3394 data samples. This will make the results of this strategy on the SEED-IV dataset worse than expected.

**Table 2.** Ablation study for cross-subject classification accuracy (mean/std) on SEED and SEED-IV.

| Model | SEED | SEED-IV |
|---|---|---|
| Source domain data selection strategy | **85.92(07.44)*** | 62.02(09.45)* |
| - Source domain data selection strategy | 83.66(06.81) | **67.74(08.78)** |
| - Entropy regularizer | 83.97(07.14)* | 66.86(08.76) |
| - Global domain discriminator | 83.83(07.38)* | 66.31(09.37) |

* Denotes the experiment results obtained are based on the source domain data selection strategy.
- Indicates the following component is removed.

## 4 Conclusion

Based on the asymmetric difference of brain, this paper proposes a Bi-CapsNet method to improve the cross-subject EEG emotion recognition performance. Furthermore, we propose a regularization method to reduce the prediction uncertainty of the target domain data to increase the stability of the model. The experimental results on SEED and SEED-IV indicate that the proposed Bi-CapsNet model achieves better separability than most methods. In addition, the proposed model is easy to implement and does not require additional calculated features. In the future work, we plan to use more neuroscience findings of emotion as the prior knowledge to extract features or develop models.

**Acknowledgements.** This work was supported by the National Natural Science Foundation of China (No.61876126 and 61503278).

## References

1. Quan, X., Zeng, Z., Jiang, J., Zhang, Y., Lu, B., Wu, D.: Physiological signals based affective computing: a systematic review. IEEE/CAA J. Autom. Sinica **8** (2021)
2. Recio, G., Schacht, A., Sommer, W.: Recognizing dynamic facial expressions of emotion: specificity and intensity effects in event-related brain potentials. Biol. Psychol. **96**, 111–125 (2014)
3. Gunes, H., Piccardi, M.: Bi-modal emotion recognition from expressive face and body gestures. J. Netw. Comput. Appl. **30**(4), 1334–1345 (2007)
4. Han, K., Yu, D., Tashev, I.: Speech emotion recognition using deep neural network and extreme learning machine. In: Interspeech 2014 (2014)
5. Wang, X.W., Nie, D., Lu, B.L.: Emotional state classification from EEG data using machine learning approach. Neurocomputing **129**, 94–106 (2014)
6. Alarcao, S.M., Fonseca, M.J.: Emotions recognition using EEG signals: a survey. IEEE Trans. Affect. Comput. **10**(3), 374–393 (2017)
7. Li, H., Jin, Y.-M., Zheng, W.-L., Lu, B.-L.: Cross-subject emotion recognition using deep adaptation networks. In: Cheng, L., Leung, A.C.S., Ozawa, S. (eds.) ICONIP 2018. LNCS, vol. 11305, pp. 403–413. Springer, Cham (2018). https://doi.org/10.1007/978-3-030-04221-9_36
8. Zheng, W.L., Lu, B.L.: Investigating critical frequency bands and channels for EEG-based emotion recognition with deep neural networks. IEEE Trans. Auton. Ment. Dev. **7**(3), 162–175 (2015)
9. Zheng, W.L., Liu, W., Lu, Y., Lu, B.L., Cichocki, A.: Emotionmeter: a multimodal framework for recognizing human emotions. IEEE Trans. Cybern. **49**(3), 1110–1122 (2018)
10. Li, Y., Zheng, W., Zong, Y., Cui, Z., Zhang, T., Zhou, X.: A bi-hemisphere domain adversarial neural network model for EEG emotion recognition. IEEE Trans. Affect. Comput. **12**(2), 494–504 (2018)
11. Li, Y., et al.: A novel bi-hemispheric discrepancy model for EEG emotion recognition. IEEE Trans. Cogn. Develop. Syst. **13**(2), 354–367 (2020)
12. Zhong, P., Wang, D., Miao, C.: EEG-based emotion recognition using regularized graph neural networks. IEEE Trans. Affect. Comput. **14**, 1290– 1301 (2020)
13. Hochreiter, S., Schmidhuber, J.: Long short-term memory. Neural Comput. **9**(8), 1735–1780 (1997)
14. Jain, A., Singh, A., Koppula, H.S., Soh, S., Saxena, A.: Recurrent neural networks for driver activity anticipation via sensory-fusion architecture. In: 2016 IEEE International Conference on Robotics and Automation (ICRA), pp. 3118–3125. IEEE (2016)
15. Liu, Y., et al.: Multi-channel EEG-based emotion recognition via a multi-level features guided capsule network. Comput. Biol. Med. **123**, 103927 (2020)

16. Sabour, S., Frosst, N., Hinton, G.E.: Dynamic routing between capsules. In: Advances in Neural Information Processing Systems 30 (2017)
17. Fleuret, F., et al.: Uncertainty reduction for model adaptation in semantic segmentation. In: Proceedings of the IEEE/CVF Conference on Computer Vision and Pattern Recognition, pp. 9613–9623 (2021)
18. Zheng, W.L., Lu, B.L.: Personalizing EEG-based affective models with transfer learning. In: Proceedings of the Twenty-fifth International Joint Conference on Artificial Intelligence, pp. 2732–2738 (2016)
19. Suykens, J.A., Vandewalle, J.: Least squares support vector machine classifiers. Neural Process. Lett. **9**(3), 293–300 (1999)
20. Pan, S.J., Tsang, I.W., Kwok, J.T., Yang, Q.: Domain adaptation via transfer component analysis. IEEE Trans. Neural Networks **22**(2), 199–210 (2010)
21. Song, T., Zheng, W., Lu, C., Zong, Y., Zhang, X., Cui, Z.: MPED: a multi-modal physiological emotion database for discrete emotion recognition. IEEE Access **7**, 12177–12191 (2019)
22. Song, T., Zheng, W., Song, P., Cui, Z.: EEG emotion recognition using dynamical graph convolutional neural networks. IEEE Trans. Affect. Comput. **11**(3), 532–541 (2018)

# Attention 3D Fully Convolutional Neural Network for False Positive Reduction of Lung Nodule Detection

Guitao Cao[1(✉)], Qi Yang[1], Beichen Zheng[1], Kai Hou[2], and Jiawei Zhang[3,4]

[1] East China Normal University, 3663 North Zhongshan Road, Shanghai 200062, China
gtcao@sei.ecnu.edu.cn

[2] Zhongshan Hospital Affiliated Fudan University, 1474 Yan'an Road, Shanghai 200032, China

[3] Shanghai MicroPort EP MedTech Co., Ltd., #19 Building, 588 Tianxiong Road, Shanghai 201318, China

[4] Shanghai Open University, 288 Guoshun Road, Shanghai 200433, China

**Abstract.** Deep Learning based lung nodule detection is rapidly growing. It is one of the most challenging tasks to increase the true positive while decreasing the false positive. In this paper, we propose a novel attention 3D fully Convolutional Neural Network for lung nodule detection to tackle this problem. It performs automatic suspect localization by a new channel-spatial attention U-Network with Squeeze and Excitation Blocks (U-SENet) for candidate nodules segmentation, following by a Fully Convolutional C3D (FC-C3D) network to reduce the false positives. The weights of spatial units and channels for U-SENet can be adjusted to focus on the regions related to the lung nodules. These candidate nodules are input to FC-C3D network, where the convolutional layers are re-placed by the fully connected layers, so that the size of the input feature map is no longer limited. In addition, voting fusion and weighted average fusion are adopted to improve the efficiency of the network. The experiments we implement demonstrate our model outperforms the other methods in the effectiveness, with the sensitivity up to 93.3%.

**Keywords:** Deep Learning · Fully Convolutional Neural Network · Attention · Lung Nodule Detection · False Positive

## 1 Introduction

Lung cancer has the highest mortality among all types of cancers worldwide [1]. In 2022, a report from the American Cancer Society showed that lung cancer was the leading type of cancer, making up almost 25% of all cancer deaths [2]. The morphological characteristics of lung nodules are closely associated with the incidence of lung cancer. If the early diagnosis rate can be improved before the patient's condition deteriorates, the mortality rate of high-risk patients can be significantly reduced.


M. Tanveer et al. (Eds.): ICONIP 2022, CCIS 1793, pp. 337–350, 2023.
https://doi.org/10.1007/978-981-99-1645-0_28

With the successful application of DNN in natural images, researchers have recently attempted to use DNN to detect lung nodules [3,5]. Based on DCNNs, Ding et al. developed a new method to detect pulmonary nodules by incorporating the deconvolution structure into the CNN based Faster R-CNN [12]. The experimental results in LUNA16 Challenge have proved the superior performance of this method in pulmonary nodule detection. Wang et al. proposed a new Semi-Supervised Learning framework, FocalMix, which utilized less-annotated raw medical images to improve the diagnostic accuracy of deep learning models [19]. Through a large number of experiments, researchers demonstrated that the SSL method can significantly improve the performance of the supervised learning method. To further improve the segmentation precision, Hao et al. designed DAU-Net based on 3D U-Net by replacing every two adjacent convolutional layers with a residual structure, and adding the parallelly-arranged position and channel attention modules between the contraction and expansion paths, thus improving the segmentation precision [14].

DNNs make it possible to use lung nodules detection as a precise tool for auxiliary diagnosis of lung cancer. How to reduce the false positive rate and improve the diagnosis precision while maintaining a high segmentation or classification accuracy has become a key research topic [7]. Liu et al. developed 3DFPN-HS2 by using a FPN-based 3D feature pyramid network and incorporating a High Sensitivity and Specificity (HS2) network [16], which can track the appearance changes in consecutive CT slices of each candidate nodule, resulting in the sensitivity of 90.4%. We designed a 3D convolutional network to automatically learn multi-scale strong semantic features from CT images and distinguish lung nodules from similar ones in complex environments [6].

There are three steps in our research including candidate nodule detection, high-precision nodule refinement, and model fusion to reduce the false positive of lung nodule detection. The main contributions of this paper are as follows:

- We propose a channel-spatial attention based lung nodule segmentation network, U-Net with Squeeze and Excitation blocks (U-SENet) for candidate nodule selection. It can focus on the regions associated with lung nodule by spatial feature extraction and adjustment of the weights of spatial units and channels.
- We develop a high-sensitivity Fully Convolutional C3D (FC-C3D) network to induce the false positives. The fully connected layer is replaced by the fully convolution layer so that the size of the input feature image is no longer limited and the efficiency of network forward propagation is improved.
- We adopt model fusion to improve the sensitivity. The fused model is equivalent to four experienced professional radiologists who conduct the clinical diagnosis, improving the sensitivity of lung nodule detection to 93.3%.

## 2 Methods

Our proposed lung nodule detection framework is shown in Fig. 1. The lung nodules are preprocessed and input into the U-Net with Squeeze and Excitation blocks (U-SENet) for extracting candidate nodules. U-SENet screens out the candidate nodules with a confidence level higher than 0.3 as the false positive samples which enter the highly sensitive Fully Convolutional C3D (FC-C3D) network together with the morphologically preprocessed lung nodule samples to eliminate the false positives. Then, model fusion is applied to improve the sensitivity. The fused model is equivalent to four experienced radiologists to do a clinical diagnosis, which increases the sensitivity of lung nodule detection to 93.3%.

### 2.1 U-Net with Squeeze and Excitation Blocks (U-SENet)

U-SENet in this research is a U-net network structure to carry out tasks of 3D semantic segmentation. The traditional U-net network consists of two parts, encoder and decoder, mirror-symmetrical in structure (Fig. 1-I). Inspired by ResNet, skip connections are introduced into the relative parts of the encoder and decoder amid U-Net, and the encoding features of corresponding layers are fused during decoding.

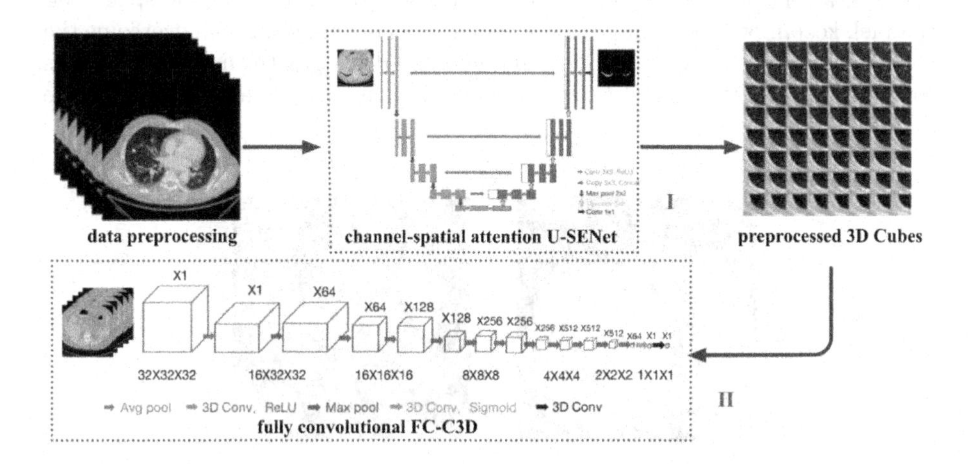


**Fig. 1.** Our proposed Lung Nodule Detection framework

The input of U-SENet is not a single 2D picture, but a 2D picture with multiple frames. And the output becomes a multi-frame 2D labeled picture treated as 3D data. The input also changes from 4D data (i.e., number of pictures, number of channels, width, and height) to 5D data (i.e., number of pictures, number of channels, length, width, and height). Accordingly, all of the operations in U-net should be altered to corresponding 3D operation.

### 2.2 Channel-Spatial Attention Based Fully CNN

To focus on the more important information of lung nodule in images, channel-based and space-based attention mechanisms are introduced into U-SENet, respectively. The channel attention mechanism is to distinguish the importance levels of the information provided by different channels and adjust the values of each group of channels through a set of weights. The spatial attention mechanism centers on the more important ones among spatial features. This mechanism distinguishes the importance levels of different spatial location features, and then trains a set of weights through supervised learning to act on spatial features and adjust the values of spatial features.

**Spatial Attention.** Figure 2 shows the activation probability map of the spatial attention mechanism, by which the spatial information of input images can be trans-formed, with their key information remained.

**Channel Attention.** The "squeeze" operation is to utilize all the global spatial feature of each channel as the representation of the channel and use the global average pooling to generate the statistics of each channel. The "excitation" operation is to learn the degree of dependence of each channel by adjusting different feature maps based on the degree of dependence [11].

**Channel-Spatial Attention Mechanism.** In this paper, we introduce Spatial-Channel attention mechanism to simultaneously concentrate on the space and channel, keeping the diversity of models in different channels. It can overcome the harm of ignoring the information in channels and dealing with the image features equivalently in each channel by spatial attention only. It can also handle a large number of feature channels.

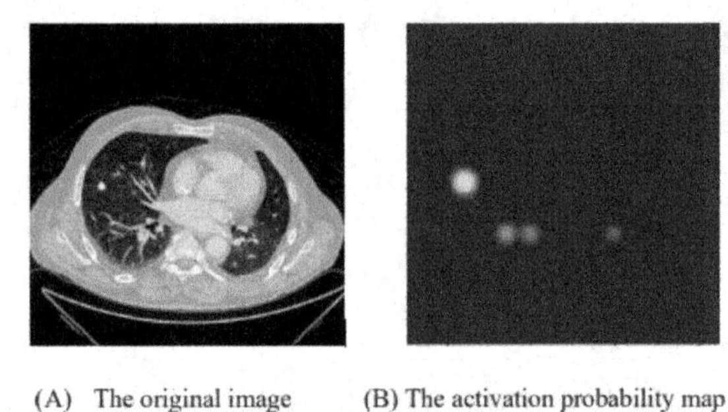

(A) The original image (B) The activation probability map

**Fig. 2.** The distribution of activation probability map of spatial attention.

### 2.3 FC-C3D Network

As illustrated in Fig. 1-II, the proposed FC-C3D network in this research contains 14 layers. The main process of FC-C3D is as follows:

1. Down-sample the z-axis through a $2 \times 1 \times 1$ pooling kernel and stride, using the average pooling operation. The target is to average the z-axis to 2 mm per voxel, making the network lighter for better performance.

2. Execute two consecutive convolution and max pooling operations on the network. The kernel size of the two convolution operations is adjusted to $3 \times 3 \times 3$ structure. According to the contrastive experiment of three convolution kernels of $3 \times 3 \times 3$, $5 \times 3 \times 3$ and $7 \times 3 \times 3$, the network with a convolution kernel size of $3 \times 3 \times 3$ starts to perform significantly better than others after training for 4 epochs. Moreover, the use of smaller convolution kernels has advantages over larger receptive fields in terms of parameters and FLOPs, which can speed up network convergence, reduce training time, and reduce model size, thereby shortening prediction time and improving real-time performance. Besides, in order to improve the expressive ability of the network, a rectified linear unit is added after the convolutional layer.
3. Down-sample the network while keeping the z-axis scale unchanged, through the max-pooling layer with $1 \times 2 \times 2$ receptive field and stride. After the pooling operation, each voxel in the feature map represents a $1 \times 2 \times 2$ pixel in the input feature map, reducing dimensionality in depth. For the reduced input to the subsequent convolutional layer, the FLOPs and parameters of the model after pooling operation are also reduced. As such, the network can extract wider features. To transform the size of the input 3D Cubes to $1/16$ of the original, the pooling kernel sizes of the following three max-pooling layers are set to $2 \times 2 \times 2$.
4. Ignore half of the feature detectors in each batch of training. This adjustment significantly reduces overfitting. A Dropout layer is added after each max-pooling layer in the FC-C3D network, thereby reducing feature detectors and the interactions between hidden nodes.

## 3 Experimental Results and Analysis

### 3.1 Dataset

**LUNA16 Dataset.** The dataset consists of 888 CT scans and corresponding diagnosis results of lung nodules. This dataset roots in the Lung Image Database Consortium (LIDC-IDRI), an lung nodules dataset containing 1018 CT scans.

**Ndsb Dataset.** The Ndsb dataset originates from the National Data Science Bowl 2017 supported by Kaggle. In this challenge, more than 60 GB of CT data are provided. This is an epoch-making milestone in the of lung cancer detection algorithm developed by the Data Science Bowl and the medical community.

### 3.2 Data Preprocessing

The lung image preprocessing mainly contains three-dimensional CT image resampling and segmentation of the pulmonary region.

**Three-Dimensional CT Image Resampling.** When each voxel of CT represents a volume of 1mm, the precision and FLOPs are the best. Therefore, bilinear interpolation is used in the z-axis, and the number of slice samples is increased to set the slice thickness evenly to 1mm.

**Pulmonary Region Segmentation.** The lung region segmentation of CT images is accomplished via 8 steps: image binarization, clearing lung boundaries, labeling connected regions, deleting small regions, erosion operations, closing operations, and hole filling, shown in Fig. 3.

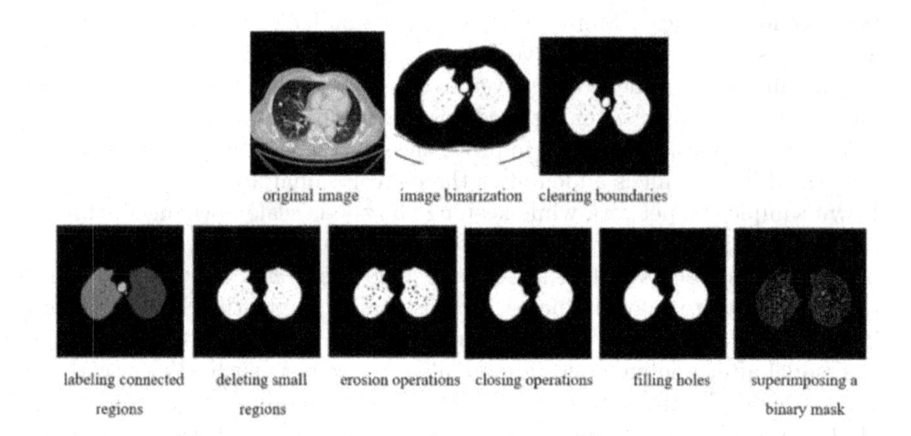


**Fig. 3.** The procedure of pulmonary region segmentation.

### 3.3 Data Augmentation

Due to the relatively small quantity of data in LUNA16 dataset, six data augmentation methods are used in this study, including flipping, rotating, scaling, cropping, translation and adding Gaussian noise.

The augmented data are used for training. The six data augmentation methods act on training data with a probability of 0.5.

### 3.4 Candidate Nodules Extraction Based on U-SENet

In this research, we use the U-SENet to extract candidate nodules in the LUNA16, with 750 cases for training, and 138 cases for testing. Our environment is an NVIDIA Titan X 11G graphics card, and a system environment of ubuntu16.04. The network structure applies a 3D U-net network with a channel-spatial attention mechanism. The image input size is scaled to $64 \times 64 \times 64$, using Adam optimizing, where the initial learning rate is set to 0.01, which decays to 0.001 and 0.0001 at epochs of 200 and 250, respectively. We conduct four sets of comparative experiments, namely 2D U-net, 3D U-net, 3D U-net with channel attention mechanism, and 3D U-net with spatial attention mechanism. The results after training for 300 rounds are shown in Table 1.

**Table 1.** Comparison of different attention mechanisms

| Network | Loss | Dice | IoU | Training Time (h) |
|---|---|---|---|---|
| 2D U-net | 0.072 | 0.928 | 0.866 | 104 |
| 3D U-net | 0.059 | 0.941 | 0.888 | 75 |
| Channel Attention 3D U-net | 0.051 | 0.949 | 0.902 | 80 |
| Spatial Attention 3D U-net | 0.046 | 0.954 | 0.912 | 78 |
| Channel-Spatial Attention U-SENet | 0.032 | 0.968 | 0.937 | 83 |

**IoU Evaluation Metric.** In this paper, we use the Intersection over Union (IoU) as an evaluation metric for semantic segmentation.

**Loss Function.** Dice coefficient is another commonly used loss function in the segmentation of medical images, which can be calculated by:

$$Dice = \frac{2\left|A \cap B\right|}{\left|A\right| + \left|B\right|} \tag{1}$$

where, A and B are two sets, the numerator represents the norm of the intersection of A and B, and the denominator is the sum of norms of the two sets. The larger the value of Dice, the better the segmentation results.

We use 1-Dice as the loss function of U-SENet to optimize the deep learning:

$$loss = 1 - \frac{2\left|GT * Pred\right|}{\left|GT\right| + \left|Pred\right|} \tag{2}$$

where, GT represents the real mask label of the image, Pred represents the predicted mask label of the image, "*" represents element-wise multiplication, "+" represents element-wise addition, and $|\bullet|$ represents the 1-norm of the matrix.

**Comparison of Different Attention Mechanism.** We carry out 5 comparative training for the identification of candidate lung nodule regions in this study, and the experimental results are shown in Table 2.

As can be seen from Table 2, the segmentation result of 3D U-net is better than that of 2D U-net, and the IoU value is increased by 2.2%. The IoU value of 3D U-net is further improved to varying degrees by adding the attention mechanisms, i.e., the IoU values are increased by 1.4%, 2.4%, and 3.3% with the addition of channel attention mechanism, spatial attention mechanism or the channel-spatial mixed domain attention mechanism, respectively. The addition of attention mechanism has little influence on the training time of 3D U-net, which is far less than that of 2D U-net.

Figure 4 shows the images after segmentations, including the original image and processed counterparts through 2D U-net, 3D-Unet, 3D U-net with channel attention, 3D U-net with spatial attention, and U-SENet with channel-spatial attention. The white circular areas are the circles drawn according to the coordinates and diameters of the lung nodule label (x, y, z), and the red areas are generated by using the segmentation network. The greater overlap between the red and white are-as, the better segmentation effect obtained. From (6), we can see that the red area segmented by the channel-spatial attention mechanism U-SENet has the highest overlap with the white, achieving the best result.

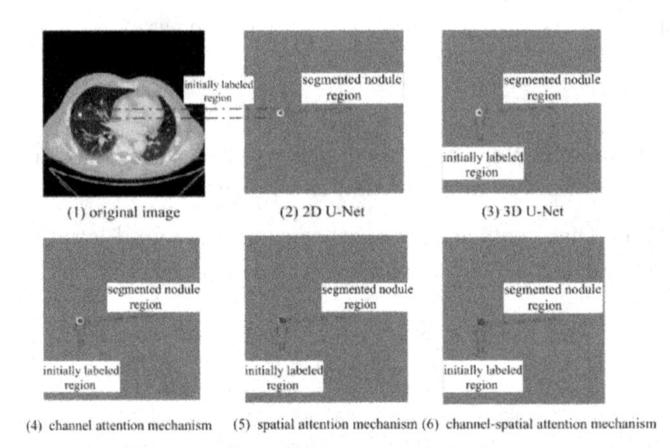


**Fig. 4.** Comparison of segmentation images with different attentions. White circles are the initially labeled nodules, and the red regions are the segmented nodules, respectively. (Color figure online)

### 3.5 False Positive Reduction Based on FC-C3D Network

The nodules detected by U-SENet still have a high false positive rate, so we construct the FC-C3D network model to reduce the false positives. Both the U-SENet outputs with confidence greater than 0.3, and the morphologically preprocessed lung nodules, are input into FC-C3D network.

**Dealing with Sample Imbalance.** To solve the problem of sample imbalance, optimization is carried out on data and algorithm. Up-sampling and down-sampling are done on positive and negative samples, respectively. The amplification ratios of positive samples for each batch of training are set to 1:10, 1:5, and 1:2 for comparative experiments. The experimental results demonstrate that the model works the best when the amplification ratio is 1:10. There are positive, false positive, and negative samples shown in Fig. 5, which are extracted by U-SENet on the LUNA16 and Ndsb datasets. In order to construct the 3D cube, the slices are tiled to generate 88 plane graphics in sequence for graphical exhibition, making full use of the z-axis spatial features of lung CT. Some of the amplified images of the nodule input to the FC-C3D network are also shown.

**Optimization of Network Training Process.** In this research, we utilize Lecun normal distribution initialization, an internal method of tensorflow. The weight parameters are initialized to a normal distribution with 0 as the mean and square root of 1/n as the standard deviation, where n is the number of input units in the weight tensor. The distribution can be expressed as:

$$W = \sum_{l=1}^{n} w_i \sim N(0, \sqrt{\tfrac{1}{n}}\ ) \tag{3}$$

where, W is the overall parameter of the network, and l (1, 2, ..., n) is the sequence number of network layers. Given the occupancy of GPU memory, the batch size is set to 12 in this research. The Mini-batch Gradient Descent method (MBGD) is used based on the size of the lung nodule samples, the complexity of the time and the accuracy of the algorithm. In doing so, iterative training of large data sets runs smoothly, while the small batch can avoid the bias of a single sample. And the momentum is used to optimize the MBGD, speeding up model learn-ing and preventing parameters from going into gradient-sensitive regions. In this research, the momentum and the initial learning rate are set to 0.95 and 0.001, respectively. The learning rate is reduced to 10% of the original value after every 30 iterations. Since this research does not use any existing pre-trained models, a relatively high learning rate is used at the beginning of the iteration in order to let the model quickly learn the 3D network from scratch. After the network gradually converges, a lower learning rate is used for refine learning.

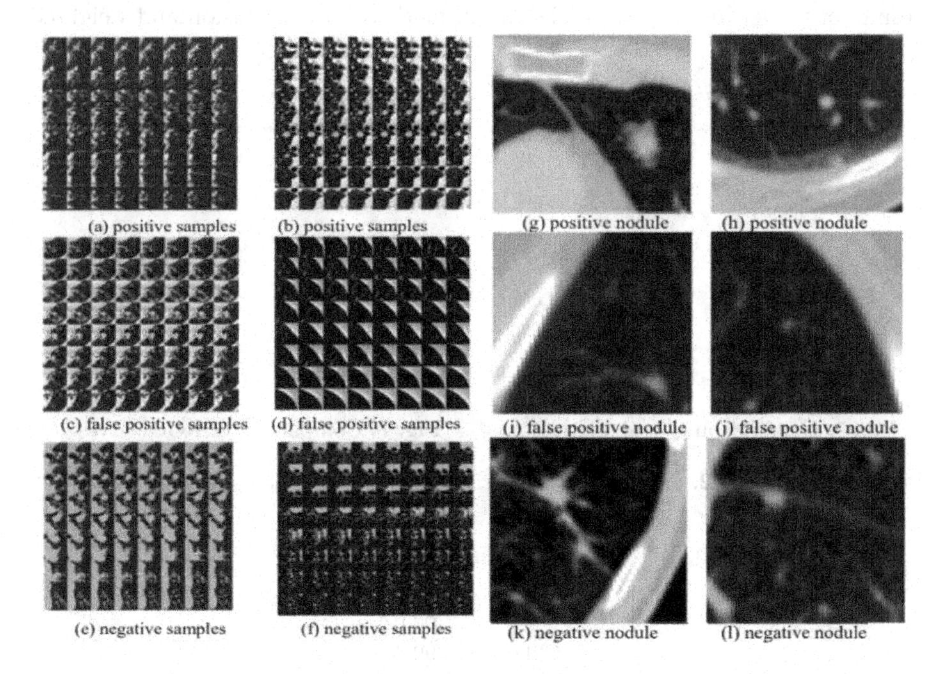


**Fig. 5.** The samples of 3D Cubes construction and nodules.

In addition, we also adopt the Dropout operation to prevent overfitting, coupled with Max-Norm regularization to limit the size of the network weights[18]. Besides, k-fold cross-validation is used. In this study, 10-fold cross-validation is chosen, i.e., k = 10.

## Experimental Results of False Positive Reduction

*Comparison Among Three Models.* The network trained with Luna16 dataset is FC-C3D-V1, the network trained with Ndsb dataset is FC-C3D-V2, and the network trained with both Luna16 and Ndsb datasets is Fusion-FC-C3D. The models in this research are trained for 200 Epochs. The sensitivities of the four single models of 3DResnet, C3D, FC-C3D-V1, and FC-C3D-V2 are shown in Table 2. As shown in Fig. 6, the training process of the fusion model is stabler, and the loss of convergence is also smaller (Fig. 6).

Using the same testing set, the Receiver Operating Characteristic Curves (ROC) of the testing results of the three models are shown in Fig. 7. The abscissa and ordinate of the ROC curve are specificity and sensitivity, respectively, and the area of the ROC curve and the abscissa is called Area Under Curve (AUC). The AUCs of the FC-C3D-V1, FC-C3D-V2, and Fusion-FC-C3D models are 0.897, 0.921 and 0.939, respectively.

*Comparison Among Model Fusion Methods.* In order to verify the experimental results of model fusion, two model fusion methods, voting fusion and weighted

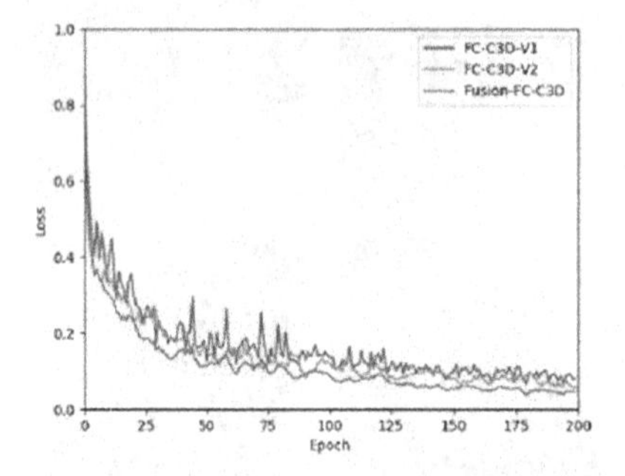


Fig. 6. Comparison of training loss.

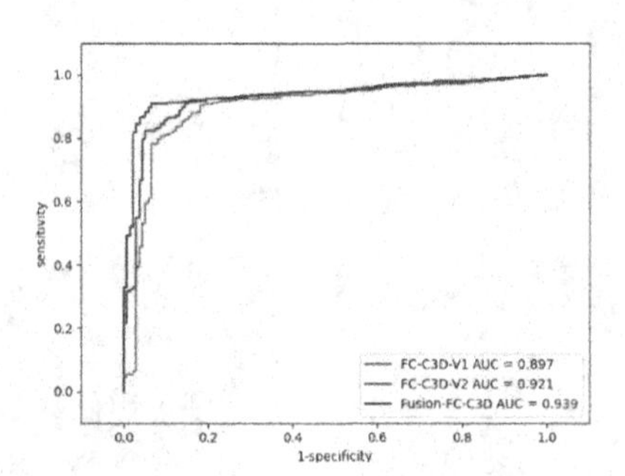


Fig. 7. Comparison of ROC curves.

**Table 2.** Sensitivity Comparison of Different Single Model

| Model | Sensitivity |
|---|---|
| 3DResnet | 90.2% |
| C3D | 89.6% |
| FC-C3D-V1 | 90.4% |
| FC-C3D-V2 | **90.9%** |

average fusion, are adopted. The sensitivities of them are shown in Tables 3 and 4, respectively.

In this research, weighted model fusion is executed in combinations of different weights. Table 4 shows the 10 groups of weighted models. The sensitivity reaches the highest of 93.33% when the fusion weights of models 3Dresnet, C3D, FC-C3D-V1 and FC-C3D-V2 are 0.25, 0.2, 0.275 and 0.275, respectively.

*Comparison with Other Methods.* The optimal experimental results of model fusion in this study (Fig. 8) are compared with other methods of lung nodule detection. The comparison results are shown in Table 5. From Table 5, we can see that our method is the best one with the sensitivity of 93.3%.

**Table 3.** Sensitivity Comparison of Voting Fusion Models

| Model Weights | Sensitivity |
|---|---|
| 3Dresnet+ C3D | 90.5% |
| 3Dresnet+ FC-C3D-V1 | 90.7% |
| 3Dresnet+ FC-C3D-V2 | 91.3% |
| C3D+ FC-C3D-V1 | 90.8% |
| C3D+ FC-C3D-V2 | 91.0% |
| FC-C3D-V1+ FC-C3D-V2 | 93.2 % |
| 3Dresnet+ C3D+ FC-C3D-V1 | 92.2 % |
| 3Dresnet+ C3D+ FC-C3D-V2 | 92.6 % |
| C3D+ FC-C3D-V1+ FC-C3D-V2 | 92.4% |
| **3Dresnet+C3D+FC-C3D-V1+FC-C3D-V2** | **93.2%** |

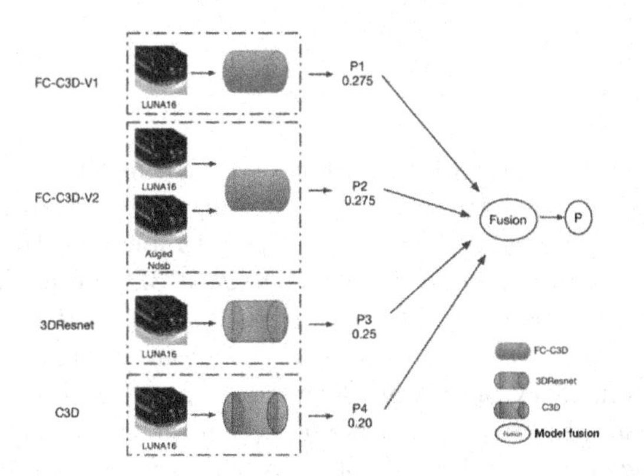


**Fig. 8.** The best model fusion scheme in this research.

**Table 4.** Sensitivity Comparison of Weighted Fusion Model

| Model Weights | | | | Sensitivity |
|---|---|---|---|---|
| 3Dresnet | C3D | FC-C3D-V1 | FC-C3D-V2 | |
| 0.25 | 0.2 | 0.275 | 0.275 | 93.33% |
| 0.25 | 0.22 | 0.26 | 0.27 | 93.30% |
| 0.24 | 0.23 | 0.25 | 0.28 | 93.27% |
| 0.28 | 0.23 | 0.25 | 0.24 | 93.26% |
| 0.25 | 0.25 | 0.25 | 0.25 | 93.23% |
| 0.24 | 0.24 | 0.26 | 0.26 | 93.22% |
| 0.26 | 0.24 | 0.25 | 0.25 | 93.20% |
| 0.25 | 0.26 | 0.24 | 0.25 | 93.01% |
| 0.24 | 0.26 | 0.25 | 0.25 | 93.00% |
| 0.23 | 0.24 | 0.27 | 0.26 | 92.99% |

**Table 5.** Comparison with other Methods

| Model | Sensitivity |
|---|---|
| k-nearest-neighbour-2D [17] | 80% |
| Multi-View Convolutional Networks-2D [4] | 90.1% |
| DCNN-3D [12] | 89.3% |
| 3DFPN-HS2-3D [16] | 90.4% |
| knowledge-infused-3D [15] | 88.0% |
| 3D-Resnet (ours) | 90.2% |
| Fusion(ours) | 93.3% |

## 4 Conclusions

In this research, we propose a lung nodule detection method based on attention 3D fully convolutional neural network. After lung nodule segmentation network named U-SENet with channel-spatial attention to focus on the nodule regions, a high-sensitivity Fully Convolutional C3D (FC-C3D) network is proposed to re-move the false positives. The fully convolutional layer is used instead of the fully connected layer, so that the size of the input feature map is no longer limited and the efficiency of network forward propagation is improved. And the model fusion is adopted to improve the detection sensitivity, which is equivalent to four experienced professional radiologists who conduct the clinical diagnosis. The experi-mental result demonstrates the effectiveness of our methods with the sensitivity of lung nodule detection up to 93.3%.

## References


1. Siegel, R.L., Miller, K.D., Fuchs, H.E., Jemal, A.: Cancer statistics, 2022. CA Cancer J. Clin. (2022). https://doi.org/10.3322/caac.21708
2. American Cancer Society.https://www.cancer.org/cancer/lung-cancer/about/key-statistics.html. Accessed 1 June 2022
3. Arindra, A., Setio, A., Traverso, A.: Validation, comparison, and combination of algorithms for automatic detection of pulmonary nodules in computed tomography images: the LUNA16 challenge. Med. Image Anal. **42**, 1–13 (2017) https://doi.org/10.1016/j.media.2017.06.015
4. Arindra, A., Setio, A., Ciompi, F., et al.: Pulmonary nodule detection in CT images: false positive reduction using multi-view convolutional networks. IEEE **35**(5), 1160–1169 (2016). https://doi.org/10.1109/TMI.2016.2536809
5. Qu, K., Chai, X., Liu, T., Zhang, Y., Leng, B., Xiong, Z.: Computer-aided diagnosis in chest radiography with deep multi-instance learning. In: Liu, D., Xie, S., Li, Y., Zhao, D., El-Alfy, ES. (eds.) ICONIP 2017. LNCS, vol. 10637, pp. 723–731. Springer, Cham (2017). https://doi.org/10.1007/978-3-319-70093-9_77
6. Cao, G.T., et al.: 3D convolutional neural networks fusion model for lung nodule detection on clinical CT scans. In: IEEE International Conference on Bioinformatics and Bio-medicine. Madrid, Spain, pp. 973–978 (2018). https://doi.org/10.1109/BIBM.2018.8621468
7. Yang, F., Zhang, H., Tao, S., et al.: Graph representation learning via simple jumping knowledge networks. Appl. Intell. **52**, 11324–11342 (2018). https://doi.org/10.1007/s10489-021-02889-z
8. Chen, W.Q., et al.: Cancer incidence and mortality in China. Chin. J. Cancer Res. **27**(1), 1004–0242 (2018). https://doi.org/10.21147/J.ISSN.1000-9604.2018.01.01
9. Song, J., et al.: Human action recognition with 3D convolution skip-connections and RNNs. In: Cheng, L., Leung, A.C.S., Ozawa, S. (eds.) ICONIP 2018. LNCS, vol. 11301, pp. 319–331. Springer, Cham (2018). https://doi.org/10.1007/978-3-030-04167-0_29
10. Wu, C., Liu, X., Li, S., Long, C.: Coordinate attention residual deformable U-net for vessel segmentation. In: Mantoro, T., Lee, M., Ayu, M.A., Wong, K.W., Hidayanto, A.N. (eds.) ICONIP 2021. LNCS, vol. 13110, pp. 345–356. Springer, Cham (2021). https://doi.org/10.1007/978-3-030-92238-2_29
11. Cheng, D.C., et al.: SeNet: structured edge network for sea-land segmentation. IEEE Geosci. Remote Sens. Lett. **14**(2), 247–251 (2017). https://doi.org/10.1109/LGRS.2016.2637439
12. Ding J., Li, A., Hu, Z.Q., Wang, L.W.: Accurate pulmonary nodule detection in computed tomography images using deep convolutional neural networks. In: Medical Image Computing and Computer-Assisted Intervention, Quebec City, Quebec, Canada, pp. 559–567 (2017). https://doi.org/10.1007/978-3-319-66179-7_64
13. AbdelMaksoud, E., Barakat, S., Elmogy, M.: A computer-aided diagnosis system for detecting various diabetic retinopathy grades based on a hybrid deep learning technique. Med. Biol. Eng. Comput. **60**, 2015–2038 (2022). https://doi.org/10.1007/s11517-022-02564-6
14. Hao, X.Y., Xiong, J.F., Xue, X.D., et al.: 3D U-net with dual attention mechanism for lung tumor segmentation. J. Image Graph. **25**(10), 2119–2127 (2020). https://doi.org/10.11834/jig.200282
15. Tan, J.X., Huo, Y.M., et al.: Expert knowledge-infused deep learning for automatic lung nodule detection. J. Xray Sci. Technol. **27**(1), 17–35 (2018). https://doi.org/10.3233/XST-180426

16. Liu, J., Cao, L., Akin, O., Tian, Y.: 3DFPN-HS$^2$: 3D feature pyramid network based high sensitivity and specificity pulmonary nodule detection. In: Shen, D., et al. (eds.) MICCAI 2019. LNCS, vol. 11769, pp. 513–521. Springer, Cham (2019). https://doi.org/10.1007/978-3-030-32226-7_57
17. Murphy, K., et al.: A large-scale evaluation of automatic pulmonary nodule detection in chest CT using local image features and k-nearest-neighbour classification. **13**(5), 757–770 (2009). https://doi.org/10.1016/J.MEDIA.2009.07.001
18. Srivastava, N., Hinton, G., et al.: Dropout: a simple way to prevent neural networks from overfitting. J. Mach. Learn. Res. **15**(1), 1929–1958 (2013)
19. Wang, D., Zhang, Y., Zhang, K.X., et al.: FocalMix: semi-supervised learning for 3D medical image detection. In: CVPR: Computer Vision and Pattern Recognition. Seattle, WA, USA, pp. 3951–3960 (2020). https://doi.org/10.1109/CVPR42600.2020.00401

# A Novel Optimized Context-Based Deep Architecture for Scene Parsing

Ranju Mandal[1(✉)], Brijesh Verma[1], Basim Azam[1], and Henry Selvaraj[2]

[1] Centre for Intelligent Systems, School of Engineering and Technology, Central Queensland University, Brisbane, Australia
{r.mandal,b.verma,b.azam}@cqu.edu.au

[2] Department of Electrical and Computer Engineering, University of Nevada, Las Vegas, USA
henry.selvaraj@unlv.edu

**Abstract.** Determining the optimal parameter values for a scene parsing network architecture is an important task as a network with optimal parameters produces the best performance. A manual selection of parameter values requires expert domain knowledge of the intrinsic structure and extensive trial and error, which does not guarantee optimal results. An automatic search of the optimal parameters is desirable to harness the full potential of a scene parsing framework. The network architecture needs to be evaluated with various combinations for several parameters to achieve optimum performance. We propose a stacked three-level deep context-adaptive end-to-end network. The end-to-end network architecture extracts visual features, and contextual features, and optimally integrates both features in three logical levels for scene parsing. Particle Swarm Optimization (PSO) algorithm has been used to efficiently search for optimal solutions. Using PSO, we set an optimal set of parameters in all three levels to obtain optimum performance. The PSO aims to optimize the network globally by considering all hyperparameters in the network to achieve the best performance.

**Keywords:** Image Parsing · Semantic Segmentation · Scene Parsing · PSO · Deep Learning

## 1 Introduction

The main aim of semantic segmentation is to classify a given image into one of the predefined semantic classes, including objects (e.g., vehicle, animal, person) and backgrounds (e.g., sky, road) at the pixel level. Semantic segmentation can be described as a dense prediction task where output maintains the matching resolution as input. Image understanding is a key task in computer vision research that has a wide range of applications. The partitioning of an image into semantically meaningful classes and finding interaction patterns among classes are some primary challenges. The reliability of a few high-impact and demanding intelligent vision applications, such as hazard detection, image compression, augmented reality, robotic perception, AI-enabled video surveillance, and self-driving vehicle navigation are dependent upon estimating the co-occurrence of predefined objects within a given environment. Visual variability among


M. Tanveer et al. (Eds.): ICONIP 2022, CCIS 1793, pp. 351–364, 2023.
https://doi.org/10.1007/978-981-99-1645-0_29

structured and unstructured objects [1] in complex natural scenes make accurate pixel labeling into object categories a challenging task. In Fig. 1, a few sample images with corresponding labels from our experimental datasets are presented to show some existing challenges. There are several variations in the appearance of class objects along with other factors such as occlusion, illumination, angle, and objects' magnitude. Three implicit tasks such as classification, localization, and edge delineation are performed object-wise in a robust segmentation network. The network needs to accomplish all the above-implied tasks effectively to achieve acceptable accuracy.

Recently, the remarkable performance [2] of multi-layered CNN on computer vision tasks has enabled the researchers to achieve encouraging results for semantic (i.e., pixel-wise) labeling, yet CNN architectures are unable to capture rich context representations fully, eventually leaving significant scope for improvement of the semantic segmentation architectures. A substantial number of works are aimed at building scene parsing deep network architecture because the CNN-based models outperformed many scenes parsing benchmark datasets [2, 3]. However, there are significant notable limitations of deep CNN when applied directly to dense prediction tasks like semantic segmentation [4, 5].

**Fig. 1.** Scene parsing accomplishes dense prediction of semantic categories from natural images captured in an unconstrained environment. Original images are presented in the top row, while labeled images (i.e. ground truth) are presented in the bottom row.

Semantic segmentation consists of a few subtasks (e.g., superpixel extraction, superpixel-wise feature computation, training and testing of classifiers, modeling of pixel-level context information, and the integration of contextual and visual properties for final class estimation), making it a considerably challenging problem. However, deep ConvNets can capture abstract feature representation of an image globally and hence capture features of salient objects of an image, enabling such networks to perform impressively on classification and detection tasks. However, for dense prediction, the network needs to go further to capture crucial pixel-wise discriminative features and spatial information. Deep convolutional network-based architectures [5, 6] have labeled images pixel-wise for the segmentation of natural images to a great extent. However, incorporating a lack of distinct contextual properties and natural scene segmentation tasks remains challenging thus far. Recently developed CNN-based multi-layered models with an end-to-end pipeline have emerged for numerous computer vision projects including scene parsing [7]. The concept of dense predictions by using only visual features limits the performance unless consideration of objects' contextual properties (appearing in the image samples) is used for training [8, 9]. The proposed method produces class labels for each pixel, regardless of the pixel distribution appearing in the full image. Image or scene parsing becomes a challenging task due to some prominent attributes which are the intra-class variation of information for class representation and invariably inter-class objects hold identical attributes and have homogeneous representations. A good example of such a scenario is the appearance of motor vehicles and roads together, while a motor vehicle is most unlikely to appear in the sky or in an ocean scene.

To get optimum performance, the proposed deep architecture needs to be evaluated with different configurations for various numbers of superpixels, feature-length, the number of nodes for each fully connected network, etc. The trial and error method for selecting parameters fails to attain optimum performance and demands through observation for execution and performance analysis for a combination of parameters. We employed the PSO method independently for altering the configuration of our context-adaptive architecture to obtain optimal performance and record the best cost among all iterations. The proposed model introduces a novel three-level deep architecture optimized by PSO which harnesses patterns by modeling visual properties and the contextual relationship of superpixel patches existing in an image to attain the final class label for object classes. The final integration layer integrates three sets of probability scores acquired from the superpixel classification, global and local context-adaptive contextual features. We summarize our primary contributions to this work as follows:

a) The proposed deep architecture optimized by PSO captures optimal parameter values across three levels of our architecture (i.e. visual feature layer, context feature layer, and integration layer). To improve the overall accuracy, we find PSO algorithm-based parameters play important roles.
b) The middle layer of our model successfully extracts relationships among object classes from the image samples assigned for the training process. This layer harnesses the existing pattern from the spatial information of objects globally and locally (block-wise).
c) Robust evaluation of widely used benchmark Stanford Background Dataset (SBD) [10] and CamVid [5] datasets show the remarkable performance of our model. We found the obtained accuracy to be comparable with earlier approaches on the SBD and the CamVid street scene dataset.

## 2 Related Works

### 2.1 Optimization Algorithm

A broad spectrum of hyper-parameter optimization and search problems has been solved by evolutionary algorithms. PSO [11, 12] and Genetic Algorithm (GA) [13, 14] are widely used in a plethora of optimization problems. In previous works, GA was employed decades ago to learn the network structure of artificial neural networks [15, 16] or to optimize the weights of artificial neural networks [17, 18]. PSO is a very popular and nature-inspired evolutionary optimization technique in the hyper-parameter solution space. These derivation-free techniques are categorized into meta-heuristics and avoid being trapped into 'local minima'. PSO is a metaheuristic global optimization technique that can be applied easily in unsupervised, complex multi-dimensional problems where the traditional deterministic method fails. Junior et al. [12] applied their PSO method to update particles depending on the layer rather than the layer's parameters. Li et al. [19] employed the bias-variance model to improve the accuracy and stability of their proposed multi-objective PSO. Gao et al. [20] created a gradient-priority particle swarm optimization technique to address concerns such as PSO's low convergence efficiency when there are many hyperparameters to tune. They anticipated that the particle will first obtain the locally optimal solution before moving on to the global optimal solution. A hybrid PSO-GA approach was proposed by Wang et al. [21]. The PSO was applied to steer the evolution of the decimal-encoded parameters within every block. In the meantime, the use of GA to guide the evolution of shortcut connections is represented in binary notation. This hybrid model can extensively search structures since PSO works well on continuous optimization and GA is ideal for optimization with binary data.

## 2.2 Scene Parsing

A detailed review was done to analyze the pros and cons of the various CNN-based scene parsing architectures and their performance on the publicly available scene parsing datasets. Many network models based on CNN have performed remarkably on various advanced semantic segmentation models. Recent scene or image-parsing models discussed in our literature survey used convolution-based visual features with either implicit global [22] or local context [23] and were unable to integrate both relative and absolute contextual information with visual features simultaneously. In contrast, the proposed model is based on explicit context integration. Published articles on scene parsing contain two major approaches, either multi-scale context-based or variants of the CNN model.

Early proposed approaches secure pixels' class labels by extracting visual features pixel-wise [24] or patch-wise surrounding every image pixel [25]. Scene parsing architectures developed with feature hierarchies and region proposal-based [28] technique use region proposals to secure class labels in a few early proposed scene parsing tasks. However, the global context-based feature outperforms pixel-level features as pixel-level features are unable to capture robust neighboring region statistics, and the patch-wise features extraction technique is liable to be influenced by background noise from objects. UPSNet [26] is a panoptic segmentation network that includes a residual network, semantic segmentation head, and Mask R-CNN-based segmentation head. The network controls subtasks concurrently, and eventually, the panoptic segmentation is performed by a panoptic head using pixel-wise classification. PSPNet [6] enhance scene parsing accuracy by exploiting global contextual information using spatial pyramid-based pooling.

A segmentation method based on a hybrid Deep Learning-Gaussian Process (DL-GP) network [23] was developed to segment lane and background regions from an image. An encoder-decoder network was integrated with a hierarchical GP classifier into a single architecture, which outperforms previous works such as SegNet [39] both qualitatively and quantitatively on pedestrian lane segmentation (single class). Zhang et al. [7] proposed a model that is a weakly-supervised deep network making use of the descriptive sentences of the images using a semantic tree approach to understand the training set image configurations. The model addresses scene understanding by parsing images into a structured layout or configuration. It contains two networks; pixel-wise object labeling was performed by a CNN for image representation, and an RNN was deployed to discover the hierarchical object representations and the inter-object relations. Residual Atrous Pyramid Network (RAPNet) [27] incorporates the importance of various object classes to address street scene parsing. The salient features are selected for label predictions, and the RASP module sequentially aggregates global-to-local context to enhance the labelling performance.

## 3 Proposed Method

We propose a PSO-based optimization for carrying out iterations to obtain optimal parameter values for the proposed segmentation architecture (see Fig. 2). PSO follows a swarm strategy for searching to find a set of optimal parameters from a solution space (see Fig. 3). It does so by defining parameters of a problem in terms of bit string (for logic problems) or a set of real numbers called chromosomes. The proposed model extracts visual features from segmented superpixels at the initial stage. The extracted features are then used for a class-semantic supervised classifier labelled visual feature prediction layer which produces probabilities of all superpixels for predefined classes as a class-wise probability matrix. The contextual voting layer contains two components to extract two variants of contextual features (local and global region-based) from each superpixel present in an image.

The most probable class information obtained at the first layer along with the corresponding Object Co-occurrence Priors (OCPs) is used to compute the contextual features. Finally, we use a multi-layer perceptron (MLP) or a fully connected network with a single hidden layer as an integration layer to concatenate visual and contextual features. The integration layer produces a final class label for each superpixel.

### 3.1 Visual Features

An input image is over-segmented into groups of pixels called superpixels. Our feature extraction algorithm extracts visual features from each superpixel present in the image followed by a feature selection process to obtain the most significant feature set. Next, we trained our classifier using the features computed from the training image samples in the dataset. The visual feature classification layer accepts superpixels as input, obtains visual features, and uses a classifier to distinguish among object classes. The selected feature subset helps the training process by enhancing the classifier performance.

### 3.2 Context Sensitive Features

In the proposed architecture the contextual properties of the object play a significant role. The relationship among superpixels is modeled carefully by computing neighboring probabilities in regard to a local region of the image as well as a global region. We use the adjacent superpixel values during the computation of local probabilities and utilize the superpixels present in a spatial block for global probabilities. We consider both types of information (i.e., local and global) to obtain the probability value.

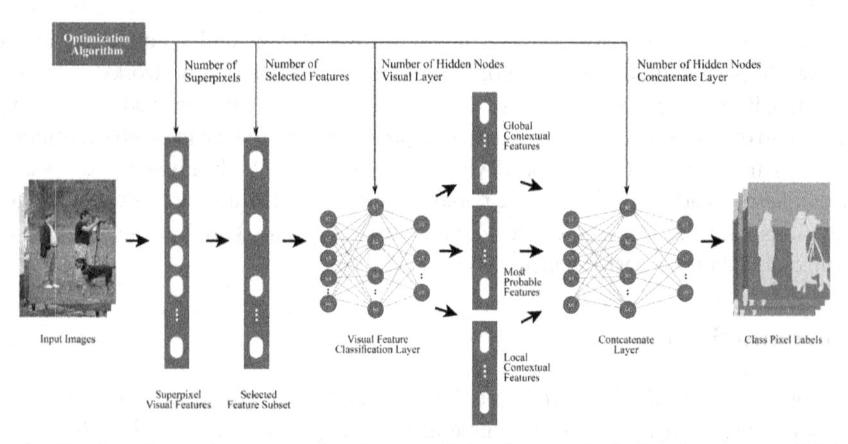


**Fig. 2.** The three-layered scene parsing framework has been optimized using a state-of-the-art optimization algorithm. The visual Features Layer computes visual features based on probability in the first layer. The second layer computes context features using superpixel blocks probability and votes from neighboring superpixels to predict the probability for classes. The final integration layer determines the final class label for each superpixel by optimally fusing the probability vector obtained from the visual feature classification (layer 1) and the contextual properties (layer 2).

**Adjoining Superpixel Votes:** The adjacent superpixel information is rich in context and our local contextual information utilizes adjacent superpixel information fully. We compute the probability vector of the neighboring superpixels. Every superpixel casts votes for its surrounding superpixel in a predefined class label. As a result, we obtain a matrix containing an adjacent superpixel probability vector computed only from the training set of images. The matrix is represented in such a way that for each superpixel we can retrieve class probability information of the surrounding superpixels.

**Block-Wise Superpixel Votes:** Block-wise votes for superpixel also play a significant role in the exploitation of long-range superpixel dependencies. This block-wise partition is done to take account of both relative spatial offsets and the absolute location of objects. The block-wise representation retains a good trade-off between absolute and relative location. Spatial relationships between blocks help to encode the relative location offsets of objects while the spatial coordinates of each block preserve the absolute location. In addition, the directional spatial relationships between blocks (e.g., left and right spatial relationships) are also preserved by the spatial distributions of all blocks.

**Integration Layer:** The final layer aims to produce a class label for each superpixel through learning optimal weights. The optimal weights integrate probability values computed from visual feature prediction and the context-adaptive voting process. The seamless integration is performed using a fully connected network (FCN) to approximate their correlation, and a majority voting strategy is used to assign a class label to each superpixel.

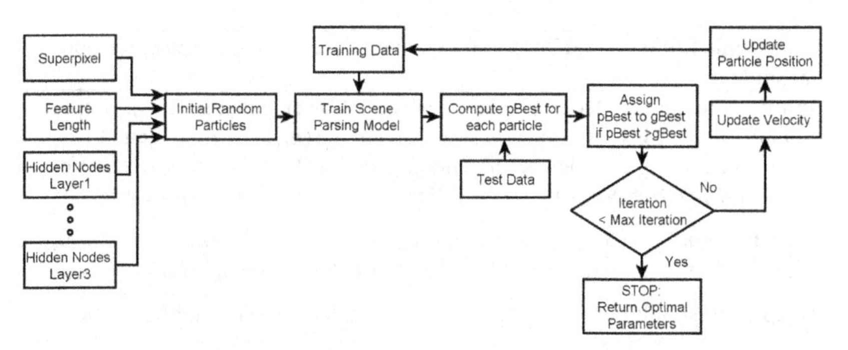


**Fig. 3.** Flowchart of the PSO-based optimization system for scene parsing.

## 3.3 Optimization

The feature dimension, the nodes in each layer, and other hyperparameters are included in our suggested network architecture. Parameter optimization is a non-differentiable and non-convex problem, and gradient-based optimization algorithms are ineffective [29] in such a scenario. The PSO approach of initializing the initial population chromosomes or vector with parameter range values was followed and several offspring are derived based on this initial model (see Algorithm 1). The fitness evaluation checks compute the cost to determine which of the models performs best. Some models outperform others, because of the way the models are randomly initialized. The proposed network determines an optimized set of hyperparameters for all its layers. The scene parsing network was optimized with a mutation rate of 0.01, a population size of 50, and a crossover rate ranging between 0.05 and 0.1. The accuracy improved over the iterations and its best value was achieved when the crossover rate was 0.1. In our experiment, the convergence was accelerated during the optimization when the crossover rate was high. The proposed network learns parameters in each layer to generate an approximation.

The algorithm has been implemented using MATLAB and makes use of Neural Network Toolbox, Image Processing Toolbox and Global Optimization Toolbox (for PSO Algorithm). Experiments have been carried over SBD [10] and CamVid dataset [5]. A novel PSO-based approach is also applied to optimize the proposed context-based scene parsing architecture. The optimized PSO-based scene parsing architecture can automatically determine the optimal combination of parameters by navigating the solution space. In the experimental setup the parsing network was optimized using inertia weight $(w) = 1$, personal learning coefficient $(c_1) = 1.5$, $(c_2) = 2.0$, and population size $(n_{pop})$ was 50.

**Algorithm 1.** PSO-based Process for Context-based Deep Architecture Optimization

**Input:** Number of Decision Variables ($n_V$), Lower Bounds ($v_{Min}$), Upper Bounds ($v_{Max}$), Max Iterations ($max_{It}$), Populations($n_{Pop}$), Inertia Weight ($w$), Personal Learning Coefficient ($c_1$), Global Learning Coefficient ($c_2$), Max Velocity ($vel_{Max}$), Min Velocity ($vel_{Min}$), Best Particle ($p_{Best}$), Global Best ($g_{Best}$)

**Initialization:** $n_V = 3$, $v_{Min} = \{50, 16, 16\}$, $v_{Max} = \{100, 32, 32\}$, $max_{It} = 100$, $n_{pop} = 50$, $w = 1$, $c_1 = 1.5$, $c_2 = 2.0$, $vel_{Max} = 0.1*(v_{Max} - v_{Min})$, $vel_{Min} = -vel_{Max}$

**Output:** Best Solution values ranging between $v_{Min}$ and $v_{Max}$ and Best Cost

```
** Compute the first population
for i = 1 to n_pop do
  Initialize particle (i) position and velocity
  Evaluate cost of particle (i)
  Update personal best of particle(i) (p_Best)
  Update Global Best (g_Best);
end for
** compute Global Best
for it = 1 to Max_It
  for i = 1 to n_pop
     Update velocity
     Update position of particle(i)
     Evaluation of particle(i) cost
     Update Personal Best (p_Best)
     Update Global Best (g_Best)
  end for
end for
Output: Best Solution and Best Cost
```

# 4 Results and Discussions

In this section, the experimental outcomes obtained from the evaluations of our proposed approach are analyzed and discussed. Two widely used benchmark scene parsing datasets are considered for our experiments. A detailed comparative study was conducted with recently proposed state-of-art published works in the literature on the scene parsing algorithms, and here we draw a comparison from a performance perspective.

## 4.1 Datasets

We use the SBD [10] and the Cambridge video dataset (CamVid) [5] to validate the proposed methodology. The SBD contains 715 photos of outdoor sceneries compiled from publicly available databases. Using Amazon Mechanical Turk, image pixels are labelled into one of eight categories or undefined categories. In each fold, 572 photos are chosen at random for training and 143 photos for training. The CamVid dataset [5] provides ground truth labels for 32 object categories. The dataset provides qualitative experimental data for the quantitative evaluation of semantic segmentation algorithms. The image frames are extracted from video captured from the viewpoint of a car. The specific scenario improves the heterogeneity of the perceived object classes. The CamVid dataset is popular among computer vision researchers due to its relevant contributions. The original resolution of images is 960 × 720, and the images are down-sampled to 480 × 360 for our experiments to follow previous works. We implemented our model with the Python and MATLAB environment. All evaluations are done on a university HPC cluster facility. Limited computing nodes (12 to 16 CPUs) and 170 GB of memory were used during the experiments.

**Table 1.** Confusion Matrix on **Stanford Background Dataset**, Accuracy = 92.42%

| | Sky | Tree | Road | Grass | Water | Bldng | Mtn | Fgnd |
|---|---|---|---|---|---|---|---|---|
| Sky | 0.95 | 0.02 | 0 | 0 | 0 | 0.02 | 0 | 0.01 |
| Tree | 0.01 | 0.89 | 0.01 | 0.01 | 0 | 0.1 | 0 | 0.01 |
| Road | 0 | 0 | 0.95 | 0 | 0 | 0.02 | 0 | 0.05 |
| Grass | 0 | 0.03 | 0.01 | 0.95 | 0 | 0.01 | 0 | 0.05 |
| Water | 0.01 | 0 | 0 | 0.01 | 0.96 | 0.01 | 0 | 0.1 |
| Bldg | 0.01 | 0.06 | 0.01 | 0 | 0 | 0.94 | 0 | 0.04 |
| Mtn | 0.02 | 0.01 | 0 | 0.03 | 0.01 | 0.01 | 0.9 | 0.04 |
| Fgnd | 0.02 | 0.05 | 0.06 | 0.01 | 0.01 | 0 | 0 | 0.85 |

**Table 2.** Comparative study on segmentation performance (%) with various approaches on **Stanford Background dataset**

| Method | Pixel Acc. | Class Acc. |
|---|---|---|
| Gould et al. [10] | 76.4 | NA |
| Kumar et al. [30] | 79.4 | NA |
| Lempitsky et al. [31] | 81.9 | 72.4 |
| Farabet et al. [24] | 81.4 | 76.0 |
| Sharma et al. [25] | 82.3 | 79.1 |
| Luc et al. [32] | 75.2 | 68.7 |
| Chen et al. [33] | 87.0 | 75.9 |
| Zhu et al. [34] | 87.7 | 79.0 |
| Proposed Approach (PS0-50) | 92.4 | 92.0 |

### 4.2 Evaluation of Stanford Dataset

As a confusion matrix, Table 1 illustrates the acquired results on the Stanford dataset. On the Stanford dataset, the optimized deep architecture achieves a score of 92.4%. Table 2 compares the accuracy given by prior methodologies with the accuracy achieved by the proposed PSO optimized model. Class accuracies of 92.0% were attained using the network model when the population size was 50. Figure 4 illustrates the qualitative results achieved on the Stanford dataset. The findings reveal that the proposed solution correctly predicts object pixels with a high degree of accuracy.

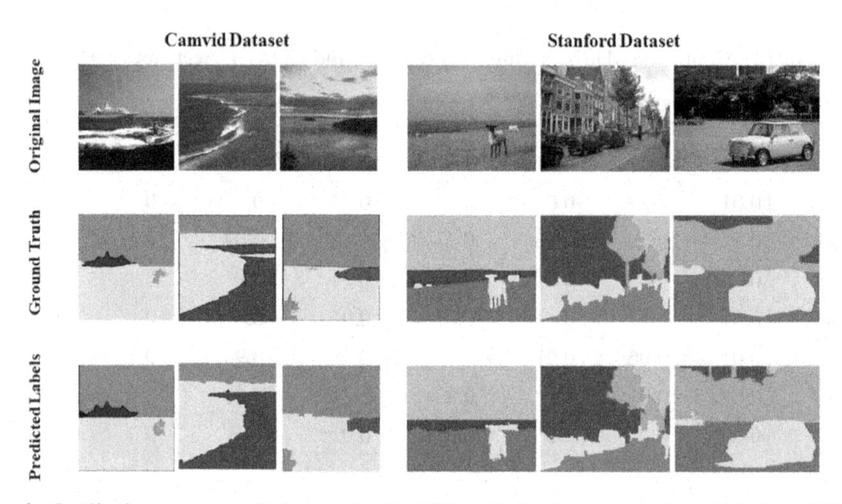


**Fig. 4.** Qualitative accuracy obtained using the PSO optimized model on Camvid (left) and Stanford Background dataset (right) is presented. The original image, ground truth, and predicted labels are shown column-wise in top to bottom approach.

### 4.3 Segmentation Results on CamVid Dataset

Using the commonly applied Intersection-Over-Union (IoU), often known as the Jaccard Index, we present the experimental results achieved from the CamVid [5] test dataset. The mean IoU metric (the weighted Jaccard Index) was adopted to analyze the model's performance on the CamVid dataset, as it is commonly used by the best-performing algorithms in the literature. In semantic segmentation, the IoU metric is a highly successful evaluation metric. The proposed optimized 3-level network model achieved 89.78% mIoU without using the pre-trained model weights. On the CamVid dataset, a comparison of mean IoU with prior techniques is shown in Table 3. We found that by using the best parameter choices, the network model was able to outperform existing techniques. Figure 4 shows the visual results of the suggested approach on the CamVid Dataset. The top row contains photos from the dataset, the center row contains ground truth, and the bottom row interprets the output of the trained model.

### 4.4 Efficacy of the Proposed Optimization

The network makes use of the proposed optimization algorithm to choose the optimal set of hyperparameters for the framework. It can be argued that the proposed optimization technique enhances global accuracy. Over the iterations (see Fig. 5) cost is reduced to 0.0767 and accuracy is increased to 92.33% when the population is set to 25. The cost was reduced to 0.0758 and accuracy was 92.42% when the population was set to 50. Iteration vs. cost obtained from PSO-based optimization process on CamVid dataset are plotted after 100 iterations. Over the iterations cost is reduced to 0.103 and accuracy is increased to 89.77% when the population is set to 25. The cost was reduced to 0.102 and the accuracy was 89.78% when the population was set to 50. Table 4 presents the optimal set of parameters, and accuracy scores on the benchmark datasets, produced by the proposed optimization-based image parsing architecture.

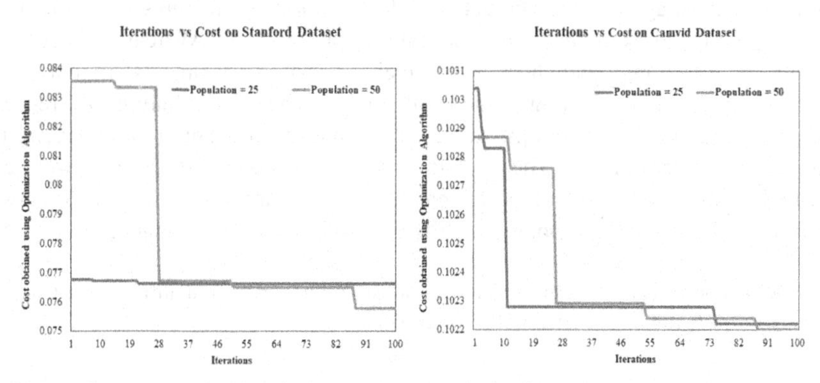


**Fig. 5.** Iteration vs cost obtained from PSO-based optimization process on Stanford Background dataset (left) and CamVid dataset (right) are plotted after 100 iterations.

**Table 3.** Comparative study on segmentation performance (%) on **CamVid** dataset

| Method | Model Pre-trained | Encoder | mIoU (%) |
|---|---|---|---|
| Yu and Koltun [22] | ImageNet | Dilate | 65.3 |
| Zhao et al. [6] | ImageNet | ResNet50 | 69.1 |
| Huang et al. [35] | ImageNet | VGG16 | 62.5 |
| Yu et al. [36] | ImageNet | ResNet18 | 68.7 |
| Bilinski and Prisacariu [37] | ImageNet | Res-NeXt101 | 70.9 |
| Chandra et al. [38] | Cityscapes | Res-Net101 | 75.2 |
| Badrinarayanan et al. [39] | ImageNet | VGG16 | 60.1 |
| Liu et al. [40] | ImageNet | Mo-bileNetV2 | 76.3 |
| Proposed Approach (PSO-50) | – | – | 89.7 |

**Table 4.** Efficacy of the proposed optimization on segmentation performance (%) population size of 25 and 50 on the **Stanford Background** and **CamVid** dataset.

| Population Size | Superpixel | Feature Dimension | Node 1st Layer | Node3rd Layer | Stanford | CamVid |
|---|---|---|---|---|---|---|
| 25 | 512 | 88 | 31 | 21 | 92.33 | 89.77 |
| 50 | 512 | 77 | 26 | 20 | 92.42 | 89.78 |

## 5 Conclusion

We present a deep architecture optimized by PSO algorithms for scene parsing in natural scene images. A novel context-adaptive deep network is created and evaluated by incorporating spatial image blocks. The network derives the CAV features using the object co-occurrence priors computed from the training phase. The CAV feature is designed to obtain the long-range and short-range label correlations of objects simultaneously in a scene image while adapting to the local context. The goal of nature-inspired optimization was to determine the optimal set of parameters in the entire architecture that would improve performance. The experiments on two benchmark datasets were conducted which demonstrated high performance. In our future research, we will focus on further optimization of the proposed architecture to improve performance.

**Acknowledgments.** This research project was supported under Australian Research Council's Discovery Projects funding scheme (ARC-DP200102252).

## References

1. Yang, M., Yu, K., Zhang, C., Li, Z., Yang, K.: DenseASPP for Semantic Segmentation in Street Scenes. In: Proceedings of the IEEE Conference on Computer Vision and Pattern Recognition, pp. 3684–3692 (2018)
2. Qi, M., Wang, Y., Li, A., Luo, J.: STC-GAN: spatio-temporally coupled generative adversarial networks for predictive scene parsing. IEEE Trans. Image Process. **29**, 5420–5430 (2020)
3. Zhou, L., Zhang, H., Long, Y., Shao, L., Yang, J.: Depth embedded recurrent predictive parsing network for video scenes. IEEE Trans. Intell. Transp. Syst. **20**(12), 4643–4654 (2019)
4. Zhang, R., Tang, S., Zhang, Y., Li, J., Yan, S.: Perspective-adaptive convolutions for scene parsing. IEEE Trans. Pattern Anal. Mach. Intell. **42**(4), 909–924 (2019)
5. Brostow, G.J., Fauqueur, J., Cipolla, R.: Semantic object classes in video: a high-definition ground truth database. Pattern Recogn. Lett. **30**(2), 88–97 (2009)
6. Zhao, H., Shi, J., Qi, X., Wang, X., Jia, J.: Pyramid scene parsing network. In: Proceedings of the IEEE Conference on Computer Vision and Pattern Recognition, pp. 2881–2890 (2017)
7. Zhang, R., Lin, L., Wang, G., Wang, M., Zuo, W.: Hierarchical scene parsing by weakly supervised learning with image descriptions. IEEE Trans. Pattern Anal. Mach. Intell. **41**(3), 596–610 (2019)
8. Mandal, R., Azam, B., Verma, B., Zhang, M.: Deep learning model with GA-based visual feature selection and context integration. In: IEEE Congress on Evolutionary Computation (CEC), pp. 288–295 (2021)

9. Mandal, R., Azam, B., Verma, B.: Context-based deep learning architecture with optimal integration layer for image parsing. In: Mantoro, T., Lee, M., Ayu, M.A., Wong, K.W., Hidayanto, A.N. (eds.) ICONIP 2021. LNCS, vol. 13109, pp. 285–296. Springer, Cham (2021). https://doi.org/10.1007/978-3-030-92270-2_25
10. Gould, S., Fulton, R., Koller, D.: Decomposing a scene into geometric and semantically consistent regions. In: Proceedings of IEEE Conference on Computer Vision and Pattern Recognition, pp. 1–8 (2009)
11. Wang, B., Sun, Y., Xue, B., Zhang, M.: Evolving deep convolutional neural networks by variable-length particle swarm optimization for image classification. In: Proceedings of IEEE Congress on Evolutionary Computation (CEC), pp. 1–8 (2018)
12. Junior, F.E.F., Yen, G.G.: Particle swarm optimization of deep neural networks architectures for image classification. Swarm Evol. Comput. **49**, 62–74 (2019)
13. Houck, C.R., Joines, J., Kay, M.G.: A genetic algorithm for function optimization: a Matlab implementation. Ncsuie tr **95**(09), 1–10 (1995)
14. Deb, K., Pratap, A., Agarwal, S., Meyarivan, T.: A fast and elitist multiobjective genetic algorithm: NSGA-II. IEEE Trans. Evol. Comput. **6**(2), 182–197 (2002)
15. Bayer, J., Wierstra, D., Togelius, J., Schmidhuber, J.: Evolving memory cell structures for sequence learning. In: International Conference on Artificial Neural Networks, pp. 755–764 (2009)
16. Stanley, K.O., Miikkulainen, R.: Evolving neural networks through augmenting topologies. Evol. Comput. **10**(2), 99–127 (2002)
17. Ding, S., Li, H., Su, C., Yu, J., Jin, F.: Evolutionary artificial neural networks: a review. Artif. Intell. Rev. **39**(3), 251–260 (2013)
18. Yao, X.: Evolving artificial neural networks. Proc. IEEE **87**(9), 1423–1447 (1999)
19. Li, L., Qin, L., Qu, X., Zhang, J., Wang, Y., Ran, B.: Day-ahead traffic flow forecasting based on a deep belief network optimized by the multi-objective particle swarm algorithm. Knowl.-Based Syst. 1–14 (2019)
20. Gao, Z., Li, Y., Yang, Y., Wang, X., Dong, N., Chiang, H.-D.: A GPSO-optimized convolutional neural networks for EEG-based emotion recognition. Neurocomputing **380**, 225–235 (2020)
21. Wang, B., Sun, Y., Xue, B., Zhang, M.: A hybrid GA-PSO method for evolving architecture and short connections of deep convolutional neural networks. In: Nayak, A.C., Sharma, A. (eds.) PRICAI 2019. LNCS (LNAI), vol. 11672, pp. 650–663. Springer, Cham (2019). https://doi.org/10.1007/978-3-030-29894-4_52
22. Yu, F., Koltun, V.: Multi-scale context aggregation by dilated convolutions. arXiv Preprint arXiv:1511.07122 (2015)
23. Choi, S., Kim, J.T., Choo, J.: Cars can't fly up in the sky: improving urban-scene segmentation via height-driven attention networks. In: Proceedings of the IEEE Conference on Computer Vision and Pattern Recognition, pp. 9373–9383 (2020)
24. Farabet, C., Couprie, C., Najman, L., LeCun, Y.: Learning hierarchical features for scene labeling. IEEE Trans. Pattern Anal. Mach. Intell. **35**(8), 1915–1929 (2013)
25. Sharma, A., Tuzel, O., Jacobs, D.W.: Deep hierarchical parsing for semantic segmentation. In: Proceedings of the IEEE Conference on Computer Vision and Pattern Recognition, pp. 530–538 (2015)
26. Xiong, Y., et al.: UPSNet: a unified panoptic segmentation network. In: Proceedings of the IEEE Conference on Computer Vision and Pattern Recognition, pp. 8818–8826 (2019)
27. Zhang, P., Liu, W., Lei, Y., Wang, H., Lu, H.: RAPNet: residual atrous pyramid network for importance-aware street scene parsing. IEEE Trans. Image Process. **29**, 5010–5021 (2020)
28. Girshick, R., Donahue, J., Darrell, T., Malik, J.: Rich feature hierarchies for accurate object detection and semantic segmentation. In: Proceedings of the IEEE Conference on Computer Vision and Pattern Recognition, pp. 580–587 (2014)

29. Sun, Y., Xue, B., Zhang, M., Yen, G.G.: An experimental study on hyper-parameter optimization for stacked auto-encoders. In: Proceedings of IEEE Congress on Evolutionary Computation (CEC), pp. 1–8 (2018)
30. Kumar, M.P., Koller, D.: Efficiently selecting regions for scene understanding. In: Proceedings of the IEEE Conference on Computer Vision and Pattern Recognition, pp. 3217–3224 (2010)
31. Lempitsky, V., Vedaldi, A., Zisserman, A.: Pylon model for semantic segmentation. In: Advances in Neural Information Processing System, pp. 1485–1493 (2011)
32. Luc, P., Couprie, C., Chintala, S., Verbeek, J.: Semantic segmentation using adversarial networks. arXiv Preprint arXiv:1611.08408 (2016)
33. Chen, L.-C., Papandreou, G., Kokkinos, I., Murphy, K., Yuille, A.L.: DeepLab: semantic image segmentation with deep convolutional nets, atrous convolution, and fully connected CRFs. IEEE Trans. Pattern Anal. Mach. Intell. **40**(4), 834–848 (2017)
34. Zhu, X., Zhang, X., Zhang, X.-Y., Xue, Z., Wang, L.: A novel framework for se-mantic segmentation with generative adversarial network. J. Vis. Commun. Image Represent. **58**, 532–543 (2019)
35. Huang, P.-Y., Hsu, W.-T., Chiu, C.-Y., Wu, T.-F., Sun, M.: Efficient uncertainty estimation for semantic segmentation in videos. In: Ferrari, V., Hebert, M., Sminchisescu, C., Weiss, Y. (eds.) ECCV 2018. LNCS, vol. 11205, pp. 536–552. Springer, Cham (2018). https://doi.org/10.1007/978-3-030-01246-5_32
36. Yu, C., Wang, J., Peng, C., Gao, C., Yu, G., Sang, N.: BiSeNet: bilateral segmentation network for real-time semantic segmentation. In: Ferrari, V., Hebert, M., Sminchisescu, C., Weiss, Y. (eds.) ECCV 2018. LNCS, vol. 11217, pp. 334–349. Springer, Cham (2018). https://doi.org/10.1007/978-3-030-01261-8_20
37. Bilinski, P., Prisacariu, V.: Dense decoder shortcut connections for single-pass semantic segmentation. In: Proceedings of the IEEE Conference on Computer Vision and Pattern Recognition, pp. 6596–6605 (2018)
38. Chandra, S., Couprie, C., Kokkinos, I.: Deep spatio-temporal random fields for efficient video segmentation. In: Proceedings of IEEE Conference on Computer Vision and Pattern Recognition, pp. 8915–8924 (2018)
39. Badrinarayanan, V., Kendall, A., Cipolla, R.: SegNet: a deep convolutional encoder-decoder architecture for image segmentation. IEEE Trans. Pattern Anal. Mach. Intell. **39**(12), 2481–2495 (2017)
40. Liu, Y., Shen, C., Yu, C., Wang, J.: Efficient semantic video segmentation with per-frame inference. In: Vedaldi, A., Bischof, H., Brox, T., Frahm, J.-M. (eds.) ECCV 2020. LNCS, vol. 12355, pp. 352–368. Springer, Cham (2020). https://doi.org/10.1007/978-3-030-58607-2_21

# Resnet-2D-ConvLSTM: A Means to Extract Features from Hyperspectral Image

Anasua Banerjee(✉) and Debajyoty Banik(✉)

Kalinga Institute of Industrial Technology, Bhubaneswar, India
anasua123.banerjee@gmail.com, debajyoty.banik@gmail.com

**Abstract.** There are many spectral bands of different wavelengths present in Hyperspectral Image containing a huge amount of information that helps to detect and identify various objects. Many challenges are faced at the time of analyzing a hyperspectral image like information loss, hindrances posed by redundant information lingering on input data and the presence of high dimensions, etc. In this paper, we proposed a Resnet-2D-ConvLSTM model which is composed of a 2D Convolution Neural Network together with Batch Normalization and it helps to minimize the computational complexity and to extract features from Hyperspectral Image. At the same time, we added shortcut connections to eliminate the vanishing gradient problem, being followed by the Long Short Term Memory layer to remove redundant information from an input image. We implemented our model on three different types of hyperspectral data sets and also on three different types of time series data sets. Our model produced better accuracy than others' proposed models reaching the levels of 0.07%, 0.01%, 0.56% more in the "Indian Pines", "Pavia University", and "Botswana" data set respectively. The commitment of our errors decreased in time series datasets by 0.44, 0.08, and 0.5 in "Electricity production", "International Airline Passenger" and "Production of shampoo over three years" respectively.

**Keywords:** Hyperspectral Imaging (HSI) · Feature extraction · Classification and Dimension Reduction

## 1 Introduction

Hyperspectral imaging (HSI) is also acquainted with us as imaging spectroscopy the sensors collect spectral vector with innumerable elements from each and every pixel. This technology i.e HSI, has the ability to identify different objects and features. As a result whenever images are captured in high-resolution satellite sensors with the help of this technique, monitoring and the tracking of changes that are being noticed in the urban land cover area can be done effectively [1]. Earlier hand based feature extraction technique had come up. Afterwards the

A. Banerjee and D. Banik—These authors contributed equally to this work.


M. Tanveer et al. (Eds.): ICONIP 2022, CCIS 1793, pp. 365–376, 2023.
https://doi.org/10.1007/978-981-99-1645-0_30

linear and non-linear technique came up for the purpose of classification. These techniques are Support Vector Machine (SVM) and K Nearest Neighbor (KNN), etc. all these are supervised models [2]. The very structure of Hyperspectral image is there in the form of sequence-based data structure. For this reason, if the supervised model is used which is vector based methodology for the classification of Hyperspectral image, information loss will occur [13]. So to overcome this problem learning-based technique has come up. To overcome these challenges, as discussed earlier, many techniques have come recently like Convolution Neural Network (CNN) [6], Recurrent Neural Network (RNN) [9] and Boltzmann band selection [4]. By using learning-based technique any object can be detected, image can be classified and segmentation of image can be done.

Authors in [15] found out that when only the 3D Convolution Neural Network (3DCNN) model is applied on HSI, then it faces the problem of high computational complexity. As a result, they chose 2D Convolution Neural Network (2DCNN) over 3DCNN but still, information loss problem is not resolved here. CNN fails to extract intra layer information from HSI due to the lack of information interaction between each layer according to authors [7]. Liu et al. proposed Bidirectional convolutional long short term memory (Bi-CLSTM) in [8] to get rid of carrying redundant information to the next cell. Information loss problem was also reduced at the time of HSI classification. However, the LSTM model only processes one-dimensional data, and HSI is not one-dimensional, so intrinsic structure will be destroyed here. Hu et al. developed the ConvLSTM model to resolve issues that had previously arisen at the time of using the CNN or LSTM models exclusively on HSI. But the issue of vanishing gradient problem remains unsolved here. The Resnet-2D-ConvLSTM (RCL) model, on the other hand, helps in the elimination of vanishing gradient, information loss, and computational complexity. RCL also extracts the intra layer information from HSI data. The combined effect of the significance of 2DCNN, Resnet and LSTM models can be found here.

## 2 Challenges and Objectives

Many challenges are faced at the time of analyzing a hyperspectral image. These challenges relate to high volumes of data; redundant information [11], vanishing gradient problem, limited availability with training samples, presence of mixed pixels in the data, high interclass similarity, high intraclass variability, high computational complexity and information loss problem.

In this paper, our main aim is to reduce the vanishing gradient problem, time complexity, and information loss problem also decreased.

## 3 Model Description

Hybrid-Spectral-Net [15] beats the state-of-the-art approach; however, choosing it is difficult due to poor time management. In this section, we tried to describe our proposed architecture briefly. Here we applied a dimensional reduction technique (PCA) to an input image because HSIs are volumetric data and have a spectral dimension [15]. Hence, PCA helps to reduce the enormous number of bands. 2DCNN layers are applied to reduce the complexity of the model. Any DNN model is required to solve the vanishing gradient and over-fitting problems. It is anticipated that the model will need to be straightforward and effective with fewer trainable parameters to address the overfitting issue, which prevents fully adapting the training data when the number of CNN layers increases. The vanishing gradient problem also increases in a deep neural network. Batch Normalization with skip connections is applied to the 2DCNN layers to reduce the vanishing gradient problem [14]. According to the authors, [7] CNN cannot extract intra-layer information from HSI since there is no information interaction between each layer. For this reason, we applied an LSTM layer to get rid of carrying redundant information to the next cell [8]. First of all as an input a hyperspectral image was taken, which is denoted by L $\epsilon$ $E^{W \times H \times Y}$ Furthermore the width and height of an input image are denoted by W, H. The presence of intraclass variability and interclass similarity is to be had in L. V is preparing one hot label vector V=($V_1$,$V_2$,............,$V_K$). Whereas Y denotes number of band in hyperspectral image and K represents the number of land cover categories. Now to remove the redundant spectral information from L the dimension reduction technique has to be used. With this end in view Principal Component Analysis (PCA) was used here. For this reason modified input was achieved, which is denoted as N $\epsilon$ $E^{W \times H \times R}$. PCA helps to reduce the number of spectral bands from Y to R although it keeps spatial dimensions as it is. This modified HSI data is split into 2D patches. After that, we generated 2D adjacent patches from the modified input image N. At this stage, we transported the obtained data into the 2D CNN layer together with Batch Normalization in $(M+1)^{th}$ layer.

$$N_K^{M+1} = \phi \left( \sum_{l=1}^{N^M} F_{norm}(N_l^M) \# q_K^{M+1} + b_K^{M+1} \right) \tag{1}$$

In this Eq. 1 $F_{norm}$ is the batch normalization function for $l^{th}$ feature map of $M^{th}$ layer; $\phi$ represents the ReLU activation function; The bias of the K filter bank of the $M^{th}$ layer is denoted by $q_K^{M+1}$ and $b_K^{M+1}$kernel parameters; 2DCNN operations is denoted by #.

$$F_{norm}(N_l^M) = \left( \frac{N_l^M - \omega(N_l^M)}{\sqrt{\gamma^2(N_l^M) + \epsilon}} \right) * \theta + \beta \tag{2}$$

In this Eq. 2 $F_{norm}(N_l^M)$ is the batch normalization function for $l^{th}$ feature map of $M^{th}$ layer; Mean and variance are denoted by $\omega$ and $\gamma^2$; $\theta$ and $\beta$ are learnable vector parameters. 2DCNN operations are accomplished by convolving two dimensional kernels. With the help of the sum dot product between the kernel and the input data, the entire convolution process is possible. The kernel starts passing over the whole input data. The activation value at spatial position (c, d) in the $K^{th}$ layer of $l^{th}$ feature map, is denoted as $M_{K,l}^{c,d}$, created in 2-D convolution, which is shown in Eq. 3.

$$M_{K,l}^{c,d} = \phi\left(b_{K,l} + \sum_{Y=1}^{Y_{K-1}} \sum_{n=-\theta}^{\theta} \sum_{\alpha=-z}^{z} w_{K,l,Y}^{\alpha,n} M_{(K-1),Y}^{(c+\alpha)(d+n)}\right) \tag{3}$$

In this Eq. 3 $Y_{K-1}$ is the number of feature maps present in $K-1^{th}$ layer; $w_{K,l}$ is the depth of the kernel of $K^{th}$ layer for $l^{th}$ feature map; weight of the parameter is denoted by $w_{K,l}$ for $l^{th}$ feature map of $K^{th}$ layer.

The output of the 2DCNN was fed into a one-dimensional Long Short Term Memory (LSTM). Only useful information is carried to the next cell by the LSTM model.

$$M_t = sigmoid\left(W_M\left[h_{(t-1)}, P_t\right] + b_M\right) \tag{4}$$

In this Eq. 4 $M_t$ stands for input gate; $W_M$ is the input gate's weight; $h_{(t-1)}$ denotes the previous hidden state; $b_M$ is the bias for the input gate.

$$Q_t = sigmoid\left(W_Q\left[h_{(t-1)}, P_t\right] + b_Q\right) \tag{5}$$

In this Eq. 5 $Q_t$ is the forget gate; $W_Q$ is denoted as the weight for the forget gate; $b_Q$ is denoted as bias for the input gate.

$$G_t = sigmoid\left(W_G\left[h_{(t-1)}, P_t\right] + b_G\right) \tag{6}$$

In this Eq. 6 $G_t$ denotes output gate of the LSTM model; $W_G$ represents the weight for the forget gate; $b_G$ is the bias for the input gate.

$$\overline{R_t} = tanh\left(W_R\left[h_{(t-1)}, P_t\right] + b_R\right) \tag{7}$$

$$R_t = Q_t * R_{(t-1)} + M_t \times \overline{R_t} \tag{8}$$

From the above Eqs. 7, 8 $R_t$ represents the cell state; $W_R$ is the weight of the cell state; $\overline{R_t}$ is the candidate hidden state, which is evaluated on the basis of previous hidden state and current input.

$$H_t = U_t \times tanh(C_t) \tag{9}$$

In this Eq. 9 $C_t$ denotes the cell state of LSTM model. Internal memory, $U_t$, is calculated by multiplying forget gate by previous memory. The result of this operation is added, and the hidden state is now multiplied by the input gate.

The RCL model which has been used in this paper has the following sizes: At first we took an input image. The dimension of this input image is $(W \times H \times Y)$. Here W, H shows the spatial dimension of an input image, and Y shows number of spectral bands present in a hyperspectral image. Here we have applied dimension reduction technique on input images so that the number of spectral bands gets reduced. Now the modified input image in the reduced form of bands. In the next phase a 2DCNN layer has been applied kernel size is $10 \times 10$, filter size is 64 and the activation function ReLU is used here. Then comes the next phase, here we applied another 2DCNN layer. Filter size is 64 and kernel size is $1 \times 1$. We repeated this process ten times with same filter and kernel size. As well as we added a 2DCNN shortcut connection (filter size being 64, kernel size being $1\times1$) which retrieved the output from the modified image and thereafter it fed the obtained result into the Average Pooling layer and its size was ($2\times2$). In the next phase we used LSTM layer. After that we applied dense layer of size (1*256). In the last phase, we applied a dense layer with the size depending on the number of classes in a data set. In this model, 150 batch size, 100 epochs, Adam optimizer, and categorical cross-entropy loss have been applied.

## 4 Data Set Description and Experimental Setup

With the help of satellite NASA EO-1, the capturing of the data set Botswana (BW) can be effectively had. In the BW data set, there are 242 bands and 14 identified classes present. The types of the land cover present here are occasional swamps, seasonal swaps. The remaining 145 bands are kept here. With this AVIRIS sensor, the capturing of the Indian Pines (IP) data set can be obtained. Within this, there are 224 spectral bands the range of which varies from 0.4 to 2.5. This data set covers the area of 2/3 agriculture and 1/3 forest. There are 16 classes present. ROSIS sensor is of great help to capture Pavia University (PU) data set. This data set contains 103 spectral bands, 9 identified classes, and $610\times610$ pixel. Three more time-series data sets have been used here. These are "Electricity production", "International Airline Passengers", and "Production of shampoo over three years". We got Time Series data sets from the kaggle website: https://www.kaggle.com/shenba/time-series-datasets. We downloaded the hyperspectral datasets [16] from here: http://www.ehu.eus/ccwintco/index.php?title=Hyperspectral_Remote_Sensing_Scenes#Indian_Pines.

The entire experiment has been done in Google Colab. Here a programming language has been used the name of the language is python the version used is 3.6. The version of used Tensorflow is 2.3.0 and the version of used Keras is 2.4.3. The size of the RAM is 25GB. Four data sets have been used here for the purpose of the classification of hyperspectral images. The names of which are Indian Pines, Pavia University and Botswana. The learning rate of the model is 0.001.

## 5 Result Analysis

To measure the degree of performance of this RCL model, some methods have been taken namely Overall Accuracy (OA), Average Accuracy (AA), and Kappa (KAA). The functions of these methods are described thus: Overall Accuracy (OA) gives the result after dividing the number of correctly classified pixels of an image by total number of test samples. Average accuracy (AA) is the classification accuracy of different classes. Kappa (KAA): It produces statistical measurement accuracy of the classification between the final output and the ground truth map. We used three different hyperspectral data sets in this paper. Each and every hyperspectral data set has been divided here into two parts. One part contains thirty percent training data and another part contains seventy percent testing data. This Fig. 1 shows the accuracy of different data sets by using the variation in input patches like 19 × 19, 25 × 25, 23 × 23, and 21 × 21. The empirical analysis has shown that each and every data set produces best results among all results when the input patch size is 25 × 25 due to increasing the window size. After that we tried to increase the window size but due to the limitation of RAM we were unable to obtained the results. From this we concluded that 25 × 25 spatial window size gives us the best results.

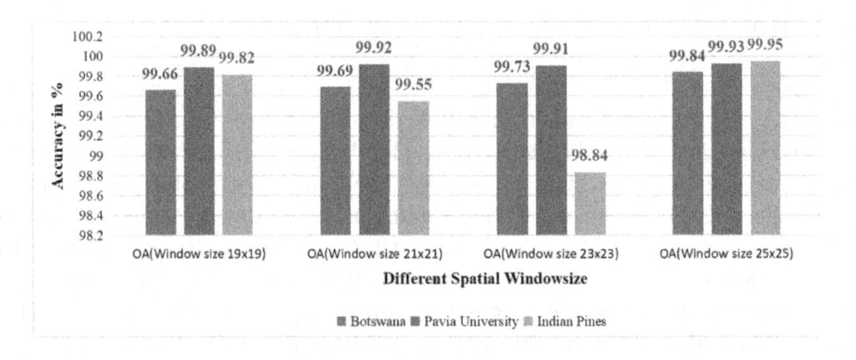


**Fig. 1.** Overall Accuracy of hyperspectral data sets on different window size

From the empirical analysis it is observed that our model gives better results when the training ratio is thirty percent. In the proposed model it is also shown that when the training ratio is reduced from thirty percent to ten percent by fixing the window size at 25 × 25, the accuracy does not decrease very much and this is shown in Fig. 2. We decreased the training ratio from thirty percent to ten percent. Because authors in [15] also decreased the training ratio.

Mean Square Error (MSE) finds out the mean value of n number of elements of the squared difference between the observed value and predicted value. In this Eq. 10 $Y_i$ denotes predicted value whereas $\overline{Y_i}$ denotes observed value.

$$MSE = 1/n \sum_{i=1}^{i=n} (Y_i - \overline{Y_i})^2 \tag{10}$$

Mean Absolute Percentage Error (MAPE) is an evaluation of the predicted accuracy of forecasting methods. The mathematical operation of MAPE is being shown in Eq. 11. $A_t$ denotes actual value whereas $F_t$ denotes forecast value.

$$MAPE = 1/n \sum_{i=1}^{i=n} |A_t - F_t/A_t| \tag{11}$$

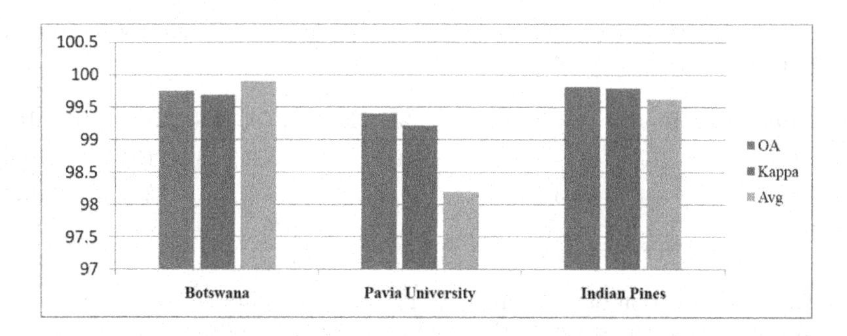


**Fig. 2.** Accuracy assessment on different hyperspectral data sets by with training ratio of 10%

Symmetric Mean Absolute Percentage Error (SMAPE) evaluates the absolute value of the forest and also the average of the absolute value of the actual parameter. The mathematical operation of SMAPE is being shown in Eq. 12.

$$SMAPE = 1/n \sum_{i=1}^{i=n} |A_t - F_t/(A_t + F_t)^2| \tag{12}$$

We took three different types of time-series data sets. Initially, we applied the preprocessing technique to each and every time-series data set. Then we broke the data set into two parts, namely training and testing ratio, seventy percent and thirty percent respectively. Next, we applied our proposed RCL model and compared it with others' the state of art of the methods.

Authors in ([3,12,15,17]) did not use their model on time-series data set, as they did not apply LSTM model, which is required to process on time-series data set.

We applied different types of supervised models published in ( [3,12,15,17]) on the "Electricity Production", "International-airline-passenger" and "Production of a shampoo over three years" time-series data set. We did simulation after using their models in our data set. Afterwards we compared our model results with their model results. The results we obtained are being shown in Table 1, 2 and 3.

**Table 1.** On Electricity Production data set the accuracy value obtained using proposed as well as the state-of-the-art methods.

| **Performance Estimator** | **SVM** | **LSTM** | **2DCNN** | **3DCNN** | **HybridSN** | **RCL** |
|---|---|---|---|---|---|---|
| MSE | 0.43 | 0.09 | 0.21 | 0.13 | 0.51 | **0.07** |
| MAPE | 0.6 | 0.1 | 0.1 | 0.2 | 0.5 | **0.1** |
| SMAPE | 0.7 | 0.2 | 0.3 | 0.3 | 0.7 | **0.2** |

**Table 2.** On International-airline-passengers data set the accuracy value obtained using proposed as well as the state-of-the-art method.

| Performance Estimator | SVM | LSTM | 2DCNN | 3DCNN | HybridSN | RCL |
|---|---|---|---|---|---|---|
| MSE | 0.45 | 0.25 | 0.48 | 0.55 | 0.17 | **0.09** |
| MAPE | 0.3 | 0.1 | 0.5 | 0.5 | 0.2 | **0.1** |
| SMAPE | 0.4 | 0.3 | 0.6 | 0.7 | 0.3 | **0.2** |

**Table 3.** On Production of a shampoo over three years data set the accuracy value obtained using proposed as well as the state-of-the-art methods.

| Performance Estimator | SVM | LSTM | 2DCNN | 3DCNN | HybridSN | RCL |
|---|---|---|---|---|---|---|
| MSE | 1.27 | 0.35 | 0.45 | 0.40 | 0.71 | **0.21** |
| MAPE | 0.8 | 0.5 | 0.5 | 0.7 | 0.7 | **0.5** |
| SMAPE | 0.8 | 0.7 | 0.7 | 0.7 | 0.75 | **0.5** |

In this Fig. 3 we showed the comparison between our RCL model's overall accuracy obtained with others ( [3,12,15,17]) supervised models OA on three different types of time series data sets namely Electricity production, International Airline Passengers and Production of shampoo over three years.

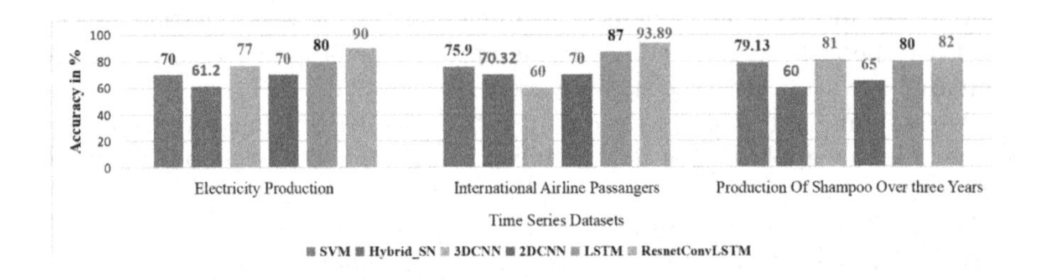


**Fig. 3.** Overall Accuracy of three different Time-Series Data sets

Among all these, it has been noticed that the results obtained from the proposed RCL model are better than the results obtained by other models. The reason behind getting better results is RCL model has the power to carry only the important information to the next cell after discarding the redundant information.

In this Table 7 we showed training time and testing time obtained from different deep learning models like 3DCNN [5], 2DCNN [12], HybridSN [15] along with our proposed RCL model on HSI datasets. From this table it is clearly noticed that our model takes less time for training and testing. All this is possible owing to the presence of fewer number of trainable parameters (Table 7).

**Table 4.** Proposed and state–of–the–art method accuracy reaped through Indian Pines Hyperspectral data set.

| Accuracy | SVM | 2DCNN | 3DCNN | SSRN | Bi-CLSTM | HybridSN | CSMS-SSRN | RCL |
|---|---|---|---|---|---|---|---|---|
| OA | 81.67 ± 0.65 | 89.48 ± 0.2 | 98.53 ± 0.29 | 99.19 ± 0.26 | 96.78 ± 0.35 | 99.75 ± 0.1 | 95.58 | 99.82 ± 0.0 |
| AA | 79.84 ± 3.37 | 86.14 ± 0.8 | 99.50 ± 0.80 | 98.18 ± 0.4 | 94.47 ± 0.83 | 99.63 ± 0.2 | 95.85 | 99.81 ± 0.1 |
| Kappa | 78.76 ± 0.77 | 87.96 ± 0.5 | 98.20 ± 0.35 | 99.03 ± 0.30 | 96.33 ± 0.40 | 99.63 ± 0.1 | 95.58 | 99.75 ± 0.5 |

**Table 5.** Proposed and state–of–the–art method accuracy reaped through Pavia University Hyperspectral data set.

| Accuracy | SVM | 2DCNN | 3DCNN | SSRN | Bi-CLSTM | HybridSN | CSMS-SSRN | RCL |
|---|---|---|---|---|---|---|---|---|
| OA | 90.58 ± 0.47 | 97.86 ± 0.2 | 99.66 ± 0.03 | 99.79 ± 0.0 | 99.10 ± 0.16 | 99.98 ± 0.09 | 99.29 | 99.99 ± 0.0 |
| AA | 92.99 ± 0.39 | 96.55 ± 0.0 | 99.77 ± 0.08 | 99.66 ± 0.17 | 99.20 ± 0.17 | 99.97 ± 0.0 | 99.30 | 99.97 ± 0.1 |
| Kappa | 87.21 ± 0.70 | 97.16 ± 0.5 | 99.56 ± 0.04 | 99.72 ± 0.12 | 98.77 ± 0.21 | 99.63 ± 0.1 | 99.05 | 99.77 ± 0.1 |

**Table 6.** Proposed and state–of–the–art method accuracy reaped through Botswana Hyperspectral data set.

| Accuracy | SVM | 2DCNN | 3DCNN | SSRN | Bi-CLSTM | HybridSN | RCL |
|---|---|---|---|---|---|---|---|
| OA | 82.10 ± 0.4 | 96.08 ± 0.47 | 98.40 ± 0.2 | 98.66 ± 0.2 | 97.55 ± 0.5 | 99.28 ± 0.4 | 99.81 ± 0.4 |
| AA | 67.64 ± 0.82 | 96.00 ± 0.8 | 97.1 ± 0.45 | 98.67 ± 0.57 | 97.85 ± 0.1 | 99.38 ± 0.4 | 99.89 ± 0.0 |
| Kappa | 78.01 ± 0.6 | 96.00 ± 0.43 | 97.52 ± 0.53 | 98.67 ± 0.21 | 98.1 ± 0.2 | 99.38 ± 0.4 | 99.75 ± 0.2 |

Their model have been applied on "Botswana" and "Pavia University" hyperspectral data set. We want to mention their best portion of accuracy achievement face to face our RCL model's accuracy achievement. For this reason, here we applied their models proposed by authors in ([3,8,12,15,17,18]) and did simulation in our "Botswana" and "Pavia University" data set. All this is shown in Table 6.

In this paper we compared our RCL model's results with others supervised model's results published in [3,8,10,12,15,17,18] namely 2DCNN, 3DCNN, Support Vector Machine (SVM), Bidirectional-Convolutional LSTM (Bi-CLSTM), Spectral-Spatial Residual Network (SSRN), 3D channel and spatial attention-based multi-scale spatial-spectral residual network (CSMS-SSRN), and HybridSN on different types of Hyperspectral datasets. In this paper [15] authors shows that by applying input patch size 25 × 25 on the "Indian Pines" Hyperspectral data set they obtained overall accuracy that is shown in Table 4. The best portion of very accuracy of achievement here has been shown here. From the above Tables it has been noticed that the RCL model's performance is better than the performance of all other models. Here we combine the LSTM and CNN models in one place, it can extract data from the inter layer as well as from the intra layer. Still, one more problem exists here and that is the vanishing gradient problem. The problem of vanishing gradient was also solved by combining the Resnet model with LSTM and CNN. Furthermore, we applied Batch Normalization after every CNN layer, which is very helpful to stabilize the model. Moreover, we applied the LSTM model, which helps to reduce the

**Table 7.** Training and Testing time (in minutes) using different models on HSI.

| Datasets | Training Time | | | | Testing Time | | | |
|---|---|---|---|---|---|---|---|---|
| | 3DCNN | 2DCNN | HybridSN | RCL | 3DCNN | 2DCNN | HybridSN | RCL |
| IP | 15.2 | 1.9 | 14.1 | 4.8 | 4.3 | 1.1 | 4.8 | 2.0 |
| BW | 10.5 | 1.0 | 2.1 | 1.5 | 3.0 | 0.5 | 1.0 | 0.7 |
| PU | 58.0 | 1.8 | 20.3 | 19.0 | 10.6 | 1.3 | 6.6 | 5.0 |

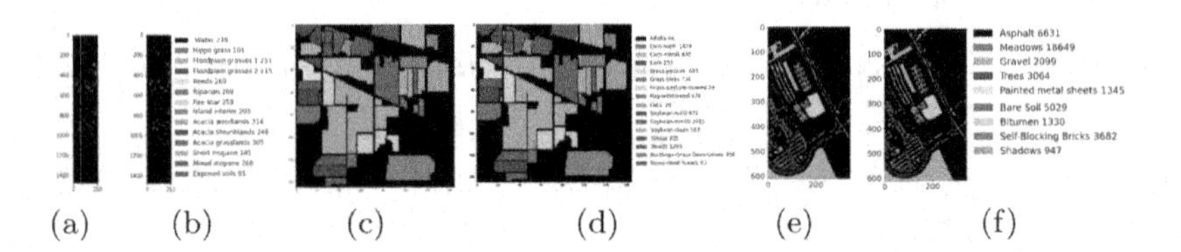


**Fig. 4.** Ground truth image cum predicted image of three different types of hyperspectral data set:(a) Ground truth image of BW (b) Predicted image of BW (c) Ground truth image of IP (d) Predicted image of IP (e) Ground truth image of PU and (f) Predicted image of PU.

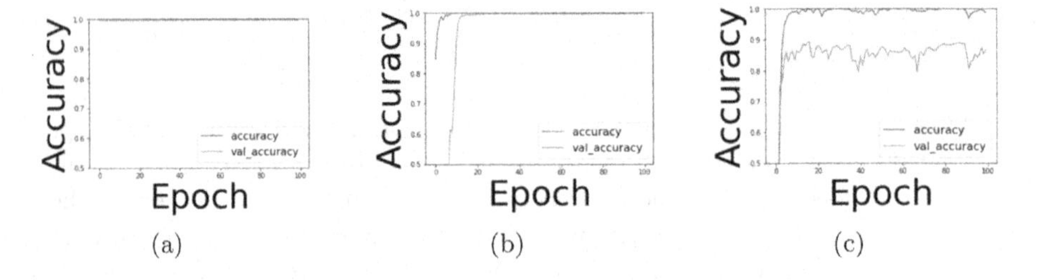


**Fig. 5.** Convergence graph over "IP", "BW", "PU" hyperspectral image datasets.

number of trainable parameters. The ground truth and predicted images of three different HSI dataset are shown in Fig. 4 and the convergence graph are shown in Fig. 5.

## 6 Conclusion

We have implemented the Resnet ConvLstm model to extract features from the hyperspectral image. We noticed that the RCL model outperform on the hyperspectral datasets but also overcame other issues. These issues relate to the problem we had faced when we applied exclusively the LSTM, Resnet, or CNN models on the same datasets. Problems that may arise include the inability to extract the intralayer feature from HSI by using only the CNN model. LSTM is a one-dimensional input data but the HSI is not single dimensional. For this reason, the intrinsic structure of HSI will be affected. We used 2DCNN to reduce the computational complexity. Furthermore, our proposed model took less time to train and has fewer trainable parameters. The model may also be useful in the fields of medical analysis. So we will apply the proposed approach on the top of medical image in the recent future to understand their behavior.

## References


1. Cai, Y., Zhang, Z., Liu, X., Cai, Z.: Efficient graph convolutional self-representation for band selection of hyperspectral image. IEEE J. Sel. Top. Appl. Earth Observations Remote Sens. **13**, 4869–4880 (2020)
2. Camps-Valls, G., Gomez-Chova, L., Muñoz-Marí, J., Vila-Francés, J., Calpe-Maravilla, J.: Composite kernels for hyperspectral image classification. IEEE Geosci. Remote Sens. Lett. **3**(1), 93–97 (2006)
3. Chen, Y., Jiang, H., Li, C., Jia, X., Ghamisi, P.: Deep feature extraction and classification of hyperspectral images based on convolutional neural networks. IEEE Trans. Geosci. Remote Sens. **54**(10), 6232–6251 (2016)
4. Gao, P., Zhang, H., Jia, D., Song, C., Cheng, C., Shen, S.: Efficient approach for computing the discrimination ratio-based variant of information entropy for image processing, 1-1 IEEE Access (2020)
5. Hamida, A.B., Benoit, A., Lambert, P., Amar, C.B.: 3-D deep learning approach for remote sensing image classification. IEEE Trans. Geosci. Remote Sens. **56**(8), 4420–4434 (2018)
6. He, K., Zhang, X., Ren, S., Sun, J.: Deep residual learning for image recognition. In: Proceedings of the IEEE conference on computer vision and pattern recognition, pp. 770–778 (2016)
7. Hu, W.S., Li, H.C., Pan, L., Li, W., Tao, R., Du, Q.: Feature extraction and classification based on spatial-spectral convlstm neural network for hyperspectral images. (2019) arXiv preprint arXiv:1905.03577
8. Liu, Q., Zhou, F., Hang, R., Yuan, X.: Bidirectional-convolutional LSTM based spectral-spatial feature learning for hyperspectral image classification. Remote Sens. **9**(12), 1330 (2017)
9. Lowe, A., Harrison, N., French, A.P.: Hyperspectral image analysis techniques for the detection and classification of the early onset of plant disease and stress. Plant Methods **13**(1), 80 (2017)
10. Lu, Z., Xu, B., Sun, L., Zhan, T., Tang, S.: 3-D channel and spatial attention based multiscale spatial-spectral residual network for hyperspectral image classification. IEEE J. Sel. Top. Appl. Earth Observations Remote Sens. **13**, 4311–4324 (2020). https://doi.org/10.1109/JSTARS.2020.3011992
11. Lv, W., Wang, X.: Overview of hyperspectral image classification. J. Sens. 2020, 1-13 (2020)
12. Makantasis, K., Karantzalos, K., Doulamis, A., Doulamis, N.: Deep supervised learning for hyperspectral data classification through convolutional neural networks. In: 2015 IEEE International Geoscience and Remote Sensing Symposium (IGARSS), pp. 4959–4962 (2015). https://doi.org/10.1109/IGARSS.2015.7326945
13. Mou, L., Ghamisi, P., Zhu, X.X.: Deep recurrent neural networks for hyperspectral image classification. IEEE Trans. Geosci. Remote Sens. **55**(7), 3639–3655 (2017). https://doi.org/10.1109/TGRS.2016.2636241
14. Noh, S.H.: Performance comparison of CNN models using gradient flow analysis. In: Informatics. vol. 8, p. 53. MDPI (2021)

15. Roy, S.K., Krishna, G., Dubey, S.R., Chaudhuri, B.B.: HybridSN: Exploring 3-D-2-D CNN feature hierarchy for hyperspectral image classification. IEEE Geosci. Remote Sens. Lett. **17**(2), 277–281 (2019)
16. Wang, W., Dou, S., Wang, S.: Alternately updated spectral-spatial convolution network for the classification of hyperspectral images. Remote Sens. **11**(15), 1794 (2019)
17. Waske, B., van der Linden, S., Benediktsson, J.A., Rabe, A., Hostert, P.: Sensitivity of support vector machines to random feature selection in classification of hyperspectral data. IEEE Trans. Geosci. Remote Sens. **48**(7), 2880–2889 (2010)
18. Zhong, Z., Li, J., Luo, Z., Chapman, M.: Spectral-spatial residual network for hyperspectral image classification: A 3-D deep learning framework. IEEE Trans. Geosci. Remote Sens. **56**(2), 847–858 (2018). https://doi.org/10.1109/TGRS.2017.2755542

# An Application of MCDA Methods in Sustainable Information Systems

Jakub Więckowski[1], Bartosz Paradowski[2], Bartłomiej Kizielewicz[1], Andrii Shekhovtsov[1], and Wojciech Sałabun[1,2](✉)

[1] National Institute of Telecommunications, Szachowa 1, 04-894 Warsaw, Poland
{j.wieckowski,b.kizielewicz,a.shekhovtsov,w.salabun}@il-pib.pl
[2] Research Team on Intelligent Decision Support Systems, Department of Artificial Intelligence and Applied Mathematics, Faculty of Computer Science and Information Technology, West Pomeranian University of Technology in Szczecin, ul. Żołnierska 49, 71-210 Szczecin, Poland
{bartosz-paradowski,wojciech.salabun}@zut.edu.pl

**Abstract.** The importance of information system is rising every year due to the ever-increasing amount of data. In many cases, data can be accessed through various types of software. For example, web browsers are the most used medium in everyday life. Therefore, it was necessary to check whether they were sustainable. For this purpose, multi-criteria decision-making methods were applied. Namely, the EDAS method was used to identify the multi-criteria model, and the COMET method was used to evaluate the analyzed alternatives. The analysis was carried out in two ways: the first, in which a given version was given multiple points and then summed. Finally, the different steps were summed to obtain a final score in the second case. An investigation has shown that it is necessary to raise awareness about sustainability in software development, as this can lead to much better results. However, this is not given much importance, as can be seen by significant fluctuations in the evaluation of different versions.

**Keywords:** MCDA · Browser assessments · Sustainability · Power management · Mozilla Firefox

## 1 Introduction

Sustainability is an essential topic on many current environmental issues. Due to the growing impact of negative impacts of human activity on the environment, it is necessary to make responsible decisions that ensure the minimization of damage [8]. One of the fundamental areas of sustainable development is using information systems that adapt operations to the latest requirements of functioning [13,16]. Meeting the assumptions of green systems, these solutions should ensure optimal energy management. The broad area of application means that the amount of energy used by these information systems consumes significant


M. Tanveer et al. (Eds.): ICONIP 2022, CCIS 1793, pp. 377–388, 2023.
https://doi.org/10.1007/978-981-99-1645-0_31

resources every day [27]. In addition, the amount of heat produced during operation and other aspects that have a negative impact on the environment directly translate into the need to select the solutions used carefully. It makes it essential that the current systems reflect the spirit of green systems as closely as possible.

When considering the possible decision variants, the most efficient performance with the least amount of environmental degradation should be selected [6,22]. The choice of eco-systems should be conditioned by a thorough analysis of the mode of operation and statistical data characterizing the energy management [3]. Responsible information systems should meet sustainability requirements and minimize their harmful effects on the environment [26]. Due to the number of available information systems, it is necessary to conduct a reliable analysis that, based on many factors, can indicate which of the available alternatives are the most recommended as green systems.

Multi-Criteria Decision Analysis (MCDA) methods can be used for this purpose. Multi-criteria analysis is an appropriate tool for evaluating the analyzed set of alternatives based on many criteria characterizing the problem's specificity [29]. Thus, such a model indicates the most preferred solutions, which is necessary when considering available information systems. The performance of these methods has been repeatedly verified in a wide area of research [10,20], thus confirming their effectiveness in complex problems. Furthermore, it shows that these techniques can be a reliable tool for assessing the suitability of information systems for sustainable development [2,5].

In green systems, it is worthwhile to look at the problem of evaluating the quality of web browser versions in terms of how they manage energy and how they adapt to sustainability goals. The Mozilla Firefox browser is one of information systems users' most widely used solutions [18]. With successive versions, the characteristics of the browser have changed in the way it manages energy [21]. How the developers managed the information system and the changes introduced affected the quality of functioning of the solution within many parameters concerning energy management [31]. It makes it necessary to identify which options offered are the most efficient and least damaging to the environment.

In this paper, we determine the quality of available versions of the Mozilla Firefox browser using MCDA methods. For this purpose, a combination of Evaluation based on Distance from Average Solution (EDAS) methods was used to identify a multi-criteria model and Characteristic Objects Method (COMET) to evaluate the analyzed alternative. The purpose of assessing the offered solutions' quality is to ensure a reliable ranking of available decision-making alternatives. To compare different approaches in rankings calculation, it is valuable to provide a similarity assessment to examine their coherence. Furthermore, the problem of selecting optimal decisions in the process of green systems management is essential for meeting the requirements of sustainable development and reducing the negative impact of used solutions on the environment.

The rest of the paper is organized as follows. In Sect. 2, we present the preliminaries of the Multi-Criteria Decision Analysis methods used in research, namely the COMET method combined with the EDAS method, and the correlation coefficients which enables to compare similarity of the results. Section 3 includes the problem description concerning the Firefox browser version assessment and its effectiveness in power management. In Sect. 4, the results are included, and their practical usage is described. Finally, in Sect. 5, we present the summary of the results and conclusions drawn from the conducted research.

## 2 Preliminaries

### 2.1 The COMET Method

The Characteristic Objects Method method stands out from other MCDA methods in that it is completely free of ranking reversal phenomenon [25]. It allows for creating a multi-criteria decision analysis model, which will serve unequivocal results regardless of the number of alternatives used in the system [11,12]. This method is more and more willingly used to solve multi-criteria problems, and its effectiveness has already been confirmed many times [7,24]. What is more, it is worth to introduce basic assumptions that accompany the work of this method.

**Step 1.** Define the Space of the Problem – the expert determines the dimensionality of the problem by selecting the number $r$ of criteria, $C_1, C_2, ..., C_r$. Then, the set of fuzzy numbers for each criterion $C_i$ is selected (1):

$$C_r = \{\tilde{C}_{r1}, \tilde{C}_{r2}, ..., \tilde{C}_{rc_r}\} \tag{1}$$

where $c_1, c_2, ..., c_r$ are numbers of the fuzzy numbers for all criteria.

**Step 2.** Generate Characteristic Objects – The characteristic objects ($CO$) are obtained by using the Cartesian Product of fuzzy numbers cores for each criteria as follows (2):

$$CO = C(C_1) \times C(C_2) \times ... \times C(C_r) \tag{2}$$

**Step 3.** Rank the Characteristic Objects – the expert determines the Matrix of Expert Judgment ($MEJ$). It is a result of pairwise comparison of the COs by the problem expert. The $MEJ$ matrix contains results of comparing characteristic objects by the expert, where $\alpha_{ij}$ is the result of comparing $CO_i$ and $CO_j$ by the expert. The function $f_{exp}$ denotes the mental function of the expert. It depends solely on the knowledge of the expert and can be presented as (3). Afterwards, the vertical vector of the Summed Judgments ($SJ$) is obtained as follows (4).

$$\alpha_{ij} = \begin{cases} 0.0, & f_{exp}(CO_i) < f_{exp}(CO_j) \\ 0.5, & f_{exp}(CO_i) = f_{exp}(CO_j) \\ 1.0, & f_{exp}(CO_i) > f_{exp}(CO_j) \end{cases} \tag{3}$$

$$SJ_i = \sum_{j=1}^{t} \alpha_{ij} \tag{4}$$

Finally, values of preference are approximated for each characteristic object. As a result, the vertical vector $P$ is obtained, where $i-th$ row contains the approximate value of preference for $CO_i$.

**Step 4.** The Rule Base – each characteristic object and value of preference is converted to a fuzzy rule as follows (5):

$$IF\ C(\tilde{C}_{1i})\ AND\ C(\tilde{C}_{2i})\ AND\ ...\ THEN\ P_i \tag{5}$$

In this way, the complete fuzzy rule base is obtained.

**Step 5.** Inference and Final Ranking – each alternative is presented as a set of crisp numbers (e.g., $A_i = \{a_{1i}, a_{2i}, ..., a_{ri}\}$). This set corresponds to criteria $C_1, C_2, ..., C_r$. Mamdani's fuzzy inference method is used to compute preference of $i-th$ alternative. The rule base guarantees that the obtained results are unequivocal. The bijection makes the COMET a completely rank reversal free.

## 2.2 The EDAS Method

In Evaluation based on Distance from Average Solution method, the Positive Distance from Average solution (PDA) and Negative Distance from Average solution (NDA) solutions are calculated to obtain the alternative preference value [23]. The optimal alternative has the higher distance from the worst solution and lowest distance from the ideal solution [1,14]. Its advantage is that it needs fewer calculations to obtain the results [30]. EDAS is designed to evaluate decision alternatives according to the following steps:

**Step 1.** Define a decision matrix of dimension $n \times m$, where $n$ is the number of alternatives, and $m$ is the number of criteria (6).

$$X_{ij} = \begin{bmatrix} x_{11} & x_{12} & \dots & x_{1m} \\ x_{21} & x_{22} & \dots & x_{2m} \\ \dots & \dots & \dots & \dots \\ x_{n1} & x_{n2} & \dots & x_{nm} \end{bmatrix} \tag{6}$$

**Step 2.** Calculate the average solution for each criterion according to the formula (7).

$$AV_j = \frac{\sum_{i=1}^{n} X_{ij}}{n} \tag{7}$$

**Step 3.** Calculating the positive distance from the mean solution and the negative distance from the mean solution for the alternatives. When the criterion is of profit type, the negative distance and the positive distance are calculated using equations (8) and (9), while when the criterion is of cost type, the distances are calculated using formulas (10) and (11).

$$NDA_{ij} = \frac{\max(0, (AV_j - X_{ij}))}{AV_j} \tag{8}$$

$$PDA_{ij} = \frac{\max\left(0, (X_{ij} - AV_j)\right)}{AV_j} \tag{9}$$

$$NDA_{ij} = \frac{\max\left(0, (X_{ij} - AV_j)\right)}{AV_j} \tag{10}$$

$$PDA_{ij} = \frac{\max\left(0, (AV_j - X_{ij})\right)}{AV_j} \tag{11}$$

**Step 4.** Calculate the weighted sums of $PDA$ and $NDA$ for each decision variant using equations (12) and (13).

$$ASP_i = \sum_{j=1}^{m} w_j PDA_{ij} \tag{12}$$

$$SN_i = \sum_{j=1}^{m} w_j NDA_{ij} \tag{13}$$

**Step 5.** Normalize the weighted sums of negative and positive distances using equations (14) and (15).

$$NSN_i = 1 - \frac{SN_i}{\max_i (SN_i)} \tag{14}$$

$$NSP_i = \frac{SP_i}{\max_i (SP_i)} \tag{15}$$

**Step 6.** Calculate the evaluation score ($AS$) for each alternative using the formula (16). A higher point value determines a higher ranking alternative.

$$AS_i = \frac{1}{2} \left(NSP_i + NSN_i\right) \tag{16}$$

### 2.3 Rank Similarity Coefficients

Similarity coefficients are the most commonly used measure to determine how two rankings differ. Their frequent use in the literature shows that they are a proven measure that is supported by many studies [4,32]. In our study, we used two coefficients, namely weighted Spearman's correlation coefficient (17) and WS ranking similarity coefficient (18). The formulas for calculation both coefficients are presented below.

$$r_w = 1 - \frac{6 \cdot \sum_{i=1}^{n} (x_i - y_i)^2 \left((N - x_i + 1) + (N - y_i + 1)\right)}{n \cdot (n^3 + n^2 - n - 1)} \tag{17}$$

$$WS = 1 - \sum_{i=1}^{n} \left(2^{-x_i} \frac{|x_i - y_i|}{\max\left\{|x_i - 1|, |x_i - N|\right\}}\right) \tag{18}$$

## 3 Practical Problem

Web browsers are one of the most widely used programs found on personal computers today [15,28]. They allow access to an unlimited amount of information, which is very important in today's ever-changing world. The need to keep up to date with events, new discoveries or research is growing unimaginably every year. In addition, browsers are an important development tool in the current market where they are responsible for the correct display of more than two million different applications [19]. Because of their widespread use, it is important to ensure the sustainability not only of their use, but of all software on personal computers and smartphones [9].

By developing software to consume less power and therefore be more efficient, we can allow for the development of a better green future [17]. For this purpose, we will use data obtained from a study [33]. The authors in their study created a benchmark that allowed to obtain certain parameters of different types of software depending on its version. From the available data, we focused on the Mozilla Firefox browser. From the available parameters, we selected eleven criteria which are presented in Table 1.

**Table 1.** Decision matrix criteria for model identification.

| $C_i$ | Criterion | Description |
|---|---|---|
| $C_1$ | Power.amps | Amperage measured |
| $C_2$ | Power.volts | Voltage measured |
| $C_3$ | Power.watts | Watts measured |
| $C_4$ | Power.watt hours | Energy consumed in Kw |
| $C_5$ | SAR.%idle | % CPU time spent idle |
| $C_6$ | SAR.%memused | the percentage of used memory |
| $C_7$ | SAR.%system | % CPU time spent in kernel |
| $C_8$ | SAR.%user | % CPU time spent in userspace |
| $C_9$ | SAR.%iowait | % CPU time spent waiting on I/O operations to complete |
| $C_{10}$ | SAR.fault/s | Page Faults / Second |
| $C_{11}$ | SAR.tps | Transfers to disk / Second |

Based on the considered set of criteria, the preferences of subsequent versions of the Mozilla Firefox browser need to be calculated. Due to the significant length of time the application has been on the market, many versions have been created. Therefore, we select the 1032 versions with approximately 350 variants based on the available set. It brings the entire set of alternatives to 1337923 decision options. Because of the large dimensionality of the problem, we have chosen to include the results of the multi-criteria analysis for the last 75 browser versions considered. However, it should be mentioned that identifying the multi-criteria model through EDAS and COMET methods was carried out for 1337923 decision variants in the case of average ratings and 1032 alternatives in the study taking into account the average decision matrices for individual versions.

## 4 Results

Taking into account the considered set of alternatives and the criteria presented in Table 1, we obtained preference values defining the quality of the proposed solutions offered by the different versions of the Mozilla Firefox browser. Visualizations of the calculated results are shown in Fig. 1 and Fig. 2, where the approaches considering the averaged scores for 75 alternatives and the averaged decision matrix for 75 versions are presented, respectively.

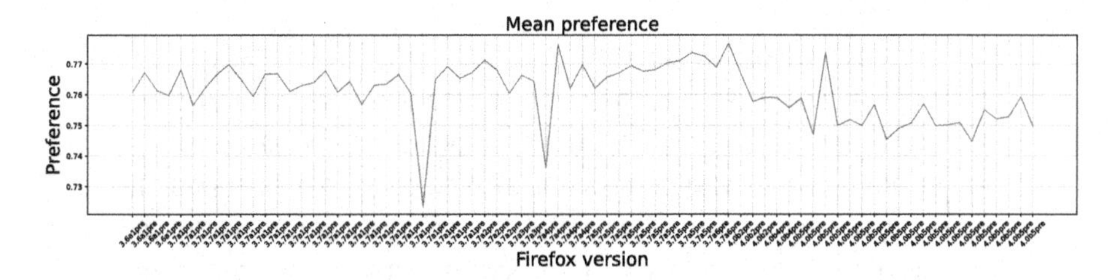


**Fig. 1.** Averaged ratings for the 75 most current variants of the Firefox browser.

The results presented in Fig. 1 show that successive browser variants offer the varying quality of solutions. Over the analyzed range, it can be seen that some of the modifications significantly resulted in worse performance compared to earlier versions. That is especially important for sustainable performance, so changes that produce worse results than previously used should be avoided.

However, the dynamics of change in the world of the Internet and the need to support ever newer and more demanding features introduced to applications sometimes require significant changes that decrease the quality of performance, especially when analyzing power management. However, the fact that successive versions of the Mozilla browser quickly return to ratings at a similar or slightly better level than earlier variants is encouraging and shows that the deterioration in the quality of the solution is temporary.

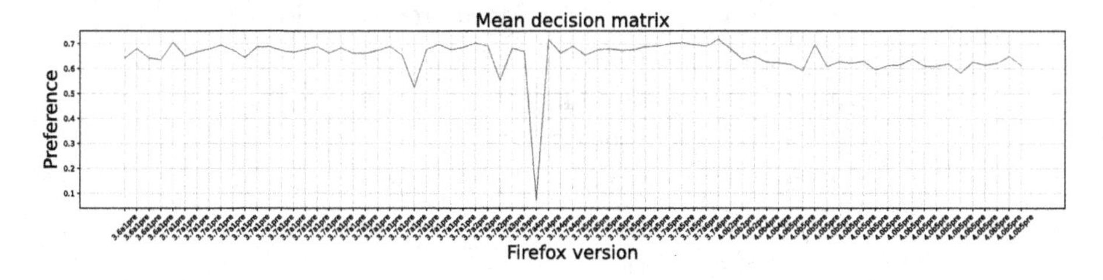


**Fig. 2.** Evaluation of the averaged decision matrix for 75 recent versions of the Firefox browser.

In contrast, the run of average decision matrix ratings for the last 75 versions shown in Fig. 2 indicates that older browser versions performed at a similar or slightly better level than current versions. It is definitely because browsers are now responsible for much more information processing, displaying animations, or handling more users than was previously the case. Thus, significant demands are placed on these applications, with the result that a trade-off has to be made between performance and a way of adapting the performance to the objectives of sustainable information systems.

**Table 2.** Decision matrix for the latest 10 variants of the Firefox browser.

| $A_i$ | $C_1$ | $C_2$ | $C_3$ | $C_4$ | $C_5$ | $C_6$ | $C_7$ | $C_8$ | $C_9$ | $C_{10}$ | $C_{11}$ |
|---|---|---|---|---|---|---|---|---|---|---|---|
| $A_1$ | 24.68 | 122.64 | 23.89 | 1.17 | 70.29 | 95.84 | 4.72 | 24.16 | 828.44 | 4.94 | 0.18 |
| $A_2$ | 23.31 | 122.71 | 23.43 | 1.15 | 71.16 | 96.01 | 4.84 | 23.05 | 796.99 | 5.30 | 0.30 |
| $A_3$ | 25.90 | 122.82 | 24.47 | 1.16 | 66.54 | 96.37 | 5.50 | 27.17 | 504.68 | 5.84 | 0.17 |
| $A_4$ | 23.19 | 122.88 | 23.08 | 0.98 | 72.97 | 95.97 | 4.77 | 21.54 | 861.00 | 4.47 | 0.09 |
| $A_5$ | 24.79 | 122.76 | 23.84 | 1.04 | 70.46 | 95.85 | 4.71 | 23.98 | 801.22 | 5.12 | 0.18 |
| $A_6$ | 25.48 | 122.82 | 23.93 | 1.17 | 69.64 | 95.88 | 4.79 | 24.65 | 821.50 | 5.85 | 0.26 |
| $A_7$ | 26.84 | 122.90 | 24.82 | 1.23 | 65.18 | 96.35 | 5.41 | 28.60 | 505.73 | 5.84 | 0.16 |
| $A_8$ | 24.84 | 122.94 | 23.70 | 1.16 | 68.90 | 96.53 | 5.34 | 24.65 | 565.11 | 6.01 | 0.44 |
| $A_9$ | 25.57 | 122.62 | 23.94 | 1.18 | 69.38 | 96.29 | 4.72 | 24.92 | 553.81 | 5.43 | 0.32 |
| $A_{10}$ | 26.91 | 123.47 | 24.73 | 1.24 | 64.68 | 96.61 | 5.69 | 28.06 | 503.40 | 6.19 | 0.35 |

To demonstrate in detail how the preference values were calculated and the results obtained, we selected the last ten versions of the Mozilla Firefox browser, and the data describing their characteristics concerning a set of eleven criteria are shown in Table 2. It is noticeable that there are subtle differences in the values describing the different alternatives for specific criteria, which directly translates into a better or worse rating from a sustainability study perspective.

**Table 3.** Preferences and rankings obtained for the latest 10 versions of the Firefox browser.

| $A_i$ | Mean pref | Mean dec | Mean pref | Mean dec |
|---|---|---|---|---|
| $A_1$ | 0.756981 | 0.638435 | 2 | 2 |
| $A_2$ | 0.749842 | 0.609149 | 8 | 9 |
| $A_3$ | 0.750191 | 0.609947 | 7 | 8 |
| $A_4$ | 0.750866 | 0.618392 | 6 | 5 |
| $A_5$ | 0.744806 | 0.581833 | 10 | 10 |
| $A_6$ | 0.755033 | 0.626543 | 3 | 3 |
| $A_7$ | 0.752083 | 0.613332 | 5 | 6 |
| $A_8$ | 0.752877 | 0.621158 | 4 | 4 |
| $A_9$ | 0.759274 | 0.648755 | 1 | 1 |
| $A_{10}$ | 0.749790 | 0.613169 | 9 | 7 |

Table 3 presents the obtained preference values and calculated rankings based on the multi-criteria model identified for all available Firefox browser variants. The calculated results are further broken down into averaged preference values and averaged decision matrices to investigate the effect of the approach on the rankings' consistency. Differences can be seen in the ranking order of the alternatives in positions 7, 8, and 9. The calculated correlation values were 0.9680 for the weighted Spearman correlation coefficient ($r_w$) and 0.9882 for the rank similarity coefficient ($WS$). It indicates the high correlation of the results and shows that both approaches can be used interchangeably without significantly affecting the rankings.

Analyzing the results in practical dimensions shows the superiority of earlier versions of browsers over current solutions. It is directly related to the workload and the number of tasks the browsers performs to ensure the available applications' efficient operation. Increasing demands mean that more resources are needed to maintain the process of continuous execution. However, it is worth noting that despite the need for more resources, development work to improve browser performance has resulted in variants within versions that are more responsive to sustainable conditions and have better power management.

Moreover, the older versions included much less demanding information processing while maintaining similar power consumption parameters. Still, the ratings for one and the other are not significantly different. It shows that the available capabilities of the newer versions of the Firefox browser are managing to perform very well, despite the increased functional workload.

## 5 Conclusions

Energy management and efficiency control are crucial when analyzing the performance of information system solutions. To meet the assumptions of sustainable development, it is necessary to strive for solutions offering high efficiency and relatively low energy consumption. Furthermore, wise planning of the resources used in information systems allows for limiting the harmful influence on the environment and reducing the produced heat or carbon dioxide accompanying the operation of these systems. However, to choose the most rational options, it is necessary to use appropriate tools enabling the analysis of available decision variants.

A multi-criteria analysis to assess the quality of the Mozilla Firefox browser versions showed that older solutions performed slightly better than newer variants. However, the evaluation was influenced by the requirements set for these applications, so it is worth noting that despite the much higher expectations for the performance of web browsers, newer solutions perform well in combining power management and effects. The conducted analysis showed which parameters should be paid attention to achieve more beneficial solutions that meet the assumptions of green information systems.

For further directions, it is worth considering conducting the sensitivity analysis, which could indicate the direction of changes to receive more well-managed solutions. Moreover, it would be meaningful to provide a comprehensive ranking of available browsers and their versions to promote the most rational choice regarding the sustainable goals for information systems.

## References


1. Bączkiewicz, A., Kizielewicz, B., Shekhovtsov, A., Wątróbski, J., Sałabun, W.: Methodical aspects of MCDM based E-commerce recommender system. J. Theor. Appl. Electron. Commer. Res. **16**(6), 2192–2229 (2021)
2. Bączkiewicz, A., Watrobski, J., Sałabun, W.: Towards MCDA based decision support system addressing sustainable assessment, pp. 1–12 (2021)
3. Capehart, B.L., Kennedy, W.J., Turner, W.C.: Guide to energy management: international Version. River Publishers (2020)
4. Chok, N.S.: Pearson's versus Spearman's and Kendall's correlation coefficients for continuous data. Ph.D. thesis, University of Pittsburgh (2010)
5. Devi, P., Kizielewicz, B., Guleria, A., Shekhovtsov, A., Wątróbski, J., Królikowski, T., Więckowski, J., Sałabun, W.: Decision Support in Selecting a Reliable Strategy for Sustainable Urban Transport Based on Laplacian Energy of T-Spherical Fuzzy Graphs. Energies **15**(14), 4970 (2022)
6. Ding, G.K.: Sustainable construction-The role of environmental assessment tools. J. Environ. Manage. **86**(3), 451–464 (2008)
7. Faizi, S., Sałabun, W., Rashid, T., Wątróbski, J., Zafar, S.: Group decision-making for hesitant fuzzy sets based on characteristic objects method. Symmetry **9**(8), 136 (2017)
8. Gholami, R., Watson, R.T., Hasan, H., Molla, A., Bjorn-Andersen, N.: Information systems solutions for environmental sustainability: How can we do more? J. Assoc. Inf. Syst. **17**(8), 2 (2016)
9. Johann, T., Dick, M., Kern, E., Naumann, S.: Sustainable development, sustainable software, and sustainable software engineering: an integrated approach. In: 2011 International Symposium on Humanities, Science and Engineering Research, pp. 34–39. IEEE (2011)
10. Kizielewicz, B., Sałabun, W.: A New Approach to Identifying a Multi-Criteria Decision Model Based on Stochastic Optimization Techniques. Symmetry **12**(9), 1551 (2020)
11. Kizielewicz, B., Shekhovtsov, A., Sałabun, W.: A New Approach to Eliminate Rank Reversal in the MCDA Problems. In: Paszynski, M., Kranzlmüller, D., Krzhizhanovskaya, V.V., Dongarra, J.J., Sloot, P.M.A. (eds.) ICCS 2021. LNCS, vol. 12742, pp. 338–351. Springer, Cham (2021). https://doi.org/10.1007/978-3-030-77961-0_29
12. Kizielewicz, B., Shekhovtsov, A., Sałabun, W.: How to Make Decisions with Uncertainty Using Hesitant Fuzzy Sets? In: International Conference on Intelligent and Fuzzy Systems, pp. 763–771. Springer (2022) https://doi.org/10.1007/978-3-031-09176-6_84
13. Kizielewicz, B., Dobryakova, L.: How to choose the optimal single-track vehicle to move in the city? Electric scooters study case. Procedia Comput. Sci. **176**, 2243–2253 (2020)
14. Kundakcı, N.: An integrated method using MACBETH and EDAS methods for evaluating steam boiler alternatives. J. Multi-Criteria Decis. Anal. **26**(1–2), 27–34 (2019)

15. Kurt, S.: Moving toward a universally accessible web: web accessibility and education. Assistive Technol. **31**, 199–208 (2018)
16. Melville, N.P.: Information systems innovation for environmental sustainability. MIS quarterly, pp. 1–21 (2010)
17. Naumann, S., Dick, M., Kern, E., Johann, T.: The greensoft model: A reference model for green and sustainable software and its engineering. Sustain. Comput. Inf. Syst. **1**(4), 294–304 (2011)
18. Nelson, R., Shukla, A., Smith, C.: Web Browser Forensics in Google Chrome, Mozilla Firefox, and the Tor Browser Bundle. In: Zhang, X., Choo, K.-K.R. (eds.) Digital Forensic Education. SBD, vol. 61, pp. 219–241. Springer, Cham (2020). https://doi.org/10.1007/978-3-030-23547-5_12
19. Oh, J., Lee, S., Lee, S.: Advanced evidence collection and analysis of web browser activity. Digital Invest. **8**, S62–S70 (2011)
20. Oliveira, M.D., Mataloto, I., Kanavos, P.: Multi-criteria decision analysis for health technology assessment: addressing methodological challenges to improve the state of the art. Eur. J. Health Econ. **20**(6), 891–918 (2019). https://doi.org/10.1007/s10198-019-01052-3
21. Peters, N., et al.: Phase-aware web browser power management on HMP platforms. In: Proceedings of the 2018 International Conference on Supercomputing, pp. 274–283 (2018)
22. Raisinghani, M.S., Idemudia, E.C.: Green information systems for sustainability. In: Green business: Concepts, methodologies, tools, and applications, pp. 565–579. IGI Global (2019)
23. Rashid, T., Ali, A., Chu, Y.M.: Hybrid BW-EDAS MCDM methodology for optimal industrial robot selection. PLoS ONE **16**(2), e0246738 (2021)
24. Sałabun, W., Karczmarczyk, A., Wątróbski, J.: Decision-making using the hesitant fuzzy sets COMET method: an empirical study of the electric city buses selection. In: 2018 IEEE Symposium Series on Computational Intelligence (SSCI), pp. 1485–1492. IEEE (2018)
25. Sałabun, W., Piegat, A., Wątróbski, J., Karczmarczyk, A., Jankowski, J.: The COMET method: the first MCDA method completely resistant to rank reversal paradox. Eur. Working Group Ser. 3 (2019)
26. Seidel, s, et al.: The sustainability imperative in information systems research. Commun. Assoc. Inf. Syst. **40**(1), 3 (2017)
27. Shrouf, F., Miragliotta, G.: Energy management based on Internet of Things: practices and framework for adoption in production management. J. Cleaner Prod. **100**, 235–246 (2015)
28. Snyder, P., Ansari, L., Taylor, C., Kanich, C.: Browser feature usage on the modern web. In: Proceedings of the 2016 Internet Measurement Conference, pp. 97–110 (2016)
29. Stewart, T.J., Durbach, I.: Dealing with Uncertainties in MCDA. In: Greco, S., Ehrgott, M., Figueira, J.R. (eds.) Multiple Criteria Decision Analysis. ISORMS, vol. 233, pp. 467–496. Springer, New York (2016). https://doi.org/10.1007/978-1-4939-3094-4_12
30. Torkayesh, S.E., Amiri, A., Iranizad, A., Torkayesh, A.E.: Entropy based EDAS decision making model for neighborhood selection: a case study in Istanbul. J. Ind. Eng. Decis. Making **1**(1), 1–11 (2020)
31. Yee, G., Webster, T.: State of practice of energy management, control, and information systems. Web Based Energy Information and Control Systems: Case Studies and Applications, p. 275 (2021)

32. Zar, J.H.: Spearman rank correlation. Encyclopedia of biostatistics, 7 (2005)
33. Zhang, C., Hindle, A.: A green miner's dataset: mining the impact of software change on energy consumption. In: Proceedings of the 11th working conference on mining software repositories, pp. 400–403 (2014)

# Decision Support System for Sustainable Transport Development

Jakub Więckowski[1], Jarosław Wątróbski[2], Bartosz Paradowski[3], Bartłomiej Kizielewicz[1], Andrii Shekhovtsov[1], and Wojciech Sałabun[1,3](✉)

[1] National Institute of Telecommunications, Szachowa 1, 04-894 Warsaw, Poland
{j.wieckowski,b.kizielewicz,a.shekhovtsov,w.salabun}@il-pib.pl

[2] Institute of Management, University of Szczecin, ul. Cukrowa 8, 71-004 Szczecin, Poland
jaroslaw.watrobski@usz.edu.pl

[3] Research Team on Intelligent Decision Support Systems, Department of Artificial Intelligence and Applied Mathematics, Faculty of Computer Science and Information Technology, West Pomeranian University of Technology in Szczecin, ul. Żołnierska 49, 71-210 Szczecin, Poland
{bartosz-paradowski,wojciech.salabun}@zut.edu.pl

**Abstract.** Information systems prove to be useful tools aiming to support the operation of the user. One of the areas where these systems provide significant support is decision-making. Complex problems are solved effectively with the help of decision support systems. Based on this knowledge, in this paper, we determine the dedicated information system aiming to assess the quality of electric vehicles regarding sustainable development. The importance of reducing negative human impact on the environment makes it necessary to maximize the quality of the decisions made, especially in transport. We used the Interval TOPSIS method, which assures quality assessment. Moreover, the indication of preferential values is provided, which could be used to increase the attractiveness of alternatives. The proposed solution shows that information systems could be effectively used in complex assessments, influencing the quality of the decision variants selected.

**Keywords:** IT systems · Sustainability · Electric vehicles · MCDA

## 1 Introduction

Decision-making is a daily part of everyone's life. However, some problems require us to think more or less carefully to identify the preferred choice [17]. To avoid difficulties and to support the decision-making process, Multi-Criteria Decision Analysis (MCDA) has been developed [15]. They work by evaluating a set of alternatives and calculating their positional ranking. Thanks to the flexibility of operation and the possibility of configuring parameters, they are


M. Tanveer et al. (Eds.): ICONIP 2022, CCIS 1793, pp. 389–397, 2023.
https://doi.org/10.1007/978-981-99-1645-0_32

effectively used in many areas [19,25], and their effectiveness has been verified many times [24,30].

The rankings obtained from MCDA methods can be obtained by evaluating a decision matrix represented as crisp or fuzzy numbers [7]. The first approach has the advantage of simple implementation and ease of interpretation of results. Hence, it is used for problems where all the alternatives' values are known to the decision-maker. However, performing calculations on fuzzy numbers is more related to real problems where there is often little change in the values of input parameters, or not all values are known exactly [16,22].

The evaluation of alternatives using MCDA methods allows creating positional rankings based on preference values. However, slight differences between the alternatives are often noticeable [23]. In addition, these differences are so subtle that a slight change in the values of the input parameters could affect the change in the ranking hierarchy [4]. Therefore, it is worth investigating whether it is possible to determine the magnitude of these changes needed to increase the ranking position of selected alternatives.

One of the available techniques based on interval arithmetic is the Interval Technique for Order Preference by Similarity to an Ideal Solution (TOPSIS) method [9,11]. It has been applied in a wide range of research areas, and its operation has been repeatedly verified [5]. The capabilities offered by the assumptions of interval arithmetic make it possible to study the change of parameter values on the achieved results [31]. Furthermore, it makes it possible to examine a specific range of parameters to determine the influence of modifications applied to the results.

The presented approach could have meaningful usage considering many practical fields [26]. One of them is sustainable development. The growing consciousness of societies causes taking up the more responsible decisions for saving the planet [6]. However, many possible decision variants and changing characteristics make it challenging to indicate the most rational choice unanimously. Information systems come with help [2,32]. Based on the various techniques, they aim to support the decision-maker and provide him the preferential values, which could help him select the alternative that best fits the sustainable development goals [1].

In this paper, we want to address the problem of selecting the most rational choice from the considered set of electric vehicles (EV) regarding their impact on the environment. The popularity of passenger cars makes them the most eagerly used type of transport nowadays. Therefore, choosing the most green-friendly solutions for reducing carbon dioxide emissions is necessary. We want to highlight the significance of the raised problem and show that dedicated information systems could be an effective tool to support the decision-making process. Furthermore, the determination of the system based on the Interval TOPSIS method precisely for the assessment of the electric vehicle allows equipping the decision-maker with additional preferential values, which could affect the attractiveness of regarding solutions.

The rest of the paper is organized as follows. In Sect. 2, we present the main assumptions of the Interval TOPSIS method used as the main technique in proposed system. Then, in Sect. 3, we describe the problem of the assessment of electric vehicles using the multi-criteria approach. Finally, Sect. 4 summarizes the examined research and presents its conclusions.

## 2 Interval TOPSIS Method

One of the most eagerly used extensions of the crisp TOPSIS method is the Interval TOPSIS [21]. It is based on interval arithmetic [13]. The main advantage of this method is the possibility to perform evaluations in an uncertain environment. Moreover, its effectiveness was examined many times in various fields. Therefore, the subsequent steps of the method should be shortly recalled.

**Step 1.** The calculation of the interval lower ($r^L$) and upper ($r^U$) bounds is performed based on formula presented below (1).

$$\begin{aligned} r_{ij}^L &= \frac{x_{ij}^L}{\left(\sum_{k=1}^m\left(\left(x_{kj}^L\right)^2+\left(x_{kj}^U\right)^2\right)\right)^{\frac{1}{2}}}, \quad i=1,\ldots,m, j=1,\ldots,n \\ r_{ij}^U &= \frac{x_{ij}^U}{\left(\sum_{k=1}^m\left(\left(x_{kj}^L\right)^2+\left(x_{kj}^U\right)^2\right)\right)^{\frac{1}{2}},} \quad i=1,\ldots,m_i, j=1,\ldots,n \end{aligned} \tag{1}$$

**Step 2.** The next step is to calculate the weighted values for each interval value (2):

$$\nu_{ij}^L = w_j \times r_{ij}^L, \quad \nu_{ij}^U = w_j \times r_{ij}^U, \quad i=1,\ldots,m; j=1,\ldots,n \tag{2}$$

**Step 3.** Subsequently, the next step is to determine a positive and negative value for the ideal solution as follows (3):

$$\begin{aligned} A^+ &= \left\{v_1^+, v_2^+, \ldots, v_n^+\right\} = \left\{\left(\max_i v_{ij}^U | j \in K_b\right), \left(\min_i v_{ij}^L | j \in K_c\right)\right\} \\ A^- &= \left\{v_1^-, v_2^-, \ldots, v_n^-\right\} = \left\{\left(\min_i v_{ij}^L | j_i \in K_b\right), \left(\max_i v_{ij}^U | j \in K_c\right)\right\} \end{aligned} \tag{3}$$

**Step 4.** The calculation of the distance from the ideal solution is described as (4):

$$\begin{aligned} S_i^+ &= \left\{\sum_{j\in K_b}\left(v_{ij}^L - v_j^+\right)^2 + \sum_{j\in K_c}\left(v_{ij}^U - v_j^+\right)^2\right\}^{\frac{1}{2}}, \quad i=1,\ldots,m \\ S_i^- &= \left\{\sum_{j\in K_b}\left(v_{ij}^U - v_j^-\right)^2 + \sum_{j\in K_c}\left(v_{ij}^L - v_j^-\right)^2\right\}^{\frac{1}{2}}, \quad i=1,\ldots,m \end{aligned} \tag{4}$$

**Step 5.** The final ranking and preferences of the alternatives are determined by (5):

$$RC_i = \frac{S_i^-}{S_i^+ + S_i^-}, \quad i=1,2,\ldots,m, 0 \leqslant RC_i \leqslant 1 \tag{5}$$

## 3 Practical Problem

With the increasing impact of human activity on ecosystems and the growing emission of pollutants, it is essential that the solutions used and technologies developed fit in as well as possible with the assumptions of sustainable development [14]. One of the most significant areas devoted to research on improving functioning and reducing the production of harmful substances is transport [3]. The rising number of passenger cars on urban streets means that the amount of carbon dioxide emitted due to fuel combustion is increasing [10]. Therefore, significant requirements are attached to car manufacturers producing new cars fuelled by oil or petrol to reduce the average production of harmful substances during the use of these vehicles. However, despite meeting these conditions, this is not a sufficient step towards reducing the damaging environmental impact of this area. Furthermore, it was decided to develop quality solutions for electric cars, with the main advantage being producing zero carbon dioxide while driving.

It offers promising opportunities to reduce the amount of carbon dioxide emitted when using cars with combustion engines. Consequently, there has been a definite increase in the popularity of electric cars [33]. It, in turn, has resulted in a more extensive set of alternatives to choose from, making it difficult to know which car might be the most rational decision. Cars have different characteristics, depending on the version or class of vehicle the consumer chooses [29]. Based on this assumption, it is possible to examine the impact of changing the value of selected parameters on their attractiveness. For example, the possibility to negotiate the price of a car and lower the required amount may make it a better choice than the others, even though the initial price suggested that this alternative was less attractive.

The amount of factors translates into difficulties with the analogical comparison of available alternatives by experts. However, the dedicated information systems could be used as the decision support systems to help the decision-maker make the most rational choice [8]. These IT systems could have a more or less significant impact on the environment, depending on the field. Nevertheless, even small changes in the decision-making process directed toward sustainability could positively reduce the pollutants [27]. Providing the information systems that help to make the most rational choices from a considered set of alternatives is valuable in implementing eco-friendly solutions in everyday situations.

In this study, we decided to propose a dedicated information system for assessing the quality of the electric vehicle regarding sustainable performance. Based on the Interval TOPSIS method, we perform a comprehensive analysis aiming to provide preferential values suggesting the desirable characteristics of alternatives regarding specific criteria. We want to address that information systems could be used as decision support systems indicating the green-friendly solutions, which selection will reduce the damaging effect of human activity. Moreover, these systems can also be used as a tool for performing the sensitivity analysis, managing the indications of possible changes in alternative characteristics increasing their quality.

We used the problem of assessing the quality of electric cars, which is a crucial topic due to the growing importance of sustainable transport. Furthermore, the number of offers on the electric car market makes it necessary to make a reliable assessment of them based on several criteria. Based on the factors considered in the literature for the evaluation of electric cars, we selected a set of criteria presented below [12, 18].

- $C_1$ - Price
- $C_2$ - Acceleration
- $C_3$ - Battery range
- $C_4$ - Charging time
- $C_5$ - Fast charging time
- $C_6$ - Maximum speed
- $C_7$ - Average energy use
- $C_8$ - Battery capacity
- $C_9$ - Engine power
- $C_{10}$ - Vehicle weight
- $C_{11}$ - Maximum torque

To determine the autonomous IT system, it was necessary to establish the weights for the criteria. We used the entropy method for this reason. It uses the decision matrix to calculate the importance of each factor [34]. The more significant value is assigned to a particular criterion, the more important it is in the considered problem. In addition, because of the characteristics of the MCDA methods, the types of criteria also had to be defined [28]. We present both of the mentioned factors in Table 1.

**Table 1.** Determined types and weights for criteria set.

| $C_i$ | $C_1$ | $C_2$ | $C_3$ | $C_4$ | $C_5$ | $C_6$ | $C_7$ | $C_8$ | $C_9$ | $C_{10}$ | $C_{11}$ |
|---|---|---|---|---|---|---|---|---|---|---|---|
| Type | Cost | Cost | Profit | Cost | Cost | Profit | Cost | Profit | Profit | Cost | Profit |
| Weight | 0.1152 | 0.0742 | 0.0488 | 0.0792 | 0.1465 | 0.0364 | 0.0243 | 0.1051 | 0.1844 | 0.0494 | 0.1365 |

Based on the determined set of criteria, we collect the data which contributes to the decision matrix in problem of assessing the quality of electric vehicles. It is presented in Table 2. We chose three alternatives, subjected to sensitivity analysis of solutions, namely Skoda Citigo-e iV ($A_1$), Renault Zoe Zen r110 ($A_2$), and Nissan Leaf N-Connecta ($A_3$). The selection of electric cars was dictated by their popularity and the number of cars sold in Poland in 2021 [20]. The values describing each decision variant were assigned based on the data provided by EV manufacturers. The positional ranking of alternatives calculated by crisp TOPSIS method was presented to indicate which alternative proved to be most rational choice regarding the base set of values.

The sensitivity analysis study carried out with the Interval TOPSIS method consisted in using interval arithmetic to change the value of the selected criterion for a single alternative and determining whether this change affects the

**Table 2.** Decision matrix for the electric cars assessment.

| $C_i$ | $C_1$ | $C_2$ | $C_3$ | $C_4$ | $C_5$ | $C_6$ | $C_7$ | $C_8$ | $C_9$ | $C_{10}$ | $C_{11}$ | Rank |
|---|---|---|---|---|---|---|---|---|---|---|---|---|
| $A_1$ | 82500 | 12 | 260 | 240 | 60 | 130 | 16.4 | 36 | 83 | 1265 | 212 | 1 |
| $A_2$ | 145400 | 9.5 | 395 | 565 | 65 | 135 | 13.0 | 52 | 110 | 1502 | 225 | 2 |
| $A_3$ | 161000 | 6.9 | 385 | 690 | 90 | 157 | 17.8 | 62 | 217 | 1727 | 340 | 3 |

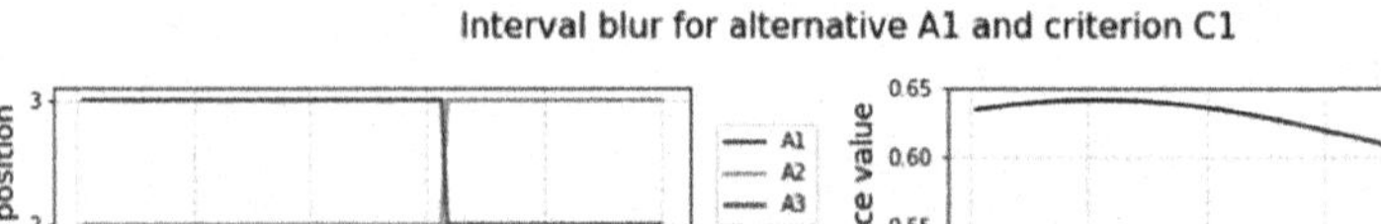
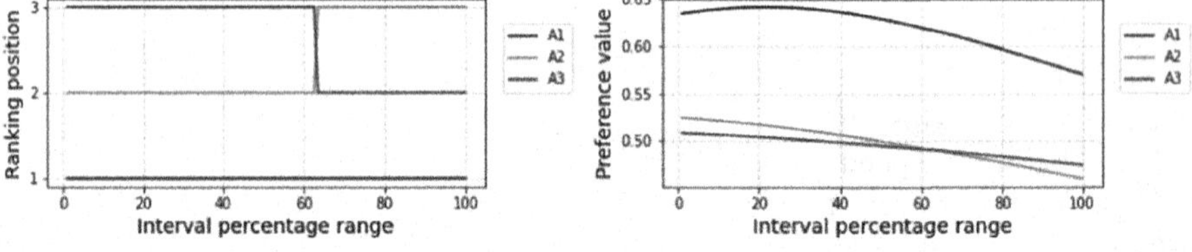


**Fig. 1.** Sensitivity analysis of alternative $A_1$ (Skoda Citigo-e iV) and criterion $C_1$ (price).

calculated preference values and positional ranking. The approach used took into account modifications of the initial values starting from 1% up to 100% of the magnitude of change. Selected results and visualizations of the obtained rankings and preference values are shown in Figs. 1, 2 and 3.

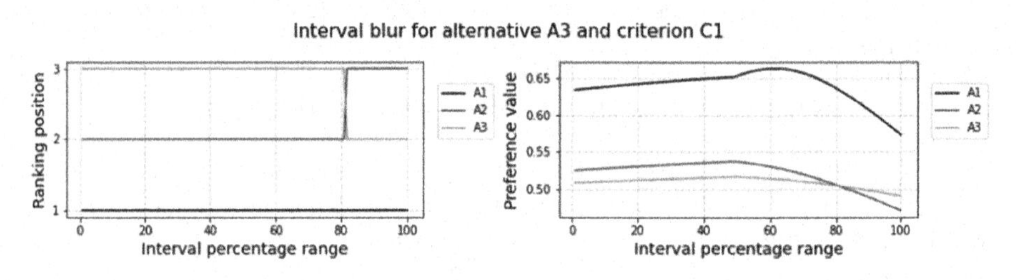


**Fig. 2.** Sensitivity analysis of alternative $A_3$ (Nissan Leaf N-Connecta) and criterion $C_1$ (price).

The results show that slight differences in the evaluation of the alternatives can be compensated by modifying the selected input parameters. The rating runs shown in the figures demonstrate how great a change needs to occur from the initial value for a criterion for the considered alternative to be a more favorable choice compared to the initial rating.

It can be noted that the $A_1$ alternative (Skoda Citigo-e iV) was ranked significantly better than the other two electric cars in each example. In contrast, the change in price and charging time of the $A_3$ alternative (Nissan Leaf N-Connecta) improved its ranking position relative to the reference ranking calculated for the crisp values. It is worth noting that the study showed that not every change in

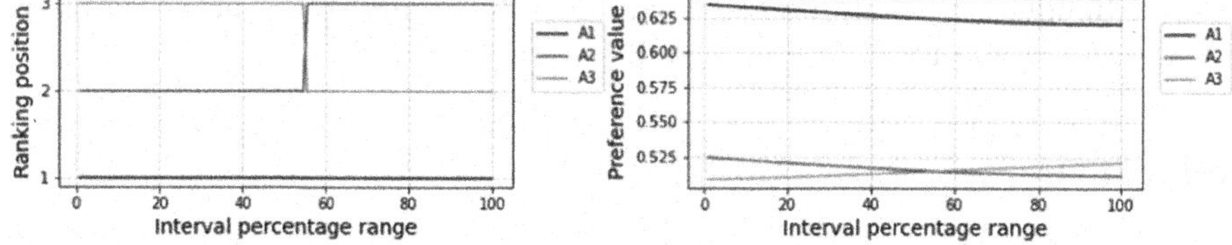


**Fig. 3.** Sensitivity analysis of alternative $A_3$ (Nissan Leaf N-Connecta) and criterion $C_4$ (charge time).

value made using intervals results in a change in ranking. It means that not all parameters are adjustable enough to ensure the greater attractiveness of selected electric cars.

From a practical point of view, it can be concluded from the results indicated that an 80% reduction in the price of the $A_3$ alternative would result in it being placed in 2nd position in the ranking. Reducing the charging time by 55% would have a similar effect. The sensitivity analysis showed that the obtained initial solutions are stable, and only significant modifications of the initial values can reverse the order in the rankings of the selected alternatives.

Furthermore, the provided information system based on the Interval TOPSIS method shows that it could be used as a electric vehicles quality assessment model. Its suggestions indicates the most green-friendly solutions and equip decision-maker in additional data, which shows direction of changes for particular car to raise its attractiveness and sustainable performance.

## 4 Conclusions

MCDA methods provide wide applications due to their flexibility and configurability. So that they could be used as a core of the decision support systems, which contributes to spreading the usability of information systems. It is crucial to develop models aiming to propose green-friendly solutions and reduce damaging human activity.

Sensitivity analysis of solutions is a crucial issue, as it makes it possible to determine the impact of changes in values on results. Its use in sustainable development may be beneficial, as it makes it possible to define the boundary values affecting the advantage of selected solutions. Many information systems lack the mentioned approach, which makes them more limited in practical usage. The proposed model proves to be an effective technique for assessing the quality of the electric vehicle regarding sustainable development. Moreover, it equips the expert with additional data indicating desirable preferential values and strives to increase decision variants' attractiveness.

For further directions, it is worth considering extending the set of alternatives for the comprehensive electric cars benchmark. Moreover, the other MCDA methods could be used to assess the similarity of obtained results regarding different techniques. It would be meaningful for the quality of the proposed information system and increase the reliability of rankings.

## References

1. Allaoui, H., Guo, Y., Sarkis, J.: Decision support for collaboration planning in sustainable supply chains. J. Clean. Prod. **229**, 761–774 (2019)
2. Bączkiewicz, A., Kizielewicz, B., Wątróbski, J., Shekhovtsov, A., Więckowski, J., Salabun, W.: Towards innovative MCDM-based sustainable consumer choices system: automotive evaluation case study. In: 2021 Innovations in Intelligent Systems and Applications Conference (ASYU), pp. 1–6. IEEE (2021)
3. Banister, D., Button, K.: Transport, the environment and sustainable development. Routledge (2015)
4. Baumann, M., Weil, M., Peters, J.F., Chibeles-Martins, N., Moniz, A.B.: A review of multi-criteria decision making approaches for evaluating energy storage systems for grid applications. Renew. Sustain. Energy Rev. **107**, 516–534 (2019)
5. Celik, E., Gul, M., Aydin, N., Gumus, A.T., Guneri, A.F.: A comprehensive review of multi criteria decision making approaches based on interval type-2 fuzzy sets. Knowl.-Based Syst. **85**, 329–341 (2015)
6. Chomsky, N., Pollin, R.: Climate crisis and the global green new deal: the political economy of saving the planet. Verso Books (2020)
7. De Montis, A., De Toro, P., Droste-Franke, B., Omann, I., Stagl, S.: Assessing the quality of different MCDA methods. In: Alternatives for Environmental Valuation, pp. 115–149. Routledge (2004)
8. Dell'Ovo, M., Capolongo, S., Oppio, A.: Combining spatial analysis with MCDA for the siting of healthcare facilities. Land Use Policy **76**, 634–644 (2018)
9. Dymova, L., Sevastjanov, P., Tikhonenko, A.: A direct interval extension of TOPSIS method. Expert Syst. Appl. **40**(12), 4841–4847 (2013)
10. Figueres, C., Schellnhuber, H.J., Whiteman, G., Rockström, J., Hobley, A., Rahmstorf, S.: Three years to safeguard our climate. Nature **546**(7660), 593–595 (2017)
11. Fu, Y., Xiangtianrui, K., Luo, H., Yu, L.: Constructing composite indicators with collective choice and interval-valued TOPSIS: the case of value measure. Soc. Indic. Res. **152**(1), 117–135 (2020)
12. Giansoldati, M., Monte, A., Scorrano, M.: Barriers to the adoption of electric cars: evidence from an Italian survey. Energy Policy **146**, 111812 (2020)
13. Giove, S.: Interval TOPSIS for multicriteria decision making. In: Marinaro, M., Tagliaferri, R. (eds.) WIRN 2002. LNCS, vol. 2486, pp. 56–63. Springer, Heidelberg (2002). https://doi.org/10.1007/3-540-45808-5_5
14. Holmberg, J., Sandbrook, R.: Sustainable development: what is to be done? In: Policies for a Small Planet, pp. 19–38. Routledge (2019)
15. Huang, I.B., Keisler, J., Linkov, I.: Multi-criteria decision analysis in environmental sciences: ten years of applications and trends. Sci. Total Environ. **409**(19), 3578–3594 (2011)
16. Kizielewicz, B., Bączkiewicz, A.: Comparison of Fuzzy TOPSIS, Fuzzy VIKOR, Fuzzy WASPAS and Fuzzy MMOORA methods in the housing selection problem. Procedia Comput. Sci. **192**, 4578–4591 (2021)

17. Kizielewicz, B., Więckowski, J., Paradowski, B., Sałabun, W.: Dealing with non-monotonic criteria in decision-making problems using fuzzy normalization. In: International Conference on Intelligent and Fuzzy Systems, pp. 27–35. Springer, Cham (2022). https://doi.org/10.1007/978-3-031-09173-5_5
18. Kubiczek, J., Hadasik, B.: Segmentation of passenger electric cars market in Poland. World Electric Veh. J. **12**(1), 23 (2021)
19. Marttunen, M., Lienert, J., Belton, V.: Structuring problems for multi-criteria decision analysis in practice: a literature review of method combinations. Eur. J. Oper. Res. **263**(1), 1–17 (2017)
20. OptimalEnergy: najpopularniejsze samochody elektryczne. https://optimalenergy.pl/aktualnosci/samochody-elektryczne/najpopularniejsze-samochody-elektryczne/ (2021). Accessed 25 Sept 2021
21. Palczewski, K., Sałabun, W.: The fuzzy TOPSIS applications in the last decade. Procedia Comput. Sci. **159**, 2294–2303 (2019)
22. Pamucar, D., Ecer, F.: Prioritizing the weights of the evaluation criteria under fuzziness: the fuzzy full consistency method-FUCOM-F. Facta Universitatis. Ser.: Mech. Eng. **18**(3), 419–437 (2020)
23. Paradowski, B., Sałabun, W.: Are the results of MCDA methods reliable? Selection of materials for thermal energy storage. Procedia Comput. Sci. **192**, 1313–1322 (2021)
24. Sałabun, W., Wątróbski, J., Shekhovtsov, A.: Are MCDA methods benchmarkable? A comparative study of TOPSIS, VIKOR, COPRAS, and PROMETHEE II methods. Symmetry **12**(9), 1549 (2020)
25. Sałabun, W., Wickowski, J., Wątróbski, J.: Swimmer Assessment Model (SWAM): expert system supporting sport potential measurement. IEEE Access **10**, 5051–5068 (2022)
26. Saltelli, A., et al.: Why so many published sensitivity analyses are false: a systematic review of sensitivity analysis practices. Environ. Model. Softw. **114**, 29–39 (2019)
27. Tian, H., et al.: Optimizing resource use efficiencies in the food-energy-water nexus for sustainable agriculture: from conceptual model to decision support system. Curr. Opinion Environ. Sustain. **33**, 104–113 (2018)
28. Triantaphyllou, E., Baig, K.: The impact of aggregating benefit and cost criteria in four MCDA methods. IEEE Trans. Eng. Manag. **52**(2), 213–226 (2005)
29. Van Vliet, O.P., Kruithof, T., Turkenburg, W.C., Faaij, A.P.: Techno-economic comparison of series hybrid, plug-in hybrid, fuel cell and regular cars. J. Power Sources **195**(19), 6570–6585 (2010)
30. Velasquez, M., Hester, P.T.: An analysis of multi-criteria decision making methods. Int. J. Oper. Res. **10**(2), 56–66 (2013)
31. Xu, J., Wu, Z., Yu, X., Hu, Q., Dou, X.: An interval arithmetic-based state estimation framework for power distribution networks. IEEE Trans. Ind. Electron. **66**(11), 8509–8520 (2019)
32. Ye, J., Zhan, J., Xu, Z.: A novel decision-making approach based on three-way decisions in fuzzy information systems. Inf. Sci. **541**, 362–390 (2020)
33. Zhao, X., Ke, Y., Zuo, J., Xiong, W., Wu, P.: Evaluation of sustainable transport research in 2000–2019. J. Clean. Prod. **256**, 120404 (2020)
34. Zhu, Y., Tian, D., Yan, F.: Effectiveness of entropy weight method in decision-making. Mathematical Problems in Engineering 2020 (2020)

# Image Anomaly Detection and Localization Using Masked Autoencoder

Xiaohuo Yu, Jiahao Guo, and Lu Wang(✉)

School of Computer Engineering and Science, Shanghai University, Shanghai, China
{xiaohuoyu,g747173965,luwang}@shu.edu.cn

**Abstract.** Generally speaking, abnormal images are distinguished from normal images in terms of content or semantics. Image anomaly detection is the task of identifying anomalous images that deviate from normal images. Reconstruction based methods detect anomaly using the difference between the original image and the reconstructed image. These methods assume that the model will be unable to properly reconstruct anomalous images. But in practice, anomalous regions are often reconstructed well due to the network's generalization ability. Recent methods propose to decrease this effect by turning the generative task to an inpainting problem. By conditioning on the neighborhood of the masked part, small anomalies will not contribute to the reconstrued image. However, it is hard to reconstruct the masked regions when neighborhood exists much anomalous information. We suggest that it should include more useful information of the image when doing inpainting. Inspired by masked autoencoder (MAE), we propose a new anomaly detection method, which called MAE-AD. The architecture of the method can learn global information of the image, and it can avoid being affected by the large anomalous region. We evaluate our method on the MVTec AD dataset, and the results outperform the previous inpainting based approach. In comparison with the methods which use pre-trained models, MAE-AD also has a competitive performance.

**Keywords:** Anomaly detection · Masked autoencoder · Image inpainting

## 1 Introduction

The purpose of anomaly detection and localization in computer vision is to identify anomalous images that different from those seen in normal images and locate anomalous regions. Anomaly detection and localization have wide applications in industrial defect detection [1], medical image analysis [6], security check [21], etc. Due to the problem that anomalies occur rarely and appear in different types, it is often hard to collect and label a large amount of anomalous images [26]. Current approaches of anomaly detection and localization usually try to model the normal data. At the time of prediction, an anomaly score is given


M. Tanveer et al. (Eds.): ICONIP 2022, CCIS 1793, pp. 398–409, 2023.
https://doi.org/10.1007/978-981-99-1645-0_33

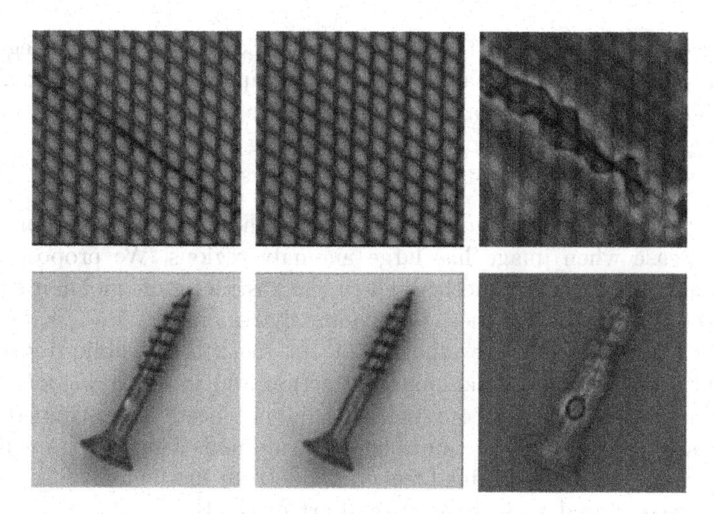

**Fig. 1.** Examples: By inpainting all patches of an input image(left), a full inpainting is obtained(middle). Comparison of original and inpainting yields a pixel-wise anomaly score(right).

to each image that indicates how much it deviates from normal samples. For anomaly localization, a similar score is assigned to each pixel of the image [14].

According to the idea mentioned above, one class of methods use the reconstruction models such as autoencoder to learn the manifold of normal training data. The difference between the input and reconstructed image is used to compute the anomaly scores. As the model is trained on normal data only, it should reconstruct normal images well but not be able to properly reconstruct anomalous images. But because autoencoders sometimes having a high generalization capacity, the anomalies are reconstructed with a high fidelity [13]. This violates the core assumption and makes anomalous regions indistinguishable from anomaly-free regions based on the reconstruction error alone [27].

Recent approaches based on inpainting propose to alleviate this problem by masking image patches and training a model to inpaint them [2,9,13,27]. The reason why the inpainting methods work is that they restrict the network's capability to reconstruct anomalies by hiding anomalous regions. RIAD [27] inpaints masked patches by convolutional neural network(CNN). Due to the limited receptive field of CNN, the masked patches are inpainted only by their immediate neighbourhood. It is hard to inpaint a masked patch to a normal one when the neighborhood includes many anomalous regions, so we need to acquire more useful information while inpainting.

Motivated by the recent success of Vision Transformer(ViT) [5] and masked autoencoder(MAE) [10] in image classification task, we propose a new method based on inpainting for anomaly detection and localization. In the inpainting process, compared to inpainting only relies on the neighbourhood, our method can extract more valid information from the whole image by self-attention.

The masked images are inputted into our network and the masked regions are inpainted to normal using global information. Compared to the previous method, our method can get better inpainting ability due to the usage of more valid information. Figure 1 shows the inpainting result using our method.

Our contribution can be summarized as follows:

- The performance of traditional inpainting based anomaly detection method will decrease when image has large anomaly regions. We propose a MAE based anomaly detection method which can abstract more global information and avoid be affected by these large anomaly regions.
- We did a comprehensive evaluation of our method in public dataset. The method's performance on anomaly detection and localization both surpass previous work. Moreover, our method did not need a pre-trained feature model, so it is more flexible compared to the methods using pre-trained model. Despite not using a pre-trained feature model, our method also has close performance compared with these state-of-art methods.

## 2 Related Work

Recent approaches for anomaly detection can be broadly classified into three categories: one-class classification, probabilistic density estimation and reconstruction models [17]. Deep SVDD [18] and Patch SVDD [25] train a network to form a hypersphere with minimum volume using normal data only. The anomalies are determined based on the distance between the predicted data and the centre of the hypersphere. The probability density based approaches model all normal samples with probability density, generally at the feature level. At the prediction time, the likelihood value of the predicted image in the modelled distribution is used as the anomaly score. [3,8,15,16,26] achieve great performance in anomaly detection. But these methods need to have a great feature extractor like pre-trained ResNet and not easy to transfer to some specific data and tasks. The methods based on autoencoder and generative adversarial networks compute the reconstructed distance between input and reconstructed images, and we do anomaly detection task according to the reconstructed distance, like [7,19,20].

In the reconstruction models such as autoencoder, the anomalies are often reconstructed with a high fidelity due to the networks have a high generalization capacity. RIAD [27] proposes that viewing the reconstruct task as an inpainting problem. The anomalous patches are masked and then inpainted to normal ones can alleviate the problem that occurs in the autoencoder. It uses a U-Net architecture to inpaint the masked patch and gets a great performance in anomaly detection. Due to the limited receptive field of CNN, the restricted information can be used to inpaint masked patches in RIAD, and it vastly influences the effect for anomaly detection. In another similar work, InTra [14] masks a small patch in a large patch and executes self-attention in the large patch, and the masked patch can be inpainted by all patches within the large patch. Regularly slide the large patch in the image and inpaint the masked patch in the large patch, and it can inpaint all patches in the image eventually. The method can

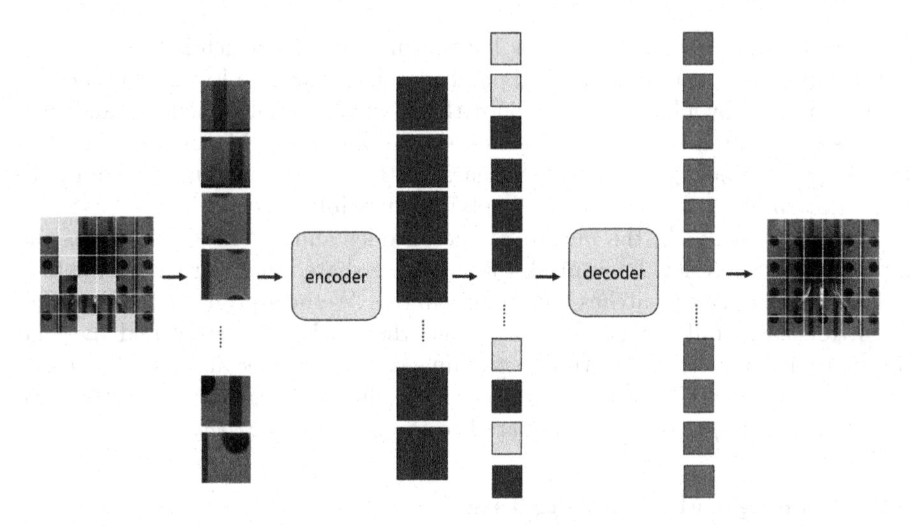


**Fig. 2.** Our method architecture. The encoder is applied to the unmasked patches. Mask tokens are introduced after the encoder, and the full set of encoded patches and mask tokens is processed by a decoder that reconstructs the original images in pixels.

improve inpainting's ability by expanding local information and can also achieve better performance than the aforementioned method.

Transformer [22] and BERT [4] architecture have already achieved success in natural language processing(NLP) and sequence models. ViT [5] migrates Transformer to the image field and gets good performance in image classification and other tasks. Compared to CNN, the transformer can get global information by self-attention. Recently, He [10] proposes masked autoencoder(MAE) in the image classification task. It mainly learns the image feature through the self-supervised task of inpainting image patches like in BERT, and it gets a great performance in the image classification task. The inpainting task in anomaly detection and localization is similar to MAE, so we can attempt to improve the ability of inpainting by the architecture of MAE.

## 3 Approach

Our method is designed based on MAE [10], and it can make full use of global information in inpainting and improve the performance of anomaly detection and localization. A detailed architecture of the method is shown in Fig. 2.

### 3.1 Network Architecture

We divide an image into square non-overlapping patches and then random mask some patches with a ratio within all patches. The remaining unmasked patches will be inputted into the network.

The encoder of our network is a Transformer encoder which is from ViT [5], but it only applied on the unmasked patches [10], as shown in Fig. 2. Our encoder embeds patches by a linear projection with added positional embedding and then processes the patch embedding via a series of Transformer blocks. Compared to the encoder, which processes all patches in the image, our encoder can save computation and memory by only applying on visible patches.

As shown in Fig. 2, the output of the encoder and the mask tokens as the input into the decoder. Each mask token [4] is a shared and learned vector that indicates the masked patches can be inpainted. We add positional embedding to all tokens in full patches set, otherwise the masked tokens would have no information about their location in the image. The decoder also has a series of Transformer blocks. In order to get a high-resolution image for the output, we would have a deeper decoder compared to the decoder in MAE.

### 3.2 Training and Inpainting Target

The network is trained by randomly masking patches on normal images only. Let $x \in \mathbb{R}^{H \times W \times C}$ be an input image, where the parameters of $H, W, C$ denote the height-size, width-size, and number of channels of the image respectively. We set the parameter of $K$ to be the desired side length of a patch. We can describe the image patches as

$$x_p \in \mathbb{R}^{(N \times M) \times (K^2 \cdot C)}, \tag{1}$$

where $N = \frac{H}{K}$, $M = \frac{W}{K}$, and the value of $M \times N$ is the number of patches, each patch has $K^2$ pixels.

In the training phase, we randomly choose some patches $x_p^1, x_p^2, ..., x_p^i, ..., x_p^n$ to mask that the mask ratio is 25%, where $x_p^i \in \mathbb{R}^{K \times K \times C}$ and n is the number of masked patches. In the original work of MAE, the conclusion is that the mask ratio is higher within a certain range, and the performance of the image classification task is better. But in our anomaly detection task, the best performance occurs when the mask ratio is low. A detailed comparison can see Ablation 5 that we discuss the difference in different mask ratio.

The remaining unmasked patches are inputted to our network, and we hope the network can inpaint the masked patches that the inpainted patches $\hat{x}_p$ as similar to the original ones $x_p$ as possible.

In order to better measure the similarity between a patch before masked $x_p^i$ and the inpainted one $\hat{x}_p^i$. Besides the pixel-wise $L_2$ loss between $x_p^i$ and $\hat{x}_p^i$, we also compute other two similarity measures which used in RIAD [27]: structural similarity(SSIM) [23] and gradient magnitude similarity(GMS) [24].

Each loss function is given by

$$\mathcal{L}_2(x_p^i, \hat{x}_p^i) = \frac{1}{K^2} \sum_{(m,n) \in K \times K} (x_p^i - \hat{x}_p^i)^2_{(m,n)}, \tag{2}$$

$$\mathcal{L}_{GMS}(x_p^i, \hat{x}_p^i) = \frac{1}{K^2} \sum_{(m,n) \in K \times K} (1 - GMS(x_p^i, \hat{x}_p^i))_{(m,n)}, \tag{3}$$

$$\mathcal{L}_{SSIM}(x_p^i, \hat{x}_p^i) = \frac{1}{K^2} \sum_{(m,n)\in K\times K} (1 - SSIM(x_p^i, \hat{x}_p^i))_{(m,n)}, \tag{4}$$

where (m,n) is the location of a pixel within a $K \times K$ patch. $GMS(x_p^i, \hat{x}_p^i)$ and $SSIM(x_p^i, \hat{x}_p^i)$ denote the structural similarity and gradient magnitude similarity between $x_p^i$ and $\hat{x}_p^i$.

We can get the final loss function as

$$\mathcal{L}(x_p^i, \hat{x}_p^i) = \mathcal{L}_2(x_p^i, \hat{x}_p^i) + \alpha\mathcal{L}_{GMS}(x_p^i, \hat{x}_p^i) + \beta\mathcal{L}_{SSIM}(x_p^i, \hat{x}_p^i), \tag{5}$$

where $\alpha, \beta$ are individual hyper-parameters.

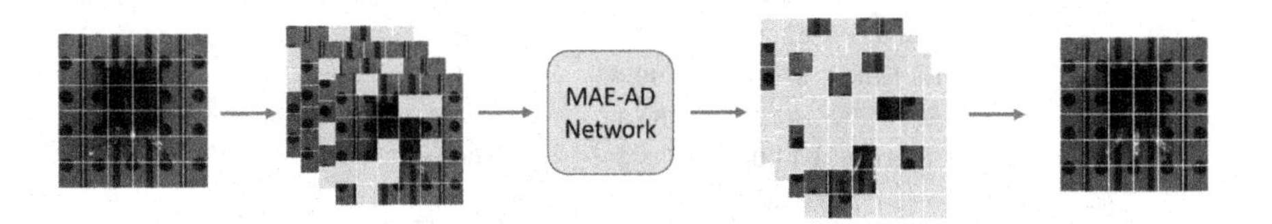


**Fig. 3.** The detailed operation in the testing phase.

### 3.3 Inference and Anomaly Detection

In the testing phase, we can get an inpainted image for each test image. The difference between the inpainted and original image is used to compute a pixel-wise anomaly map.

Let $x \in \mathbb{R}^{H\times W\times C}$ be a test image, and it also has $N \times M$ patches. We divide the full patches set $s$ into $S$ subsets, where the intersection of these subsets is empty and the union of these subsets is $s$. It can be described as $s_1 \cap ... s_j \cap ... \cap s_S = \emptyset$ and $s_1 \cup ... s_j \cup ... \cup s_S = s$.

As shown in Fig. 3, we successively mask the patches in each subset $s_j$ within the original image and input the remaining ones to the network, and we can get the inpainted patches of the subset $\hat{s}_j$. When all patches have been inpainted, we can get an inpainted image $\hat{x}$, where $\hat{x} = \hat{s} = \hat{s}_1 \cup ... \cup \hat{s}_j ... \cup \hat{s}_S$. Then we compute the difference based on GMS between inpainted image $\hat{x}$ and original image x as the pixel-wise anomaly map. The anomaly map can be computed as

$$anomap(x) = blur(\mathbf{1} - GMS(x, \hat{x})), \tag{6}$$

where $\mathbf{1}$ is a $H \times W$ matrix that each element of it is 1. The operation of blur is to smooth the anomaly map by a Gaussian, just similar in [27]. The anomaly map can be used to do the anomaly localization task in each image.

The maximal value in the anomaly map can be used as an anomaly score for anomaly detection. The anomaly score can be formulated as

$$score(x) = max(anomap(x)). \tag{7}$$

**Table 1.** The comparison between different methods on anomaly detection AUC-ROC.

| Category | RIAD [27] | CutPaste [11] | MAE-AD(Ours) |
|---|---|---|---|
| Bottle | **99.9** | 98.3 | 99.8 |
| Capsule | 88.4 | **96.2** | 86.4 |
| Grid | 99.6 | 99.9 | **100.0** |
| Leather | **100.0** | **100.0** | 99.7 |
| Tile | **98.7** | 93.4 | 98.4 |
| Transistor | 90.9 | 95.5 | **97.2** |
| Zipper | 98.1 | 99.4 | **99.8** |
| Cable | 81.9 | 80.6 | **84.7** |
| Carpet | 84.2 | **93.1** | 82.7 |
| Hazelnut | 83.3 | **97.3** | 96.3 |
| Metal Nut | 88.5 | **99.3** | 88.7 |
| Screw | 84.5 | 86.3 | **94.3** |
| Toothbrush | **100.0** | 98.3 | **100.0** |
| Wood | 93.0 | **98.6** | 98.5 |
| $Avg_{tex}$ | 95.1 | **97.0** | 95.9 |
| $Avg_{obj}$ | 89.9 | **94.3** | 93.5 |
| $Avg_{all}$ | 91.7 | **95.2** | 94.3 |

# 4 Experiments

We evaluate our method on the MVTec AD [1] dataset. The MVTec AD is a real-world anomaly detection dataset, it contains 5354 high-resolution color images and includes 10 types of object and 5 types of texture [12]. It has 73 different types of industrial product anomalies in the objects and texture [1]. For each anomalous image, it provides pixel-accurate ground truth regions that allow to evaluate methods for both anomaly detection and localization. It is a good way to evaluate the performance of our method.

The performance of the proposed method and all comparable methods is measured by the area under the receiver operating characteristic curve(AUC-ROC) at image-level and pixel-level [26]. For the anomaly detection task, we need to get an image-level anomaly score for each test image. In the anomaly localization task, the pixel-level anomaly score is used to locate the anomalous areas.

We compare the results of our method to the previous method RIAD [27] and the state-of-the-art method that without using pre-trained models.

## 4.1 Implementation Details

We train the network on normal data only and evaluate the network on the data that include normal and anomalies images. For each category, an independent model can be trained and then evaluate on the testing data. For all images, we resize the image resolution to $224 \times 224$, i.e. $H = 224, W = 224$. In the capsule, carpet, and tile category we set the side length of patch $K = 4$, the

**Table 2.** The comparison between different methods on anomaly localization AUC-ROC.

| Category | RIAD [27] | CutPaste [11] | MAE-AD(Ours) |
|---|---|---|---|
| Bottle | **98.4** | 97.6 | 96.7 |
| Capsule | 92.8 | 97.4 | **97.8** |
| Grid | **98.8** | 97.5 | 98.6 |
| Leather | 99.4 | **99.5** | **99.5** |
| Tile | **95.7** | 90.5 | 95.4 |
| Transistor | 87.7 | 93.0 | **97.5** |
| Zipper | 97.8 | **99.3** | 99.1 |
| Cable | 84.2 | 90.0 | **95.1** |
| Carpet | 96.3 | **98.3** | 96.8 |
| Hazelnut | 96.1 | 97.3 | **98.3** |
| Metal Nut | 92.5 | 93.1 | **95.3** |
| Screw | **98.8** | 96.7 | 98.4 |
| Toothbrush | **98.9** | 98.1 | **98.9** |
| Wood | 93.0 | **95.5** | 91.4 |
| $Avg_{tex}$ | 93.9 | **96.3** | **96.3** |
| $Avg_{obj}$ | 94.3 | 95.8 | **97.4** |
| $Avg_{all}$ | 94.2 | 96.0 | **97.0** |

other categories we set $K = 8$. The mask ratio is 25% in training as we desire to have a high-resolution output. Meanwhile, we set the number of patch subsets $S = 4$ in testing, like Fig. 3 shows. It means the mask ratio in the testing phase is also 25%. The encoder and decoder of the model both have 8 Transformer blocks with 8 attention heads each and a latent of $D = 1024$. In the training phase, the optimizer is AdamW with a learning rate of 0.0001 and a batch size of 4. The hyper-parameters of $\alpha, \beta$ are both set to 0.01, and for each category we trained 600 epochs.

### 4.2 Results and Discussion

The previous method RIAD [27] uses a U-Net architecture and gets a great performance. It confirms that the method based on inpainting is effective for anomaly detection and localization. Compared to RIAD, we extract the global information for the inpainting task instead of neighbour information. For test the improvement of our method, we compare our result to RIAD. We also have a comparison with CutPaste [11]. CutPaste uses a special data segmentation strategy to train a one-class classifier in a self-supervised way [14], and it gets a state-of-the-art performance in anomaly detection. We compare the results without using pre-trained models in CutPaste, which conform to our training procedures.

The results of ROC-AUC on anomaly detection and anomaly localization are shown in Table 1 and Table 2. From Table 1 and Table 2, our results surpass RAID both on detection and localization. Compared to CutPaste, although our method is inferior to it on detection, we can improve the localization result by 1%.

Compared to inpaint anomalous areas only rely on neighbour information, our method can easily inpaint the large anomalous regions to normal with the usage of the global information. The improvement of inpainting ability improves the capacity that detects the anomaly. From Table 1, the results of capsule and carpet are not good. It demonstrates that our method has few effect in some categories, especially in some texture categories. The visual results of all categories are shown in Fig. 4 and Fig. 5.

**Table 3.** The comparison between different mask ratios on anomaly detection and anomaly localization AUC-ROC for our method.

| Category | Mask ratio | | |
|---|---|---|---|
| | 75% | 50% | 25%(Ours) |
| | Det./Loc. | Det./Loc. | Det./Loc. |
| Bottle | 99.0/94.2 | 99.7/95.8 | 99.8/96.7 |
| Capsule | 82.1/96.3 | 83.4/97.1 | 86.4/97.8 |
| Grid | 99.7/98.3 | 99.7/98.5 | 100.0/98.6 |
| Leather | 99.5/99.4 | 99.7/99.4 | 99.7/99.5 |
| Tile | 97.4/90.1 | 98.4/95.1 | 98.4/95.4 |
| Transistor | 93.9/96.9 | 96.0/97.1 | 97.2/97.5 |
| Zipper | 95.8/97.6 | 96.6/98.7 | 99.8/99.1 |
| Cable | 80.0/87.3 | 80.9/91.3 | 84.7/95.1 |
| Carpet | 73.4/93.4 | 74.4/94.8 | 82.7/96.8 |
| Hazelnut | 95.4/98.2 | 96.1/98.3 | 96.398.3 |
| Metal Nut | 73.7/84.9 | 85.6/95.2 | 88.7/95.3 |
| Screw | 76.5/96.1 | 85.4/98.4 | 94.3/98.4 |
| Toothbrush | 85.6/96.8 | 100.0/98.8 | 100.0/98.9 |
| Wood | 94.3/89.6 | 95.6/89.8 | 98.5/91.4 |
| $Avg_{tex}$ | 92.9/94.2 | 93.6/95.5 | **95.9/96.3** |
| $Avg_{obj}$ | 86.3/94.3 | 91.0/96.7 | **93.5/97.4** |
| $Avg_{all}$ | 88.5/94.3 | 91.9/96.3 | **94.3/97.0** |

## 5 Ablation

To test the validity of our ideas, we repeat our experiment on different mask ratio and the results on the mask ratio of 25%, 50%, 75% are shown in Table 3. The encoder can learn the image feature best at the mask ratio of 75% in MAE, and the best performance of image classification task achieved at this the time. But for anomaly detection and localization, we can get the best performance at the mask ratio of 25%.

We can observe from Table 3, that although the mask ratio is as high as 75%, the anomaly localization result of our approach can exceed the result in RIAD.

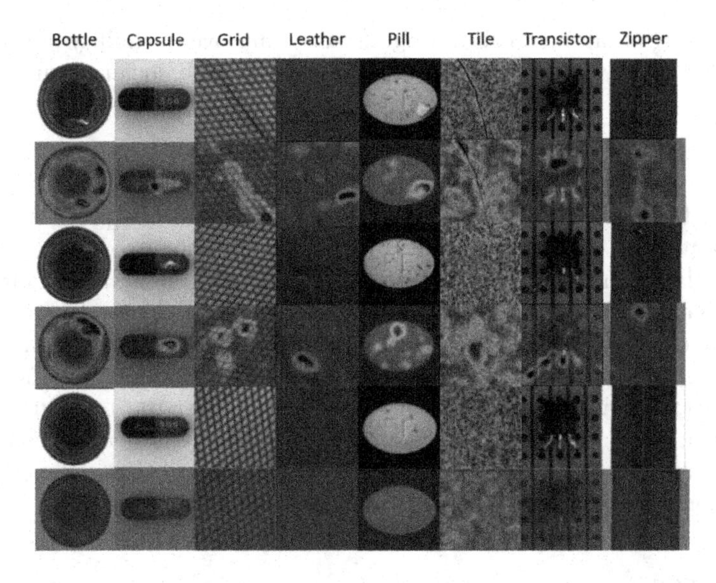


**Fig. 4.** Visual results of MAE-AD on the MVTec AD dataset containing anomalous images (row 1 and 3) and overlaid anomaly maps produced by MAE-AD (row 2 and 4). Row 5 contains non-anomalous images and row 6 contains the corresponding anomaly maps.

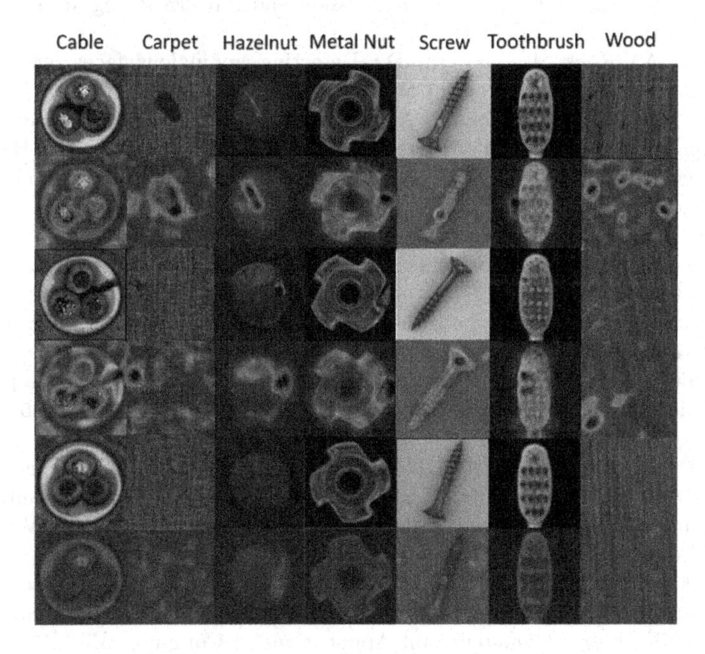


**Fig. 5.** Visual results of MAE AD on the MVTec AD dataset containing anomalous images (row 1 and 3) and overlaid anomaly maps produced by MAE-AD (row 2 and 4). Row 5 contains non-anomalous images and row 6 contains the corresponding anomaly maps.

It demonstrates that our work can significantly improve the ability of inpainting and achieve excellent performance, especially in anomaly localization.

## 6 Conclusion

Inpainting based anomaly detection method can mitigate the problem of traditional reconstruction based method. But it has performance reduction when the image has large anomalous regions. In this paper, we propose a new inpainting based anomaly detection and localization method. This method uses a MAE-like architecture to extract global information, which will avoid be affected by large anomalous regions. With the comprehensive evaluations on the MVTec AD dataset, our method has a great improvement over the previous method. In addition, our method is more flexible compare with other methods which use pre-trained feature model. But the experiment showed that our method had close performance compare with these state-of-the-art methods.

## References

1. Bergmann, P., Fauser, M., Sattlegger, D., Steger, C.: Mvtec ad-a comprehensive real-world dataset for unsupervised anomaly detection. In: Proceedings of the IEEE/CVF Conference on Computer Vision and Pattern Recognition, pp. 9592–9600 (2019)
2. Bhattad, A., Rock, J., Forsyth, D.: Detecting anomalous faces with'no peeking'autoencoders. arXiv preprint arXiv:1802.05798 (2018)
3. Defard, T., Setkov, A., Loesch, A., Audigier, R.: PaDiM: a patch distribution modeling framework for anomaly detection and localization. In: Del Bimbo, A., et al. (eds.) ICPR 2021. LNCS, vol. 12664, pp. 475–489. Springer, Cham (2021). https://doi.org/10.1007/978-3-030-68799-1_35
4. Devlin, J., Chang, M.W., Lee, K., Toutanova, K.: BERT: pre-training of deep bidirectional transformers for language understanding. arXiv preprint arXiv:1810.04805 (2018)
5. Dosovitskiy, A., et al.: An image is worth $16 \times 16$ words: transformers for image recognition at scale. arXiv preprint arXiv:2010.11929 (2020)
6. Fernando, T., Gammulle, H., Denman, S., Sridharan, S., Fookes, C.: Deep learning for medical anomaly detection-a survey. ACM Comput. Surv. (CSUR) **54**(7), 1–37 (2021)
7. Gong, D., Liu, L., Le, V., Saha, B., Mansour, M.R., Venkatesh, S., Hengel, A.V.D.: Memorizing normality to detect anomaly: memory-augmented deep autoencoder for unsupervised anomaly detection. In: Proceedings of the IEEE/CVF International Conference on Computer Vision, pp. 1705–1714 (2019)
8. Gudovskiy, D., Ishizaka, S., Kozuka, K.: Cflow-ad: real-time unsupervised anomaly detection with localization via conditional normalizing flows. In: Proceedings of the IEEE/CVF Winter Conference on Applications of Computer Vision, pp. 98–107 (2022)
9. Haselmann, M., Gruber, D.P., Tabatabai, P.: Anomaly detection using deep learning based image completion. In: 2018 17th IEEE International Conference on Machine Learning and Applications (ICMLA), pp. 1237–1242. IEEE (2018)
10. He, K., Chen, X., Xie, S., Li, Y., Dollár, P., Girshick, R.: Masked autoencoders are scalable vision learners. arXiv preprint arXiv:2111.06377 (2021)

11. Li, C.L., Sohn, K., Yoon, J., Pfister, T.: Cutpaste: self-supervised learning for anomaly detection and localization. In: Proceedings of the IEEE/CVF Conference on Computer Vision and Pattern Recognition, pp. 9664–9674 (2021)
12. Mishra, P., Verk, R., Fornasier, D., Piciarelli, C., Foresti, G.L.: VT-ADL: a vision transformer network for image anomaly detection and localization. In: 2021 IEEE 30th International Symposium on Industrial Electronics (ISIE), pp. 01–06. IEEE (2021)
13. Nguyen, B., Feldman, A., Bethapudi, S., Jennings, A., Willcocks, C.G.: Unsupervised region-based anomaly detection in brain MRI with adversarial image inpainting. In: 2021 IEEE 18th International Symposium on Biomedical Imaging (ISBI), pp. 1127–1131. IEEE (2021)
14. Pirnay, J., Chai, K.: Inpainting transformer for anomaly detection. arXiv preprint arXiv:2104.13897 (2021)
15. Rippel, O., Mertens, P., Merhof, D.: Modeling the distribution of normal data in pre-trained deep features for anomaly detection. In: 2020 25th International Conference on Pattern Recognition (ICPR), pp. 6726–6733. IEEE (2021)
16. Rudolph, M., Wehrbein, T., Rosenhahn, B., Wandt, B.: Fully convolutional cross-scale-flows for image-based defect detection. In: Proceedings of the IEEE/CVF Winter Conference on Applications of Computer Vision, pp. 1088–1097 (2022)
17. Ruff, L., et al.: A unifying review of deep and shallow anomaly detection. In: Proceedings of the IEEE (2021)
18. Ruff, L., et al.: Deep one-class classification. In: International Conference on Machine Learning, pp. 4393–4402. PMLR (2018)
19. Schlegl, T., Seeböck, P., Waldstein, S.M., Langs, G., Schmidt-Erfurth, U.: f-AnoGan: fast unsupervised anomaly detection with generative adversarial networks. Med. Image Anal. **54**, 30–44 (2019)
20. Schlegl, T., Seeböck, P., Waldstein, S.M., Schmidt-Erfurth, U., Langs, G.: Unsupervised anomaly detection with generative adversarial networks to guide marker discovery. In: Niethammer, M., et al. (eds.) IPMI 2017. LNCS, vol. 10265, pp. 146–157. Springer, Cham (2017). https://doi.org/10.1007/978-3-319-59050-9_12
21. Sultani, W., Chen, C., Shah, M.: Real-world anomaly detection in surveillance videos. In: Proceedings of the IEEE Conference on Computer Vision and Pattern Recognition, pp. 6479–6488 (2018)
22. Vaswani, A., et al.: Attention is all you need. In: Advances in Neural Information Processing Systems, vol. 30 (2017)
23. Wang, Z., Bovik, A.C., Sheikh, H.R., Simoncelli, E.P.: Image quality assessment: from error visibility to structural similarity. IEEE Trans. Image Process. **13**(4), 600–612 (2004)
24. Xue, W., Zhang, L., Mou, X., Bovik, A.C.: Gradient magnitude similarity deviation: a highly efficient perceptual image quality index. IEEE Trans. Image Process. **23**(2), 684–695 (2013)
25. Yi, J., Yoon, S.: Patch SVDD: patch-level SVDD for anomaly detection and segmentation. In: Proceedings of the Asian Conference on Computer Vision (2020)
26. Yu, J., et al.: Fastflow: unsupervised anomaly detection and localization via 2D normalizing flows. arXiv preprint arXiv:2111.07677 (2021)
27. Zavrtanik, V., Kristan, M., Skočaj, D.: Reconstruction by inpainting for visual anomaly detection. Pattern Recogn. **112**, 107706 (2021)

# Cross-domain Object Detection Model via Contrastive Learning with Style Transfer

Ming Zhao[1], Xing Wei[1,2,3](✉), Yang Lu[1], Ting Bai[1], Chong Zhao[1,2], Lei Chen[4], and Di Hu[2]

[1] School of Computer and Information, Hefei University of Technology, Hefei, China
zhaomingpc@mail.hfut.edu.cn
[2] Intelligent Manufacturing Technology Research Institute, Hefei University of Technology, Hefei, China
[3] Intelligent Interconnected Systems Laboratory of Anhui Province, Hefei University of Technology, Hefei, China
[4] Institute of Intelligent Machines, HFIPS, Chinese Academy of Sciences, Hefei, China

**Abstract.** Cross-domain object detection usually solves the problem of domain transfer by reducing the difference between the source domain and target domain. However, existing solutions do not effectively solve the performance degradation caused by cross-domain differences. To address this problem, we present the Cross-domain Object Detection Model via Contrastive Learning with Style Transfer(COCS). Our model is based on generating new samples with source domain information and target domain style. In addition, the importance of new samples feature information are aimed to match positive and negative samples for comparative learning better. So, we transfer source domain with labeled to get new samples with style of target domain. Then we employ momentum contrast learning method to maximize the similarly between positive sample pairs representations and minimize the loss function. Moreover, our model can be adapted to different style domains, which further expands the application scenarios. Experiments on a benchmark dataset demonstrate that our model achieves or matches the state-of-the-art approaches.

**Keywords:** Cross-domain Object Detection · Style Transfer · Contrastive learning

## 1 Introduction

Cross-domain adaptive object detection aims to learn feature-relevant representations in the case of domain shifting, where the training data (source domain) is richly labeled with bounding box annotations, and the testing data (target domain) is less or no labels. Due to the difference of feature distribution between source domain and target domain, the trained model has poor universality. Various approaches have been proposed to address the problem of domain


M. Tanveer et al. (Eds.): ICONIP 2022, CCIS 1793, pp. 410–421, 2023.
https://doi.org/10.1007/978-981-99-1645-0_34

shift [4,9,22]. The first class of methods focuses on feature distribution. The second class of methods are pseudo-label based methods and the third class of methods is to use generative models to convert the source domain image into an image similar to the target domain image. However, the existing algorithms have the following disadvantages: on the one hand, they cannot reduce the negative migration caused by non-shared categories in the source domain. On the other hand, it fails to fully promote forward migration and realize effective feature migration in shared category space.

Therefore, we introduce a method for cross-domain object detection. First, we adapt the source domain image to the target domain using a style transfer method. Second, we propose to send the transformed source domain images into a comparative learning task to match the positive and negative samples and label the pseudo-label information. Finally, the pseudo-label samples and target domain samples are used for training.

Our main contributions are as follows:

1. To solve the problem of reducing domain differences, we introduce a novel cross-domain object detection method, the stylization is embedded into contrast learning by constructing an embedded stylization network to minimize contrast loss and the difference between source domain and target domain.
2. The mechanism of momentum contrastive learning method is constructed to make up for the deficiency of feature extraction ability of object detection model and it has higher memory efficient.
3. We use multiple datasets to conduct a series of experiments to evaluate the effect of our domain-adaptive model embedding stylized contrastive learning. Experiments show that our proposed method can improve the detection effect in various scenarios with relative field differences.

## 2 Related Work

**Cross-Domain Object Detection:** Cross-domain object detection usually solves the problem of domain transition by aligning the features or region proposals of source domain and target domain [25]. Alignment is typically achieved through adversarial training, while a detection model aims to fool the classifier. Some difference-based methods explicitly compute the maximum mean difference (MMD) [10] between the source and target domains to align the two domains. Another class of approaches [24] is to iteratively train the model to generate pseudo-bounding box labels for the target image and update the model with the generated pseudo-labels. Our work will revolve around and extend these methods to discuss the possibility of reducing domain differences.

**Style Transfer:** Style transfer is a generalization of texture synthesis that constrains the output to resemble some content image while maintaining the style to a given style image. Early work on style transfer was nonparametric, [5] can only capture low-level style features. With the advent of deep neural networks trained on ImageNet, Gatys et al. [7] introduced CNNs as a new method for texture synthesis. After that, they proposed a neural style transfer algorithm for image formulation, which defined two input images namely content image and style image and output a composite image, preserving the semantic information of the content image. Hertzman et al. [16] use image analogy to transfer texture from an already trained image to a target image and they found that the matrix of deep features represents the artistic style of the image, of which future work has also found applications in a wide variety of tasks.

**Contrastive Representation Learning:** DIM [12] maximizes the mutual information between regions input to the encoder and other outputs, MoCo [11] contrastive learning is a method of constructing a dynamic discrete dictionary on high-dimensional continuous input such as images, while SimCLR [2] does not use a memory bank, which introduces nonlinear transformations between representation and loss function, PIRL [18] learns similar representations for different transformations of images, while SWAV [1] avoids explicit pairwise feature comparisons, choosing to contrast multiple image views by comparing their cluster assignments. In this paper, we use momentum contrast to carry on the stylized samples.

## 3 Proposed Method

Here we present the definition of the domain adaptive task. Firstly, we get labeled samples $D_s = \{(x_s^i, y_s^i)\}_{i=1}^{N_s}$ from the source domain that have undergone style transfer and unlabeled samples $D_t = \{x_t^i\}_{i=1}^{N_t}$ from the target domain, every $\{x_s^i\}$ and $\{x_t^i\}$ belong to the same set of predefined categories $M$. Here $N_s$ and $N_t$ are the numbers of samples from two domains. Then, we use $y_s^i \in \{0, 1, ..., M-1\}$ to denote the labels of M classes of source domain samples but the sample label $y_t^i \in \{0, 1, ..., M-1\}$ of the target domain is unknown at training time. The goal of domain adaptation is to predict labels in the target domain and assuming that the source and target tasks are the same. The constructed model is trained on $D_s \cup D_t$. Figure 1 shows the architecture of our approach.

### 3.1 Stylize Source Domain

Transferring a picture scene style from one image to another can be thought of as a problem of texture transfer. Previous texture transfer algorithms [7] mostly rely on these nonparametric methods for texture synthesis, while using different methods to preserve the structure of the target image. In the style transfer process based on convolutional neural network, firstly, we initialize the synthetic image $I_p \in \mathbb{R}^{3\times H\times W}$ to be the content image. The synthetic image is the only variable that needs to be updated in the style transfer process, that is, the iterative model parameters required for style transfer. Texture synthesis is to generate texture samples from a given style image $I_s \in \mathbb{R}^{3\times H\times W}$, and then minimize the style loss formula Eq. 1 by the pixel value $I_p \in \mathbb{R}^{3\times H\times W}$ of the source domain image. The style loss is defined as the difference between the Gram matrices of the synthetic image $G_p^l = \Phi_l(I_p)^T \Phi_l(I_p)$ and the style image $G_s^l = \Phi_l(I_s)^T \Phi_l(I_s)$. It is defined that $\Phi_l(\cdot)$ represents the output of the

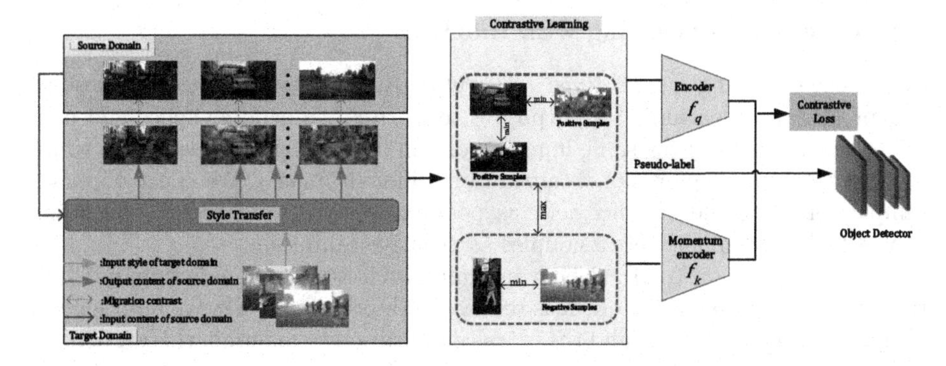


**Fig. 1.** Overview of the Cross-domain Object Detection Model via Contrastive Learning with Style Transfer: (Left part) Style transfer network enables source domain to stylize target domain to form source domain data samples of target domain style. (Right part) Select positive sample pairs that are close to each other and mark pseudo-label information, while negative sample pairs are far away from each other. The positive sample pairs can be composed of samples from the source domain, or can be composed of samples from the source domain and the target domain, which is random. In the acquisition of feature loss methods, momentum contrastive learning is used, and a queue and a moving average encoder are dynamically maintained to achieve a large and consistent dictionary to improve the overall effect of the upstream part of the input.

ImageNet pretrained VGG-19 network at layer $l$, and $\gamma^l$ is a coefficient factor representing a hyperparameter control layer. Gram matrix $G$ features extracted from the VGG-19 network to capture style-related perceptual features such as color, texture, and pattern.

$$\mathcal{L}_{style} = \sum_{l} \gamma^l \|G_s^l - G_p^l\|_2^2 \tag{1}$$

Our goal is to obtain features that are close to the source domain and target domain, it can better compare the learned features and reduce the domain difference. Therefore, in addition to minimizing the style loss Eq. 1, the synthetic image $I_p$ also needs to be calculated to satisfy the characteristics of the source domain image $I_c \in \mathbb{R}^{3\times H\times W}$. The resulting synthetic image is similar in style to $I_s \in \mathbb{R}^{3\times H\times W}$ , while maintaining the structural features of image $I_c$ in content. The content loss formula Eq. 2 is introduced to define the difference between the feature $f_c^l = \Phi_l(I_c)$ of the content image and the feature of the style synthesis image $f_p^l = \Phi_l(I_p)$.

$$\mathcal{L}_{content} = \|f_c^l - f_p^l\|_2^2 \tag{2}$$

The total loss for style transfer is a linear combination between content and style losses. Through the style transfer network, we generate the source domain dataset with the target domain style, where $\lambda_{content}$ and $\lambda_{style}$ respectively control the trade-off of the two terms. Experiments shows that style total loss is optimized when $\lambda_{content} = 0.6$ and $\lambda_{style} = 0.4$.

$$\mathcal{L}_{neural} = \lambda_{content}\mathcal{L}_{content} + \lambda_{style}\mathcal{L}_{style} \tag{3}$$

### 3.2 Contrastive Learning with InfoNCE

Contrastive learning has achieved relatively advanced performance and results in representation learning. Its purpose is to learn an embedding space and pair the enhanced parts of the same image to obtain a pair of positive samples, while the negative sample pairs are obtained from different images. We select $N$ image samples of the same enhancement as positive samples, and regard the other $2(N-1)$ image enhancement samples as negative samples.

According to MoCo [11], the current min-batch is put into a dictionary while the oldest min-batch in the queue is removed. The dictionary always represents a sampled subset of all data, and the extra computation to maintain this dictionary is manageable.

An encoder $f_q$ with parameter $\theta_q$, and a momentum update encoder $f_k$ with parameter $\theta_k$ are used for feature transformation. Generally solution is to copy key encoder $\theta_k$ from query encoder $\theta_q$. $\theta_k$ is provided by $\theta_k \leftarrow m\theta_k + (1-m)\theta_q$ updates, where $m \in [0, 1)$ is the momentum coefficient, and is only updated when the parameter $\theta_q$ is back-propagated. The momentum update of the formula makes $\theta_k$ evolve more smoothly than $\theta_q$, although the keys in the queue are encoded by different encoders (in different mini-batches), the differences between these encoders are generally small. The contrastive loss function InfoNCE [19] is used as an unsupervised objective function to train the encoder network representing the query and key:

$$\mathcal{L}_{NCE} = -log\frac{exp(q \cdot k^{+}/\tau)}{exp(q \cdot k^{+}/\tau) + \sum\limits_{k^{-}} exp(q \cdot k^{-}/\tau)} \tag{4}$$

where $q$ is the source domain sample, $k^{+}$ is the positive sample, $k^{-}$ is the negative sample, and $\tau$ is the hyperparameter, known as the temperature coefficient. In this article, we use a variant of momentum contrast. For feature pairs of InfoNCE loss, there is an encoding queue $w_q = f_q(x)$ and a key encoding queue $w_k = f_k(x)$, $w_q$ is an informative feature from the source domain, $w_k^{+}$ is a feature with the same label as $w_q$, and $w_k^{-}$ is a set of features with a different label from $w_q$.

$$\mathcal{L}_c = -log\frac{exp(w_q \cdot w_k^{+}/\tau)}{exp(w_q \cdot w_k^{+}/\tau) + \sum_{j=1}^{N-1} exp(w_q \cdot w_k^{-}/\tau)} \tag{5}$$

### 3.3 Algorithm

Algorithm 1 represents the process of entering an embedding stylized contrastive learning architecture from an input image. First, we stylize the source domain dataset $D_s$ into a dataset with the style of the target domain and denote it as $D_t$. Then the obtained images are fed to a comparative learning framework, where the image enhancements used by the contrastive learning method are cropping, horizontal flipping, color distortion, Gaussian blurring, and grayscale. Finally, the momentum encoder $f_k$ is updated.

```
Algorithm 1: Pseudo code of COCS
Input:Source Data:S = {(x_1^s, y_1^s), ..., (x_N^s, y_N^s)}
      Target Data:T = {x_1^t, ..., x_N^t}
      Max epoch E, iterations per epoch K, encoder parameters f_q = f_k and parameter of I_p
Output:parameters of encoder f_q and momentum encoder f_k
Procedure:
1 for sampled minibatch do
2     Calculate the style parameters of I_p
3     Calculate the content parameters of I_p
4     Calculate L_neural using Eq. 3 and update parameters of stylized network
5 end for
6 #put target samples and stylized samples into momentum contrastive learning
7 for sampled minibatch do
8     Make two augmentations per samples
9     Update encoder f_q
10    Update encoder f_k = m · f_k + (1 − m) · f_q
11    Calculate L_c using Eq. 5
12 end for
```

# 4 Experiments

In this section, we conduct experiments on multiple domain adaptation benchmarks to verify the effectiveness of our method. SSD with the VGG16 backbone pre-trained on the ImageNet is employed as the base detector. We evaluate our method on six public datasets and select labeled images from the source domain based on previous work.

## 4.1 Datasets and Experiments Settings

Cityscapes dataset [4] - Cityscapes dataset with 5000 images of driving scenes in urban environments (2975 train, 500 val, 1525 test). The Foggy Cityscapes dataset [22] is rendered using image and depth features from Cityscapes and is ideal for studying domain shifts caused by weather conditions. KITTI [8] is another real-world dataset containing 7481 images of real traffic conditions. SIM10K [15] is a collection of synthetic images that contains 1000 images and their corresponding bounding box annotations. The Watercolor [14] dataset contains six categories of watercolor-style artistic images. PASCAL VOC [6] is a dataset containing 20 classes of real images.

We train the network using mini-batch stochastic gradient descent (SGD) with momentum of 0.9 and weight decay of 0.0005. We fine-tune the network with 50 iterations each, and each batch consists of two images, one from the source domain and one from the target domain. We follow the same learning rate time, i.e. the initial learning rate $\eta_p = \eta_0$ is adjusted accordingly according to formula $\eta_p = \frac{\eta_0}{(1+ap)^b}$. $\eta_p$ increases linearly from 0 to 1. The initial learning rate is, which is 0.001 for the convolutional layer. We use a 12GB GTX TITAN Xp for experiments.

### 4.2 Use Cityscapes Datasets Adapt to Foggy Cityscapes Datasets

The biggest beneficiary group from the popularity of object detection lies in traffic scenarios. When weather conditions change, visual data is also affected to some extent. When we train the dataset, usually not all weather conditions can be collected for training, so the model must adapt to different weather conditions. Here, we evaluate our method and demonstrate its superiority over some current methods.

**Table 1.** Using the Cityscape dataset and the Foggy Cityscape dataset, the performance is evaluated using the mean precision (mAP) of 8 classes. We recorded average precision on the Foggy Cityscapes dataset. The last row shows the performance of the base detector when labeled data for the target domain is available.

| Cityscapes → Foggy Cityscapes | | | | | | | | | |
|---|---|---|---|---|---|---|---|---|---|
| Method | person | rider | car | truck | bus | train | motorcycle | bicycle | mAP |
| Source-only | 29.7 | 32.2 | 44.6 | 16.2 | 27.0 | 9.1 | 20.7 | 29.7 | 26.2 |
| Faster RCNN [20] | 23.3 | 29.4 | 36.9 | 7.1 | 17.9 | 2.4 | 13.9 | 25.7 | 19.6 |
| FRCNN in the wild [3] | 25.0 | 31.0 | 40.5 | 22.1 | 35.3 | 20.2 | 20.0 | 27.1 | 27.6 |
| Selective Align [25] | 33.5 | 38.0 | 48.5 | 26.5 | 39.0 | 23.3 | 28.0 | 33.6 | 33.8 |
| Progress Domain [13] | 36.0 | 45.5 | 54.4 | 24.3 | 44.1 | 25.8 | 29.1 | 35.9 | 36.9 |
| Ours | 37.2 | 44.6 | 56.5 | 24.0 | 47.7 | 23.9 | 31.6 | 35.2 | 37.6 |
| Oracle | 37.8 | 48.4 | 58.8 | 25.2 | 53.3 | 15.8 | 35.4 | 39.0 | 39.2 |

Table 1 compares our method with baselines for multi-label domain adaptation. In this experiment, the classes included are, person, rider, car, truck, bus, train, motorcycle, bicycle. The mean precision for each class and the mean precision (mAP) for all subjects are recorded in the table. Our method improves the mAP of Faster RCNN by +18% compared to the state-of-the-art method by +0.7%.

### 4.3 Domain Transfer Using Other Datasets

We conduct several sets of experiments, training from synthetic data and real data, respectively, to evaluate the performance of the model. In the first two sets of experiments, we only train on cars with sticky notes. Table 2 investigates the ability to transfer from a synthetic dataset to a real dataset and compares the average accuracy of the car class, which improves by 1.68% on the same baseline. Table 3 investigates the ability to adapt from one real dataset to another and compares the average precision for the car class. The latter set of experiments, Table 4 studies the transfer of different styles of scenes and adapts to the data under different styles.

**Table 2.** The performance is evaluated using the average precision of the car class between the SIM10K and Cityscapes domains.

| SIM10K → Cityscapes | |
|---|---|
| Method | AP |
| Faster R-CNN [20] | 31.08 |
| FRCNN in the wild [3] | 35.47 |
| Progressive Domain [13] | 42.56 |
| Ours | 44.24 |
| Oracle | 52.19 |

**Table 3.** The performance is evaluated using the average precision of the car class between the two domains, KITTI and Cityscapes.

| KITTI → Cityscapes | |
|---|---|
| Method | AP |
| Faster R-CNN [20] | 28.8 |
| FRCNN in the wild [3] | 35.8 |
| Progressive Domain [13] | 43.9 |
| Ours | 46.5 |
| Oracle | 55.8 |

**Table 4.** Use PASCAL VOC dataset and Watercolor dataset to adapt data under different styles respectively. The performance is evaluated using the mean precision (mAP) of the 6 classes.

| PASCAL VOC → Watercolor | | | | | | | |
|---|---|---|---|---|---|---|---|
| Method | person | bike | car | bird | cat | dog | mAP |
| Source-only | 53.7 | 83.7 | 35.6 | 44.8 | 37.0 | 29.5 | 47.4 |
| SWDA [21] | 60.0 | 75.2 | 48.0 | 40.6 | 31.5 | 20.6 | 46.0 |
| SOAP [23] | 55.4 | 77.7 | 40.1 | 43.2 | 48.2 | 38.8 | 50.6 |
| Ours | 62.1 | 79.4 | 50.2 | 41.6 | 43.1 | 35.6 | 52.0 |

### 4.4 Ablation Experiments

In this subsection, to verify the effectiveness of the method proposed, we conduct ablation experiments based on the Cityscapes→Foggy Cityscapes experiments to demonstrate the effectiveness of different parts of the components for the entire model, and discuss them.The results are shown in Table 5. All experiments are based on the adaptation of Cityscapes→Foggy Cityscapes. Hyperparameter settings for all model variants remain as described in Sect. 4.1. As shown in the table, adding the style transfer module to the only object detection SSD model can bring a significant performance gain of 7.2%, increasing the AP from 37.9% to 45.1%. This shows that style transfer can effectively reduce the differences between domains, thereby improving the domain adaptability. Adding the self-supervised contrastive learning method to the only object detection SSD model can also bring about a significant performance gain of 11.3%, proving the effectiveness of contrastive learning for accuracy improvement. However, it is worth noting that when adding both the style transfer module and the contrastive learning method, our entire model achieves an average accuracy of 56.5%. In general, compared with a single module, both the style transfer module and the contrastive learning module make their own contributions, and the combination of the two improves the effect better.

**Fig. 2.** Detection results from our three domain adaptation tasks. The first column is Cityscapes→Foggy Cityscapes, the second column is SIM10K→Cityscapes, and the last column is KITTI→Cityscapes. We show the detection results of different detection methods in the target domain.

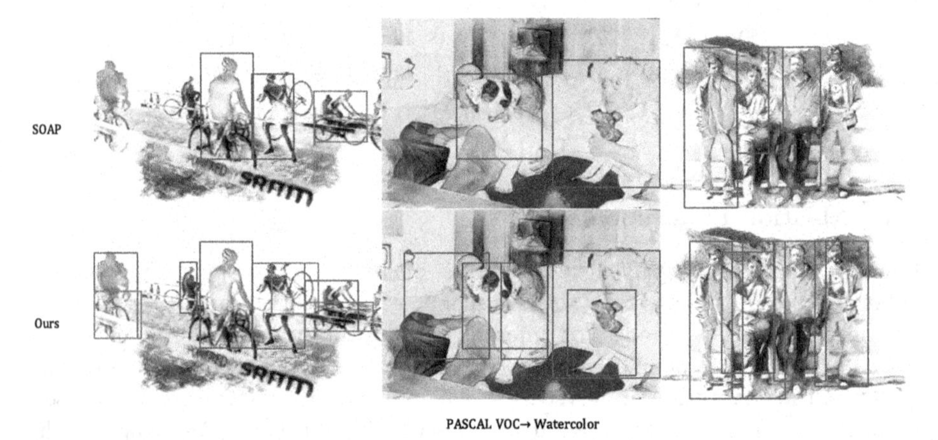


**Fig. 3.** Visualization of detection results. The first row is the detection result of the SOAP model, and the second row is the detection result of our model.

**Table 5.** Average precision (AP) on the car class measured with variants of different component methods on the Cityscapes→Foggy Cityscapes datasets to verify the effectiveness of the proposed algorithm. SN stands for Style Transfer Network and CL stands for Contrastive Learning.

| SN | CL | SSD | AP on car |
|---|---|---|---|
| | | √ | 37.9 |
| √ | | √ | 45.1 |
| | √ | √ | 49.2 |
| √ | √ | √ | 56.5 |

### 4.5 Qualitative Analysis

Figure 2 shows the qualitative results of cross-domain target detection on three different datasets, which are mainly experiments conducted under specific weather conditions such as light intensity in different scenes, rainy and foggy days. In experiments, our method mostly succeeds in correcting the size and position of bounding boxes, and detects vehicles and pedestrians missed by Faster RCNN and Progress Domain. Although the probability of detection in the small target area is not high, further research is needed in the small target area, but in general, the proposed framework has improved the accuracy of target detection to a certain extent, confirming the feasibility of our method.

Figure 3 shows the qualitative results of two different styles of cross-domain target detection. Compared with SOAP, our method can still achieve high accuracy in different styles. More experiments with different styles will be conducted in the future to verify our overall effect.

T-SNE visualization [17]. As shown, we show the category feature distribution from Cityscapes→Foggy Cityscapes, we randomly select a certain number of object features for each category from the source domain (Cityscapes) and the target domain (Foggy Cityscapes), respectively. It can be seen that with our method, the classification distributions of the source and target domains can be well aligned (Fig. 4).

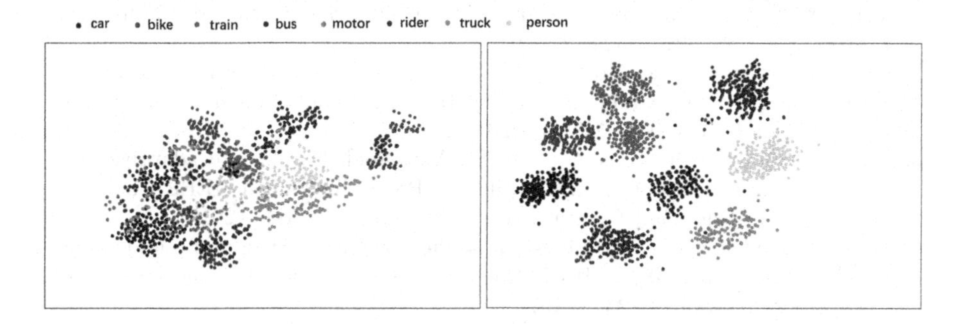


**Fig. 4.** T-SNE visualization of object features from different classes in the dataset. We denote different categories with circles of different colors. The feature distribution on the left is the result obtained with the source domain model only, and the feature distribution on the right is the result obtained with our proposed method.

## 5 Conclusion

In this paper, we propose a style-embedded contrastive learning to align source and target domains to solve the domain-adaptive object detection problem. Specifically, we explore the possibility of optimizing the model upstream of object detection. By optimizing the input image of object detection, the style information with target domain features is transferred to the source domain to form a new dataset. This dataset can be viewed as an enhancement of the source domain dataset and further reduces the differences between the two domains. As a kind of tool, contrastive learning connects the data work from style transfer. This part focuses on the data structure and information correlation of pictures, so that pictures with similar features are close to each other, and pictures with dissimilar features are kept away from each other, so as to maximize the similarity between identical images. Finally, we show the experimental procedures and ablation experiments, which achieve relatively advanced results in a large number of datasets and settings.

**Acknowledgements.** This work was supported by Joint Fund of Natural Science Foundation of Anhui Province in 2020 (2008085UD08), Anhui Provincial Key R&D Program (202004a05020004), Open fund of Intelligent Interconnected Systems Laboratory of Anhui Province (PA2021AKSK0107), Intelligent Networking and New Energy Vehicle Special Project of Intelligent Manufacturing Institute of HFUT (IMIWL2019003, IMIDC2019002).

## References

1. Caron, M., Misra, I., Mairal, J., Goyal, P., Bojanowski, P., Joulin, A.: Unsupervised learning of visual features by contrasting cluster assignments. Adv. Neural. Inf. Process. Syst. **33**, 9912–9924 (2020)
2. Chen, T., Kornblith, S., Norouzi, M., Hinton, G.: A simple framework for contrastive learning of visual representations. In: International Conference on Machine Learning, pp. 1597–1607. PMLR (2020)
3. Chen, Y., Li, W., Sakaridis, C., Dai, D., Van Gool, L.: Domain adaptive faster R-CNN for object detection in the wild. In: Proceedings of the IEEE Conference on Computer Vision and Pattern Recognition, pp. 3339–3348 (2018)
4. Cordts, M., et al.: The cityscapes dataset for semantic urban scene understanding. In: Proceedings of the IEEE Conference on Computer Vision and Pattern Recognition, pp. 3213–3223 (2016)
5. Efros, A.A., Freeman, W.T.: Image quilting for texture synthesis and transfer. In: Proceedings of the 28th Annual Conference on Computer Graphics and Interactive Techniques, pp. 341–346 (2001)
6. Everingham, M., Van Gool, L., Williams, C.K., Winn, J., Zisserman, A.: The pascal visual object classes (voc) challenge. Int. J. Comput. Vision **88**(2), 303–338 (2010)
7. Gatys, L.A., Ecker, A.S., Bethge, M.: Image style transfer using convolutional neural networks. In: Proceedings of the IEEE Conference on Computer Vision and Pattern Recognition, pp. 2414–2423 (2016)
8. Geiger, A., Lenz, P., Stiller, C., Urtasun, R.: Vision meets robotics: The KITTI dataset. Int. J. Robot. Res. **32**(11), 1231–1237 (2013)
9. Girshick, R.: Fast R-CNN. In: Proceedings of the IEEE International Conference on Computer Vision, pp. 1440–1448 (2015)
10. Gretton, A., Borgwardt, K.M., Rasch, M.J., Schölkopf, B., Smola, A.: A kernel two-sample test. J. Mach. Learn. Res. **13**(1), 723–773 (2012)

11. He, K., Fan, H., Wu, Y., Xie, S., Girshick, R.: Momentum contrast for unsupervised visual representation learning. In: Proceedings of the IEEE/CVF Conference On Computer Vision And Pattern Recognition, pp. 9729–9738 (2020)
12. Hjelm, R.D., et al.: Learning deep representations by mutual information estimation and maximization. arXiv preprint arXiv:1808.06670 (2018)
13. Hsu, H.K., et al.: Progressive domain adaptation for object detection. In: Proceedings of the IEEE/CVF Winter Conference on Applications of Computer Vision, pp. 749–757 (2020)
14. Inoue, N., Furuta, R., Yamasaki, T., Aizawa, K.: Cross-domain weakly-supervised object detection through progressive domain adaptation. In: Proceedings of the IEEE Conference on Computer Vision and Pattern Recognition, pp. 5001–5009 (2018)
15. Johnson-Roberson, M., Barto, C., Mehta, R., Sridhar, S.N., Rosaen, K., Vasudevan, R.: Driving in the matrix: can virtual worlds replace human-generated annotations for real world tasks? arXiv preprint arXiv:1610.01983 (2016)
16. Liu, Z., Qi, X., Torr, P.H.: Global texture enhancement for fake face detection in the wild. In: Proceedings of the IEEE/CVF Conference on Computer Vision and Pattern Recognition, pp. 8060–8069 (2020)
17. Van der Maaten, L., Hinton, G.: Visualizing data using t-SNE. J. Mach. Learn. Res. **9**(11), 2579–2605 (2008)
18. Misra, I., van der Maaten, L.: Self-supervised learning of pretext-invariant representations. In: Proceedings of the IEEE/CVF Conference on Computer Vision and Pattern Recognition, pp. 6707–6717 (2020)
19. Van den Oord, A., Li, Y., Vinyals, O., et al.: Representation learning with contrastive predictive coding. arXiv preprint arXiv:1807.03748 (2018)
20. Ren, S., He, K., Girshick, R., Sun, J.: Faster R-CNN: towards real-time object detection with region proposal networks. In: Advances in Neural Information Processing Systems 28 (2015)
21. Saito, K., Ushiku, Y., Harada, T., Saenko, K.: Strong-weak distribution alignment for adaptive object detection. In: Proceedings of the IEEE/CVF Conference on Computer Vision and Pattern Recognition, pp. 6956–6965 (2019)
22. Sakaridis, C., Dai, D., Van Gool, L.: Semantic foggy scene understanding with synthetic data. Int. J. Comput. Vision **126**(9), 973–992 (2018)
23. Xiong, L., Ye, M., Zhang, D., Gan, Y., Li, X., Zhu, Y.: Source data-free domain adaptation of object detector through domain-specific perturbation. Int. J. Intell. Syst. **36**(8), 3746–3766 (2021)
24. Zhu, J.Y., Park, T., Isola, P., Efros, A.A.: Unpaired image-to-image translation using cycle-consistent adversarial networks. In: Proceedings of the IEEE International Conference on Computer Vision, pp. 2223–2232 (2017)
25. Zhu, X., Pang, J., Yang, C., Shi, J., Lin, D.: Adapting object detectors via selective cross-domain alignment. In: Proceedings of the IEEE/CVF Conference on Computer Vision and Pattern Recognition, pp. 687–696 (2019)

# A Spatio-Temporal Event Data Augmentation Method for Dynamic Vision Sensor

Xun Xiao, Xiaofan Chen, Ziyang Kang, Shasha Guo, and Lei Wang(✉)

College of Computer Science and Technology, National University of Defense Technology, Changsha 410073, China
{xiaoxun520,chenxiaofan19,kangziyang14,guoshasha13,leiwang}@nudt.edu.cn

**Abstract.** The advantages of Dynamic Vision Sensor (DVS) camera and Spiking Neuron Networks (SNNs) have attracted much attention in the field of computer vision. However, just as many deep learning models, SNNs also have the problem of overfitting. Especially on DVS datasets, this situation is more severe because DVS datasets are usually smaller than traditional datasets. This paper proposes a data augmentation method for event camera, which augments asynchronous events through random translation and time scaling. The proposed method can effectively improve the diversity of event datasets and thus enhance the generalization ability of models. We use a Liquid State Machine (LSM) model to evaluate our method on two DVS datasets recorded in real scenes, namely DVS128 Gesture Dataset and SL-Animals-DVS Dataset. The experimental results show that our proposed method improves baseline accuracy without any augmentation by 3.99% and 7.3%, respectively.

**Keywords:** Data Augmentation · Dynamic Vision Sensor · Liquid State Machine

## 1 Introduction

Dynamic Vision Sensor (DVS) is one of the most widely applied neuromorphic sensors [8,17]. Compared with traditional cameras, event cameras have the advantages of high time resolution, low power consumption, less motion blur, and low data redundancy [9]. These advantages make DVS very suitable for computer vision tasks. Events recorded by an event camera are organized in the form of $[x, y, t, p]$. However, for such unique format of event data from DVS [2], the existing computer vision algorithm can not process the event sequences directly. Spiking Neural Networks (SNNs) are artificial neural networks more similar to biological neural systems. [15]. It processes data by simulating biological neurons to receive and emit spikes. Unsurprisingly, SNNs perform very well on neuromorphic sensors because of their event-driven characteristics.

Deep learning models can abstract high-level semantic information from data through training with a large amount of samples. Great potential has been seen for deep learning techniques in multiple fields, including image recognition, data mining, natural language processing, and so on [6,19]. However, due to the


M. Tanveer et al. (Eds.): ICONIP 2022, CCIS 1793, pp. 422–433, 2023.
https://doi.org/10.1007/978-981-99-1645-0_35

lack of training data, overfitting is still a challenging problem in deep learning, whose models perform excellently just on the training data. SNN is no exception. Increasing the number of sample is a common method to solve overfitting. However, it is not easy to implement on DVS datasets. Different from traditional image datasets, such as ImageNet [5], a large number of samples can be easily obtained from the Internet. Since the event camera is not widely applied in our daily life yet, a DVS dataset with high quality is relatively difficult to acquire. The sample size of DVS dataset is relatively smaller, and the overfitting problem is severer [1]. Data augmentation is effective in alleviating the overfitting problems of deep learning models [22] by mainly transforming existing samples. More robust model generalization can be obtained with such augmented data [7,27–29]. For image data, augmentation methods being commonly used include translation, rotation, flipping, clipping, contrast, sharpness, clipping, and so on [4]. For event data, which have fundamentally different formats compared with RGB images, data augmentation technology has not been widely used.

In this paper, we propose a data augmentation method for event data. First, we augment the spatial position of event data through random translation, since event data still contain abundant information of relative position. Second, we scale the timestamp of events to increase the data diversity in the time dimension. Furthermore, we evaluate our method on the DVS128 Gesture Dataset and SL-Animals-DVS Dataset. The experimental results show that our augmentation method improves accuracy by 3.99% and 7.9%, respectively, compared with baseline accuracy without data augmentation. To the best of our knowledge, our work is the first to augment asynchronous event data in the time dimension.

In summary, our main contributions are as follows:

- We propose a data augmentation strategy based on event data. We propose to augment event data in the dimension of time for the first time. Our method is easy to implement and has low computational overhead. It can be widely used in all kinds of DVS datasets.
- We evaluate our work on the DVS128 Gesture Dataset and SL-Animals-DVS Dataset. The Liquid State Machine (LSM), a purely event-driven spiking neural network, is employed to be the feature extractor, while a three-layer Multilayer Perceptron (MLP) to be the classifier. The experimental results show that our augmentation strategies bring a significant accuracy boost.

## 2 Background

### 2.1 DVS

Dynamic Vision Sensor (DVS) is a neuromorphic vision sensor. DVS can simulate the biological retina by generating asynchronous events when the brightness change of each pixel exceeds a preset threshold. Compared with traditional cameras, the way of recording active pixels greatly reduces data redundancy. Usually, we organize the data collected by DVS in Address Event Representation (AER) format [3]. The event data generated by DVS tends to be sparse in many scenarios. Traditional image representations with matrix cause a considerable memory overhead. An event is generated when the luminosity change on any pixel exceeds the threshold. The output of DVS is a sequence of tuples. An event can be represented by a quad of $[x, y, p, t]$ where $[x, y]$ is the pixel coordinates, the $t$ represents the timestamp of the event, and the $p$ signifies the polarity of luminosity change.

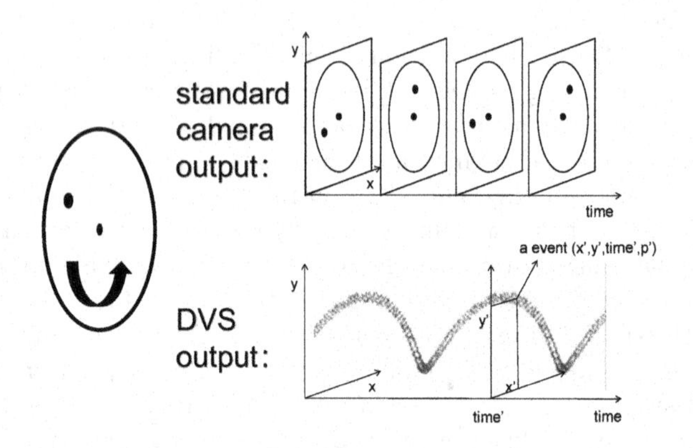


**Fig. 1.** Output of traditional camera and DVS camera

Figure 1 compares the output data of a traditional camera and a DVS camera. The traditional camera can only output each frame at a fixed speed. The output of DVS cameras is a continuous stream of events in Spatio-temporal space, with each point in the space representing an event. Compared with traditional cameras, DVS cameras can obtain more detailed information about moving objects.

### 2.2 LSM

The Liquid State Machine (LSM) is a type of reservoir computing that uses spiking neural networks [16]. As shown in Fig. 2, LSM mainly consists of three parts: input layer, liquid layer, and readout layer. The input layer receives the external spike sequence and sends the input spikes to the corresponding liquid neuron. The core of LSM is the liquid layer, which consists of a large number of interconnected spiking neurons. Each liquid neuron receives time-varying input from the input layer as well as from the liquid layer [13]. Liquid neurons are randomly connected to each other. The recurrent nature of the connections turns the time-varying input into a Spatio-temporal pattern of activations in the network. The readout layer reads out the Spatio-temporal patterns of activation for each time step.

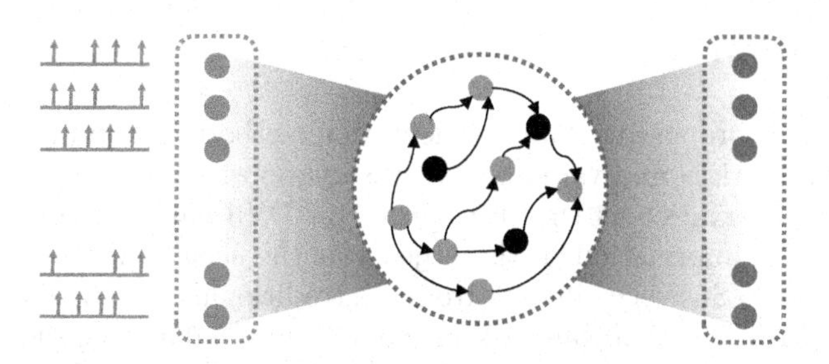

**Fig. 2.** The structure of LSM

# 3 Relatedwork

## 3.1 Data Augmentation for Frame-based Data

Data augmentation can significantly upgrade generalization performance. For image data, there are very mature augmentation methods [22]. Traditional transformations, such as random flipping, rotation, clipping, deformation scaling, adding noise, color disturbance, and so on, can effectively increase the number of samples and train networks with better generalization. In addition, there are some new image data augmentation methods. Zhu et al. [29] used the Generative Adversarial Network to realize image style migration, which can produce new samples with very high quality. Cutout [7] implements data augmentation by randomly cropping the part of the picture. Mixup [28] provides a smoother uncertainty estimation by superimposing two pictures, resulting in a linear transition of the decision boundary from one class to another. Cutmix [27] cuts and pastes some pictures in the training images, and the sample labels are also mixed according to the area ratio. Augmix [11] utilizes stochasticity and diverse augmentations, a Jensen-Shannon Divergence consistency loss, and a formulation to mix multiple augmented images to achieve state-of-the-art performance. These works have achieved excellent results on Cifar-10 and Cifar-100 datasets [12].

## 3.2 Data Augmentation for Event-based Data

Gu et al. [10] propose an augmentation method for event-based datasets, which is called EventDrop. Three augmentation strategies are used in EventDrop, namely Random Drop, Drop by Time, and Drop by Area. Experiments on two event datasets (N-Caltech101 and N-Cars) demonstrate that EventDrop can notably improve generalization across various deep networks [23]. Li et al. [14] proposed an augmentation strategy called NDA, which is a series of geometric augmentation strategies designed for the event-based dataset (including Horizontal Flipping, Rolling, Rotation, Cutout, Shear, and Mixup). On the N-Caltech 101 dataset [18], their strategy improved the absolute accuracy by 15.8%. On the N-MNIST dataset [18], NDA improved the accuracy of the model by 0.12%.

## 4 Motivation

In this paper, we proposed a data augmentation method for event data generated by DVS. This work is motivated by two observations.

The first observation is that the output of DVS camera still contains the relative position information of the objects in the scene, although the output data format of DVS is very different from the traditional camera. The traditional data augmentation method can also be applied to DVS datasets, such as random translation. The sensitivity of the model to location information can be reduced by random translation.

The second observation is that the subjects' gestures generally move at different speeds. Some subjects may be fast, while others may be slow for the same gesture. However, there are limited speed differences in the available datasets. Thus, the models trained based on the limited datasets may have poor generalization. The event representation of DVS output contains time tags, which enable us to proceed with time scaling conveniently and generate samples at different speeds. The model is better to adapt more subjects by generating new samples simulating different speeds.

## 5 Method

In order to solve the problem that there is much few sample in DVS dataset, we propose two strategies to increase event samples in the Spatio-temporal space, namely random translation and time scaling. The first strategy is set for the reason that the subject may be located in different positions in the DVS camera scene. The second strategy takes into consideration that the gesture speeds of different subjects may be at different levels.

In this section, we describe the implementation of the augmentation strategy. Algorithm 1 gives the procedures of augmenting event data. The augmentation level controls the augmentation degree, specifically the maximum offset in random translation and the maximum scaling ratio in time scaling. Overall, our strategy is computationally low-cost and easy to implement.

**Algorithm 1.** Procedures of augmenting event data

**Input:** event sequence $\varepsilon$, pixel resolution $(W, H)$, *augmentation_method*, augmentation level $C$

**Output:** augmented data $\varepsilon^*$

1: $\varepsilon = \{e_i\}_{i\in[1,N]}, e_i = \{x_i, y_i, t_i, p_i\}$
2: **if** $augmentation\_method == random\_translation$ **then**
3: $\delta_x \leftarrow random(-C, C) \times W$
4: $\delta_y \leftarrow random(-C, C) \times H$
5: **for** $e_i$ in $\varepsilon$ **do**
6: $x_i \leftarrow x_i + \delta_x$
7: $y_i \leftarrow y_i + \delta_y$
8: **if** $(x_i \in [1, W] \& y_i \in [1, H])$ **then**

9: *add* $e_i$ *to* $\varepsilon^*$
10: **end if**
11: **end for**
12: **end if**
13: **if** *augmentation_method* == *time_scaling* **then**
14: $\tau \leftarrow random(1 - C, 1 + C)$
15: **for** $e_i$ in $\varepsilon$ **do**
16: $t_i \leftarrow t_i \times \tau$
17: *add* $e_i$ *to* $\varepsilon^*$
18: **end for**
19: **end if**

### 5.1 Random Translation

We randomly generate relative offsets $\delta_x$ and $\delta_y$ in the X and Y directions within the augmentation level in the random translation method. Then we add this relative offset to all events. Offset is valid only if the newly generated event does not exceed the original resolution range. As shown in Fig. 3(a), we translate all events to the left and down. We delete events that exceed the original pixel range after translation.

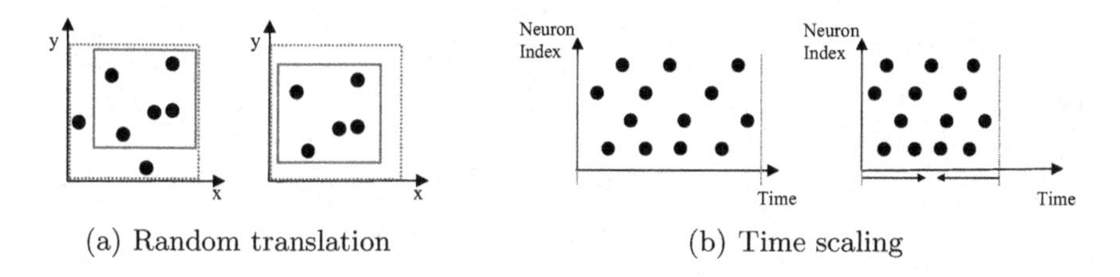


(a) Random translation (b) Time scaling

**Fig. 3.** Data augmentation strategies.

### 5.2 Time Scaling

In the time scaling method, we generate a random scaling factor $\tau$ within the augmentation level. Then the timestamp of all events is multiplied by $\tau$, as shown in Fig. 3(b). The events are becoming denser within the same duration of timesteps, which means the speed information of targeted motions has been diversified.

## 6 Experiments and Results

### 6.1 Datasets

We evaluate our augmentation strategy on the DVS128 Gesture Dataset [1] and the SL-Animals-DVS Dataset [25]. These datasets were recorded by DVS cameras in real scenes with a 128×128 resolution [20,21]. The DVS128 Gesture Dataset contains eleven different gestures of 29 subjects, including 1342 gesture samples, under three different lighting conditions. The 11 classes of gestures include ten fixed gestures and a random gesture that is different from the previous ten gestures. The first ten classes are shown in Fig. 4. The SL-Animals-DVS Dataset was a Sign Language dataset recorded by a DVS, composed of more than 1100 samples of 58 subjects performing 19 signs in isolation corresponding to animals. The 58 subjects were divided into four groups and recorded gestures in different scenes and lighting conditions.

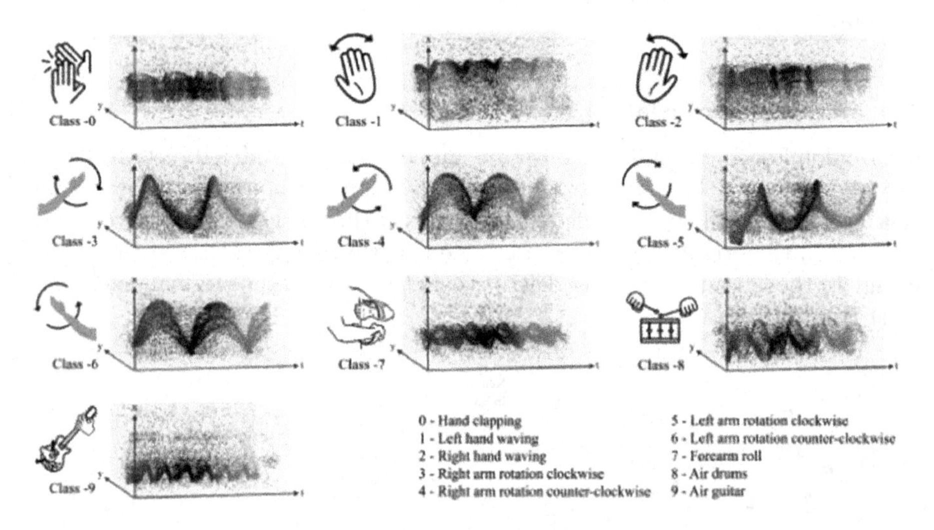


**Fig. 4.** The ten main gestures [26]

### 6.2 Experiment Setup

We use LSM to evaluate our proposed method. The LSM model used in our experiment contains 343 spiking neurons, which are divided into inhibitory neurons and excitatory neurons according to the proportion of 20% and 80%. The connection probability between neurons is calculated according to the spatial distance. The LSM runs on the SNN simulator Brian2 of version 2.4 [24]. Brian2 is an efficient and powerful python-based SNN simulator. We use an MLP classifier with only one hidden layer, and the number of neurons in each layer is (1029,400,11) or (1029,400,19). We use SGD to train the MLP classifier, with learning rate attenuation coefficient and batch size set at 0.01, 1e-6, and 128, respectively. We also use a dropout parameter of 0.5 during training. For the DVS128 Gesture Dataset, we take the first 1000 samples of DVS128 Gesture Dataset as the training set while the remaining 341 samples as the test set. For the SL-Animals-DVS Dataset, We selected a total of 14 subjects in four scenes

**Table 1.** The hardware and software configuration of Experimental platform

| Hardware | Software |
|---|---|
| Intel(R) Core(TM) i7-10700K CPU @3.8GHz | Python 3.6.12, Brian 2.4 |
| NVIDIA GeForce RTX 2060 | Keras 2.3.1, CUDA 10.0, cuDNN 7.6.5 |

as the test set and the remaining subjects as the training set. The configuration of software and hardware is shown in Table 1.

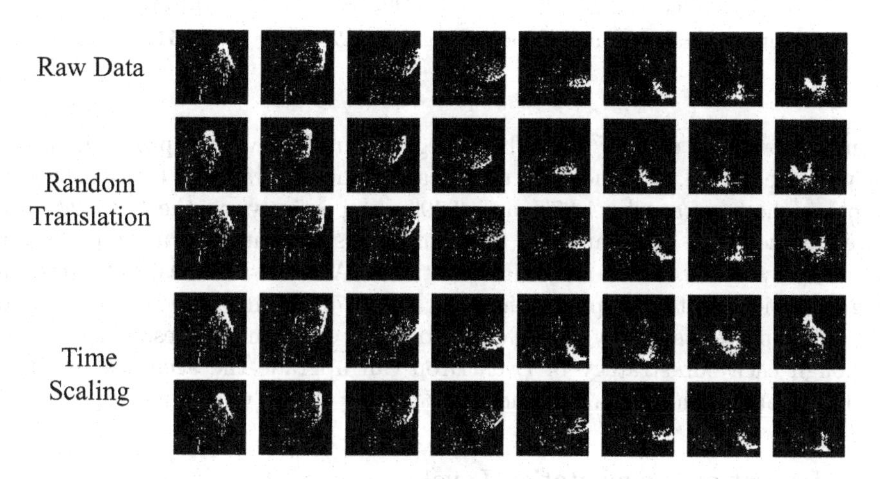


**Fig. 5.** Visualisation of data augmentation on DVS128 Gesture Dataset

### 6.3 Experiment Results

Figure 5 shows the visualization of our augmentation results. We convert the event stream into video frames according to the cumulative time of 20 ms. At the same time, in order to show the gesture completely, we structure video frames at the time intervals of 100 ms. In Fig. 5, the gesture action we visualized is left hand counter clockwise. The first line is the original data. The second two rows show the effect of random translation, with the data shifting to the left and up, respectively. The third two rows show the effect of time scaling, with the scaling factor set at 0.8 and 1.2, respectively. After enhancement with time scaling, the speed of the gesture changed in the same time interval.

Table 2 shows the experimental results of our data augmentation strategy. We used the accuracy of aforementioned LSM model on the dataset without data augmentation being as the baseline. Meanwhile, we also compare the accuracy improvement of our method with that of EventDrop [10]. The experimental

**Table 2.** Results on DVS128 Gesture Dataset and SL-Animals-DVS Dataset

| No. | Method | Accuracy | |
|---|---|---|---|
| | | DVS128 Gesture | SL-Animals-DVS |
| 1 | Baseline | 93.45% | 62.21% |
| 2 | Drop by Time | 94.66% | 63.31% |
| 3 | Drop by Area | 94.75% | 65.86% |
| 4 | Random Drop | 94.13% | 65.45% |
| 5 | Event Drop | 95.71% | 66.20% |
| 6 | Random translation | 96.94% | 67.67% |
| 7 | Time scaling | 95.42% | 67.51% |
| 8 | Random translation + Time scaling | **97.44%** | **69.51%** |

results show that each of our sub-strategies can effectively improve the accuracy of the model, in which the time scaling and the random translation can improve the accuracy by 1.97% and 3.49%, respectively, on the DVS128 Gesture Dataset. The application of the combined sub-strategies can improve the accuracy by 3.99%. At the same time, on the SL-Animals-DVS Dataset, our data augmentation strategy improves the accuracy by 7.3%. Finally, we also compared our augmentation strategy with EventDrop. The experimental results show that although each sub-strategy of EventDrop can upgrade the accuracy by more than 1%, our method is 1.73% and 3.31% higher than EventDrop, respectively.

### 6.4 Impact of Augmentation Level

In this section, we analyze the effect of different augmentation levels. In algorithm 1, the augmentation levels control the maximum translation ratio of random translation and the maximum scaling ratio of time scaling. Different datasets probably have different optimal augmentation levels. On these two datasets, the recording position and gesture speed are relatively fixed. So we only experiment on the results of different augmentation levels in the range of 0.05–0.3. We fix the parameters and structure of the model and then compare their performance on different augmentation levels.

As shown in Fig. 6, it presents the experimental results of different augmentation levels. What's more, different augmentation levels significantly impact the classification accuracy and they all improve the model performance compared to baseline. On the DVS128 Gesture Dataset, when the maximum relative translation ratio is 0.2, the random translation achieves the best effect, and the best maximum scaling ratio of time scale is 0.15. The optimal augmentation levels of random translation and time scaling are 0.15 and 0.2, respectively.

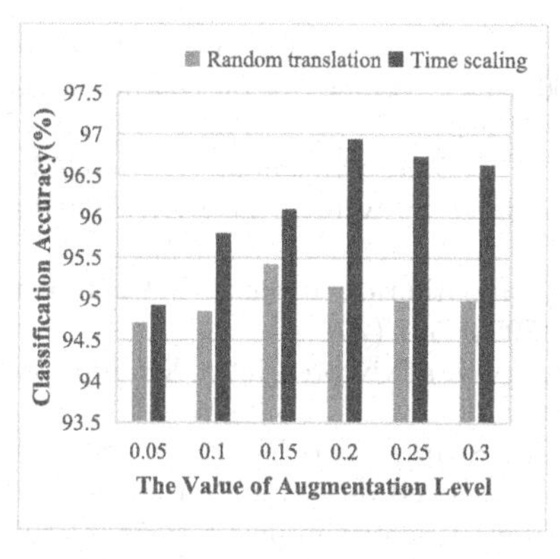


(a) DVS128 Gesture Dataset

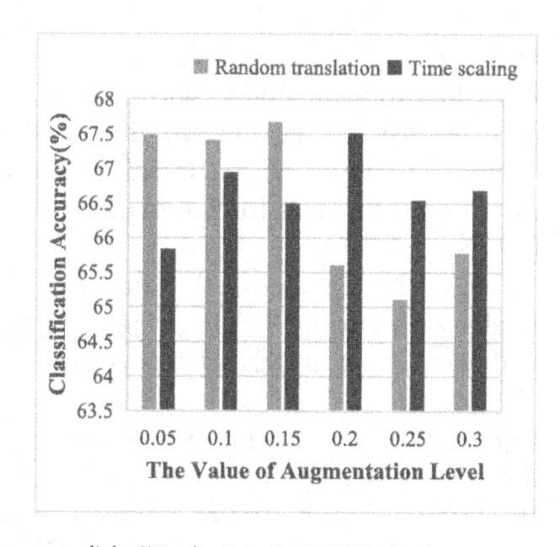


(b) SL-Animals-DVS Dataset

**Fig. 6.** Impact of Augmentation Level on DVS128 Gesture Dataset and SL-Animals-DVS Dataset

## 7 Conclusion

In this paper, we propose a data augmentation method for event data generated by DVS. It can efficiently generate high-quality event samples and solve the problem of insufficient samples in DVS datasets. We propose a method to augment the data output from DVS in terms of temporal and spatial information, which greatly enriches the relative speed, distance and direction features for targeted gestures. The experimental results show that our augmentation method can effectively improve the accuracy and generalization ability of the model. In the future, we will focus on applying more data augmentation strategies to event data and experimenting with our strategy on more datasets and models.

**Acknowledgements.** This work is supported by the National Natural Science Foundation of China (61902408).

## References

1. Amir, A., et al.: A low power, fully event-based gesture recognition system. In: Proceedings of the IEEE conference on computer vision and pattern recognition. pp. 7243–7252 (2017)
2. Camunas-Mesa, L., Acosta-Jimenez, A., Serrano-Gotarredona, T., Linares-Barranco, B.: A digital pixel cell for address event representation image convolution processing. In: Bioengineered and Bioinspired Systems II. vol. 5839, pp. 160–171. SPIE (2005)
3. Chan, V., Liu, S.C., van Schaik, A.: Aer ear: A matched silicon cochlea pair with address event representation interface. IEEE Transactions on Circuits and Systems I: Regular Papers **54**(1), 48–59 (2007)
4. Cubuk, E.D., Zoph, B., Mane, D., Vasudevan, V., Le, Q.V.: Autoaugment: Learning augmentation strategies from data. In: Proceedings of the IEEE/CVF Conference on Computer Vision and Pattern Recognition. pp. 113–123 (2019)

5. Deng, J., Dong, W., Socher, R., Li, L.J., Li, K., Fei-Fei, L.: Imagenet: A large-scale hierarchical image database. In: 2009 IEEE conference on computer vision and pattern recognition. pp. 248–255. Ieee (2009)
6. Devlin, J., Chang, M.W., Lee, K., Toutanova, K.: Bert: Pre-training of deep bidirectional transformers for language understanding. arXiv preprint arXiv:1810.04805 (2018)
7. DeVries, T., Taylor, G.W.: Improved regularization of convolutional neural networks with cutout. arXiv preprint arXiv:1708.04552 (2017)
8. Gallego, G., et al.: Event-based vision: a survey. arXiv preprint arXiv:1904.08405 (2019)
9. Gehrig, D., Loquercio, A., Derpanis, K.G., Scaramuzza, D.: End-to-end learning of representations for asynchronous event-based data. In: Proceedings of the IEEE/CVF International Conference on Computer Vision. pp. 5633–5643 (2019)
10. Gu, F., Sng, W., Hu, X., Yu, F.: Eventdrop: data augmentation for event-based learning. arXiv preprint arXiv:2106.05836 (2021)
11. Hendrycks, D., Mu, N., Cubuk, E.D., Zoph, B., Gilmer, J., Lakshminarayanan, B.: Augmix: A simple data processing method to improve robustness and uncertainty. arXiv preprint arXiv:1912.02781 (2019)
12. Krizhevsky, A., Hinton, G., et al.: Learning multiple layers of features from tiny images (2009)
13. Lapique, L.: Recherches quantitatives sur l'excitation electrique des nerfs traitee comme une polarization. Journal of Physiology and Pathololgy **9**, 620–635 (1907)
14. Li, Y., Kim, Y., Park, H., Geller, T., Panda, P.: Neuromorphic data augmentation for training spiking neural networks. arXiv preprint arXiv:2203.06145 (2022)
15. Maass, W.: Networks of spiking neurons: the third generation of neural network models. Neural networks **10**(9), 1659–1671 (1997)
16. Maass, W., Natschläger, T., Markram, H.: Real-time computing without stable states: A new framework for neural computation based on perturbations. Neural computation **14**(11), 2531–2560 (2002)
17. Mahowald, M.A.: Vlsi analogs of neuronal visual processing: a synthesis of form and function (1992)
18. Orchard, G., Jayawant, A., Cohen, G.K., Thakor, N.: Converting static image datasets to spiking neuromorphic datasets using saccades. Frontiers in neuroscience **9**, 437 (2015)
19. Pak, M., Kim, S.: A review of deep learning in image recognition. In: 2017 4th International Conference on Computer Applications and Information Processing Technology (CAIPT). pp. 1–3 (2017). https://doi.org/10.1109/CAIPT.2017.8320684
20. Patrick, L., Posch, C., Delbruck, T.: A 128 $\times$ 128 120 db 15$\mu$ s latency asynchronous temporal contrast vision sensor. IEEE journal of solid-state circuits **43**, 566–576 (2008)
21. Serrano-Gotarredona, T., Linares-Barranco, B.: A 128 $\times$ 128 1.5% contrast sensitivity 0.9% fpn 3 μs latency 4 mw asynchronous frame-free dynamic vision sensor using transimpedance preamplifiers. IEEE Journal of Solid-State Circuits 48(3), 827–838 (2013). https://doi.org/10.1109/JSSC.2012.2230553
22. Shorten, C., Khoshgoftaar, T.M.: A survey on image data augmentation for deep learning. Journal of big data **6**(1), 1–48 (2019)
23. Sironi, A., Brambilla, M., Bourdis, N., Lagorce, X., Benosman, R.: Hats: Histograms of averaged time surfaces for robust event-based object classification. In: Proceedings of the IEEE Conference on Computer Vision and Pattern Recognition. pp. 1731–1740 (2018)

24. Stimberg, M., Brette, R., Goodman, D.F.: Brian 2, an intuitive and efficient neural simulator. Elife **8**, e47314 (2019)
25. Vasudevan, A., Negri, P., Linares-Barranco, B., Serrano-Gotarredona, T.: Introduction and analysis of an event-based sign language dataset. In: 2020 15th IEEE International Conference on Automatic Face and Gesture Recognition (FG 2020). pp. 675–682. IEEE (2020)
26. Wang, Q., Zhang, Y., Yuan, J., Lu, Y.: Space-time event clouds for gesture recognition: From rgb cameras to event cameras. In: 2019 IEEE Winter Conference on Applications of Computer Vision (WACV). pp. 1826–1835. IEEE (2019)
27. Yun, S., Han, D., Oh, S.J., Chun, S., Choe, J., Yoo, Y.: Cutmix: Regularization strategy to train strong classifiers with localizable features. In: Proceedings of the IEEE/CVF international conference on computer vision. pp. 6023–6032 (2019)
28. Zhang, H., Cisse, M., Dauphin, Y.N., Lopez-Paz, D.: mixup: Beyond empirical risk minimization. arXiv preprint arXiv:1710.09412 (2017)
29. Zhu, J.Y., Park, T., Isola, P., Efros, A.A.: Unpaired image-to-image translation using cycle-consistent adversarial networks. In: Proceedings of the IEEE international conference on computer vision. pp.eps

# FCFNet: A Network Fusing Color Features and Focal Loss for Diabetic Foot Ulcer Image Classification

Chuantao Xie(✉)

School of Computer Engineering and Science, Shanghai University, Shanghai, China
ctxie@shu.edu.cn

**Abstract.** Diabetic foot ulcers (DFU) are one of the most common and severe complications of diabetes. More than one million diabetics face amputation due to failure to recognize and treat DFU properly every year, and it is very important to early recognize the condition of DFU. As deep learning has achieved outstanding results in computer vision tasks, some deep learning methods have been used for diabetic foot ulcer image classification. However, these models do not pay attention to the learning of color information in images and may not effectively learn the characteristics of hard examples, which are difficult to recognize by models trained by common loss functions. In addition, they may be time-consuming or not suitable for small datasets. According to DFU medical classification systems, the presence of infection and ischemia has important clinical implications for DFU assessment. In our work, based on the characteristics of gangrene and wound, which are significantly different from the surrounding skin and the background, we use the K-Means algorithm to segment the features and further build the classification model with the segmentation module. Next, we use focal loss to train EfficientNet B3, which can make this model better learn the characteristics of hard examples. We finally use the two powerful networks for testing. The experimental results demonstrate that compared with other excellent classification models, our model has better performance with a macro-average F1-score of 0.6334, which has great potential in medical assisted diagnosis.

**Keywords:** Diabetic foot ulcers · Deep learning · Segmentation · Classification

## 1 Introduction

Diabetes has become a common disease. According to the global report on diabetes, 422 million people were living with diabetes mellitus in 2014, compared to 108 million people in 1980 [30]. It is estimated that the number of global diabetes in 2019 is 463 million, rising to 578 million by 2030 and 700 million by 2045 [31]. Worldwide annual incidence of diabetic foot ulcers ranges from


M. Tanveer et al. (Eds.): ICONIP 2022, CCIS 1793, pp. 434–445, 2023.
https://doi.org/10.1007/978-981-99-1645-0_36

9.1 million to 26.1 million [4]. There is about 15%-25% chance that a diabetic patient will eventually develop diabetic foot ulcers (DFU). DFU are one of the most common and severe complications of diabetes, and if proper care is not taken, it may result in lower limb amputation [1]. Every year, more than one million diabetics face amputation due to failure to properly recognize and treat DFU [5].

The treatment of diabetic foot is a global concern, which will bring a heavy burden to society. Recognition of infection and ischemia is significant to determine factors that predict the healing progress of DFU and the risk of amputation. Lower limb ischemia is a lack of blood supply. As the disease progresses, it can lead to resting pain, gangrene at the toe, and ulcers at the heel or metatarsal joint and even lead to the amputation of toes or legs if not recognized and treated early. In previous studies, it is estimated that patients with critical ischemia have a three-year limb loss rate of about 40% [2]. Approximately, 56% of DFU become infected and 20% of DFU infections lead to amputation of a foot or limb [20,22,27]. And this paper focuses on early detection of ischemia and infection of diabetic foot ulcers.

Insufficient blood supply affects DFU healing, and the manifestations of lower limb ischemia are skin malnutrition, muscle atrophy, dry skin with poor elasticity, decreased skin temperature, pigmentation, and weakened or disappearance of acral artery pulsation. Patients may have intermittent claudication of lower limbs. As the disease progresses, there may be resting pain, gangrene at the toes, ulcers at the pressure of the heel or metatarsophalangeal joint, and infection of the limb in some patients. From the view of computer vision, visual features may indicate the presence of ischemia. Infection is defined as bacterial soft tissue or bone infection in the DFU, which is based on at least two classic findings of inflammation or purulence. Infective diabetic foot is usually characterized by swelling of foot lesions. If the infection affects a wide range, swelling of the whole foot may occur. And if there are ulcers in the foot, the swelling around the ulcers is the most obvious. Infection is often associated with necrosis. From the perspective of computer vision, visual features such as redness and swelling may indicate the presence of infection.

Recognition of infection and ischemia is essential to determine factors that predict the healing progress of DFU and the risk of amputation. High risks of infection and ischemia in DFU can lead to hospital admission and amputation [25]. However, accurate diagnosis of ischemia and infection requires establishing a good clinical history, physical examination, blood tests, bacteriological study, and Doppler study of leg blood vessels [15]. But it also comes with high costs and cumbersome tests. Experts working in the field of diabetic foot ulcers have good experience of predicting the presence of underlying ischemia or infection simply by looking at the ulcers. As deep learning has achieved excellent results in computer vision tasks, we can construct a model based on some labeled diabetic foot ulcer images to predict the presence of ischemia and infection in diabetic foot ulcers. Recognition of infection and ischemia in DFU with cost-effective deep learning methods is a significant step towards developing a complete computerized DFU assessment system for remote monitoring in the future.

The main contributions of this paper are as follows: (1) We design a novel model with the segmentation module. Based on the prior that gangrene and wound are significantly different from the surrounding skin and the background, the paper uses the K-Means algorithm to segment the features and further build the classification model with the segmentation module. (2) EfficientNet B3 [32] trained by focal loss [21] and the classification model with the segmentation module are used simultaneously to predict ischemia and infection in diabetic foot ulcers. (3) The results on the large Diabetic Foot Ulcer Challenge (DFUC) 2021 dataset [36] show that our method achieves better performance compared with other state-of-the-art methods. (4) It is very possible to put our research into clinical practice to aid clinicians in making the correct diagnosis.

The rest of the paper is organized as follows: Sect. 2 introduces the related work. Section 3 describes the proposed methods in detail. Section 4 mainly presents the dataset, experimental datails and results. Lastly, Sect. 5 draws conclusions and indicate future work.

## 2 Related Work

In the beginning, researchers generally use machine learning methods to analyse DFU. Vardasca et al. [33] used the k-Nearest Neighbour algorithm to perform the classification of infrared thermal images. Patel et al. [26] used Gabor filter and k-means methods to identify and label three types of tissue images of diabetic foot ulcers. With deep learning rapidly developing and the increased amount of data collected by researchers, many researchers have used deep learning methods to recognize and analyze DFU. Goyal et al. [14] proposed the DFUNet, which classified standard skin images and the diabetic foot (DF) images. Alzubaidi et al. [3] proposed the DFU_QUTNet, which also classified standard skin images and the DF images. Goyal et al. [16]proposed an object detection method based on the Faster-RCNN [29] to detect DFU. What's more, the model is implemented in a mobile phone terminal. There is still other work about the detection of DFU found in the papers [8,9]. Goyal et al. [17] used a transfer learning method to train the fully convolutional network (FCN) [24] to perform the segmentation of DFU and surrounding skin on the full foot images. Rania et al. [28] used three models to perform the segmentation of DFU, respectively.

The next work is about infection and ischaemia classification in diabetic foot ulcers. Goyal et al. [15] used an ensemble convolutional neural network and support vector machine(SVM) to recognize ischemia and infection of DFU. Although they achieved high accuracy in ischemia recognition, there were some shortcomings. The dataset was small and highly imbalanced, so the generalization ability of models trained on the dataset may not be strong. The recognition rate of infection was 73%, which was low, which may require lots of work to improve accuracy. Xu et al. also [35] train models on this dataset [15], and there are some similar problems. Yap et al. [36] want to address these problems, who

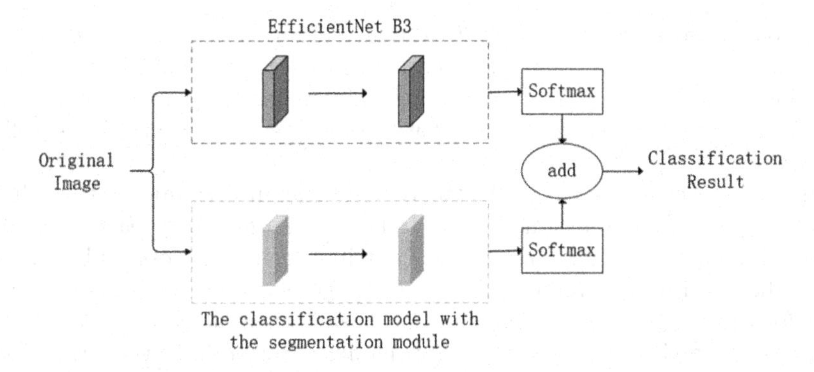


**Fig. 1.** The overall testing process for an input image.

introduced the Diabetic Foot Ulcer Challenge (DFUC) 2021 dataset, which consists of 4474 clinically annotated images, together with DFU patches with the label of infection, ischemia, both infection and ischemia, and none of those conditions (control). In this dataset, there is some people's work. Bloch et al. [6] used an ensemble of EfficientNets [32] with a semi-supervised training strategy. Galdran et al. [13] used an ensemble network based on Resnext50 [34] and EfficientNet B3. There is still other work found in the paper [7]. However, these models do not pay attention to the learning of color information in images and may not effectively learn the characteristics of a few hard examples. In addition, they may be time-consuming or not suitable for small datasets.

## 3 Proposed Method

To accurately recognize whether diabetic foot ulcers are ischemic or infected, we train two powerful models separately and use them for testing. Specifically, we add the outputs of the two models, which are EfficientNet B3 and the model based on ResNeXt50 with the segmentation module. Figure 1 illustrates the overall testing process for an input image.

### 3.1 Preprocessing

In order to enhance the generalization ability of models, we use a two-stage data augmentation method to expand the dataset. The ways in the first stage include horizontal flip, vertical flip, random rotation, and random clipping and then scaling to a fixed size. The data augmentation methods in the second stage include contrast change, brightness change, and saturation change.

### 3.2 EfficientNet B3 and Focal Loss

EfficientNet B3 contains approximately 12M parameters, designed by neural architecture search. EfficientNet B3 introduces the attention idea of squeeze and excitation network (SENet). SENet adds attention to the channel dimension of the feature map. This attention mechanism allows the model to pay more attention to the most information channel features and suppress those unimportant channel features. The model has fewer parameters, but it achieves excellent performance.

In the dataset, there may be a lot of easily classified samples, of which the cumulative loss function value will be relatively large. Many easily classified samples will play a significant role in contributing to the loss and dominate the gradient's update direction. As a result, the classifier has a poor learning effect for those hard samples and can not classify them accurately. These hard samples may be difficult to distinguish for models when training them with cross-entropy loss function, so when training EfficientNet B3, we use focal loss [21] as the optimized loss function. The specific focal loss function which we use is defined as Eq. 1. We set the $\gamma$ to 1, and $p$ is the probability of predicting the correct classification. In this way, for easy samples, $p$ will be relatively large, so the weight will naturally decrease. If $p$ is small for hard examples, the weight is considerable, so the network tends to use such samples to update parameters, and the weights change dynamically. If hard samples gradually become easy samples, their influence will gradually decrease.

$$L(p) = -(1-p)^{\gamma} \log(p) \tag{1}$$

### 3.3 The Classification Network with the Segmentation Module

We design the classification network with the segmentation module based on ResNeXt50, which contains approximately 26M parameters and implements convolutional layers with Weight Standardization.

We first scale the images to 28 * 28 and then gray them. We can still see the location and the outline of ulcers in the gray images. Finally, we use the K-Means algorithm to cluster the gray images, and these images will be used as the masks of the feature maps predicted by the segmentation module. We want to segment the essential features of diabetic foot ulcers, classify them as a class, and classify unimportant information as another class.

Patients with ischemic diabetic foot ulcers may have skin pigmentation and gangrene on the toes. We use the K-Means clustering algorithm to segment these features based on these characteristics. If the color of the surrounding skin is similar to the color of the background in the gray images, the gangrenous part will probably not be classified as the class of most of the surrounding skin. Conversely, most of the surrounding skin may be classified as the class of the gangrenous part. And we also use the K-Means algorithm to segment the wound area. If there are black or red wound areas in the original image, the areas will be black in the gray image corresponding to the original image, which are easy to segment. In short, if there are gangrene, black wound, or red wound in the original image, their color is darker than the color of the surrounding skin and background environment in the gray image, which are easy to be segmented from the gray image. In essence, our method focuses on the black parts of the

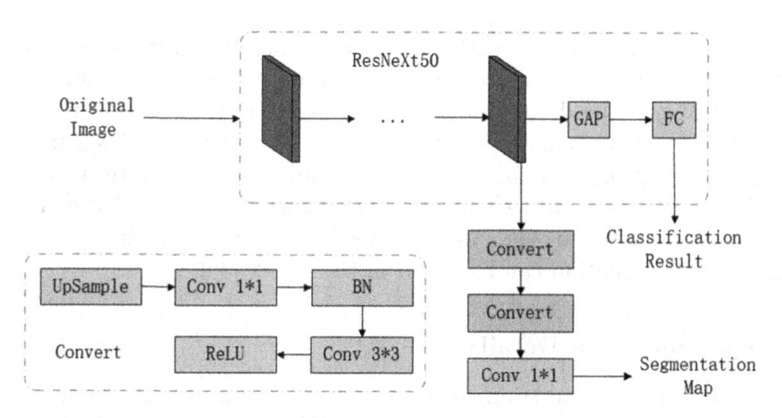


**Fig. 2.** The classification model with the segmentation module. In this figure, "GAP" indicates the global average pooling layer. "FC" indicates the fully connected layer. "UpSample" refers to the upsampling layer. "Conv" means the convolutional layer. "BN" refers to batch normalization. "ReLU" means rectified linear unit. "Convert" is an information conversion module.

gray images corresponding to the original images, in which there will likely be important features of diabetic foot ulcers.

We use the K-Means algorithm to segment the 28 * 28 gray images, and the resulting images are used as the masks of the predicted feature maps generated by the segmentation module. Compared with using masks whose size is 224*224, it will greatly reduce our training time, and we can know the general location and outline of ulcers and learn most of the important features in an image. We use ResNeXt50 as our primary network. Since the mask size is 28 * 28, we need to upsample the output feature map before the last global average pooling layer to make the output size 28 * 28. We make full use of the segmentation mask of important information of an image as prior information to guide the network to extract essential features better. We use the mean square error (MSE) loss as the loss function of our segmentation module $L_{seg}$ and use the cross-entropy function as our classification loss function $L_{cls}$. Finally, the sum of the segmentation loss function and the classification loss function is our final optimization objective function $L_{sum}$, as shown in Eq. 2. The network with the segmentation module is shown in Fig. 2.

$$L_{sum} = L_{cls} + L_{seg} \tag{2}$$

## 4 Experiments

### 4.1 Dataset

The dataset we used came from DIABETIC FOOT ULCER CHALLENGE (DFUC) 2021, which was held in conjunction with the Medical Image Computing and Computer-Assisted Intervention (MICCAI) 2021 conference. Foot images with DFU were collected from the Lancashire Teaching Hospital over the past few years. Images were acquired by a podiatrist and a consultant physician specializing in the diabetic foot, with more than five years of professional experience. The total images for DFUC 2021 are 15683, with 5955 images for the training set (2555 infection only, 227 ischemia only, 621 both infection and ischemia, and 2552 without ischemia and infection), 3994 unlabeled images, and 5734 images for the testing set. And we only use the labeled training set to train models. More information can be found in the paper [36].

### 4.2 Implementation Details

We use 10% of the official training set as our validation set and the rest as our training set. We train our models on the NVIDIA Tesla P100 GPU device with a learning rate of 0.01 and a momentum of 0.9. For all models, we initialize them using the pretraining weights on the Imagenet dataset [10]. Macro-average F1-score is monitored on an independent validation set, and we save all models with the highest Macro-average F1-scores during training. Also, We train all models with a batch size of 8.

General Neural network optimizers only seek to minimize the training loss. Although they can converge to a good local optimum, it is often not smooth near the local optimum. In the evaluation, it is easy to achieve the training error close to 0, but the generalization error is high. These prompt us to find an optimizer that can be aware of the steepness, making the loss smoother in the training process. Sharpness Aware Optimization (SAM) [12,19], instead of focusing on a single point, try to find parameters lying in neighborhoods with low loss. The goal of SAM is to improve model generalization and enforce robustness against label noise, so we optimize all models with Sharpness-Aware loss minimization with SGD.

For testing of every model, we generate four different versions of each image by horizontal flipping and vertical flipping, predict each of them, and average the results. We finally average the output results of EfficientNet B3 and the model based on ResNeXt50 with the segmentation module.

### 4.3 Results and Analysis

We use macro-average F1-score as the first metric to measure the performance of a model which is the average value of F1-scores of four classes, taking into account the classes with few samples. Next, we will compare our model with five models with good performance, which are ResNet50 [18], ResNeXt50, EfficientNet B3, Vision Transformer (ViT) [11] and Swin Transformer (SwinT) [23]. And we will also compare our model with three models of the paper [7], whose authors are Galdran et al., Bloch et al., and Ahmed et al.

It can be seen from the Table 1 that our macro-average F1-score, macro-average AUC, macro-average precision, micro-average F1-score, micro-average AUC, and accuracy all achieved the highest scores. And our macro-average recall ranked second. Specially, we completed the best macro-average F1-score

**Table 1.** Overall results of models. Note that "Macro" means macro-average and "Micro" means micro-average.

| Models/Authors | Macro Precision | Macro Recall | Macro F1-score | Macro AUC | Micro F1-score | Micro AUC | Accuracy |
|---|---|---|---|---|---|---|---|
| Ours | **0.6344** | 0.6469 | **0.6334** | **0.8915** | **0.6961** | **0.9143** | **0.6997** |
| Galdran et al. [7] | 0.6140 | **0.6522** | 0.6216 | 0.8855 | 0.6801 | 0.9071 | 0.6856 |
| Bloch et al. [7] | 0.6207 | 0.6246 | 0.6077 | 0.8616 | 0.6532 | 0.8734 | 0.6657 |
| ResNeXt50 | 0.6296 | 0.6152 | 0.6057 | 0.8787 | 0.6703 | 0.8935 | 0.6817 |
| EfficientNet B3 | 0.6184 | 0.6008 | 0.5996 | 0.8699 | 0.6633 | 0.8809 | 0.6723 |
| Ahmed et al. [7] | 0.5984 | 0.5979 | 0.5959 | 0.8644 | 0.6714 | 0.8935 | 0.6711 |
| SwinT | 0.5916 | 0.6279 | 0.5921 | 0.8698 | 0.6618 | 0.8840 | 0.6716 |
| ViT | 0.5792 | 0.6118 | 0.5800 | 0.8765 | 0.6678 | 0.8972 | 0.6711 |
| ResNet50 | 0.5736 | 0.5979 | 0.5687 | 0.8664 | 0.6286 | 0.8895 | 0.6416 |

(0.6334), an improvement of 1.18% on the result of the model of agaldran et al.(0.6216), and an improvement of 2.77% on the RexNeXt50 result (0.6057). And we achieved the best accuracy (0.6997), an improvement of 1.41% on the outcome of the model of agaldran et al. (0.6856).

It can be seen from the Table 2 that we achieved the best F1-score for infection (0.6782) and the best F1-score for ischemia (0.5769). ResNeXt50 achieved the best F1-score for control (0.7600) which shows an improvement of 0.21% on the result of our model (0.7579), and our score ranks second in the measure. Galdran achieved the best F1-score for infection and ischemia (0.5619). All in all, for the four per class F1-scores, our scores are at the top and more balanced.

**Table 2.** per class F1-scores of models.

| Models/Authors | Control | Infection | Ischaemia | Both |
|---|---|---|---|---|
| Ours | 0.7579 | **0.6782** | **0.5769** | 0.5205 |
| Galdran et al. [7] | 0.7574 | 0.6388 | 0.5282 | **0.5619** |
| Bloch et al. [7] | 0.7453 | 0.5917 | 0.5580 | 0.5359 |
| ResNeXt50 | **0.7600** | 0.6262 | 0.5510 | 0.4855 |
| EfficientNet B3 | 0.7440 | 0.6233 | 0.5095 | 0.5215 |
| Ahmed et al. [7] | 0.7157 | 0.6714 | 0.4574 | 0.5390 |
| SwinT | 0.7368 | 0.6298 | 0.4825 | 0.5192 |
| ViT | 0.7421 | 0.6525 | 0.4814 | 0.4442 |
| ResNet50 | 0.7341 | 0.5688 | 0.5471 | 0.4250 |

In general, our model has achieved good results. There may be a lot of easily classified samples and a few hard examples, so we use focal loss to train EfficientNet B3. Next, we use the model based on ResNeXt50 with the segmentation module. Specially, we use the K-Means algorithm to segment gangrene and wound. The segmented images are used as the masks of the feature maps predicted by our segmentation module, which makes the model focus on the disease features of the foot and learn useful features more easily. We use these two powerful models for testing so that our model can achieve outstanding results.

### 4.4 Ablation Study

In this section, we evaluate the contribution of focal loss and the segmentation module.

**Table 3.** overall results of models in ablation study. Note that "Macro" means macro-average and "Micro" means micro-average.

| Models | Macro Precision | Macro Recall | Macro F1-score | Macro AUC | Micro F1-score | Micro AUC | Accuracy |
|---|---|---|---|---|---|---|---|
| Ours | 0.6344 | **0.6469** | **0.6334** | **0.8915** | **0.6961** | **0.9143** | **0.6997** |
| E3_f | 0.6192 | 0.6306 | 0.6196 | 0.8762 | 0.6811 | 0.9044 | 0.6838 |
| R50_s | 0.5911 | 0.6131 | 0.5960 | 0.8826 | 0.6806 | 0.9073 | 0.6828 |
| R50__E3 | 0.6328 | 0.6125 | 0.6094 | 0.8837 | 0.6736 | 0.8959 | 0.6840 |
| R50__E3_f | **0.6376** | 0.6343 | 0.6236 | 0.8866 | 0.6883 | 0.9083 | 0.6964 |
| R50_s__E3 | 0.6277 | 0.6141 | 0.6121 | 0.8906 | 0.6785 | 0.9105 | 0.6864 |

"E3" indicates EfficientNet B3, "R50" means ResNeXt50, "_s" refers to adding the segmentation module, "_f" indicates using focal loss to train a model, and "__" means ensembling two models. It can be seen from the Table 3 that the performance of a single model is significantly lower than that of the integrated model, indicating the two models are complementary. Next, we will analyze a primary integrated network based on ResNeXt50 and EfficientNet B3. When we use focal loss to train EfficientNet B3 of the integrated network, the performance of "R50__E3_f" is significantly higher than "R50_E3", and the performance of our final model also substantially surpasses that of "R50_s__E3". Significantly, the macro-average F1-score of our last network has an improvement of 2.13% than the result of "R50_s__E3". When we add the segmentation module to ResNeXt50 of the integrated network, the performance of "R50_s_E3" is higher than that of "R50_E3," and the performance of our final network also surpass that of "R50_E3_f". Significantly, the macro-average F1-score of our last network has an improvement of 0.98% than the result of "R50_E3_f". Finally, when we use both the segmentation module and focal loss, the macro-average F1-score of our last network has an improvement of 2.4% than the result of "R50__E3". In short, adding focal loss and the segmentation module can improve the performance of models.

### 4.5 Some Deficiencies

The F1-score for ischemia and infection classification (0.52) and the F1-score for ischemia classification (0.58) are relatively low, compared to other F1-scores, which is a possible further reflection of the class imbalance within the dataset and some similarities in the two classes. At the same time, the K-Means algorithm can affect the generated masks and further affect classification results to a certain extent. Maybe better clustering algorithms can be used in the future.

## 5 Conclusion and Future Work

In this paper, we propose an integrated model for classifying ischemia and infection in diabetic foot ulcers. This model trained on a large dataset of diabetic foot ulcers has achieved good results in the test set. And compared with other models, our model achieves better performance, so we believe the model will help to early identify DFU complications to guide treatment and help prevent further complications.

In the future, we plan to use our model to construct an auxiliary diagnostic application of DFU, which can be used by patients and alleviate the pressure on the world's medical services. And it is of great significance to the development of computer-aided diagnosis systems in the future. Next, we will do the semi-supervised segmentation of DFU images because it is costly to label all ulcer areas in many DFU images. At the same time, we plan to collect the foot images of patients over a period of time, construct a model to analyze wound changes, and provide helpful information to doctors.

**Acknowledgements.** We have obtained the authorization of the organizers of DFUC 2021 for research, who have received approval from the UK National Health Service (NHS) Research Ethics Committee (REC). The computation is supported by the School of Computer Engineering and Science of Shanghai University. This work is supported by the NSFCs of China (No. 61902234).

## References

1. Aguiree, F., et al.: IDF diabetes atlas (2013)
2. Albers, M., Fratezi, A.C., De Luccia, N.: Assessment of quality of life of patients with severe ischemia as a result of infrainguinal arterial occlusive disease. J. Vasc. Surg. **16**(1), 54–59 (1992)
3. Alzubaidi, L., Fadhel, M.A., Oleiwi, S.R., Al-Shamma, O., Zhang, J.: DFU_QUTNet: diabetic foot ulcer classification using novel deep convolutional neural network. Multimedia Tools Appl. **79**(21), 15655–15677 (2020)
4. Armstrong, D.G., Boulton, A.J., Bus, S.A.: Diabetic foot ulcers and their recurrence. New England J. Med. **376**(24), 2367–2375 (2017)
5. Armstrong, D.G., Lavery, L.A., Harkless, L.B.: Validation of a diabetic wound classification system: the contribution of depth, infection, and ischemia to risk of amputation. Diabetes Care **21**(5), 855–859 (1998)
6. Bloch, L., Brüngel, R., Friedrich, C.M.: Boosting efficientnets ensemble performance via pseudo-labels and synthetic images by pix2pixHD for infection and ischaemia classification in diabetic foot ulcers. arXiv preprint arXiv:2112.00065 (2021). https://doi.org/10.1007/978-3-030-94907-5_3

7. Cassidy, B., et al.: Diabetic foot ulcer grand challenge 2021: evaluation and summary. arXiv preprint arXiv:2111.10376 (2021). https://doi.org/10.1007/978-3-030-94907-5_7
8. Cassidy, B., et al.: DFUC 2020: analysis towards diabetic foot ulcer detection. arXiv abs/2004.11853 (2021)
9. Cassidy, B., et al.: The DFUC 2020 dataset: analysis towards diabetic foot ulcer detection. TouchREV. Endocrinol. **17**(1), 5–11 (2021)
10. Deng, J., Dong, W., Socher, R., Li, L.J., Li, K., Fei-Fei, L.: ImageNet: a large-scale hierarchical image database. In: 2009 IEEE Conference on Computer Vision and Pattern Recognition, pp. 248–255. IEEE (2009)
11. Dosovitskiy, A., et al.: An image is worth $16 \times 16$ words: Transformers for image recognition at scale. arXiv preprint arXiv:2010.11929 (2020)
12. Foret, P., Kleiner, A., Mobahi, H., Neyshabur, B.: Sharpness-aware minimization for efficiently improving generalization. arXiv preprint arXiv:2010.01412 (2020)
13. Galdran, A., Carneiro, G., Ballester, M.A.G.: Convolutional nets versus vision transformers for diabetic foot ulcer classification. arXiv preprint arXiv:2111.06894 (2021)
14. Goyal, M., Reeves, N.D., Davison, A.K., Rajbhandari, S., Spragg, J., Yap, M.H.: DFUNet: convolutional neural networks for diabetic foot ulcer classification. IEEE Trans. Emerg. Top. Comput. Intell. **4**(5), 728–739 (2018)
15. Goyal, M., Reeves, N.D., Rajbhandari, S., Ahmad, N., Wang, C., Yap, M.H.: Recognition of ischaemia and infection in diabetic foot ulcers: dataset and techniques. Comput. Biol. Med. **117**, 103616 (2020)
16. Goyal, M., Reeves, N.D., Rajbhandari, S., Yap, M.H.: Robust methods for real-time diabetic foot ulcer detection and localization on mobile devices. IEEE J. Biomed. Health Inform. **23**(4), 1730–1741 (2018)
17. Goyal, M., Yap, M.H., Reeves, N.D., Rajbhandari, S., Spragg, J.: Fully convolutional networks for diabetic foot ulcer segmentation. In: 2017 IEEE International Conference on Systems, Man, and Cybernetics (SMC), pp. 618–623. IEEE (2017)
18. He, K., Zhang, X., Ren, S., Sun, J.: Deep residual learning for image recognition. In: Proceedings of the IEEE Conference on Computer Vision and Pattern Recognition, pp. 770–778 (2016)
19. Korpelevich, G.M.: The extragradient method for finding saddle points and other problems. Matecon **12**, 747–756 (1976)
20. Lavery, L.A., Armstrong, D.G., Wunderlich, R.P., Tredwell, J., Boulton, A.J.: Diabetic foot syndrome: evaluating the prevalence and incidence of foot pathology in Mexican Americans and non-Hispanic whites from a diabetes disease management cohort. Diabetes Care **26**(5), 1435–1438 (2003)
21. Lin, T.Y., Goyal, P., Girshick, R., He, K., Dollár, P.: Focal loss for dense object detection. In: Proceedings of the IEEE International Conference on Computer Vision, pp. 2980–2988 (2017)
22. Lipsky, B.A., et al.: 2012 infectious diseases society of America clinical practice guideline for the diagnosis and treatment of diabetic foot infections. Clin. Infect. Dis. **54**(12), e132–e173 (2012)
23. Liu, Z., et al.: Swin transformer: hierarchical vision transformer using shifted windows. arXiv preprint arXiv:2103.14030 (2021)
24. Long, J., Shelhamer, E., Darrell, T.: Fully convolutional networks for semantic segmentation. In: Proceedings of the IEEE Conference on Computer Vision and Pattern Recognition, pp. 3431–3440 (2015)

25. Mills Sr, J.L., et al.: Society for Vascular Surgery Lower Extremity Guidelines Committee. The society for vascular surgery lower extremity threatened limb classification system: risk stratification based on wound, ischemia, and foot infection (WIfI). J. Vasc. Surg. **59**(1), 220–234 (2014)
26. Patel, S., Patel, R., Desai, D.: Diabetic foot ulcer wound tissue detection and classification. In: 2017 International Conference on Innovations in Information, Embedded and Communication Systems (ICIIECS). pp. 1–5. IEEE (2017)
27. Prompers, L., et al.: High prevalence of ischaemia, infection and serious comorbidity in patients with diabetic foot disease in Europe. Baseline results from the Eurodiale study. Diabetologia **50**(1), 18–25 (2007)
28. Rania, N., Douzi, H., Yves, L., Sylvie, T.: Semantic segmentation of diabetic foot ulcer images: dealing with small dataset in DL approaches. In: El Moataz, A., Mammass, D., Mansouri, A., Nouboud, F. (eds.) ICISP 2020. LNCS, vol. 12119, pp. 162–169. Springer, Cham (2020). https://doi.org/10.1007/978-3-030-51935-3_17
29. Ren, S., He, K., Girshick, R., Sun, J.: Faster R-CNN: towards real-time object detection with region proposal networks. IEEE Trans. Pattern Anal. Mach. Intell. **39**(6), 1137–1149 (2016)
30. Roglic, G., et al.: WHO Global report on diabetes: a summary. Int. J. Noncommun. Dis. **1**(1), 3 (2016)
31. Saeedi, P., et al.: Global and regional diabetes prevalence estimates for 2019 and projections for 2030 and 2045: results from the international diabetes federation diabetes atlas. Diabetes Res. Clin. Pract. **157**, 107843 (2019)
32. Tan, M., Le, Q.: EfficientNet: rethinking model scaling for convolutional neural networks. In: International Conference on Machine Learning, pp. 6105–6114. PMLR (2019)
33. Vardasca, R., Magalhaes, C., Seixas, A., Carvalho, R., Mendes, J.: Diabetic foot monitoring using dynamic thermography and AI classifiers. In: Proceedings of the 3rd Quantitative InfraRed Thermography Asia Conference (QIRT Asia 2019), Tokyo, Japan. pp. 1–5 (2019)
34. Xie, S., Girshick, R., Dollár, P., Tu, Z., He, K.: Aggregated residual transformations for deep neural networks. In: Proceedings of the IEEE Conference on Computer Vision and Pattern Recognition, pp. 1492–1500 (2017)
35. Xu, Y., Han, K., Zhou, Y., Wu, J., Xie, X., Xiang, W.: Classification of diabetic foot ulcers using class knowledge banks. Front. Bioeng. Biotechnol. 9 (2021)
36. Yap, M.H., Cassidy, B., Pappachan, J.M., O'Shea, C., Gillespie, D., Reeves, N.: Analysis towards classification of infection and ischaemia of diabetic foot ulcers. arXiv preprint arXiv:2104.03068 (2021)

# ClusterUDA: Latent Space Clustering in Unsupervised Domain Adaption for Pulmonary Nodule Detection

Mengjie Wang[1], Yuxin Zhu[1], Xiaoyu Wei[1], Kecheng Chen[4], Xiaorong Pu[1,3(✉)], Chao Li[2], and Yazhou Ren[1]

[1] School of Computer Science and Engineering, University of Electronic Science and Technology of China, Chengdu, China
puxiaor@uestc.edu.cn

[2] Cancer Hospital of University of Electronic Science and Technology School of Medicine, Chengdu, China

[3] NHC Key Laboratory of Nuclear Technology Medical Transformation (Mianyang Central Hospital), Mianyang, China

[4] Department of Electrical Engineering, City University of Hong Kong, Hong Kong, China

**Abstract.** Deep learning has achieved notable performance in pulmonary nodule (PN) detection. However, existing detection methods typically assume that training and testing CT images are drawn from a similar distribution, which may not always hold in clinic due to the variety of device vendors and patient population. Hence, the idea of domain adaption is introduced to address this domain shift problem. Although various approaches have been proposed to tackle this issue, the characteristics of samples are ignored in specific usage scenarios, especially in clinic. To this end, a novel unsupervised domain adaption method (namely ClusterUDA) for PN detection is proposed by considering characteristics of medical images. Specifically, a convenient and effective extraction strategy is firstly introduced to obtain the Histogram of Oriented Gradient (HOG) features. Then, we estimate the similarity between source domain and target one by clustering latent space. Finally, an adaptive PN detection network can be learned by utilizing distribution similarity information. Extensive experiments show that, by introducing a domain adaption method, our proposed ClusterUDA detection model achieves impressive cross-domain performance in terms of quantitative detection evaluation on multiple datasets.

**Keywords:** Domain adaption · pulmonary nodules detection · clustering

Supported in part by Open Foundation of Nuclear Medicine Laboratory of Mianyang Central Hospital (No. 2021HYX017), Sichuan Science and Technology Program (Nos. 2021YFS0172, 2022YFS0047, 2022YFS0055), Clinical Research Incubation Project, West China Hospital, Sichuan University (No. 2021HXFH004), Guangdong Basic and Applied Basic Research Foundation (No. 2020A1515011002), and Fundamental Research Fund for the Central Universities of China (No. ZYGX2021YGLH022).



M. Tanveer et al. (Eds.): ICONIP 2022, CCIS 1793, pp. 446–457, 2023.
https://doi.org/10.1007/978-981-99-1645-0_37

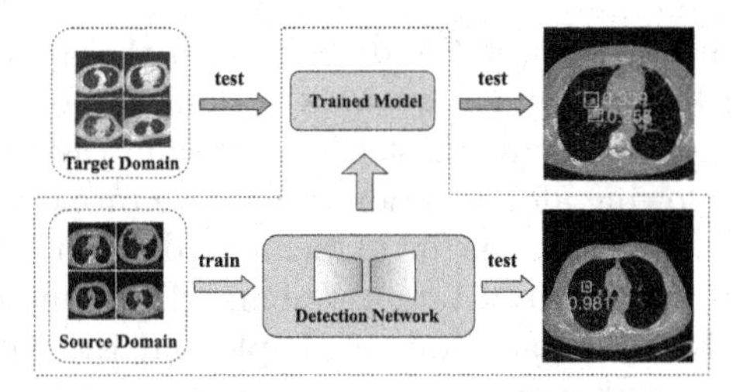


**Fig. 1.** Existing PN detection pipeline without domain adaption encounters a performance drop when testing on target domain. The red boxes outline the predicted pulmonary nodules. The yellow numbers represent the Intersection over Union (IoU) of the hitted nodule candidates. For IoU, the higher the better. (Color figure online)

# 1 Introduction

Lung cancer, one of the most common malignant tumors, is a leading cause of cancer deaths. The 5-year relative survival rate of lung cancer is only 4% at terminal stage [18]. However, early diagnosis of pulmonary nodule (PN) could decrease the mortality rate of patients for high-risk population effectively. As reported by [9], a pulmonary nodule is defined as an approximately round or irregularly shaped opacity lesion with diameter no more than 3 cm. On chest CT scans, a pulmonary nodule usually refers to a small block that occupies a few pixels.

In clinic, it is a troublesome task for radiologists to analyze the chest CT images and make the decision of potential abnormalities due to exhaustion, inattention and the lack of experience. Computer-assisted detection methods thus have been widely adopted for preliminary screening in PN detection, which benefit from their less time consumption and competitive detection performance. In recent years, automatic PN detection networks using convolutional neural network (CNN) have achieved impressive performance thanks to the development of deep learning. Overall, quite a few models aim to improve comprehensive detection accuracy on their adopted datasets. However, existing deep learning-based PN detection approaches still have some limitations that need to be considered.

The biggest issue is that most existing PN detection models ignore the potential discrepancy between training data (source domain) and test one (target domain), resulting from various patients, device vendors, and scan parameters (e.g. slice thickness). The model trained on source domain thus may suffer from performance degradation on target domain. For example, we observed an unaccepted decrease of detected effect when the detection model was trained on LUNA16 [24] dataset and tested on LNDb [20] dataset as shown in Fig. 1. It is no doubt that this phenomenon would increase diagnosed risk in practical clinic scenarios.

To tackle this issue, an intuitive idea is to alleviate distribution mismatch between source and target domains. By doing so, the training process on source pulmonary images can fuse the information from the target one and thereby benefit the performance of target PN detection. In this paper, we concentrate on PN detection task and pulmonary images domain adaption task, on consideration of the characteristic of chest CT scans and distribution information between domains. Following aforementioned idea, we propose a simple yet effective network for PN detection via introducing a domain adaption module to model distribution information of target domain. To be more specific, we estimate the similarity between each source sample and the entire target domain in latent feature space. Subsequently, the similarity is integrated into the training process for PN detection on source samples. To this end, the trained PN detection network is used to detect target samples, which can achieve a significant improvement on target domain. In summary, the major contributions in our paper are three-fold as follows:

1. To the best of our knowledge, this is the first attempt to take performance degradation between domains into account for PN detection by introducing a domain adaptive method.
2. On consideration of the characteristic between domains, we estimate similarity between source and target samples to alleviate distribution mismatch.
3. Our method can significantly alleviate performance degradation between domains, which would promote data sharing on multi-center clinical scenarios.

## 2 Related Work

### 2.1 PN Detection

Pulmonary nodule (PN) detection has been studied for many years. The mainstream PN detection methods are drawn from the object detection task for natural images in the field of computer vision. Faster R-CNN [21] and Retinanet [13] are widely adopted object detection frameworks. Many detection methods make some improvements based on Faster R-CNN to adapt medical imaging features [2,6,16,25,27]. Ding et al. [6] applied the two-dimensional Faster R-CNN to detect candidates for each slices, and the three-dimensional deep convolutional neural network for false positive reduction. Tang et al. [27] developed a end-to-end 3D deep convolutional neural network to solve nodule detection, false positive reduction and nodule segmentation jointly in a multi-task fashion by incorporating several useful tricks. To speed up training process and improve PN detection performance, Shao et al. [25] integrated slice context information in two-dimensional detection network and combined attention mechanism. All the methods mentioned above assume that train and test samples are drawn from similar distribution, which is not the case in clinic. Therefore, introducing a domain adaption method may be a feasible solution.

### 2.2 Domain Adaption

Domain adaption is proposed to mitigate the performance drop caused by variant distribution between domains. According to the availability of labels on target samples, domain adaption methods can be divided into three categories: supervised, semi-supervised and unsupervised methods. Our work focuses on unsupervised domain adaption (UDA) methods due to the bottleneck on high-quality annotated data for medical object detection task. The main challenge of the UDA methods [1,3,8,22,29] is the domain shift. DA-faster [3] is a Domain-adversarial Neural Network (DANN) [8] based framework for object detection task proposed to reduce domain shift by aligning features on image level and instance level. Inspired by [3], Saito et al. [22] learned domain-invariant features, including the local patch level and the global scene level. All these methods overcome the domain shift in adversarial training manner, which learn shared representation of domains by training a domain discriminator to reduce the feature distribution divergence. In addition, many researches adopt generative models to enforce image-level feature alignment on two domains. Bousmalis et al. [1] directly generated synthetic target images to mitigate domain shift. However, these methods were designed on specific application scenarios without consideration of similarity between source and target samples. Our methods attend to learn information according to similarity between source and target samples to boost the performance of PN detection.

## 3 Method

### 3.1 Preliminary

For existing PN detection tasks, quite a few efforts have been made to improve performance on a certain domain, the cross-domain discrepancy yet may be ignored. As discussed in Sect. 1, the detection performance would degrade on a cross-domain scenario. In this paper, we aim to tackle this issue by introducing a domain adaption method.

Note that the source domain $\mathcal{D}_s$ is fully labeled and the target domain $\mathcal{D}_t$ is completely unlabeled. $X$ and $B$ represent the pulmonary images and bounding boxes (which consists of 4 coordinates and can be regarded as the labels), respectively. We denote the source samples on a joint space $X_s \times B_s$ as $\mathcal{D}_s = \{x_s^i, b_s^i\}_i^{n_s}$, where $n_s$ denotes the number of source domain images. The unsupervised domain adaption is assumed in this paper, $B_t = \{b_t^j\}_j^{n_t}$ is thus invisible and $\mathcal{D}_t = \{x_t^j\}_j^{n_t}$ where $n_t$ denotes the number of target images. Note that subscript $s$ and $t$ represent the data from $\mathcal{D}_s$ and $\mathcal{D}_t$ respectively for abovementioned declarations.

### 3.2 Motivation

Before introducing our method, we analyze the differences between medical images and natural images. On the one hand, there are big differences between source domain and target one in natural images. For real pictures and cartoons (from PASCAL [7] and Clipart [11] datasets respectively), their position, geometric shapes and detailed textures are quite different. As for more similar pictures of original and foggy urban scene (from Cityscape [4] and FoggyCityscape [23]

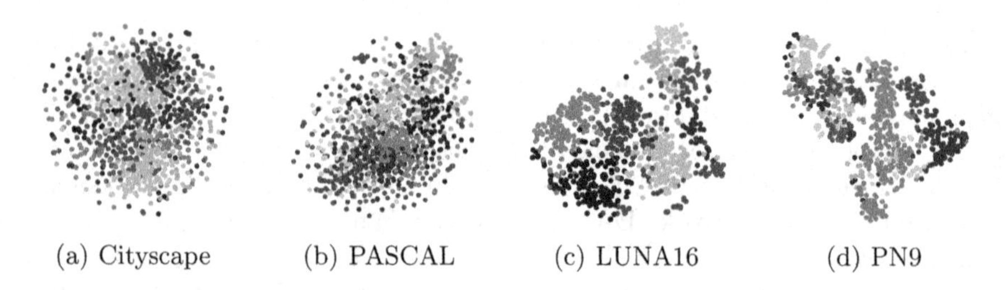

(a) Cityscape (b) PASCAL (c) LUNA16 (d) PN9

**Fig. 2.** Comparison of clustering results of features between natural images and medical images using t-SNE.

respectively), the fog will blur or cover part of the objects, and thereby widen the differences between domains. However, the contents of medical images are generally a certain part of the human body, which exist little variation between people, and thereby cause high degree similarity between domains. On the other hand, natural images contain rich colors, categories and scenes, whereas medical images contain monotonous scenes and simple semantic information.

To validate our analysis, we visualize the clustering results of features. As shown in Fig. 2, the natural datasets of Cityscape and PASCAL are cluttered with the dots of different colors mixed together, while medical datasets of LUNA16 and PN9 can be effectively discriminated. Intuitively, we can utilize the characteristic of medical images to design a special and effective domain adaption model for PN detection.

### 3.3 Cluster-based Domain Adaption

For pulmonary images between source domain and target one, some intrinsic characteristics may be leveraged as following

- Similarity: Source domain and target one exist highly semantic similarity such as similar background and the distribution of CT intensity value, as they are collected from the pulmonary region.
- Discrepancy: The distribution between domains is also slightly different, which usually results from various patients, device vendors, and scan parameters.

To this end, a ClusterUDA model is proposed to estimate the similarity between domains on consideration of aforementioned prior knowledge.

**Distribution Information Extraction.** In this subsection, we utilize a convenient traditional machine learning feature extraction method, which can maintain good invariance to the geometric and optical deformation of the images. As shown in Fig. 2(c) and Fig. 2(d), this convenient strategy is effective enough for medical images.

Given a pulmonary image $x$, we impose a feature extractor to obtain Histogram of Oriented Gradient (HOG) features [5] after resizing the image to a

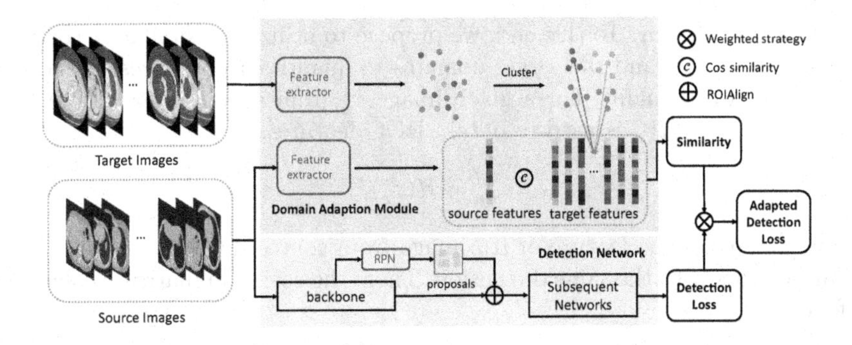


**Fig. 3.** The framework of CluserUDA PN detection network.

fixed size. The process of the HOG method can be simply described as: a) Divide input images to several small cell units. b) Calculate and compute the gradient direction histogram for each cell unit. c) Combine these histograms into a vector as the extracted feature. The process can be formulated as follow:

$$f_t^j = \mathcal{H}(x_t^j), j = 1, 2, \ldots, n_t, \tag{1}$$

where $f_t^j$ denotes the feature from the $j$th target image. $\mathcal{H}(\cdot)$ denotes the HOG feature extractor while ignoring image preprocessing. Following the same step, all HOG features can be extracted from the entire target domain.

$$\mathcal{F}_t = \{f_t^1, f_t^2, \ldots, f_t^{n_t}\}, \tag{2}$$

where $\mathcal{F}_t$ denotes extracted HOG features from all target samples. In order to further model the distribution information from target domain, we propose to cluster latent space on extracted features $\mathcal{F}_t$, which is inspired by [15] (who constructed subspaces by clustering the samples from the same class for exploring distribution of attribute vectors). Specially, pulmonary images can be regarded as the same class due to monotone colors, single category and similar contents.

Note that we introduce k-means clustering algorithm with k-means++ [28] initialization method to divide images into $k$ sets and obtain features of $k$ center points finally. The clustering step of domain adaption component is formulated as below:

$$\mathcal{F}^* = \{f_c^1, f_c^2, \ldots, f_c^k\} = \mathcal{C}(\mathcal{F}_t), \tag{3}$$

where $\mathcal{F}^*$ denotes target features shown in the dashed box of Domain Adaption Module in Fig. 3, $f_c^m$ denotes the feature of $i$th center point and $m = 1, 2, \ldots, k$, $\mathcal{C}(\cdot)$ denotes the clustering method. Specially, we utilize $\mathcal{F}^*$ to represent modeled distribution information.

**Image Similarity.** In this subsection, we will describe how to further encode aforementioned distribution information $\mathcal{F}^*$. [3] assumed that if the image distribution between source and target domains is similar, the distribution of detection objects is also similar. To this end, we propose to utilize the similarity (between each source sample and the target domain) to quantitatively represent the relationship of two domains. For a given image $x_s^i$ from source domain, the same feature extractor $\mathcal{H}(\cdot)$ is used to obtain HOG features,

$$f_s^i = \mathcal{H}(x_s^i), \tag{4}$$

where $f_s^i$ denotes the feature of $i$th image from source domain. $f_s^i$ and $f_c^m$ are feature vectors with the same dimension. Then, the cosine similarity is calculated as follow:

$$cs_{i\_m} = \cos(f_c^m, f_s^i) = \frac{f_c^m \cdot f_s^i}{\|f_c^m\| \|f_s^i\|}, m = 1, 2, \ldots, k, \tag{5}$$

where $cs_{i\_m}$ denotes cosine similarity between $x_s^i$ and $m$th center point. $\|\cdot\|$ denotes Euclidean norm. Finally, we leverage the maximum value of $k$ cosine similarities to represent the image similarity with target domain.

$$\delta_i = \max_{1 \le m \le k} cs_{i\_m}, m = 1, 2, \ldots, k, \tag{6}$$

where $\delta_i$ denotes the similarity of $x_s^i$, which is a quantitative representation to measure how similar the source and target samples are. Following the same steps, we can obtain the similarity for each source sample.

The $\delta_i$ will be used to optimize the training loss of detection network, which will be discussed in next subsection.

### 3.4 PN Detection Network

**Faster R-CNN.** Before introducing our method, we review the plain detection framework Faster R-CNN [21], which is an end-to-end detection network including the backbone, the RPN, the ROI Pooling layer, and the subsequent networks (the classifier). The pipeline consists of two main steps. An input image is firstly fed into the backbone network to extract feature maps. Then, the RPN is used to generate candidate region proposals based on extracted feature maps. The ROI Pooling layer collects proposals on shared feature maps and outputs proposal feature maps with fixed size. Finally, the classifier will complete box prediction and classification. In our method, we utilize a slightly modified Faster R-CNN (which replaces the ROI Pooling layer with the ROI Align layer). By the way, the ROI Align method uses floating point number to precisely represent the coordinates of the candidate box before the pooling operation, which achieves more accurate detection and segmentation performance than the ROI Pooling strategy. The total detection loss of Faster R-CNN is represented as $L_{det}$.

**Image Weighting.** Here, we will describe how to integrate the similarity into modified Faster R-CNN. In order to eliminate the domain distribution mismatch, we propose to focus on the source domain images that are similar to target domain during the training process. According to the instance weighting method,

**Table 1.** The main parameters of used datasets.

| Dataset | LUNA16 | LNDb | PN9 |
|---|---|---|---|
| Modality | CT | CT | CT |
| Nodule size | $\geq$ 3 mm | $\geq$ 3 mm | 0–5 mm |
| Slice interval | 0.45–2.50 mm | 1.00 mm | 0.40–2.50 mm |
| Pixel spacing | 0.46–0.98 mm | 0.43–0.90 mm | 0.31–1.09 mm |
| Slice total | 1171 | 621 | 4650 |
| Nodule total | 1186 | 629 | 4690 |

we introduce the adapted detection loss to weight original detection loss with the computed similarity. The adapted detection loss $L$ can be expressed as following:

$$L = (1 + \theta \cdot \delta) \cdot L_{det}, \tag{7}$$

where $\delta$ denotes the similarity between each source domain sample and the entire target domain. $\theta$ is a trade-off parameter to adjust the effect of $\delta$ and balance the original detection loss and our weighting loss. It is worth mentioned that the existing detection networks will be learned better via a target-oriented loss. By doing so, our strategy can be performed on other detection models easily.

### 3.5 Network Overview

In this subsection, the proposed ClusterUDA model is illustrated in Fig. 3. Our network consists of domain adaption module and detection network, respectively. In the domain adaption module, the distribution information of target domain is encoded. In addition, a quantitative representation of similarity for each pulmonary image from source domain is computed. In detection network, we utilize obtained similarity to weight the training loss in the Faster R-CNN.

## 4 Experiment

### 4.1 Dataset

Our experiments are conducted on three datasets, including LUNA16 [24], LNDb [20], and PN9 [17] (more details can be found in Table 1). For better performance and convenience, several preprocessing steps are performed. For PN9, we only utilize a subset of the given dataset (it is already enough for the experiments) after the processing steps demonstrated in [17]. For LUNA16 and LNDb, the slice spacing between two slices and pixel spacing of each slice are all normalized to 1 mm. Then, we adjust window width to 1800 Hounsfield Unit (HU) and window level to -300 HU. These settings would best observe pulmonary nodules in clinic. Finally, all labels of three datasets are transferred to COCO [14] format. Note that only key slices of nodules are extracted as the input of detection framework in our experiments. We use average precision (AP) at 0.5 Intersection over Union (IoU) threshold (namely AP50 in [14]) as the evaluation metric.

**Table 2.** The performance comparison between ClusterUDA (Ours) and other methods on LUNA16 (A), LNDb (B) and PN9 (P) datasets, in terms of AP50 (%)

| Method | A→B | P→B | B→A | P→A | A→P | B→P | Average |
|---|---|---|---|---|---|---|---|
| Source Only | 39.93 | 38.35 | 52.82 | 21.42 | 13.63 | 16.56 | 30.45 |
| DA-Faster | 22.79 | 14.42 | 37.26 | 18.42 | 24.45 | 20.25 | 22.93 |
| Strong-Weak | 20.60 | 14.60 | 36.72 | 19.76 | 25.89 | 20.95 | 23.09 |
| SCL | 23.18 | 14.26 | 43.51 | 19.91 | **25.96** | 20.56 | 24.56 |
| DA-RetinaNet | 35.92 | 36.73 | 47.13 | 32.70 | 18.78 | 19.01 | 31.71 |
| Ours | **44.21** | **52.70** | **57.11** | **51.18** | 19.26 | **27.16** | **41.94** |

**Table 3.** The influence of different strategies, including calculation method ($\theta$), trade-off parameter ($\delta$) and feature extractor ($\mathcal{H}$), in terms of AP50 (%)

| | | A→B | P→B | B→A | P→A | A→P | B→P | Average |
|---|---|---|---|---|---|---|---|---|
| $\delta$ | max | 44.97 | **51.78** | 59.17 | **55.55** | 20.67 | **25.49** | **42.94** |
| | min | **48.15** | 49.22 | 57.76 | 48.03 | 19.97 | 23.08 | 41.03 |
| | avg | 42.19 | 48.71 | **60.11** | 46.30 | **20.70** | 23.93 | 40.32 |
| $\theta$ | 1 | **44.97** | **51.78** | **59.17** | **55.55** | 20.67 | **25.49** | **42.94** |
| | 2 | 43.34 | 49.22 | 56.66 | 50.27 | **20.70** | 23.31 | 40.58 |
| | 5 | 40.16 | 47.28 | 56.92 | 47.19 | 19.53 | 25.24 | 39.39 |
| $\mathcal{H}$ | hog | 44.97 | **51.78** | 59.17 | **55.55** | 20.67 | **25.49** | **42.94** |
| | ResNet50 | **47.18** | 50.29 | 60.20 | 51.82 | **22.29** | 25.37 | 42.86 |
| | ResNet101 | 45.58 | 51.62 | **60.37** | 52.22 | 21.91 | 22.32 | 42.33 |

### 4.2 Implementation Details

We conduct experiments of our method on detectron2 benchmark [30] developed by Facebook AI Research. Pre-trained ResNet50 [10] and ResNet101 [10] with Feature Pyramid Networks (FPN) [12] are employed as backbones in our experiments. The training pipeline is optimized by stochastic gradient descent with batch size as 4, max iteration as 25000, base learning rate as 0.01 and iteration number to decrease learning rate between 10000 to 20000. The size of all pulmonary images is stretched to 512×512. Other detection parameters are consistent with Faster R-CNN in detectron2 benchmark.

### 4.3 Results

In this section, we report the results of comparison experiments. To the best of our knowledge, the proposed ClusterUDA model is the first attempt to take the cross-domain pulmonary detection into account. We thus compared our method with existing domain adaption detection methods, such as DA-Faster [3], Strong-Weak [22], SCL [26] and DA-RetinaNet [19]. In order to be consistent with these models, we use the ResNet101 as backbone in the comparative experiments. Table 2 shows the comparing quantitative results.

As we can see, our method can achieve better performance than Faster R-CNN (Source Only), which is reasonable as our proposed ClusterUDA inserts the domain adaption module to reduce the discrepancy between domains. We can also observe that the performance of baseline methods degrade significantly compared with the model trained on source domain only on most datasets. The phenomena may result from possible reasons: First, as reported by [22], complete feature matching can disturb the training process for detection and thereby degrade the performance on target domain. Second, these baseline methods may not be adaptive to medical images well, although we try our best to adjust their settings. Note that the experiments are conducted on different benchmarks, the detectron2 used by source only model has made some optimizations (e.g. initialization, code implementation) to improve the original detection performance. Here, our method achieves slightly lower performance than other UDA models on LUNA16 $\rightarrow$ PN9. The mismatch between data and labels may lead to this performance degradation of cross-domain detection for our proposed method. As shown in Table 1, LNDA16 and LNDb only mark nodules $\geq$ 3 mm, but PN9 marks nodules in 0–5 mm (e.g. a 2 mm nodule would be labeled in PN9 but ignored in LNDb and LUNA16), which may cause quite a few nodules to be missed. Besides, we can see that our method performs best on average among the source only model and domain adaption methods for detection. Overall, the proposed ClusterUDA achieves better comprehensive detection performance.

## 4.4 Model Analysis

In order to verify the effectiveness of each key module, we perform parametric analysis experiments and ablation study (which utilize ResNet50 as backbone) on our network. In this section, we mainly explored the calculation method $\delta$ to compute similarity, the trade-off parameter $\theta$ and the feature extractor $\mathcal{H}$. As shown in Table 3, the first column denotes parameters and the second column denotes different strategies of these parameters. For each strategy on certain cross-domain datasets, the best results are highlighted in Bold fonts.

The similarity is used to weight the detection loss. We are thus interested in the calculation method of estimating the similarity. The quantitative results of different methods to calculate similarity is shown in Table 3 at the first sub-block. The *min*, *max*, *avg* denote minimum, maximum, average value of $k$ (the number of center points obtained by elbow method) cosine similarity respectively. We can observe that the calculation method of *max* achieves best performance in five cases. In addition, the average effect of *max* operation is also the best. To this end, we adopt the calculation method of *max* in our network.

The trade-off parameter $\theta$ is used to weight similarity and thereby adjust weighted loss in detection network. At the second sub-block in Table 3, it is shown that lower values achieve better effect. We set the value of $\theta = 1$ finally in our method on account of the slight advantage comparing with others.

The extracted feature in domain adaption module will affect subsequent clustering effects, we therefore wonder the performance of different feature extractor $\mathcal{H}$. As shown in Table 3 at the last sub-block, we compare ResNet (pre-trained on natural images) and HOG. It is shown that the HOG has achieved better results than ResNet overall. Intuitively, the pre-trained ResNet unable to extract suitable features due to a modal gap between medical and natural images. However, HOG (which extracts features according to the histogram) is decoupled from the image mode. Overall, we utilize HOG as featuer extractor to benefit PN detection effects.

## 5 Conclusion and Future Work

In this paper, we propose a ClusterUDA detection network for pulmonary images to mitigate the domain distribution mismatch. A domain adaption module is conducted to model the distribution information of target domain by clustering latent space. A novel target-oriented objective is further introduced to alleviate the performance degradation in the detection network. The experimental results show that our proposed method achieved an impressive improvement for PN detection via comprehensive evaluations. In the future, we will do further research on more effective methods for similarity evaluation.

## References

1. Bousmalis, K., Silberman, N., Dohan, D., Erhan, D., Krishnan, D.: Unsupervised pixel-level domain adaptation with generative adversarial networks. In: CVPR, pp. 95–104 (2017)
2. Chen, K., Long, K., Ren, Y., Sun, J., Pu, X.: Lesion-inspired denoising network: connecting medical image denoising and lesion detection. In: ACM MM, pp. 3283–3292 (2021)
3. Chen, Y., Li, W., Sakaridis, C., Dai, D., Van Gool, L.: Domain adaptive faster R-CNN for object detection in the wild. In: CVPR, pp. 3339–3348 (2018)
4. Cordts, M., et al.: The cityscapes dataset for semantic urban scene understanding. In: CVPR, pp. 3213–3223 (2016)
5. Dalal, N., Triggs, B.: Histograms of oriented gradients for human detection. In: CVPR, pp. 886–893 (2005)
6. Ding, J., Li, A., Hu, Z., Wang, L.: Accurate pulmonary nodule detection in computed tomography images using deep convolutional neural networks. In: MICCAI, pp. 559–567 (2017)
7. Everingham, M., Van Gool, L., Williams, C.K., Winn, J., Zisserman, A.: The pascal visual object classes (VOC) challenge. IJCV, 303–338 (2010)
8. Ganin, Y., et al.: Domain-adversarial training of neural networks. JMLR **17**, 1–35 (2016)
9. Hansell, D.M., Bankier, A.A., MacMahon, H., McLoud, T.C., Muller, N.L., Remy, J.: Fleischner society: glossary of terms for thoracic imaging. Radiology, 697–722 (2008)
10. He, K., Zhang, X., Ren, S., Sun, J.: Deep residual learning for image recognition. In: CVPR, pp. 770–778 (2016)
11. Inoue, N., Furuta, R., Yamasaki, T., Aizawa, K.: Cross-domain weakly-supervised object detection through progressive domain adaptation. In: CVPR, pp. 5001–5009 (2018)

12. Lin, T.Y., Dollár, P., Girshick, R., He, K., Hariharan, B., Belongie, S.: Feature pyramid networks for object detection. In: CVPR (2017)
13. Lin, T.Y., Goyal, P., Girshick, R., He, K., Dollár, P.: Focal loss for dense object detection. In: ICCV, pp. 2980–2988 (2017)
14. Lin, T.Y., et al.: Microsoft COCO: common objects in context. In: ECCV, pp. 740–755 (2014)
15. Liu, X., et al.: VoxelHop: successive subspace learning for ALS disease classification using structural MRI. arXiv preprint arXiv:2101.05131 (2021)
16. Long, K., et al.: Probability-based mask R-CNN for pulmonary embolism detection. Neurocomputing, pp. 345–353 (2021)
17. Mei, J., Cheng, M.M., Xu, G., Wan, L.R., Zhang, H.: SANet: a slice-aware network for pulmonary nodule detection. TPAMI (2021)
18. Miller, K.D., et al.: Cancer treatment and survivorship statistics, 2019. CA Cancer J. Clin., 363–385 (2019)
19. Pasqualino, G., Furnari, A., Signorello, G., Farinella, G.M.: An unsupervised domain adaptation scheme for single-stage artwork recognition in cultural sites. Image Vis. Comput., 104098 (2021)
20. Pedrosa, J., et al.: LNDb: a lung nodule database on computed tomography. arXiv preprint arXiv:1911.08434 (2019)
21. Ren, S., He, K., Girshick, R., Sun, J.: Faster R-CNN: towards real-time object detection with region proposal networks. In: NIPS (2015)
22. Saito, K., Ushiku, Y., Harada, T., Saenko, K.: Strong-weak distribution alignment for adaptive object detection. In: CVPR, pp. 6956–6965 (2019)
23. Sakaridis, C., Dai, D., Van Gool, L.: Semantic foggy scene understanding with synthetic data. IJCV, 973–992 (2018)
24. Setio, A.A.A., et al.: Validation, comparison, and combination of algorithms for automatic detection of pulmonary nodules in computed tomography images: the luna16 challenge. Med. Image Anal., 1–13 (2021)
25. Shao, Q., Gong, L., Ma, K., Liu, H., Zheng, Y.: Attentive CT lesion detection using deep pyramid inference with multi-scale booster. In: MICCAI (2019)
26. Shen, Z., Maheshwari, H., Yao, W., Savvides, M.: SCL: towards accurate domain adaptive object detection via gradient detach based stacked complementary losses. arXiv preprint arXiv:1911.02559 (2019)
27. Tang, H., Zhang, C., Xie, X.: NoduleNet: decoupled false positive reduction for pulmonary nodule detection and segmentation. In: MICCAI, pp. 266–274 (2019)
28. Vassilvitskii, S., Arthur, D.: K-means++: the advantages of careful seeding. In: SODA (2007)
29. Wei, P., Zhang, C., Li, Z., Tang, Y., Wang, Z.: Domain-adaptation person re-identification via style translation and clustering. In: Mantoro, T., Lee, M., Ayu, M.A., Wong, K.W., Hidayanto, A.N. (eds.) ICONIP 2021. LNCS, vol. 13108, pp. 464–475. Springer, Cham (2021). https://doi.org/10.1007/978-3-030-92185-9_38
30. Wu, Y., Kirillov, A., Massa, F., Lo, W.-Y., Girshick, R.: Detectron2. https://github.com/facebookresearch/detectron2 (2019)

# Image Captioning with Local-Global Visual Interaction Network

Changzhi Wang and Xiaodong Gu(✉)

Department of Electronic Engineering, Fudan University, Shanghai 200438, China
xdgu@fudan.edu.cn

**Abstract.** Existing attention based image captioning approaches treat local feature and global feature in the image individually, neglecting the intrinsic interaction between them that provides important guidance for generating caption. To alleviate above issue, in this paper we propose a novel Local-Global Visual Interaction Network (LGVIN) that novelly explores the interactions between local feature and global feature. Specifically, we devise a new visual interaction graph network that mainly consists of visual interaction encoding module and visual interaction fusion module. The former implicitly encodes the visual relationships between local feature and global feature to obtain an enhanced visual representation containing rich local-global feature relationship. The latter fuses the previously obtained multiple relationship features to further enrich different-level relationship attribute information. In addition, we introduce a new relationship attention based LSTM module to guide the word generation by dynamically focusing on the previously output fusion relationship information. Extensive experimental results show that the superiority of our LGVIN approach, and our model obviously outperforms the current similar relationship based image captioning methods.

**Keywords:** Image captioning · Visual interaction · Graph network

## 1 Introduction

Image captioning aims to automatically predict a meaningful and grammatically correct natural language sentence that can precisely and accurately describe the main content of a given image [17].

Some related visual attention based methods [2,6] achieved remarkable progress for image captioning. However, these methods treat the image local feature and global feature individually, which ignore the intrinsic interactions between them. That makes the model difficult to generate an accurate or appropriate description that can correctly describe the interactions between local feature and global feature in the image. For example, as illustrated in Fig. 1(Left), in the image caption "*A dog standing in a small boat in the water.*", the verb

This work was supported by the National Natural Science Foundation of China under grant 62176062.


M. Tanveer et al. (Eds.): ICONIP 2022, CCIS 1793, pp. 458–469, 2023.
https://doi.org/10.1007/978-981-99-1645-0_38

**Caption:**
A dog standing in a small boat in the water.

**GLA (base):** Three boys are running on the playground.
**LGVIA (ours):** Three children are playing baseball in a field.

**Fig. 1.** (Left) Illustration of interaction information in generated sentence. An interaction word/phase often appears between two nouns, which is often a verb word/phrase (e.g., "standing in") or preposition (e.g., "in"). (Right) Illustration GLA (base) baselines and our LGVIN method for image captioning. Blue denotes the image local information, red means the global information, and underline denotes their interaction words.

word/phrase "standing in" denotes an interaction between the words "dog" and "boat", and the preposition word/phrase "in" indicates an interaction between the words "boat" and "water". Further, as illustrated in Fig. 1(right), GLA (base) [9] method directly takes the image local feature and global feature as the input of language model, ignoring the visual interactions between local feature and global feature in the image. Therefore, GLA (base) generates an inappropriate caption (i.e., *"Three boys are running on the playground."*), which falsely predicts the interactive word "playing" as the "running on". Contrarily, the caption *"Three children are playing baseball outside in a field."* generated by our LGVIN accurately describes the input image main content. The word "playing" in the caption appropriately represents the interaction between the two local regions "children" and "baseball", and the word "in" accurately describes the interaction between the local region "children" and the global concept "field". More importantly, generating such an interactive word "in" (or "playing") requires the relationship encoded on local region "children" and global concept "field" (or local regions "children" and "baseball". ). Therefore, considering the relationships between local feature and global feature in the image is needed by the caption decoder (e.g., LSTM) to generate relationship-focused words.

Based on the above observations, different from existing relationship based methods [10,18,23] (See Fig. 2 ) that explore the relationships between local feature or global feature separately, this work proposes a novel local-global visual interaction network which novelly leverages the improved Graph AtTention network (GAT) to automatically learns the visual relationships between local feature and global feature in the image, as presented in Fig. 3.

The main contributions of our work are as follows:

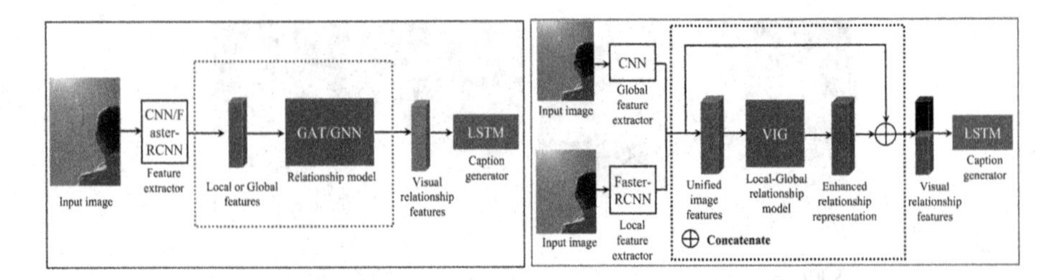


**Fig. 2.** (Left) Relationship based methods [10,18,23] usually use GAT/GNN (Graph Neural Network) to explore the relationships between local feature or global feature separately. (Right) Our LGVIN method can learn the visual relationships between local feature and global feature. In addition, the visual relationship features effectively retains the original image local-global features.

- We propose a novel Local-Global Visual Interaction Network (LGVIN) for image captioning, which fully takes into account the joint local and global feature relationships.
- We devise a new visual interaction graph network which consists of a visual interaction encoding module and a visual interaction fusion module. The former automatically encodes the relationships between local feature and global feature, and latter fuses previously obtained interaction information.
- Extensive experiments comprehensively indicate that our model achieves superior performance and outperforms the similar relationship based image captioning methods.

## 2 Related Work

**Relationship Based Methods:** Currently, the relationship based methods can effectively boost the performance of image captioning model. For example, Wang et al. [18] exploited a Graph Neural Network (GNN) to establish the visual relationship between image salient regions in which each visual region is regarded as a graph node, and all the nodes are fully connected in the graph. However, the key adjacency matrix in GNN needs to be constructed in advance. [23] explored the semantic and spatial relationship between the detected image regions of objects by using graph convolutional networks plus long short-term memory. Li et al. [10] explored the visual relationship, which explicitly uses the semantic relationship triples (scene graph) as additional inputs, and then introduce the hierarchical visual attention. In our work, different from these similar relationship based methods that only explore the relationships between local feature or global feature separately, we propose a LGVIN framework, which utilizes a new visual interaction encoding module to learn the intrinsic relationships between local feature and global feature in the image, and devise a new visual interaction fusion module to fusion the output local-global feature relationships from previous multiple visual interaction encoding layers.

# 3 Proposed Approach

The main purpose of this work is to learn an accurate interactive information between image local feature and global feature for generating sentence. In the following contents, some key elements in the LGVIN framework are introduced.

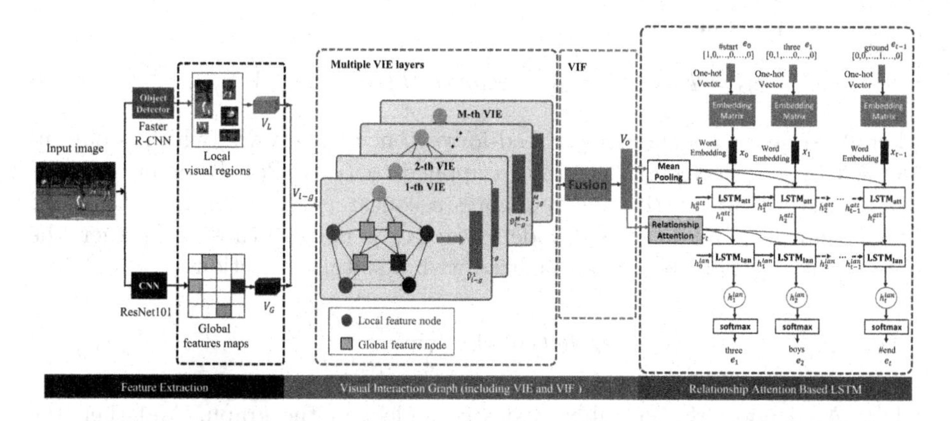


**Fig. 3.** The framework of the proposed method. It consists of three modules, namely Feature Extraction module, Visual Interaction Graph (VIG) network (including Visual Interaction Encoding (VIE) and Visual Interaction Fusion (VIF) modules) and Relationship Attention Based LSTM module. We first utilize Feature Extraction module to extract local features $V_L$ and global features $V_G$ from the input image. These features are fed into VIE module to capture the visual relationships between the image features. After that, an enhanced local-global visual interaction representation $V_f = \{\widetilde{V}_{l-g}^1, ..., \widetilde{V}_{l-g}^M\}$ from multiple VIE layers are fused by VIF module. Further, the obtained fused feature $V_o$ is input into the Relationship Attention based LSTM module for word generation.

## 3.1 Feature Extraction

We first utilize an object detector to obtain the local features $L$, as $L = \{l_1, ..., l_k\}$, $l_k \in \mathbb{R}^{D_l}$, where $D_l$ is the dimension of each local features, and $k$ is the number of local visual regions. Then, we use CNN to extract the global feature $G$. Further, the feature maps of the last layer is used and reshaped into a set of features $G = \{g_1, ..., g_n\}$, $g_n \in \mathbb{R}^{D_p}$, where $D_p$ is the each pixel dimension, and $n$ is the reshaped size of the feature maps. In addition, to embed them into the same dimensional space, a single layer perceptron with rectifier activation function is utilized to transform these feature vectors into new vectors with dimension $D_e$,

$$V_L = ReLU(W_r L + b_r), \quad (1)$$

$$V_G = ReLU(W_g G + b_g), \quad (2)$$

where $W$ and $b$ are model trainable parameters. $V_L \in \mathbb{R}^{D_e}$ and $V_G \in \mathbb{R}^{D_e}$ are the final extracted image local feature and global feature.

### 3.2 Visual Interaction Graph

**Graph AtTention Network:** Given a fully-connected graph $G = (V, E)$, where $V = \{v_1, ..., v_i, ..., v_N\}$, $(v_i \in \mathbb{R}^{D_e})$ is the node feature set, and $E = \{e_{1j}, ..., e_{ij}, ..., e_{nj}\}$ denotes the edge set. Following the previous work [16], for each pair nodes $(v_i, v_j)$, we calculate the correlation score $e_{ij}$ by performing the self-attention operation,

$$e_{ij} = \Psi(Wv_i, Wv_j) = LeakyReLU(\beta^T[Wv_i||Wv_j]), \tag{3}$$

where $\Psi(\cdot)$ denotes a single-layer feed-forward neural network, which is parameterized by weight vector $\beta$ and then applied to *LeakyReLU* nonlinearity, $W$ is a shared weight matrix, and $||$ denotes a concatenation operation. Equation (3) indicates the importance of node feature $v_i$ to node feature $v_j$. Further, the obtained correlation score is normalized with a softmax function,

$$\alpha_{ij} = softmax(e_{ij}) = \frac{exp(e_{ij})}{\sum_{z \in N_i} exp(e_{iz})}, \tag{4}$$

where $N_i$ denotes the neighbor nodes of node $i$ in the graph. And then the multi-head attention is used to stabilize the learning process of self-attention as,

$$v_i' = ||_{h=1}^{H} \delta(\sum_{j \in N_i} \alpha_{ij}^h W_{N_i}^h v_j), \tag{5}$$

where $W_{N_i}^h$ is a learnable matrix and $\delta$ is a activation function.

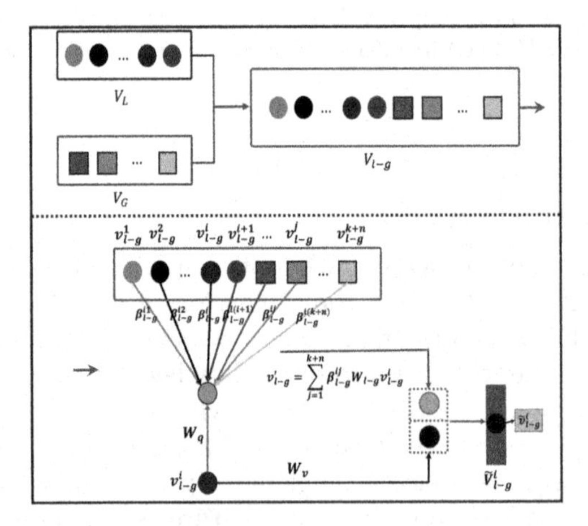


**Fig. 4.** Illustration of the proposed visual interaction encoding module. The representation of the aggregated feature $v_{l-g}'$ is reconstructed by aggregating information from all neighboring nodes with attention mechanism.

**Visual Interaction Encoding:** To obtain an enhanced visual representation containing rich local-global relationship information, we devise the new VIE module to learn the intrinsic relationships between the local feature and complementary global feature. As illustrated in Fig. 4, firstly, to generate a relationship graph, for the global feature maps, we define each $1 \times 1$ x $c$ grid as a global graph node $v_{gn}$, and $c$ is the number of global feature channel; for the local feature maps, we define each detected object as a local graph node $v_{ln}$. Then, we further construct a visual interaction graph $G_{l-g} = (V_{l-g}, E_{l-g})$ to model the intrinsic relationship between local feature and global feature, where $V_{l-g} = \{v_{l-g}^1, ..., v_{l-g}^{k+n}\}$, $v_{l-g}^i \in \mathbb{R}^{D_e}$ is a new local-global feature node set which is concatenated by the defined local graph node $V_{ln} = \{v_{ln}^1, ..., v_{ln}^k\}$ and global graph node $V_{gn} = \{v_{gn}^1, ..., v_{gn}^n\}$, and $E_{l-g}$ is the edge set of VIG. Different to the Eq. (3), to adaptively learn a better visual representation between the feature nodes, we first apply the linear transformations to the node features, and then replace the single-layer feed-forward neural network with the inner product operation to compute the correlation score as follows,

$$e_{l-g}^{ij} = \Psi(W_q v_{l-g}^i, W_k v_{l-g}^j), \tag{6}$$

$$\Psi(W_q v_{l-g}^i, W_k v_{l-g}^j) = (W_q v_{l-g}^i)^T (W_k v_{l-g}^j), \tag{7}$$

where $W$ are the linear transformation matrices. Equation (7) measures the similarity of the two feature nodes $v_{l-g}^i$ and $v_{l-g}^j$, i.e., the higher the correlation between local feature and global feature, the higher the edges affinity scores.

Similar to Eq. (4), we further normalize $e_{l-g}^{ij}$ with a softmax function,

$$\beta_{l-g}^{ij} = \frac{exp(e_{l-g}^{ij})}{\sum_{z \in N_i} exp(e_{l-g}^{iz})}, \tag{8}$$

Intuitively, $\beta_{l-g}^{ij}$ measures how much attention the $i$-th local feature should pay to the $j$-th global feature.

We then calculate the aggregated information for feature node $i$ with,

$$v_{l-g}' = \sum\nolimits_{j \in (k+n)} \beta_{l-g}^{ij} W_{l-g} v_{l-g}^j, \tag{9}$$

where $W_{l-g}$ denotes a linear transformation matrix.

Notably, unlike Eq. (5), we further concatenate the aggregated feature $v_{l-g}'$ with the original local-global node feature $v_{l-g}^i$ and then exploit the ReLU non-linear activation to obtain a more powerful feature representation,

$$\tilde{v}_{l-g}^i = ReLU(v_{l-g}' || (W_v v_{l-g}^i)), \tag{10}$$

where $W_v$ is a linear transformation matrix. The obtained visual interaction encoding features $\tilde{V}_{l-g} = \{\tilde{v}_{l-g}^1, ..., \tilde{v}_{l-g}^{k+n}\}$ will be used for subsequent visual interaction fusion.

**Visual Interaction Fusion:** We input the concatenated feature $V_{l-g}$ into $M$ different VIE layers, and the output is represented as $V_f = \{\tilde{V}_{l-g}^1, ..., \tilde{V}_{l-g}^M\}$, where $\tilde{V}_{l-g}^i$ is the $i$-th feature vector of $\tilde{V}_{l-g}$. In addition, we apply a mean-pooling operation on $\tilde{V}_{l-g}^i$, $i \in [1, M]$,

$$\tilde{V}_{l-g}^i = Mean(\tilde{V}_{l-g}^i), \tag{11}$$

Further, to enrich different-level relationship attribute features, we design a new VIF module to fuse the output multi-head feature $V_f$. Specifically, we take two feature vectors $\tilde{V}_{l-g}^i$ and $\tilde{V}_{l-g}^j$ as the input and output a fused relationship feature representation, the corresponding fusion process is as follows,

$$\tilde{V}_1 = W_1\tilde{V}_{l-g}^i, \tilde{V}_2 = W_2\tilde{V}_{l-g}^j, \ell = \gamma(\tilde{V}_1 W_{\tilde{V}_1} + \tilde{V}_2 W_{\tilde{V}_2}) \tag{12}$$

$$\tilde{V} = F(\tilde{V}_1, \tilde{V}_2) = \ell\tilde{V}_1 + (1-\ell)\tilde{V}_2, \tag{13}$$

where $W$ are the model learnable parameters, $\gamma$ denotes a sigmoid function that projects the fusion coefficient $\ell$ to the value (0,1), and $F(\cdot)$ is the fused function.

### 3.3 Relationship Attention Based LSTM

We introduce a relationship attention based LSTM module to generate sentence by feeding the obtained visual representation $V_o = \{v_o^1, ..., v_o^{n+k}\}$ into a two-layer LSTM. As presented in the right part of Fig. 3, it mainly includes attention LSTM layer, relationship attention mechanism and language LSTM layer.

**Attention LSTM Layer:** To generate attention information, attention LSTM (i.e., $LSTM_{att}$) takes the concatenation of the input word $x_t = W_e e_t$, $t \in [1, l]$, ($W_e$ is an embedding matrix, $e_t$ is a corresponding one-hot vector of each word) the previous output $h_{t-1}^{lan}$ of language LSTM, and the mean-pooled image fused feature $\bar{u} = \dfrac{1}{n+k}\sum_{i=1}^{n+k} v_o^i$ as input. Thus, the updating procedure of attention LSTM at time step $t$ is as

$$h_t^{att} = f_{att}^{lstm}(h_{t-1}^{att}, [x_t; h_{t-1}^{lan}; \bar{u}]), \tag{14}$$

where $h_t^{att}$ is the current output of attention LSTM, and $f_{att}^{lstm}$ is the updating function of attention LSTM.

**Relationship Attention Mechanism:** In many cases, a generated word in the sentence is only related to some of the features contained in the image. Therefore, the relationship attention ensures that model can adaptively focus on the specific relationship information related to the predicted word at each time step. Specifically, we generate a normalized attention distribution over the fused visual interactive feature $v_o^i$ depending on the previous hidden state $h_{t-1}^{att}$ of attention LSTM, as

$$\eta_t^i = W_\eta tanh(W_v v_o^i + W_h h_{t-1}^{att}), \quad (15)$$

$$\alpha_t^i = softmax(W_\alpha \eta_t^i + b_\alpha), \quad (16)$$

where $W_\eta$, $W_v$, $W_h$, $W_\alpha$ and $b_\alpha$ are parameters to be learned.

Based on the above attention weight distribution $\alpha_t^i$, we compute the relationship attention context $c_t$ as

$$c_t = \sum_{i=1}^{n+k} \alpha_t^i v_o^i, \quad (17)$$

where the $i$-th element $(\alpha_t^i)$ of $\alpha_t$ denotes the attention probability of the fused visual interactive feature $v_o^i$.

**Language LSTM Layer:** We employ language LSTM to predict the image caption word by word. The concatenation of the relationship attention context $c_t$ and $h_t^{att}$ is fed into the language LSTM (i.e., $LSTM_{lan}$), and the current output $h_t^{lan}$ of language LSTM is obtained by,

$$h_t^{lan} = f_{lan}^{lstm}(h_{t-1}^{lan}, [c_t; h_t^{att}]). \quad (18)$$

where $f_{lan}^{lstm}$ is the updating function of language LSTM. Finally, $h_t^{lan}$ is used to generate the next word vector $e_{t+1}$ by feeding it into a fully connected layer followed by a softmax,

$$P(X = e_{t+1}) = softmax(W_h h_t^{lan}). \quad (19)$$

where $P(\cdot)$ denotes the generation probability of the $(t+1)$-th word vector, and $W_h$ is the model learnable parameter.

# 4 Experiments

**Dataset and Evaluation Metrics:** We evaluate our model on the standard MSCOCO dataset, which consists of 82,783, 40,775 and 40,504 images for training, testing and validation respectively. Each image includes 5 description sentences. In addition, several popular evaluation metrics: BLEU (B-1∼B-4) [14], ROUGE-L (R) [11], METEOR (M) [3], CIDEr (C) [15] and SPICE (S) [1] are used, and coco-caption code[1] is utilized to compute these metrics.

## 4.1 Ablation Analysis

In the subsection, we perform some ablation experiments to clarify the effectiveness of following modules: **1)** Local Feature (L Fea.), **2)** Global Features

**Table 1.** Ablation performance of LGVIN model on MSCOCO dataset. "✓" means that this component is integrated into the proposed model.

| Num. | Model Settings | | | | | Model Metrics | | | | |
|---|---|---|---|---|---|---|---|---|---|---|
| | **L Fea.** | **G Fea.** | **VIE** | **VIF** | **R Att.** | B-1 | B-4 | M | R | C |
| 1 | ✓ | ✓ | | | | 78.8 | 37.4 | 27.4 | 57.2 | 124.4 |
| 2 | ✓ | ✓ | ✓ | | | 79.2 | 37.9 | 27.8 | 57.6 | 125.7 |
| 3 | ✓ | ✓ | ✓ | | ✓ | 79.9 | 38.2 | 28.1 | 57.8 | 127.8 |
| 4 | ✓ | ✓ | ✓ | ✓ | | 81.1 | 38.3 | 28.2 | 58.2 | 128.3 |
| 5 | ✓ | ✓ | | ✓ | ✓ | 80.1 | 37.6 | 27.9 | 57.7 | 127.7 |
| 6 | ✓ | ✓ | ✓ | ✓ | ✓ | **81.4** | **38.9** | **28.5** | **58.6** | **128.6** |

(G Fea.), **3)** Visual Interaction Encoding (VIE) module, **4)** Visual Interaction Fusion (VIF) module, **5)** Relationship Attention (R Att.) module.

From Table 1, we have the following conclusions. **1)** The scores in line 2 is higher than line 1, which shows that VIE module can effectively learn the two feature relationships. **2)** The model in line 6 outperforms line 5, which further shows the effectiveness of VIE. **3)** The performance of model in line 3 is better than line 2, it indicates that the visual attention is helpful to boost the evaluation metric scores. **4)** Our full model in line 6 achieves the best performance, which further demonstrates the overall effectiveness of our proposed model.

### 4.2 Comparison with State-of-the-Arts

*Note that* since our model only uses the image visual information, these state-of-the-art models [5,8,12,13,24] using external resources such as semantic information are excluded in the comparison. Table 2 presents the experimental results

**Table 2.** Comparison results on MSCOCO dataset. GLA (base) and TDA (base) are our baselines. We use Beam Search (BS) strategy for predicting caption.

| | $B-1\uparrow$ | $B-2\uparrow$ | $B-3\uparrow$ | $B-4\uparrow$ | $M\uparrow$ | $R\uparrow$ | $C\uparrow$ | $S\uparrow$ |
|---|---|---|---|---|---|---|---|---|
| SCST [7] | 77.9 | 61.5 | 46.8 | 35.0 | 26.9 | 56.3 | 115.2 | 20.42 |
| DHEDN [21] | 80.8 | 63.7 | 48.8 | 36.7 | 27.2 | 57.2 | 117.0 | – |
| IPSG [25] | 78.5 | 62.2 | 47.5 | 36.4 | 27.2 | **62.1** | 102.6 | – |
| SDCD [6] | 74.8 | 52.5 | 36.5 | 23.5 | 23.5 | 50.5 | 104.1 | – |
| SCA-CNN [4] | 71.9 | 54.8 | 41.1 | 31.1 | 25.0 | 53.1 | 95.2 | – |
| Trans+KG [26] | 76.24 | – | – | 34.39 | 27.71 | – | 112.60 | 21.12 |
| TDA+GLD [19] | 78.8 | 62.6 | 48.0 | 36.1 | 27.8 | 57.1 | 121.1 | 21.6 |
| cLSTM-RA [22] | **81.7** | 64.5 | 49.4 | 37.2 | 28.0 | 57.9 | 121.5 | – |
| NADGCN-rl [20] | 80.4 | – | – | 38.3 | 28.5 | 58.2 | 126.4 | 21.8 |
| GLA (base) | 72.3 | 55.3 | 41.4 | 31.4 | 25.2 | 53.5 | 98.7 | 20.3 |
| TDA (base) | 79.7 | 63.3 | 48.3 | 36.4 | 27.9 | 57.1 | 121.1 | 21.6 |
| LGVIN (BS=4)(ours) | 81.1 | 64.9 | 50.3 | 38.8 | **28.6** | 58.7 | **128.8** | 22.2 |
| **LGVIN (BS=5)(ours)** | 81.4 | **65.1** | **50.5** | **38.9** | 28.5 | 58.6 | 128.6 | **22.3** |
| LGVIN (BS=6)(ours) | 81.2 | 64.7 | 50.1 | 38.6 | 28.3 | 58.4 | 128.4 | 22.1 |

[1] Available: https://github.com/tylin/coco-caption.

on MSCOCO dataset. As can be observed, our model outperforms all of them across the most metrics. Specifically, compared with the baselines TDA (base) [2], the score of metric BLEU4 is increased by 6.87% from 36.4 to 38.9, and the CIDEr score is increased by 6.19% from 121.1 to 128.6. That is a significant improvement obviously.

**Table 3.** Comparisons of LGVIN with relationship based methods on MSCOCO.

| | $B-1\uparrow$ | $B-2\uparrow$ | $B-3\uparrow$ | $B-4\uparrow$ | $M\uparrow$ | $R\uparrow$ | $C\uparrow$ | $S\uparrow$ |
|---|---|---|---|---|---|---|---|---|
| ARL [18] | 75.9 | 60.3 | 46.5 | 35.8 | 27.8 | 56.4 | 111.3 | – |
| KMSL [10] | 79.2 | 63.2 | 48.3 | 36.3 | 27.6 | 56.8 | 120.2 | 21.4 |
| **LGVIN (ours)** | **81.4** | **65.1** | **50.5** | **38.9** | **28.5** | **58.6** | **128.6** | **22.3** |

In addition, unlike these similar relationship based methods, our proposed model does not use any semantic information input and resorts to the visual interaction graph to automatically explore the visual relationship between local feature and global feature in the image. As showed in Table 3, we can see that our method is significantly outperforms them across all metrics, which indicates the potentials of our local-global visual interaction attention model quantitatively.

## 5 Conclusion

In the paper, a Local-Global Visual Interaction Attention (LGVIN) model is proposed, which fully explores the intrinsic interaction between local feature and global feature for boosting image captioning. To obtain feature interaction information, we develop a visual interaction encoding module to automatically encode the visual relationships between local feature and global feature in the image. To further enrich the relationship representation with different attributes, we devise a visual interaction fusion module to fuse multi-layer outputs from previous visual interaction encoding. Different from the previous similar relationship based methods that only explore the image local or global feature relationship separately, our LGVIN effectively addresses the issue of the inaccurate interaction information between local feature and global feature in the image by implicitly encoding their relationship. The extensive ablation and comparative experiment results demonstrate that LGVIN achieves a promising performance.

## References


1. Anderson, P., Fernando, B., Johnson, M., Gould, S.: SPICE: semantic propositional image caption evaluation. In: Leibe, B., Matas, J., Sebe, N., Welling, M. (eds.) ECCV 2016. LNCS, vol. 9909, pp. 382–398. Springer, Cham (2016). https://doi.org/10.1007/978-3-319-46454-1_24
2. Anderson, P., et al.: Bottom-up and top-down attention for image captioning and visual question answering. In: Proceedings of IEEE Conference on Computer Vision and Pattern Recognition, pp. 6077–6086 (2018)

3. Banerjee, S., Lavie, A.: Meteor: An automatic metric for MT evaluation with improved correlation with human judgments. In: Proceedings of Meeting of the Association for Computational Linguistics, pp. 65–72 (2005)
4. Chen, L., et al.: SCA-CNN: spatial and channel-wise attention in convolutional networks for image captioning. In: Proceedings of IEEE Conference on Computer Vision and Pattern Recognition, pp. 5659–5667 (2017)
5. Cornia, M., Stefanini, M., Baraldi, L., Cucchiara, R.: Meshed-memory transformer for image captioning. In: Proceedings of IEEE Conference on Computer Vision and Pattern Recognition, pp. 10578–10587 (2020)
6. Ding, X., Li, Q., Cheng, Y., Wang, J., Bian, W., Jie, B.: Local keypoint-based faster R-CNN. Appl. Intell. **50**(10), 3007–3022 (2020)
7. Gao, J., Wang, S., Wang, S., Ma, S., Gao, W.: Self-critical n-step training for image captioning. In: Proceedings of IEEE Conference on Computer Vision and Pattern Recognition, pp. 6300–6308 (2019)
8. Li, G., Zhu, L., Liu, P., Yang, Y.: Entangled transformer for image captioning. In: Proceedings of the IEEE International Conference on Computer Vision, pp. 8928–8937 (2019)
9. Li, L., Tang, S., Zhang, Y., Deng, L., Tian, Q.: GLA: global-local attention for image description. IEEE Trans. Multimedia **20**(3), 726–737 (2018)
10. Li, X., Jiang, S.: Know more say less: image captioning based on scene graphs. IEEE Trans. Multimedia **20**(8), 2117–2130 (2020)
11. Lin, C.Y.: Rouge: A package for automatic evaluation of summaries. In: Proceedings of Meeting of the Association for Computational Linguistics, pp. 74–81 (2004)
12. Luo, Y., et al.: Dual-level collaborative transformer for image captioning. In: Proceedings of the AAAI Conference on Artificial Intelligence (2021)
13. Pan, Y., Yao, T., Li, Y., Mei, T.: X-linear attention networks for image captioning. In: Proceedings of IEEE Conference on Computer Vision and Pattern Recognition, pp. 10971–10980 (2020)
14. Papineni, K., Roukos, S., Ward, T., Zhu, W.J.: BLEU: a method for automatic evaluation of machine translation. In: Proceedings of Meeting of the Association for Computational Linguistics, pp. 311–318 (2002)
15. Vedantam, R., Zitnick, C.L., Parikh, D.: CIDEr: consensus-based image description evaluation. In: Proceedings of IEEE Conference on Computer Vision and Pattern Recognition, pp. 4566–4575 (2015)
16. Velickovic, P., Cucurull, G., Casanova, A., Romero, et al. : Graph attention networks. arXiv:1710.10903 (2017)
17. Wang, C., Gu, X.: Image captioning with adaptive incremental global context attention. Appl. Intell. **52**(6), 6575–6597 (2021). https://doi.org/10.1007/s10489-021-02734-3
18. Wang, J., Wang, W., Wang, L., Wang, Z., Feng, D.D., Tan, T.: Learning visual relationship and context-aware attention for image captioning. Pattern Recogn. **98**, 107075 (2020)
19. Wu, J., Chen, T., Wu, H., Yang, Z., Luo, G., Lin, L.: Fine-grained image captioning with global-local discriminative objective. IEEE Trans. Multimedia **23**, 2413–2427 (2021)
20. Wu, L., Xu, M., Sang, L., Yao, T., Mei, T.: Noise augmented double-stream graph convolutional networks for image captioning. IEEE Trans. Circuits Syst. Video Technol. **31**(8), 3118–3127 (2021)
21. Xiao, X., Wang, L., Ding, K., Xiang, S., Pan, C.: Deep hierarchical encoder-decoder network for image captioning. IEEE Trans. Multimedia **21**(11), 2942–2956 (2019)

22. Yang, L., Hu, H., Xing, S., Lu, X.: Constrained LSTM and residual attention for image captioning. ACM Trans. Multimed. Comput. Commun. Appl. **16**(3), 1–18 (2020)
23. Yao, T., Pan, Y., Li, Y., Mei, T.: Exploring visual relationship for image captioning. In: Ferrari, V., Hebert, M., Sminchisescu, C., Weiss, Y. (eds.) Computer Vision – ECCV 2018. LNCS, vol. 11218, pp. 711–727. Springer, Cham (2018). https://doi.org/10.1007/978-3-030-01264-9_42
24. Yao, T., Pan, Y., Li, Y., Mei, T.: Hierarchy parsing for image captioning. In: Proceedings of the IEEE International Conference on Computer Vision, pp. 2621–2629 (2019)
25. Zhang, J., Mei, K., Zheng, Y., Fan, J.: Integrating part of speech guidance for image captioning. IEEE Trans. Multimedia **23**, 92–104 (2021)
26. Zhang, Y., Shi, X., Mi, S., Yang, X.: Image captioning with transformer and knowledge graph. Pattern Recogn. Lett. **143**, 43–49 (2021)

# Rethinking Voxelization and Classification for 3D Object Detection

Youshaa Murhij[1(✉)], Alexander Golodkov[1], and Dmitry Yudin[1,2]

[1] Moscow Institute of Physics and Technology, Dolgoprudny, Moscow Region, Russia
{yosha.morheg,golodkov.ao}@phystech.edu, yudin.da@mipt.ru
[2] Artificial Intelligence Research Institute, Moscow, Russia

**Abstract.** The main challenge in 3D object detection from LiDAR point clouds is achieving real-time performance without affecting the reliability of the network. In other words, the detecting network must be confident enough about its predictions. In this paper, we present a solution to improve network inference speed and precision at the same time by implementing a fast dynamic voxelizer that works on fast pillar-based models in the same way a voxelizer works on slow voxel-based models. In addition, we propose a lightweight detection sub-head model for classifying predicted objects and filter out false detected objects that significantly improves model precision in a negligible time and computing cost. The developed code is publicly available at: https://github.com/YoushaaMurhij/RVCDet.

**Keywords:** 3D Object Detection · Voxelization · Classification · LiDAR Point Clouds

## 1 Introduction

Current LiDAR-based 3D object detection approaches follow a standard scheme in their pipelines. Most of 3D detection pipelines consists of reading module that prepares point clouds for voxelization stage that converts raw points into a fixed size 2D or 3D grid which can be fed to a detection neural network. Most common grid formats use voxels or pillars in this stage. Compared to other methods like 2D projected images and raw Lidar point, voxel representation can be processed efficiently using 3D sparse convolution [2] and preserve approximately similar information to raw point cloud and make feature sparsity learnable with position-wise importance prediction. A Pillar represents multiple voxels which are vertically stacked and treated as a one tall voxel. Voxels are generally used in 3D backbones such as VoxelNet [24], while pillars are used in 2D backbones such as PointPillars [7]. In this paper, we are going to discuss the differences between the two representations and introduce a new voxelization approach to benefit from voxel features in pillars representations by implementing a fast dynamic voxel encoder for 2D backbones like PointPillars.

This work was supported by the Russian Science Foundation (Project No. 21-71-00131).


M. Tanveer et al. (Eds.): ICONIP 2022, CCIS 1793, pp. 470–481, 2023.
https://doi.org/10.1007/978-981-99-1645-0_39

Current 3D detection neural nets suffer from noisy outputs which can be seen as false positive objects in the network predictions. This problem can be reduced by filtering the network output based on each object score (How much the network is confident that this object is corresponding to a certain class). But this will reduce the network precision as it could filter out a true positive object with low confidence score. To address this problem, we present an auxiliary module that can be merged to the detection head in the network and works as classification sub-head. Classification sub-head learns to distinguish true and false objects base on further processing intermediate features generated by neck module in the detection pipeline.

## 2 Related Work

Different forms of point cloud representation have been explored in the context of 3D object detection. The main idea is to form a structured representation where standard convolution operation can be applied.

Existing representations are mainly divided into two types: 3D voxel grids and 2D projections. A 3D voxel grid transforms the point cloud into a regularly spaced 3D grid, where each voxel cell can contain a scalar value (e.g., occupancy) or vector data (e.g., hand-crafted statistics computed from the points within that voxel cell). 3D convolution is typically applied to extract high-order representation from the voxel grid [4]. However, since point clouds are sparse by nature, the voxel grid is very sparse and therefore a large proportion of computation is redundant and unnecessary. As a result, typical systems [4,9] only run at 1-2 FPS.

Several Point cloud based 3D object detection methods utilize a voxel grid representation. [4] encode each non-empty voxel with 6 statistical quantities that are derived from all the points contained within the voxel. [14] fuses multiple local statistics to represent each voxel. [16] computes the truncated signed distance on the voxel grid. [8] uses binary encoding for the 3D voxel grid. [1] introduces a multi-view representation for a LiDAR point cloud by computing a multi-channel feature map in the bird's eye view and the cylindral coordinates in the frontal view. Several other studies project point clouds onto a perspective view and then use image-based feature encoding schemes [9,11]. While Center-based 3D object detecion and tracking method [10,21] are designed based on [3,25] to represent, detect, and track 3D objects as points. CenterPoint framework, first detects centers of objects using a keypoint detector and regresses to other attributes, including 3D size, 3D orientation, and velocity. In a second stage, CenterPoint refines these estimates using additional point features on the object. In CenterPoint, 3D object tracking simplifies to greedy closest-point matching. An attempt to synergize the birds-eye view and the perspective view was done in [23] through a novel end-to-end multiview fusion (MVF) algorithm, which can learn to utilize the complementary information from both.

A pipeline for multi-label object recognition, detection and semantic segmentation was introduced in [5] benefiting from classification approach on 2D image data. In this pipeline, They tried to obtain intermediate object localization and pixel labeling results for the training data, and then use such results to train task-specific deep networks in a fully supervised manner. The entire process consists of four stages, including object localization in the training images, filtering and fusing object instances, pixel labeling for the training images, and task-specific network training. To obtain clean object instances in the training images, they proposed an algorithm for filtering, fusing and classifying predicted object instances.

In our work, we will also show that adding an auxiliary classifier to 3D detection head will improve the model precision as this sub-head will learn not from positive ground truth examples but from negative examples (false positive), too. Several approaches are available to reduce the number of false positives in 3D object detection. One of them is adding a point cloud classifier. As an input, the classifier takes the points clouds which are included in the predicted 3D boxes by the detector, and returns which class the given box belongs to: true or false predictions. The detector predictions contain many false positives, and, accordingly, by reducing their number, one can significantly improve the accuracy of the approach used. At the moment, there are three main approaches for point cloud classification.

Algorithms based on voxels [13], which represent the classified point cloud as a set of voxels. Further, 3D grid convolutions are usually used for classification. The downside of this approach is that important information about the geometry of points can be lost, while voxels complicate calculations and such algorithms require more memory. Projective approaches [17]. These algorithms usually classify not the point cloud itself, but its projections on planes, and convolutional networks are usually used for classification. The disadvantage of such approaches is that data on the spatial relative position of the points is lost, which can be critical if there are few points in the classified cloud. Algorithms that process points directly [12]. Usually these are algorithms that apply convolutions to points directly and take into account the density of points. Such algorithms are invariant with respect to spatial displacements of objects. All this makes this type of algorithm the most interesting for use in the problem of classifying lidar point clouds.

PAConv [20] from the third group, and one of the most effective classification models was trained, and evaluated on ModelNet40 [19] dataset. PAConv shows a prediction accuracy of 93.9 on objects with 1024 points. But, this approach was not used in conjunction with our detection pipeline due to the large inference time, which made the pipeline not real-time. An alternative to such a classifier is a lighter network, which is an MLP, possibly with the addition of a small number of convolutional layers. It is proposed to classify intermediate tensors in the sub-heads of the detector, containing information about the detected boxes. This topic will be discussed in this article in details.

# 3 Method

Our 3D detection pipeline consists of a fast dynamic voxelizer (FDV), where we implemented our fast voxelization method for pillar-based models based on scatter operations, a 2D backbone for real-time performance (RV Backbone), which is adapted to take multiple-channel FDV features as an input, regional proposal network (RPN module) as neck for feature map generation and a Center-based detection head to predict 3D bounding boxes in the scene with our additional sub-head classifier to filter out false detected objects. Next, in this section, we discuss the proposed FDV voxelizer in Sect. 3.1, our adapted RV backbone that accepts FDV output in Sect. 3.2 and our proposed classifier sub-head in Sect. 3.3.

## 3.1 Fast Dynamic Voxelizer

We have implemented a fast dynamic voxelizer (FDV) for PointPillars model, which works similarly to the dynamic 3D voxelizer for the VoxelNet model. The proposed voxelizer is based on scatter operations on sparse point clouds. FDV voxelizer runs with $O(N)$ time complexity faster than most current similar approaches. FDV implementation does not require to sample a predefined number of points per voxel. This means that every point can be used by the model, minimizing information loss. Furthermore, no need to pad voxels to a predefined size, even when they have significantly fewer points. This can greatly reduce the extra space and compute overhead compared to regular methods (hard voxelization), especially at longer ranges where the point cloud becomes very sparse.

As an input, we feed a point cloud in the format: $Batch\_id, x, y, z$. We determine the size of the grid by the voxel size and the range of the input point cloud after filtering not in-range points. $gridsize_i = cloudrange_i/voxelsize_i$, where $i$ represents size along $x, y, z$ axis. After that, we determine the coordinates of the voxels on the grid from the point cloud range and remove the redundant points (this may happen in the center of the coordinates). Next, we calculate the mean of all points in each non-empty pillar using the scatter mean method and find the distance $x, y$ and $z$ from the center of the voxel to the calculated mean point. Scatter mean averages all values from the $src$ tensor into $out$ at the indices specified in the index tensor along a given axis dimension as seen in Fig. 1 For one-dimensional tensors, the operation computes:

$$out_i = out_i + \frac{1}{N_i} \cdot \sum_j src_j \tag{1}$$

where $\sum_j$ is over $j$ such that $indexj = i.N_i$ indicates referencing $i$.

Finally, we combine all the features together and pass them to the RV backbone module. The features include $x_{pt}, y_{pt}, z_{pt}$ for each point in the voxel, $x_{center}$, $y_{center}$, $z_{center}$ the distance for each point to the voxel center and $x_{mean}$, $y_{mean}$, $z_{mean}$ the distance from each point to the mean point.

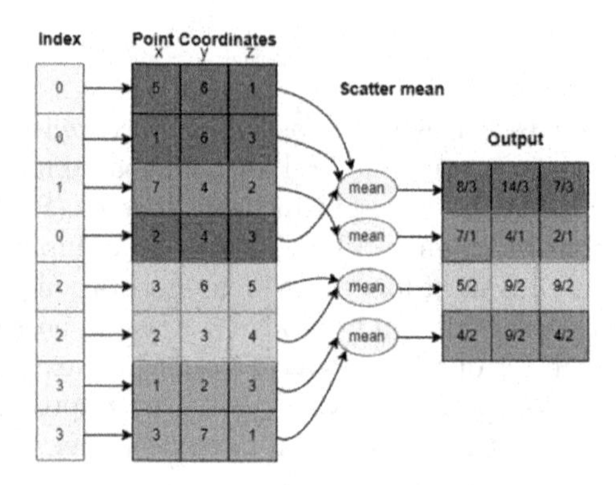


**Fig. 1.** Scatter Mean operation over 2D/3D voxel grid

### 3.2 RV Backbone Module

After concatenating all the features from FDV module, we feed them to our proposed RV backbone. RV backbone takes advantage from the gathered scatter data as it consists of multiple modified Pillar Feature Nets (PFN) for further feature extraction. The main difference between our RV backbone and original FPN [7] is that RV backbone accepts input features from multiple channels in a similar way voxel feature encoder (VFE) [24] works in voxel-based models. In our implementation, RV backbone includes two Pillar-like feature nets with additional scatter layer to get the features maximum values (scatter-max) from each sub-module. Next, we re-concatenate the new features, again calculate their final scatter-max and rearrange them to generate initial feature maps which we feed to RPN module.

### 3.3 Classification Sub-head Module

Mean average precision metric does not provide rich information about the model performance in terms of true and false positive detections. We noticed that even if a certain model gives a high mAP metric, it still detects enough amount of false positive objects (FP), which is a major issue in self-driving real-world scenarios. To tackle this problem, we propose an additional refining module in our detection pipeline to classify and filter FP predictions based on their features though a classification sub-head in our Center-based detection head by additional processing of the corresponding middle-features learned in the neck module.

The main reason to add such a sub-head in our model structure is that after voxelizing the input point cloud into 2d grid and generating spatial features though a scatter backbone, we feed these middle-features into a regional proposal network (RPN) that calculate the final features and provide a set of heat-maps (HM) that can be directly used to regress all the 3D attributes of the predicted output (position, scale, orientation, class and score). So, these temporary features are sufficient to achieve this task by further prepossessing and training.

In addition to regressing to all these attributes, we add NN Sub-Head module that takes the crops of HM tensors corresponding to the predicted objects and refine them to False/True positive predictions depending on the number of classes we train on.

The proposed method pipeline is illustrated in Fig. 2.

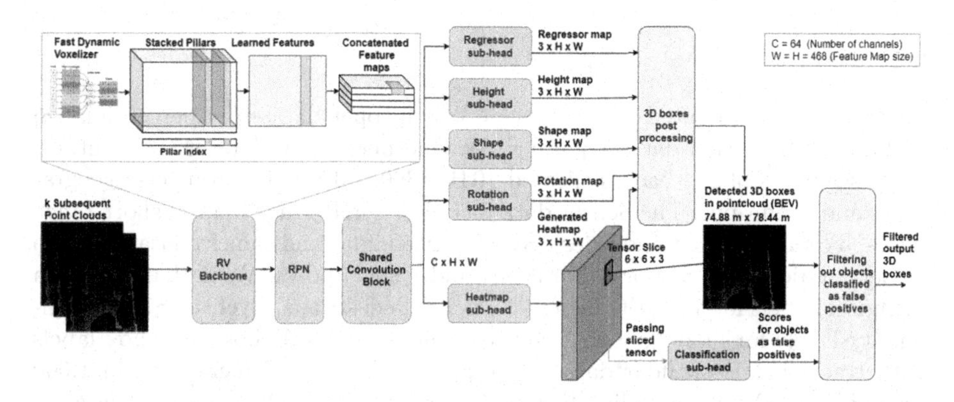


**Fig. 2.** Scheme of our 3D detection pipeline named as RVDet including classification sub-head

As seen in Fig. 2, we are interested in HM sub-head as it represents the Gaussian distribution map and contains all the required information to locate the objects in the scene. According to the proposed scheme the classifier takes the predicted output of the detection pipeline as an input in addition to middle features (HM) generated in the RPN module. Our sub-head classifier crops the input HM tensors to multiple small windows including the predicted objects based on their predicted coordinates in 3D space and maps them to their 2D plane coordinates in the feature map so, it can focus only on the objects of interest in the generated map.

Next, our classification module refines these predictions based on their initialized predicted classes into two categories: true positive class and false negative class. Where, Class could be either Vehicle, Pedestrian or Cyclist in Waymo or KITTI datasets.

Based on the predicted TP/FP classes it refines the initial pipeline predictions and returns the final predictions. Figure 2 shows more detailed scheme illustrating the tensor cropping process and classification.

Regarding the classifier's architecture, we have implemented several models based on multi-layer perceptrons (MLP) and convolutional layers (Conv). Our motivation while choosing the architecture was to provide simple and lightweight module that can achieve this task considering real-time performance of the pipeline.

To train the classification sub-head, we prepared a custom dataset for this purpose by running our detection model in inference mode on Waymo and KITTI datasets. As an input data, we stored the feature heat map generated after neck module. We classified the network predictions into: False Vehicle, False Pedestrian, False Cyclist according to the ground-true data from the original dataset. During training, we used cross entropy as loss function, Adam as optimizer.

### 3.4 Datasets

We trained and evaluated our pipeline on Waymo open dataset for perception [18] and on KITTI vision dataset [6]. The Waymo perception dataset [18] contains 1,950 segments of 20 s each, collected 10 Hz (390,000 frames) in diverse geographies and conditions. The Sensor data includes: 1 mid-range lidar, 4 short-range lidars, 5 cameras (front and sides), Synchronized lidar and camera data, Lidar to camera projections, Sensor calibrations and vehicle poses. While Labeled data includes: Labels for 4 object classes - Vehicles, Pedestrians, Cyclists, Signs, High-quality labels for lidar data in 1,200 segments, 12.6 M 3D bounding box labels with tracking IDs on lidar data, High-quality labels for camera data in 1,000 segments, 11.8M 2D bounding box labels with tracking IDs on camera data.

The Kitti 3D object detection benchmark [6] consists of 7481 training images and 7518 test images as well as the corresponding point clouds, comprising a total of 80.256 labeled objects.

## 4 Experimental Results

We mainly tested our voxelizer and classifier on two popular datasets for 3D object detection: Waymo Open Dataset [18] and KITTI dataset [6]. We compared our pipeline performance with other current open real-time approaches on Waymo leaderboard. Our implementation with PointPillars backbone added +3% $mAPHL2$ on Waymo testset outperforming all other PointPillars-based models while running +18FPS on Tesla-V100. This implementation is now comparable with Heavy VoxelNet-based models that run under 11FPS on Tesla-V100. Table 1 shows our pipeline mean average precision metrics with/without heading (both Level 1 and 2) on Waymo testset compared to real-time PointPillars models and VoxelNet models. Table 2 shows the performance of our pipeline on KITTI testset. Our RVDet model including FDV and RV backbone achieved +1 mAP on Car class, +2.4 mAP on Pedestrian class and +1.4 mAP on Cyclist class. Again on KITTI testset our real-time PointPillars implementation is comparable with CenterPoint VoxelNet method. Figure 3 shows detection pipeline performance example on Waymo before and after adding an classifier subhead. Red boxes refer to ground true data while black ones are network predictions. We show more detailed metrics related to classifier's performance in Sect. 5.

**Table 1.** Dynamic voxelizer impact on 3D detection metrics on Waymo test dataset

| Model | Range | mAP/L1 | mAPH/L1 | mAP/L2 | mAPH/L2 |
|---|---|---|---|---|---|
| PointPillars [7] | PerType | 0.4406 | 0.3985 | 0.3933 | 0.3552 |
| | [0, 30) | 0.5322 | 0.4901 | 0.5201 | 0.4792 |
| | [30, 50) | 0.4225 | 0.3792 | 0.3859 | 0.3458 |
| | [50, +inf) | 0.3010 | 0.2594 | 0.2376 | 0.2038 |
| SECOND- [24] VoxelNet | PerType | 0.4364 | 0.3943 | 0.3914 | 0.3530 |
| | [0, 30) | 0.5291 | 0.4865 | 0.5134 | 0.4721 |
| | [30, 50) | 0.4211 | 0.3771 | 0.3846 | 0.3440 |
| | [50, +inf) | 0.2981 | 0.2571 | 0.2366 | 0.2030 |
| CIA-SSD [22] | PerType | 0.6148 | 0.5794 | 0.5605 | 0.5281 |
| | [0, 30) | 0.8002 | 0.7630 | 0.7873 | 0.7509 |
| | [30, 50) | 0.5946 | 0.5549 | 0.5520 | 0.5153 |
| | [50, +inf) | 0.3085 | 0.2763 | 0.2482 | 0.2222 |
| QuickDet | PerType | 0.7055 | 0.6489 | 0.6510 | 0.5985 |
| | [0, 30) | 0.7995 | 0.7418 | 0.7841 | 0.7280 |
| | [30, 50) | 0.6905 | 0.6385 | 0.6457 | 0.5969 |
| | [50, +inf) | 0.5484 | 0.4863 | 0.4620 | 0.4089 |
| CenterPoint-PointPillars [21] | PerType | 0.7093 | 0.6877 | 0.6608 | 0.6404 |
| | [0, 30) | 0.7766 | 0.7501 | 0.7627 | 0.7367 |
| | [30, 50) | 0.7095 | 0.6892 | 0.6692 | 0.6497 |
| | [50, +inf) | 0.5858 | 0.5698 | 0.5042 | 0.4899 |
| **RVDet (ours*)** | **PerType** | **0.7488** | **0.7280** | **0.6986** | **0.6789** |
| | [0, 30) | 0.8503 | 0.8268 | 0.8357 | 0.8127 |
| | [30, 50) | 0.7195 | 0.7001 | 0.6787 | 0.6602 |
| | [50, +inf) | 0.5873 | 0.5706 | 0.5061 | 0.4911 |
| CenterPoint-VoxelNet [21] | PerType | 0.7871 | 0.7718 | 0.7338 | 0.7193 |
| | [0, 30) | 0.8766 | 0.8616 | 0.8621 | 0.8474 |
| | [30, 50) | 0.7643 | 0.7492 | 0.7199 | 0.7055 |
| | [50, +inf) | 0.6404 | 0.6248 | 0.5522 | 0.5382 |
| CenterPoint++ VoxelNet [21] | PerType | 0.7941 | 0.7796 | 0.7422 | 0.7282 |
| | [0, 30) | 0.8714 | 0.8568 | 0.8566 | 0.8422 |
| | [30, 50) | 0.7743 | 0.7605 | 0.7322 | 0.7189 |
| | [50, +inf) | 0.6610 | 0.6469 | 0.5713 | 0.5586 |

## 5 Ablation Study

To inspect our voxelizer impact on model performance, we validate our trained model on Waymo and KITTI validation sets and report the results in Table 3 for Waymo main metrics and in Table 4 for KITTI metrics.

To choose our classifier architecture, we have carefully run multiple experiments to get the most efficient model in terms of time complexity and accuracy. Among the parameters of the classifier architecture that affect the metrics, the following are considered: types of network layers, the number of layers in the network, as well as the size of the layers, depending on the size of the sliced section of the tensor. To determine the most suitable architecture, experiments were carried out on Waymo Validation set considering parameters that determine the network architecture. Based on the results of CenterPoint PointPillars detector on Train set, a new dataset was assembled from objects of six classes: True Vehicle, False Vehicle, True Pedestrian, False Pedestrian, True Cyclist, False Cyclist. Each class has 5000 objects. Table 6 considers MLP models with

**Table 2.** Dynamic voxelizer main results on 3D detection KITTI test dataset

| | | Car | | Pedestrian | | Cyclist | |
|---|---|---|---|---|---|---|---|
| **Model** | **Difficulty** | **3D Det** | **BEV** | **3D Det** | **BEV** | **3D Det** | **BEV** |
| CenterPoint-PointPillars [21] | Easy | 82.58 | 90.52 | 41.31 | 48.21 | 68.78 | 72.16 |
| | Moderate | 72.71 | 86.60 | 35.32 | 41.53 | 52.52 | 57.43 |
| | Hard | 67.92 | 82.69 | 33.35 | 39.82 | 46.77 | 51.28 |
| RVDet (ours) | Easy | 82.37 | 89.90 | **45.93** | **52.91** | **69.50** | **76.11** |
| | Moderate | **73.50** | 86.21 | **38.54** | **45.18** | 51.78 | **58.40** |
| | Hard | **68.79** | 82.63 | **36.26** | **42.99** | 46.05 | **51.93** |
| CenterPoint-VoxelNet [21] | Easy | 81.17 | 88.47 | 47.25 | 51.76 | 72.16 | 76.38 |
| | Moderate | 73.96 | 85.05 | 39.28 | 44.08 | 56.67 | 61.25 |
| | Hard | 69.48 | 81.19 | 36.78 | 41.80 | 50.60 | 54.68 |
| PV-RCNN [15] | Easy | 90.25 | 94.98 | 52.17 | 59.86 | 78.60 | 82.49 |
| | Moderate | 81.43 | 90.65 | 43.29 | 50.57 | 63.71 | 68.89 |
| | Hard | 76.82 | 86.14 | 40.29 | 46.74 | 57.65 | 62.41 |

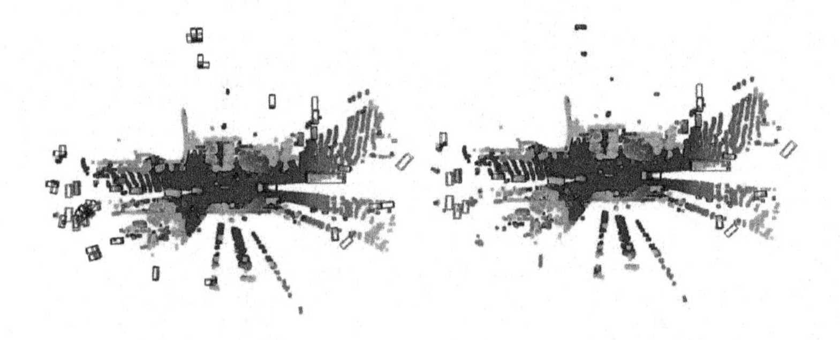

**Fig. 3.** Detection pipeline performance example on Waymo before (left) and after (right) adding classification subhead

different number of layers. The MLP-1-layer model has only 1 linear layer. The size of the input layer of all the linear models listed below is determined by the size of the supplied tensor multiplied by number of channels in the tensor (3 channels in our case). The output of this model is 3 neurons. In the MLP-2-layer model, there are two linear layers, between which the ReLU activation function is applied. At the output of the first layer, the number of neurons is twice as large as input. This number of neurons goes as input size of second layer. There are also 3 neurons at the output of the second layer. In the MLP-3-layer model, the difference is that, only a layer with 24 output neurons and a ReLU activation function is added between the two layers. In the MLP-4-layer model, a layer with 6 output neurons and a ReLU activation function is added before the last layer. In the classifier models with convolutional layers, the size of the MLP input was adjusted to the output of the convolutional layers in front of them. In the 1Conv convolution layer, the parameters: $Number\ of\ in\ channels = 3$, $number\ of\ out\ channels = 6$, $kernel\ size = (2, 2)$, $stride = (1, 1)$ were used. One more convolutional layer is added to 2Conv with parameters: $Number\ of\ in\ channels = 6$, $number\ of\ out\ channels = 12$, $kernel\ size = (2, 2)$, $stride = (1, 1)$

Table 6 shows the validation accuracy during Linear-convolutional classifier training on the prepared dataset depending on the number of layers and the size

**Table 3.** Dynamic voxelizer impact on 3D detection metrics on Waymo validation dataset. Latency measured on RTX-3060Ti

| Model | Range | mAP/L1 | mAPH/L1 | mAP/L2 | mAPH/L2 | Latency |
|---|---|---|---|---|---|---|
| CenterPoint-PointPillars | PerType | 0.6920 | 0.6716 | 0.6338 | 0.6147 | 56 ms |
| | [0, 30) | 0.8137 | 0.7936 | 0.7946 | 0.7749 | |
| | [30, 50) | 0.6612 | 0.6424 | 0.6098 | 0.5922 | |
| | [50, +inf) | 0.5061 | 0.4834 | 0.4235 | 0.4036 | |
| RVDet (ours)-with removed ground points | PerType | 0.7287 | 0.7095 | 0.6721 | 0.6541 | 43 ms |
| | [0, 30) | 0.8266 | 0.8075 | 0.8075 | 0.7888 | |
| | [30, 50) | 0.6969 | 0.6775 | 0.6465 | 0.6283 | |
| | [50, +inf) | 0.5721 | 0.5530 | 0.4863 | 0.4694 | |
| RVDet(ours) | PerType | **0.7374** | **0.7176** | **0.6807** | **0.6621** | 47 ms |
| | [0, 30) | **0.8323** | **0.8129** | **0.8134** | **0.7945** | |
| | [30, 50) | **0.7065** | **0.6875** | **0.6558** | **0.6379** | |
| | [50, +inf) | **0.5840** | **0.5612** | **0.4974** | **0.4772** | |

**Table 4.** Dynamic voxelizer detailed metrics on KITTI validation dataset compared with CenterPoint (Diff: difficulty, Mod: Moderate)

| | | Car | | | | Pedestrian | | | | Cyclist | | | |
|---|---|---|---|---|---|---|---|---|---|---|---|---|---|
| **Model** | **Diff.** | **BBox** | **BEV** | **3D** | **AOS** | **BBox** | **BEV** | **3D** | **AOS** | **BBox** | **BEV** | **3D** | **AOS** |
| CenterPoint-PointPillars | Easy | 90.62 | 89.66 | 86.60 | 90.61 | 63.45 | 56.07 | 50.84 | 58.76 | 86.83 | 81.26 | 80.21 | 86.56 |
| | Mod | 89.06 | 86.05 | 76.69 | 88.96 | 61.40 | 52.80 | 47.96 | 55.64 | 72.62 | 64.39 | 62.68 | 71.91 |
| | Hard | 87.40 | 84.42 | 74.33 | 87.23 | 60.22 | 50.66 | 45.22 | 54.13 | 68.39 | 61.42 | 59.87 | 67.70 |
| **RVDet (ours)** | Easy | **91.87** | 89.58 | **86.76** | **91.85** | **68.63** | 55.60 | **51.58** | **63.62** | **90.29** | **85.61** | **81.49** | **90.14** |
| | Mod | 89.05 | 86.21 | 76.87 | 88.86 | **66.36** | **53.04** | **48.27** | **60.92** | **74.02** | **68.51** | **63.63** | **73.62** |
| | Hard | **88.09** | **84.53** | **75.26** | **87.80** | **64.70** | 49.97 | 45.41 | **58.66** | **71.56** | **64.33** | **61.32** | **71.16** |

of the input tensor. It can be seen from the tested architectures that, MLP with $8 \times 8$ input tensor shows the highest accuracy. For the most detailed study of MLP with such a number of layers, the experiments reflected in Table 7 were carried out. In this case, the architecture was examined on a larger number of possible input tensor sizes. We also show Recall of each class for different input sizes. From the experiments, we conclude that a classifier having two linear layers with $9 \times 9$ input tensor has the best accuracy. Table 5 shows experiments demonstrating the effect of the classifier on the precision metric. In this case, we set IoU threshold when calculating the metric to 0.4. To compare the metrics, experiments were carried out with the method of box filtering by the number of LiDAR points that fell inside the predicted box, and an additional filtering of all predictions with a confidence score less than 0.3.

**Table 5.** Classifier architecture impact on precision of CenterPoint PointPillars on Waymo validation dataset

| **Pipeline** | **Overall** | **Vehicle** | **Pedestrian** | **Cyclist** |
|---|---|---|---|---|
| CenterPoint PP | 31.70% | 53.44% | 10.77% | 4.02% |
| CenterPoint PP + Point Filtering (Threshold 5) | 77.48% | 90.02% | 36.28% | 10.04% |
| CenterPoint PP + Score Filtering (Threshold 0.3) | 78.91% | 90.02% | 38.30% | 10.38% |
| RVCDet (MLP-2-layers) | 85.94% | **94.24%** | 41.29% | **11.61%** |
| RVCDet (2Conv + MLP-2-layers) | **86.95%** | 92.56% | **44.39%** | 11.4% |

**Table 6.** Accuracy for different classifier architectures on Waymo validation data

| Number of layers | 2 × 2 input | 4 × 4 input | 6 × 6 input | 8 × 8 input | 10 × 10 input |
|---|---|---|---|---|---|
| MLP-1-layer | 89.01% | 89.23% | 89.61% | 89.67% | 90.11% |
| MLP-2-layers | **89.92%** | 90.08% | 90.82% | **90.91%** | 90.47% |
| MLP-3-layers | 89.48% | **90.29%** | 90.02% | 90.56% | 90.31% |
| MLP-4-layers | 89.40% | 90.02% | 90.40% | 90.05% | 90.37% |
| 1Conv+MLP-2-layers | – | – | 89.70% | 90.01% | 90.43% |
| 2Conv+MLP-2-layers | – | – | **91.20%** | **90.91%** | **90.91%** |

**Table 7.** Validation accuracy and recall (for false predictions) for MLP-2-layers with different input sizes on Waymo dataset

| Input | Accuracy | Recall (Vehicle) | Recall (Pedestrian) | Recall (Cyclist) |
|---|---|---|---|---|
| 1 × 1 | 85.27% | 87.22% | 84.08% | 86.64% |
| 3 × 3 | 89.79% | 90.75% | 86.43% | 91.23% |
| 5 × 5 | **91.49%** | 87.67% | **87.81%** | 93.04% |
| 7 × 7 | 90.91% | 91.31% | 86.33% | 95.12% |
| 9 × 9 | 91.43% | **93.29%** | 87.31% | **95.52%** |

## 6 Conclusions

Achieving real-time performance without affecting the reliability of the network is still a challenge in 3D object detection from LiDAR point clouds.In our work, we addressed this problem and proposed a solution to improve network inference speed and precision at the same time by implementing a fast dynamic voxelizer that works in a $O(N)$ time complexity faster than other current methods. In addition, we presented a lightweight detection sub-head model for classifying predicted objects and filter out false detected objects that significantly improves model precision in a negligible time and computing cost.

## References

1. Chen, X., Ma, H., Wan, J., Li, B., Xia, T.: Multi-view 3D object detection network for autonomous driving. In: Proceedings of the IEEE conference on Computer Vision and Pattern Recognition, pp. 1907–1915 (2017)
2. Chen, Y., Li, Y., Zhang, X., Sun, J., Jia, J.: Focal sparse convolutional networks for 3D object detection. In: Proceedings of the IEEE/CVF Conference on Computer Vision and Pattern Recognition, pp. 5428–5437 (2022)
3. Duan, K., Bai, S., Xie, L., Qi, H., Huang, Q., Tian, Q.: CenterNet: keypoint triplets for object detection (2019)
4. Engelcke, M., Rao, D., Wang, D.Z., Tong, C.H., Posner, I.: Vote3Deep: fast object detection in 3D point clouds using efficient convolutional neural networks. In: 2017 IEEE International Conference on Robotics and Automation (ICRA), pp. 1355–1361. IEEE (2017)
5. Ge, W., Yang, S., Yu, Y.: Multi-evidence filtering and fusion for multi-label classification, object detection and semantic segmentation based on weakly supervised learning. In: Proceedings of the IEEE Conference on Computer Vision and Pattern Recognition (CVPR) (2018)

6. Geiger, A., Lenz, P., Urtasun, R.: Are we ready for autonomous driving? The kitti vision benchmark suite. In: Conference on Computer Vision and Pattern Recognition (CVPR) (2012)
7. Lang, A.H., Vora, S., Caesar, H., Zhou, L., Yang, J., Beijbom, O.: PointPillars: fast encoders for object detection from point clouds. In: Proceedings of the IEEE/CVF Conference on Computer Vision and Pattern Recognition, pp. 12697–12705 (2019)
8. Li, B.: 3D fully convolutional network for vehicle detection in point cloud. In: 2017 IEEE/RSJ International Conference on Intelligent Robots and Systems (IROS), pp. 1513–1518. IEEE (2017)
9. Li, B., Zhang, T., Xia, T.: Vehicle detection from 3D lidar using fully convolutional network. arXiv preprint arXiv:1608.07916 (2016)
10. Murhij, Y., Yudin, D.: FMFNet: improve the 3D object detection and tracking via feature map flow. In: Proceedings of the IEEE International Joint Conference on Neural Network (IJCNN) (2022)
11. Premebida, C., Carreira, J., Batista, J., Nunes, U.: Pedestrian detection combining rgb and dense lidar data. In: 2014 IEEE/RSJ International Conference on Intelligent Robots and Systems, pp. 4112–4117. IEEE (2014)
12. Qi, C.R., Su, H., Mo, K., Guibas, L.J.: PointNet: deep learning on point sets for 3D classification and segmentation. In: IEEE (2017)
13. Riegler, G., Ulusoy, A.O., Geiger, A.: OctNet: learning deep 3D representations at high resolutions. In: IEEE (2017)
14. Schwarz, K., Sauer, A., Niemeyer, M., Liao, Y., Geiger, A.: VoxGRAF: fast 3D-aware image synthesis with sparse voxel grids. arXiv preprint arXiv:2206.07695 (2022)
15. Shi, S., Guo, C., Jiang, L., Wang, Z., Shi, J., Wang, X., Li, H.: PV-RCNN: point-voxel feature set abstraction for 3D object detection. In: CVPR (2020)
16. Song, S., Xiao, J.: Deep sliding shapes for amodal 3D object detection in RGB-D images. In: Proceedings of the IEEE Conference on Computer Vision and Pattern Recognition, pp. 808–816 (2016)
17. Su, H., Maji, S., Kalogerakis, E., Learned-Miller, E.: Multi-view convolutional neural networks for 3D shape recognition. In: IEEE (2015)
18. Sun, P., et al.: Scalability in perception for autonomous driving: Waymo open dataset. In: Proceedings of the IEEE/CVF Conference on Computer Vision and Pattern Recognition, pp. 2446–2454 (2020)
19. Wu, Z., Song, S., Khosla, A., Yu, F., Zhang, L., Xiao, X.T.J.: 3D shapeNets: a deep representation for volumetric shapes. In: IEEE (2015)
20. Xu, M., Ding, R., Zhao, H., Qi, X.: PAConv: position adaptive convolution with dynamic kernel assembling on point clouds. In: IEEE/CVF (2021)
21. Yin, T., Zhou, X., Krahenbuhl, P.: Center-based 3d object detection and tracking. In: Proceedings of the IEEE/CVF Conference on Computer Vision and Pattern Recognition, pp. 11784–11793 (2021)
22. Zheng, W., Tang, W., Chen, S., Jiang, L., Fu, C.W.: CIA-SSD: Confident IoU-aware single-stage object detector from point cloud. In: AAAI (2021)
23. Zhou, Y., et al.: End-to-end multi-view fusion for 3D object detection in lidar point clouds (2019)
24. Zhou, Y., Tuzel, O.: VoxelNet: end-to-end learning for point cloud based 3D object detection. In: Proceedings of the IEEE Conference on Computer Vision and Pattern Recognition, pp. 4490–4499 (2018)
25. Zhu, B., Jiang, Z., Zhou, X., Li, Z., Yu, G.: Class-balanced grouping and sampling for point cloud 3D object detection (2019)

# GhostVec: Directly Extracting Speaker Embedding from End-to-End Speech Recognition Model Using Adversarial Examples

Xiaojiao Chen[1], Sheng Li[2], and Hao Huang[1](✉)

[1] Xinjiang University, Urumqi, China
hwanghao@gmail.com
[2] National Institute of Information and Communications Technology, Kyoto, Japan
sheng.li@nict.go.jp

**Abstract.** Obtaining excellent speaker embedding representations can leverage the performance of a series of tasks, such as speaker/speech recognition, multi-speaker dialogue, and translation systems. The automatic speech recognition (ASR) system is trained with massive speech data and contains many speaker information. There are no existing attempts to protect the speaker embedding space of ASR from adversarial attacks. This paper proposes GhostVec, a novel method to export the speaker space from the ASR system without any external speaker verification system or real human voice as reference. More specifically, we extract speaker embedding from a transformer-based ASR system. Two kinds of targeted adversarial embedding (GhostVec) are proposed from features-level and embedding-level, respectively. The similarities are evaluated between GhostVecs and corresponding speakers randomly selected from Librispeech. Experiment results show that the proposed methods have superior performance in generating a similar embedding of the target speaker. We hope the preliminary discovery in this study to catalyze future downstream research speaker recognition-related topics.

**Keywords:** speaker recognition · speaker embedding · adversarial examples · transformer · speech recognition

## 1 Introduction

The speaker information has an important impact on the recognition rate since different speakers have substantial differences in pronunciation, speed, stress, pronunciation, etc. The speaker information is represented as speaker embedding (a.k.a. voiceprint) as features of each vocal cavity, which can fully express the differences of voices. The existing speaker representation approaches, which is GMM-based, include Gaussian Mixture Model (GMM) super-vector [1], the joint factor analysis (JFA) [12], and the i-vector [7]. And those GMM-based approaches are


M. Tanveer et al. (Eds.): ICONIP 2022, CCIS 1793, pp. 482–492, 2023.
https://doi.org/10.1007/978-981-99-1645-0_40

replace by the deep neural network (DNN), such as d-vector [26] and x-vector [24], which is the current state-of-the-art speaker representation technique. Obtaining excellent speaker embedding representations can boost the performance of a series of tasks, such as speaker/speech recognition, multi-speaker dialogue, and translation systems. However, heavy reliance on high-quality speaker recognition datasets makes the system development cost essentially high.

Previous studies have also shown deep neural network (DNN) to be vulnerable to adversarial perturbations [2,4,25,30], and adding some small perturbations to the original input can mislead the ASR system to get erroneous recognition results. The misleading perturbed example is often denoted as adversarial example and the procedure of using adversarial example to obtain erroneous recognition results is called adversarial attack. The research on adversarial attack has drawn lots of attentions in recent speech research community. Similarly, the speaker recognition system has also security concerns because of adopting DNNs as the back-bone in speaker recognition system [10,13,17,20,28,29].

We notice in [14] that the automatic speech recognition (ASR) system is trained with massive speech data, and the model contains a lot of speaker information. Inspired by these, this paper proposes GhostVec, a novel method to effectively export the speaker space from ASR system without any external speaker verification system [5,11] or real human voice as reference. We generate the specific speaker embedding by using the adversarial example from the speaker-adapted Transformer-based ASR system [16]. The main contributions of this paper can be summarized as follows:

- We investigate effective adversarial perturbation to the speaker space of a speaker-adapted transformer-based ASR system.
- We propose two different GhostVec generation methods. Both methods can obtain speaker embedding matching with exact target speaker information.

The remainder of this paper is organized as follows: Sect. 2 introduces the technical background of this paper. Sections 3 and 4 describe our proposed methods and experimental evaluations. Conclusions and plans for future works are presented in Sect. 5.

## 2 Background

### 2.1 Transformer-Based E2E Speaker-Adapted ASR Systems

End-to-End (E2E) speech recognition has been widely used in speech recognition. The most crucial component is the encoder, which can convert the input waveform or feature into a high-dimensional feature representation. Transformer-based [8] seq2seq speech recognition architecture generally includes two parts: an encoder and a decoder, where the encoder is responsible for encoding the input speech feature sequence, and the decoder predicts the output sequence according to the encoding information of the encoder.

Though the DNN based ASR system has shown great success in recent years, however, the ASR system performance degrades significantly when testing conditions are different from training, and the adaptive method came into being. Adaptation algorithms aim to seek balance between the test data and training data in ASR. Speaker adaptation, which adapt the system to a target speaker, is the most common form of adaptation and generally augment the input speech with speaker-level variations [15,23,33]. In [14], researchers observe gradual removal of speaker variability and noise as ASR model's layer goes deeper. In [15], it is shown that the output of the acoustic features by the encoder of the transformer can effectively show the classification characteristics of the speaker. Therefore, this paper proposes to effectively export the speaker space from ASR system without any external speaker verification system.

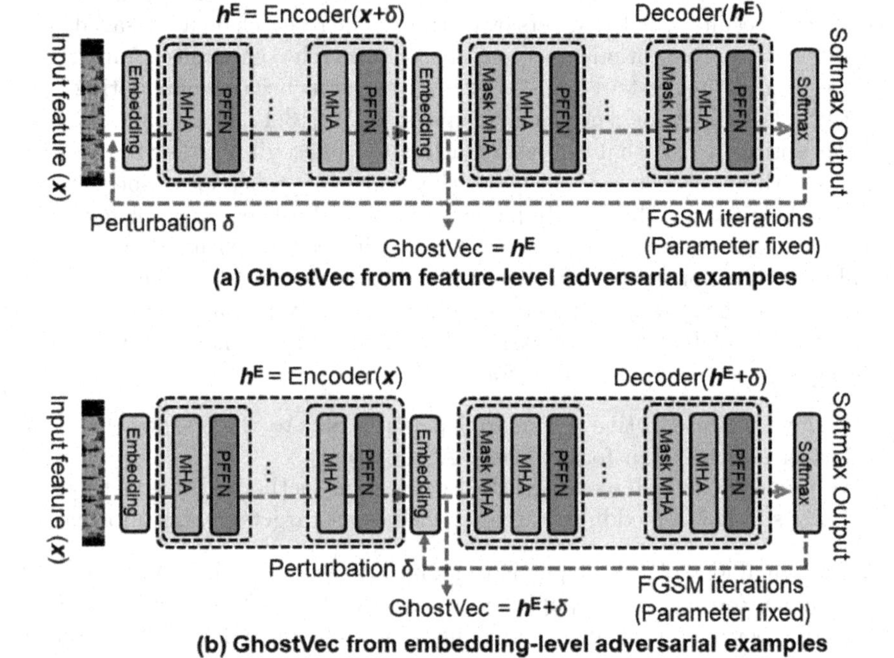


**Fig. 1.** The flowchart of the proposed method (a: GhostVec from feature-level adversarial perturbation, b: GhostVec from embedding-level adversarial perturbation, multi-head self-attention: MHA, and position-wise feed-forward networks: PFFN)

### 2.2 Generating Adversarial Examples

DNN, as a technology in the field of machine learning, has shown unique advantages in various areas in recent years. And it has achieved state-of-the-art performance in various artificial intelligence research fields. Unfortunately, DNNs are vulnerable to small perturbation, specifically designed to mislead the network.

Adversarial examples, the most critical technology in adversarial attack, can be defined as:

$$\tilde{x} = x + \delta \quad \text{such that} \quad ||\delta||_p < \epsilon, \tag{1}$$

where $x$ is the input of a neural network $f$, $\tilde{x}$ is adversarial example. $||\cdot||_p$ is the p-norm operation. $\epsilon$ is chosen to be sufficiently small so as to undetectable. And we want to find a small malicious perturbation $\delta$: to force the network to produce erroneous output when input is adversarial examples $\tilde{x}$.

According to Eq. (1), the method based on gradient optimization is usually used to optimize the adversarial perturbation when the model parameters are transparent. For the optimization methods of adversarial perturbation, there are mainly methods, such as fast gradient sign method (FGSM) [9], Projected Gradient Descent Method [18], etc. The genetic algorithm is often used in the black-box model to craft adversarial examples. In recent research, [34] proposed prepending perturbation in ASR system. In this paper, considering iteration speed, we adopt a one-step optimized FGSM and prepending perturbation as the main method to generate GhostVec.

**FGSM Optimization.** This paper uses the FGSM, which is optimized for $L_\infty$ distance measurement. Given an input $x$, the adversarial examples can be formulated:

$$\tilde{x} = x + \epsilon \cdot \text{sign}(\nabla_x J(\theta, x, y)) \tag{2}$$

where, the adversarial perturbation is $\delta = \epsilon \cdot \text{sign}(\nabla_x J(\theta, x, y))$ ; $y$ satisfies $y = f(x)$ and is the correct output of $x$; $J(\theta, x, y)$ is the loss function used in this deep neural network.

**Prepending Perturbation.** In addition, we also adopted the new adversarial example method proposed by Lu et al. [34]. Compared with the traditional way of adding adversarial perturbations to the entire audio, the attack success rate of prepending perturbation on the LAS system [3] is very high. The prepending adversarial perturbation has a more robust attack performance due to relatively larger receptive fields from the attention mechanism. The formulated expression is as follows:

$$adv_\delta = [\delta_1, \delta_2, \delta_3 ... x_1, x_2, x_3 ....], \tag{3}$$

where $\delta$ is concatenated before $x$. This method can get a good attack effect in speech recognition based on the attention mechanism, so we mainly use this method to generate embedding of the target speaker. The details will be shown in Sect. 5.1.

## 3 Proposed Methods

This section discusses details of the proposed GhostVec, in which we use adversarial examples to generate similar embedding of specific speakers. Moreover, this paper adopts two GhostVec generation methods from the feature-level and embedding-level adversarial examples, which is shown in Fig. 1.

The proposed method has there steps: First prepare the model, then train to generate adversarial samples, and finally extract GhostVec. In generating adversarial samples step and extracting GhostVec steps, we define $x$ to represent the input feature of the encoder; $y$ is the recognition result of the target speaker, which contains the speaker-id; $y_t$ is the output token of the target speaker's sentence ($t$ stands for target). The detailed description is in the following subsections.

### 3.1 GhostVec from Feature-Level Adversarial Examples

In the conventional method of generating adversarial examples, we consider the encoder-decoder structure of the transformer as a whole and create the target label from softmax output. The entire GhostVec generation process can be expressed as follows:

$$\mathrm{h}^{E} = \mathrm{Encoder}(\mathrm{x} + \delta) \tag{4}$$

$$\hat{\mathrm{h}}_{l}^{D} = \mathrm{Decoder}(\mathrm{h}^{\mathrm{E}}, \hat{\mathrm{y}}_{1:\mathrm{l}-1}) \tag{5}$$

$$P(y_t|\hat{y}_{1:l-1}, \mathrm{h}^{E}) = \mathrm{softmax}(\hat{\mathrm{h}}_{\mathrm{l}}^{\mathrm{D}}) \tag{6}$$

In this method, as shown in Fig. 1 (a), we obtain an ideal recognition result containing the speaker-id for the input $x$ through the FGSM optimization iterations with the adversarial perturbation $\delta$ at the feature input side. Finally, we extract the output of the encoder $\mathrm{h}^{E}$ as the GhostVec of target speakers.

### 3.2 GhostVec from Embedding-Level Adversarial Examples

As shown in Fig. 1 (b), this method is aimed at only the decoder of the transformer. Since we need a high-level representation for the speaker, it is more direct to add the adversarial perturbation to the embedding output of the encoder.

$$\mathrm{h}^{E} = \mathrm{Encoder}(\mathrm{x}) \tag{7}$$

$$\mathrm{h}^{*E} = \mathrm{h}^{E} + \delta \tag{8}$$

$$\hat{\mathrm{h}}_{l}^{D} = \mathrm{Decoder}(\mathrm{h}^{*\mathrm{E}}, \hat{\mathrm{y}}_{1:l-1}) \tag{9}$$

$$P(y_t|\hat{y}_{1:\mathrm{l}-1}, \mathrm{h}^{*E}) = \mathrm{Softmax}(\hat{\mathrm{h}}_{\mathrm{l}}^{\mathrm{D}}) \tag{10}$$

where $h^{*E}$ is the embedding representation of the target speaker, we also use the FGSM method to add adversarial perturbation $\delta$ to the embedding extraction of the original sentence $h^{E}$, and obtain a related embedding of the target speaker.

## 4 Experiment Setup

### 4.1 Datasets

The primary model used in experiments is the speaker adaptation system based on the transformer trained by the Librispeech dataset [22] (train-clean-100).

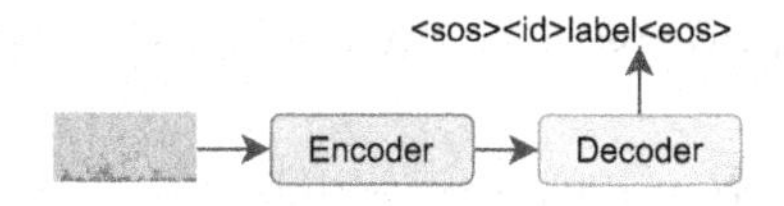


**Fig. 2.** Speaker-adapted ASR model using speaker-id in the label

During the experiment, we randomly selected six speakers. Based on these six speakers, experiments were carried out.

We randomly selected six speakers from train-clean-100 and test-clean, in total 239 sentences for experiments.

### 4.2 ASR Models

Recently, the self-attention [6,27,31,32] Transformer architecture has been widely adopted for sequence modeling due to its ability to capture long-distance interactions and high training efficiency. Also, the transformer has been applied to E2E speech recognition systems and has achieved promising results. ASR-transformer can map an input sequence to a sequence of intermediate representations by the encoder. The biggest difference between the transformer-based ASR model and commonly used E2E models is that transformer-based ASR relies on attention and feedforward components.

ASR model required for embedding extraction is trained on the LibriSpeech train-clean-100 but based on multitask training method following [15,27] with the speaker-id and label. The WER% was approximately 9.0%.

As shown in Fig. 2, the speaker-ID is explicitly added as a label during training [15,27]. And we inculcated speaker instinct as ground truth in training. The training labels are organized as "<SOS> <speaker-id> labels <EOS>". The trained network can automatically output these attributes at the beginning of decoding, so we do not have to prepare classifiers for these attributes.

### 4.3 Evaluation Metrics

To evaluate the similarity between speakers, we use cosine similarity. For a more intuitive display, we use the uniform manifold approximation and projection (UAMP) [19] tool to reflect the relationship between speakers. The cosine similarity between two speakers$x_1$ and $x_2$ can be expressed as follows:

$$S(x_1, x_2) = \frac{x_1 \cdot x_2}{||x_1|| \cdot ||x_2||} \tag{11}$$

It can be seen that the range of similarity is between -1 and 1, and the closer the absolute value of similarity is to 1, the higher the similarity is, and vice versa.

Secondly, Word Error Rate (WER) is used to evaluate the accuracy in recognizing speech. The WER calculation is based on a measurement called the "Levenshtein distance" [21]. The Levenshtein distance is a measurement of the differences between two "strings". We compute WER with the Levenshtein distance $\ell$:

$$WER = 100\% \cdot \frac{\ell}{N} = 100\% \cdot \frac{S+D+I}{N} \tag{12}$$

where the sum of word substitution is $S$, word insertions $I$, and word deletion $D$. It is worth noting that since the methods proposed in this paper adopt the targeted adversarial example generation method, the obtained WER is almost equal to 1.

Moreover, the accuracy of speaker-id is also an essential metric in evaluating our proposed method, which we define as:

$$SIER = 100\% \cdot \frac{N_{error}}{N_{sum}} \tag{13}$$

where $N_{error}$ is the total number of mispredicted speaker-id; $N_{sum}$ is the total speaker utterances.

## 5 Experimental Results

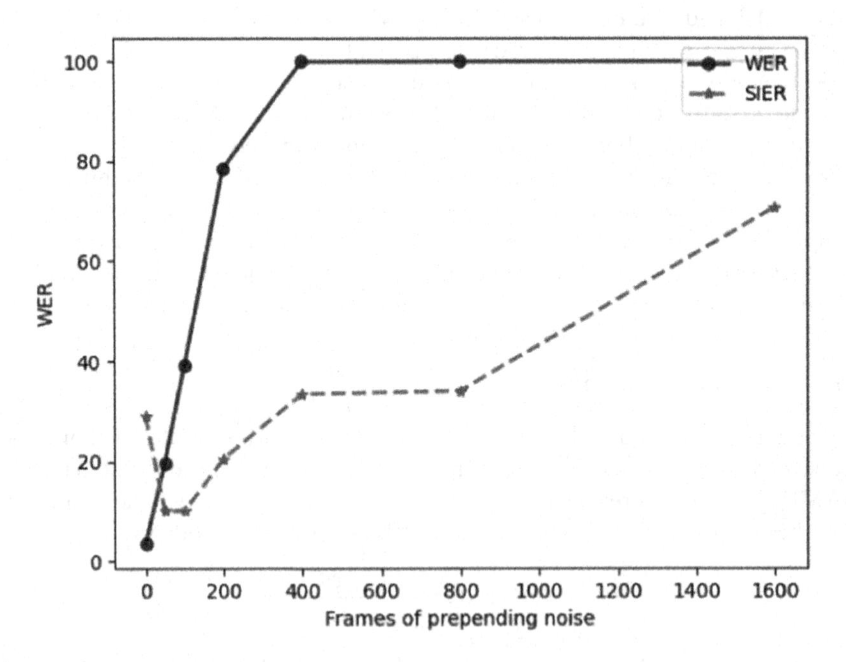


**Fig. 3.** WER and SIER trends of adversarial perturbations prepending at the beginning of the speech with different lengths

### 5.1 Effectiveness of Adversarial Examples

Before proposing our method, we obtained an interesting phenomenon during speaker analysis. The arbitrary adversarial examples pretended to noise without any prior speaker information can be more helpful for generating targeting GhostVec.

Since the transformer-based speech recognition system is auto-regressive, we assume the adversarial perturbations prepending at the beginning of the speech will highly influence the successive recognition. As we expected, the recognition results in Fig. 3 change dramatically when we use prepending adversarial perturbation method. In Fig. 3, using the prepending perturbation method, there is a clear trend of attack success rate creasing as the adversarial noise length increases. This phenomenon is also consistent with the auto-regressive nature of the model.

### 5.2 Effectiveness of GhostVec

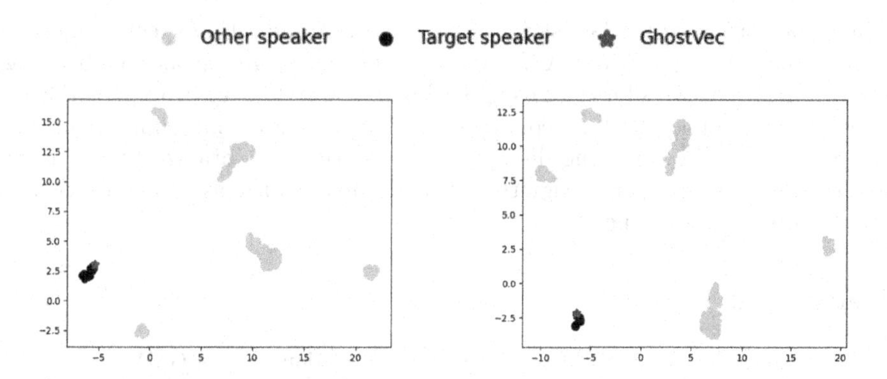


**Fig. 4.** UMAP similarity between target speakers' embeddings and GhostVecs (a:GhostVects from feature-level adversarial examples, b:GhostVects from embedding-level adversarial examples).

This section mainly shows the results of the method proposed in Sect. 3. Figure 4 is the embedding distribution of six speakers and GhostVec from two different types of adversarial examples, and the red ones in (a), (b) are generated GhostVec of the target speaker by two proposed methods, and green-colored one is the real target speaker, and pink-colored ones are the other five non-target speakers.

We can see that both generated adversarial examples can get good speaker embedding matching with the target speaker. Experimental results once again confirmed the feasibility of using the feature or embedding-level adversarial perturbations to get GhostVec.

Figure 4 (a), (b) are the GhostVecs from two types of adversarial examples seeded from background noise. We confirmed in Fig. 4 that the background noise itself has no language information, but through the transformer's FGSM optimization, embedding with the target speaker information can be obtained.

### 5.3 Further Discussions

The meaning of our work can be summaries as follows:

- Our proposed method can easily export speaker embedding with the similar behavior of arbitrary target speaker without external resources (system and utterances).
- Moreover, it is also kind of alert to all community developers, and the voice-privacy can be extracted from a well-constructed ASR system if we rely too much on the opensource toolkit.

## 6 Conclusions

This paper proposes the GhostVec, which utilizes the existing speaker information in the well-constructed ASR system to obtain target speaker embedding. We confirmed the feasibility of using pretending-based adversarial perturbation on feature or embedding-level could get speaker embedding matching with exact target speaker. We hope the discovery in this study to catalyze future downstream research speaker recognition-related topics, such as low-resource and speech privacy-preserving.

## References

1. Campbell, W., et al.: SVM based speaker verification using a GMM supervector Kernel and NAP variability compensation. In: Proceedings IEEE-ICASSP (2006)
2. Carlini, N., Wagner, D.: Audio adversarial examples: targeted attacks on speech-to-text. in Proceedings IEEE Security and Privacy Workshops (SPW), pp. 1–7 (2018)
3. Chan, W., Jaitly, N., Le, Q., Vinyals, O.: Listen, attend and spell: a neural network for large vocabulary conversational speech recognition. In: 2016 IEEE international conference on acoustics, speech and signal processing (ICASSP), pp. 4960–4964. IEEE (2016)
4. Chen, G., et al.: Who is real bob? adversarial attacks on speaker recognition systems. arXiv preprint arXiv:1911.01840 (2019)
5. Cooper, E., Lai, C.I., Yasuda, Y., Fang, F., Wang, X., Chen, N., Yamagishi, J.: Zero-shot multi-speaker text-to-speech with state-of-the-art neural speaker embeddings. In: Proc. IEEE-ICASSP, pp. 6184–6188 (2020)
6. Dalmia, S., Liu, Y., Ronanki, S., Kirchhoff, K.: Transformer-transducers for code-switched speech recognition. In: ICASSP 2021–2021 IEEE International Conference on Acoustics, Speech and Signal Processing (ICASSP), pp. 5859–5863. IEEE (2021)
7. Dehak, N., et al.: Front-end factor analysis for speaker verification. IEEE Trans. ASLP **19**, 788–798 (2011)

8. Dong, L., Xu, S., Xu, B.: Speech-transformer: a no-recurrence sequence-to-sequence model for speech recognition. In: 2018 IEEE International Conference on Acoustics, Speech and Signal Processing (ICASSP), pp. 5884–5888 (2018). https://doi.org/10.1109/ICASSP.2018.8462506
9. Goodfellow, I.J., Shlens, J., Szegedy, C.: Explaining and harnessing adversarial examples. arXiv preprint arXiv:1412.6572 (2014)
10. Jati, A., Hsu, C.C., Pal, M., Peri, R., AbdAlmageed, W., Narayanan, S.: Adversarial attack and defense strategies for deep speaker recognition systems. Comput. Speech Lang. **68**, 101199 (2021)
11. Jia, Y., et al.: Transfer learning from speaker verification to multispeaker text-to-speech synthesis. In: Advances in Neural Information Processing Systems, pp. 4480–4490 (2018)
12. Kenny, P., et al.: A study of inter-speaker variability in speaker verification. IEEE Trans. ASLP **16**(5), 980–988 (2008)
13. Kreuk, F., Adi, Y., Cisse, M., Keshet, J.: Fooling end-to-end speaker verification with adversarial examples. In: Proceedings IEEE-ICASSP, pp. 1962–1966 (2018)
14. Li, C.Y., Yuan, P.C., Lee, H.Y.: What does a network layer hear? Analyzing hidden representations of end-to-end ASR through speech synthesis. In: Proceedings IEEE-ICASSP, pp. 6434–6438 (2020)
15. Li, S., Dabre, R., Lu, X., Shen, P., Kawahara, T., Kawai, H.: Improving transformer-based speech recognition systems with compressed structure and speech attributes augmentation. In: Proceedings INTERSPEECH (2019)
16. Li, S., lu, X., Dabre, R., Shen, P., Kawai, H.: Joint training end-to-end speech recognition systems with speaker attributes, pp. 385–390 (2020). https://doi.org/10.21437/Odyssey
17. Li, X., Zhong, J., Wu, X., Yu, J., Liu, X., Meng, H.: Adversarial attacks on GMM i-vector based speaker verification systems. In: Proceedings IEEE-ICASSP, pp. 6579–6583 (2020)
18. Madry, A., Makelov, A., Schmidt, L., Tsipras, D., Vladu, A.: Towards deep learning models resistant to adversarial attacks. arXiv preprint arXiv:1706.06083 (2017)
19. McInnes, L., Healy, J., Melville, J.: UMAP: uniform manifold approximation and projection for dimension reduction. arXiv preprint arXiv:1802.03426 (2018)
20. Miyato, T., Maeda, S., Koyama, M., Nakae, K., Ishii, S.: Distributional smoothing with virtual adversarial training. arXiv preprint arXiv:1507.00677 (2015)
21. Navarro, G.: A guided tour to approximate string matching. ACM Comput. Surv. (CSUR) **33**(1), 31–88 (2001)
22. Panayotov, V., Chen, G., Povey, D., Khudanpur, S.: Librispeech: an ASR corpus based on public domain audio books. In: 2015 IEEE International Conference on Acoustics, Speech and Signal Processing (ICASSP), pp. 5206–5210 (2015). https://doi.org/10.1109/ICASSP.2015.7178964
23. Sarı, L., Moritz, N., Hori, T., Le Roux, J.: Unsupervised speaker adaptation using attention-based speaker memory for end-to-end ASR. In: ICASSP 2020–2020 IEEE International Conference on Acoustics, Speech and Signal Processing (ICASSP), pp. 7384–7388. IEEE (2020)
24. Snyder, D., et al.: X-vectors: Robust DNN embeddings for speaker recognition. In: Proceedings IEEE-ICASSP, pp. 5329–5333 (2018)
25. Szegedy, C., et al.: Intriguing properties of neural networks. arXiv preprint arXiv:1312.6199 (2013)
26. Variani, E., et al.: Deep neural networks for small footprint text-dependent speaker verification, pp. 4052–4056 (2014)

27. Vaswani, A., et al.: Attention is all you need. CoRR abs/1706.03762 (2017)
28. Wang, Q., Guo, P., Sun, S., Xie, L., Hansen, J.: Adversarial regularization for end-to-end robust speaker verification. In: Proceedings INTERSPEECH, pp. 4010–4014 (2019)
29. Wang, Q., Guo, P., Xie, L.: Inaudible adversarial perturbations for targeted attack in speaker recognition. arXiv preprint arXiv:2005.10637 (2020)
30. Yuan, X., et al.: CommanderSong: a systematic approach for practical adversarial voice recognition. In: Proceedings 27th {USENIX} Security Symposium ({USENIX} Security 18), pp. 49–64 (2018)
31. Zeyer, A., Bahar, P., Irie, K., Schlüter, R., Ney, H.: A comparison of transformer and LSTM encoder decoder models for ASR. In: 2019 IEEE Automatic Speech Recognition and Understanding Workshop (ASRU), pp. 8–15. IEEE (2019)
32. Zhang, Q., et al.: Transformer transducer: a streamable speech recognition model with transformer encoders and RNN-T loss. In: ICASSP 2020–2020 IEEE International Conference on Acoustics, Speech and Signal Processing (ICASSP), pp. 7829–7833. IEEE (2020)
33. Zhao, Y., Ni, C., Leung, C.C., Joty, S.R., Chng, E.S., Ma, B.: Speech transformer with speaker aware persistent memory. In: INTERSPEECH. pp. 1261–1265 (2020)
34. Zong, W., Chow, Y.-W., Susilo, W., Rana, S., Venkatesh, S.: Targeted universal adversarial perturbations for automatic speech recognition. In: Liu, J.K., Katsikas, S., Meng, W., Susilo, W., Intan, R. (eds.) ISC 2021. LNCS, vol. 13118, pp. 358–373. Springer, Cham (2021). https://doi.org/10.1007/978-3-030-91356-4_19

# An End-to-End Chinese and Japanese Bilingual Speech Recognition Systems with Shared Character Decomposition

Sheng Li[1(✉)], Jiyi Li[2], Qianying Liu[3], and Zhuo Gong[4]

[1] National Institute of Information and Communications Technology, Kyoto, Japan
sheng.li@nict.go.jp
[2] University of Yamanashi, Kofu, Japan
jyli@yamanashi.ac.jp
[3] Kyoto University, Kyoto, Japan
ying@nlp.ist.i.kyoto-u.ac.jp
[4] University of Tokyo, Tokyo, Japan
gongzhuo@gavo.t.u-tokyo.ac.jp

**Abstract.** The rising number of tourists in most areas in East Asia has increased the requirement for East Asian speech recognition systems (e.g., Chinese and Japanese). However, the large existing character vocabulary limits the performance of Chinese-Japanese bilingual speech recognition systems. In this study, we propose a novel End-to-End ASR system for Chinese-Japanese bilingual circumstances based on a smaller granularity and quantity modeling unit set. We apply the decomposition strategy to both Chinese character and Japanese Kanji, so that two languages can also benefit from the similar representations of sub-character level during the joint training. For Chinese-Japanese bilingual speech, our proposed bilingual model outperformed our monolingual baseline models for both Japanese and Chinese. This proves the feasibility of joint modeling among similar languages, while decreasing the modeling units.

**Keywords:** Bilingual speech recognition · modeling units · end-to-end model · radicals · transformer

## 1 Introduction

With the constantly increasing number of East Asian tourists in Japanese, particularly Chinese tourists, the demand for general Chinese-Japanese bilingual speech recognition systems is rising. Early studies use GMM-HMM and DNN-HMM to build multilingual ASR systems, which cannot handle the problem of sequence labeling between the variable-length speech frame input and label output. Additionally, they rely on either dictionaries that map words to sequences of subwords that require heavy expert knowledge or a universal phone set that can only achieve limited performance. Recent advances in end-to-end (E2E) speech recognition architectures that encapsulate an acoustic, pronunciation, and language model


M. Tanveer et al. (Eds.): ICONIP 2022, CCIS 1793, pp. 493–503, 2023.
https://doi.org/10.1007/978-981-99-1645-0_41

jointly in a single network does not only improve the ASR performance, but also solve the sequence labeling problem and allows one input frame to predict multiple output tokens. Meanwhile, the grapheme has been used as an alternative modeling unit to build multilingual ASR systems [20]. The use of graphemes does not need extra expert knowledge and also achieves better performance than phonetic systems.

While E2E models and the grapheme modeling unit boost the ASR system performance, however, in East Asian languages such as Chinese, Japanese, and Korean, the character vocabulary size could reach tens of thousands. CJK characters include traditional Chinese characters ("zhengti" or "fanti"), simplified Chinese characters ("jianti"), Kanji, and Hanja, which comprise more than 20,000 characters in total [13]. Chunom, formerly used in Vietnamese, is sometimes also included in the set, which is then abbreviated as CJKV, thereby increasing the number of characters even further [15]. The large size of character vocabulary results in the label sparsity problem. It also increases the complexity of character-based speech processing for eastern languages.

Several character sets coexist in the Chinese, Japanese, and Korean languages. Traditional Chinese characters were used as the official script in ancient China and Japan. At the present time, traditional characters are being gradually replaced in mainland China with Simplified Chinese characters, while the modern Japanese writing system uses a combination of syllabic kana and logographic Kanji, which are adopted Chinese characters, and syllables [1,5]. Such linguistic characteristics allows us to merge such Chinese character tokens and compress the vocabulary size. Instead of directly mapping the Chinese characters and Kanji, we consider sub-character level tokens as the unit tokens following [14], which decomposed Chinese characters into a compact set of basic radicals assigned to three levels to effectively reduce the vocabulary. In this study, we apply the decomposition strategy to Japanese Kanji and propose a novel radical-based Chinese-Japanese bilingual speech recognition system under the limited data resource. Not only the vocabulary size is compressed, the two languages can also benefit from the similar representations of sub-character level during the joint training and further improve the performance. We visualize the embeddings of Chinese and Japanese characters, radicals and strokes to further analysis the similarity between Japanese and Chinese.

The rest of this paper is structured as follows. The related works are overviewed in Sect. 2. The proposed method is Sect. 3. Section 4 evaluates benchmark systems with our dataset. This paper concludes in Sect. 5.

## 2 Related Work

### 2.1 Multilingual ASR Systems

When building multilingual automatic speech recognition (ASR) systems for East Asian languages, the conventional ASR system based on GMM-HMM and DNN-HMM cannot handle the problem of sequence labeling between the variable-length speech frame input and label output. Additionally, conventional

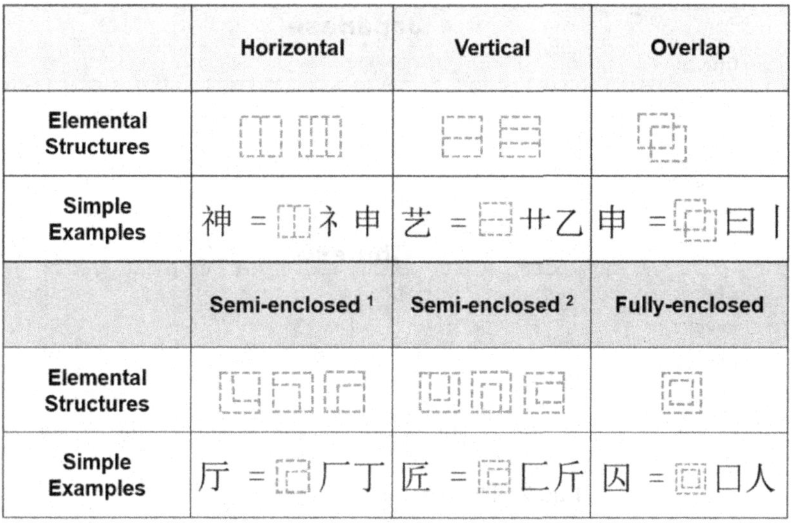

| | Horizontal | Vertical | Overlap |
|---|---|---|---|
| Elemental Structures | | | |
| Simple Examples | 神 = 礻 申 | 艺 = 艹 乙 | 申 = 曰 丨 |
| | Semi-enclosed 1 | Semi-enclosed 2 | Fully-enclosed |
| Elemental Structures | | | |
| Simple Examples | 厅 = 厂 丁 | 匠 = 匚 斤 | 囚 = 口 人 |

**Fig. 1.** Structural analysis of characters.

multilingual ASR systems require expert knowledge of each language and pronunciation dictionaries, which provide a mapping from words to sequences of sub-word units [18]. To solve the language dependency problem, the universal phone set has been proposed [9]. However, the performance of phonetic-based ASR systems is always challenging when merging different languages.

Because of advances in end-to-end (E2E) speech recognition architectures that encapsulate an acoustic, pronunciation, and language model jointly in a single network, the sequence labeling problem has been solved to some extent [10,21,23]. E2E-based multilingual ASR systems have also improved. Meanwhile, the grapheme has been used as an alternative modeling unit to build multilingual ASR systems [20]. The use of graphemes enables multilingual ASR task optimization.

## 2.2 The Decomposition Strategy for Chinese

Because the large vocabulary of existing characters, we were inspired by the wubi input method [14]. It can encode Chinese characters completely based on strokes and font features, and decompose each character into four radicals, or even less; hence, the wubi input method[1] is regarded as the most efficient input method in China and Singapore. However, using only this approach, we cannot select the desired character from a list of homophonic possibilities, which is different from traditional phonetic input methods.

In Fig. 1, we summarize the following 12 basic structures for characters: two "**horizontal**" structures, two "**vertical**" structures, one "**overlap**" structure,

[1] http://wubi.free.fr.

<table>
<tr><th></th><th>Japanese</th></tr>
<tr><td>Char.</td><td>青と空で青空のように</td></tr>
<tr><td>Level-1</td><td>龶月|と|穴工|で|龶月|穴工|のように</td></tr>
<tr><td>Level-2</td><td>龶月|と|宀八工|で|龶月|宀八工|のように</td></tr>
<tr><td>Level-3</td><td>一一丨一丿𠃌一一|と|丶丶乛丿乀一丨一|で|<br>一一丨一丿𠃌一一|丶丶乛丿乀一丨一|のように</td></tr>
</table>

<table>
<tr><th></th><th>Chinese</th></tr>
<tr><td>Char.</td><td>状况稳定</td></tr>
<tr><td>Level-1</td><td>丬犬|冫兄|禾急|宀疋</td></tr>
<tr><td>Level-2</td><td>丬犬|冫口丿乚|丿木ク彐心|宀一|一丿乀</td></tr>
<tr><td>Level-3</td><td>丶丿丨一丿乀丶|丶丿丨𠃍一丿乚|丿一丨丿乀丿フ<br>𠃍一一丶乚丶丶|丶丶乛一|一丿乀</td></tr>
</table>

**Fig. 2.** Three decomposition levels.

three "**semi-enclosed** [1]" structures, three "**semi-enclosed** [2]" structures, and one "**fully enclosed**" structure. Using these structures, we can decompose the characters into particles. We refer to these particles using the technical term "radical" [14]. The decomposition rules are all adopted from an existing open-source project[2].

To ensure that the model can learn the different structures of characters, we input the structure marks as the blue forms shown in Fig. 1 and propose three levels of decomposition strategy, as shown in Fig. 2 [14]. We evaluated these methods on the Beijing dialect (also known as Mandarin, MA) corpus, which includes 35.5 h of clean recordings on the traveling topic and uses simplified Chinese characters. As we introduced in [14], the level 1 (L1) decomposition strategy without structural marks performed best. We reason that the structural marks may increase the probability of conflict appearance (the sharing of the same radical sequence by different characters), and the L1 decomposition is a moderately aggressive approach and preserves the significant information correlated with the characters called the "meaning radical" and "sound radical", which indicate the meaning and sound of the character, respectively.

## 3 Proposed Method

In this section, we describe our proposed method. Our method is consisted of the decomposition strategy for Chinese and Japanese and an attention-based multilingual ASR transformer.

### 3.1 Decomposition Strategy

In [14], we decomposed Chinese characters into radicals assigned to three levels of complexity; meanwhile, we found that the L1-decomposition performed best,

[2] https://github.com/amake/cjk-decomp.

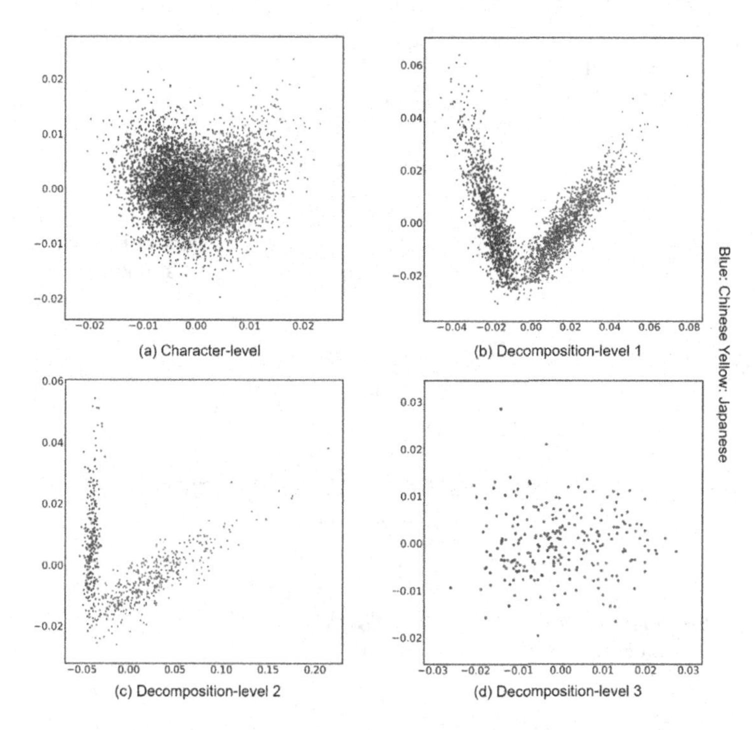


**Fig. 3.** **a** to **d** are the distributions of vectors of Chinese and Japanese characters and decomposed radicals or strokes in three different levels. The blue dots represent Chinese characters and radicals while the red ones are Japanese counterpart. (Color figure online)

as we mentioned in Sect. 2. In this paper, we firstly visualized the embeddings of characters and three levels of decomposed radicals or strokes, as shown in Fig. 3 to further prove that the similarity between Chinese and Japanese increases with the granularity of modeling unit reducing. Based on the FastText tool[3], we extracted the 100 dimensional embeddings of Chinese and Japanese modeling units and using Principal Component Analysis (PCA) to reduce the dimension. Then we visualized the vectors with dots to show the different distributions. In Fig. 3, the blue dots represents Chinese modeling units and the yellow ones are Japanese modeling units.

Based on this knowledge, we applied the decomposition strategy to Japanese Kanji. The Japanese writing system includes Kanji, Katakana, and Hiragana. Among them, Kanji is the most similar to Chinese characters and more decomposable, whereas the other two are more indecomposable in font and their limited in number of characters (both are about 48). After decomposing Kanji, the size of Japanese vocabulary has been reduced from 2651 to 1276. When Kanji characters are decomposed into two radicals, it is worth mentioning that

[3] http://github.com/facebookresearch/fastText.

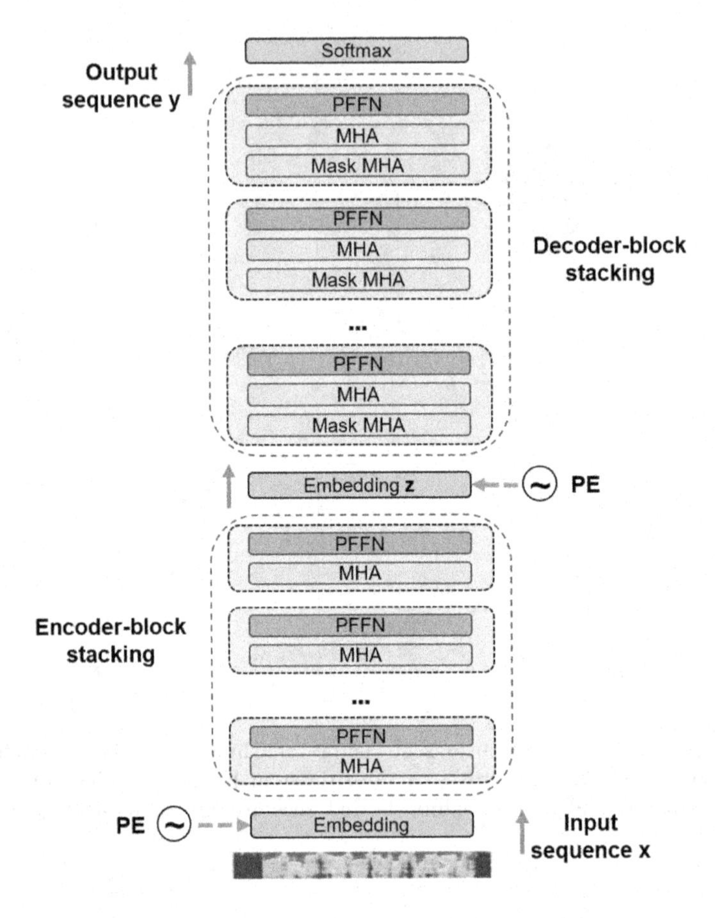


**Fig. 4.** Structure of the Transformer model.

the L1-decomposed Japanese radicals are probably similar to simplified Chinese characters. Furthermore, we jointly modeled the Chinese radicals and Japanese radicals to build a bilingual ASR system, which is described in Sect. 4.3 [3].

## 3.2 Multilingual ASR Transformer

We develop a multilingual ASR system based on the attention-based Transformer. Figure 4 depicts the structure of the model. The multihead attention (MHA) and positionwise (PE), completely connected layers of the ASR transformer's design are used for both the encoder and the decoder. The transformer encoder and decoder are both a stack of blocks.

The prior monolingual transformer type and the multilingual transformer are highly similar. It still models the acoustic feature sequence using the position feedforward network and a stack of multilayer encoder-decoders with multihead self-attention. The sole difference between the two models is the Softmax layer in the decoder. The final output node in the monolingual transformer model is monolingual; in contrast, the final output node in the multilingual model with mixed modeling units of several languages is multilingual. The multilingual transformer model is more adaptable since it has one Softmax layer, as opposed to the multilingual DNN model's several Softmax layers for various languages.

# 4 Experiments

For experiment, a set of systems were constructed based on the proposed method.

## 4.1 Model Structure and Data Description

In this study, ASR tasks are based on the attention-based Transformer model (ASR-Transformer) in the tensor2tensor[4] library [2,4,22], with the output target changed from characters to radicals. Table 1 shows the settings of the training and testing, which are similar to [19].

In this paper, we use Chinese and Japanese to build a bilingual ASR system. Firstly, we chose 34 h of the Beijing dialect corpus (known as Mandarin, MA), which includes clean recordings about travel and the corresponding simplified Chinese characters, as the training set and used the remainder as the test set. To build the Japanese ASR system, we selected 27.6 h of data from the Corpus of Spontaneous Japanese (CSJ) [16] as the training set to build the character-based Japanese monolingual baseline, and evaluated it on two official evaluation sets, **CSJ-Eval01** (E01), and **CSJ-Eval02** (E02) [6–8,17], each of which contains 10 academic lecture recordings [8][5]. The development set (**CSJ-Dev**) was ten selected lecture recordings.

## 4.2 Monolingual Baseline Systems

To build the Chinese monolingual baseline system, we chose 34 h of the Beijing dialect corpus (known as Mandarin, MA) as we mentioned above. To build the Japanese monolingual baseline system, we selected 27.6 h of data from the Corpus of Spontaneous Japanese (CSJ) [16] as the training set to build the character-based Japanese monolingual baseline, and evaluated it on two official evaluation sets as we mentioned above. We used 72-dimensional filterbank features (24-dimensional static $+\Delta$ $+\Delta\Delta$) and did not use speed perturbation [11] to save training time. Based on the ASR-Transformer model mentioned above, our character-based Chinese baseline system (CH(**c**)) yielded 14% character error rates (CER) on the MA test set, whereas the character-based Japanese baseline system (JA(**c**)) yielded 28.0%, and 25.5% CERs on the two evaluation sets shown in Table 3.

[4] https://github.com/tensorflow/tensor2tensor.

[5] There are three test sets, and we only use real academic lecture recordings. The other simulated test set is not used.

**Table 1.** Major experimental settings.

| Model structure | | | |
|---|---|---|---|
| Attention-heads | 8 | Decoder-blocks | 6 |
| Hidden-units | 512 | Residual-drop | 0.3 |
| Encoder-blocks | 6 | Attention-drop | 0.0 |
| Training settings | | | |
| Max-length | 5000 | GPUs (K40 m) | 4 |
| Tokens/batch | 10000 | Warmup-steps | 12000 |
| Epochs | 30 | Steps | 300000 |
| Label-smooth | 0.1 | Optimizer | Adam |
| Testing settings | | | |
| Ave. chkpoints | last 20 | Batch-size | 100 |
| Length-penalty | 0.6 | Beam-size | 13 |
| Max-length | 200 | GPUs (K40 m) | 4 |

**Table 2.** Character error rate (CER%) of the recognition result of the systems trained using different modeling units (Improvement with statistical significance are highlighted with bold color ($p$-value $< 0.05$)).

| Network | #unit | E01 | E02 |
|---|---|---|---|
| char | 3178 | 8.2 | 5.9 |
| word | 98245 | 10.2 | 8.6 |
| WPM | 3000 | 8.4 | 6.1 |
| WPM | 8000 | **7.8** | **6.0** |
| L1+kana | 1171 | **8.0** | **5.8** |

### 4.3 Chinese-Japanese Bilingual Speech Recognition Task

As we mentioned in Sect. 1, East Asian characters (called "Kanji" in Japanese) have been deeply embedded in the Japanese writing system since ancient times. Additionally, the rising number of tourists who may read Kanji in Chinese from China puts the requirement of Chinese-Japanese bilingual ASR tasks on the map. In [14], we built a series of Japanese monolingual models using approximately 577 h of lecture recordings as the training set (**CSJ-Train**), following [6–8,17]. In this work, several modeling units were compared including words, a word-piece model (WPM) [12], and characters, as shown in Table 2. The ASR-Transformer model trained with 1,000 radicals and 171 kanas yielded a 2% relative CER reduction on average on the two evaluation sets. The results inspired us to challenge the bilingual ASR tasks in this way.

To train the bilingual models using a small dataset, we selected 27.6 h of data from the CSJ Japanese corpus and 34 h of spontaneous Mandarin speech recordings (simplified Chinese), as we mentioned in Sect. 4.2. We used the CSJ

**Table 3.** ASR performance (CER%) with different settings on Japanese (**JA**) and Mandarin (**CH**) ("**c**" indicates a character-based model, and "**r**" indicates a L1 radical-based model)(The improvement of results with statistical significance (two-tailed $t$-test at a significance of $p$-value $< 0.05$) are shown in bold font.).

| Models | | CER% (Bilingual Testset) | |
|---|---|---|---|
| Training Sets | #units | JA (E01/02) | CH |
| CH (**c**) | 2569 | / | 14.3 |
| JA (**c**) | 2651 | 28.0/25.5 | / |
| JA+CH (**c**) | 3583 | **14.1/10.6** | **11.3** |
| CH (**r**) | 1072 | / | 17.3 |
| JA (**r**) | 1276 | 31.4/29.7 | / |
| JA+CH (**r**) | 1474 | **14.1/10.7** | **13.7** |

evaluation sets and 1.4 h of Mandarin data to evaluate the bilingual ASR results, as shown in Table 3. From the results, the two multilingual models (JA+CH (**c**, **r**)) obtained a dramatic leap in performance and reduced the CER% by half compared with the two baseline systems. This proves the validity of the joint-modeling method for challenging ASR tasks for similar languages. By contrast, for these two multilingual models (JA+CH (**c**, **r**)), the Mandarin ASR results of the radical-based multilingual model (JA+CH (**r**)) were 2% worse, whereas the Japanese ASR results were very close. We reason that the L1-decomposition radicals in Kanji are similar not only to Chinese characters but also radicals. Thus, compared with joint-modeling with Chinese characters (JA+CH (**c**)), it is possible to decrease the accuracy of the ASR system when joint-modeling with Chinese radicals (JA+CH (**r**)). Despite this, the radical-based method can obtain essentially equivalent performance with less than half of the modeling units.

## 5 Conclusions and Future Work

To overcome the confusion that occurs in ASR tools when Chinese tourists read Japanese words and solve the large vocabulary problem, in this study, we proposed a novel Chinese-Japanese bilingual E2E system based on the decomposition strategy, which can decompose a character into radicals. Our experiments demonstrated that the radical-based E2E ASR system achieved great performance and effectively reduced the vocabulary by sharing basic radicals across Chinese and Japanese. We also proved that the joint-modeling method probably can improve the precision of E2E ASR systems in a low-resource scenario. In future work, we may expand our method to more East Asian languages, such as Korean, Kazakh, and Tibetan, some of which may have limited resources.

**Acknowledgements.** The work is partially supported by JSPS KAKENHI No. 21K17837, JST SPRING Grant No. JPMJSP2110 and NICT tenure-track startup funding.

## References

1. Akahori, K., et al.: Learning Japanese in the network society. University of Calgary Press (2002)
2. Chan, W., Jaitly, N., Le, Q., Vinyals, O.: Listen, attend and spell: a neural network for large vocabulary conversational speech recognition. In: Proceedings IEEE-ICASSP (2016)
3. Chen, Z., Jain, M., Wang, Y.: Joint grapheme and phoneme embeddings for contextual end-to-end ASR. In: Proceedings INTERSPEECH, pp. 3490–3494 (2019)
4. Chorowski, J., Bahdanau, D., Serdyuk, D., Cho, K., Bengio, Y.: Attention-based models for speech recognition. In: Proceedings NIPS (2015)
5. Columbus, A.: Advances in psychology research, vol. 45. Nova Publishers (2006)
6. Hori, T., Watanabe, S., Zhang, Y., Chan, W.: Advances in joint CTC-attention based end-to-end speech recognition with a deep CNN Encoder and RNN-LM. In: Proceedings INTERSPEECH (2017)
7. Kanda, N., Lu, X., Kawai, H.: Maximum a posteriori based decoding for CTC acoustic models. In: Proceedings INTERSPEECH, pp. 1868–1872 (2016)
8. Kawahara, T., Nanjo, H., Shinozaki, T., Furui, S.: Benchmark test for speech recognition using the corpus of spontaneous Japanese. In: Proceedings ISCA & IEEE Workshop on Spontaneous Speech Processing and Recognition (2003)
9. Köhler, J.: Multilingual phone models for vocabulary-independent speech recognition task. Speech Commun. **35**(1), 21–30 (2001)
10. Kim, S., Hori, T., Watanabe, S.: Joint CTC-attention based end-to-end speech recognition using multi-task learning. In: Proceedings IEEE International Conference on Acoustics, Speech and Signal Processing (ICASSP), pp. 4835–4839 (2017)
11. Ko, T., Peddinti, V., Povey, D., Khudanpur, S.: Audio augmentation for speech recognition. In: Proceedings INTERSPEECH (2015)
12. Kudo, T.: Subword regularization: Improving neural network translation models with multiple subword candidates. In: CoRR abs/1804.10959 (2018)
13. Leban, C.: Automated Orthographic Systems for East Asian Languages. (Chinese, Japanese, Korean), State-of-the-art Report, Prepared for the Board of Directors. Association for Asian Studies (1971)
14. Li, S., Lu, X., Ding, C., shen, P., Kawahara, T., Kawai, H.: Investigating radical-based end-to-end speech recognition systems for Chinese dialects and Japanese. In: Proceedings INTERSPEECH), pp. 2200–2204 (2019)
15. Lunde, K.: CJKV Information Processing, 2nd Edition, Chinese. Korean, and Vietnamese Computing. O'Reilly & Associates, Japanese (2009)
16. Maekawa, K.: Corpus of spontaneous Japanese: its design and evaluation. In: Proceedings ISCA & IEEE Workshop on Spontaneous Speech Processing and Recognition (2003)
17. Moriya, T., Shinozaki, T., Watanabe, S.: Kaldi recipe for Japanese spontaneous speech recognition and its evaluation. In: Autumn Meeting of ASJ (2015)
18. Schultz, T., Waibel, A.: Multilingual and crosslingual speech recognition. In: Proceedings DARPA Workshop on Broadcast News Transcription and Understanding, pp. 259–262 (1998)

19. Zhou, S., Dong, L., Xu, S., Xu, B.: Syllable-based sequence-to-sequence speech recognition with the transformer in mandarin Chinese. In: Proceedings INTERSPEECH (2018)
20. Toshniwal, S., Sainath, T., Weiss, R.: Multilingual speech recognition with a single end-to-end model. In: Proceedings IEEE International Conference on Acoustics, Speech and Signal Processing (ICASSP), pp. 4904–4908 (2018)
21. Vaswani, A., Shazeer, N., Parmar, N.: Attention is all you need. In: Proceedings Advances in Neural Information Processing Systems, pp. 5998–6008 (2017)
22. Vaswani, A., et al.: Attention is all you need. In: CoRR abs/1706.03762 (2017)
23. Zhang, Y., Chan, W., Jaitly, N.: Very deep convolutional networks for end-to-end speech recognition. In: Proceedings IEEE International Conference on Acoustics, Speech and Signal Processing (ICASSP), pp. 4845–4849 (2017)

# An Unsupervised Short- and Long-Term Mask Representation for Multivariate Time Series Anomaly Detection

Qiucheng Miao, Chuanfu Xu, Jun Zhan, Dong Zhu, and Chengkun Wu(✉)

Institute for Quantum Information and State Key Laboratory of High Performance Computing, College of Computer Science and Technology, National University of Defense Technology, Changsha, Hunan, People's Republic of China
{miaoqiucheng19,chengkun_wu}@nudt.edu.cn

**Abstract.** Anomaly detection of multivariate time series is meaningful for system behavior monitoring. This paper proposes an anomaly detection method based on unsupervised **S**hort- and **L**ong-term **M**ask **R**epresentation learning(SLMR). The main idea is to extract short-term local dependency patterns and long-term global trend patterns by using multi-scale residual dilated convolution and Gated Recurrent Unit(GRU) respectively. Furthermore, our approach can comprehend temporal contexts and feature correlations by combining spatial-temporal masked self-supervised representation learning and sequence split. It considers the importance of features is different, and we introduce the attention mechanism to adjust the contribution of each feature. Finally, a forecasting-based model and a reconstruction-based model are integrated to focus on single timestamp prediction and latent representation of time series. Experiments show that the performance of our method outperforms other state-of-the-art models on three real-world datasets. Further analysis shows that our method is good at anomaly localization.

**Keywords:** Anomaly detection · Multivariate time series · Representation learning · Multi-scale convolution

## 1 Introduction

Cyber-physical systems are generally used to control and manage industrial processes in some significant infrastructures. Active monitoring of sensor readings and actuator status is crucial for early system behavior detection [1]. Anomaly detection is widely studied in different fields to find significant deviations data

The work is supported by National Natural Science Foundation of China (62006236), NUDT Research Project (ZK20-10), National Key Research and Development Program of China (2020YFA0709803), Hunan Provincial Natural Science Foundation (2020JJ5673), National Science Foundation of China (U1811462), National Key R&D project by Ministry of Science and Technology of China (2018YFB1003203), and Autonomous Project of HPCL (201901–11, 202101–15).


M. Tanveer et al. (Eds.): ICONIP 2022, CCIS 1793, pp. 504–516, 2023.
https://doi.org/10.1007/978-981-99-1645-0_42

from normal observations [2], such as images and time series. This paper focuses on the anomaly detection of multivariate time series data (MTS). Nevertheless, the data is collected from interconnected sensor networks, so there are usually few labeled anomalies. Therefore, unsupervised learning seems to be the ideal choice for anomaly detection.

Many kinds of research based on machine learning have been proposed to concern MTS anomaly detection [5,8], such as One-class and isolation forest [3]. However, a more generally used strategy is residual-error based anomaly detection. Specifically, the residual-error based anomaly detection relies on a forecasting-based model to predict future sensor measurements [4], or a reconstruction-based model (such as autoencoder) to capture a lower-dimensional representation of sensor measurements [4]. Then the forecasting or reconstruction measurements are compared with the ground-truth measurements to yield a residual error. A system is considered abnormal if the residual error exceeds a threshold.

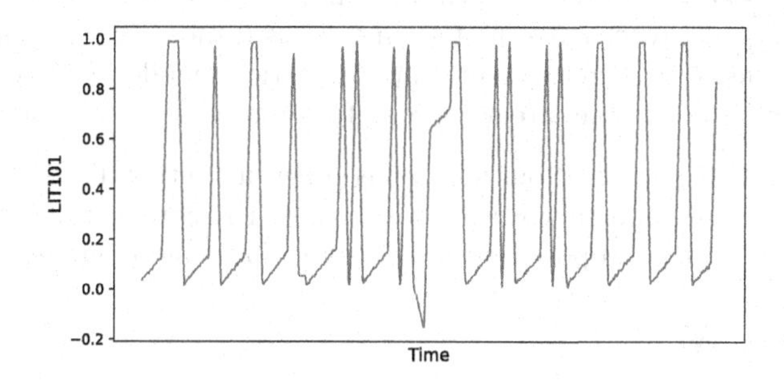


**Fig. 1.** LIT101 state values(The dataset of SWaT)

However, the existing methods often fail to consider that time series have different time scales, namely short-term and long-term patterns. Figure 1 shows the state values of LIT101 in the real-world dataset [15]. Obviously, there are two patterns, which are the long-term overall trends and short-term fluctuations. A robust anomaly detection model should be able to capture series correlations on time scales. The long-term patterns reflect the overall trend, and the short-term patterns capture the subtle changes in local regions. Furthermore, MTS is a special kind of sequence data [6]. Most working condition information can be retained when downsampling the sequence. Sequence split guarantees the model learns an efficient representation with different resolutions. Finally, MTS is composed of a collection of univariate time series, each of which describes an attribute of the system. Therefore, MTS not only has time dependence within the feature, which characterizes the temporal pattern but has inter-feature dependence within the system, which characterizes the linear or non-linear relationship between the features of a system in each period [9]. A key concern in anomaly detection is effectively extracting the temporal context and features correlation.

Therefore, we propose a jointly optimized anomaly detection method based on short- and long-term mask representation. The main contributions are the following:

(1) Random mask and sequence split: The input data is masked randomly and then the original data is used for reconstruction and forecasting representation. Mask can promote the model to understand temporal contexts and learn the dynamic information between features. In addition, the input data is split to obtain odd subsequences and even subsequences. Different convolution filters are used to extract the features, maintain heterogeneity information and ensure that the model learns different sequence resolutions.
(2) Short- and long-term patterns extract: We perform multi-scale residual dilated convolution to extract short-term spatial-temporal local dependency patterns and filter irrelevant information. Also, an attention mechanism is applied to different channels, adjusting the contribution of different feature weights. Finally, the jointly optimized method based on GRU is introduced to identify long-term patterns for time series trends. The forecasting-based model focuses on single timestamp prediction, while the reconstruction-based model learns the latent representation of the entire time series.

The results illustrate that our method is generally better than other advanced methods, and more importantly, achieves the best F1 score on the three datasets. It also shows that the proposed method enables locating anomalies.

## 2 Related Work

There is plenty of literature on time-series anomaly detection, which can be classified into two categories. One is the forecasting-based model and another is the reconstruction-based model.

The forecasting-based model implements the prediction of the next timestamp, which is compared with the ground-truth value to generate a residual error, to decide whether an abnormality occurs according to the threshold. Long Short-Term Memory (LSTM) is one of the most popular methods to predict the next time sequence in anomaly detection [2]. The attention mechanism and its variants have also obtained widespread application, such as Dsanet [8]. Building a model, the complex dependencies of features, is a great challenge in traditional machine learning, so GNN-based anomaly detection has become a hot research topic [4,9]. CNN can automatically extract and learn the different components of the signal on time scales, which yield unusually brilliant results in multivariate time series anomaly detection [10].

The reconstruction-based model learns the latent representation of the entire time series by reconstructing the original data to generate the observed value. Then, the deviation between the observed value and the ground-truth value is evaluated to identify anomalies. The current state-of-the-art deep reconstruction anomaly detection models mainly include DAGMM [11], AE [13], GAN [14]. In addition, convolutional neural networks perform well in feature extraction, especially in noisy environments [5,17].

## 3 Methodology

**Problem definition:** The dataset of MTS anomaly detection can be expressed as $\boldsymbol{S} \in \mathbb{R}^{n \times k}$, where $n$ is the length of the timestamp, and $k$ is the number of input features. A fixed input $\boldsymbol{X} \in \mathbb{R}^{w \times k}$ is generated from long time series by a sliding window of length $w$. The target of our algorithm is to produce a set of binary labels $\boldsymbol{y} \in \mathbb{R}^{n-w}$, where $y(t) = 1$ indicates the $t_{th}$ timestamp is abnormal.

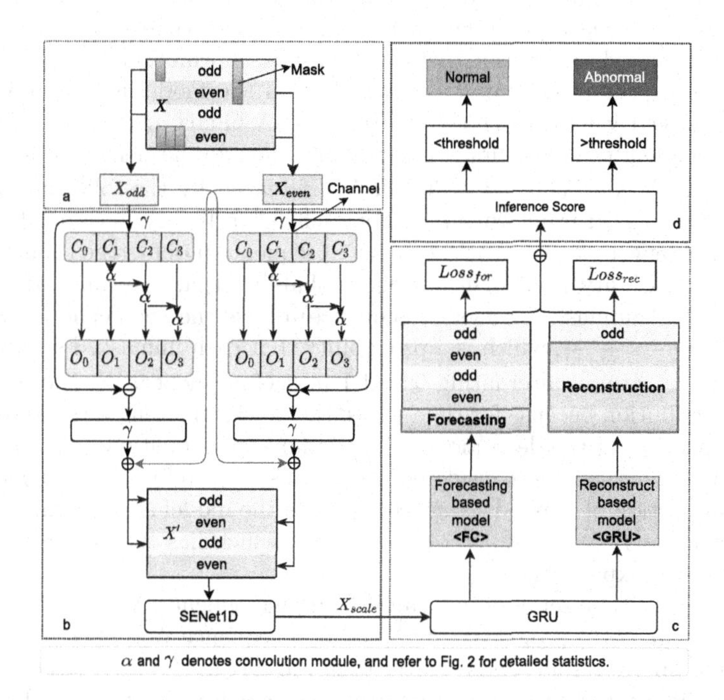


**Fig. 2.** An overview of our method for anomaly detection. a) We randomly set part of the input data to 0, then divide the input into odd and even subsequences according to the time dimension. b) After odd and even subsequences are generated, firstly, we realize the channel dimension transformation with a $1 \times 1$ convolution (i.e., $\gamma$, see Fig. 3), and then perform multi-scale residual convolution on different channels (i.e., $\alpha$, see Fig. 3). The output channel is inverse transformation through a $1 \times 1$ convolution, merging odd and even sequences. Then all of them are input into SENet1D (see Fig. 4), and the contribution weights of different channels are adjusted. c) GRU-based sequence joint optimization, and balance the scores of both. d) Determine whether an abnormality occurs according to the threshold and the final inference score.

### 3.1 Mask and Sequence Split

As a task for spatial-temporal masked self-supervised representation, the mask prediction explores the data structure to understand the temporal context and features correlation. We will randomly mask part of the original sequence before we input it into the model, specifically, we will set part of the input to 0. The training goal is to learn the latent representation of the masked signal and then predict the masked value. The mask prediction task involves the basic assumption that there is a correlation between temporal context and features. The benefits are twofold: First, masked values do not change the size of the time series, which is essential for time series anomaly detection. Second, they also improve the robustness of the learned representation by forcing each timestamp to reconstruct itself in a different context. Intuitively, masked values enable the network to infer the representation in an incomplete time series, which helps predict missing values in incomplete surrounding information.

Expressly, part of the input is set to 0, and the model is required to predict the masked values, as shown in Fig. 2(a). A binary mask $\boldsymbol{M} \in \mathbb{R}^{w \times k}$ for each independent sample is firstly created, and then do the element-wise product with the input $\boldsymbol{X}$ : $\widetilde{\boldsymbol{X}} = \boldsymbol{M} \odot \boldsymbol{X}$. $\boldsymbol{M}$ is referenced by the following method, a transformer-based framework [18]. For each $\boldsymbol{M}$, a binary mask is alternately set, and the length of each mask segment (0 sequences) obeys the geometric distribution probability $1/s_m$, which is the probability that each masked sequence will stop. And unmask segment (1 sequence) obeys the geometric distribution probability $\frac{1}{s_m} \times \frac{r}{(1-r)}$, which is a probability that each unmasked sequence will stop. $r$ is the ratio of alternating 0 and 1 in each sequence. We chose the geometric distribution because for time series, the powerful latent variable representation capability of the deep model allows adaptive estimation of short masked sequences (which can be considered as outliers). Therefore, we need to set a higher proportion of masked sequences to force the model actively mine contextual associations. We set $s_m = 3$ and $r = 0.1$, which has the best performance with extensive experiments.

The original time-series is split, and the input sequence $\boldsymbol{X} \in \mathbb{R}^{w \times k}$ is divided into even sequence $\boldsymbol{X}_{even} \in \mathbb{R}^{(w/2) \times k}$ and odd sequence $\boldsymbol{X}_{odd} \in \mathbb{R}^{(w/2) \times k}$, each of which temporal resolution is coarser than the original input. The subsequence only contains part of the original information, which well preserves the heterogeneous information of the original sequence. Feature extraction is performed on the odd-sequence and even-sequence, respectively. Interactive learning between features is added to each sequence to compensate for the representation loss during downsampling. As shown in Fig. 2(a), after projecting $\boldsymbol{X}_{even}$ and $\boldsymbol{X}_{odd}$ to two different one-dimensional convolutions $\gamma$ and $\alpha$, the $\boldsymbol{X}'_{even}$ and $\boldsymbol{X}'_{odd}$ are obtained by the residuals connection. After all downsampling—convolution—interaction, all low-resolution components are rearranged and connected to a new sequence representation.

$$\boldsymbol{X}'_{even} = \boldsymbol{X}_{odd} \oplus Conv1d(\boldsymbol{X}_{even}) \tag{1}$$

$$\boldsymbol{X}'_{odd} = \boldsymbol{X}_{even} \oplus Conv1d(\boldsymbol{X}_{odd}) \tag{2}$$

Conv1d denotes the multi-scale residual convolution operation, and $\oplus$ is the residual connection. Time series downsampling can retain most information and exchange information with different time resolutions. In addition, the designed sequence sampling does not require domain knowledge and can be easily generalized to various time-series data.

### 3.2 Multi-scale residual Convolution

CNN uses the learnable convolution kernels to automatically extract features from different scales to obtain a better representation [19]. Therefore, we propose a simple and effective module for multi-scale residual convolution. Unlike existing multi-layer and multi-scale methods, we have improved the multi-scale representation ability of CNN at a more fine-grained level. To achieve this goal, firstly, realizing the feature channels dimension($n$) transformation through $1 \times 1$ convolution ($\gamma$, as shown in Fig. 3). Then, the feature channels will be divided into $s = 4$ subsets on average, each subset can be represented by $\boldsymbol{C}_i$, $(i = 1, 2, .., s)$, and each subset of the segmented has $w$ channels, i.e., $\boldsymbol{C}_i \in \mathbb{R}^{\frac{n}{2} \times w}$, where $w$ denotes the number of feature map subgroups. A group of $3 \times 3$ kernels extracts features from a set of input feature maps, denoted by $\alpha$, as shown in Fig. 3. Finally, the output features of the previous group are connected with another group in a residual manner, and the kernel size of this group is $2i + 1, i = 0, 1, ..., s - 1$. This process is repeated several times until the feature maps of all groups are connected and concatenate all outputs together.

The output $\boldsymbol{O}_i$ can be expressed as:

$$\boldsymbol{O}_i = \begin{cases} \boldsymbol{C}_i, i = 1 \\ \alpha\left(\boldsymbol{C}_i\right), i = 2 \\ \alpha\left(\boldsymbol{C}_i + \boldsymbol{O}_{i-1}\right), 2 < i \leq s \end{cases} \tag{3}$$

$\alpha$ is a convolution module, as shown in Fig. 3. The multi-scale residual can capture a larger receptive field. To effectively extract local and global information, different convolution scales are integrated. The $1 \times 1$ kernel convolution ($\gamma$) adjusts the output data to the same size as the input, as shown in Fig. 3.

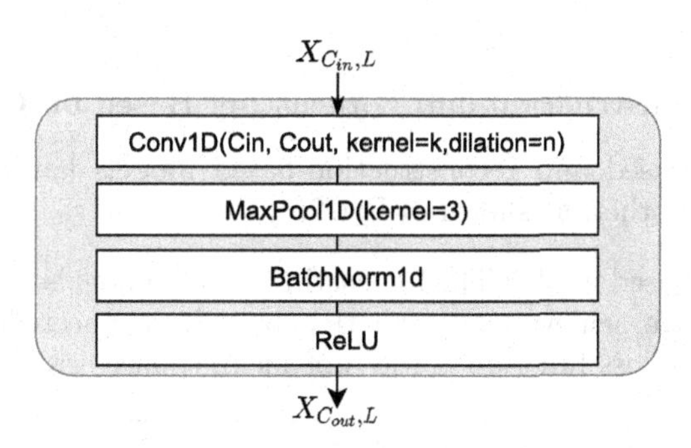


**Fig. 3.** Convolution modules, where $\alpha$: k=2i+1 (i=1,...,s-1), dilation=2, $\gamma$: k=1, dilation=1, $\beta$: k=3, dilation=1.

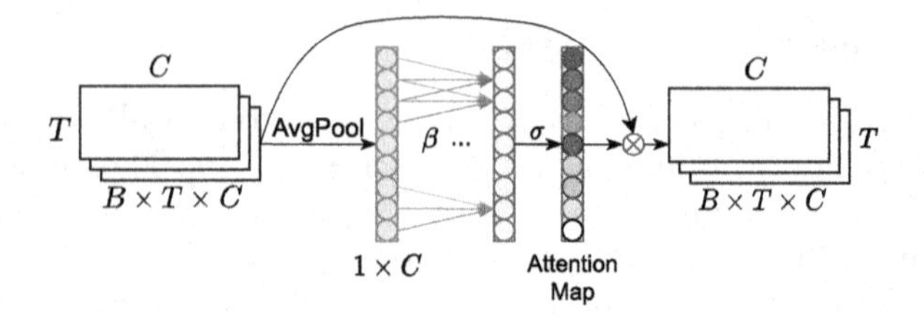


**Fig. 4.** SENet1D module. $\beta$ denotes convolution with a kernel size of 3, $\sigma$ is the Sigmoid activation function, $\otimes$ is element-wise broadcast dot multiplication.

In different feature channels, the contribution in the anomaly detection is often different, and SENet1D is used to adjust the weight contribution of different features, as shown in Fig. 4. The specific path to obtain weights is "Global Average Pooling (AvgPool) → Conv1d → ReLU Function → Conv1d → Sigmoid Function ($\sigma$)", as shown in Fig. 4.

$$\boldsymbol{z} = \sigma(Conv1d(\delta(Conv1d(Avgpool(\boldsymbol{X}'))))) \tag{4}$$

$$\boldsymbol{X}_{scale} = \boldsymbol{z} \otimes \boldsymbol{X}' \tag{5}$$

where $\boldsymbol{X}'$ is the original sequence sample, $\delta$ represents the ReLU function, Conv1d is a one-dimensional convolution with a kernel size of 3, which implements a nonlinear gating mechanism. Avgpool is a global average pooling, which compresses the global spatial information to the channel descriptors to generate channel-oriented statistical information. Finally, through $\sigma$, the attention weights of different channels are compressed to [0,1], realizing the channel weight contribution adjustment, $\boldsymbol{X}_{scale}$ inputs into the GRU to capture the long-term sequence trends.

### 3.3 Joint Reconstruction and Forecasting Based on GRU

The forecasting-based and reconstruction-based models have advantages and complement each other, as shown in Fig. 2(c).

(1) Forecasting-based model: The forecasting-based model is to predict the next timestamp. We achieve the prediction by fully connected layers after the GRU, and the loss function is the root mean square error (RMSE):

$$Loss_f = \sqrt{\sum_{m=1}^{k} (\hat{x}_{t+1,m} - x_{t+1,m})^2} \tag{6}$$

$Loss_f$ represents the loss function of the forecasting-based model. Where $\hat{\boldsymbol{x}}_{t+1,m}$ denotes the predicted value of node $m$ at $t+1$, $\boldsymbol{x}_{t+1,m}$ represents the groundtruth value of node $k$ at time $t+1$. By calculating the residual between the predicted value and the ground truth value, it is determined whether this point is abnormal.

(2) Reconstruction-based model: Reconstruction-based model learns low-dimensional representations and reconstructs the "normal patterns" of data. For the input $\boldsymbol{X}_{t-w:t}$, GRU decodes the input to obtain the reconstruction matrix $\hat{\boldsymbol{X}}_{t-w:t}$, $w$ is window sizes, and the RMSE is:

$$Loss_r = \sqrt{\sum (\boldsymbol{X}_{t-w:t} - \hat{\boldsymbol{X}}_{t-w:t})^2} \tag{7}$$

$Loss_r$ represents the loss function of the reconstruction-based model. The total loss is the sum of $Loss_f$ and $Loss_r$:

$$Loss_{total} = Loss_f + Loss_r \tag{8}$$

### 3.4 Anomaly Detection

For the joint optimization, we have two inference results for each timestamp. One is the predicted value $\hat{x}_i|i = 1, 2, \ldots, k$ computed forecasting-based model, and the other is the reconstructed value $\hat{x}_j|j = 1, 2, \ldots, k$, which was obtained from reconstruction-based model. The final inference score balances their benefits to maximize the overall effectiveness of anomaly detection, as shown in Fig. 2(d). We use the mean of all features to compute the final inference score and to determine whether the current timestamp is abnormal with the corresponding inference score.

For the forecasting-based model, the input data is $\boldsymbol{X}_{t-w:t}$, and inference score is $\sqrt{\sum_{m=1}^{k} (\hat{x}_{t+1,k} - x_{t+1,k})^2}$. For the reconstruction-based model, the input data is $\boldsymbol{X}_{t-w+1:t+1}$, using GRU decodes the input to obtain the reconstruction matrix $\hat{\boldsymbol{X}}_{t-w+1:t+1}$, and inference score is $\sqrt{\sum (\hat{\boldsymbol{X}}_{t:t+1} - \boldsymbol{X}_{t:t+1})^2}$. The final inference score can be calculated by:

$$score = \frac{1}{k}\left(\sqrt{\sum_{m=1}^{k} (\hat{x}_{t+1,m} - x_{t+1,m})^2} + \gamma\sqrt{\sum (\hat{\boldsymbol{X}}_{t:t+1} - \boldsymbol{X}_{t:t+1})^2}\right) \tag{9}$$

The optimal global threshold selection is similar to previous work, we also use best-F1 [7,9,13,16]. It is considered an anomaly when the inference score exceeds the threshold.

## 4 Experiment and Analysis

### 4.1 Datasets and Evaluation Metrics

**a) Datasets:** To verify the effectiveness of our model, we conduct experiments on three datasets, SMAP (Soil Moisture Active Passive satellite) [2], MSL (Mars Science Laboratory rover) [2], and SWaT (The Secure Water Treatment) [15]. Table 1 is the detail statistic information of the three datasets. Our code with Pytorch1.8 and data are released at https://github.com/qiumiao30/SLMR.

**Table 1.** Statistical details of datasets. (%) is the percentage of abnormal data.

| Dataset | Train | Test | Dimensions | Anomalies (%) |
|---|---|---|---|---|
| SWaT | 480599[a] | 449919 | 51 | 11.98 |
| SMAP | 135183 | 427617 | 55*25[b] | 13.13 |
| MSL | 58317 | 73729 | 27*55 | 10.72 |

[a] Remove the first four hours of data [14].
[b] 55 is the dimension, 25 is the number of entities

**b) Metrics:** We use precision, recall, and F1 score to measure the performance of our model. In a continuous anomaly segment, the entire segment is correctly predicted if at least one moment is detected to be anomalous. The point-adjust method is applied to evaluate the performance of models according to the evaluation mechanism in [9,13,16]. The model is trained with the Adam optimizer, the learning rate is initialized to 0.001, the batch size is 256, 10% of the training data is used as the validation set, and the window size is set to 100 or 80 (Table 2).

**Table 2.** Performance of our model and baselines.

| Method | SWaT | | | SMAP | | | MSL | | |
|---|---|---|---|---|---|---|---|---|---|
| | Prec. | Rec. | F1 | Prec. | Rec. | F1 | Prec. | Rec. | F1 |
| IF | 0.962 | 0.731 | 0.831 | 0.442 | 0.510 | 0.467 | 0.568 | 0.674 | 0.598 |
| AE | 0.991 | 0.704 | 0.823 | 0.721 | 0.979 | 0.777 | 0.853 | 0.974 | 0.879 |
| LSTM-VAE | 0.979 | 0.766 | 0.860 | 0.855 | 0.636 | 0.729 | 0.525 | 0.954 | 0.670 |
| MAD-GAN | 0.942 | 0.746 | 0.833 | 0.671 | 0.870 | 0.757 | 0.710 | 0.870 | 0.782 |
| DAGMM | 0.829 | 0.767 | 0.797 | 0.633 | 0.998 | 0.775 | 0.756 | 0.980 | 0.853 |
| LSTM-NDT | 0.990 | 0.707 | 0.825 | 0.896 | 0.884 | 0.890 | 0.593 | 0.537 | 0.560 |
| OmniAnomaly | 0.722 | 0.983 | 0.832 | 0.758 | 0.975 | 0.853 | 0.914 | 0.889 | 0.901 |
| USAD | 0.987 | 0.740 | 0.846 | 0.769 | 0.983 | 0.863 | 0.881 | 0.978 | 0.927 |
| MTAD-GAT | 0.903 | 0.821 | 0.860 | 0.809 | 0.912 | 0.901 | 0.875 | 0.944 | 0.908 |
| GTA | 0.948 | 0.881 | 0.910 | 0.891 | 0.917 | 0.904 | 0.910 | 0.911 | 0.911 |
| **SLMR** | 0.963 | 0.874 | **0.916** | 0.915 | 0.992 | **0.952** | 0.965 | 0.967 | **0.966** |

### 4.2 Performance and Analysis

We compare our method with other 10 advanced models that deal with multivariate time series anomaly detection, including Isolation Forest(IF) [3], DAGMM [11], basic Autoencoder, LSTM-VAE [12], MAD-GAN [14], LSTM-NDT [7], USAD [16], OmniAnomaly [13], MTAD-GAT [9], GTA [20]. The results illustrate that our method generally achieves the highest F1 score on the three datasets. We can also observe that our method achieves a great balance between precision and recall.

**Overview Performance:** Overall, the non-parametric methods, Isolation Forest and DAGMM, perform the worst. Because they are poor at capturing temporal information. As a generative model, DAGMM has high recall scores on the SMAP and MSL datasets, but it still does not solve the time correlation problem, resulting in poor performance. For time series, temporal information is necessary because observations are dependent and historical data helps reconstruct/predict current observations. AutoEncoder(AE) uses an encoder and decoder to complete the reconstruction of time series. USAD adds adversarial training based on AE to generate better representations for downstream tasks. However, the above two methods reconstruct the time series point by point without capturing the time correlation, limiting the model's detection performance. Generative models based on VAE or GAN, such as LSTM-VAE, MAD-GAN, and OmniAnomaly, can effectively capture temporal information but do not consider feature-level correlation. MTAD-GAT and GTA capture feature dependence through graph attention networks, but to a certain extent, the temporal information is ignored. The method proposed in this paper firstly combines mask spatial-temporal representation, which facilitates the model to comprehend temporal contexts and feature correlations. In addition, multi-scale convolution extracts short-term dependency patterns, catches rich information on time scales, filters the original data, and reduces the impact of irrelevant information on the results. Finally, GRU acquires long-term dependencies. Therefore, our method achieves the highest F1 score and the best performance in balancing recall and precision.

### 4.3 Ablation Study

We perform ablation studies using several variants of SLMR to further point out the validity of the module described in Sect. 3, as shown in Table 3.

**Table 3.** Performance of our method and its variants(F1).

| Method | SWaT | SMAP | MSL |
|---|---|---|---|
| **SLMR** | **0.916** | **0.952** | **0.966** |
| W/o mask | 0.854 | 0.942 | **0.966** |
| W/o odd/even | 0.873 | 0.911 | 0.953 |
| W/o multi_CNN | 0.849 | 0.929 | 0.951 |
| W/o SENet1D | 0.870 | 0.903 | 0.950 |
| W/o forecast | 0.852 | 0.923 | 0.952 |
| W/o reconstruct | 0.847 | 0.935 | 0.941 |

**Self-supervised mask representation.** The SWaT and SMAP have been dramatically improved, but the MSL has poor performance. It is proved that when datasets size is small, the performance is not improved. The main reason is that the context learning ability of the model is poor on small-scale datasets, and

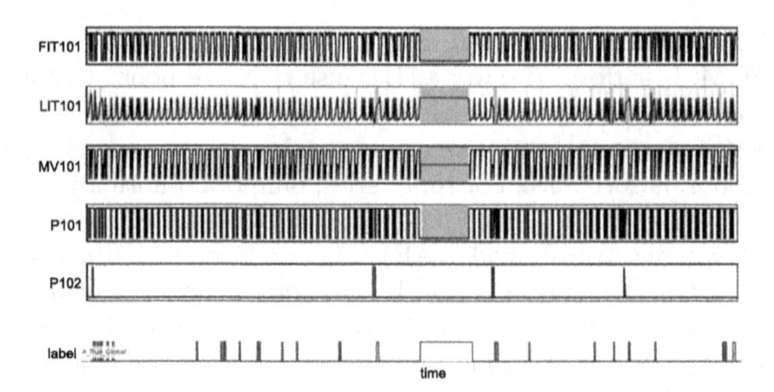


**Fig. 5.** Anomalies localization(the first 5 features). The shadow is the abnormal segment detected by the model, and the 'label' is the true global abnormal segment.

the masked value cannot be effectively filled. **Sequence split.** The data downsampling is beneficial to improving performance. This method can ensure that the model learns the effective representation of the sequence between different levels. **Multi-scale residual dilation convolution.** The proposed method can learn more latent knowledge than a basic convolution layers, and it can effectively reduce the impact of irrelevant information on the results. **Joint optimization.** The forecasting-based model is sensitive to the randomness of the time series, while the reconstruction-based model alleviates it by learning the distribution of random variables. Besides, the reconstruction-based model can capture the global data distribution well, but it may ignore abrupt perturbations, thereby destroying the periodicity of the time series. In contrast, the forecasting-based model can effectively compensate for this drawback.

### 4.4 Localization Abnormal

Figure 5 shows the abnormal location information of the SWaT. The blue part is the detected abnormal event. It can be seen that the intermediate abnormal information can be wholly detected. To other shorter abnormal segments, most of the abnormal information is detected. As shown in Fig. 6, the abnormal detection of LIT101 (the green line of (b)) corresponds to the actual measured abnormal state, which fully proves that our model can effectively detect abnormal information. Therefore, accurate anomaly location interpretation is convenient for practitioners to find the abnormal parts early, and timely mitigation measures can be taken to avoid equipment downtime or damage due to major failures and reduce potential economic and environmental losses.

'A_score_1' in Fig. 6(b) represents the actual anomaly score inferred from LIT101, which fully proves that our model can effectively detect abnormal information. Therefore, accurate anomaly location interpretation is convenient for the engineer to find the abnormal parts early. They can timely take mitigation measures to avoid equipment outages or damage due to major failures and reduce potential economic and environmental losses.

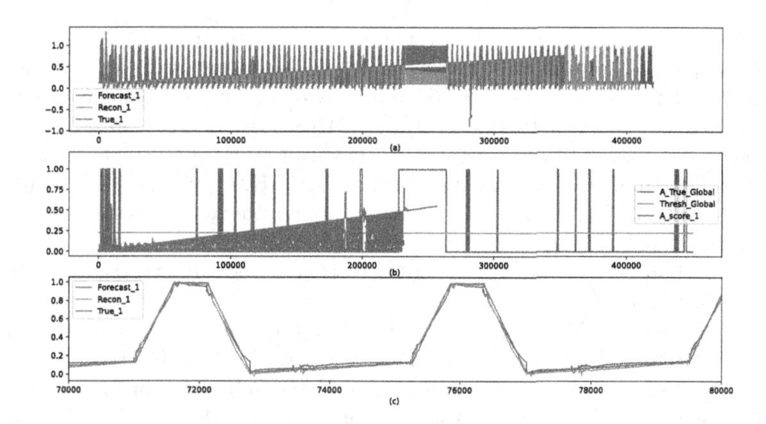


**Fig. 6.** Visualization of LIT101. (a): True_1 is ground-truth values, and Forecast_1 and Recon_1 is generated value by forecasting and reconstruction; (b): A_True_Global is the true global anomaly state, and A_score_1 is inferred anomaly scores, and Thres_Global is the global threshold; (c): Partial subset of (a).

# 5 Conclusion

This paper proposes anomaly detection methods based on MTS. Firstly, we use mask-based self-supervised representation learning to enable the model to understand temporal contexts and features correlation, and sequence split enhances the representation capacity of different resolutions of sequences. Then, multi-scale residual convolution effectively improves the capability to capture short-term dynamic changes, and filters existing irrelevant information. GRU identifies the long-term trend patterns by joint optimization of the forecasting and reconstruction-based model. It is observed that our method effectively captures the temporal-spatial correlation, and generally outperforms the other advanced methods.

# References

1. Feng, C., Tian, P.: Time series anomaly detection for cyber-physical systems via neural system identification and bayesian filtering, In: Proc. of SIGKDD, ser. KDD '21, New York, NY, USA, p. 2858–2867 (2021)
2. Hundman, K., Constantinou, V., Laporte, C., Colwell, I., Soderstrom, T.: Detecting spacecraft anomalies using lSTMS and nonparametric dynamic thresholding, In: Proc of SIGKDD, pp. 387–395 (2018)
3. Liu, F., Ting, K., Zhou, Z.: Isolation forest In: Proc of ICDM, pp. 413–422. Institute of Electrical and Electronics Engineers, IEEE (2008)
4. Deng, A., Hooi, B.: Graph neural network-based anomaly detection in multivariate time series. Proc. AAAI **35**(5), 4027–4035 (2021)
5. Zhang, C., et al.: A deep neural network for unsupervised anomaly detection and diagnosis in multivariate time series data. Proc AAAI **33**(01), 1409–1416 (2019)

6. Liu, M., Zeng, A., Xu, Z., Lai, Q.: Time series is a special sequence: Forecasting with sample convolution and interaction, (2021) arXiv preprint arXiv:2106.09305
7. Goh, J., Adepu, S., Tan, M., Lee, Z.: Anomaly detection in cyber physical systems using recurrent neural networks, In: Proc of HASE. IEEE, 2017, pp. 140–145 (2017)
8. Huang, S., Wang, D., Wu, X., Tang, A.: DSANet: Dual self-attention network for multivariate time series forecasting, In: Proc of CIKM, pp. 2129–2132 (2019)
9. Zhao, H., et al.: Multivariate time-series anomaly detection via graph attention network, In: ICDM. IEEE, 2020, pp. 841–850 (2020)
10. Ren, H., et al.: Time-series anomaly detection service at microsoft, In: Proc of SIGKDD, pp. 3009–3017 (2019)
11. Zong, B., Song, Q., Min, M., Cheng, W., Lumezanu, C.: Deep autoencoding gaussian mixture model for unsupervised anomaly detection In: ICLR, (2018)
12. Park, D., Hoshi, Y., Kemp, C.: A multimodal anomaly detector for robot-assisted feeding using an LSTM-based variational autoencoder. IEEE Robot. Autom. Lett. **3**(3), 1544–1551 (2018)
13. Su, Y., Zhao, Y., Niu, C., Liu, R., Sun, W., Pei, D.: Robust anomaly detection for multivariate time series through stochastic recurrent neural network, In: Proc of SIGKDD, pp. 2828–2837 (2019)
14. Li, D., Chen, D., Jin, B., Shi, L., Goh, J., Ng, S.-K.: MAD-GAN: Multivariate Anomaly Detection for Time Series Data with Generative Adversarial Networks. In: Tetko, I.V., Kůrková, V., Karpov, P., Theis, F. (eds.) ICANN 2019. LNCS, vol. 11730, pp. 703–716. Springer, Cham (2019). https://doi.org/10.1007/978-3-030-30490-4_56
15. Goh, J., Adepu, S., Junejo, K.N., Mathur, A.: A Dataset to Support Research in the Design of Secure Water Treatment Systems. In: Havarneanu, G., Setola, R., Nassopoulos, H., Wolthusen, S. (eds.) CRITIS 2016. LNCS, vol. 10242, pp. 88–99. Springer, Cham (2017). https://doi.org/10.1007/978-3-319-71368-7_8
16. Audibert, J., Michiardi, P., Guyard, F.: USAD: unsupervised anomaly detection on multivariate time series, In: Proc of SIGKDD. ACM, pp. 3395–3404 (2020)
17. Zhang, Y., Chen, Y., Wang, J., Pan, Z.: Unsupervised deep anomaly detection for multi-sensor time-series signals, In: IEEE Trans. Knowl. Data Eng., p. 1, (2021)
18. Zerveas, G., Jayaraman, S., Patel, D., Bhamidipaty, A., Eickhoff, C.: A transformer-based framework for multivariate time series representation learning, In: Proc. of SIGKDD, ser. KDD '21, NY, USA, p. 2114–2124 (2021)
19. Gao S.Z. K., Cheng, M.: Res2net: a new multi-scale backbone architecture, In: IEEE Trans. Pattern Anal. Mach. Intell. **43**(2), pp. 652–662, (2021)
20. Chen, Z., Chen, D., Zhang, X., Yuan, Z., Cheng, X.: Learning graph structures with transformer for multivariate time-series anomaly detection in IoT. IEEE Internet Things J. **9**(12), 9179–9189 (2022)

# Investigating Effective Domain Adaptation Method for Speaker Verification Task

Guangxing Li[1], Wangjin Zhou[2], Sheng Li[3], Yi Zhao[4], Jichen Yang[5], and Hao Huang[1,6(✉)]

[1] School of Information Science and Engineering, Xinjiang University, Urumqi, China
ligx2022@gmail.com, huanghao@xju.edu.cn

[2] Graduate School of Informatics, Kyoto University, Sakyo-ku, Kyoto, Japan
zhou.wangjin.54r@st.kyoto-u.ac.jp

[3] National Institute of Information and Communications Technology (NICT), Kyoto, Japan
sheng.li@nict.go.jp

[4] Kuaishou Technology, Beijing, China
zhaoyi07@kuaishou.com

[5] School of Cyberspace Security, Guangdong Polytechnic Normal University, Guangzhou, China
NisonYoung@163.com

[6] Xinjiang Key Laboratory of Multi-lingual Information Technology, Urumqi, China

**Abstract.** CN-Celeb data set is a popular Chinese data set used for speaker recognition. In order to improve the speaker recognition performance on CN-Celeb, this paper mainly makes efforts in following two aspects: applying scenario information and using the self-supervised learning (SSL) model to the speaker recognition system. Compared to the SSL model, scenario information, named domain embedding, is more effective. We integrate speaker embedding and speaker scenario information abstracted as word embedding and puts forward several back-end processing methods. These methods are more suitable for each registered speaker with multiple enroll utterances. Moreover, data augmentation is also investigated. Experiments show that concatenating and averaging each utterance's domain embedding dimensions of multiple registered speakers with data augmentation shows better performance. Our proposed method effectively reduces 6.0%, 21.7% and 13.5% on EER, minDCF (P = 0.01) and minDCF (P = 0.001), respectively, compared with the baseline system. We did not observe improvement by using SSL as reported in previous works, and the possible reasons and analysis are also given in this paper.

**Keywords:** speaker recognition · speaker verification · speaker scenarios · multiple enroll utterances

G. Li and W. Zhou—Equal contribution.


M. Tanveer et al. (Eds.): ICONIP 2022, CCIS 1793, pp. 517–527, 2023.
https://doi.org/10.1007/978-981-99-1645-0_43

## 1 Introduction

CN-Celeb [7,13] data set currently has two versions of Chinese data set, which is the most popular open source Chinese speaker recognition data set. At present, the effect of using CN-Celeb [7,13] for speaker recognition is different from that of Vox-Celeb [6,15] data set, which generally has an EER of less than 1%. The main reason is that CN-Celeb [7,13] is more complex than Vox-Celeb, including more and more complex scenarios, statements of different lengths and overlapping background speakers. Therefore, using CN-Celeb [7,13] for speaker recognition is a challenging task.

In order to improve the efficiency of speaker recognition in CN-Celeb [7,13] data set. We only used CN-Celeb-T as the training set, and non-speech data set for data augmentation. According to the baseline speaker recognition system provided in ASV-Subtools [25], a speaker classifier model is first trained using the CN-Celeb-T, and this model output speakers in the last layer. We use the CN-Celeb-E to extract speaker embedding before the test stage's full connection layer.

In addition, cross domain is one of the key factors that lead to the unsatisfactory effect of speaker recognition, such as cross channel (8kHz vs 16KHz), cross language or cross dialect, cross device, cross environment, etc. In order to solve the impact of cross domain problems on speaker recognition, domain adaptation came into being. At present, there are many domain adaptive solutions for cross domain speaker recognition. [12,24] starting with the alignment covariance, find a transformation matrix to minimize the Frobenius norm with in domain and out of domain. [14] is an optimization method of [12,24], which mainly makes z-norm for data in domain, and adopts up sampling on this basis. [5] uses DA [2,12] as a speaker recognition model that is robust to speech style changes for the first time. Different from the above adaptive methods, we mainly aim at the problem that CN-Celeb [7,13] data set has complex scenarios to solve the speaker recognition between different scenarios, so as to avoid the same speaker being recognized as different speakers in different scenarios and different speakers being recognized as the same speaker in the same scenario. We try to recognize speakers in the same scenario by introducing scenario information.

Besides, using self-supervised learning to learn more abundant speech representation has shown surprising results in automatic speaker recognition task. Similarly, the application of self-supervised learning in speaker recognition task should also achieve good results. Some recent self-supervised learning speaker recognition tasks [3,8,16] also illustrate the feasibility of this method, and even [4] specifically analyzes why self-supervised learning of speech recognition is conducive to speaker recognition. The next chapter also introduces the use of self-supervised learning.

In the rest of this paper, we will briefly introduce the system which we improved. We will describe the method we proposed in Sect. 2 and describe the specific process of the experiment in Sect. 3, including the back-end test method we proposed. In addition, we will summarize the experimental results in Sect. 4.

## 2 Proposed Method

Different from other data sets, the CN-Celeb [7,13] has many scenarios, such as speech, singing, and interview. We referred to the different scenarios of CN-Celeb [7,13] as domains, and there are 11 different domains in the CN-Celeb [7,13] data set . Based on this, we tried to abstract the domain labels of each utterance into different word embeddings in jointly training SE-ResNet34 [9]. We concatenate the word embedding and speaker embedding and name the concatenated embedding as domain embedding. The benefit of domain embedding for the speaker recognition task was found in subsequent tests.

Another feature of the CN-Celeb [7,13] data set compared to other data sets is that each speaker can enroll multiple utterances. However, the general processing method is to concatenate multiple utterances of a speaker into one utterance. And then, the concatenated utterance is used as the enrollment. We used various testing methods when testing the effect of domain embedding on speaker recognition. We found that enrolling multiple utterances for a single speaker significantly influences recognition. Moreover, we verified that the universality of enrolling multiple utterances for a single speaker could effectively improve the baseline system's recognition.

More detail speaking, we use the learnability of word embedding to abstract these 11 domains into different 32-dimensional embeddings. Then we concatenate them after the 256-dimensional embedding calculated by SE-ResNet34 [9] during training. We first try to maximize the same-speaker similarity and minimize the different-speaker similarity scores in the testing phase. The specific method is to find the rules by analyzing the cosine scores of test utterances and the corresponding registered utterances. We want to filter out some test-enrollment pairs with high scores as positives and others as negatives.

Besides, in order to make the training data more diverse, we also investigate the effectiveness of data augmentation using the non-speech data sets.

SSL can capture rich features because it is trained in large-scale data sets. There are some precedents that using SSL for speaker recognition, [8] fine tune in wav2vec 2.0 [1,21] based on Vox-Celeb [6,15] data set, [16] fine tune in wav2vec 2.0 [1,21] based on NIST SRE [18,19] series data sets, Vox-Celeb [6,15] and several Russian data sets, and [3] has a number of state-of-the-art results in SUPERB, which has surprising results in speaker recognition. However, none of them is fine tuned on the Chinese data set, so we want to verify the effect of various pre-training models on the speaker recognition task on the Chinese data set. We investigate using SSL models to extract speaker embeddings, inspired by the methods of the top teams [20,27] in the VoiceMOS Challenge 2022.

## 3 Experimental Settings

### 3.1 Data

All experiments are conducted on the CN-Celeb [7,13] data set.

- **Training set**: The CN-Celeb 1 & 2 [7,13] train sets, which contain more than 600,000 audio files from 3000 speakers, are used for training.
- **Enrollment set**: The enrollment set is from CN-Celeb1 [7], which contain 799 audio files from 196 speakers.
- **Testing set**: The testing set is from CN-Celeb1 [7], which 17777 audio files from 200 speakers.

Compared with enrollment speakers, four speakers in the test set are out-of-set speakers. Note that both the enrollment and test sets do not intersect with the training set. In the experiments, we focus on speaker recording scenarios. The CN-Celeb [7,13] data set annotate the recording scenarios for each speech, which provides an essential guarantee for our experiments. There are the 11 domains in CN-Celeb [7,13], and the 11 domains in CN-Celeb [7,13] is shown in Table 1.

**Table 1.** Distribution of 11 scenarios in CN-Celeb1 & 2 [7,13]

| Domains | #Spks | #Utters | #Hours |
|---|---|---|---|
| Entertainment | 1099 | 54046 | 94.51 |
| Interview | 1299 | 93341 | 217.05 |
| Singing | 712 | 54708 | 104.12 |
| Play | 196 | 19237 | 26.99 |
| Movie | 195 | 7198 | 7.97 |
| Vlog | 529 | 126187 | 181.15 |
| Live Broadcast | 617 | 175766 | 456.30 |
| Speech | 516 | 35081 | 118.80 |
| Drama | 428 | 20363 | 22.75 |
| Recitation | 259 | 60978 | 134.16 |
| Advertisement | 83 | 1662 | 4.04 |
| Overall | 3000 | 659594 | 1363.74 |

### 3.2 Baseline System

Our baseline system refers to the implementation of speaker verification provided by ASV-Subtools [25]. For the input features, 81-dimensional filter banks are extracted within a 25ms sliding window for every 10ms, and then we used Voice Activity Detection(VAD) to remove silence frames. SE-ResNet34 [9] is used in baseline system to train a classification network, which AM-Softmax [26] is used as loss function. And then we extract 256-dimensional speaker embedding for enrollment data set and test data set respectively after training.

In addition, enroll utterances belonging to the same speaker are concatenated into one utterance for enrollment, and all of our experiments are based on cosine similarity as the back-end scoring criterion.

### 3.3 Domain Embedding

It is a major feature to introduce the scenario information of the utterances in our system. The speaker's scenario information is absolutely non-negligible. The voiceprint characteristics of the same speaker in the case of speech and singing are quite different, which is also confirmed by the scoring results in the baseline system. In considering this, our goal is to introduce the speaker scenarios information and make the model learn the scenario features in the training process. In the test process, we hope to make the similarity score of speakers in the same scenario as their score as far as possible. We can reduce the recognition errors caused by the differences in speaker scenarios in this way.

The specific implementation method is as follows. Input information of the model includes 81-dimensional FBank acoustic features and scenario information corresponding to the input utterance. With the preprocessing of VAD, 256-dimensional speaker embedding is calculated through the input acoustic features. Then, the corresponding 32-dimensional word embedding is calculated through the incoming speaker scenario information. Finally, the 256-dimensional speaker embedding and 32-dimensional word embedding are concatenated in series to form a 288-dimensional domain embedding, which we can extract embeddings for the speakers. The model structure is shown in Fig. 1.

**Table 2.** Multi enroll for Baseline systems and Domain Embedding. The column of Augmentation indicate that non-speech data sets are used or not for data augmentation.

| Systems | Augmentation | EER(%) | minDCF (P = 0.01) | minDCF (P = 0.001) |
|---|---|---|---|---|
| Baselines | ✘ | 17.611 | 0.7335 | 0.8130 |
| | ✔ | 15.067 | 0.6996 | 0.7877 |
| Domain Embeddings | ✘ | 16.542 | 0.7129 | 0.7909 |
| | ✔ | **14.784** | **0.6785** | **0.7716** |

### 3.4 Different Back-ends

With the scenario information is introduced into the training, the back-end test method will be different from the baseline system. Because the baseline system concatenates multiple enrollment utterances into single utterance as the enrollment, the registered voice loses the speaker scenario information, which is inconsistent with our test settings. In order to retain the scenario information of enrollment, we will register multiple utterances of the speakers in the enrollment set, and each test utterance will make cosine similarity score with 799 utterances of 196 enroll speakers during the testing. However, the results showed that there is no significant improvement, because the test method is not the comparison of the same scenario. We propose the following four solutions to solve this problem.

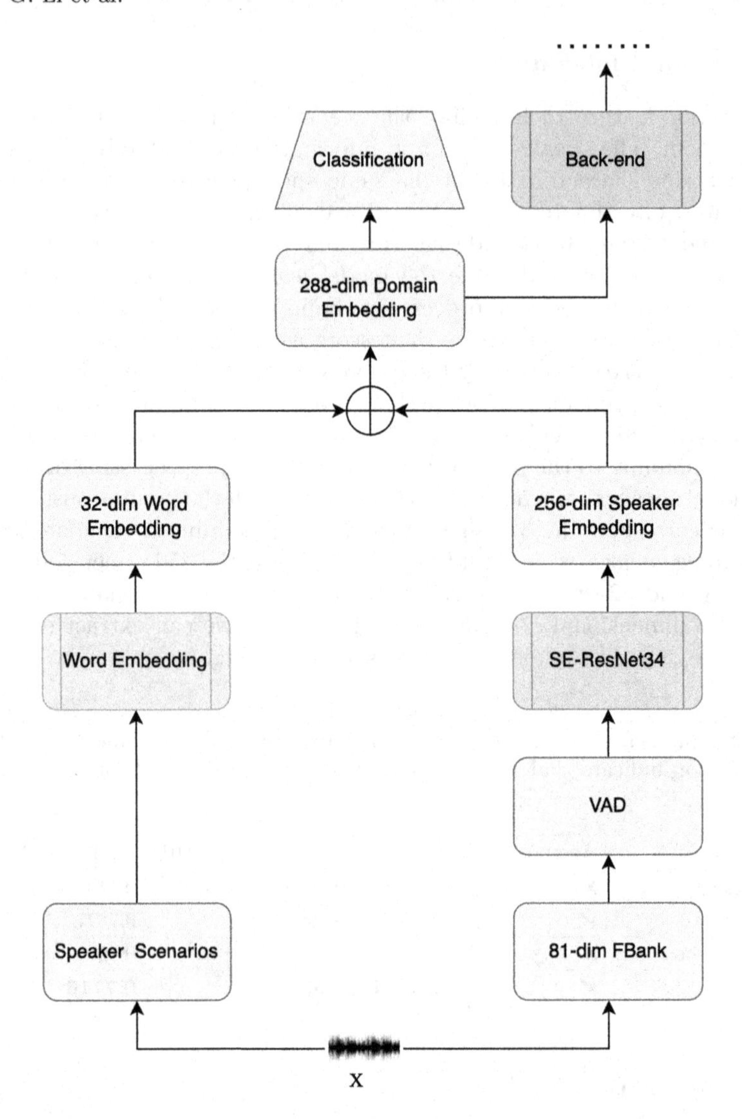


**Fig. 1.** The structure of Domain Embedding. **x** is the input utterance, which extracted the information of speaker scenarios and acoustic features, respectively. The Back-end approaches are showed in Subsect. 3.4.

- **Max_All**: We want to obtain the maximum score of each test utterance to the utterance of the enroll speakers as the score of this speaker and each enroll speaker when testing.
- **Max_1_Min**: Based on the first method, we take the maximum score of the test utterance and the enroll utterance as the score of the speaker corresponding to the test utterance and the enrollment utterance, and select the minimum score of the test utterance and other utterance in the enrollment set as their similarity score.

- **Max_10_Min** and **Max_20_Min**: Through a simple example analysis of the score histogram of the test utterances and the enrollment utterances, which shown in Fig. 2, we maximize the scores of 10 pairs and 20 pairs as the scores of the test speech and the enrollment speech respectively, and minimize the other score pairs as the scores of the test set speaker corresponding to each registered speaker.
- **Concate_Ave**: For multiple enroll utterance of the speaker in the enrollment set, we concatenate the speaker embeddings of multiple utterances by dimensions, and then average it as the enroll speaker embedding.

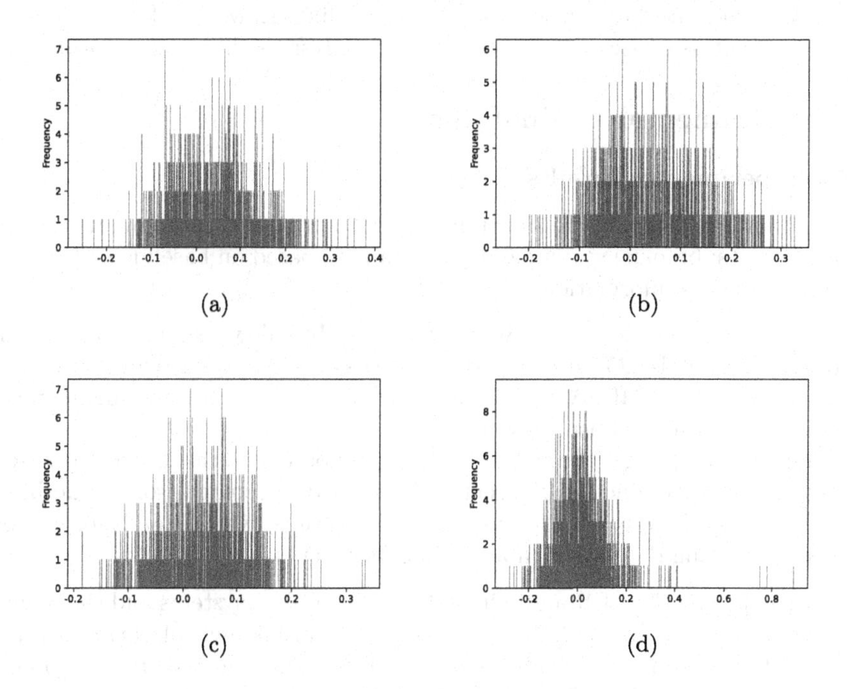


**Fig. 2.** Score histogram of the test utterances and the enrollment utterances. a, b, c and d are the score histogram examples of 4 test utterances to enrollment utterances. The horizontal axis is the score, and the vertical axis is the number of enrollment utterances corresponding to the score.

### 3.5 Data Augmentation

In order to make the training data more diverse, we used the non-speech datasets (MUSAN [23] and RIRS NOISE [11]) for data augmentation, which finally doubled the data size.

### 3.6 Self-Supervised Learning

In order to obtain advanced effects, we spent a lot of time for SSL, mainly using four pre-training models in the experiment, which including *wav2vec_small*, *wav2vec_vox_new*, *hubert_large* [10] and *wavLM_base*, using CN-Celeb-T's random 100 h and full volume data for fine tuning respectively. In these fine-tuning processes, we used the same loss function as the baseline system, AM-Softmax [26], to obtain better classification results, and all of them trained with the fairseq [17].

Because SSL models can extract robust speaker embedding, we do not design complex downstream tasks but add a statistical pooling layer and some linear layers to obtain fixed dimension speaker embedding. However, the experimental results are not as expected, and the specific analysis is shown in Subsect. 4.2.

## 4 Experimental Evaluations

### 4.1 Experimental Results

Our experimental results are shown in Tables 2 and 3, which are mainly divided into two parts: baseline system and joint training based on baseline system with speaker scenarios information.

- The baselines include two systems. One totally follows the baseline training of ASV-Subtools [25], and the other is the Base2 system, which uses non-speech data sets, MUSAN [23], and RIRS NOISE [11] for data augmentation compared with the Base1 system.
- Domain embedding is the system that we proposed, and the different back-end approaches mentioned in Subsect. 3.4 are used for comparison. In addition, each back-end approach corresponds to two results, which stands for without and with using data augmentation in experiments, respectively.

By comparing the EER and minDCF of the baseline systems and the domain embedding systems in Table 2, when a speaker enrolls multiple utterances, we can find that adding the speaker scenarios information in the training process reduces both the EER and minDCF significantly.

Besides, we can find the following conclusions through the comparisons in Table 3. Firstly, whether comparing the Base1 system with the domain embedding system without data augmentation or comparing the Base2 system with the domain embedding system augmented, we can find that these back-end approaches significantly improve the effectiveness of the task. The last three back-end testing methods of domain embedding, *Max_10_Min*, *Max_20_Min*, and *Concat_Ave*, all significantly affect EER and minDCF, especially in the system without data augmentation, which relative maximum reduction of 13.1% in EER scale. In addition, *Max_1_Min* has the best effect in minDCF scale, whether data augmentation is used or not (Without data augmentation, minDCF (P = 0.01 and P = 0.001) decreased by 20.2% and 14.4% respectively. And minDCF (P = 0.01 and P = 0.001) decreased by 21.8% and 13.5% respectively when data augmentation is used).

**Table 3.** EER and minDCF (P = 0.01 and P = 0.001) of baselines and our systems with different back-ends. There are mainly two systems which include Baseline systems and joint training with domain embedding, which use different back-end approaches mentioned in Subsect. 3.4. The column of Augmentation indicate that non-speech data sets are used or not for data augmentation.

| Systems | Augmentation | EER (%) | minDCF (P = 0.01) | minDCF (P = 0.001) |
|---|---|---|---|---|
| Baselines | ✗ | 11.450 | 0.5357 | 0.6527 |
| | ✓ | 9.580 | 0.4919 | 0.6027 |
| Domain Embeddings | – | – | – | – |
| +Max_All | ✗ | 11.000 | 0.4711 | 0.5803 |
| | ✓ | 10.270 | 0.4398 | 0.5538 |
| +Max_1_Min | ✗ | 11.631 | 0.4273 | 0.5590 |
| | ✓ | 10.696 | **0.3847** | **0.5213** |
| +Max_10_Min | ✗ | 9.952 | 0.4711 | 0.5803 |
| | ✓ | 9.378 | 0.4397 | 0.5538 |
| +Max_20_Min | ✗ | 9.655 | 0.4711 | 0.5803 |
| | ✓ | 9.085 | 0.4398 | 0.5538 |
| +Concate_Ave | ✗ | 9.784 | 0.4765 | 0.6096 |
| | ✓ | **9.006** | 0.4248 | 0.5627 |

### 4.2 Further Discussions

It is worth mentioning that with the rise of SSL in speech recognition, we hope that using SSL can also improve the speaker recognition task. Both [8,16] use wav2vec 2.0 fine-tuned with their data sets for English-speaking speaker recognition. Moreover, [8] almost entirely follows the wav2vec 2.0 model structure to realize speaker recognition and language recognition, while [16] uses wav2vec 2.0 to extract better features and access TDNN [22] twice to filter out better speaker embeddings. The experiments of both methods show that the first several transformer layers can better classify speakers. We had hoped to fine-tune multiple models on the CN-Celeb [7,13] data set and fuse the fine-tuning results to improve the effect. Unfortunately, the fine-tuning results of these four models are not satisfactory, and the EER of the best result is still greater than 14%. Based on this situation, we have the following three assumptions:

- **Different languages**: The pre-training process of the above models are based on English, which the fine-tuning effect on the Chinese data set is poor.
- **Simple classification network**: The classification network is too sample to achieve the desired effect after extracting speaker features through SSL.
- **Complex model**: It can be seen from [8,16] that the overly complex model plays an excessive role in the task of speaker recognition.

Based on the above analysis, the most likely reason for less-perfect classification results might be that SSL is overly strong in capturing too much unnecessary information, but not the essential voiceprint features for speaker recognition. We will focus on improving it in future research.

In addition, although we did not mention it in Sect. 3, CN-Celeb [7,13] does have a long tail problem in each speaker scenario. The unbalanced data distribution must lead to biased results inconsistent with our expectations in the training process. Therefore, we can use resampling to obtain a more balanced data set for training in the subsequent experiments.

## 5 Conclusion

In this paper, to effectively test under different scenarios, we propose domain embedding, the speaker scenario word embedding combined with the speaker embedding. From the experiments, the proposed domain embedding outperforms the conventional speaker embedding. In addition, to test in the same scenario as much as possible, we investigate several back-ends with data augmentation strategies and select the most effective one. Although the SSL model did not show its superiority in our experiment, it should be a promising direction in the future, and we will continue to work on it. Another observation is that it is very challenging to use a general speaker recognition system to recognize some specific scenarios, such as singing and drama, the voiceprint of which is unclear. Therefore, the data processing method has much room for improvement in the future.

**Acknowledgements.** This work was supported by the Opening Project of Key Laboratory of Xinjiang, China (2020D04047), National Key R&D Program of China (2020AAA0107902).

## References

1. Baevski, A., Zhou, Y., Mohamed, A., Auli, M.: wav2vec 2.0: a framework for self-supervised learning of speech representations. In: Advances in Neural Information Processing Systems 33, pp. 12449–12460 (2020)
2. Bahmaninezhad, F., Zhang, C., Hansen, J.H.: An investigation of domain adaptation in speaker embedding space for speaker recognition. Speech Commun. **129**, 7–16 (2021)
3. Chen, S., et al.: WavLm: large-scale self-supervised pre-training for full stack speech processing. arXiv preprint arXiv:2110.13900 (2021)
4. Chen, S., et al.: Why does self-supervised learning for speech recognition benefit speaker recognition? arXiv preprint arXiv:2204.12765 (2022)
5. Chowdhury, A., Cozzo, A., Ross, A.: Domain adaptation for speaker recognition in singing and spoken voice. In: ICASSP 2022–2022 IEEE International Conference on Acoustics, Speech and Signal Processing (ICASSP), pp. 7192–7196. IEEE (2022)
6. Chung, J.S., Nagrani, A., Zisserman, A.: VoxCeleb2: deep speaker recognition. In: Interspeech 2018 (2018)
7. Fan, Y., et al.: CN-Celeb: a challenging Chinese speaker recognition dataset. In: ICASSP 2020–2020 IEEE International Conference on Acoustics, Speech and Signal Processing (ICASSP), pp. 7604–7608 (2020). https://doi.org/10.1109/ICASSP40776.2020.9054017
8. Fan, Z., Li, M., Zhou, S., Xu, B.: Exploring wav2vec 2.0 on speaker verification and language identification. arXiv preprint arXiv:2012.06185 (2020)

9. He, K., Zhang, X., Ren, S., Sun, J.: Deep residual learning for image recognition. In: Proceedings of the IEEE Conference on Computer Vision and Pattern Recognition, pp. 770–778 (2016)
10. Hsu, W.N., Bolte, B., Tsai, Y.H.H., Lakhotia, K., Salakhutdinov, R., Mohamed, A.: HuBERT: self-supervised speech representation learning by masked prediction of hidden units. IEEE/ACM Trans. Audio Speech Lang. Process. **29**, 3451–3460 (2021)
11. Ko, T., Peddinti, V., Povey, D., Seltzer, M.L., Khudanpur, S.: A study on data augmentation of reverberant speech for robust speech recognition. In: 2017 IEEE International Conference on Acoustics, Speech and Signal Processing (ICASSP), pp. 5220–5224. IEEE (2017)
12. Lee, K.A., Wang, Q., Koshinaka, T.: The CORAL+ algorithm for unsupervised domain adaptation of PLDA. In: ICASSP 2019–2019 IEEE International Conference on Acoustics, Speech and Signal Processing (ICASSP), pp. 5821–5825 (2019). https://doi.org/10.1109/ICASSP.2019.8682852
13. Li, L., et al.: CN-Celeb: multi-genre speaker recognition. Speech Communication (2022)
14. Li, R., Zhang, W., Chen, D.: The CORAL++ algorithm for unsupervised domain adaptation of speaker recognition. In: ICASSP 2022–2022 IEEE International Conference on Acoustics, Speech and Signal Processing (ICASSP), pp. 7172–7176. IEEE (2022)
15. Nagrani, A., Chung, J.S., Zisserman, A.: VoxCeleb: a large-scale speaker identification dataset. arXiv preprint arXiv:1706.08612 (2017)
16. Novoselov, S., Lavrentyeva, G., Avdeeva, A., Volokhov, V., Gusev, A.: Robust speaker recognition with transformers using wav2vec 2.0. arXiv preprint arXiv:2203.15095 (2022)
17. Ott, M., et al.: FAIRSEQ: a fast, extensible toolkit for sequence modeling. arXiv preprint arXiv:1904.01038 (2019)
18. Reynolds, D., et al.: The 2016 NIST speaker recognition evaluation. Tech. Rep., MIT Lincoln Laboratory Lexington United States (2017)
19. Sadjadi, S.O., Greenberg, C., Singer, E., Reynolds, D., Hernandez-Cordero, J.: The 2018 NIST speaker recognition evaluation. In: Interspeech 2019 (2019)
20. Saeki, T., Xin, D., Nakata, W., Koriyama, T., Takamichi, S., Saruwatari, H.: UTMOS: utokyo-sarulab system for voicemos challenge 2022. arXiv e-prints arXiv:2204.02152 (2022)
21. Schneider, S., Baevski, A., Collobert, R., Auli, M.: wav2vec: unsupervised pre-training for speech recognition. arXiv preprint arXiv:1904.05862 (2019)
22. Snyder, D., Garcia-Romero, D., Povey, D., Khudanpur, S.: Deep neural network embeddings for text-independent speaker verification. In: Interspeech 2017 (2017)
23. Snyder, D., Chen, G., Povey, D.: MUSAN: a music, speech, and noise corpus. arXiv preprint arXiv:1510.08484 (2015)
24. Sun, B., Feng, J., Saenko, K.: Return of frustratingly easy domain adaptation. In: Proceedings of the AAAI Conference on Artificial Intelligence, vol. 30 (2016)
25. Tong, F., Zhao, M., Zhou, J., Lu, H., Li, Z., Li, L., Hong, Q.: ASV-SUBTOOLS: open source toolkit for automatic speaker verification. In: ICASSP 2021–2021 IEEE International Conference on Acoustics, Speech and Signal Processing (ICASSP), pp. 6184–6188. IEEE (2021)
26. Wang, F., Cheng, J., Liu, W., Liu, H.: Additive Margin Softmax for Face Verification. IEEE Signal Process. Lett. **25**(7), 926–930 (2018). https://doi.org/10.1109/LSP.2018.2822810
27. Yang, Z., et al.: Fusion of self-supervised learned models for MOS prediction. arXiv e-prints, pp. arXiv–2204 (2022)

# Real-Time Inertial Foot-Ground Contact Detection Based on SVM

Di Xia, YeQing Zhu, and Heng Zhang(✉)

College of Computer and Information Science College of Software, SouthWest University, Chongqing, China
{xiadi20,zyq0923}@email.swu.edu.cn, dahaizhangheng@swu.edu.cn

**Abstract.** At present, inertial motion capture systems are widely used by consumer users due to their advantages of easy deployment and low price. The inertial sensor itself does not accurately locate itself in 3D space. The most accurate localization method is currently represented by the algorithm of forward kinematic estimation of crotch position, and the accurate estimation of this algorithm needs to be based on the accurate judgment of foot-ground contact. In this work, we implemented plain Bayesian, decision tree, random forest, SVM, and GBDT models to find the model with the highest recognition rate of classified foot-ground contact states. This work evaluates the quality of each algorithm in terms of computational speed and accuracy, achieving SOTA under the condition of wearing only crotch, left and right thigh, and left and right calf IMUs compared to most people in the field using 17 IMUs for the whole body. Finally, this paper uses SVM models to classify foot-ground contact states for captured poses, yielding an average foot-ground contact accuracy of 97% for various motions. The method in this paper is applicable to any inertial motion capture system and satisfies the accuracy of an inertial motion capture system for foot-ground contact detection, providing kinematic constraints for pose estimation and global position estimation.

**Keywords:** Inertial motion capture system · 3D positioning · SVM · Foot-ground detection

## 1 Introduction

Motion capture technology is now widely used in human motion measurement, film and television production [1,2]. Capture devices are usually divided into vision-based motion capture technology and inertial sensor-based motion capture. IMU is widely used because of its small size, low cost, and independence from space volume. Traditional inertial motion capture combines multiple inertial measurement units (IMUs) with acceleration and poses measured using an extended Kalman filter (EKF) to obtain the position of the human body [3]. However, inertial measurement units contain only accelerometers, magnetometers, and gyroscopes and lack position measurement. If only using acceleration as


M. Tanveer et al. (Eds.): ICONIP 2022, CCIS 1793, pp. 528–539, 2023.
https://doi.org/10.1007/978-981-99-1645-0_44

the basis for displacement calculation, using double integration of acceleration as inertial displacement estimation, there is inevitably a cumulative error in position if it is not adequately processed [4]. The unreliability of simple, direct double integration of acceleration is widely recognized in the field of inertial sensors and position estimation. On the other hand, the ZUPT [5] approach requires the inertial sensor to be foot-mounted to obtain the required fixed period for integration and drift reset. This method is mainly used for pedestrian heading estimation and cannot be directly used for human displacement estimation [6]. The current more flexible and accurate displacement estimation algorithm obtains the overall displacement of the human body by estimating the foot-ground contact state and calculating the position difference of the foot in contact between adjacent frames [7]. The accuracy of this method depends on the judgment of the foot-ground contact state during the motion. Although many contact detection methods have been developed, Miguel et.al. [8] proposes to use motion data from a single inertial measurement unit (IMU) placed on the instep, for which angular velocity and linear acceleration signals are extracted. The pressure pattern from a force sensitive resistor (FSR) on a custom insole is used as a reference value. During the treadmill walking task, the performance of the threshold-based (TB) algorithm and the Hidden Markov Model (HMM)-based algorithm is compared. Using the acceleration signals, Tao Zhen et.al. [9] proposes a long short-term memory deep neural network (LSTM-DNN) algorithm for gate detection. Compared with the traditional thresholding algorithm and LSTM, the algorithm has higher detection accuracy for different walking speeds and different test objects. Experimental results show that the gait recognition accuracy and F-score of the LSTM-DNN algorithm exceed 91.8% and 92%, respectively. However, both require foot-mounted IMU, which is extremely inconvenient for users to foot-mounted IMU for motion. Since the task of identifying contact states does not currently have a large dataset with access to raw inertial data and GRF (Ground Reaction Forces). Therefore, we used our own collected data as our data source, using the Perception Legacy [2] full-body motion capture system from Noitom and a pair of homemade pressure insoles to collect foot-ground contact data containing various motions.

Our contributions:

1. We collected a dataset containing inertial data and pressure data that can be used to determine the foot-ground contact state under motion.
2. Without using foot-mounted IMUs, the lower body can accurately identify the foot contact state using only 5 IMUs.
3. We build a high-speed and high-precision foot-ground contact judgment model using five conventional machine learning models modelled on this self-built dataset.
4. The proposed method is applicable to any inertial motion capture system.

## 2 Dataset

In this section, we present the motion capture dataset. Since the task of identifying contact states does not currently have a large dataset provided to raw inertial data, GRF. Therefore, we used our own collected data as our data source, using the Perception Legacy full-body motion capture system from Noitom and a pair of homemade compression insoles to capture a contact dataset containing various motions.

## 2.1 Data Collection Equipment

Here we introduce the above two devices in detail. At present, most of the algorithms proposed in this field use the calibrated attitude, joint position, acceleration and other information provided by the inertial motion capture system manufacturer, but this information has been processed by the manufacturer's algorithm [8,10,11]. And then using self-developed algorithms to analyze it is unfair. Since our goal is to generalize the algorithm to any inertial sensor and eliminate the unfairness of this secondary processing, it is not possible to use the manufacturer's own calibration algorithm and the joint data calculated by their algorithm. For Perception Legacy we only use the IMU and use our own calibration algorithm to map each joint of the subject to the joints of the body model.

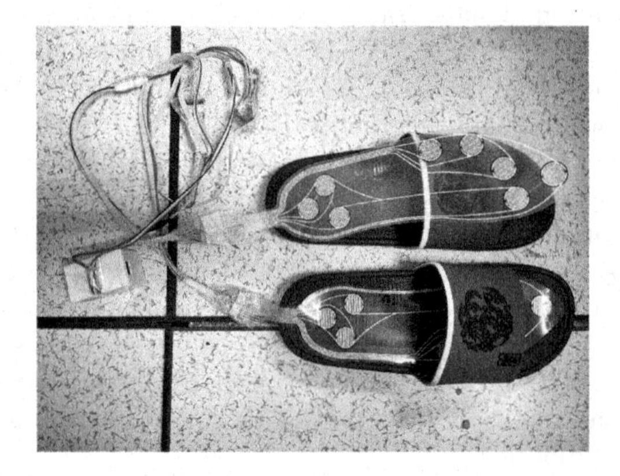

**Fig. 1.** Pressure insoles that carry 8 FSR sensors on each insole.

Each pressure insole (see Fig. 1) consists of eight force sensing resistors (FSRs) [12] mounted at different locations that are well to capture the pressure of the foot on the ground. Current commercially available force measuring stations such as Xsens' devices can measure forces in three directions, including vertical, lateral and anterior-posterior forces, as well as moments and centres of pressure. However, it is difficult to do research outside the laboratory because of its high price and small measurement range. Although our homemade pressure insoles placed on the soles of the feet are subject to wear and tear of the insoles, translation of the insoles relative to the soles of the feet, and measurement errors due to residual stresses, our pressure insoles are wireless and inexpensive, so they can solve the problems of small measurement range and high cost of the force measuring station. Both the IMU and the pressure insole collect data at a frequency of 100 HZ, and the data is synchronized through the ESP-NOW [14] data transmission protocol to ensure that our data is reliable.

### 2.2 Data Collection

We recruited ten volunteers to participate in the data collection experiment and performed four regular movements and one arbitrary form of movement. Four common movements included fast running, jogging, slow walking, and brisk walking, and the arbitrary form of movement was decided by the subjects, and each movement was collected twice for 5 min each, which was approximately 3,000,000 frames of inertial data and GRF data. These data will be used as our dataset to train and test the model. After that, we will use GRF to calculate the support leg labels. Our task will change to predicting the support leg labels from inertial data. Since we use only a single data frame for training and testing, the 3 million samples collected can extract rich features and generalizability.

### 2.3 Data Pre-processing

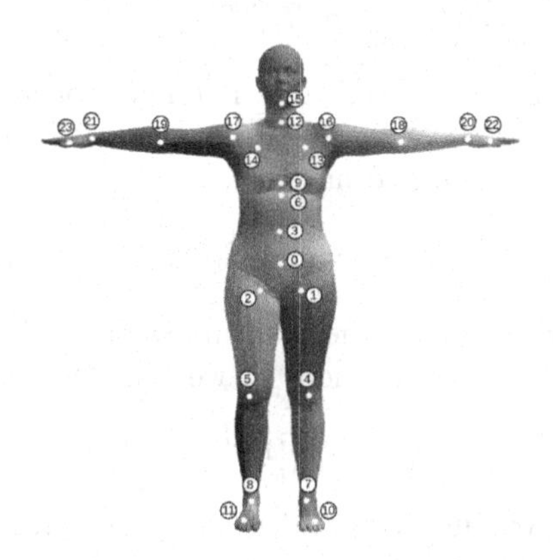

**Fig. 2.** The pose of Tpose in the SMPL model

**Inertial Sensor Calibration.** Here we are using the same calibration method as DIP [18]. We set the inertial sensor to output acceleration data relative to the sensor's own coordinate system $P^S$ and rotation matrix data relative to the global coordinate system $P^I$. Here we perform a two-step calibration operation. The first step calibrates any specified IMU so that its axes coincide with the coordinate system of the SMPL model, which yields a conversion from coordinate system $P^I$ to coordinate system $P^M$ with $P^{IM}$.

$$P^M = P^I P^{IM} \tag{1}$$

Since we apply the algorithm to the foot-ground contact detection task, in the second calibration step, we only place five IMUs on the body segments of the hips, left and right thighs and left and right calves corresponding to the joint subscripts 0, 1, 2, 4, and 5 of SMPL, respectively, and execute Tpose standing still for 3 s to obtain the average acceleration $a_S$ and rotation matrix $R_I$. Since the IMUs placed on the body segments may be at the arbitrary position on the body segment, i.e., controlling the rotation $R_I^{sensor}$ in any direction, this results in a rotational deviation from the joint rotation $R_I^{bone}$ of the standard human model, and here we assume that this deviation $R_I^{offset}$ remains constant in the motion capture experiment, we have:

$$R_I^{offset} = R_I^{sensor-1} R_I^{bone} \tag{2}$$

Since we use Tpose as the calibrated pose, we set $R_M^{bone}$ to the unit matrix for any given pose we have:

$$P^M R_M^{bone} = P^I R_I^{bone} \tag{3}$$

Combining Eqs. (1) (3) yields:

$$R_I^{offset} = R_I^{sensor-1} P^{IM} \tag{4}$$

For acceleration, since the IMU does not move once it is mounted on the subject, so $R_I^{sensor} = R_I^{bone}$. We then convert the IMU's own coordinate system to a global coordinate system and have:

$$a_I^{sensor} = R_I^{sensor} a_S^{sensor} \tag{5}$$

Due to the sensor error and the inaccuracy of $P^{IM}$, a constant deviation is produced in the global acceleration, so we take this error into account and obtain:

$$P^M (a_M^{bone} + a_M^{offset}) = P^I a_I^{bone} \tag{6}$$

and the calibration state, since the subject is stationary, so $a_M^{bone} = 0$, and have:

$$a_M^{offset} = (P^{IM})^{-1} R_I^{sensor} a_S^{sensor} \tag{7}$$

Finally, we calculate $a_M^{bone}$ and $R_M^{bone}$ in each frame of the action capture and record $R_M^{bone}$ by replacing it with a quaternion.

### 2.4 Pressure Sensor Calibration

Since our homemade pressure insoles based on FSR sensors cannot accurately measure the pressure applied, we define the pressure data of each pressure sensor on foot in the Tpose state as the baseline data as $F_s^i$. Here, we assume that the FSR sensors on each pressure insole can be fully pressed by the foot when the subject does Tpose, and average the body weight to each foot. And in order to obtain the stable pressure data under Tpose, we ask the subject to do Tpose and stand in place for 3 s to obtain $N$ frames of data. We obtain the average of these 3 s of pressure data as the benchmark data $F_s^i$. In this way, and we obtain the benchmark pressure of 16 sensors of a pair of shoes.

$$F_s^i = \frac{1}{N}\sum_{j=0}^{N} F^i, i = 1, 2, \ldots, 15, 16 \tag{8}$$

Although the data collected by the FSR sensor are ADC values, they are not pressure data that can be used directly. However, there is a linear relationship between the body weight and ADC values in the FSR sensor, so we use the body weight to convert the reference pressure ADC data to convert the ADC values to pressure data. The GRF per foot can simply be considered as the sum of the data collected by 8 FSR sensors per foot [13] and defining the subject's weight as $W_s$, from which the scaling factor $S_f^i$ corresponds to the weight, and the pressure $F^i$ applied to each FSR sensor can be calculated.

$$S_f = \frac{W_s}{2 * \sum_{i=1}^{N=16} F_s^i} \tag{9}$$

$$F^i = F_s^i * S_f \tag{10}$$

Finally, the pressure data converted to weight are recorded for post-process analysis.

### 2.5 Contact Label Generation

Considering that the contact state between two feet and the ground has a very strong correlation with the GRF, we use the GRF data to automatically set the foot-ground contact state for each frame of data. Before setting the foot-ground contact state automatically, we need to do pre-processing on the collected GRF data. Firstly, due to the nature of FSR sensors, we will collect some outliers that are not in line with common sense. Here we use box plots to find the outliers and replace this outlier with the nearby average value. After that, since the GRF values collected for the two insoles are too large, we normalize their features to make them more suitable for model training. Finally, for different models and materials of FSR, the characteristic curves are different. We need to scale the data collected for both insoles to the same scale uniformly. Here we use the sigmoid function to scale them; the advantage of using this function is to be able to suppress very large values to around 1, as shown in Fig. 3. According to the human kinetic properties, when there is no contact between the foot and the ground, the GRF at this moment should be the smallest. Therefore, we follow this theory and select the smallest value of GRF in the whole action sequence as the threshold for automatically determining the foot-ground contact status, as long as the GRF greater than this threshold is considered a contact.

Also, taking into account the sensor error and generalization, an empirical value of 0.1 is added to this threshold to prevent the lowest point in the GRF will also be considered a contact state because of the sensor error. In this paper, only set the left foot touching the ground and the right foot touching the ground in these two states if the contact state is set to 1 of the ground state set to 0. If the left foot GRF is greater than the right foot GRF both as the left foot in the contact state and vice versa is the right foot for the contact state. Identify the foot ground contact state as shown in Fig. 3.

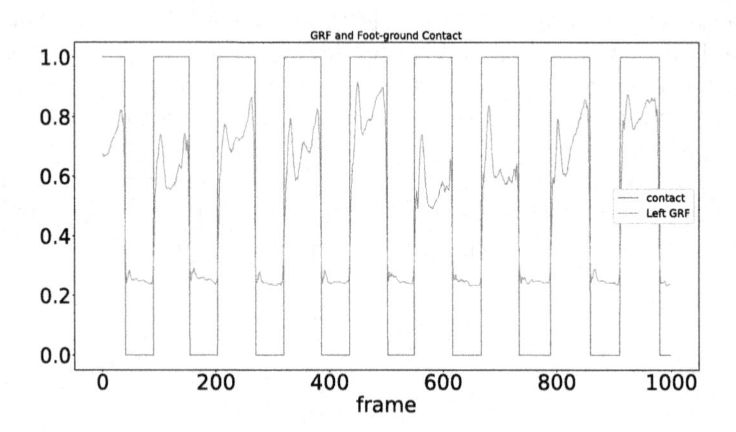


**Fig. 3.** Contact status of GRF and autoclassification of the left foot after preprocessing.

We use the method described above to process our collected pressure insole data and synthesize our exposure dataset for subsequent experiments.

## 3 Method

In this section, we present the specific workflow of the foot-ground contact algorithm. We consider the uncertainty and real-time nature of the inertial sensor's own pose solution. This paper uses a traditional machine learning model to model the foot-ground contact state. The model we design integrates data-based inference and physical a priori knowledge.

### 3.1 Feature Engineering

First, we consider the two characteristics of acceleration and joint position. Acceleration: The ideal condition is when the foot touches the ground, at which time the acceleration of the foot should be 0. Joint position: Since this paper only considers the foot-ground contact state, that is, one foot must touch the ground, the ideal condition is that when the foot touches the ground, the foot at the lowest point should be in the contact state. Due to the uncertainty of

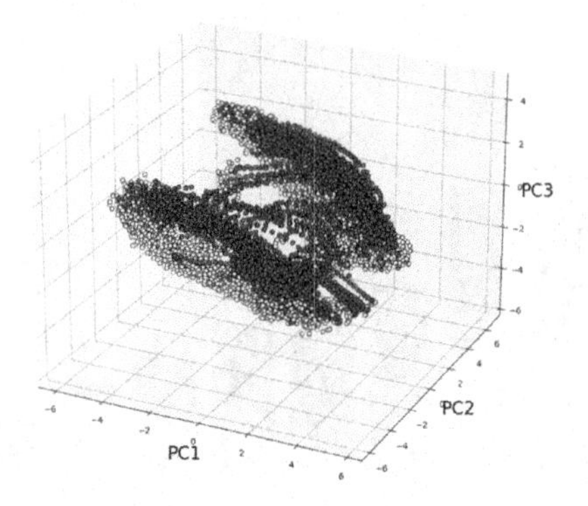


**Fig. 4.** Scatter plots created by PC1, PC2 and PC3.

the inertial sensor itself, the ideal result is often not achieved, which requires the use of probabilistic modelling to calculate the contact probability for the foot-ground contact state by combining all kinematic features.

To obtain the feature values, we use forward kinematics to calculate the joint position from the joint rotation, which is used as an input feature together with the acceleration. This gives us a total of 5 joint eigenvalues, each with six eigenvalues of triaxial acceleration and position in 3d space, for a total of 30 eigenvalues.

From the above analysis, we can simply assume that the classification can be done by the height of the foot joint position, but there may be cases when human sports have similar heights for both feet, and then other features need to be introduced to assist in the judgment. In order to predict the dataset with 30 features as a whole, we introduce PCA [19]. PCA is used to reduce the number of features in the training model. It does this by constructing what is called a principal component (PC) from multiple features. The PC is constructed in such a way that the PC1 direction explains as many of your features as possible on the maximum variation. Then PC2 explains the remaining features as much as possible on the maximum variation, and the other PCs do the same, and generally speaking, PC1 and PC2 explain a large fraction of the overall feature variation. Some features will have a higher variance because they are not normalized. So we normalize the original data. After the scaling is done, we fit the PCA model and convert our features to PCs. since we have 30 features, we can have up to 30 PCs. but for visualization, since we can't draw a four-dimensional image, we only pick the first three PCs. Then we use PC1, PC2 and PC3 to create a scatter plot, as shown in Fig. 4.

In Fig. 5, we can clearly see the two different clusters. Although there is some overlap, it is still possible to classify the majority of the data. And we only used the first three PCs, which means that the figure does not capture all the feature variations. This means that the model trained with all the features is able to classify the foot-ground contact state very well. We can determine the predictive power of the data by looking at the PCA-scree plot.

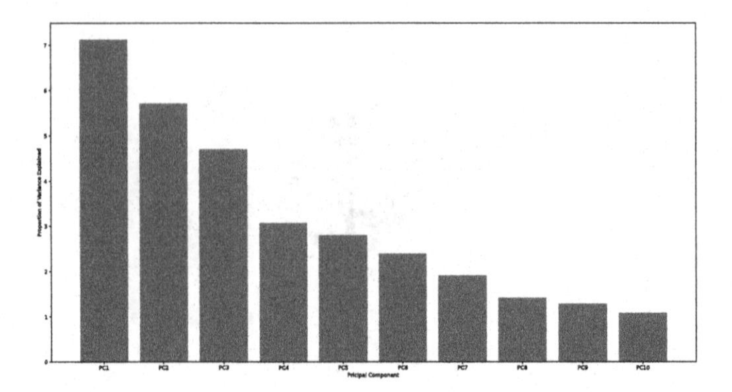


**Fig. 5.** Scatter plots created by PC1, PC2 and PC3.

The height of each bar in Fig. 4 represents the percentage of variance explained by the PCs. Only about 17% of the characteristic variance of the three PCs we selected is explained by PC1 and PC2, but still, two clusters are clearly obtained. This also emphasizes the predictive power of the data.

### 3.2 Machine Learning Models

On our exposure dataset, salient features are extracted by feature engineering, after which the features are selectively dimensioned down using PCA, and these dimensioned-down features are modelled using five classification models (plain Bayesian, decision tree, random forest, SVM, GBDT). This section will describe these machine learning classification models.

1. Naive Bayes [10] The Naive Bayes algorithm originated from classical mathematical theory and has a well-studied theory as well as a stable classification efficiency. The Naive Bayes model requires very few parameters for training and is insensitive to miss data. The Naive Bayes algorithm requires that the individual data be independent of each other, but this often does not hold true in practice. If the attributes in the dataset are not independent of each other, i.e., they are correlated, the classification accuracy is greatly reduced.
2. Decision tree [15] The decision tree algorithm is essentially a binary classification regression tree. For a binary tree, each parent node has two branches, one for True and the other for False, and the branch that satisfies the conditions is selected according to the decision conditions set by the parent node. Each node is recursive until it reaches a leaf node. The leaf nodes have decision rights, and any data from the root node will eventually reach the only leaf node. Decision trees are able to obtain rules with explanatory power, but they tend to produce overfitting problems when applied in practice. There is often noise in the training set, and the decision tree will use the noisy data as the segmentation criteria, resulting in decisions that cannot be generalized to the test set.

3. Random Forest [16] Random forest algorithm, i.e. consists of several different decision trees. When constructing a decision tree, we need to randomly select a portion of samples from the training set with playback, and also do not use all the features of the data but randomly select some features for training. Each tree uses different samples and features, and the trained results are not the same. The model of random forest is random and not easy to overfitting; it is resistant to noise and insensitive to the outlier of anomalies; it is faster than a decision tree in processing high-dimensional data sets.
4. GBDT [17] The GBDT algorithm uses the negative gradient of the loss function as an approximation of the residuals, iterates and fits the regression tree with the residuals continuously, and finally generates a strong learner. GBDT can easily obtain the importance ranking of the features and is very explanatory, and GBDT can ensure low bias and low variance, making the flush generalization ability very strong.
5) SVM [20] SVM itself is a linear model, and its nonlinear fitting ability is better when a nonlinear kernel, such as radial basis or Gaussian kernel, is introduced. On small and medium-sized datasets, SVM has a strong classification effect. Compared with GBDT, it can better control overfitting, but it is sensitive to missing values and requires normalization or normalization operations on the data.

### 3.3 Experimental Setup

The training and testing phases use the five machine learning models described above. We selected the data of the first eight individuals from the collected inertial and GRF datasets as the training set and the data of the last two individuals as the test set on which to test our trained models. The experiment was run on a personal computer running python 3.7 with an Intel Core i5 10th generation CPU and 48 GB RAM. The experiment was modelled and evaluated using third-party libraries such as sklearn, pandas, numpy, and matplotlib. Next we describe in detail the training parameters of each model.

1. Naive Bayes with default hyperparameters.
2. DecisionTree with default hyperparameters.
3. RandomForest with default hyperparameters.
4. GBDT(learning_rate = 0.1, n_estimators = 80, min_samples_split = 800, min_samples_leaf = 60, max_depth = 13, max_features = 'sqrt', subsample = 0.8).
5. SVM(kernel = 'rbf', C = 4, gamma = 'auto').

### 3.4 Experimental Results

We used the grid search algorithm to search the hyperparameters of the following models and obtained the optimal model. To be fair, each model uses the same random seed and the same dataset and training set. The following results were obtained after experimentation, as shown in Table 1:

**Table 1.** Evaluation results of the 5 models

| Models | Accuracy (%) | Precision (%) | Recall (%) | F1-score |
|---|---|---|---|---|
| Naive Bayes | 92.0% | 91.8% | 91.6% | 91.7% |
| DecisionTree | 95.7% | 95.7% | 95.3% | 95.5% |
| RandomForest | 96.8% | 96.9% | 96.4% | 96.6% |
| GBDT | 96.7% | 96.9% | 96.3% | 96.6% |
| SVM | 97.0% | 97.2% | 96.6% | 96.9% |

The experiments prove that SVM has higher accuracy and more powerful generalization performance compared to the other four machine learning models. The current state-of-the-art [10] method achieves 95.8% accuracy using plain Bayes, but he wears as many as 17 IMUs. Our algorithm, on the other hand, only needs 5 IMUs to surpass his results.

## 4 Conclusion

Most of the current work in this field uses foot-mounted IMUs to estimate the foot-ground contact state, but little consideration has been given to estimating the foot-ground contact state using only five IMUs in the calf, thigh, and crotch. Machine learning algorithms are used to analyze whether only five IMUs are also capable of estimating the foot-ground contact state. The main goal of this work is to find an optimally performing classifier for foot-ground contact detection, which can give reliable constraints on global position estimation. This work applies five machine learning algorithms DT, WNB, GBDT, SVM, and RF, to predict the foot-ground contact state on a self-built dataset. After implementing the models, this work achieves the best accuracy of 97% on the SVM. We compared these findings with previous research and discovery work, and we found that this approach performed the best. This study relied heavily on the dataset and preprocessing methods. However, as far as the self-harvested dataset is concerned, the results are satisfactory to us. We should use more efficient training and preprocessing methods to obtain better results. Increasing the sample size of the dataset in the future will ensure the accuracy and validity of this study.

## References

1. XsensMVN (2019). https://www.xsens.com/products/xsens-mvn-analyze/
2. Perception Neuron (2019). http://www.neuronmocap.com/
3. Roetenberg, D., Luinge, H., Slycke, P.: Xsens MVN: full 6DOF human motion tracking using miniature inertial sensors. Xsens Motion Technologies BV (2009)
4. Yan, H., Shan, Q., Furukawa, Y.: RIDI: robust IMU double integration. In: Ferrari, V., Hebert, M., Sminchisescu, C., Weiss, Y. (eds.) ECCV 2018. LNCS, vol. 11217, pp. 641–656. Springer, Cham (2018). https://doi.org/10.1007/978-3-030-01261-8_38

5. Suresh, R., Sridhar, V., Pramod, J., Talasila, V.: Zero Velocity Potential Update (ZUPT) as a Correction Technique. In: 2018 3rd International Conference On Internet Of Things: Smart Innovation And Usages (IoT-SIU), pp. 1–8 (2018)
6. Zampella, F., Jiménez, A., Seco, F., Prieto, J., Guevara, J.: Simulation of foot-mounted IMU signals for the evaluation of PDR algorithms. In: 2011 International Conference On Indoor Positioning And Indoor Navigation, pp. 1–7 (2011)
7. Wang, Q., Guo, Z., Sun, Z., Cui, X., Liu, K.: Research on the forward and reverse calculation based on the adaptive zero-velocity interval adjustment for the foot-mounted inertial pedestrian-positioning system. Sensors **18**, 1642 (2018). https://www.mdpi.com/1424-8220/18/5/1642
8. Manchola, M., Bernal, M., Múnera, M., Cifuentes, C.: Gait phase detection for lower-limb exoskeletons using foot motion data from a single inertial measurement unit in hemiparetic individuals. Sensors **19**, 2988 (2019)
9. Zhen, T., Yan, L., Yuan, P.: Walking gait phase detection based on acceleration signals using LSTM-DNN algorithm. Algorithms **12**, 253 (2019)
10. Ma, H., Yan, W., Yang, Z., Liu, H.: Real-time foot-ground contact detection for inertial motion capture based on an adaptive weighted Naive Bayes model. IEEE Access **7**, 130312–130326 (2019)
11. Miezal, M., Taetz, B., Bleser, G.: Real-time inertial lower body kinematics and ground contact estimation at anatomical foot points for agile human locomotion. In: 2017 IEEE International Conference On Robotics And Automation (ICRA), pp. 3256–3263 (2017)
12. Hu, J., Cao, H., Zhang, Y., Zhang, Y.: Wearable plantar pressure detecting system based on FSR. In: 2018 2nd IEEE Advanced Information Management, Communicates, Electronic And Automation Control Conference (IMCEC), pp. 1687–1691 (2018)
13. Ma, H., Liao, W.: Human gait modeling and analysis using a semi-Markov process with ground reaction forces. IEEE Trans. Neural Syst. Rehabilit. Eng. **25**, 597–607 (2017)
14. ESP-Now (2022). https://www.espressif.com.cn/en/products/software/esp-now/overview
15. Breiman, L., Friedman, J., Olshen, R., Stone, C.: Classification and regression trees. (Routledge 2017)
16. Ho, T.: Random decision forests. In: Proceedings Of 3rd International Conference On Document Analysis And Recognition, vol. 1, pp. 278–282 (1995)
17. Zhang, Z., Jung, C.: GBDT-MO: gradient-boosted decision trees for multiple outputs. IEEE Trans. Neural Netw. Learn. Syst. **32**, 3156–3167 (2021)
18. Huang, Y., Kaufmann, M., Aksan, E., Black, M., Hilliges, O., Pons-Moll, G.: Deep inertial poser learning to reconstruct human pose from SparseInertial measurements in real time. ACM Trans. Graph. (Proc. SIGGRAPH Asia) **37**, 1–15 (2018)
19. Karl Pearson, F.R.S.: LIII. On lines and planes of closest fit to systems of points in space. London Edinburgh Dublin Philosop. Mag. J. Sci. **2**, 559–572 (1901)
20. Chang, C., Lin, C.: LIBSVM: a library for support vector machines. ACM Trans. Intell. Syst. Technol. (TIST) **2**, 1–27 (2011)

# Channel Spatial Collaborative Attention Network for Fine-Grained Classification of Cervical Cells

Peng Jiang[1], Juan Liu[1(✉)], Hua Chen[1], Cheng Li[2], Baochuan Pang[2], and Dehua Cao[2]

[1] Institute of Artificial Intelligence, National Engineering Research Center for Multimedia Software, School of Computer Science, Wuhan University, Wuhan, China
{pelenjiang,liujuan,chen_hua}@whu.edu.cn
[2] Landing Artificial Intelligence Center for Pathological Diagnosis, Wuhan, China
{pangbaochuan,caodehua}@landing-med.com

**Abstract.** Accurately classifying cervical cells based on the commonly used TBS (The Bethesda System) standard is critical for building the automatic cytology diagnosing system. However, the existing two publicly available datasets (Herlev and SIPaKMeD) do not provide the TBS labels for the cervical cells, leading to the lack of practicability of the models trained based on them. In addition, most deep learning-based methods for cervical cell classification do not explore the attention spots of the model, resulting in the limit of further focusing on discriminative parts of tiny cells. In this paper, we first build a cervical cell image dataset consisting of 10,294 cell image patches with TBS labels from 238 specimens, named LTCDD (Landing TBS Cervical Cell Dataset). Then we propose a novel visual attention module, channel spatial collaborative attention (CSCA), based on the integration of the channel and spatial attention. Two kinds of comparing experiments on LTCDD and two public datasets show that the CSCA embedded network can achieve the best classification performance. Moreover, the CSCA module performs better than other attention modules when integrated into ResNet. The visualization results illustrate that CSCA embedded networks can pay more attention to the nucleus part of cell images so as to focus on learning discriminative features that are beneficial to the classification of the cells.

**Keywords:** Cervical cell identification · cervical cancer screening · attention network · deep learning

## 1 Introduction

Cervical cancer is one of the most prevalent malignancies that threaten the lives of women and ranks fourth in both morbidity and mortality among all cancers. Globally, about 600,000 new cases of cervical cancer were diagnosed and more


M. Tanveer et al. (Eds.): ICONIP 2022, CCIS 1793, pp. 540–551, 2023.
https://doi.org/10.1007/978-981-99-1645-0_45

than 340,000 people died of this disease in 2020 [1]. Since cervical cancer has a long precancerous stage, it is possible to detect and treat cervical cancer timely via annual screening programs. Thus, accurate identification of precancerous cervical cells is vital for effective prevention measures and treatments.

Traditionally, cervical cytology screening is conducted by cytologists with the naked eyes under a microscope, which is laborious, inefficient and error-prone [2]. Therefore, many advanced technologies and devices have been developed to automatically digitalize the stained sections as images. Since machine learning techniques have the advantages of objective results and high precision, computer-aided diagnosis (CAD) methods based on machine learning have been studied for cervical cytology screening [3,4].

Nevertheless, conventional machine learning approaches for cervical cell identification have complex image preprocessing and feature selection steps, which need precise segmentation algorithms to detect and segment the contours of cytoplasm and nuclei in general [5,6]. In recent years, deep learning methods have made an essential breakthrough in computer vision and image processing field [7–9]. As a result, deep learning-based methods for cervical cell classification have gradually become a promising research direction [10–13].

Compared with natural images, the size of cell images is tiny. Moreover, the nucleus containing the most essential information for cell classification only accounts for a small part of the whole cell. As a result, conventional deep learning methods are easily disturbed by the information of non-cellular parts, which reduces their ability to learn useful distinguishing features and affects the final classification accuracy of the models. Nowadays, the visual attention mechanism has been proposed to let the learning process pay more attention to meaningful information while inhibiting irrelevant information and has been successfully applied in many areas of medical image analysis [14–18]. Therefore, visual attention provides a solution that allows the model to focus on learning valuable features from cells in the image patches.

However, there are few studies applying visual attention to cervical cell classification. In addition, general visual attention modules incline to focus on channel attention of deep CNN models while paying less attention to the efficient combination of channel attention and spatial attention. To better focalize discriminative features of small cervical cells, we make full use of cross-channel information and spatial position to generate a 3D attention. Besides, most existing studies for cervical cell classification are based on two public datasets, Herlev [19] and SIPaKMeD [20], where the cell images are not labeled according to the TBS (The Bethesda System) criteria widely used in clinical diagnosis nowadays, leading to difficult application in practical cervical cancer screening programs.

To tackle these issues, in this paper, we first construct a cervical cell dataset for cervical cytology screening based on the TBS clinical diagnosis criteria. Then we propose a novel visual attention approach, which can be easily embedded into deep CNN models so that the models can focus on the features beneficial to cell classification. The main contributions of this paper are summarized as follows:

- We build a dataset of cervical cells named LTCCD (Landing TBS Cervical Cell Dataset), for cervical cancer screening. In LTCCD, the cells are annotated according to the TBS standard so that the models trained with this dataset can be directly used for the follow-up cytology diagnosis.
- We propose a novel visual attention module, CSCA (Channel Spatial Collaborative Attention), to enhance the feature extraction capability from small and limited cells, so as to realize high-performance cervical cell identification.
- We validate the CSCA module on two public datasets and the LTCCD dataset, and we perform visualized analysis from CSCA embedded deep networks. The experimental results demonstrate that the proposed CSCA module is effective for cervical cell classification and helps the model concentrate on discriminative parts.

## 2 Materials and Methods

### 2.1 Construction of LTCCD

**Background of TBS.** In the past, the Pap-smear test was the most crucial and successful method to prevent cervical cancer in medical practice. However, in the 1980s, lots of tracking studies and news reported the severe problem of Pap-smear tests about lax practices and inaccuracies in medical laboratories [21]. As a result, TBS was established to provide a more standardized framework for reporting cervical cytology. TBS adopted a two-tier grading system (low and high-grade intraepithelial lesions) to take the place of nonreproducible reporting terms such as Pap test five-level classification or three-tiers dysplasia (mild/ moderate/ severe, or carcinoma in situ) [22]. TBS has been updated three times to meet the evolving cervical cytology. According to the newest TBS criterion [23], squamous epithelial cell abnormalities involve five categories: low-grade squamous intraepithelial lesion (LSIL), high-grade squamous intraepithelial lesion (HSIL), atypical squamous cells-undetermined significance (ASC-US), atypical squamous cells-cannot exclude an HSIL (ASC-H) and squamous cell carcinoma (SCC). Among them, The first four categories suggest precancerous lesions while SCC has already developed into cancer. With the development of more advanced sample preparation techniques, liquid-based preparations (LBPs) have gradually become a satisfactory alternative to conventional smears. In recent years, the Liquid-based cytology test, together with the TBS diagnosis criterion, has dramatically improved the detection rate of precancerous and cancerous lesions, thus playing a vital role in cervical cancer prevention and is increasingly applied all over the world.

**Image Acquisition.** We constructed a cervical cell dataset, LTCCD, based on the TBS standard for better application in practical cervical cancer screening. Figure 1 illustrates the generation process of LTCCD. We collected 238 specimens from Wuhan University-Landing Pathology Research Center. Since this work is anonymous and retrospective, the need for informed consent is waived. All specimens were produced into liquid-based preparations using Thinprep Cytological

Technology (TCT) with the H&E staining method. A specimen's whole slide image (WSI) was automatically scanned by a digital slide scanner (LD-Cyto-1004 adapted from Olympus BX43 microscope) in 20× objective lenses. After scanning, a total of 10,294 normal or precancerous cervical cells were manually annotated by experienced cytopathologists into five types: normal (3,350), ASC-US (1,744), ASC-H (1,600), LSIL (2,150) and HSIL (1,450). We divided LTCCD according to the ratio of 8:1:1 for model training, validation, and testing. The validation set is used in the training process to check the overfitting problems and whether the model converges normally. The independent testing set is used for the final performance evaluation of trained models.

**Data Augmentation.** Data augmentation has been widely used to reduce overfitting problems and help build more accurate and generalizable models. Considering of imaging diversity of different pathological scanners and stain variance during sample preparation, we adopt color disturbance (50% enhancement and 50% decrease) including disturbances of brightness, sharpness, and contrast. In actual observation, cells appear from various angles. Thus, the raw image is also augmented by rotation transformations (90°, 180°, 270°) and horizontal and vertical flips. We utilize Gaussian blur to simulate the unclarity of images caused by scanning too fast without proper focusing. Via the above operations, a raw image can be expanded into 12 images, as exhibited in Fig. 1(f).

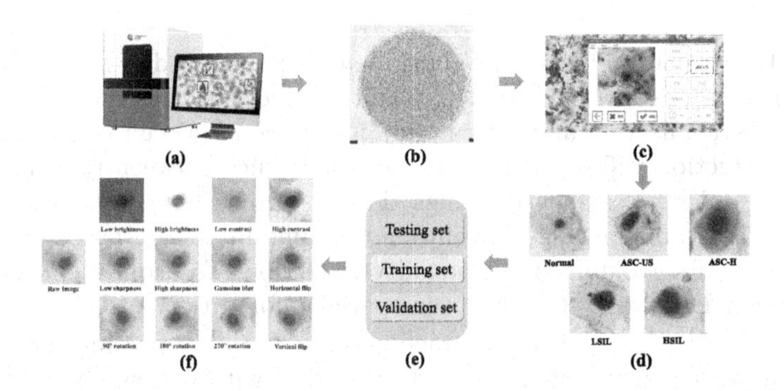


**Fig. 1.** The generation process of LTCCD: (a) Digital slide scanner, LD-Cyto-1004. (b) Auto-scanned WSI of a cervical specimen. (c) Image annotation. (d) Five types of cervical cells. (e) Dataset partitioning. (f) Data augmentation of the training set.

## 2.2 Channel Spatial Collaborative Attention Network

Many attention modules have been explored to improve the feature expression of deep CNN models [14–16]. Channel attention generates 1D weights for each channel of input feature maps, while spatial attention produces 2D weights for each spatial location. Since diagnostic features tend to gather in small nuclei of cervical cells, to obtain more comprehensive attention to limited cell features,

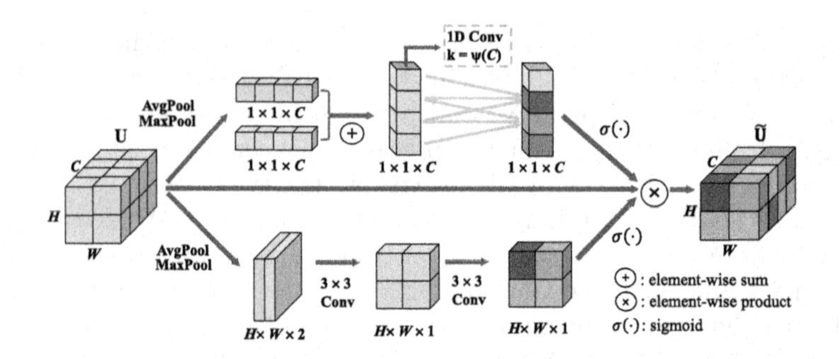


**Fig. 2.** Structure of channel spatial collaborative attention module.

we propose a CSCA module that calculates channel and spatial attention simultaneously and generates a 3D attention for every value of input feature maps. The detailed structure of the CSCA module is shown in Fig. 2. For the input feature map $\mathbf{U} \in \mathbb{R}^{H \times W \times C}$, The refined feature map generated by the CSCA module is computed as:

$$\tilde{\mathbf{U}} = \mathbf{U} \otimes \mathbf{F}_C(\mathbf{U}) \otimes \mathbf{F}_S(\mathbf{U}) \tag{1}$$

where $\otimes$ denotes element-wise product. $\mathbf{F}_C \in \mathbb{R}^{1 \times 1 \times C}$ represents channel attention, $\mathbf{F}_S \in \mathbb{R}^{H \times W \times 1}$ represents spatial attention.

**Channel Attention.** Generally, channel attention is produced with fully connected (FC) layers involving dimensionality reduction. Though FC layers can establish the connection and information interaction between channels, dimensionality reduction will destroy direct correspondence between the channel and its weight, which consequently brings side effect on channel attention prediction [16]. Thus, we adopt 1D convolution to maintain appropriate cross-channel interaction while avoiding dimensionality reduction. To acquire efficient channel attention $\mathbf{F}_C$, we first embed the spatial information as $\mathbf{U}^c_{avg} \in \mathbb{R}^{1 \times 1 \times C}$ and $\mathbf{U}^c_{max} \in \mathbb{R}^{1 \times 1 \times C}$ by using average pooling and max pooling. Further, the embedded tensor is weighted by 1D convolution with a kernel size of $k$. The whole generation process of channel attention is computed as:

$$\begin{aligned} \mathbf{F}_C(\mathbf{U}) &= \sigma(C1D(Avgpool(\mathbf{U}) + Maxpool(\mathbf{U}))) \\ &= \sigma(\mathbf{W}_0(\mathbf{U}^c_{avg} + \mathbf{U}^c_{max})) \end{aligned} \tag{2}$$

where $C1D$ indicates 1D convolution. $\sigma$ denotes a sigmoid function and $\mathbf{W}_0 \in \mathbb{R}^k$. Large channels need a large kernel size of 1D convolution to acquire longer coverage of interaction, so there may exist a mapping $\phi$ between kernel size $k$ and channel dimension $C$. Since channel dimension is often set to an integer power of 2, the mapping relationship is formulated as:

$$C = \phi(k) = 2^{\alpha k - \beta} \tag{3}$$

where the hyperparameter $\alpha$ is set to 2 and $\beta$ is set to 1 in this paper. Then, the kernel size $k$ can be calculated as follows Eq. (4), $|z|_{odd}$ represents the odd number closest to $z$.

$$k = \left| \frac{log_2(C)}{\alpha} + \frac{\beta}{\alpha} \right|_{odd} \tag{4}$$

The kernel size can be adaptively determined by the given channel dimension so that different range interaction of channels can be realized by using such non-linear mapping.

**Spatial Attention.** Different from channel attention, spatial attention focuses on informative spatial locations of the input feature maps. We first aggregate channel information to generate two maps: $\mathbf{U}^s_{avg} \in \mathbb{R}^{H\times W\times 1}$ and $\mathbf{U}^s_{max} \in \mathbb{R}^{H\times W\times 1}$. Then they are concatenated and convolved by two cascade convolution layers. The 2D spatial attention is computed as:

$$\begin{aligned} \mathbf{F}_S(\mathbf{U}) &= \sigma(f^{3\times3}(f^{3\times3}(concat(Avgpool(\mathbf{U}) + Maxpool(\mathbf{U}))))) \\ &= \sigma(\mathbf{W}_2(\mathbf{W}_1([\mathbf{U}^s_{avg}, \mathbf{U}^s_{max}]))) \end{aligned} \tag{5}$$

where $f^{3\times3}$ indicates convolution operation with the kernel size of 3×3. $\sigma$ is a sigmoid function, and $concat(\mathbf{X}, \mathbf{Y})$ denotes the concatenation of two input tensors $\mathbf{X}$ and $\mathbf{Y}$ in the channel dimension.

**Integration with Deep CNN Models.** CSCA module can be used as a plug-and-play module for deep CNN models. In this work, we incorporate the CSCA module into ResNet [7] and InceptionV3 [8] to acquire CSCA-InceptionV3 and CSCA-ResNet. Figure 3 illustrates the scheme of incorporating CSCA into Inception module and Residual block. For the CSCA-Inception module, we plug CSCA after the concatenation of multiple parallel paths in the Inception module. As for the CSCA-ResNet module, we insert CSCA into the non-identity branch of a Residual block before skip connection.

## 3 Experiments

### 3.1 Datasets

To evaluate the performance of cervical cell identification, we train and test the proposed model on LTCCD and two public datasets, Herlev and SIPaKMeD. These three datasets are acquired by different classification criteria, sample preparation methods and imaging conditions. Table 1 lists the detailed cell distributions of experimental datasets. For Herlev and SIPaKMeD datasets, the same data augmentation techniques with LTCCD (mentioned in Sect. 2.1) are performed only on the training set. The raw validation set and testing set remain the same. Since the Herlev dataset only consists of 917 cells in total, we used a five-fold cross-validation strategy to train and evaluate the performance of the

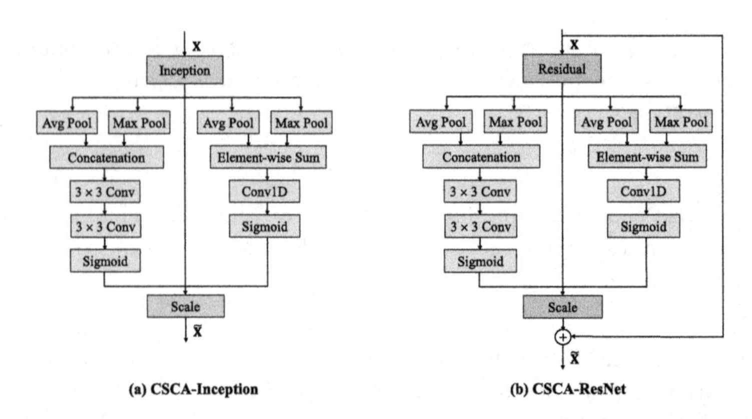


**Fig. 3.** Integrating CSCA module into standard CNN models: (a) CSCA-Inception Module. (b) CSCA-ResNet Module.

CNN models, and the average value is used to evaluate the final performance of the models. We performed 7-class and 2-class (classes 1-3 are normal, and classes 4-7 are abnormal) classification experiments on this dataset. For the other two datasets, the samples are partitioned according to the ratio of 8:1:1 for training, validation, and testing.

**Table 1.** Cell Distribution of Three Datasets.

| Dataset | Class | Cell Type | Total Number |
|---|---|---|---|
| LTCCD | 1 | Epithelial Normal | 3350 |
| | 2 | ASC-US | 1744 |
| | 3 | ASC-H | 1600 |
| | 4 | LSIL | 2150 |
| | 5 | HSIL | 1450 |
| Herlev | 1 | Superficial Squamous Epithelial | 74 |
| | 2 | Intermediate Squamous Epithelial | 70 |
| | 3 | Columnar Epithelial | 98 |
| | 4 | Mild Squamous Non-Keratinizing Dysplasia | 182 |
| | 5 | Moderate Squamous Non-Keratinizing Dysplasia | 146 |
| | 6 | Severe Squamous Non-Keratinizing Dysplasia | 197 |
| | 7 | Squamous Cell Carcinoma in Situ Intermediate | 150 |
| SIPaKMeD | 1 | Supercial-intermediate | 831 |
| | 2 | Parabasal | 787 |
| | 3 | Metaplastic | 793 |
| | 4 | Dyskeratotic | 813 |
| | 5 | Koilocytotic | 825 |

### 3.2 Implementation Details and Evaluation Metrics

This experiment was performed on an Ubuntu 18.04.5 LTS server with 2.60 GHz Intel® Xeon® Gold 6240 CPU, 256 GB memory, and eight NVIDIA GeForce RTX 2080 Ti GPU cards. In addition, all deep CNN models were implemented with the PyTorch library. Adam optimizer with an initial learning rate of 0.001 was adopted in this work. The batch size was set to 32 and the max number of training iterations for all models was 80 epochs. We adopted the four most standard metrics to evaluate the classification performance, including precision, recall, accuracy, and F1 score.

### 3.3 Performance Comparison of Different CNN Architectures

In order to investigate how much effect visual attention takes for deep CNN models, We first install the CSCA module into InceptionV3 and ResNet50 to evaluate the promotion of performance. Detailed results list in Table 2. On four different types of datasets, CSCA improves the classification accuracy of InceptionV3 by 8.6%, 2.71%, 2.97% and 1.36%, respectively. CSCA also brings 9.14%, 2.18%, 3.46%, 6.41% promotion for ResNet. Then we compare the attention-based model with two advanced networks, ResNeXt50 [24] and DenseNet121 [9], which improve ResNet in two different ways. The experimental results demonstrate that the performance improvement brought by the CSCA module surpasses the refinement methods of ResNeXt and DenseNet. CSCA module presents excellent feature extraction capacity in cervical cell identification.

**Table 2.** Classification performance comparison of different CNN architectures.

| Dataset | CNN Model | Precision | Recall | F1 Score | Accuracy |
|---|---|---|---|---|---|
| 7-class Herlev | InceptionV3 | 72.72 | 68.45 | 69.84 | 67.74 |
| | CSCA-InceptionV3 | 79.73 | 79.39 | 78.65 | 76.34 |
| | ResNet50 | 69.93 | 68.32 | 68.94 | 66.67 |
| | CSCA-ResNet50 | 77.66 | 78.62 | 77.81 | 75.81 |
| | ResNeXt50 | 71.08 | 67.51 | 68.56 | 65.05 |
| | DenseNet121 | 76.08 | 73.27 | 73.67 | 72.58 |
| 2-class Herlev | InceptionV3 | 94.15 | 91.75 | 92.86 | 94.57 |
| | CSCA-InceptionV3 | 96.81 | 96.20 | 96.50 | 97.28 |
| | ResNet50 | 92.91 | 88.68 | 90.51 | 92.93 |
| | CSCA-ResNet50 | 95.23 | 92.12 | 93.52 | 95.11 |
| | ResNeXt50 | 91.80 | 88.31 | 89.86 | 92.39 |
| | DenseNet121 | 93.32 | 89.71 | 91.30 | 93.48 |
| 5-class SIPaKMeD | InceptionV3 | 95.37 | 95.31 | 95.32 | 95.30 |
| | CSCA-InceptionV3 | 98.29 | 98.29 | 98.26 | 98.27 |
| | ResNet50 | 95.39 | 95.31 | 95.31 | 95.30 |
| | CSCA-ResNet50 | 98.77 | 98.77 | 98.76 | 98.76 |
| | ResNeXt50 | 96.05 | 96.05 | 96.03 | 96.04 |
| | DenseNet121 | 97.30 | 97.29 | 97.27 | 97.28 |
| 5-class LTCCD | InceptionV3 | 82.69 | 82.29 | 82.47 | 84.44 |
| | CSCA-InceptionV3 | 85.22 | 84.43 | 84.76 | 86.80 |
| | ResNet50 | 82.11 | 81.63 | 81.77 | 84.08 |
| | CSCA-ResNet50 | 89.65 | 88.77 | 89.15 | 90.49 |
| | ResNeXt50 | 83.02 | 82.32 | 82.58 | 84.56 |
| | DenseNet121 | 85.58 | 85.29 | 85.38 | 87.28 |

### 3.4 Performance Comparison Using Different Attention Modules

To further investigate the effect of different visual attention modules, based on ResNet, we compared the proposed CSCA module with another three attention modules, SE block [14], CBAM [15]and ECA Module [16] on SIPaKMeD and LTCCD datasets. Table 3 summarizes the experimental results. CSCA module brings the biggest performance improvement for ResNet on both datasets. On SIPaKMeD, CSCA-ResNet50 achieved the highest overall accuracy score of 98.76%. We also notice that CBAM using spatial attention performs better than the ECA module or SE block, which uses channel attention only. As for LTCCD, though the ECA module achieves better performance than CBAM or SE block, the CSCA module still performs the best. CSCA module utilizes channel and spatial attention synergistically, thus making up for the deficiency of single attention to realize more precise cervical cell identification. To intuitively estimate the performance, the confusion matrix of CSCA-ResNet50 results is illustrated in Fig. 4. The vast majority of the cells are correctly identified that fully verifying the effectiveness and robustness of the CSCA module.

**Table 3.** Classification performance comparison of different visual attention modules.

| Dataset | CNN Model | Precision | Recall | F1 Score | Accuracy (%) |
|---|---|---|---|---|---|
| 5-class SIPaKMeD | ResNet34 | 94.33 | 94.33 | 94.29 | 94.31 |
| | SE-ResNet34 | 96.10 | 96.06 | 96.03 | 96.04 |
| | CBAM-ResNet34 | 97.06 | 97.04 | 97.04 | 97.03 |
| | ECA-ResNet34 | 96.68 | 96.57 | 96.54 | 96.53 |
| | CSCA-ResNet34 | **98.30** | **98.28** | **98.28** | **98.27** |
| | ResNet50 | 95.39 | 95.31 | 95.31 | 95.30 |
| | SE-ResNet50 | 96.54 | 96.53 | 96.53 | 96.53 |
| | CBAM-ResNet50 | 97.77 | 97.76 | 97.76 | 97.77 |
| | ECA-ResNet50 | 96.30 | 96.31 | 96.29 | 96.29 |
| | CSCA-ResNet50 | **98.77** | **98.77** | **98.76** | **98.76** |
| | ResNet101 | 94.86 | 94.84 | 94.79 | 94.80 |
| | SE-ResNet101 | 96.29 | 96.30 | 96.29 | 96.29 |
| | CBAM-ResNet101 | 97.27 | 97.28 | 97.27 | 97.28 |
| | ECA-ResNet101 | 95.86 | 95.81 | 95.79 | 95.79 |
| | CSCA-ResNet101 | **98.07** | **98.04** | **98.03** | **98.02** |
| 5-class LTCCD | ResNet34 | 79.88 | 79.63 | 79.72 | 82.33 |
| | SE-ResNet34 | 81.37 | 80.57 | 80.91 | 83.01 |
| | CBAM-ResNet34 | 82.46 | 81.75 | 81.96 | 84.37 |
| | ECA-ResNet34 | 82.17 | 81.36 | 81.70 | 83.79 |
| | CSCA-ResNet34 | **85.26** | **84.95** | **85.09** | **86.87** |
| | ResNet50 | 82.11 | 81.63 | 81.77 | 84.08 |
| | SE-ResNet50 | 84.68 | 84.39 | 84.50 | 86.21 |
| | CBAM-ResNet50 | 86.20 | 85.53 | 85.74 | 87.38 |
| | ECA-ResNet50 | 87.11 | 86.77 | 86.80 | 88.74 |
| | CSCA-ResNet50 | **89.65** | **88.77** | **89.15** | **90.49** |
| | ResNet101 | 80.82 | 79.87 | 80.26 | 82.62 |
| | SE-ResNet101 | 83.61 | 82.64 | 83.02 | 85.05 |
| | CBAM-ResNet101 | 83.64 | 82.87 | 83.16 | 85.34 |
| | ECA-ResNet101 | 86.07 | 85.68 | 85.79 | 87.67 |
| | CSCA-ResNet101 | **88.00** | **87.48** | **87.67** | **89.22** |

### 3.5 Class Activation Visualization

To figure out where a trained model mainly focuses on for final classification. We use Grad-CAM [25] to show the visualization results of ResNet50 and CSCA-ResNet50. Figure 5 displays the activation mapping results of five different types of cervical cells in the LTCCD dataset, in which the red highlight represents the central mapping location responsible for the classification results. We observe that with the integration of the CSCA module, the CNN model pays more attention to the nucleus part, which has implicitly learned the diagnostic rules of cytologists. Besides, compared to the original ResNet50, the area of red highlight is relatively more minor in CSCA-ResNet50, which demonstrates that the CSCA module makes the network acquire more centralized and precise features.

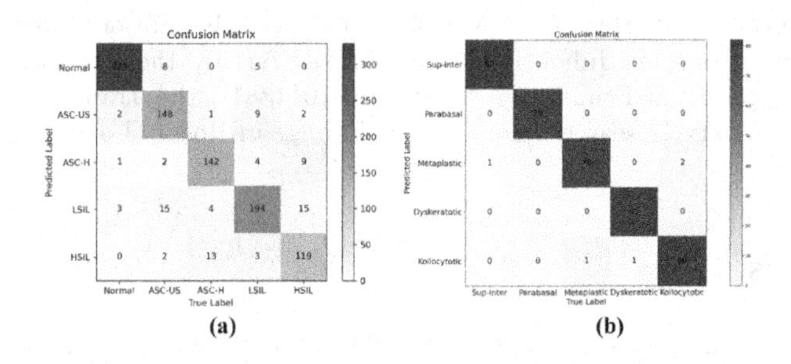


**Fig. 4.** Confusion matrix of CSCA-ResNet50 classification results on: (a) LTCCD dataset. (b) SIPaKMeD dataset.

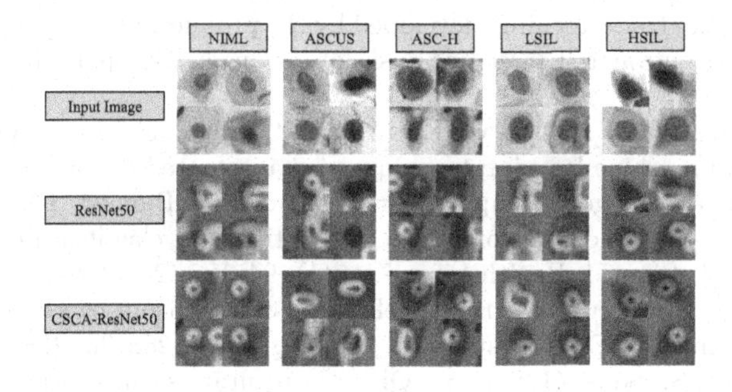


**Fig. 5.** Class activation mapping results of LTCCD dataset.

## 4 Conclusion and Future Works

In this paper, we collected a TBS-based cervical cell dataset, LTCCD. Via image processing techniques like data augmentation, the proposed deep learning method was able to be more robust and generalizable. The images of LTCCD were derived from practical screening programs, thus making it convenient to migrate the trained models to hospitals or medical laboratories. Moreover, in order to extract as many useful features as possible from a small cervical cell and its tiny nucleus, we proposed a channel spatial collaborative attention network that generates 3D attention for feature maps and focuses on valid information with high weights. Experimental results demonstrated that the proposed method provides a feasible solution for the fine-grained identification of cervical cells.

We observed a characteristic from the confusion matrix (Fig. 4(a)) that ASC-US and LSIL cells were easily misrecognized with each other, and ASC-H and HSIL cells inclined to be confused. This is due to the mimics in morphology and biomedical characteristics between them. Even experienced experts tend to make mistakes in these categories. However, distinguishing them precisely is crucial for implementing different treatments. Thus, we will conduct further research to mine helpful information about confusing categories in the near future.

**Acknowledgements.** This work was supported by the Major Projects of Technological Innovation in Hubei Province (2019AEA170), the Frontier Projects of Wuhan for Application Foundation (2019010701011381), the Translational Medicine and Interdisciplinary Research Joint Fund of Zhongnan Hospital of Wuhan University (ZNJC202226).

## References

1. Sung, H., et al.: Global cancer statistics 2020: GLOBOCAN estimates of incidence and mortality worldwide for 36 cancers in 185 countries. CA Cancer J. Clin. **71**(3), 209–249 (2021)
2. Elsheikh, T.M., Austin, R.M., Chhieng, D.F., Miller, F.S., Moriarty, A.T., Renshaw, A.A.: American society of cytopathology workload recommendations for automated pap test screening: developed by the productivity and quality assurance in the era of automated screening task force. Diagn. Cytopathol. **41**(2), 174–178 (2013)
3. William, W., Ware, A., Basaza-Ejiri, A.H., Obungoloch, J.: A review of image analysis and machine learning techniques for automated cervical cancer screening from pap-smear images. Comput. Methods Programs Biomed. **164**, 15–22 (2018)
4. Chen, Y.F., et al.: Semi-automatic segmentation and classification of pap smear cells. IEEE J. Biomed. Health Inform. **18**(1), 94–108 (2013)
5. Song, Y., et al.: Accurate cervical cell segmentation from overlapping clumps in pap smear images. IEEE Trans. Med. Imaging **36**(1), 288–300 (2016)
6. Wan, T., Xu, S., Sang, C., Jin, Y., Qin, Z.: Accurate segmentation of overlapping cells in cervical cytology with deep convolutional neural networks. Neurocomputing **365**, 157–170 (2019)
7. He, K., Zhang, X., Ren, S., Sun, J.: Deep residual learning for image recognition. In: Proceedings of the IEEE Conference on Computer Vision and Pattern Recognition, pp. 770–778 (2016)
8. Szegedy, C., Vanhoucke, V., Ioffe, S., Shlens, J., Wojna, Z.: Rethinking the inception architecture for computer vision. In: Proceedings of the IEEE Conference on Computer Vision and Pattern Recognition, pp. 2818–2826 (2016)
9. Huang, G., Liu, Z., Van Der Maaten, L., Weinberger, K.Q.: Densely connected convolutional networks. In: Proceedings of the IEEE Conference on Computer Vision and Pattern Recognition, pp. 4700–4708 (2017)

10. Zhang, L., Lu, L., Nogues, I., Summers, R.M., Liu, S., Yao, J.: DeepPap: deep convolutional networks for cervical cell classification. IEEE J. Biomed. Health Inform. **21**(6), 1633–1643 (2017)
11. Dong, N., Zhao, L., Wu, C.H., Chang, J.F.: Inception v3 based cervical cell classification combined with artificially extracted features. Appl. Soft Comput. **93**, 106311 (2020)
12. Chen, H., et al.: CytoBrain: cervical cancer screening system based on deep learning technology. J. Comput. Sci. Technol. **36**(2), 347–360 (2021)
13. Sabeena, K., Gopakumar, C.: A hybrid model for efficient cervical cell classification. Biomed. Signal Process. Control **72**, 103288 (2022)
14. Hu, J., Shen, L., Sun, G.: Squeeze-and-excitation networks. In: Proceedings of the IEEE Conference on Computer Vision and Pattern Recognition, pp. 7132–7141 (2018)
15. Woo, S., Park, J., Lee, J.-Y., Kweon, I.S.: CBAM: convolutional block attention module. In: Ferrari, V., Hebert, M., Sminchisescu, C., Weiss, Y. (eds.) ECCV 2018. LNCS, vol. 11211, pp. 3–19. Springer, Cham (2018). https://doi.org/10.1007/978-3-030-01234-2_1
16. Wang, Q., Wu, B., Zhu, P., Li, P., Zuo, W., Hu, Q.: ECA-Net: efficient channel attention for deep convolutional neural networks. In: Proceedings of the IEEE/CVF Conference on Computer Vision and Pattern Recognition (CVPR) (2020)
17. Zhang, J., Xie, Y., Xia, Y., Shen, C.: Attention residual learning for skin lesion classification. IEEE Trans. Med. Imaging **38**(9), 2092–2103 (2019)
18. He, X., Deng, Y., Fang, L., Peng, Q.: Multi-modal retinal image classification with modality-specific attention network. IEEE Trans. Med. Imaging **40**(6), 1591–1602 (2021)
19. Marinakis, Y., Dounias, G., Jantzen, J.: Pap smear diagnosis using a hybrid intelligent scheme focusing on genetic algorithm based feature selection and nearest neighbor classification. Comput. Biol. Med. **39**(1), 69–78 (2009)
20. Plissiti, M.E., Dimitrakopoulos, P., Sfikas, G., Nikou, C., Krikoni, O., Charchanti, A.: Sipakmed: a new dataset for feature and image based classification of normal and pathological cervical cells in pap smear images. In: 2018 25th IEEE International Conference on Image Processing (ICIP), pp. 3144–3148. IEEE (2018)
21. Nayar, R., Wilbur, D.C.: The Bethesda system for reporting cervical cytology: a historical perspective. Acta Cytol. **61**(4–5), 359–372 (2017)
22. Reyes, M.C., Cooper, K., et al.: Cervical cancer biopsy reporting: a review. Indian J. Pathol. Microbiol. **57**(3), 364 (2014)
23. Nayar, R., Wilbur, D.C. (eds.): Springer, Cham (2015). https://doi.org/10.1007/978-3-319-11074-5
24. Xie, S., Girshick, R., Doll'ar, P., Tu, Z., He, K.: Aggregated residual transformations for deep neural networks. In: Proceedings of the IEEE Conference on Computer Vision and Pattern Recognition, pp. 1492–1500 (2017)
25. Selvaraju, R.R., Cogswell, M., Das, A., Vedantam, R., Parikh, D., Batra, D.: Grad-CAM: visual explanations from deep networks via gradient-based localization. Int. J. Comput. Vision **128**(2), 336–359 (2020)

# Multimodal Learning of Audio-Visual Speech Recognition with Liquid State Machine

Xuhu Yu[1], Lei Wang[1(✉)], Changhao Chen[2], Junbo Tie[1], and Shasha Guo[1]

[1] College of Computer Science and Technology, National University of Defense Technology, Changsha 410073, China
{leiwang,tiejunbo11,guoshasha13}@nudt.edu.cn
[2] College of Intelligence Science, National University of Defense Technology, Changsha 410073, China

**Abstract.** Audio-visual speech recognition is to solve the multimodal lip-reading task using audio and visual information, which is an important way to improve the performance of speech recognition in noisy conditions. Deep learning methods have achieved promising results in this regard. However, these methods have complex network architecture and are computationally intensive. Recently, Spiking Neural Networks (SNNs) have attracted attention due to their being event-driven and can enable low-power computing. SNNs can capture richer motion information and have been successful in work such as gesture recognition. But it has not been widely used in lipreading tasks. Liquid State Machines (LSMs) have been recognized in SNNs due to their low training costs and are well suited for spatiotemporal sequence problems of event streams. Multimodal lipreading based on Dynamic Vision Sensors (DVS) is also such a problem. Hence, we propose a soft fusion framework with LSM. The framework fuses visual and audio information to achieve the effect of higher reliability lip recognition. On the well-known public LRW dataset, our fusion network achieves a recognition accuracy of 86.8%. Compared with single modality recognition, the accuracy of the fusion method is improved by 5% to 6%. In addition, we add extra noise to the raw data, and the experimental results show that the fusion model outperforms the audio-only model significantly, proving the robustness of our model.

**Keywords:** Liquid State Machine · Multimodal Fusion · Audio-visual Speech Recognition

## 1 Introduction

Human-to-human communication is not only reflected in the sound but also includes observing mouth shape, body posture, and movement. It is essentially a multimodal fusion process. The integration of multimodal complementary information in the brain enables efficient communication. The task of lipreading is


M. Tanveer et al. (Eds.): ICONIP 2022, CCIS 1793, pp. 552–563, 2023.
https://doi.org/10.1007/978-981-99-1645-0_46

extended to the problem of audio-visual speech recognition, automatically extracting features from both visual and audio information for speech recognition.

The studies on audio-visual speech recognition can be mainly divided into two categories, namely hand-crafted methods and end-to-end deep neural network approaches. Hand-crafted methods mainly use hand-detected visual and audio features to establish a concrete mathematical model, e.g., Hidden Markov Model [1]. However, these methods are poorly generalized in new domains, limiting their real-world usage. The end-to-end deep neural network approaches, such as Sequence-to-Sequence [2] models and Bidirectional Gated Recurrent Units (BGRUs) [3], learn to automatically extract features and solve the task in a data-driven manner, achieving good performance in audio-visual speech recognition tasks. But these models are at the cost of large computational consumption.

In recent years, the emerging neuromorphic computing that implements brain-inspired intelligence has gained great attention [4]. Brain-inspired computing uses spiking neural networks (SNNs) that are similar to the architecture of biological brains, to work in an asynchronous and event-driven way. SNNs are more suitable for processing unstructured data, e.g., multi-sensory and cross-modal data [5]. More importantly, SNNs can be deployed on neuromorphic processors, which can realize intelligent computing tasks with lower power consumption, overcoming the energy consumption bottleneck of current methods to a certain extent. This also helps to move audio-visual speech recognition models to edge devices (e.g. smart glasses with lipreading capability, etc.). However, there is a lack of research on SNNs-based visual-audio recognition. To this end, we have a first trial on applying spiking neural networks to solve the task of audio-visual speech recognition. *To the best of our knowledge, this work is the first one to apply spiking neural network to solve a multimodal lipreading task.*

In this work, we propose a fusion framework based on spiking neural networks, which can effectively fuse the information from different modalities. Specifically, our model is mainly composed of two parts, namely a single-modal feature encoder based on LSM and a soft fusion network based on attention mechanism. The single-modal feature encoder is used to extract the feature information of different modalities, and then the extracted features are fused by the soft fusion network, so as to realize the integration of different modal information.

In summary, the main contributions of this work are as follows:

- We propose the first SNN framework for audio-visual speech recognition, that uses a soft fusion mechanism to effectively integrate multimodal information.
- We design different spiking network models for visual and audio modalities to achieve effective extraction of different modal features.
- On the public LRW dataset, our proposed network achieves a recognition accuracy of 86.8%. Moreover, we validate the robustness of our proposed multimodal lipreading model in noisy environments.

# 2 Background

## 2.1 Liquid State Machine

Liquid State Machine (LSM) [6] is a major type of reservoir computing. The structure of LSM is shown in Fig. 1, which is mainly composed of three parts: the input layer, the liquid layer, and the readout layer, the core of which is the liquid layer. Input layer neurons are sparsely connected to neurons in the liquid layer, depending on their application. The liquid layer transforms the input stream into non-linear patterns in higher dimensions, acting as a filter. The liquid state is then analyzed and interpreted by a readout layer, which consists of memoryless artificial neurons or spiking neurons. Only the connections between the readout layer and the liquid layer need to be trained, and the reservoir synapses are fixed to ease the training challenge. For a given input spike train, the state $x^M(t)$ of the liquid at time t can be represented by Eq. 1.

$$x^M(t) = L^M(I(s)) \tag{1}$$

$I(s)$ represents the spike input from $0 - t$, and $L^M$ represents the response of the liquid layer neuron to the input spike train.

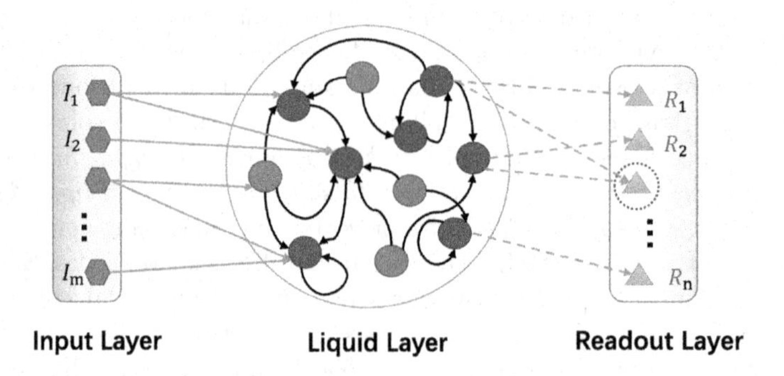


**Fig. 1.** Illustration of Liquid State Machine.

Each input produces a response in the liquid layer, and this response of the liquid layer is called the liquid state. The readout layer can be viewed as a function that transforms this liquid state into a feature vector. Hence, the output feature vector $y(t)$ can be written as a function of the liquid state $x^M(t)$ in Eq. 2.

$$y(t) = F^M(x^M(t)) \tag{2}$$

## 2.2 Dynamic Vision Sensor

Dynamic Vision Sensor (DVS) is a neuromorphic sensor [7]. It records changing objects in the scene using an event-driven approach. When objects in the real scene change, DVS will generate a series of pixel-level event outputs. Each event is represented as (x, y, t, P), where (x, y) is the pixel coordinates of the event, and t is the timestamp of the event, which indicates when the event occurred. P is the polarity of the event, which represents whether the pixel is brighter or darker than before. This kind of data stream generated by DVS is naturally suitable for processing by spiking neural networks.

Compared with traditional cameras that output each frame of pictures at a fixed rate, DVS cameras are more suitable for acquiring moving target information. In addition, in a recent study, Rebecq et al. [8] proposed that a software simulator can convert the data captured by a traditional camera into this event data stream by imitating the DVS principle.

### 2.3 Related Works

**Lipreading.** When deep learning techniques were not yet popular, lipreading tasks mainly extracted features by hand-crafted methods, such as optical flow, lip landmark tracking [9], etc. The classification was mainly completed by methods such as Support Vector Machine and Hidden Markov Models (HMMs) [1]. These methods drove the development of early lipreading. With the rapid development of deep learning, more and more researchers use deep neural network methods to complete the lipreading task. Petridis et al. [10] proposed an end-to-end lipreading model consisting of spatiotemporal convolutions, residuals, and bidirectional long short-term memory networks. Specifically, 3D convolution and 2D ResNet are used as the front end to extract lipreading video features, and a bidirectional LSTM is used as the back end to realize the classification and recognition of English words. Martinez et al. [11]proposed Multi-Scale Temporal Convolutional Network (MS-TCN) to improve temporal encoding and boosted word-level lipreading performance. Assael et al. [12] proposed an end-to-end deep learning model using the loss function of Connectionist Temporal Classification (CTC).

**Audio-Visual Speech Recognition.** Audio-visual speech recognition (AVSR) is tightly coupled with lipreading, which is essentially a multimodal fusion problem. Petridis et al. [13] used ResNet networks and BGRU networks to learn from raw pixels and waveforms. Then, the learned feature is concatenated and fed into another two-layer BGRU network to get the most likely class of a word. Afouras et al. [14] adopted the Transformer model combined with the CTC loss function to implement sentence-level audio-visual speech recognition tasks.

**Lipreading Based on Event Data.** Neil et al. [15] used dynamic visual sensors and dynamic auditory sensors to transcribe an existing lipreading dataset into event data streams, thereby bringing the lipreading recognition problem into the event domain. But the network they designed is still based on DNN to process event data. It's a shame they didn't explore it with spiking neural networks that are better suited for processing event data.

# 3 Methodology

In this section, we introduce end-to-end multimodal lipreading (also known as audio-visual speech recognition) architecture using a liquid state machine. Figure 2 shows an overview of our proposed architecture, consisting of three parts, i.e. visual and audio feature extraction, feature fusion, and word recognition.

SNNs differ from an artificial neural network in that their input needs to be encoded with spikes. The frame-based lipreading datasets need to be transformed into the spike train, and only event data format can be processed by our model. In the visual and audio feature encoding modules, we use LSM to extract semantic features of different modalities respectively. In the feature fusion component, we propose a soft fusion method to fuse features extracted from the different modalities. Finally, the fused features are fed into a classifier to achieve recognition.

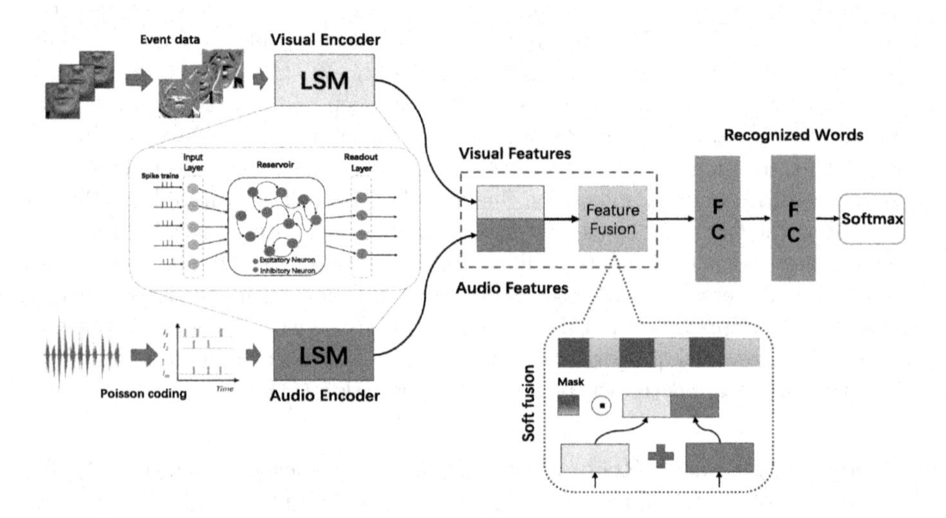


**Fig. 2.** An overview of our multimodal audio-visual model structure with our soft fusion methods, including visual and audio feature encoders, feature fusion and recognition.

## 3.1 LSM Based Feature Extraction

We construct the feature encoder of the spiking neural network using LSM. First, we transform the raw dataset into a stream of events data [8], which is similar to the data generated by a DVS camera. The feature encoder then encodes this stream of event data into the liquid layer of the LSM, which is displayed as the liquid state. Finally, We use the time window division sampling technique to read out the liquid state as a feature vector.

There are two types of neurons in the liquid layer: excitatory neurons and inhibitory neurons. In the liquid layer, we have 80% of them as excitatory neurons and 20% as inhibitory neurons [16]. Both the excitatory and inhibitory neuron models choose the leaky integrate-and-fire (LIF) neuron, whose dynamics are modeled by

$$\tau_m \frac{\partial V}{\partial t} = -V + I_{ext} \cdot R \tag{3}$$

where the membrane potential $V$ represents the charge accumulated through the incoming current ($I_{ext}$) over time. The membrane potential leaks are proportional to the time constant $\tau_m$.

**Visual Feature Encoder.** The Visual Feature Encoder ($f_{vision}$) uses a total of 1000 neurons, which forms a neuron model with a $10 \times 10 \times 10$ cubic structure. This structure simplifies the space complexity of the model. The event data converted from visual image information ($x_v$) is fed into our visual feature encoder, and the liquid state machine encodes each sample into a liquid state. Finally, the liquid state is converted into a feature vector by the readout layer, and we use this feature as our visual feature.

$$V_{lip} = f_{vision}(x_v). \tag{4}$$

**Audio Feature Encoder.** The processing complexity of speech signals is lower than that of visual signals. Therefore, for the Audio Feature Encoder($f_{audio}$), we use a smaller number of neurons for the construction of the LSM model. Only 512 spiking neurons are used, but the neuron spatial structure is the same as the visual feature encoder. We extract Mel Frequency Cepstrum Coefficient(MFCC) features from raw audio waveforms($x_a$). Afterward, the MFCC features are converted to the spike train as the input for the $f_{audio}$. As with the visual feature encoder, we will get a set of feature vectors about the audio as our audio feature.

$$A_{lip} = f_{audio}(x_a). \tag{5}$$

### 3.2 Fusion Function

There is some redundant and complementary information between different modalities, so a fusion function is needed to effectively integrate different information in order to obtain richer feature expressions. We combine the high-level features extracted by the two encoders into a multimodal fusion function $g$, which combines information from the vision channels $V_{lip}$ and audio channels $A_{lip}$ to generate more effective features for subsequent classification and recognition tasks:

$$R = g(V_{lip}, A_{lip}). \tag{6}$$

The obvious method is to directly concatenate two features into another feature space (called direct fusion here). But this method does not take into account the characteristics and advantages of different modalities. In this paper, inspired by attention mechanisms in the field of natural language processing, we propose a soft fusion method [17]. In this method, the visual and audio features are used to generate masks, and the generated masks are used to re-weight the visual and audio features. A comparison of the two methods is shown in Fig. 3.

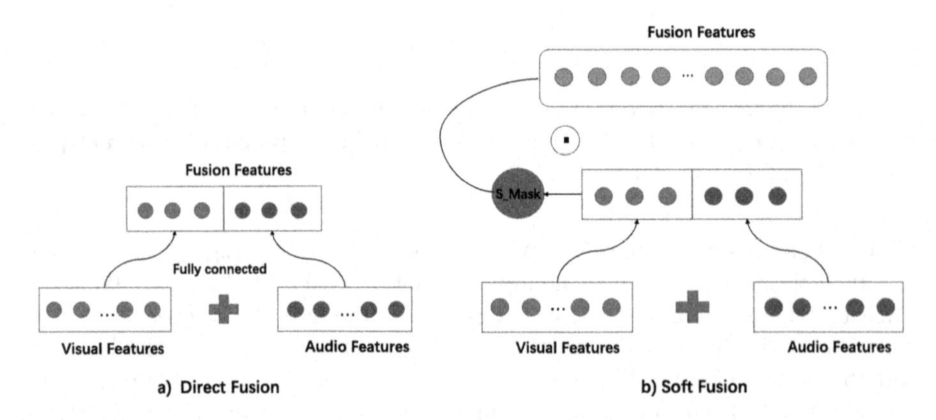


**Fig. 3.** Schematic diagram of the fusion method. a) direct fusion method. b) our soft fusion method.

**Direct Fusion.** Direct fusion can be seen as a method of indiscriminately weighting different modal information. This method uses Multi-Layer Perceptrons(MLPs) to combine features from visual and audio channels and train them end-to-end. Therefore, direct fusion is modeled as:

$$g_{direct}\left(V_{lip}, A_{lip}\right) = MLPs[V_{lip}; A_{lip}] \tag{7}$$

**Soft Fusion.** Our proposed soft fusion scheme uses the principle of attention mechanism [18], which can adjust the weights of the visual channel and audio channel features, and re-weight the two channels.

Here, the implementation principle of soft fusion is that we introduce a set of mask representations to select the features extracted by a single channel so that we can generate fusion features. This set of mask representations is a set of weights, trained from the features of both visual and audio channels.

$$S_{V,A} = Sigmoid_{V,A}\left([V_{lip}; A_{lip}]\right) \tag{8}$$

where the parameters of $S_{V,A}$ are deterministically parameterized by the neural network and depend on visual and audio features. The sigmoid function ensures that each feature is re-weighted in the range [0,1].

Then, the visual and audio features are multiplied by the previously generated mask representation, which is used as the new re-weighted vectors. This soft fusion function is modeled as:

$$g_{soft}\left(V_{lip}, A_{lip}\right) = S_{V,A} \otimes \left[V_{lip}; A_{lip}\right] \tag{9}$$

### 3.3 Recognition

This work mainly focuses on word-level lipreading task, essentially a classification problem. Hence, we transform the problem into choosing an appropriate classifier for the fused feature vector($R$). We use a three-layer fully connected network with the last layer of softmax as the classifier of this model. This recognition task can be modeled as:

$$p(y_i|r_i) = softmax(fc(r_i)) \tag{10}$$

# 4 Experiments

## 4.1 Experimental Setup

In this paper, the Brain2 simulator [19] is chosen as the realization platform of the feature encoder part of the model, which provides the construction of the neurons of the spiking neural network and the description of the synaptic behavior. The fusion network part of the model and the classification task is implemented with PyTorch. All simulation implementations of SNNs in this paper are run on the CPU. The network in the feature fusion part and the MLP classifier in the recognition part use GPU to speed up training.

In our experiments, we select a visual-only feature model, an audio-only feature model, and a directly fused visual-audio model as baselines, named V-O, A-O, and V/A-direct, respectively. The structure of the visual-only feature model and the audio-only feature model consists of two parts: a single-modal feature encoder in Fig. 2 and a single-hidden-layer MLP classifier. The directly fused visual-audio model adopts the same structure as our proposed soft fusion method, except in the feature fusion part. All of the networks including baselines were trained with a batch size of 500 using the Adam optimizer, with a learning rate $lr = 1e^{-4}$.

## 4.2 Datasets and Preprocessing

**Datasets.** We use the LRW dataset [20], which is an English lipreading dataset. The dataset has a total of 500 English words and contains over 500,000 utterances, of which 25,000 are reserved for testing. There are up to 1000 training samples per word. The content is dominated by short films (1.16 s) of BBC programs, mainly news and talk shows. In the experiments in this paper, we select ten representative words in this dataset as the original dataset samples.

**Preprocessing.** For the video sequences in the LRW dataset, we use dlib [21] to detect and track 68 facial landmarks. Then, a bounding box of $96 \times 96$ pixels is used to crop the mouth region of interest(ROI). The cropped image sequence is input to the EISM simulator [8], which is an event camera simulator. Image sequences can be converted to pixel-level event output, and the data representation format is (x, y, t, p). Such transformed event data will be used as the input of the Visual Feature Encoder in this paper.

For data preprocessing of the audio channel, we extracted the audio signal of the raw dataset. Then, we extract the MFCC features from the audio waveform. Finally, the MFCC features are transformed into the Poisson spike train [16], which is used as the input to the Audio Feature Encoder.

### 4.3 Training Strategy

The training of our model is divided into two stages: independent training of visual and audio feature encoding structures. End-to-end training of the fusion network.

**Single-Modal Training.** We connect a single-hidden-layer perception network as a classifier after the single-channel feature encoder. Then, this new single modality encoding network is trained. During training, we split the training dataset. Each part of the training data can train a set of liquid states, and finally splicing multiple sets of liquid states. The spliced liquid state is read out as the feature vector in this mode. This helps speed up training and reduces training resource consumption.

**Audio-Visual Fusion Training.** Through the above single-modal training, we obtain the synaptic connection weights between neurons of the spiking neural network in the feature encoder. During the end-to-end training process, we will fix these neuron parameters and synaptic weights.

### 4.4 Experimental Results

**Compared with Baseline Methods.** We compare the performance of our soft fusion model with baseline methods, and the difference between the direct fusion method and soft fusion method is discussed. Direct fusion does not distinguish between different modal features, it is more like a fusion method without preference. However, the soft fusion method is to make a modality fully effective in its suitable environment. Table 1 shows the comparison of the mean and best recognition accuracy between the different methods. The recognition accuracy of the visual feature model and the audio feature model in a single modality is lower than that of the fusion method, which shows that our fusion method is effective. By comparing the experimental results of the soft fusion method and the direct fusion method, we can see that our soft fusion method has about 5% improvement in the best recognition accuracy. It definitely proves that the fusion method proposed by us has better fusion ability and can better combine effective information in different modalities.

**Table 1.** Comparison of Recognition Accuracy of Different Methods.

| Model | Mean Accuracy (%) | Best Accuracy (%) |
|---|---|---|
| Visual-only (V-O) | 76.1 | 77.8 |
| Audio-only (A-O) | 80.6 | 81.3 |
| Visual-Audio (Direct Fusion) | 80.8 | 82.5 |
| Visual-Audio (Soft Fusion) | **86.3** | **86.8** |

**Comparison with State-of-the-Art Methods.** To verify the effectiveness of the proposed method, we compare word-level lipreading with some advanced methods. Table 2 shows the comparison of word-level lipreading performance on the LRW dataset. Our proposed fusion method achieves the best accuracy of 86.8%. The results show that our method can effectively combine visual representation and audio information to achieve a fusion effect. Compared with the state-of-the-art methods, our model still has a certain gap in recognition accuracy, but choosing LSM has incomparable advantages. On the one hand, the network complexity of our model is lower, and the number of trainable parameters will be much smaller than other methods, which makes our method have lower computational resource consumption. On the other hand, our method offers the possibility of subsequent work on neuromorphic processors, beneficial to marginalizing intelligent computing.

**Table 2.** Word accuracy comparison on LRW.V: visual features, A: audio features, V+A: fusion of two features.*This method only uses audio features in training, not in inference.

| Method | Backbone | ACC (%) |
|---|---|---|
| Wang (2019) | Multi-Grained + C-BiLSTM(V) | 83.3 |
| Weng and Kitani (2019) | T I3D + BiLSTM | 84.1 |
| Zhao et al. (2020) | R18 + BiGRU + LSTM(V) | 84.4 |
| Martinez et al. (2020) | R18 + MS-TCN (V) | 85.3 |
| Ma et al. (2021) | R18 + MS-TCN/LiRA | 88.1 |
| Kim et al.(2022) | R18 + MS-TCN/MVM (*) | 88.5 |
| Martinez et al. (2020) | R18 + MS-TCN (A+V) | 98 |
| **Proposed Method** | **LSM+Soft fusion** | **86.8** |

**Robustness Study.** We further investigate the robustness of our fusion method under noisy conditions. The production of audio data with noise refers to the method of Petridis et.al [13]. The noise from the NOISEX database [22] is added, and the SNR ranges from −5 dB to 15 dB. Figure 4 shows the recognition accuracy of the audio-only feature model, the visual-only feature model, the directly fused visual-audio model, and the soft fusion model under different noise levels. The visual-only feature model is not affected by noise, so its performance remains stable. Both the audio-only feature model and the fusion model have some performance degradation. And it can be found that the audio-only feature model performance drops significantly. This is because the anti-interference ability of the speech signal is poor, so it is difficult to achieve good results in speech recognition in a noisy environment. However, under the condition of high noise, the anti-interference ability of the fusion model is significantly stronger than that of the audio-only feature model. In particular, the spiking neural network model of the soft fusion method in this paper has the best anti-interference ability.

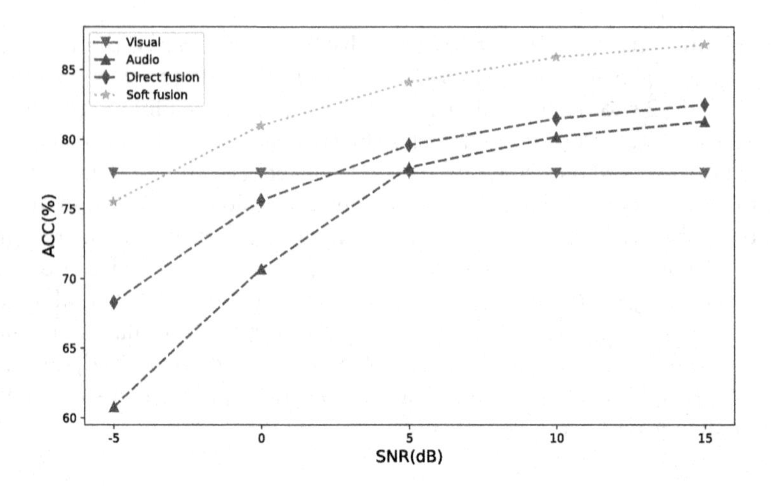


**Fig. 4.** Recognition accuracy of several models at different noise levels.

## 5 Conclusion

In this work, we presented a novel study of spiking neural networks for audio-visual speech recognition tasks. To the best of our knowledge, this is the first work that applies a spiking neural network to a multimodal lipreading task. We utilize a liquid state machine as a single-modal feature encoder and propose a soft fusion method to fuse features from different modalities, which is inspired by the attention mechanism. The effectiveness of our method has been demonstrated by conducting experiments on the LRW dataset. And experiments under noisy conditions have also shown good robustness of our fusion architecture.

## References

1. Hinton, G., Deng, L., Yu, D., Dahl, G.E., Kingsbury, B.: Deep neural networks for acoustic modeling in speech recognition: the shared views of four research groups. IEEE Signal Process. Mag. **29**(6), 82–97 (2012)
2. Vaswani, A., et al.: Attention is all you need. arXiv (2017)
3. Chung, J., Gulcehre, C., Cho, K.H., Bengio, Y.: Empirical evaluation of gated recurrent neural networks on sequence modeling. Eprint Arxiv (2014)
4. Roy, K., Jaiswal, A., Panda, P.: Towards spike-based machine intelligence with neuromorphic computing. Nature **575**(7784), 607–617 (2019)
5. Butts, D.A., et al.: Temporal precision in the neural code and the timescales of natural vision. Nature **449**(7158), 92–95 (2007)
6. Maass, W., Natschlger, T., Markram, H.: Real-time computing without stable states: a new framework for neural computation based on perturbations. Neural Comput. **14**(11), 2531–2560 (2002)
7. Gallego, G., et al.: Event-based vision: a survey. IEEE Trans. Pattern Anal. Mach. Intell. **44**(1), 154–180 (2020)

8. Rebecq, H., Gehrig, D., Scaramuzza, D.: ESIM: an open event camera simulator. In: Conference on Robot Learning, pp. 969–982. PMLR (2018)
9. Duchnowski, P., Hunke, M., Busching, D., Meier, U., Waibel, A.: Toward movement-invariant automatic lip-reading and speech recognition. In: 1995 International Conference on Acoustics, Speech, and Signal Processing, vol. 1, pp. 109–112. IEEE (1995)
10. Petridis, S., Li, Z., Pantic, M.: End-to-end visual speech recognition with LSTMs. In: ICASSP 2017–2017 IEEE International Conference on Acoustics, Speech and Signal Processing (ICASSP) (2017)
11. Martinez, B., Ma, P., Petridis, S., Pantic, M.: Lipreading using temporal convolutional networks. IEEE (2020)
12. Assael, Y.M., Shillingford, B., Whiteson, S., Freitas, N.D.: LipNet: end-to-end sentence-level lipreading. arXiv:1611.01599 (2016)
13. Petridis, S., Stafylakis, T., Ma, P., Cai, F., Pantic, M.: End-to-end audiovisual speech recognition. IEEE (2018)
14. Afouras, T., Chung, J.S., Senior, A., Vinyals, O., Zisserman, A.: Deep audio-visual speech recognition. IEEE Trans. Pattern Anal. Mach. Intell. **PP**(99), 1 (2018)
15. Li, X., Neil, D., Delbruck, T., Liu, S.C.: Lip reading deep network exploiting multi-modal spiking visual and auditory sensors. In: 2019 IEEE International Symposium on Circuits and Systems (ISCAS) (2019)
16. Tian, S., Qu, L., Wang, L., Hu, K., Li, N., Xu, W.: A neural architecture search based framework for liquid state machine design. Neurocomputing **443**, 174–182 (2021)
17. Chen, C., et al.: Selective sensor fusion for neural visual-inertial odometry. In: IEEE/CVF Conference on Computer Vision and Pattern Recognition (CVPR) (2019)
18. Mnih, V., Heess, N., Graves, A., et al.: Recurrent models of visual attention. In: Advances in Neural Information Processing Systems 27 (2014)
19. Stimberg, M., Brette, R., Goodman, D.F.: Brian 2, an intuitive and efficient neural simulator. Elife **8**, e47314 (2019)
20. Chung, J.S., Zisserman, A.: Lip reading in the wild. In: Asian Conference on Computer Vision (2016)
21. King, D.E.: Dlib-ml: a machine learning toolkit. J. Mach. Learn. Res. **10**(3), 1755–1758 (2009)
22. Andrew, V., Herman, J., Steeneken, M.: Assessment for automatic speech recognition: II. NOISEX-92: a database and an experiment to study the effect of additive noise on speech recognition systems. Speech Communication (1993)

# Identification of Fake News: A Semantic Driven Technique for Transfer Domain

Jannatul Ferdush[1(✉)], Joarder Kamruzzaman[1], Gour Karmakar[1], Iqbal Gondal[2], and Raj Das[1]

[1] Federation University, Ballarat, Australia
jannatulferdush@students.federation.edu.au,
{joarder.kamruzzaman,gour.karmakar,r.das}@federation.edu.au
[2] RMIT University, Melbourne, Australia
iqbal.gondal@rmit.edu.au

**Abstract.** Fake news spreads quickly on online social media and adversely impacts political, social, religious, and economic stability. This necessitates an efficient fake news detector which is now feasible due to advances in natural language processing and artificial intelligence. However, existing fake news detection (FND) systems are built on tokenization, embedding, and structure-based feature extraction, and fail drastically in real life because of the difference in vocabulary and its distribution across various domains. This article evaluates the effectiveness of various categories of traditional features in cross-domain FND and proposes a new method. Our proposed method shows significant improvement over recent methods in the literature for cross-domain fake news detection in terms of widely used performance metrics.

**Keywords:** Fake news · Tokenization · Bert · Transfer domain · Feature extraction

## 1 Introduction

Fake news is one of the burning issues in today's world because it directly impacts politics, businesses and society. Internet and social media have made it easy to produce and distribute any news quickly without any further check. A vast amount of fake news is produced every second that may ruin someone's reputation [1], business issue [2], while earning revenue for perpetrators by click [3]. The concept of fake news has a long history. A false explanation about earthquake by a catholic church in 1755 pressured people to speak about religious violence. When several false stories about racism were published in 1800, crime and instability increased in the United States. By the early nineteenth century, a journal called 'Sun' had become a leading publishing company by reporting false news about the existence of an alien civilisation on the moon. Creating eye-catching headlines with baseless news was also common in the 18th century, indicating that yellow journalism is an old phenomenon [4]. However, at present,


M. Tanveer et al. (Eds.): ICONIP 2022, CCIS 1793, pp. 564–575, 2023.
https://doi.org/10.1007/978-981-99-1645-0_47

its impacts are more dangerous because the propagation spread and the magnitude at which it influences people are higher than before. Another critical factor is that the post on social media does not go through a fact check like traditional media. It is always a concern for society, government, business, education, religion, and even the health sector, and therefore, it is essential to assess whether news and information items are credible and reliable. This is becoming increasingly important as news consumption has increased exponentially while people try to make sense of this rapidly evolving crisis.

A study conducted by Tel Aviv has revealed that fake news is costing the global economy 78 billion each year [5]. A 'New York Times' interview in 2017 analyzing political, economic and social fake news, noted that money is the primary motivation behind fake news creation. The creators of fake news believed that no one was physically hurt, and so there was no guilt involved [6]. However, covid-19 pandemic has shown that, without being physically hurt, people go under mental stress for the fake news seen for the covid-19. A survey that ran on Australians found that most of the covid-19 related fake news (88%) appeared on social media than on traditional media [7]. While news about the coronavirus provided an important topic of conversation (53%), people feel more anxious (52%). Overall, covid-19 has shown us that fake news can cause social unrest and make us mentally more vulnerable. As a result, detecting fake news is an increasing study area.

Efforts have been made in the literature to detect fake news. Various mathematical, machine learning, and expert system based detection schemes, or their combinations are proposed. A corpus of genuine and fake news items is needed to train a machine learning model. Mathematical models need deep domain knowledge to define mathematical equations capturing the domain characteristics which may change over time and finding accurate mathematical relations is difficult. Similarly, expert system based schemes need considerable time to develop. As new fake news are being generated almost everyday and their nature also changes, an already trained machine learning based fake news detection system may not perform at an acceptable level with new data and their performance needs to be re-evaluated. Because in real life, we are constantly facing unexpected news (such as covid-19 and the Russia-Ukraine war). There have been multiple examples in the current fake news detection methods where both supervised and unsupervised learning algorithms are used to classify text-based fake news detection. However, most of the literature focuses on specific datasets or domains, most prominently the domain of politics. Therefore, the model trained on a particular type of article's domain (i.e., with dataset(s) extracted within that domain) does not achieve optimal results when exposed to articles from other domains since articles from different domains have a unique textual and vocabulary structure. For example, technology based fake news targets to promote the brand. On the other hand, celebrity based fake news discusses celebrities' personal life. The most commonly used words in technology news are motion, launch, price, while those in celebrity news are divorce, attorney, and affair. So, developing an identification system that works well across all domains is an important topic that should be investigated. In this paper, we use a machine learning feature merging strategy to alleviate the cross-domain fake news detection problem. The following are our significant contributions:

- Cross-domain performance analysis with traditional feature extraction methods. Note that there is no comprehensive study reported on this topic in the literature.
- Improvement of the cross-domain method by merging existing features.
- Our proposed method outperforms the existing system by 13% in overall accuracy.

## 2 Literature Review

Various methods have already been proposed in the literature with the text portion of news because most of the information we obtain is on textual content. It is the primary news component, and additional metadata (title, image) appears as text enhancement data. The title is another vital piece of information that attracts people to read the news immediately. This section discusses some recent methods for FND based on text and title. FND based on text is the classification problem of natural language processing (NLP). In NLP, one of the most used feature extraction approaches is the term frequency-inverse document frequency (TF-IDF). Apart from the conventional TF-IDF, various TF-IDF variations are used where merging TF-IDF with other features showed performance improvement. Using this idea, Kumar et al. [8] proposed the TF-IDF advancement method where convolution neural network (CNN) based rumour detection is discussed. It generated a feature set using a word embedding technique called embeddings from language models (elmo), and another feature set was created by TF-IDF. Then the combined feature sets are reduced by information gain-ant colony optimization (IG-ACO) because TF-IDF and word embedding both create a huge feature set. One of the limitations of this approach is that it treats all features equally. Thus no weight is given to a particular feature depending on its importance. Another similar approach is proposed by Verma et al. [9], where an extensive dataset has been proposed by combining 4 different datasets to reduce the individual bias of the dataset. However, it does not reduce the feature set and does not give more weight to the essential features. Apart from TF-IDF and word or sentence embedding method, Balwant et al. [10] proposed a structure-based method. Here, parts of speech (POS) tags are extracted through bidirectional long short-term memory (LSTM), and speaker profile information is calculated through a convolutional neural network (CNN). The results show that the resulting hybrid architecture significantly improves the performance of the detection process for fake news. Apart from some limitations, the performance of the existing fake news detection methods is overall good. However, because of their inability to perform in cross-domain scenarios, these traditional models fail to improve real-world applications' accuracy. This failure is (1) domain-specific vocabulary and (2) domain-specific distribution of words. The current effort has been focused on cross-domain analysis for FND because the primary goal is to predict the fakeness of unseen news. FND in cross-domain is more complex than in-domain, and the major challenges are: (1) we can not predict the domain of future fake news (e.g., covid-19 came out of nowhere), (2) scarcity of training data: manual labelling of unseen news is difficult, and (3) difficult to identify domain-independent and domain-dependent features. Because of these challenges, there are only limited works that focused on cross-domain. Perez et al. [11] were one of the first to analyse FND in the cross-domain approach. In this article, the authors showed that their proposed method is good enough for the in-domain but fails for the cross-domain analysis. Their method achieved an accuracy of 78% for the in-domain, but for the cross-domain, the accuracy dropped to 65%. They also introduced two datasets: FakeNewsAMT and Celebrity. After that, Maria et al. [12] performed a broad comparative analysis where four datasets were considered for testing cross-domain for one training set, and other datasets separately used the testing set. In this article, linguistic features with an automatic readability index have been proposed. That study also shows poor performance for cross-domain where performance dropped from 90% to 20% than in-domain. So, lexical tokenize can not help so much for cross-domain even though it is better for in-domain. In this case, researchers try to go embedding methods for cross-domain analysis.

In 2020, Gautam et al. [13] proposed another method for the cross-domain analysis that used spinbot (for paraphrasing), grammarly (for grammar checking), and glove (for word-embedding) tools. But for fakenewsAMT, the accuracy is 95%, but when training with celebrity dataset, the accuracy of fakenewsAMT drops to 70%. Tanik et al. [14] also proposed elmo-based word embedding approach that demonstrated an in-domain accuracy of 83% compared to a cross-domain accuracy of 54%. It shows that the accuracy drops from previous methods [11,13].

Apart from the word embedding-based feature, Goel et al. [15] have tried RoBERTa, which is a word embedding-based transformer model that showed superior performance in in-domain detection (99% for FakeNewsAMT dataset) but suffered from significant performance degradation for cross-domain (59% when tested on Celebrity dataset). Thus, various studies in the literature showed a drastic reduction in performance for cross-domain analysis, and a solution of domain-wise training and domain-wise testing is also meaningless because we often encounter unseen domain(s). In 2021, by considering the drawbacks of cross-domain, Silva et al. [16] have proposed a potential solution using a multi-domain dataset by combining PolitiFact; (2) GossipCop; and (3) CoAID datasets. Although a multi-modal dataset is important, the domain is changed within a day (like in Covid-19, Russia-Ukraine war). So, a multi-domain dataset can not be a viable solution in real life for cross-domain. As can be seen from the literature, enhancing cross-domain performance is extremely challenging, and there exist several opportunities for improvement, which is the motivation to perform this research.

# 3 Theoretical Study

In our experiment, we have used some lexical and semantic features. For feature extraction, we used TF-IDF, count vectorizer (CV), hashing or hash vectorizer (HV), word embedding (WE), sentence embedding (SE) methods, and structure methods, and for classification, we used linear regression (LR), random forest (RF), and multilayer perceptron (MLP). These are explained below:

## 3.1 Feature Selection Methods

Lexical tokenization means dividing a document by converting into words and grouping similar words without considering any meaning. Some popular lexical tokenize methods are TF-IDF, CV and HV. The main limitations of these lexical tokenize are: (1) extensive features, and sparse feature sets, and (2) do not preserve the semantic meaning. In our experiment, we employ these lexical methods with full features set to the whole vocabulary of the training set, and we choose to stop words as the English language. On the other hand, WE and SE are more contextual-based feature selection methods. One of the main limitations of word embedding is that words with multiple meanings are grouped into a single representation (a single vector in the semantic space). In other words, polysemy (words that have the exact spelling but different meanings depending on context) is not handled correctly. However, recent word embedding methods: bidirectional encoder representations from transformers (BERT), and embeddings from language models (elmo) can handle these two limitations to some extent. However, the problem is that word meanings may change over time [17]. Pre-trained model-based word embedding can not handle this changing behavior of words. Also, the way of handling out of vocabulary (OOV) words is not so efficient, and instead, sometimes it creates irrelevant feature vectors that lead to poor accuracy. Since sentence embedding preserves the meaning of a sentence by considering the relationship between words within sentences, it is likely to perform better than word embedding. Parts of speech (POS) tagging focuses on the structure of sentences to extract features.

## 3.2 Machine Learning Techniques

We experimented with three well-known classifiers: logistic regression (LR), random forest (RF), and multilayer perceptron (MLP), to sure that the performance of our proposed method is consistent with all classifiers. LR is a statistical model that uses a logistic function to model a binary dependent variable, although many more complex extensions exist. RF is an ensemble learning method for classification, regression, and other tasks that operate by constructing many decision trees at training time. MLP is a class of feed-forward artificial neural networks (ANN). MLP consists of at least three layers of nodes: an input layer, a hidden layer, and an output layer.

## 4 Proposed Method

Figure 1 describes our system model where an extensive feature set is created by merging: (1) features are created using pos tags, and (2) features are created from the text by calculating word-level features.

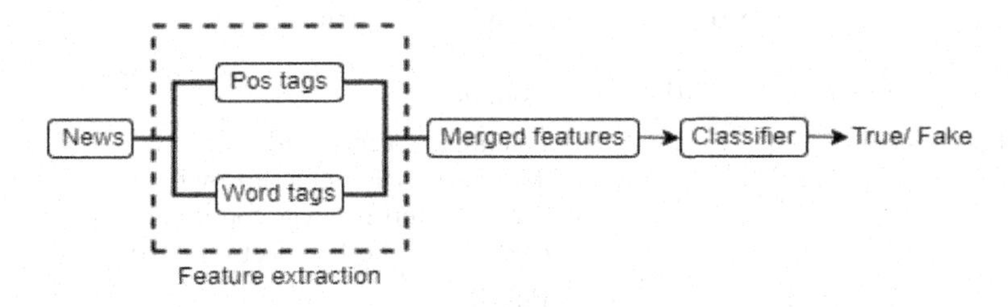


**Fig. 1.** Proposed system model.

Here, detailed pos tags are used for creating a feature set. Detailed pos tag means we not only divide a sentence from the basic pos tags such as noun, pronoun, but also a singular noun, progressive pronoun, which gives a detailed structure to a sentence.

**Table 1.** Feature set for the proposed method

| | |
|---|---|
| Word tags | Word count, char count, average word count, hashtags count, link count, number of length, user mention count |
| Pos tags | CC-Coordinating conjunction, CD-Cardinal number, DT-Determiner, EX-Existential there, FW-Foreign word, IN-Preposition, JJ-Simple adjective, JJR-Comparative Adjective, JJS- Superlative adjective, MD-Modal, NN-Singular Noun, NNP-Noun proper singular, NNPS-Noun proper plural, NNS-Noun plural, PDT-Pre determiner, POS-Possessive ending, PRP-Personal pronoun, PRP$-Possessive pronoun, RB-Adverb, RBR-Adverb comparative, RBS-Adverb superlative, RP-Particle, SYM-Symbol, TO-to, UH-Interjection, VB-Verb base form, VBD-Verb past form, VBG-Verb present or gerund particle, VBN-Verb past participle, VBP-Verb 3rd person singular, LS-List Marker, VBZ-Verb 3rd person singular, WDT, WP-Wh determiner, WP$-possessive wh pronoun, WRB-Wh adverb and other symbols |

Besides pos tags: word count, stop word count, number of mentions, etc. are also used in this article as word-based features or word tags. Pos tagging suffers from ambiguity, and word tags are used here to reduce the effect of ambiguity. Table 1 lists the feature set.

### 4.1 Fake News Datasets

In this experiment, we have worked on 4 types of datasets on different domains, namely Welfake, Isot, Pheme, Covid datasets and their brief descriptions are presented in Table 2.

**Table 2.** Datasets for the Experiment

| Dataset | Number | Title | Text | Comment |
|---|---|---|---|---|
| Welfake | Fake: 37,106<br>Real: 35,028 | yes | yes | Domains: society and politics.<br>The dataset is made up by combining four datasets and it covers political news mostly come from the US election: 2016 |
| Isot | Fake: 23,481<br>Real: 21,417 | yes | yes | Politics |
| Pheme | Fake: 71,782<br>Real: 31,430 | no | yes | Society and politics |
| Covid | Fake: 1058<br>Real: 2061 | yes | yes | Covid-19 |

Among them, Welfake and Isot are balanced datasets, but the other two are unbalanced datasets. We have taken one dataset as training and the others as test dataset(s) to measure cross-domain performance.

### 4.2 Evaluation Metrices

Evaluation metrics are used to measure the quality of the statistical or machine learning model. We selected many different types of evaluation metrics such as accuracy, sensitivity, and g-mean available to test a model (shown by Table 3).

**Table 3.** Equation for the evaluation metrices [18]

| Performance metrics | Formula |
|---|---|
| Accuracy | $\frac{(TP+TN)}{(TP+FP+TN+FN)}$ |
| Positive predictive value/ Precision | $\frac{TP}{(TP+FP)}$ |
| Negative predictive value (NPV) | $\frac{TN}{(TN+FN)}$ |
| True positive rate/Sensitivity/Recall | $\frac{TP}{(FN+TP)}$ |
| True Negative Rate/Specificity | $\frac{TN}{(FP+TN)}$ |
| F1 score | $2 * \frac{precision*recall}{precision+recall}$ |

where, TP= model correctly predicts the positive class, TN= model correctly predicts the negative class, FP= model incorrectly predicts the positive class, FN= model incorrectly predicts the negative class.

We have also considered the ROC curve, which calculates the sensitivity and specificity across a continuum of cutoffs. An area under the ROC curve (AUC) of 1 represents a perfect model, and an area of 0.5 or below 0.5 represents a worthless model.

# 5 Performance Evaluation and Analysis of Result

We have analyzed our model with respect to two aspects. Firstly, we analyze the cross-domain performance of the traditional feature extraction methods. Secondly, the performance of our proposed method for cross-domain will be discussed.

1. **Cross-domain Performance Analysis With Traditional Methods:** Table 4 illustrates the accuracy of cross-domain, where we take Welfake, Isot, and Pheme for training and covid for testing. Table 4 presents the accuracy of cross-domain, where we take Welfake, Isot, and Pheme dataset for training a model and Covid for testing the trained model. Results show that the cross-domain accuracy of three lexical tokenization methods: TF-IDF, CV, and HV is close to 0.5, demonstrating that these traditional tokenization methods do not perform well for the cross-domain fake news detection. These methods fail due to keyword mismatches from domain to domain, as seen in Fig. 2a, where the Covid dataset is compared with the other three datasets. The total number of unique keywords is compared to the matched keywords for each pair of datasets (Welfake-Covid, Isot-Covid, and Pheme-Covid). The number of matched keywords is minimal in comparison to the total keywords. Because the pheme dataset has fewer keywords matched with covid than the others, Pheme-Covid's accuracy is lower than Welfake-Covid

**Table 4.** Compare the accuracy for the existing feature extraction methods vs proposed method

| Training Dataset | Welfake | | | Isot | | | Pheme | | |
|---|---|---|---|---|---|---|---|---|---|
| Testing Dataset | Covid | | | | | | | | |
| Features | Classifier | | | | | | | | |
| | LR | RF | MLP | LR | RF | MLP | LR | RF | MLP |
| TF-IDF | 0.468 | 0.603 | 0.358 | 0.581 | 0.369 | 0.682 | 0.429 | 0.359 | 0.374 |
| Count Vectorizer | 0.544 | 0.324 | 0.471 | 0.434 | 0.637 | 0.626 | 0.531 | 0.661 | 0.406 |
| Hashing | 0.522 | 0.612 | 0.351 | 0.477 | 0.352 | 0.676 | 0.586 | 0.384 | 0.416 |
| Word Embedding | 0.471 | 0.612 | 0.366 | 0.601 | 0.401 | 0.597 | 0.368 | 0.363 | 0.342 |
| Sentence Embedding | 0.54 | 0.538 | 0.643 | 0.526 | 0.451 | 0.585 | 0.406 | 0.376 | 0.431 |
| Word Based Feature | 0.460 | 0.416 | 0.414 | 0.679 | 0.660 | 0.660 | 0.342 | 0.356 | 0.360 |
| Pos Tag | 0.526 | 0.553 | 0.545 | 0.546 | 0.539 | 0.554 | 0.494 | 0.466 | 0.432 |
| Proposed Method (word + pos tags) | 0.656 | 0.664 | 0.632 | 0.612 | 0.623 | 0.634 | 0.674 | 0.643 | 0.632 |

or Isot-Covid. In between Welfake-Covid and Isot-Covid, the more unique keywords are found on the Welfake-Covid pair. So, after matching approximately exact keywords, the performance of tokenization is slightly improved in the Isot-Covid than Welfake-Covid. But overall, their performances are not good. Embedding approaches improve the performance of lexical tokenization methods because they preserve the semantic meaning of a word or sentence. However, their technique for preserving semantic meaning is inadequate for unseen words. The ROC curve's performance is also not acceptable for traditional feature extraction techniques, as shown in Fig. 2b. The ROC-AUC $\leq$ 0.50 for all methods indicate the failure of traditional feature extractions for cross-domain.

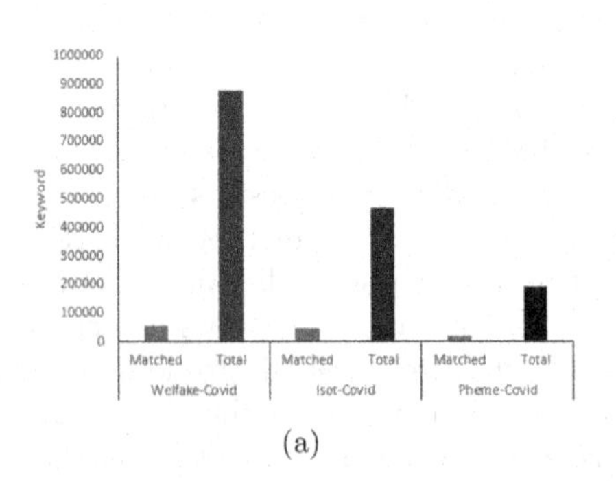


(a)

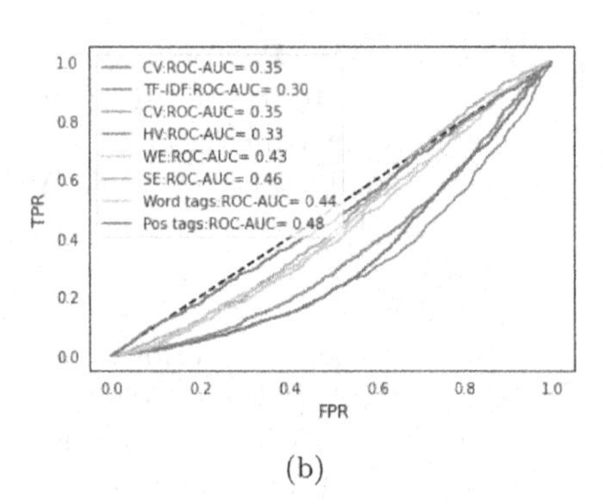


(b)

**Fig. 2.** (a) Number of matched keywords between datasets (b) ROC curve of traditional feature extraction methods

2. **Improvement of the Cross-domain with the Proposed Method:** Our proposed method performs better than the traditional methods, which is shown in the Table 4. The reason behind the success of our proposed method is that structure-based methods do not depend on the vocabulary of a domain. Our study selected two structure methods: pos tags and word tags.
   Pos tags suffer from pos ambiguity, and word tags help reduce it. So, finally, after merging, the result in terms of accuracy is improved. The Venn diagram (Fig. 3a) shows the correctly identified news from the three methods: pos tags, word tags, and the proposed method. Since our proposed method is better than these two methods, it correctly identifies more news (510-fake and real) that can not be identified by word or pos tags. The individual limitations of the two methods (pos tags and word tags) are alleviated when we merge these two features. As shown in Fig. 3b, the overall performance of our method is also improved in most of the other metrics. Our method's specificity value is lower than others, but its performance is superior in precision, recall, F1, and NPV. Notably, recall and F1 scores in our method are 0.9 and 0.75, respectively, which are much higher than other methods.

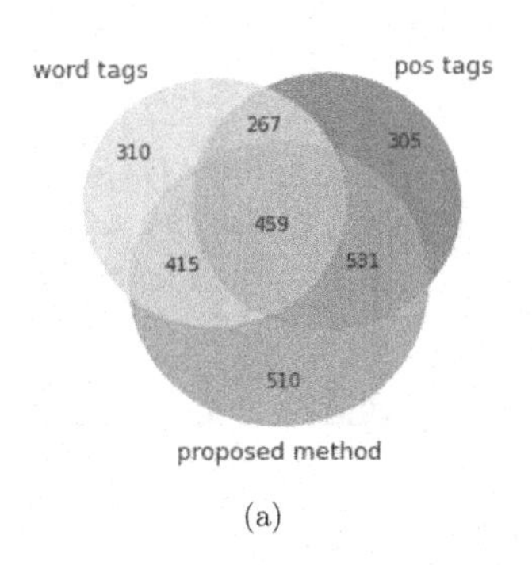


(a)

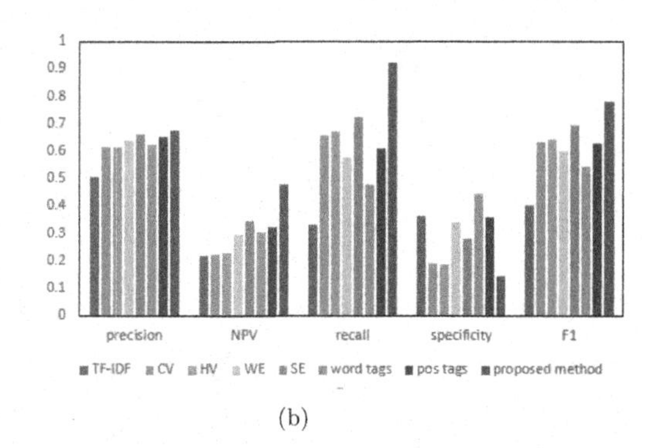


(b)

**Fig. 3.** (a) Correctly identified news by word tags, pos tags and proposed method (b) Performance metrices of various feature extraction methods including proposed method

**Table 5.** Comparison between the proposed method with the existing literature in terms of accuracy.

| Methods | Datasets | | |
|---|---|---|---|
| | Training | Celebrity | FakenewsAMT |
| | Testing | FakenewsAMT | Celebrity |
| Perez et al. [11] | | 0.64 | 0.50 |
| Gautam et al. [13] | | 0.70 | 0.56 |
| Goel et al. [15] | | 0.70 | 0.59 |
| Proposed method | | 0.79 | **0.75** |

We also performed additional experiments to compare our method with similar methods in recent literature [11,13,15] to tackle the cross-domain fake new detection problem. We take two other datasets, FakenewsAMT, and Celebrity, from two different domains to compare the existing methods. The method proposed by Goel et al. [15] shows better performance than the other two existing methods. Existing works are based on tokenization [11], word embedding based feature [13] and word embedding based transformer [15]. The existing works did not produce better results since tokenization, and word embedding-based models alone are not good enough to tackle cross-domain difficulties. On the other hand, our proposed method based on merging two structure methods improves 13%–23% (when Celebrity is used as training) and 27%–50% (when FakenewsAMT is used as training) on the accuracy, as shown in Table 5. These results demonstrate the superiority of our proposed method over others. ROC-AUC curve of our proposed method also supports this improvement, which is shown by Fig. 4a. Here ROC-AUC of our proposed method is improved **48%** with respect to [15]. The performance of other metrices is also better compared with the best literature [15]. For example, precision, NPV, specificity and F1 are better, which shows the efficiency of our proposed method.

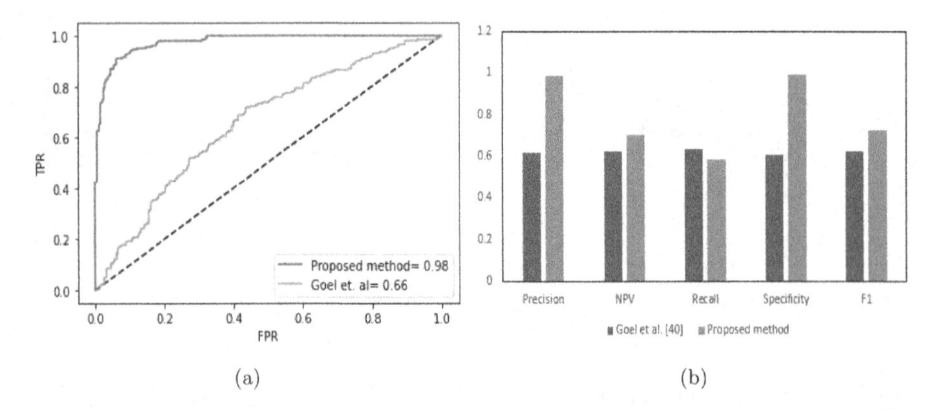


**Fig. 4.** (a) Comparison of ROC-curve between proposed method with existing literature (b) Comparison of performance metrices between proposed method with existing literature

# 6 Conclusion

In this article, the impact of traditional feature extraction techniques on detecting fake news has been studied. We conclude that traditional features do not provide an acceptable level of accuracy for cross-domain fake news detection. We propose a new method based on the merging of pos tags and word-level features in the title and text body of the news. Our method demonstrated improvement over other methods based on the existing feature extraction techniques. A comparison with recently proposed methods specifically to deal with cross-domain fake news detection further substantiate its superior performance in terms of all widely accepted performance metrics (e.g., accuracy 79% vs 70%). This work will help the broader research community in this less explored area of cross-domain fake news detection. Future work will focus on further performance improvement using innovative machine learning architecture suitable for this problem.

# References

1. Altay, S., Hacquin, A.S., Mercier, H.: Why do so few people share fake news? it hurts their reputation. New Media Soc. 24(6), 1461444820969893 (2020)
2. Parsons, D.D.: The impact of fake news on company value: evidence from tesla and galena biopharma (2020)
3. Braun, J.A., Eklund, J.L.: Fake news, real money: Ad tech platforms, profit-driven hoaxes, and the business of Journalism. Digit. J. **7**(1), 1–21 (2019)
4. A brief history of fake news. https://www.cits.ucsb.edu/fake-news/brief-history
5. Brown, E.: Online fake news is costing us 78 billion globally each year (2019). https://www.zdnet.com/article/online-fake-news-costing-us-78-billion-globally-each-year/
6. Kshetri, N., Voas, J.: The economics of "fake news". IT Profess. **19**(6), 8–12 (2017)
7. Park, S., Fisher, C., LEE, J., Mcguinness, K.: COVID-19: Australian news and misinformation. News Media Research Centre, University of Canberra (2020)

8. Kumar, A., Bhatia, M., Sangwan, S.R.: Rumour detection using deep learning and filter-wrapper feature selection in benchmark twitter dataset. Multimedia Tools Appl. **80**, 1–18 (2021)
9. Verma, P.K., Agrawal, P., Amorim, I., Prodan, R.: Welfake: word embedding over linguistic features for fake news detection. IEEE Trans. Comput. Soc. Syst. **8**(4), 881–893 (2021)
10. Balwant, M.K.: Bidirectional LSTM based on POS tags and CNN architecture for fake news detection. In: 2019 10th International Conference on Computing, Communication and Networking Technologies (ICCCNT), pp. 1–6. IEEE (2019)
11. Pérez-Rosas, V., Kleinberg, B., Lefevre, A., Mihalcea, R.: Automatic detection of fake news. arXiv preprint arXiv:1708.07104 (2017)
12. Janicka, M., Pszona, M., Wawer, A.: Cross-domain failures of fake news detection. Computación y Sistemas **23**(3), 1089–1097 (2019)
13. Gautam, A., Jerripothula, K.R.: SGG: spinbot, grammarly and glove based fake news detection. In: 2020 IEEE Sixth International Conference on Multimedia Big Data (BigMM), pp. 174–182. IEEE (2020)
14. Saikh, T., De, A., Ekbal, A., Bhattacharyya, P.: A deep learning approach for automatic detection of fake news. arXiv preprint arXiv:2005.04938 (2020)
15. Goel, P., Singhal, S., Aggarwal, S., Jain, M.: Multi domain fake news analysis using transfer learning. In: 2021 5th International Conference on Computing Methodologies and Communication (ICCMC), pp. 1230–1237. IEEE (2021)
16. Silva, A., Luo, L., Karunasekera, S., Leckie, C.: Embracing domain differences in fake news: cross-domain fake news detection using multi-modal data. In: Proceedings of the AAAI Conference on Artificial Intelligence, vol. 35, pp. 557–565 (2021)
17. Etymology- how words change over time. https://www.virtuescience.com/etymology-how-words-change-over-time.html
18. Ghori, K.M., Imran, M., Nawaz, A., Abbasi, R.A., Ullah, A., Szathmary, L.: Performance analysis of machine learning classifiers for non-technical loss detection. J. Ambient Intell. Human. Comput. **8**, 16033–16048 (2020)

# Portrait Matting Network with Essential Feature Mining and Fusion

Hang Jiang, Song Wu, Dehong He, and Guoqiang Xiao(✉)

College of Computer and Information Science, Southwest University, ChongQing, China
{jianghang,swu20201514}@email.swu.edu.cn, {songwuswu,gqxiao}@swu.edu.cn

**Abstract.** We propose an end-to-end portrait matting algorithm that emphasizes the mining and fusion of critical features to achieve higher accuracy. Previous best-performing portrait matting algorithms still have difficulties in handling difficult parts of portraits, such as hair, and often rely on increasing the network capacity to achieve better accuracy. In this paper, we argue that the effective use of critical features can help the network recover details while adding little to the network capacity and computational effort. Our proposed approach has the following two advantages: 1) it emphasizes the multiplexing of the lowest and highest level feature maps, which allows the model to recover details more effectively while locating the portrait subject more accurately. 2) we introduce a critical-feature-extraction-block to further mine this part of the feature maps, making the model parameters more effective. We evaluate the proposed method on a large portrait dataset P3M-10K, and the experimental results show that our method outperforms other portrait matting models.

**Keywords:** Portrait matting · Feature fusion · Attention mechanisms

## 1 Introduction

With the rapid development of deep learning and computer vision [14,16,23], portrait image matting has shown a wide range of applications in the real world. End-to-end portrait matting is the problem of obtaining a high-quality alpha value from a given image, i.e., predicting the transparency of each pixel point in the foreground of a given portrait image. Portrait matting is a widely used and challenging application in film and TV post-production as well as photo compositing. Mathematically, given a portrait image, this image can be represented as a linear combination of foreground and background as in the following Eq. 1:

$$I_p = \alpha_p F_p + (1 - \alpha_p) B_p, \qquad \alpha_p \in [0, 1] \tag{1}$$

where $F_p$ and $B_p$ denote the RGB values of foreground and the background at pixel $p$, respectively, and $\alpha_p$ is the desired alpha matte value. As shown in the Eq. 1,


M. Tanveer et al. (Eds.): ICONIP 2022, CCIS 1793, pp. 576–587, 2023.
https://doi.org/10.1007/978-981-99-1645-0_48

portrait matting is a difficult problem because it requires solving for seven values ($\alpha_p$, $F_p$, and $B_p$) but only three known quantities $I_p$. To simplify the problem, the previous method [6,12,15,22] introduced Trimap to simplify the process. Trimap marks the pure foreground and pure background regions in $I_p$, so only the alpha value of the uncertain region needs to be solved in the matting tasks.

Portrait matting has some tricky problems. In practice, the human pose may be quite complex, such as irregular body postures and strange gestures, and more likely to have other objects attached to the body, such as carrying a guitar bag, which is a considerable challenge for the model to locate the portrait subject. Second, the color distribution of foreground and background in portrait images may be extremely similar, which affects the accuracy of the model in predicting the alpha value of the uncertain part between foreground and background. If the portrait is accompanied by some tiny structures such as tousled hair, it is even more difficult for the model to accurately predict the alpha value.

The first problem in portrait matting is related to high-level semantic features, while the second problem is related to low-level features that contain detailed information. In recent years, deep learning-based methods have gained considerable momentum in the field of portrait matting. These methods exploit a combination of texture color information as well as semantic information of the original image. However, they usually rely on stacking the network capacity to obtain better model prediction results. Therefore, how to improve the model performance without significantly increasing the network capacity remains a problem worth optimizing.

Most current portrait matting efforts [2,7,10] employ an encoder-decoder architecture. Specifically, because the two challenges facing portrait matting are distinct in terms of the required feature maps, and neural networks are better at solving a single problem, the portrait matting task is decomposed into two sub-tasks: segmentation of portraits and uncertain region prediction. Segmenting portraits is usually considered as a task that relies on high-level semantic features, while uncertain region prediction is more considered as a low or mid-level computer vision task. Traditional Convolutional Neural Networks(CNN)-based approaches process the feature maps generated at the beginning of the network with the same weights, which makes the invalid features in them occupy the network capacity and affect the accuracy. Also, in the encoder-decoder structure, the mere interaction of features under the same level does not allow the model to notice the most useful features and even generates redundancy.

From the above points, we propose Portrait Matting Network with Essential Feature Mining and Fusion. Our contributions can be summarized as follows.

1. We propose a new view of portrait matting, arguing that current portrait matting models have a large redundancy in capacity and one way to improve model accuracy is to make the model parameters more efficient.
2. Based on 1, we introduce CFB and DFB to interactively fuse with critical features to improve the performance of the model with little increase in the number of parameters.
3. Our method reaches State-Of-the-Art on large portrait dataset P3M.

## 2 Related Work

### 2.1 Image Matting

Traditional image matting algorithms rely heavily on low-level features, such as color information. Sampling-based [3,4,20,21] or propagation-based [1,9] methods tend to have bad matting results in the case of more complex backgrounds. With the tremendous progress in deep learning, many Convolutional Neural Network (CNN)-based methods have been proposed, and they have significantly improved the matting results. Most existing matting methods use a predefined Trimap as an auxiliary input, which is a mask containing three regions: pure foreground ($\alpha = 1$), pure background ($\alpha = 0$), and unknown region ($\alpha = 0.5$). In this way, the solution space of the matting algorithm is reduced from the whole image to the unknown region.

Xu *et al.* [22] first proposed an encoder-decoder structure to estimate alpha mattes. The refinement stage of their work produces obvious bounds. They also released the Adobe Deep image Mattring dataset, a large dataset for deep learning-based matting. Hou *et al.* [6] used two encoder-decoder structures to extract local and global contextual information and perform matting. Lu *et al.* [15] argue that the parameter indexes in the un-pooling layers in the upsampling decoder can influence the matting performance, and a learnable plug-in structure is introduced, namely IndexNet, to improve the pooling and upsampling process.

While these Trimap-based matting algorithms perform well in terms of matting results, building the Trimap itself requires user-involved interaction and is not conducive to automated deployment of the model. Therefore, some approaches (including our model) try to avoid using Trimap as input, as described below.

### 2.2 End-to-End Portrait Matting

Image matting is very difficult when Trimaps are not available because foreground positioning is required before accurate alpha matting value can be predicted. Currently, Trimap-free methods always focus on specific types of foreground objects, such as portraits. Nevertheless, inputting RGB images into a single neural network to generate alpha masks is still difficult. Sengupta *et al.* [19] proposed to input an additional background image instead of Trimap to improve the model accuracy. Other works have designed pipelines containing multiple models. For example, Shen *et al.* [2] assembled a Trimap generation network before the matting network. Zhang *et al.* [24] applied a fusion network to combine the predicted foreground and background. Liu *et al.* [13] connected three networks to exploit the roughly labeled data in matting. These methods are computationally expensive and thus cannot be conveniently deployed in interactive programs. In contrast, our proposed method has better performance and superior user experience in using only one RGB image as input and the network design is very lightweight.

# 3 Proposed Method

In this section, we will describe in detail our proposed network and its internal modules.

## 3.1 Overview

Neural networks are better at handling a single task than a huge model handling a complex task. Therefore, we divide the end-to-end portrait matting task into two subtasks: portrait segmentation and detail prediction. Then the two are fused to obtain the final alpha value. The general framework of our network structure is shown in Fig. 1. Specifically, we use an encoder for feature extraction, a portrait segmentation branch to predict a rough Trimap, and a detail prediction branch to generate a fine alpha mask for the uncertain part of the Trimap. Finally, the two are fed into the fusion function $F$ for a specific fusion operation to obtain the final alpha mask value.

To enhance the effectiveness of the critical feature map in the encoder stage, we introduce the Critical feature extraction block (CFEB) to process these features, which can make the feature map extracted by the encoder more effective. Meanwhile, to enhance the effectiveness of the portrait segmentation branch and detail prediction branch in the decoder stage and improve the overall accuracy of the model, we introduce the Context Fusion block (CFB) and Detail Fusion block (DFB), respectively. We will explain the implementations of each module in detail.

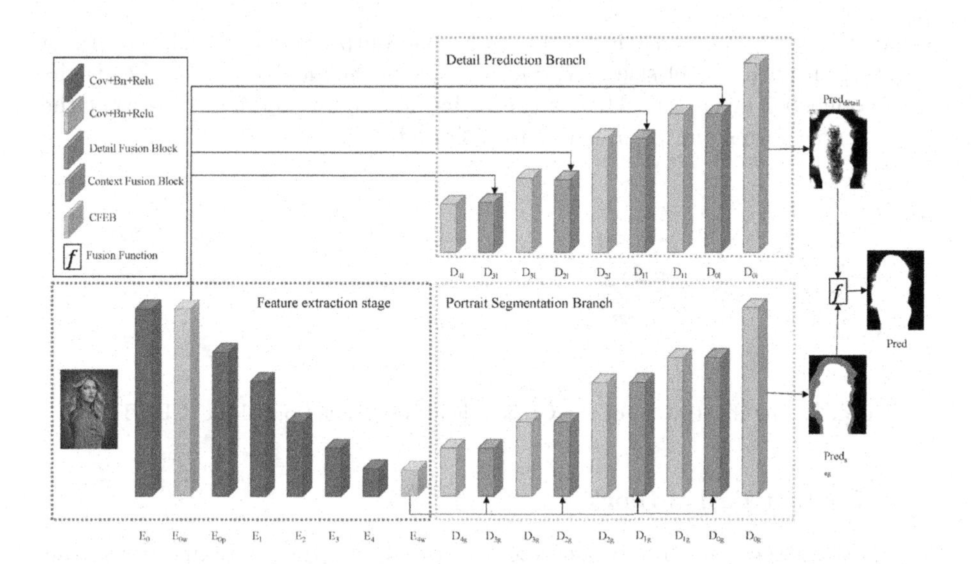


Fig. 1. Overview of our proposed portrait matting framework.

### 3.2 Critical Feature Extraction Block

Previous portrait matting methods either treat each layer of feature maps fairly [10] or do the attention mechanism only for high-level feature maps [7]. In this paper, we extend the approach of doing the attention mechanism for critical feature maps in [7]. As mentioned in Sect. 3.1, we split the matting task into two subtasks, where the portrait segmentation branch needs to utilize the high-level feature maps, and the detail prediction branch needs the assistance of the low-level feature maps. Therefore, we believe that deep mining of these two critical feature maps before information fusion can improve the accuracy of prediction for both subtasks.

Specifically, for our Critical Feature Extraction Block(CFEB), the input is the feature map $F_e$ from the encoder stage, and the output is the feature map $F_w$ with weights after the attention mechanism; the $F_e$ is fed directly into the $7\times7$ convolution for channel reduction and spatial attention extraction to obtain the spatial weighting feature map, and then multiplied by the $F_e$ directly after the sigmoid function to obtain $F_e'$. For $F_e'$, we further do global maxpooling operation by connecting a fully connected layer and multiplying it with $F_e'$ after also connecting the sigmoid function to obtain $F_w$. The internal structure of CFEB is shown in Fig. 2.

The $F_w$ obtained after $F_e$ via the CFEB can be formalized as:

$$F_e' = F_e * Sigmoid(Conv(F_e)) \tag{2}$$

$$F_w = F_e' * Sigmoid(Max(FC(F_e'))) \tag{3}$$

where $F_e$ is the input of CFEB and $F_w$ is the output after CFEB. $Conv$ is a convolution with kernel size $= 7$, $Sigmoid$ is the Sigmoid function, $FC$ is the fully connected layer, and $Max$ is the global maximum pooling. Note that the two Sigmoid forms and parameters are different.

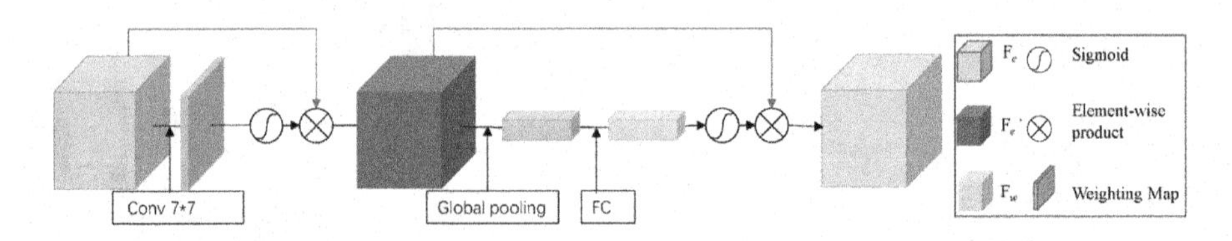


**Fig. 2.** The illustration of the Critical Feature Extraction Block (CFEB).

### 3.3 Context Fusion Block

The portrait segmentation branch requires predicting the approximate portrait location. Specifically, a Trimap with marked foreground, background and uncertain regions needs to be predicted. Therefore, high-level feature maps in the deep layer in the encoder stage may be useful because they contain rich contextual information. To exploit them to improve the portrait segmentation branch decoder, we introduce the contextual feature fusion module. As shown in Fig. 3.

For the context fusion module, we use the feature map $D_ig$ from the upper layer of the portrait segmentation branch and $E_4$ from the encoder stage as inputs. This is because $E_4$ contains rich contextual features. We think it makes sense to always use $E_4$ to guide the portrait segmentation branch. Specifically for the implementation, we first do an upsampling operation on $E_4$ with a scaling factor of $r$ to make it match the size of $D_ig$, and then $E_4'$ and $D_ig$ are fed into two $1 \times 1$ convolution control channel numbers, separately. Next, we concatenate the two sets of feature maps over the channels and feed a $3 \times 3$ convolutional layer, pick up the *Bn* layer, and *Relu* activation. Finally, we numerically sum $D_ig$ and $D_ig'$ using the idea of Resnet [5]. The context fusion module can be formalized as:

$$D_ig' = D_ig + CBR(Cat(C(D_ig), C(UP(E_4)))) \tag{4}$$

where $D_ig$ is the upper level input feature with the size of $H/r \times W/r \times C$, $UP$ is the maxupooling operation with scaling factor $r$, and $C$ is the $1 \times 1$ convolution operation. Note that the parameters of the two $C$ are different. $Cat$ is the concatenate operation on channels, $CBR$ is the sequential model of convolution, Bn, and Relu.

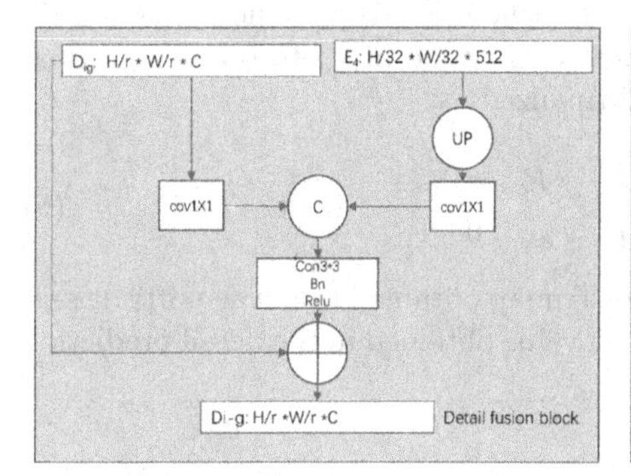


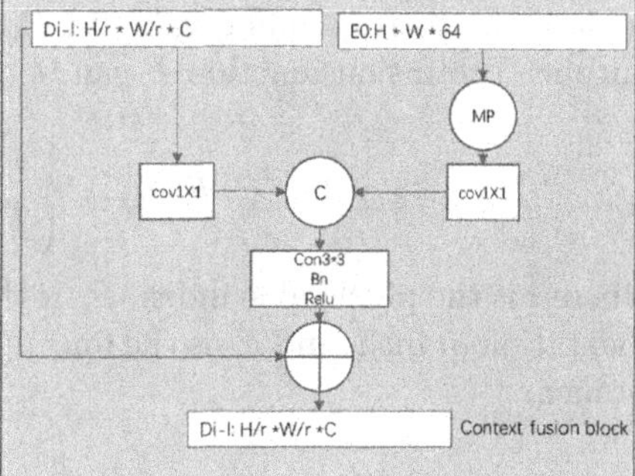


**Fig. 3.** The illustration of the Context Fusion Block (CFB) and Detail Fusion Block (DFB).

### 3.4 Detail Fusion Block

Similar to the Context Fusion Block, we believe that the shallow feature maps in the encoder stage can help the detail prediction branch to better predict details because these feature maps contain rich detail features. Therefore, we introduce a detail fusion module before upsampling each layer of the detail prediction branch in order to improve the model's ability to recover details. As shown in the Fig. 3, for the detail fusion module, we use the upper layer feature maps $D_il$

of the detail prediction branch and $E_0$ in the encoder stage as inputs. This is because $E_0$ contains rich detailed information. Specifically, we first do a max-pooling operation with a scaling factor of $r$ on $E_0$ to match the size of $D_i l$, and then we feed $E_0'$ and $D_i l$ into two 1 $\times$ 1 convolutional to control channel numbers,separately. Next, similar to the Context Fusion Block, we concatenate the two sets of feature maps over channels and then fed into the sequential model of convolution, $Bn$, and $Relu$. The same idea of Resnet is used to add up the $D_i l$ and $D_i g$ values. The detail fusion module can be formalized as:

$$D_i l' = D_i + CBR(Cat(C(D_i l), C(MP(E_0)))) \tag{5}$$

where $D_i l$ is the input feature of the previous layer with size of $H/r \times W/r \times C$, $MP$ is the maxpooling operation with scaling factor $r$, $C$ is the $1 \times 1$ operation, the parameters of the two $C$ are different, $Cat$ is the concatenate operation, and $CBR$ is the sequential model of convolution, Bn and Relu.

### 3.5 Fusion Function

The purpose of end-to-end portrait matting is to directly predict the portrait alpha mask value, so the $P_s(Pred_{seg})$ predicted by the portrait segmentation branch and the $P_d(Pred_{detail})$ predicted by the detail prediction branch need to be fused to obtain the final alpha mask $P$. Specifically, for the fusion function $F$, we trust the prediction obtained by $P_s$. The pure foreground and pure background region predicted by $P_s$ are left untouched. For the unknown region, we replace the corresponding value in $P_s$ with the value in $P_d$ to obtain the final output. The fusion function $F$ can be formalized as:

$$F(x_i) = \begin{cases} P_{ei} & P_{ei} = 0.5 \\ P_{di} & P_{ei} = 0 or 1 \end{cases} \tag{6}$$

where $i$ is the pixel value index. $P_e$ is the Trimap obtained from the portrait segmentation branch, and $P_d$ is the fine alpha value obtained in the detail prediction branch.

### 3.6 Loss Function

We design the loss function following similar ideas as in the previous work. Our loss function is made up of three parts, $L_s$ for supervising the portrait segmentation branch, $L_e$ for supervising the detail prediction branch, and $L_\alpha$ for supervising the whole picture.

For the $P_s$ of the portrait segmentation branch prediction, it contains only three types of pixel values: 0, 1, or 0.5. We use cross-entropy loss for supervision. $L_s$ is defined as:

$$L_s = CE(P_s, G_s) \tag{7}$$

where $CE$ is the cross-entropy loss function, $P_s$ is the Trimap predicted by the portrait segmentation branch, and $G_s$ is the Trimap provided in ground truth.

For the $P_d$ obtained from the detail prediction branch, it predicts the alpha value of each pixel in the uncertain part, so we use $L1$ loss for supervision. $L_d$ is defined as:

$$L_d = L_2(P_d, G_\alpha) \tag{8}$$

where $L_2$ is the $L_2$ loss, $P_d$ is the predicted detail alpha matte, and $G_\alpha$ is the ground truth. It is important to note that for $L_d$, we supervise only the uncertain part in $P_s$. Because it is determined that the pure foreground and pure background parts in $P_s$ do not need to be fused, it is meaningless to supervise them in the $P_d$ stage.

For the total model output after the fusion function $F$, we follow the practice in Deep image matting [22] and use alpha loss $L_\alpha$ for supervision. Alpha loss is the absolute difference between the grand truth in each pixel and the predicted alpha matte. $L_\alpha$ is formulated as:

$$L_\alpha = \frac{1}{N_U} \sum_{p \in U} \sqrt{(\alpha_p - \hat{\alpha_p})^2 + \epsilon^2} \tag{9}$$

where $U$ is the unknown region annotated in the Trimap, $N_U$ is the number of pixels inside region $U$. $\alpha_p$ and $\hat{\alpha_p}$ is the ground-truth and predicted alpha value of pixel $p$. $\epsilon$ is a small positive number to guarantee the full expression differentiable. In the experiment, $\epsilon$ is set to a constant $10^{-6}$.

Our total loss is defined as the weighted sum of the above three:

$$L = \lambda_s L_s + \lambda_d L_d + \lambda_\alpha L_\alpha \tag{10}$$

To make each loss valid and ensure that the three losses are in the same order of magnitude, we set $\lambda_s = 1$, $\lambda_e = 6$, and $\lambda_\alpha = 6$.

# 4 Experiments

## 4.1 Implementation Details

Our network is trained end-to-end on the P3M-10K [10] portrait dataset. The network was trained for 136 epochs and about 319,000 iterations. We used the Adam [8] optimizer and set its parameters $\beta 1 = 0.9$, $\beta 2 = 0.999$, and a fixed learning rate $lr = 0.00001$. For the training data, we first randomly cropped it to $512 \times 512$, $768 \times 768$, or $1024 \times 1024$, and then scaled it to $512 \times 512$. we used random horizontal flipping for further data augmentation. All training data is created on-the-fly. We use a single NVIDIA RTX2080 GPU for training, with a batch size of 4. It takes about 3 days to train the entire model completely.

**Table 1.** Quantitative comparisons to the baseline on P3M-10K dataset.

| Baseline | CFEB | CFB | DFB | SAD | MSE | MAD | COON | GRAD |
|---|---|---|---|---|---|---|---|---|
| ✓ | × | × | × | 14.54 | 0.0055 | 0.0084 | 14.00 | 13.87 |
| ✓ | ✓ | × | × | 13.13 | 0.0047 | 0.0076 | 12.61 | 14.15 |
| ✓ | ✓ | ✓ | × | 11.32 | 0.0037 | 0.0065 | 10.77 | 13.32 |
| ✓ | ✓ | × | ✓ | 12.67 | 0.0044 | 0.0073 | 12.14 | 14.96 |
| ✓ | ✓ | ✓ | ✓ | 10.06 | 0.0031 | 0.0058 | 9.51 | 12.05 |

### 4.2 Dataset and Evaluation Metrics

**Datasets.** We use the P3M-10k portrait dataset for training and testing the performance of our model. This portrait dataset contains 9421 portrait images for training and 1000 portrait images for testing. All of these images are annotated with high-quality alpha.

**Evaluation Metrics.** We followed previous work and used the Sum of Absolute Difference (SAD), Mean Square Error (MSE), Mean Absolute Different (MAD), Gradient Error (Grad), and Connectivity Error (Conn) [18] as metrics for the evaluation. For all end-to-end portrait matting algorithms, we compute them over the entire image.

### 4.3 Ablation Study

In this section, we report the results of ablation experiments to demonstrate the effectiveness of the modules introduced by our network. We use the metrics mentioned in the Sect. 4.2 to evaluate our model. The results of the ablation experiments are shown in Table 1. Compared to the baseline, the model using CFEB shows a performance improvement, which indicates that CFEB successfully extracts more effective feature maps. In addition, both the detail fusion module and the contextual information fusion module improve the accuracy of the model, respectively, confirming their effectiveness for both portrait segmentation and detail recovery, respectively.

**Table 2.** Testing results on the P3M-10k Dataset. The best results in each metric are emphasized in bold. For all metrics, smaller is better.

| Method | SAD | MSE | MAD | GRAD | CONN |
|---|---|---|---|---|---|
| LF [24] | 32.59 | 0.0131 | 0.0188 | 31.93 | 19.50 |
| HATT [17] | 30.53 | 0.0072 | 0.0176 | 19.88 | 27.42 |
| SHM [2] | 20.77 | 0.0093 | 0.0122 | 20.30 | 17.09 |
| MOD [7] | 28.15 | 0.010 | 0.0162 | 24.65 | 16.83 |
| GFM [11] | 15.50 | 0.0056 | 0.0091 | 14.82 | 18.03 |
| P3M [10] | 11.23 | 0.0035 | 0.0065 | 10.35 | 12.51 |
| Ours | **10.06** | **0.0031** | **0.0058** | **9.51** | **12.05** |

### 4.4 Experiment Results

We compared our model with other state-of-the-art portrait matting models, and the experimental results are shown in Table 2, from which it can be seen that our proposed model outperforms other end-to-end portrait matting models in all metrics.

The experimental visualization results are shown in Fig. 4. Our model significantly improves the problem of voids arising inside the portrait alpha matte, as a result of introducing a context fusion module to guide the model. In the case of portraits with tousled hair and a small amount of background in the foreground, our method also predicts more accurately than other models, as we also fuse detailed information to guide the model to recover them.

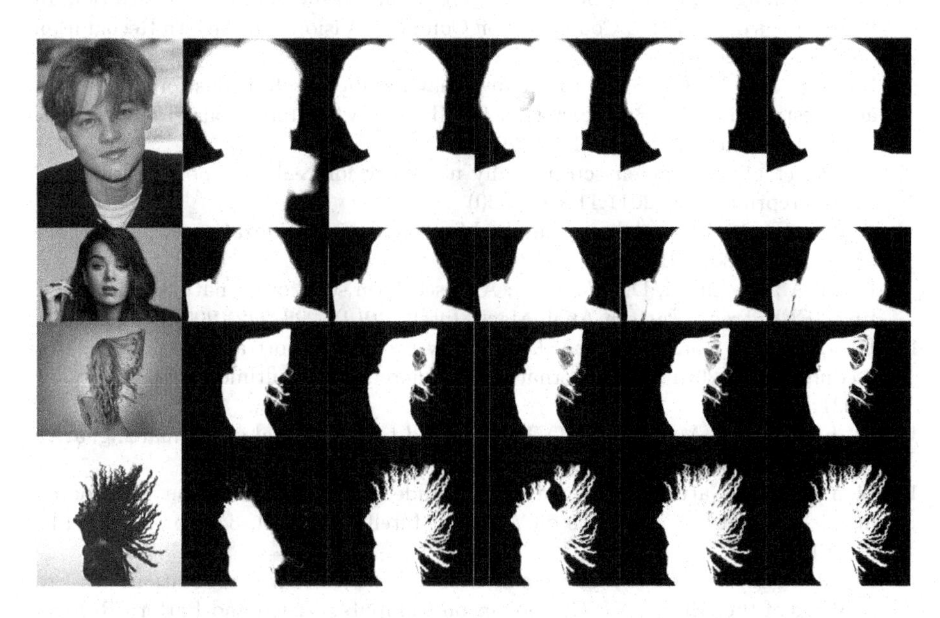

**Fig. 4.** The visual results on the P3M-10K dataset. For each row of images, from left to right: original image, MOD [7], GFM [11], P3M [10], ours, Ground-truth. Please zoom in for more details.

## 5 Conclusion

In this work, we propose a new end-to-end portrait matting architecture. We divide the portrait matting task into two subtasks and fuse the critical feature maps to guide the model, so the network can improve the model prediction without significantly increasing the network capacity. In addition, we introduce CEEB for further self-attention mining of critical features, which further strengthens the network's ability to recover details. The experimental results show that our proposed model outperforms other end-to-end portrait matting networks.

## References


1. Chen, Q., Li, D., Tang, C.K.: KNN matting. IEEE Trans. Pattern Anal. Mach. Intell. **35**(9), 2175–2188 (2013)
2. Chen, Q., Ge, T., Xu, Y., Zhang, Z., Yang, X., Gai, K.: Semantic human matting. In: Proceedings of the 26th ACM International Conference on Multimedia, pp. 618–626 (2018)
3. He, B., Wang, G., Shi, C., Yin, X., Liu, B., Lin, X.: Iterative transductive learning for alpha matting. In: 2013 IEEE International Conference on Image Processing, pp. 4282–4286. IEEE (2013)
4. He, K., Rhemann, C., Rother, C., Tang, X., Sun, J.: A global sampling method for alpha matting. In: CVPR 2011, pp. 2049–2056. IEEE (2011)
5. He, K., Zhang, X., Ren, S., Sun, J.: Deep residual learning for image recognition. In: Proceedings of the IEEE Conference on Computer Vision and Pattern Recognition, pp. 770–778 (2016)
6. Hou, Q., Liu, F.: Context-aware image matting for simultaneous foreground and alpha estimation. In: Proceedings of the IEEE/CVF International Conference on Computer Vision, pp. 4130–4139 (2019)
7. Ke, Z., et al.: Is a green screen really necessary for real-time portrait matting? arXiv preprint arXiv:2011.11961 (2020)
8. Kingma, D.P., Ba, J.: Adam: a method for stochastic optimization. arXiv preprint arXiv:1412.6980 (2014)
9. Levin, A., Lischinski, D., Weiss, Y.: A closed-form solution to natural image matting. IEEE Trans. Pattern Anal. Mach. Intell. **30**(2), 228–242 (2007)
10. Li, J., Ma, S., Zhang, J., Tao, D.: Privacy-preserving portrait matting. In: Proceedings of the 29th ACM International Conference on Multimedia, pp. 3501–3509 (2021)
11. Li, J., Zhang, J., Maybank, S.J., Tao, D.: End-to-end animal image matting. arXiv e-prints, pp. arXiv–2010 (2020)
12. Li, Y., Lu, H.: Natural image matting via guided contextual attention. In: Proceedings of the AAAI Conference on Artificial Intelligence, vol. 34, pp. 11450–11457 (2020)
13. Liu, J., et al.: Boosting semantic human matting with coarse annotations. In: Proceedings of the IEEE/CVF Conference on Computer Vision and Pattern Recognition, pp. 8563–8572 (2020)
14. Liu, S., Xiao, G., Xu, X., Wu, S.: Bi-directional normalization and color attention-guided generative adversarial network for image enhancement. In: ICASSP 2022–2022 IEEE International Conference on Acoustics, Speech and Signal Processing (ICASSP), pp. 2205–2209. IEEE (2022)
15. Lu, H., Dai, Y., Shen, C., Xu, S.: Indices matter: learning to index for deep image matting. In: Proceedings of the IEEE/CVF International Conference on Computer Vision, pp. 3266–3275 (2019)
16. Luo, P., Xiao, G., Gao, X., Wu, S.: LKD-Net: large kernel convolution network for single image dehazing. arXiv preprint arXiv:2209.01788 (2022)
17. Qiao, Y., et al.: Attention-guided hierarchical structure aggregation for image matting. In: Proceedings of the IEEE/CVF Conference on Computer Vision and Pattern Recognition, pp. 13676–13685 (2020)
18. Rhemann, C., Rother, C., Wang, J., Gelautz, M., Kohli, P., Rott, P.: A perceptually motivated online benchmark for image matting. In: 2009 IEEE Conference on Computer Vision and Pattern Recognition, pp. 1826–1833. IEEE (2009)

19. Sengupta, S., Jayaram, V., Curless, B., Seitz, S.M., Kemelmacher-Shlizerman, I.: Background matting: The world is your green screen. In: Proceedings of the IEEE/CVF Conference on Computer Vision and Pattern Recognition, pp. 2291–2300 (2020)
20. Wang, J., Cohen, M.F.: An iterative optimization approach for unified image segmentation and matting. In: Tenth IEEE International Conference on Computer Vision (ICCV2005) Volume 1, vol. 2, pp. 936–943. IEEE (2005)
21. Wang, J., Cohen, M.F.: Optimized color sampling for robust matting. In: 2007 IEEE Conference on Computer Vision and Pattern Recognition, pp. 1–8. IEEE (2007)
22. Xu, N., Price, B., Cohen, S., Huang, T.: Deep image matting. In: Proceedings of the IEEE Conference on Computer Vision and Pattern Recognition, pp. 2970–2979 (2017)
23. Xu, X., Liu, S., Zhang, N., Xiao, G., Wu, S.: Channel exchange and adversarial learning guided cross-modal person re-identification. Knowl.-Based Syst. **257**, 109883 (2022)
24. Zhang, Y., et al.: A late fusion CNN for digital matting. In: Proceedings of the IEEE/CVF Conference on Computer Vision and Pattern Recognition, pp. 7469–7478 (2019)

# Hybrid-Supervised Network for 3D Renal Tumor Segmentation in Abdominal CT

Bo Xue[1], Zhiqin Liu[1](✉), Qingfeng Wang[1], Qin Tang[1], Jun Huang[1], and Ying Zhou[2]

[1] School of Computer Science and Technology, Southwest University of Science and Technology, Mianyang, China
lzq@swust.edu.cn

[2] Radiology Department, Mianyang Central Hospital, Mianyang, China

**Abstract.** Renal cell carcinoma (RCC) is one of the top ten tumors threatening human health. Before renal tumor surgery, doctors need to understand the spatial structure and relative position of the kidney and tumor, and be familiar with the intraoperative environment in advance to improve the success rate. However, current tools for automatic contouring and localization of renal tumors are flawed. For example, the segmentation accuracy is not high and requires a lot of manual operations. This work proposes a 3D segmentation method for CT renal and tumor based on hybrid supervision. Hybrid supervision improves segmentation performance while using few labels. In the test on the public dataset KITS19 (Kidney Tumor Segmentation Challenge in 2019), the hybrid supervised method outperforms other segmentation methods, with an average Dice coefficient improvement of three percentage points under the condition of using the same amount of annotation, and the kidney and kidney tumor are well separated. Hybrid supervision provides a new intention for clinical renal tumor segmentation.

**Keywords:** Renal tumor · self-supervised · semi-supervised · Medical Segmentation

## 1 Introduction

According to the statistics [1] of the World Health Organization, kidney cancer is one of the ten most common cancers in human beings, and its morbidity and mortality [2] are constantly rising, which has aroused widespread concern from people all over the world. In the treatment and evaluation of renal cancer, the location, shape, size, spatial structure and relative position of kidney and tumors are crucial for surgical assessment and treatment plan formulation [3].

This work was supported partly by Natural Science Foundation of Sichuan Province (Nos. 2022NSFSC0940 and 2022NSFSC0894), and Doctor Foundation of Southwest University of Science and Technology (Nos. 19zx7143 and 20zx7137).


M. Tanveer et al. (Eds.): ICONIP 2022, CCIS 1793, pp. 588–597, 2023.
https://doi.org/10.1007/978-981-99-1645-0_49

While there are currently several methods to accurately detect tumors, automatically tools for delineating kidneys and tumors are still flawed. For example, the segmentation accuracy is not high and requires a lot of manual operations.

Accurate segmentation of kidney and tumors requires radiologists to draw the boundary contour of the target on each CT image [4], which is time-consuming and impressionistic and is not feasible in clinical routine. Therefore, is urgent to optimize the above manual processes and methods. Several scholars have proposed methods to segment the kidney under various imaging modalities. For CT images, traditional methods based on imaging features, deformation models, active shape models, and atlases have been proposed [5]. The widespread of deep learning in recent years has promoted the prosperity of medical image segmentation, and the field of kidney segmentation is no exception [6].

However, most of these studies focus on the automatic segmentation of kidneys, and mostly 2D networks are deficient for possible tumors due to the loss of continuous spatial information. Whats more, heterogeneity of renal tumors leads to grand challenges in segmentation [7]. It means that there may be differences in the location, size, scope, and gray image features of tumors, and sometimes tumors resemble benign renal cysts on imaging. Low contrast at the lesion boundary on CT images, and sometimes may be missing, which further increases the difficulty of differentiation. In addition, due to various rationales for patient privacy protection and differences in medical equipment, it is also hard to obtain medical image data and annotation, resulting in slow development and challenging work.

To solve above problems, This work proposes a method for CT renal tumor segmentation based on hybrid supervision. Our main contributions can be summarized as follows:

1. To solve the problem of insufficient renal annotation data in medical images, we approached a self-supervised pre-training method.
2. A hybrid supervision method is proposed according to the heterogeneity of tumors and the difference in image dimensions, and experimentally verified in 3D Data.
3. Our method is proved to be effective and efficient by numerous experiments.

## 2 Related Work

### 2.1 Self-supervision

In the history of deep learning, pre-training has always received much attentions. Many studies [8,9] have proved that effective pre-training can significantly improve the performance of the target task, but it requires a large number of labeled datasets for supervised training [10]. Another new pre-training paradigm is self-supervised learning [11], which explicitly trains the learning method of conversion network through the structure of data itself or the generated self-supervised signal, without using manual labels in the whole process. It indicates that a significant advantage of self-supervised learning is that it can rise to large datasets at a scanty cost. Mainstream self-supervised learning is mainly divided into the generation model and contrast model [11], but their fundamental goal is recovery. If some image elements need to be recovered, the network needs to learn the image pixel, texture, semantics, context features, and other information, including instances, to make an approximate prediction. It is similar to the image information learned by supervised learning classification, detection, and segmentation networks. Through self-supervised learning, downstream tasks can be performed from a higher baseline, providing beneficial improvements.

Various pretext tasks for self-supervised learning have been proposed. For example, Zhang et al. [12] proposed an image colorization task, they give the image monochrome channel L and let the model predict the values of the other two channels A and B. Similar to image inpainting proposed by Pathak et al. [13], they directly mask part of the image and let the model predict the covered part of the image based on the rest. Examples include image jigsaw [14], SIMCLR [15], Rubik's Cube restoration [16], and more.

### 2.2 Semi-supervision

Although self-supervised does not require labeling, its suitability is questionable. For example, there is the problem of pretraining cross-domain adaptation [17,18]. And when there are a large number of labeled images, the pre-training effect is limited [5,11,19]. He et al. [10] found that ImageNet pre-training only slightly improved on COCO object detection task, and Raghu et al. [19] indicated that ImageNet pre-training did not improve medical image classification task. Therefore, we introduce the concept of semi-supervised self-training. Semi-supervised means that part of the input data contains supervised signals and part does not. Semi-supervised training combines supervised and self-supervised learning to address the mismatch between the two. The self-training Noisy Student method proposed by Xie et al. [20] achieves SOTA performance in Imagenet. They predict unlabeled data through pre-training, generate pseudo labels with the highest confidence samples, and then mix pseudo labels with ground labels for self-training. And they iteratively train this step until the results converge. Unlike knowledge distillation [21], it is mainly used to compress models for deployment in industrial applications. Semi-supervised training is for better accuracy, so the trained student model is not necessarily smaller than the teacher model.

## 3 Approach

We propose a hybrid supervised learning framework combining self-supervised and semi-supervised learning, combining the advantages of both. It can be used in renal tumors and has broad application prospects in medical images lacking labeled data. The network framework and structure are shown in Fig. 1. First, we use unlabeled data for 3D image transformation. Figure 1 crops out some small voxels for demonstration. A to F on the right side of the figure are the

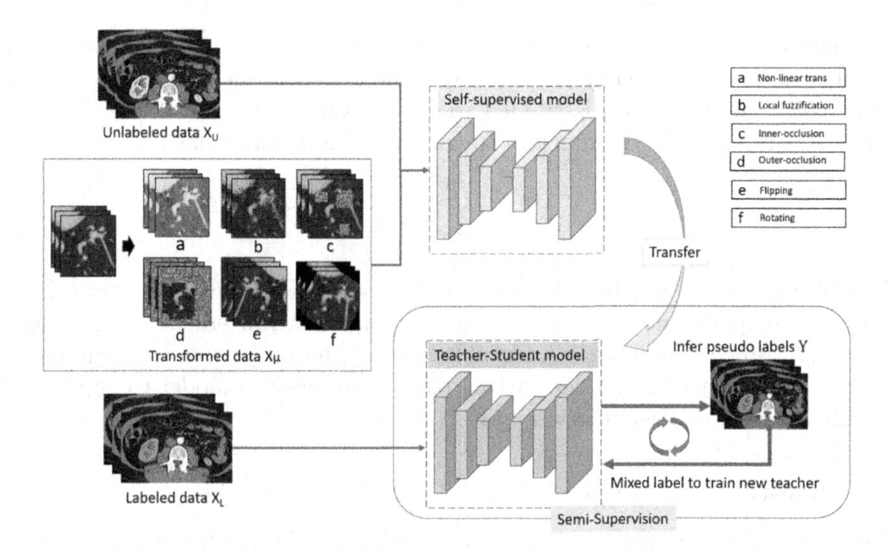


**Fig. 1.** Our proposed hybrid supervised learning framework

specific methods of transformation. Contrastive learning using unlabeled and transformed images to train a self-supervised model as a pre-training result. Then we use the labeled data to train the model to get the initial teacher model, and use teacher for inference to get more pseudo-labeled data. Mix pseudo-labels and original labels for iterative training to get the student model. The following is a details introduction to the implementation.

Our data source can divided into labeled dataset $X_L \in \{(x_1, y_1), (x_2, y_2)... (x_n, y_n)\}$ and unlabeled dataset $\overline{X}_U \in \{\overline{x}_1, \overline{x}_2, ...\overline{x}_m\}$. The hybrid supervision method can also divided into two parts: self-supervision part and semi-supervision part. Firstly, we use the unlabeled data for self-supervised pre-training and transfer weight to $X_L$ to generate the teacher model $M_t$. The teacher model $M_t$ makes semi-supervised self-training. Generates pseudo masks and mixes the pseudo mask and ground truth for interleaved training to obtain the student model $M_s$, and make the student become a new teacher to continue the above process, iterates until the training results converge. Next, each process is described in detail.

### 3.1 Pre-training Stage

For original feature map $\overline{x}_i \subset \overline{X}_U$, randomly generate subset for image transform. we apply the transformation $f(\overline{x}_i) = \mu_i$ where $f(\cdot)$ is transform function, $\mu_i$ is the scrambled image. training a fitting function $g(\cdot)$ as

$$\overline{x}_i = g(\mu_i) = f^{-1}(\mu_i) \tag{1}$$

In detail, we implement image transformations $f(\cdot)$ with a defined probability in the following six ways: 1) Non-linear transformation; 2) Local fuzzification;

3) Inner occlusion; 4) Outer occlusion; 5) Flipping; 6) Rotating; The inspiration for this came from Zhou et al. [22]. After these transformed images have been scrambled, the model $M_p$ is trained to restore the original image. We use the mean square error (MSE) function to compare the restored effect.

$$L_{MSE} = \frac{1}{m} \sum_{i=1}^{m} \left(x_i - g\left(\mu_i\right)\right)^2 \tag{2}$$

Contrastive learning by comparing signals generated by image inpainting can capture the appearance, texture, representation, and arrangement of CT images. Since this process learns a considerable part of the image features, our target task of fine-tuning the model on this basis transfers the model to supervised learning to achieve self-supervised pre-training.

### 3.2 Self-training Stage

For labeled feature map $(x_j, y_j) \subset X_L$ we initialize the teacher model $M_t$ with $M_p$ weights and use the total loss function in a combination of cross-entropy loss and Dice loss:

$$L_{Total} = \alpha L_{CE} + \beta L_{Dice} \tag{3}$$

The $\alpha$ and $\beta$ are hyperparameters to balance the terms in Eq. 3. Where cross-entropy loss describes the distance between two probability distributions. It can be used to describe whether the segmented pixels are correct. $L_{CE}$ can define as follows:

$$L_{CE} = \frac{1}{N} \sum_{k=1}^{k} l\left(y^{(i)}, f(x^{(i)}, M_t)\right) \tag{4}$$

Dice loss function can evaluate set similarity and is sensitive to the internal filling of mask. Dice Loss $L_{Dice}$ is equivalent to investigating from a global perspective. Cross-entropy loss $L_{CE}$ is zooming in pixel by pixel from a microscopic perspective. Their angles are complementary, balance two loss functions to achieve the best evaluation effect.

$$L_{Dice} = 1 - \frac{2\left|X \cap Y\right|}{\left|X\right| + \left|Y\right|} \tag{5}$$

After training the teacher model $M_t$, we use its inference to generate pseudo-labels $\hat{Y}$ by:

$$\hat{Y} = f\left(\overline{x}_i.M_t\right).\forall i = 1...m \tag{6}$$

Mix pseudo-label and real-label data, add noise [20] to train the student model $M_s$ with follows:

$$\frac{1}{n} \sum_{i=1}^{n} f_n\left(x_i, y_i\right) + \frac{1}{m} \sum_{j=1}^{m} g_n\left(\overline{x}_j, \hat{y}_j\right) \tag{7}$$

The $f_n()$ and $g_n()$ represent noisy and data augmentation methods, respectively. Iterative training is performed until the model converges. The total experimental pipeline can be described in Algorithm 1.

**Algorithm 1.** hybrid supervised training procedure

I .Pre-training

**Input:** Unlabeled images $X_U$

**Output:** Teacher model $M_t$

1: **for** i = 0 to $N_U$ **do**
2: Select unlabeled data $x_i \in X_U$
3: Generate subset $s_i$ make transform $f(x_i) = s_i$
4: Learning inpaint by minimizes the mean square error $\frac{1}{N}\sum_{i=1}^{N}\left(s_i - g^{-1}\left(x_i\right)\right)$
5: **end for**
6: **return** Teacher model $M_t$

II .Interleaved training

**Input:** Labeled images ($X_L$,$Y_L$),Teacher model $M_t$

**Output:** Student model $M_s$

1: load teacher model $M_t$
2: **repeat**
3: **for** j = 0 to $N_L$ **do**
4: Select paired data $(x_j, y_j) \in (X_L, Y_L)$
5: generate teacher model $M_t$
6: make inference $M_t(X_U) \rightarrow \overline{Y}$
7: **end for**
8: Make self-training data $(X_L + \overline{X}, Y_L + \overline{Y})$
9: minimize loss function $L_{ce} + L_{dice}$
10: **until** converged
11: **return** Student model $M_s$

## 4 Experiments

Our experiments were performed using the public kidney dataset KITS19 [23], a large scale challenge dataset organized by the Medical Image Computing and Computer-Aided Intervention(MICCAI). The dataset contains a total of 300 cases, of which 210 cases are used for training and 90 unpublished labels are used for validation. Since our hybrid supervised framework uses unlabeled data for training, continuing to use the official 90 test set may result in a shared data set for test and training. Therefore, we randomly split the 210 labeled cases 8:2 into training and test sets (168 for training and 42 for testing). And keep this setting to ensure that data is not mixed in future experiments. At the same time, we also used five-fold cross-validation in the training process to ensure data stability and avoid overfitting.

We first preprocessed the data based on preoperative arterial CT scans with resolution ranging from 0.437 mm to 1.04 mm and thickness varying from 0.5 mm to 5.0 mm. We resampled it to 3.22 * 1.62 * 1.62 mm pixel spacing (from nnU-Net [24]), maintaining isotropic is conducive to CNN learning. Then the CT value (HU) was adjusted to the range of $[-79, 304]$, and z-Score normalization was performed.

Then we test the baseline network. Since the abdominal CT is 3D data, we need to use the 3D network for baseline comparison. Which we use nnU-Net [24], v-net [25] and Cascaded-FCN [26]. Use Adam [27] as the network optimizer with hyperparameters alpha and beta set to 0.3 and 0.7. Table 1 shows their average Accuracy, Dice coefficients, and IOU for kidney and tumor in five-fold cross-validation. Although the data division is different, the results are the same as the results of the KITS19 challenge. nnU-Net outperforms other networks in terms of Dice coefficient and IOU, thus we conduct hybrid supervised experiments based on nnU-Net.

**Table 1.** Comparison between different network baselines

| | *V-Net* [25] | *Cascaded FCN* [26] | *nnU-Net* [24] |
|---|---|---|---|
| Kidney Accuracy | **99.95 ± 0.01** | 99.93 ± 0.01 | 99.94 ± 0.003 |
| Kidney Dice | 95.06 ± 0.98 | 94.55 ± 1.02 | **95.61 ± 0.95** |
| Kidney IOU | 91.86 ± 1.23 | 90.99 ± 0.97 | **92.13 ± 0.98** |
| Tumor Accuracy | 99.85 ± 0.04 | 99.82 ± 0.06 | **99.90 ± 0.03** |
| Tumor Dice | 70.30 ± 4.08 | 67.33 ± 4.36 | **71.49 ± 3.9** |
| Tumor IOU | 59.08 ± 5.43 | 54.73 ± 6.61 | **60.98 ± 3.99** |

Hybrid supervision is in two steps: self-supervised and semi-supervised stage. The improvement effect of the baseline network nnU-Net is tested with self-supervised and semi-supervised algorithms respectively, and then mixed for comparative experiments, as shown in Table 2. Base means base nnu-net, no changes. base+self means adding the self-supervision process separately, and the rest of the settings are the same as the base network. Base+Semi means joining the semi-supervised process, and Hybrid indicates a mixture of two, which is also our final proposal. This is also our ablation experiment, the role of self-supervision and semi-supervision and the degree of improvement of hybrid supervision to CNN are reflected respectively.

The experimental results show that adding the self-supervised method alone has little effect on the segmentation accuracy, and Dice and IOU are slightly improved: kidney DICE increased from 95.61 to 95.87, and tumor DICE

**Table 2.** Ablation experiment on hybrid supervision

| | *Base* | *Base+self* | *Base+semi* | *Base+Hybird (Ours)* |
|---|---|---|---|---|
| Kidney Accuracy | 99.94+0.003 | 99.94+0.003 | 99.95+0.006 | **99.96+0.003** |
| Kidney Dice | 95.61+0.95 | 95.87+0.54 | 96.46+0.87 | **96.96+0.21** |
| Kidney IOU | 92.13+0.98 | 92.26+0.84 | 93.63+0.98 | **94.21+0.38** |
| Tumor Accuracy | 99.90+0.03 | 99.89+0.02 | 99.91+0.03 | **99.92+0.01** |
| Tumor Dice | 71.49+3.90 | 72.37+1.89 | 73.83+2.37 | **74.65+3.95** |
| Tumor IOU | 60.98+3.99 | 62.09+2.17 | 62.39+4.88 | **62.61+5.31** |

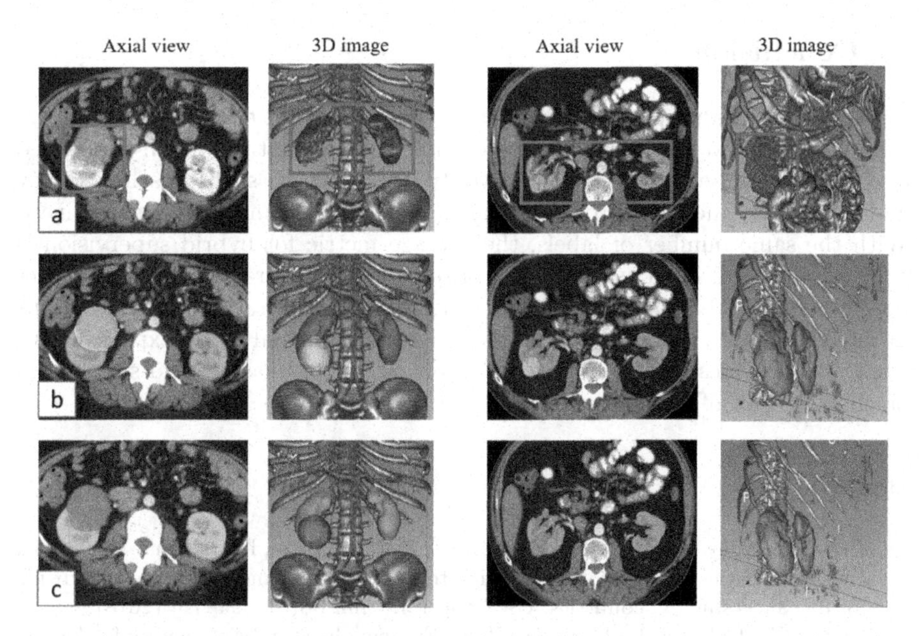


**Fig. 2.** Visualization of the representative segmentation results from our experiments

increased from 71.49 to 72.37. When semi-supervised alone is used, all metrics improve by an average of one percentage point (see the comparison of the second and fourth columns in Table 2). When using hybrid supervision, all network metrics are significantly improved, reaching the best performance. This explains their respective roles: self-supervised pre-training can obtain image information for training without the need for labels, which improves the primary performance of the model. On the other hand, semi-supervised self-training trains the student model iteratively on the self-supervised model and learns plentiful image features through the noise and data augmentation. They do not interfere much with each other, and mixing self-supervised and semi-supervised usage can maximize their advantages. The experimental results also demonstrate the effectiveness of our proposed hybrid supervision method.

Finally, we performed a 3D visualization of the segmentation results, see Fig. 2. We selected two representative patients. The axial view is a slice display of the CT in the axial direction, and his 3D reproduction is on the right. The visualization software used here is 3D Slicer. The original image is in row a, there are some interference from blood vessels and gut due to manual threshold adjustment, and the location of the kidney has been marked with a red box. Row b shows the ground truth of kidneys and tumors, marked in green and yellow, and the results of our inference are in row c, which have been marked in blue and red, respectively. Intuitively, our results are very close to those annotated by doctors. This means that our method may be more acceptable to radiologist, with certain advantages in terms of practicality.

## 5 Conclusion

In this paper, we propose a hybrid supervised approach to automatically segment kidneys and kidney tumors. We implement a mixture of self-supervised and semi-supervised supervision and verify its effectiveness: the baseline kidney and tumor Dice increased from 95.61 and 71.49 to 96.96 and 74.65, respectively. With the same number of labels, the average metric for hybrid supervision is greatly improved, demonstrating the effectiveness and practicality of our proposed method. We believe that in the context of the general lack of annotations in medical images, hybrid supervision has great potential to be explored. Our work also provides some inspiration to some medical imagers who suffer from not having a lot of labels.

## References

1. Bray, F., Ferlay, J., Soerjomataram, I., Siegel, R.L., Torre, L.A., Jemal, A.: Global cancer statistics 2018: Globocan estimates of incidence and mortality worldwide for 36 cancers in 185 countries. CA Cancer J. Clin. **68**(6), 394–424 (2018)
2. Chow, W.-H., Dong, L.M., Devesa, S.S.: Epidemiology and risk factors for kidney cancer. Nat. Rev. Urol. **7**(5), 245–257 (2010)
3. Jaffe, C.C.: Measures of response: recist, who, and new alternatives. J. Clin. Oncol. **24**(20), 3245–3251 (2006)
4. Ljungberg, B., et al.: EAU guidelines on renal cell carcinoma: 2014 update. Eur. Urol. **67**(5), 913–924 (2015)
5. Torres, H.R., Queiros, S., Morais, P., Oliveira, B., Fonseca, J.C., Vilaca, J.L.: Kidney segmentation in ultrasound, magnetic resonance and computed tomography images: a systematic review. Comput. Methods Programs Biomed. **157**, 49–67 (2018)
6. Piao, N., Kim, J.-G., Park, R.-H.: Segmentation of cysts in kidney and 3-D volume calculation from CT images. Int. J. Comput. Graph. Anim. (IJCGA) **5**(1), 1 (2015)
7. Moch, H., Cubilla, A.L., Humphrey, P.A., Reuter, V.E., Ulbright, T.M.: The 2016 who classification of tumours of the urinary system and male genital organs-part a: renal, penile, and testicular tumours. Eur. Urol. **70**(1), 93–105 (2016)
8. Schmidhuber, J.: Deep learning in neural networks: an overview. Neural Netw. **61**, 85–117 (2015)
9. Donahue, J., Jia, Y., Vinyals, O., Hoffman, J.: A deep convolutional activation feature for generic visual recognition. Proc. Mach. Learn. Res. **1** (2014)
10. He, K., Girshick, R., Dollár, P.: Rethinking imagenet pre-training. In: Proceedings of the IEEE/CVF International Conference on Computer Vision, Conference Proceedings, pp. 4918–4927 (2019)
11. Liu, X., et al.: Self-supervised learning: generative or contrastive. IEEE Trans. Knowl. Data Eng. (2021)
12. Zhang, R., Isola, P., Efros, A.A.: Colorful image colorization. In: European Conference on Computer Vision, Conference Proceedings (2016)
13. Pathak, D., Krahenbuhl, P., Donahue, J., Darrell, T., Efros, A.A.: Context encoders: feature learning by inpainting. IEEE Conference on Computer Vision and Pattern Recognition (2016)

14. Noroozi, M., Favaro, P.: Unsupervised learning of visual representations by solving Jigsaw puzzles. In: Leibe, B., Matas, J., Sebe, N., Welling, M. (eds.) ECCV 2016. LNCS, vol. 9910, pp. 69–84. Springer, Cham (2016). https://doi.org/10.1007/978-3-319-46466-4_5
15. Chen, T., Kornblith, S., Norouzi, M., Hinton, G.: A simple framework for contrastive learning of visual representations. Proc. Mach. Learn. Res. (2020)
16. Zhu, J., Li, Y., Hu, Y., Ma, K., Zhou, S.K., Zheng, Y.: Rubik's cube+: a self-supervised feature learning framework for 3D medical image analysis. Med. Image Anal. **64**, 101746 (2020). https://www.ncbi.nlm.nih.gov/pubmed/32544840
17. Ghiasi, G., Lin, T.-Y., Le, Q.V.: Dropblock: a regularization method for convolutional networks. In: Advances in Neural Information Processing Systems, vol. 31 (2018)
18. Poudel, R.P., Liwicki, S., Cipolla, R.: Fast-SCNN: fast semantic segmentation network. Comput. Vision Pattern Recogn. (2019)
19. Raghu, M., Zhang, C., Kleinberg, J., Bengio, S.: Transfusion: understanding transfer learning for medical imaging. In: Advances in Neural Information Processing Systems (NeurIPS 2019) (2019)
20. Xie, Q., Luong, M.-T., Hovy, E., Le, Q.V.: Self-training with noisy student improves imagenet classification. In: Proceedings of the IEEE/CVF Conference on Computer Vision and Pattern Recognition, Conference Proceedings, pp. 10:687–10:698 (2020)
21. Hinton, G., Vinyals, O., Dean, J., et al.: Distilling the knowledge in a neural network. arXiv preprint arXiv:1503.02531, vol. 2, no. 7 (2015)
22. Zhou, Z., Sodha, V., Pang, J., Gotway, M.B., Liang, J.: Models genesis. Med. Image Anal. (2020)
23. Heller, N., et al.: The kits19 challenge data: 300 kidney tumor cases with clinical context, CT semantic segmentations, and surgical outcomes. Quant. Biol. (2019)
24. Isensee, F., Jaeger, P.F., Kohl, S.A.A., Petersen, J., Maier-Hein, K.H., nnu-net: a self-configuring method for deep learning-based biomedical image segmentation. Nat. Methods **18**(2), 203–211 (2021). https://www.ncbi.nlm.nih.gov/pubmed/33288961
25. Milletari, F., Navab, N., Ahmadi, S.-A.: V-net: fully convolutional neural networks for volumetric medical image segmentation. In: Fourth International Conference on 3D Vision (3DV), pp. 565–571. IEEE (2016)(2016)
26. Christ, P.F., et al.: Automatic liver and lesion segmentation in CT using cascaded fully convolutional neural networks and 3D conditional random fields. In: Ourselin, S., Joskowicz, L., Sabuncu, M.R., Unal, G., Wells, W. (eds.) MICCAI 2016. LNCS, vol. 9901, pp. 415–423. Springer, Cham (2016). https://doi.org/10.1007/978-3-319-46723-8_48
27. Kingma, D.P., Ba, J.: Adam: a method for stochastic optimization. In: International Conference for Learning Representations (2014)

# Double Attention-Based Lightweight Network for Plant Pest Recognition

Janarthan Sivasubramaniam[1,2](✉), Thuseethan Selvarajah[1], Sutharshan Rajasegarar[1,2], and John Yearwood[1,2]

[1] School of Information Technology, Deakin University, Geelong, VIC 3220, Australia
{jsivasubramania,tselvarajah,srajas,john.yearwood}@deakin.edu.au
[2] Deakin-SWU Joint Research Centre on Big Data, Faculty of Science, Engineering and Built Environment, Deakin University, Geelong, VIC 3125, Australia

**Abstract.** Timely recognition of plant pests from field images is significant to avoid potential losses of crop yields. Traditional convolutional neural network-based deep learning models demand high computational capability and require large labelled samples for each pest type for training. On the other hand, the existing lightweight network-based approaches suffer in correctly classifying the pests because of common characteristics and high similarity between multiple plant pests. In this work, a novel double attention-based lightweight deep learning architecture is proposed to automatically recognize different plant pests. The lightweight network facilitates faster and small data training while the double attention module increases performance by focusing on the most pertinent information. The proposed approach achieves 96.61%, 99.08% and 91.60% on three variants of two publicly available datasets with 5869, 545 and 500 samples, respectively. Moreover, the comparison results reveal that the proposed approach outperforms existing approaches on both small and large datasets consistently.

**Keywords:** Plant Pest Recognition · Double Attention · Lightweight Network · Deep Learning

## 1 Introduction

Plant pests cause severe damage to crop yields, resulting in heavy losses in food production and to the agriculture industry. In order to reduce the risk caused by plant pests, over the years, agricultural scientists and farmers tried various techniques to diagnose the plant pests at their early stage. Although many sophisticated automatic pest recognition algorithms have been proposed in the past, farmers continue to rely on traditional methods like manual investigation of pests by human experts. This is mainly because of poor classification ability and limited in-field applicability of automatic pest recognition systems [1]. Different plant pests share common characteristics, which makes automatic pest recognition a very challenging task and hence traditional handcrafted feature


M. Tanveer et al. (Eds.): ICONIP 2022, CCIS 1793, pp. 598–611, 2023.
https://doi.org/10.1007/978-981-99-1645-0_50

extraction based approaches often failed to correctly classify pests [2]. While conventional deep learning-based techniques achieved benchmark performances in pest classification, they have limited usage with resource constraint devices due to their high computational and memory requirements [3]. Large labelled data requirement for training is another flip side of conventional deep learning techniques [4]. Recently proposed semi-supervised learning of deep networks is also not ideal for this problem as they frequently demonstrate low accuracies and produce unstable iteration results.

In order to prevent the deep model from overfitting, it is also paramount to provide sufficient data during the training phase [5]. However, constructing a large labelled data in the agriculture domain, especially for plant pests, requires not only high standard of expertise but also time-consuming. Moreover, inaccurate labelling of the training data produces deep models with reduced reliability. Few-shot learning concept is proposed simply by replicating humans' ability to recognize any objects with the help of only a few examples [6]. Few-shot learning has gained popularity across various domains as it can address the classification task with a few training samples. In few-shot learning, the classification accuracy increases as the number of shots grow. However, a major limitation of few-shot learning is that the prediction accuracy drops when the number of ways increases [7]. Directly applying the classification knowledge learned from meta-train classes to meta-test classes is mostly not feasible, which is another fundamental problem of this approach [8].

Constructing decent-performing models with the reduced number of trainable parameters by downsizing the kernel size of convolutions (e.g., from $3 \times 3$ to $1 \times 1$ as demonstrated in [9]) is a significant step towards the development of lightweight networks. In recent years, lightweight deep network architectures have gained growing popularity as an alternative to traditional deep networks [10–13]. The MobileNets [14] and EfficientNets [15] families are thus far two most widely used lightweight networks. Several lightweight deep network-based techniques have also been proposed for real-time pest recognition [16,17]. The lightweight architectures however suffer to reach the expected level of classification accuracy as they are essentially developed for faster and lighter deployment by sacrificing the performance.

Considering this issue, a novel high-performing and lightweight pest recognition approach is proposed in this study, as illustrated in Fig. 1. While preserving the lightweight characteristic of the deep network, a double attention mechanism is infused to enhance the classification performance. As the attention closely imitates the natural cognition of the human brain, the most influential regions of the pest images are enhanced to learn better feature representations. Notably, attention-aware deep networks have shown improved performances in various classification tasks [18,19]. The key contributions of this paper are three-fold:

- A novel lightweight network-based framework integrated with a double attention scheme is proposed for enhancing the in-field pest recognition, especially using small training data.
- A set of extensive experiments were conducted under diverse environments to reveal the feasibility and validate the in-field applicability of the

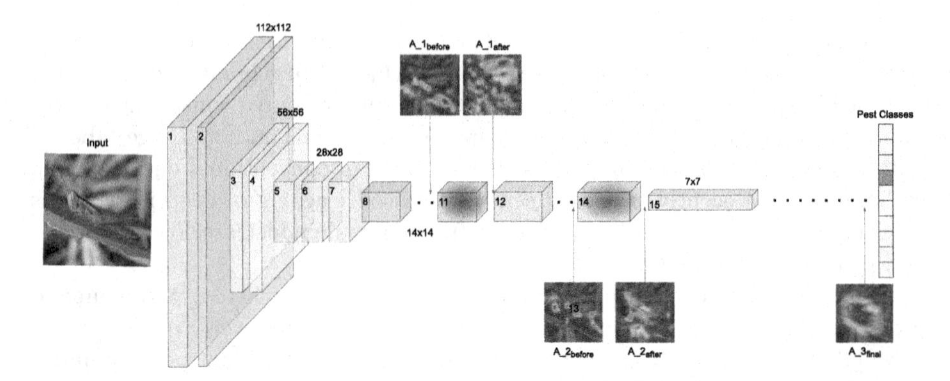


**Fig. 1.** The overall architecture of the proposed double attention-based lightweight pest recognition framework. The layers are labelled to match with the layer numbers indicated in Table 1. Some layers of the proposed architecture are avoided in this diagram for brevity. The $A\_1_{before}$, $A\_1_{after}$, $A\_2_{before}$, $A\_2_{after}$ and $A\_3_{final}$ are the activation maps generated before and after respective layers as indicated.

proposed framework. To organize diverse environments, three publicly available datasets consisting of small to large number of pest samples are utilized.
- A comparative analysis is performed to show the superiority of this framework over existing state-of-the-art lightweight networks that are often used for pest recognition methods available in the literature.

The remainder of this paper is constructed as follows. The recent advancements of deep learning based pest recognition approaches are given in Sect. 2. Section 4 provides comprehensive details on the proposed framework and Sect. 5 concludes the paper with some future directions.

## 2 Related Work

Effective pest recognition is essential for preventing the spread of crop diseases and minimizing economic losses in relation to agriculture. Over the years, researchers invested considerable effort to develop pest recognition techniques. Probe sampling, visual inspection and insect trap are some of the widely used manual pest recognition approaches that are still the farmers' favourites when it comes to in-field tasks [20]. Early studies targetted the sounds emitted by the pests to perform the classification, but the paradigm has quickly shifted to digital images in the last decade [21]. As this paper reviews a few recently proposed prominent pest recognition works, readers are recommended to read [22] and [23] for a comprehensive pest recognition literature.

With the recent development of deep learning, the automatic pest recognition techniques have become a rapidly growing agricultural research direction [24–26]. Different deep convolutional neural networks (CNNs) (i.e., state-of-the-art and tailor-made networks) and capsule networks are prominently used to develop high-performing pest recognition methods. In [27], the pre-trained VGG-16 deep network is exploited to perform tea tree pest recognition. In a similar work, transfer learning is applied on VGG-16 and Inception-ResNet-v2 to enhance the accuracy of the pest identification system [28]. An ensemble of six pre-trained state-of-the-art deep networks with majority voting showed improved performance in pest classification [29]. Zhang et al. [30] presented a compelling modified capsule network (MCapsNet) for improving the crop pest recognition. A capsule network along with a multi-scale convolution module is adapted to construct novel multi-scale convolution-capsule network (MSCCN) for fine-grained pest recognition [31]. Even though these methods showed better pest classification accuracies, they are limited in in-field environments because of their heaviness and large training data requirement. Additionally, the pest recognition methods that use CNNs and capsule networks mostly performed poorly with imbalanced data.

Recently, few-shot and semi-supervised learning mechanisms have been extensively used for plant pest and disease recognition. For instance, Li and Yang [32] introduced a few-shot pest recognition approach where a deep CNN model integrated with the triplet loss is trained and validated. According to the results, the proposed few-shot learning approach showed high generalization capability. In [33], the online semi-supervised learning is exploited for an insect pest monitoring system that achieved a substantial improvement in the accuracy. Despite the fact that the aforementioned methods alleviate large data and class imbalance problems to a certain extent, they continue to show limited classification accuracy.

With the wide adaption of lightweight architectures in image classification tasks [14], there have been many lightweight deep network-based pest recognition proposed in the last two years. It has been confirmed in the literature that the lightweight deep networks can not only be deployed in resource-constrained devices but also be trained with small data [15,34]. In [16], an improved lightweight CNN is introduced by linking low-level and high-level network features that eventually connect rich details and semantic information, respectively. Zha et al. [17] proposed a lightweight YOLOv4-based architecture with MobileNetv2 as the backbone network for pest detection. In another work, the proposed lightweight CNN model's feasibility in classifying tomato pests with imbalanced data is validated [35]. The lightweight pest recognition approaches use optimized deep models that mostly compromise the performance.

The attention mechanism has been prominently used in crop pest recognition to enhance the performance of deep learning models. A parallel attention mechanism integrated with a deep network showed improvements in the accuracy [36]. In another work, the in-field pest recognition performance is boosted by combining spatial and channel attention [37]. Attention mechanism embedded with lightweight networks obtained better classification accuracy in pest recognition domain [38].

In summary, a lightweight network-based pest recognition system is ideal for in-field utilization as it is fast and consumes less memory. However, it is noticed

**Table 1.** Layer organization of the proposed lightweight pest recognition architecture.

| # | Input | Layer | Output | Stride |
|---|---|---|---|---|
| 1 | 224 × 224 × 3 | conv 3 × 3 | 112 × 112 × 32 | 2 |
| 2 | 112 × 112 × 32 | channel reduction | 112 × 112 × 16 | 1 |
| 3 | 112 × 112 × 16 | down sampling | 56 × 56 × 24 | 2 |
| 4 | 56 × 56 × 24 | 1×inverted residual | 56 × 56 × 24 | 1 |
| 5 | 56 × 56 × 24 | down sampling | 28 × 28 × 32 | 2 |
| 6–7 | 28 × 28 × 32 | 2×inverted residual | 28 × 28 × 32 | 1 |
| 8 | 28 × 28 × 32 | down sampling | 14 × 14 × 64 | 2 |
| 9–10 | 14 × 14 × 64 | 2×inverted residual | 14 × 14 × 64 | 1 |
| 11 | 14 × 14 × 64 | double attention | 14 × 14 × 64 | 1 |
| 12 | 14 × 14 × 64 | channel expansion | 14 × 14 × 96 | 1 |
| 13 | 14 × 14 × 96 | 1×inverted residual | 14 × 14 × 96 | 1 |
| 14 | 14 × 14 × 96 | double attention | 14 × 14 × 96 | 1 |
| 15 | 14 × 14 × 96 | down sampling | 7 × 7 × 160 | 2 |
| 16-17 | 7 × 7 × 160 | 2×inverted residual | 7 × 7 × 160 | 1 |
| 18 | 7 × 7 × 160 | channel expansion | 7 × 7 × 320 | 1 |
| 19 | 7 × 7 × 320 | conv 1 × 1 | 7 × 7 × 1280 | 1 |
| 20 | 7 × 7 × 1280 | avgpool | 1 × 1 × 1280 | – |
| 21 | 1 × 1 × 1280 | dropout & softmax | 1 × 1 × $k$ | – |

that these often failed to attain expected classification accuracy. Attention can fix the performance issue of any deep learning model [19], which creates a lead for this work.

# 3 Proposed Model

This section presents the proposed double attention-based lightweight pest recognition network in detail.

## 3.1 The Lightweight Network

Figure 1 illustrates an overview of the proposed lightweight network with the example activation maps for an example input. In the figure, the $A\_1_{before}$ and $A\_1_{after}$ are the activation maps obtained before and after the first double attention layer. Similarly, the activation maps before and after the second double attention layer are given in $A\_2_{before}$ and $A\_2_{after}$, respectively. The $A\_3_{final}$ is the activation map generated by the last convolutional layer of the proposed architecture.

The comprehensive details of each layer in the proposed network, where the MobileNetv2 is used as the backbone, are individually illustrated in Table 1. Altogether, this architecture has 21 primary layers: convolution, channel reduction, down sampling, inverted residual and channel expansion, average pooling, and the final dropout and softmax layers. Inspired by the key architectural concepts

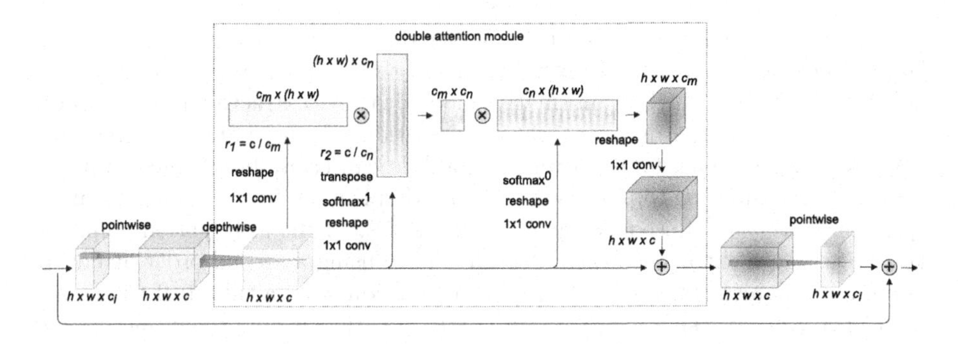


**Fig. 2.** The double attention layer used in our proposed architecture.

of the MobileNetv2 deep network, the inverted residual and depthwise separable convolution that are unique to MobileNetv2 are exploited in the proposed architecture. For example, inverted residual layers are built in a specific way which expands the channels six times and apply the depthwise separable convolution in order to learn the features efficiently. In addition, the inverted residual layers create the shortcut connection between the input and output same as ResNet [40] to enable the residual learning. This allows to build deeper networks. In depthwise separable convolution, a depthwise convolution operation is first performed and followed by a pointwise convolution operation. Notably, as demonstrated in [14], depthwise separable convolution has the benefit of computational efficiency due to the fact that the depthwise convolution can potentially reduce the spatial dimensions.

The depthwise separable convolution in the higher channel dimension with a similar structure is used to build both channel expansion and downsampling layers. However, in channel expansion layers, a higher number of pointwise filters are defined to increase the output channels. In contrast, for downsampling layers, the stride is set to 2 along with a depthwise convolution that reduces the spatial dimensions by half while the subsequent pointwise filters increasing the channels. In addition, the first standard conv $3 \times 3$ layer with the stride set to 2 reduces the spatial dimension by half and the subsequent channel reduction layer (layer 2) reduces the channels by half. This drastic reductions in dimensions makes early layers efficient.

As given in Fig. 1 and Table 1, two double attention modules are systematically affixed as layer 11 and 14 in this architecture which is explained in the next subsection.

### 3.2 Double Attention Layer

Figure 2 shows the complete structure of the double attention layer proposed in [39], which is known for capturing long-range relations. Generally, for any image classification tasks, it is confirmed that a lightweight network integrated with double attention modules obtained superior results over a much larger traditional deep network. For instance, a much smaller double attention enabled ResNet-50 model outperformed the ResNet-152 model on the ImageNet-1k dataset for a classification task. Here, the informative global features extracted from the whole image are manipulated in a two-phased attention scheme, namely aggregation and propagation. The double attention module can easily be plugged into any deep network, allowing to pass useful information learned from the entire feature space effectively to subsequent convolution layers. In the aggregation step, the key features are extracted from the entire space using a second-order attention pooling mechanism. Meanwhile, in the propagation step, a subset of the meaningful features corresponding to each spatial location is adaptively picked and dispersed using another attention mechanism.

Two hyperparameters $r_1$ and $r_2$ can be observed in this mechanism, the first one is to reduce the channels from $c$ to $c_m$ and the second one to reduce the channels from $c$ to $c_n$. Both of these parameters are fine-tuned to obtain the best performing combination. However, similar to the procedure followed in the original paper, a common hyperparameter is used to tune both of these hyperparameters. The double attention layers are built deeply integrating the double attention module into the inverted residual layers as shown in Fig. 2. The double attention module is placed next to the depthwise convolution to exploit the higher dimensional feature space (six times the number of input channels). The importance of this placement is further discussed in the ablative analysis, in Subsect. 4.4. More importantly, both of the double attention modules are placed in deeper layers since it is confirmed in the original paper that the double attention modules perform well when they are attached to deeper layers.

# 4 Experiments

## 4.1 Datasets

Two publicly available pest datasets containing images with natural backgrounds, published in [41] and [42], are used. Throughout the paper, the former [41] is referred to as the D1 dataset while the latter [42] is mentioned as the D2 dataset.

**D1 Dataset:** The D1 dataset consists of the data for the ten most prevalent crop pests with fast reproductive rates and causing significant yield losses. The images are collected using internet search through the most popular search engines, such as Google, Baidu, Yahoo, and Bing. The authors of the D1 dataset also use mobile phone cameras (i.e., Apple 7 Plus) to collect additional pest images from the environment. The published D1 dataset contains 5,869 images for ten pests, namely Cydia Pomonella (415), Gryllotalpa (508), Leafhopper (426), Locust (737), Oriental Fruit Fly (468), Pieris Rapae Linnaeus (566), Snail (1072), Spodoptera litura (437), Stink Bug (680) and Weevil (560).

**D2 Dataset:** Cheng et al. [42] constructed the D2 pest dataset by picking ten species of pests from the original dataset used in [43]. The original dataset contains high-resolution images in the size of 1280 × 960, captured using colour digital cameras (i.e., Canon and Nikon). However, in the D2 dataset, the resolution of the original images is reduced to 256 × 256. There are 545 image samples of ten species, namely Elia Sibirica (55), Cifuna Locuples (55), Cletus Punctiger (55), Cnaphalocrocis Medinalis (54), Colaphellus Bowvingi (51), Dolerus Tritici (55), Pentfaleus Major (55), Pieris Rapae (55), Sympiezomias Velatus (55) and Tettigella Viridis (55), are included in this dataset.

**$D1_{500}$ Dataset:** To evaluate the proposed network with small data, a subset of D1 dataset with 500 image samples is constructed by randomly selecting 50 samples per class.

### 4.2 Implementation and Training

The PyTorch deep learning framework and Google Colab[1] platform are utilized to implement the proposed pest recognition model and other existing state-of-the-art lightweight models used for comparison. The pre-trained ImageNet weights available with the PyTorch framework are loaded before fine-tuning all the state-of-the-art networks on pest datasets. Similarly, the proposed model is trained on the ImageNet dataset before the fine-tuning process.

During the training of all the models on pest datasets, the number of epochs, batch size, imput image size are set to 50, 8 and $224 \times 224$ with centre cropped, respectively. The stochastic gradient descent (SGD) optimizer with an initial learning rate of 0.001 is applied for all the models, except for MNasNet (0.005) and ShuffleNet (0.01) when they are trained with dataset D2 and $D1_{500}$. The momentum and weight decay of the SGD are set to 0.9 and $1e^{-4}$, respectively.

### 4.3 Results

In this section, the evaluation results obtained for the proposed double attention-based lightweight pest recognition architecture under diverse environments are discussed. The results are reported using two evaluation metrics: average accuracy and F1-measure. All the experimental results reported in this paper used 5-fold cross-validation procedure.

**Table 2.** Comparison of the average accuracy (Acc) and F1-measure (F1) obtained for the proposed model and other state-of-the-art lightweight models on D1, D2 and $D1_{500}$ pest datasets. The number of parameters (Params [Millions]) and floating point operations (FLOPs [GMAC]) are also given.

| Method | Params [Millions] | FLOPs [GMAC] | D1[5869] | | D2[545] | | $D1_{500}$[500] | |
|---|---|---|---|---|---|---|---|---|
| | | | F1 | Acc | F1 | Acc | F1 | Acc |
| ShuffleNetv2 | 1.26 | 0.15 | 93.47 | 93.90 | 97.94 | 97.94 | 85.60 | 85.60 |
| SqueezeNet | 0.74 | 0.74 | 91.40 | 91.85 | 95.19 | 95.22 | 79.11 | 78.50 |
| MobileNetv2 | 2.24 | 0.32 | 95.32 | 95.69 | 98.72 | 98.71 | 89.74 | 89.80 |
| MNasNet | 3.12 | 0.33 | 95.80 | 96.16 | 94.23 | 94.30 | 87.22 | 87.40 |
| MobileNetv3_Large | 4.21 | 0.23 | 96.04 | 96.34 | 97.60 | 97.61 | 89.02 | 89.20 |
| MobileNetv3_Small | 1.53 | 0.06 | 94.50 | 94.58 | 95.81 | 95.77 | 85.40 | 85.60 |
| **Proposed Method** | 2.56 | 0.38 | **96.37** | **96.61** | **99.09** | **99.08** | **91.55** | **91.60** |

[1] https://colab.research.google.com/.

Table 2 presents the comparison results obtained on D1, D2 and $D1_{500}$ datasets. In addition to reporting average accuracy and F1-measure, the number of parameters in millions and floating-point operations (FLOPs) of each model are also provided. The proposed double attention-based pest recognition approach achieved the average accuracy of 96.61%, 99.09% and 91.55% on D1, D2 and $D1_{500}$ datasets, respectively. Compared to other lightweight models, the proposed approach showed a comprehensive improvement. When it comes to small training data, the proposed model outperformed the compared models, at least by 0.37% on D2 and 1.8% on $D1_{500}$. This indicates that the proposed method can produce high plant pests classification capability under data constraint conditions. The proposed model achieved very high F1 measures on all datasets, especially 96.61% on the D1 dataset, which reveals the feasibility of this method in dealing with data with imbalanced class samples. While the MobileNetv3_Large model achieved the second-highest average accuracy on the D1 dataset, the MobileNetv2 outperformed it on the other two datasets. This exhibits the superiority of the MobileNetv2 model on small datasets in comparison to other state-of-the-art lightweight networks.

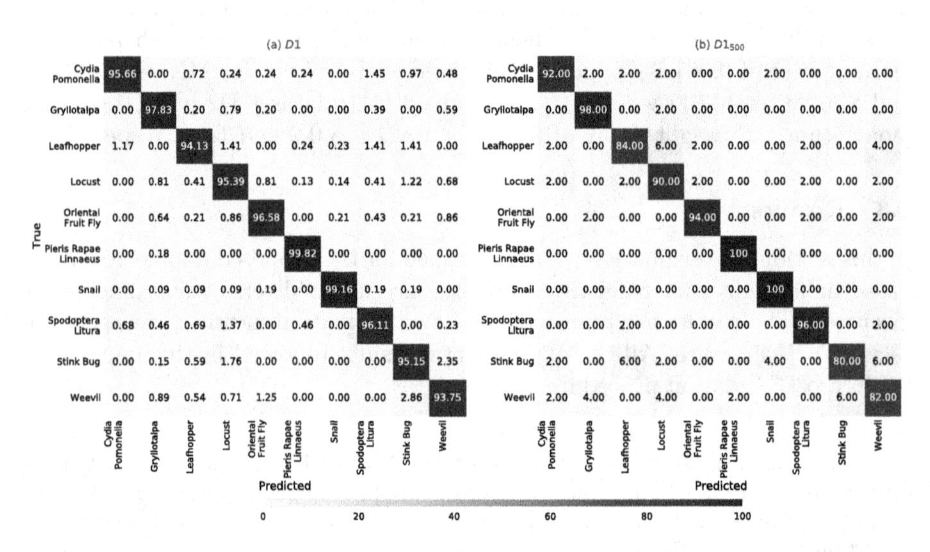


**Fig. 3.** Confusion matrix obtained for the proposed model on D1 and $D1_{500}$ pest datasets.

The number of parameters for the proposed method is 2.56 million, which is slightly more than the original MobileNetv2 model's 2.24 million parameters. However, the proposed model holds 0.56 and 1.65 million parameters lesser than MNasNet and MobileNetv3_large, respectively, which is a significant improvement. The FLOPs counts to run a single instance of all the compared models are also compared, and one can see that the proposed model manifests comparable number of FlOPs. The models with 0.4 GMAC or less are generally considered

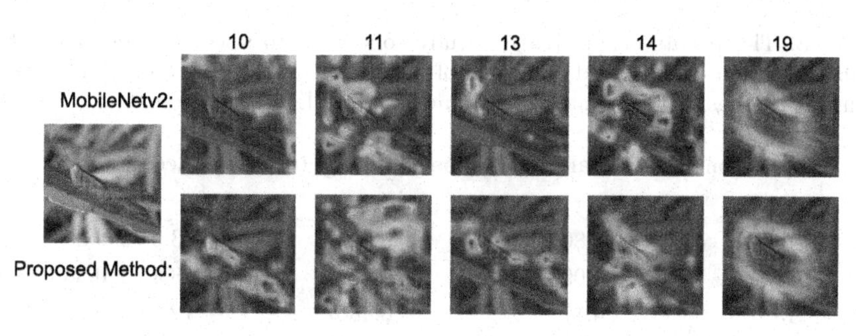


**Fig. 4.** Comparison of the activation maps generated by the proposed model and MovileNetv2 for an example pest image taken from D1 dataset.

as the mobile-size baseline [15]. Hence, it can be concluded that the proposed pest recognition model is adaptive for mobile deployment and comprehensively used in in-field settings.

In order to further emphasize the class-wise performance of the proposed model on D1 and $D1_{500}$ datasets, the confusion matrices are presented in Fig. 3. Overall, the proposed pest recognition model showed high class-wise accuracies for most of the pest classes. Compared to the other pest classes, on both datasets, the Pieris Rapae Linnaeus, Snail and Gryllotalpa pest classes consistently obtained high average accuracies because of their unique characteristics. The low average accuracies are recorded for the pest classes, such as Weevil, Leafhopper and Stink bug. The accuracies are greatly deprived for these pest classes when reducing the number of samples in the dataset (i.e., 5869 samples of D1 to 500 samples of $D1_{500}$). For example, the reduction of 15.15% for Stink bug and 11.75% for Weevil are witnessed from D1 dataset to $D1_{500}$ dataset. Further, the majority of the misclassified samples from Stink bug and Weevil pest classes are confused between those two classes due to a substantial amount of shared characteristics. For instance, 6.00% of the Stink bug pests are confused with the Weevil and vice versa.

Figure 4 provides the qualitative results of the proposed pest recognition model, where the activation maps obtained in different layers are compared. The activation maps are generated for a Locust pest sample, which is correctly classified by both the proposed method and MobileNetv2. The activation maps for the learned feature representations by the proposed model and MobileNetv2 generated on layers 10, 11, 13, 14 and 19 are compared. As can be seen, the activations obtained by the proposed model convey better meanings in respective layers, compared to MobileNetv2. Fro example, it can be observed that the pest parts are more accurately captured in the activation maps generated by the attention layers of the proposed method (layers 11 and 14). This behaviour positively contributes to the overall pest recognition performance.

**Table 3.** The results of an ablative study conducted to select the locations of the double attention modules and the values of the hyperparameters $r_1$ and $r_2$. The average accuracies alongside the number of parameters and FLOPs are presented.

| Locations | | $r_1, r_2$ | Params (Millions) | FLOPs (GMAC) | Acc |
|---|---|---|---|---|---|
| $l_1$ | $l_2$ | | | | |
| 10 | 11 | 4 | 3.80 | 0.3783 | $73.03 \pm 0.25$ |
| 11 | 14 | 4 | 3.99 | 0.4145 | $73.16 \pm 0.38$ |
| 13 | 14 | 4 | 4.17 | 0.4507 | $72.93 \pm 0.26$ |
| 10 | 11 | 6 | 3.70 | 0.3590 | $72.78 \pm 0.23$ |
| 11 | 14 | 6 | 3.83 | 0.3831 | $\mathbf{73.22} \pm 0.35$ |
| 13 | 14 | 6 | 3.95 | 0.4073 | $73.03 \pm 0.36$ |
| 10 | 11 | 8 | 3.65 | 0.3494 | $72.74 \pm 0.38$ |
| 11 | 14 | 8 | 3.75 | 0.3675 | $73.00 \pm 0.24$ |
| 13 | 14 | 8 | 3.84 | 0.3856 | $73.03 \pm 0.48$ |
| 11* | 14* | 1 | 3.56 | 0.3308 | $73.08 \pm 0.29$ |

### 4.4 Ablative Study

In this subsection, an ablative analysis is performed to explain the process behind determining the ideal location of the double attention modules in the proposed lightweight network. Table 3 gives the recorded average accuracies for various combinations of double attention location and $r_1$ and $r_2$ hyperparameter values. The selection of the positions of the double attention modules and the fine-tuning of the hyperparameters $r_1$ and $r_2$ are performed on the more sophisticated CIFAR100 dataset. The locations of the double attention modules are fixed in deeper layers as it is empirically endorsed. However, it is possible to integrate two subsequent double attention layers between 9 and 11 (set of units processing 64 channels) or 12 and 14 (processing 96 channels) or one each in those sets. Although integrating into the second set would process more channels, applying attention interleaved can help in learning the attention well. Hence, Multiple combinations of the double attention module locations (10,11), (11,14) and (13,14) and the hyperparameters $r_1$ and $r_2$ values of 4, 6 and 8 are defined to grid search the best performing tuple with the highest average accuracy. In addition, the number of parameters and FLOPs are also recorded. The best model accuracy is achieved when the double attention modules are affixed in locations 11 and 14 along with six as the value for hyperparameters $r_1$ and $r_2$, which are then utilized to build the proposed model.

The last row of Table 3 shows the accuracy obtained when the double attention modules are integrated directly to the MobileNetv2-backbone after $11^{th}$ and $14^{th}$ layers, instead of deeply integrating into the inverted residual layers. As can be seen, the proposed method achieves better accuracy of 73.22% over this, which achieves only 73.08% accuracy. The hyperparameters $r_1$ and $r_2$ were set to 1 in this case, to maintain the same feature dimensions within the attention module. This shows the attention learning is more effective in the higher dimensional feature space and importance of integrating the double attention module deeply into the inverted residual layer as shown in Fig. 2.

## 5 Conclusion

In order to enable in-field pest recognition model trained with small data, in this paper, a novel double attention-based lightweight architecture is presented. As the MobileNetv2 is utilised as the backbone, this lightweight network showed useful characteristics of less memory consumption and faster inference. The proposed architecture also exploited the benefit of increased performance derived by deploying a double attention mechanism. The evaluation results on two benchmark datasets and one reduced dataset proves the promising capability of the proposed approach in effectively classify plant pests in in-field settings with small data for training. Specifically, the recognition accuracies on small data trends to prevail, by achieving 99.08% and 91.60%. The comparative analysis demonstrates that the proposed approach is superior over existing pest recognition methods. In the future, as new pest classes appear down the line, continual learning ability can be infused to further extend the in-field applicability of the proposed approach. The proposed framework can effectively spot smaller objects and objects confusing with their background. It mainly enhances the features in the identified areas for better performance. This trait is helpful for other applications like chip defect classification, where the defects are tiny and often subtle. In addition, the lightweight aspect of this network enable edge devise deployment.

## References


1. Wang, R., et al.: An automatic system for pest recognition and forecasting. Pest Manag. Sci. **78**(2), 711–721 (2022)
2. Fina, F., Birch, P., Young, R., Obu, J., Faithpraise, B., Chatwin, C.: Automatic plant pest detection and recognition using k-means clustering algorithm and correspondence filters. Int. J. Adv. Biotechnol. Res. **4**(2), 189–199 (2013)
3. Alvarez, J.M., Salzmann, M.: Learning the number of neurons in deep networks. In: Advances in Neural Information Processing Systems, vol. 29 (2016)
4. Krizhevsky, A., Sutskever, I., Hinton, G.E.: Imagenet classification with deep convolutional neural networks. In: Advances in Neural Information Processing Systems, vol. 25 (2012)
5. Gidaris, S., Komodakis, N.: Dynamic few-shot visual learning without forgetting. In: Proceedings of the IEEE Conference on Computer Vision and Pattern Recognition, pp. 4367–4375 (2018)
6. Lake, B., Salakhutdinov, R., Gross, J., Tenenbaum, J.: One shot learning of simple visual concepts. In: Proceedings of the Annual Meeting of the Cognitive Science Society, vol. 33, no. 33 (2011)
7. Wang, Y., Yao, Q., Kwok, J.T., Ni, L.M.: Generalizing from a few examples: a survey on few-shot learning. ACM Comput. Surv. (CSUR) **53**(3), 1–34 (2020)

8. Ye, H.J., Hu, H., Zhan, D.C., Sha, F.: Few-shot learning via embedding adaptation with set-to-set functions. In: Proceedings of the IEEE/CVF Conference on Computer Vision and Pattern Recognition, pp. 8808–8817 (2020)
9. Iandola, F.N., Han, S., Moskewicz, M.W., Ashraf, K., Dally, W.J., Keutzer, K.: SqueezeNet: AlexNet-level accuracy with 50x fewer parameters and <0.5 MB model size. arXiv preprint arXiv:1602.07360 (2016)
10. Zhang, J., Zhu, H., Wang, P., Ling, X.: ATT squeeze U-net: a lightweight network for forest fire detection and recognition. IEEE Access **9**, 10858–10870 (2021)
11. Zhou, N., Liang, R., Shi, W.: A lightweight convolutional neural network for real-time facial expression detection. IEEE Access **9**, 5573–5584 (2020)
12. Zhang, J., Wang, W., Lu, C., Wang, J., Sangaiah, A.K.: Lightweight deep network for traffic sign classification. Ann. Telecommun. **75**(7), 369–379 (2020)
13. Rashid, A.H., Razzak, I., Tanveer, M., Robles-Kelly, A.: RipNet: a lightweight one-class deep neural network for the identification of RIP currents. In: Yang, H., Pasupa, K., Leung, A.C.-S., Kwok, J.T., Chan, J.H., King, I. (eds.) ICONIP 2020. CCIS, vol. 1333, pp. 172–179. Springer, Cham (2020). https://doi.org/10.1007/978-3-030-63823-8_21
14. Howard, A.G., et al.: Mobilenets: efficient convolutional neural networks for mobile vision applications. arXiv preprint arXiv:1704.04861 (2017)
15. Tan, M., Le, Q.: Efficientnet: rethinking model scaling for convolutional neural networks. In: International Conference on Machine Learning, pp. 6105–6114. PMLR (2019)
16. Yang, Z., Yang, X., Li, M., Li, W.: Automated garden-insect recognition using improved lightweight convolution network. Inf. Process. Agric. (2021)
17. Zha, M., Qian, W., Yi, W., Hua, J.: A lightweight YOLOv4-Based forestry pest detection method using coordinate attention and feature fusion. Entropy **23**(12), 1587 (2021)
18. Wang, F., et al.: Residual attention network for image classification. In: Proceedings of the IEEE Conference on Computer Vision and Pattern Recognition, pp. 3156–3164 (2017)
19. Takalkar, M.A., Thuseethan, S., Rajasegarar, S., Chaczko, Z., Xu, M., Yearwood, J.: LGAttNet: automatic micro-expression detection using dual-stream local and global attentions. Knowl.-Based Syst. **212**, 106566 (2021)
20. Banga, K.S., Kotwaliwale, N., Mohapatra, D., Giri, S.K.: Techniques for insect detection in stored food grains: an overview. Food Control **94**, 167–176 (2018)
21. Liu, H., Lee, S.H., Chahl, J.S.: A review of recent sensing technologies to detect invertebrates on crops. Precision Agric. **18**(4), 635–666 (2017)
22. Ngugi, L.C., Abelwahab, M., Abo-Zahhad, M.: Recent advances in image processing techniques for automated leaf pest and disease recognition-a review. Inf. Process. Agric. **8**(1), 27–51 (2021)
23. Nagar, H., Sharma, R.S.: A comprehensive survey on pest detection techniques using image processing. In: 2020 4th International Conference on Intelligent Computing and Control Systems (ICICCS), pp. 43–48. IEEE (2020)
24. Türkoğlu, M., Hanbay, D.: Plant disease and pest detection using deep learning-based features. Turk. J. Electr. Eng. Comput. Sci. **27**(3), 1636–1651 (2019)
25. Rustia, D.J.A., et al.: Automatic greenhouse insect pest detection and recognition based on a cascaded deep learning classification method. J. Appl. Entomol. **145**(3), 206–222 (2021)
26. Faisal, M.S.A.B.: A pest monitoring system for agriculture using deep learning. Res. Progr. Mech. Manuf. Eng. **2**(2), 1023–1034 (2021)

27. Chen, J., Liu, Q., Gao, L.: Deep convolutional neural networks for tea tree pest recognition and diagnosis. Symmetry **13**(11), 2140 (2021)
28. Liu, Y., Zhang, X., Gao, Y., Qu, T., Shi, Y.: Improved CNN method for crop pest identification based on transfer learning. Comput. Intell. Neurosci. (2022)
29. Turkoglu, M., Yanikoğlu, B., Hanbay, D.: PlantDiseaseNet: convolutional neural network ensemble for plant disease and pest detection. SIViP **16**(2), 301–309 (2022)
30. Zhang, S., Jing, R., Shi, X.: Crop pest recognition based on a modified capsule network. Syst. Sci. Control Eng. **10**(1), 552–561 (2022)
31. Xu, C., Yu, C., Zhang, S., Wang, X.: Multi-scale convolution-capsule network for crop insect pest recognition. Electronics **11**(10), 1630 (2022)
32. Li, Y., Yang, J.: Few-shot cotton pest recognition and terminal realization. Comput. Electron. Agric. **169**, 105240 (2020)
33. Rustia, D.J.A., et al.: Online semi-supervised learning applied to an automated insect pest monitoring system. Biosys. Eng. **208**, 28–44 (2021)
34. Janarthan, S., Thuseethan, S., Rajasegarar, S., Lyu, Q., Zheng, Y., Yearwood, J.: Deep metric learning based citrus disease classification with sparse data. IEEE Access **8**, 162588–162600 (2020)
35. Liang, K., Wang, Y., Sun, L.: Imbalance data set classification of tomato pest based on lightweight CNN model (2021)
36. Zhao, S., Liu, J., Bai, Z., Hu, C., Jin, Y.: Crop pest recognition in real agricultural environment using convolutional neural networks by a parallel attention mechanism. Front. Plant Sci. **13** (2022)
37. Yang, X., Luo, Y., Li, M., Yang, Z., Sun, C., Li, W.: Recognizing pests in field-based images by combining spatial and channel attention mechanism. IEEE Access **9**, 162448–162458 (2021)
38. Chen, J., Chen, W., Zeb, A., Zhang, D., Nanehkaran, Y.A.: Crop pest recognition using attention-embedded lightweight network under field conditions. Appl. Entomol. Zool. **56**(4), 427–442 (2021). https://doi.org/10.1007/s13355-021-00732-y
39. Chen, Y., Kalantidis, Y., Li, J., Yan, S., Feng, J.: $A^2$-nets: double attention networks. In: Advances in Neural Information Processing Systems, vol. 31 (2018)
40. He, K., Zhang, X., Ren, S., Sun, J.: Deep residual learning for image recognition. In: Proceedings of the IEEE Conference on Computer Vision and Pattern Recognition, pp. 770–778 (2016)
41. Li, Y., Wang, H., Dang, L.M., Sadeghi-Niaraki, A., Moon, H.: Crop pest recognition in natural scenes using convolutional neural networks. Comput. Electron. Agric. **169**, 105174 (2020)
42. Cheng, X., Zhang, Y., Chen, Y., Wu, Y., Yue, Y.: Pest identification via deep residual learning in complex background. Comput. Electron. Agric. **141**, 351–356 (2017)
43. Xie, C., et al.: Automatic classification for field crop insects via multiple-task sparse representation and multiple-kernel learning. Comput. Electron. Agric. **119**, 123–132 (2015)

# A Deep Investigation of RNN and Self-attention for the Cyrillic-Traditional Mongolian Bidirectional Conversion

Muhan Na[1,2,3], Rui Liu[1,2,3](✉), Feilong Bao[1,2,3], and Guanglai Gao[1,2,3]

[1] College of Computer Science, Inner Mongolia University, Hohhot, China
liurui_imu@163.com, {csfeilong,csggl}@imu.edu.cn
[2] National and Local Joint Engineering Research Center of Intelligent Information Processing Technology for Mongolian, Hohhot, China
[3] Inner Mongolia Key Laboratory of Mongolian Information Processing Technology, Hohhot, China

**Abstract.** Cyrillic and Traditional Mongolian are the two main members of the Mongolian writing system. The Cyrillic-Traditional Mongolian Bidirectional Conversion (CTMBC) task includes two conversion processes, including Cyrillic Mongolian to Traditional Mongolian (C2T) and Traditional Mongolian to Cyrillic Mongolian conversions (T2C). Previous researchers adopted the traditional joint sequence model, since the CTMBC task is a natural Sequence-to-Sequence (Seq2Seq) modeling problem. Recent studies have shown that Recurrent Neural Network (RNN) and Self-attention (or Transformer) based encoder-decoder models have shown significant improvement in machine translation tasks between some major languages, such as Mandarin, English, French, etc. However, an open problem remains as to whether the CTMBC quality can be improved by utilizing the RNN and Transformer models. To answer this question, this paper investigates the utility of these two powerful techniques for CTMBC task combined with agglutinative characteristics of Mongolian language. We build the encoder-decoder based CTMBC model based on RNN and Transformer respectively and compare the different network configurations deeply. The experimental results show that both RNN and Transformer models outperform the traditional joint sequence model, where the Transformer achieves the best performance. Compared with the joint sequence baseline, the word error

This research is funded by the National Key Research and Development Program of China (No. 2018YFE0122900), China National Natural Science Foundation (No. 62066033), the High-level Talents Introduction Project of Inner Mongolia University (No. 10000-22311201/002) and the Young Scientists Fund of the National Natural Science Foundation of China (No. 62206136), Applied Technology Research and Development Program of Inner Mongolia Autonomous Region (No. 2019GG372, 2020GG0046, 2021GG0158, 2020PT002), Young science and technology talents cultivation project of Inner Mongolia University (No. 21221505), The Research Program of The National Social Science Fund of China (No. 18XYY030).


M. Tanveer et al. (Eds.): ICONIP 2022, CCIS 1793, pp. 612–625, 2023.
https://doi.org/10.1007/978-981-99-1645-0_51

rate (WER) of the Transformer for C2T and T2C decreased by 5.72% and 5.06% respectively.

**Keywords:** Cyrillic Mongolian · Traditional Mongolian · Bidirectional Conversion · Recurrent Neural Network (RNN) · Self-attention

# 1 Introduction

Mongolian language belongs to the language group of the Altaic language family and is both the most widely spoken and most-known member of the Mongolic language family [1]. The number of speakers across all its dialects may be 5.2 million, including the vast majority of the residents of Mongolia and many of the ethnic Mongol residents of the Inner Mongolia Autonomous Region of the People's Republic of China. In Mongolia, the Mongolian character is currently written in Cyrillic Mongolian script [2]. In Inner Mongolia of China, the language is written in the Traditional Mongolian script [1]. The Cyrillic-Traditional Mongolian Bidirectional Conversion (CTMBC) [3] consists of Cyrillic Mongolian to Traditional Mongolian conversion (C2T) and Traditional Mongolian to Cyrillic Mongolian conversion (T2C). Therefore, the CTMBC facilitates the language communication between the compatriots of both countries and has great importance to the scientific, economic, and cultural fields of both countries.

The traditional method mainly focuses on the rule-based approach and the statistical model, such as the joint sequence model. Specifically, Bao Sarina et al. [4,5] focused on the nouns and case suffixes translation in the CTMBC task and proposed a hybrid translation method that includes bilingual dictionary, rules and N-gram language model. Gao et al. [6] also proposed a hybrid method based on dictionaries and rules. Feilong et al. [3] first adopted joint sequence model for the CTMBC task. In 2017, Feilong et al. [7] further proposed a hybrid method based on a combination of rule and joint sequence model. It adopts the rule-based approach to convert the words in the vocabulary, and uses joint sequence model to convert the out-of-vocabulary (OOV) words.

However, the conversion methods based on rule or statistical model perform some shortcomings. 1) The rule-based approach is unavailable when facing OOV words and loanwords [7]: Mongolian words are made up of stems and suffixes. Different types of suffixes are added to the same stem to form different words, which leads to a huge vocabulary; Furthermore, the Mongolian language contains many loanwords that do not follow the Mongolian word-formation rules [1]; 2) The statistical model performs poor generalization ability and limited modeling capability: the joint sequence model holds shallow architecture and does not have strong nonlinear modeling ability [8,9].

We note that the CTMBC task can be interpreted as a standard machine translation task [3] between two languages. In other words, it is also a sequence-to-sequence (Seq2Seq) [9] modeling problem. The "encoder-decoder" structure [8–10] have been successfully applied to various Seq2Seq tasks, including neural machine translation [9,11,12], speech synthesis [13–17], and the grapheme-to-phoneme (G2P) conversion [18–20], etc. Recurrent neural networks (RNN) and self-attention based "encoder-decoder" models and pre-training language model [21,22] have lately received attention from the community [8,12,23].

For the CTMBC task, we don't have enough data to train a large-scale pre-training model. Meanwhile, the RNN and Self-attention based encoder-decoder models have great performance in Seq2Seq modeling tasks. However, an open problem remains as to whether the CTMBC performance can be improved by utilizing the RNN and Transformer models. To answer this question, this paper investigates the utility of these two powerful techniques without large-scale pre-training model for CTMBC task combined with agglutinative characteristics of Mongolian language. In this study, we validate the RNN and self-attention in the CTMBC task respectively. 1) We deep study the agglutinative characteristics of the Mongolian language, including the Traditional and Cyrillic Mongolian; 2) We build RNN and self-attention based CTMBC models according to the analyzed agglutinative characteristics; 3) To identify the optimal network configurations, we investigate and compare the RNN and self-attention models with various configurations. The experimental results show that both RNN and self-attention models outperform the traditional joint sequence model, where the self-attention model achieves the best performance.

The main contributions of this paper include, 1) We conduct a deep investigation of RNN and self-attention for the CTMBC task; 2) We combine the agglutinative characteristics of Mongolian language with RNN and self-attention models to achieve outstanding performance. To our best knowledge, this is the first deep investigation of the recent powerful deep learning models, including RNN and self-attention models, for the CTMBC task.

## 2 Task Challenges

### 2.1 Data Sparseness

For machine translation tasks, a large amount of aligned sentence-level data is required. Although this paper treats CTMBC as a word-level machine translation task, there is currently no high-quality and large-scale training data for model training due to the low-resource nature of Mongolian language.

### 2.2 Agglutinative Characteristics

For the Mongolian written form, there are two styles including Cyrillic Mongolian and Traditional Mongolian scripts. The similarity and difference between them in terms of agglutinative Characteristics can be summarized as follows.

**Similarity.** The similarity between Traditional and Cyrillic Mongolian scripts are summarized as word formation, pronunciation, and grammar rules.

- **Word formation:** Cyrillic Mongolian and Traditional Mongolian is an agglutinative language in which a new word is created by joining multiple suffixes to a word stem.

- **Pronunciation:** Cyrillic Mongolian can always find at least one Traditional Mongolian with the same pronunciation.
- **Grammar rules:** Cyrillic Mongolian retains most of the grammatical features of Traditional Mongolian. They are consistent in grammar.

**Difference.** The difference between Traditional and Cyrillic Mongolian scripts are summarized as symbol systems, morphological rules and the correlation between pronunciation and spelling.

- **Different symbol systems:** Cyrillic script has 13 vowels, 20 consonants, 1 hardened character, and 1 softened character [7]. Traditional script has only 8 vowels and 27 consonants [7]. Cyrillic script performs case-sensitive and follows the rule of capitalization of the initial letter, Traditional Mongolian is not so. Traditional Mongolian have inconsistencies between code and presentation form. To avoid this phenomenon, in this work, we transform the Traditional Mongolian characters into their corresponding Latin transcriptions [24].
- **Different morphological rules:** The morphological rules of Cyrillic Mongolian has 66 categories [25,26], while Traditional Mongolian just has 4 categories [1]. Such difference lead to the fact that the characters of the two kind of Mongolian words can not correspond one by one. For example, Traditional Mongolian word ᠪᠠᠷᠢᠶᠠᠰᠤ (Latin: bariyasv) has 8 characters, and its corresponding Cyrillic Mongolian word бариас has 6 characters.
- **Different correlation between pronunciation and spelling:** The pronunciation and spelling of Cyrillic Mongolian are one-to-one correspondence, Traditional Mongolian is otherwise, occur a one-to-many relationship. In Traditional Mongolian, the vowels or consonants may be dropped, added, or changed when reading. As a result, one Cyrillic Mongolian word may correspond to more than one Traditional Mongolian word. For example, Cyrillic Mongolian for both ᠳᠠᠯᠠ (Latin: dala, means: expand) and ᠳᠠᠯᠤ (Latin: dalv, means: shoulder) is дал.

Mongolian words are constructed by successively concatenating suffixes to stems, which results in the word forms in Mongolian being very large and having a lot of out-of-vocabulary words. Meanwhile, different morphological rules make it difficult for the model to learn the mapping relationship between two Mongolian characters.

The above data sparseness and agglutinative characteristics problems bring huge challenge for our CTMBC task. The agglutinative characteristics of Traditional and Cyrillic Mongolian scripts are the knowledge we must master for the CTMBC task. In the next section, we will fully explore the above language knowledge to complete the C2T and T2C conversion.

## 3 RNN and Self-attention Based Encoder-Decoder Frameworks for C2T and T2C

In this section, we will introduce the overall Encoder-Decoder frameworks for C2T and T2C, at first. Then we will explain their workflows in detail respectively.

Last and not least, the backbone of the encoder-decoder framework, including RNN and self-attention, will be introduced.

### 3.1 Overall Framework

As shown in Fig. 1, the overall frameworks of C2T and T2C are illustrated in the upper and bottom panels respectively. We follow G2P [19] workflow to build our C2T and T2C frameworks, in which the input and output both are word-level. As mentioned in Sect. 2.2, there are some similarities and differences between Cyrillic and traditional Mongolian scripts in terms of agglutinative Characteristics. Therefore, we make necessary processing for Mongolian scripts and design the workflows for C2T and T2C specifically.

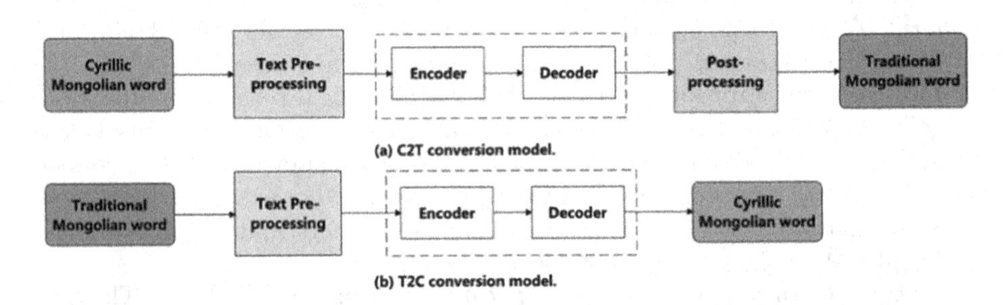


**Fig. 1.** The overall frameworks of (a) C2T conversion model, (b) T2C conversion model. Our task is similar to G2P task [19]. In the process of T2C conversions, our input is Traditional Mongolian word and output is Cyrillic Mongolian word, while vice versa in C2T conversion.

### 3.2 Workflow of C2T

As shown in Fig. 1(a), C2T conversion consists of 3 parts, including "Text Pre-processing", "Encoder-Decoder" and "Post-processing". Text Pre-processing takes the Cyrillic Mongolian word as input to output the character sequence. Then the encoder reads the character sequence to generate the high-level hidden representation, which is then fed to the decoder to predict the Latin transcriptions of Traditional Mongolian characters. At last, Post-processing module converts the Latin transcription to the Traditional Mongolian scripts as the final conversion results.

**Text Pre-processing.** Given that $C$ is the input Cyrillic Mongolian word, Text Per-processing can output the character sequence $\mathbf{c} = c_1, c_2, c_3, c_4, ..., c_n$ ($n$ means the sequence length or character number). Words and sub-words also are two possible training units. However, the complex word or sub-word generation pipeline will bring some noise and bring about side effects. In other words, incorrect tokenization or separation results will affect model performance.

**Encoder-Decoder.** The encoder summarizes the input sequence, Cyrillic Mongolian characters, into a set of vectors while the decoder conditions the encoded input sequence, and generates the output sequence, Traditional Mongolian Latin transcription characters, one token at a time. Let $\mathbf{o} = o_1, o_2, ..., o_m$ be the sequence of $m$ symbols in The encoder is simply a function of the following form: $x = Encoder(c_1, c_2, ..., c_n)$, in which $\mathbf{x} = x_1, x_2, ..., x_m$ is a list of fixed size vectors.

The decoder is often trained to predict the next word $o_t$ $(t \in [1, m])$ given all the previously predicted words $o_1, o_2, ..., o_{t-1}$. In other words, the decoder defines a probability over the conversion $\mathbf{o}$ by decomposing the joint probability into the ordered conditionals: $P(o|x) = \prod_{t=1}^{M} P(o_t|o_1, o_2, ..., o_{t-1}; x_1, x_2, ..., x_n)$.

**Post-processing.** The output of decoder is character-level Latin transcription. Therefore, Post-processing aims to restore the Latin transcription to Traditional Mongolian word $T$.

### 3.3 Workflow of T2C

As shown in Fig. 1(b), different from the C2T conversion process, T2C conversion consists of the following 2 parts, including "Text Pre-processing" and "Encoder-Decoder". Text Pre-processing module aims to convert Traditional Mongolian word to their character-level Latin transcription. The Encoder processes the character-level Latin transcription and outputs the high-level representation. Decoder module predicts the Cyrillic Mongolian as the result.

**Text Pre-processing.** There are two steps in the Text Pre-processing of T2C. First, convert the Traditional Mongolian into Latin transcription. Second, divide word-level Latin transcription into character-level Latin sequence. As mentioned in Sect. 2.2, in this work, we transform the Traditional Mongolian characters to their corresponding Latin transcription for model training. Similar with Sect. 3.2, we also divide Latin transcription into Latin characters.

**Encoder-Decoder.** In the T2C task, the encoder-decoder module is the same as that, as described in Sect. 3.2 in the C2T task. The only difference is that the input of T2C task is Traditional Mongolian Latin transcription characters and the output is Cyrillic Mongolian characters.

### 3.4 Backbone of Encoder-Decoder Framework

In this section, we use C2T as an example to introduce the backbone of Encoder-Decoder Framework. To validate the RNN and Self-attention for the CTMBC task, we employ attention based RNN and self-attention model as the backbone of Encoder-Decoder framework. We will introduce the details next.

**RNN.** Specifically, RNN-based encoder is implemented with a multi-layer bidirectional LSTM (BiLSTM), which transforms the input Cyrillic Mongolian sequence $c$ into a high-level representation $h = \{h_n\}_{n=1}^{N}$. The RNN-based decoder takes $h$ and all previously predicted outputs $o_{1:m-1}$ as input, producing the probability distribution of the token $o_m$. Specifically, at each decoder time step $t$, the posterior distribution of the predicted output is generated from the cascade of decoder states $s_t$ and context vector $a_t$. The $a_t$ is the context information produced by the attention module based on the hidden state of the encoder and decoder.

**Self-attention.** Unlike the RNN-based encoder-decoder framework, the self-attention based encoder-decoder framework, that is Transformer, replaces the RNN modules with the pure self-attention mechanism. Specifically, Transformer encoder consists of $N$ identical Transformer blocks [27]. Each block consists of two sub-layers, including the multi-head self-attention mechanism and the fully connected feed-forward network. Residual connection and normalization are added to each sub-layers.

Transformer decoder also consists of $N$ identical Transformer blocks [27], which include masked multi-head self-attention, multi-head self-attention and the fully connected feed-forward network. Residual connection and normalization are also added to each sub-layers. In addition, different from the RNN based model, the Transformer encoder and decoder take a position encoding [27] as an additional input.

## 4 Experiments and Results

### 4.1 Datasets

We report the experiments on a Cyrillic-Traditional Mongolian mapping dictionary, denoted as "*Mon_data_63668*". The *Mon_data_63668* dataset includes 63668 word pairs which collected from the New Mongolian-Chinese Dictionary [28]. Note that 58436 pairs were randomly selected as the training set, and the remaining 5232 word pairs were used as the model testing set.

### 4.2 Evaluation Setup

**Comparative Study.** We implement three frameworks for C2T and T2C in a comparative study. RNN and self-attention based encoder-decoder models are studied for the first time for CTMBC, while Joint sequence model is the baseline and re-implementation of [3].

- **Joint Sequence Model ("Joint" for short) (Baseline):** The idea of the joint sequence model [18] is to represent the relationship between an input sequence and an output sequence in terms of a common sequence of joint units composed of input and output symbols. N-gram language model is used to predict character during decoding. Note that we will seek the optimal model configuration by adjusting the value of $N$ in N-gram.We infer the model parameters through the expectation maximization and trim the evidence to avoid over-fitting. We also discount evidence for smoothing.

- **RNN based Encoder-Decoder (New):** We implement two architectures, including RNN based Encoder-Decoder without attention ("RNN" for short) and the attention based RNN Encoder-Decoder as mentioned in Sect. 3.4 ("RNN+ATT" for short), for comparison. Note that we set the RNN hidden size to 512 and 1024, and set the hidden layer to 1, 2, and 4 to explore the optimal model configuration.
  For RNN training, the input of encoder is 128-dimensional character sequence and the output of decoder is mongolinan characters. The parameters of RNN are set as following: batch size = 32, epochs = 100, learning rate = 0.0005. We decrease the learning rate every 20 epochs by a factor of 0.9. The loss function is using the cross-entropy(CE) criterion. We conduct experiments on an NVIDIA GPU (Tesla P40).
- **Self-attention based Encoder-Decoder ("Transformer" for short) (New):** "Transformer-Tiny" model [29] to build our self-attention based Encoder-Decoder model since it holds a smaller vocabulary than "Transformer-base" [29] model and therefore is more suitable for our task. Note that we set the attention head to 2 and 4, and set the hidden layer to 1, 2, 4, and 6 to explore the optimal parameter combination.

All the comparative experiments are conducted in the Tensor2Tensor [29]. We set embedding dimensional = 128, train steps = 100000, batch size = 4096, learning rate = 0.2 and the learning rate warm-up [30] steps to 8000. All word embeddings are initialized randomly and then updated with the whole model. We conduct experiments on an NVIDIA GPU (Tesla P40).

**Metrics.** The result is performed in terms of word error rate (WER) and character error rate (CER). The formulas of WER and CER are as follows: $WER = 1 - \frac{N_{corrent}}{N_{total}}$, $CER = \frac{N_{ins}+N_{del}+N_{sub}}{N_{charatertotal}}$. $N_{corrent}$ is the number of correctly predicted Mongolian words; $N_{total}$ is size of test set; $N_{ins}$ is the number of character insertion errors; $N_{del}$ is the number of character deletion errors; $N_{sub}$ is the number of character substitution errors; $N_{charatertotal}$ is the number of word characters.

Following [18,20] in the case of multiple references Mongolian, the variant with the smallest edit distance is used. Similarly, if there are multiple references of Mongolian for a word, a word error occurs only if the predicted Mongolian doesn't match any of the references.

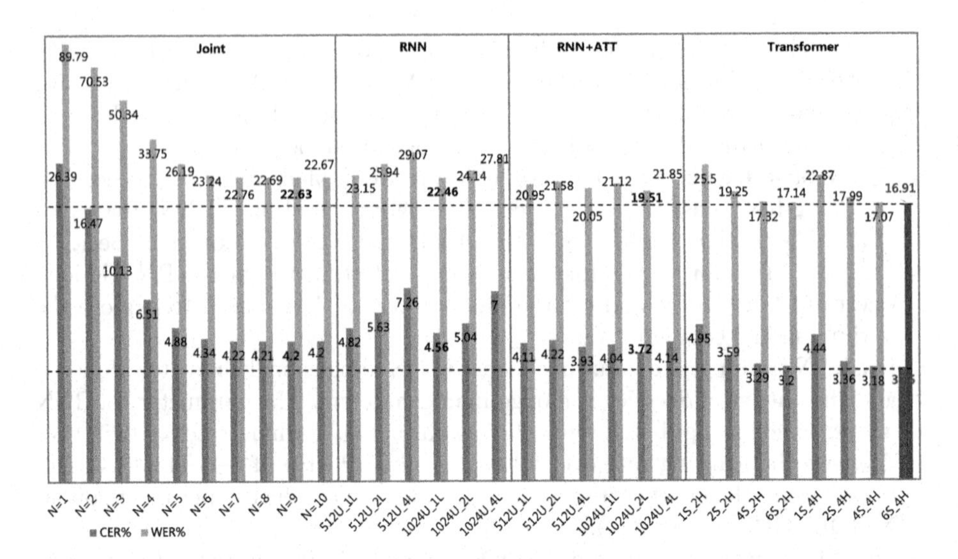


**Fig. 2.** The WER and CER results of C2T. The first panel is results of Joint sequence model and "N" is delegates N of N-gram. The second panel and third panel is results of RNN based model and RNN+ATT based model, "U" represents BiLSTM hidden units and "L" represents BiLSTM hidden layer. The fourth panel is results of Transformer model under different parameters, "S" represents the layer size of Transformer and "H" represents the attention head number.

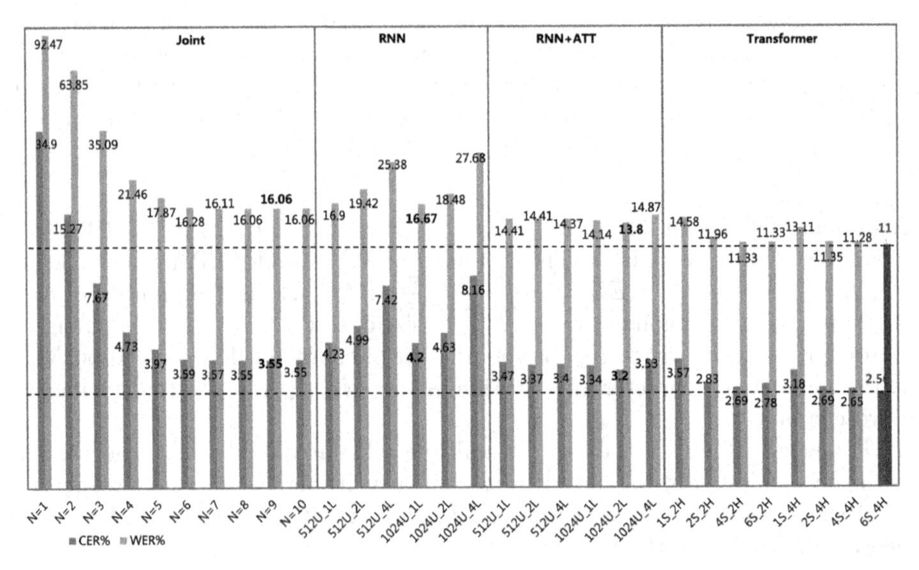


**Fig. 3.** The WER and CER results of T2C.

### 4.3 Performance Comparison for C2T and T2C

Figure 2 shows the WER and CER results of all models under different parameters for C2T. We will analysis the results next.

**Joint vs. (RNN & RNN+ATT).** As shown in the first panel of Fig. 2, by increasing $N$ from 1 to 10, we can find that the WER and CER are basically decreasing gradually. "N = 9" for optimal performance, with with WER = 22.63% and CER = 4.2%.

From the second panel of Fig. 2, the 1-layer BiLSTM with 1024 hidden units obtains the best performance, which can reach 4.56% on CER and 22.46% on WER. The optimal result, RNN(1024U_1L), is almost the same as "N = 9". We also fix hidden layer number of BiLSTM and adjust the hidden unit size to observe the performance. It can be found that the performance of the model can be improved by increasing the hidden unit size. For example, RNN(512U_1L) outperforms the RNN(1024U_1L) model. Stacking more hidden layers sometimes does not lead to significant improvements. For example, RNN(1024U_1L) beats the RNN(1024U_4L) model.

In the third panel of Fig. 2, we can find that the 2-layer BiLSTM with 1024 hidden units, RNN+ATT(1024U_2L), obtains the best performance, which can reach 3.72% in CER and 19.51% in WER. Compared with the optimal results of RNN, RNN(1024U_1L) model, the WER of RNN+ATT(1024U_2L) is reduced by 2.95%. We believe that adding the attention mechanism to the RNN based encoder-decoder model can effectively capture the internal information of input Mongolian character sequence better and improve the model performance.

T2C has similar results as C2T. As shown in Fig. 3, RNN(1024U_L1) model outperforms the "N = 8" model, RNN+ATT(1024U_L2) model beat the "N = 8" model, which consists of C2T.

**Joint vs. Transformer.** The fourth panel of Fig. 2 show the WER and CER of Transformer model under different parameters.

The results show that the Transformer framework achieves best performance when layer size is 6 and attention head is 4. "6S_4H" achieves 3.15% on CER and 16.92% on WER and gains 5.72% relative reduction compared to the "N = 9" model in terms of WER. We also find that increasing the attention head number can effectively improve the model performance. For example, the "6S_4H" and the "4S_4H" outperforms the "6S_2H" and the "4S_2H" respectively.

The result of T2C is similar to C2T. In Fig. 3, "6S_4H" model outperforms the "N = 8" model, which has identical conclusion to C2T.

**(RNN & RNN+ATT) vs. Transformer.** Comparing the optimal results of RNN and Transformer, Transformer achieves better performance. For example, the "6S_4H" achieves 16.92% WER and outperforms RNN(1024U_1L) with 5.54% WER. The self-attention mechanism has stronger ability to capture long-term dependencies within input sequence, therefore brings better performance. Comparing RNN+ATT(1024U_2L) and "6S_4H", "6S_4H" beat RNN+ATT(1024U_2L) with 2.95% WER. The reason is that the self-attention mechanism can capture global context information, it is helpful for model prediction.

The results of T2C resemble C2T. As shown in Fig. 3, "6S_4H" model beat the RNN(1024U_1L) and RNN+ATT(1024U_2L), which is consistent with C2T.

### 4.4 Case Study

To analyze the translated words produced by the various model, we use two examples extracted from test set to compare them in Fig. 4. The wrong translated characters are highlighted in blue.

| Cyrillic Mongolian Word | алгуувтар |
|---|---|
| English meaning | slower |
| Traditional Mongolian Word | [illegible] |
| Latin transcription (Ground Truth) | algavbtvr |
| Joint(N=9) | aldartan |
| RNN(1024U_1L) | algvvbtvr |
| RNN+ATT(1024U_2L) | algvbtvr |
| **Transformer(6S_4H)** | **algavbtvr** |

(a) Case study of C2T.

| Cyrillic Mongolian Word (Ground Truth) | аравтын |
|---|---|
| English meaning | ten |
| Traditional Mongolian Word | [illegible] |
| Latin transcription | arbatv-yin |
| Joint(N=8) | аравтайин |
| RNN(1024U_1L) | арватан |
| RNN+ATT(1024U_2L) | аравтуйн |
| **Transformer(6S_4H)** | **аравтын** |

(b) Case study of T2C.

**Fig. 4.** The example of translated words produced by various models. The Latin characters not correctly translated are highlighted in blue. The long vowels and suffix are highlighted in red. (Color figure online)

As shown in Fig. 4(a), we take a Cyrillic Mongolian word "алгуувтар" as an example to show the C2T conversion results. The Latin transcriptions is "alagavbtvr". Note that there is vowel substitution phenomenon in the pronunciation of this word. The traditionally Mongolian word is pronounced with the fifth vowel "a" becoming "v" and the ninth vowel "v" becoming "a". The fifth row shows the conversion result of Joint model. We can find that there are 4 wrong characters in the results, which performs poor performance, while RNN and RNN+ATT models just have 2 and 1 wrong characters respectively. We further observe that the translated word output from Transformer model exactly the same as the target Latin transcriptions "alagavbtvr" and contains no errors.

Similar with Fig. 4(a), as shown in the Fig. 4(b), we also take a traditional Mongolian word "[illegible]" as an example to show the T2C conversion results. The Latin transcriptions is "arbatv-yin". We observe the Joint, RNN, and RNN+ATT models all have some wrong conversion characters, while Transformer model contains no errors. Note that the Transformer improves the conversion accuracy of long vowels and suffix that are highlighted in red.

The case study results show that the RNN and Transformer are stronger than Joint model and Transformer achieves the best performance, which is consistent with our previous results.

## 5 Conclusion

In this paper, we introduce RNN and self-attention based models into the Cyrillic-Traditional Mongolian Bidirectional Conversion (CTMBC) task. This is the first deep investigation of the recent powerful deep learning models for the CTMBC task. Compared with the joint sequence model baseline, RNN based models and self-attention model gains significant improvements. Note that self-attention model achieves the best performance at both T2C and C2T tasks. In addition, we also compared the parameter setting of each model, such as hidden layer number, hidden unit size, attention head number, etc., in detail to determine the optimal model configuration. Note that this paper mainly studies the CTMBC task at word-level, which similar with G2P. However, word level conversion is plagued by frequent polyphonic words and lack of contextual information, which may limit model performance. In the future work, we will focus on CTMBC task at sentence-level to make full use of context information for more accurate conversion performance.

## References

1. Chinggaltai: A Grammar of the Mongolian Language. Inner Mongolia Peoples Publishing House, Hohhot (1991)
2. Chuma, S.: A Comparative Study of Mongolian and Cyrillic Orthography. Inner Mongolia Education Press, Hohhot (2010)
3. Bao, F., Gao, G., Yan, X., Wei, H.: Research on conversion approach between traditional Mongolian and Cyrillic Mongolian. Comput. Eng. Appl. 206–211 (2014)
4. Li, H., Sarina, B.: The study of comparison and conversion about traditional Mongolian and Cyrillic Mongolian. In: 2011 4th International Conference on Intelligent Networks and Intelligent Systems, pp. 199–202. IEEE (2011)
5. Sarina, B.: The research on conversion of noun and its case from Classic Mongolian into Cyrillic Mongolian. Master's thesis, Inner Mongolia University (2009)
6. Gao, H., Ma, x.: Automatic system on Cyrillic Mongolian webpage conversion to traditional Mongolian script. J. Inner Mongolia Univ. Nationalities **18**(5), 17–18 (2012)
7. Bao, F., Gao, G., Wang, H., Lu, M.: Combining of rules and statistics for Cyrillic Mongolian to traditional Mongolian conversion. J. Chin. Inf. Procession **31**(3), 156–162 (2017)
8. Cho, K., et al.: Learning phrase representations using RNN encoder-decoder for statistical machine translation. Comput. Sci. (2014)
9. Sutskever, I., Vinyals, O., Le, Q.V.: Sequence to sequence learning with neural networks. In: Advances in Neural Information Processing Systems, vol. 27 (2014)

10. Kalchbrenner, N., Blunsom, P.: Recurrent continuous translation models. In: Proceedings of the 2013 Conference on Empirical Methods in Natural Language Processing, pp. 1700–1709 (2013)
11. Bahdanau, D., Cho, K., Bengio, Y.: Neural machine translation by jointly learning to align and translate. Comput. Sci. (2014)
12. Wu, Y., et al.: Google's neural machine translation system: bridging the gap between human and machine translation. arXiv preprint arXiv:1609.08144 (2016)
13. Liu, R., Bao, F., Gao, G., Zhang, H., Wang, Y.: Improving Mongolian phrase break prediction by using syllable and morphological embeddings with BiLSTM model. In: Interspeech, pp. 57–61 (2018)
14. Liu, R., Sisman, B., Bao, F., Gao, G., Li, H.: Modeling prosodic phrasing with multi-task learning in Tacotron-based TTS. IEEE Sig. Process. Lett. **27**, 1470–1474 (2020)
15. Liu, R., Sisman, B., Bao, F., Yang, J., Gao, G., Li, H.: Exploiting morphological and phonological features to improve prosodic phrasing for Mongolian speech synthesis. IEEE/ACM Trans. Audio Speech Lang. Process. **29**, 274–285 (2020)
16. Liu, R., Sisman, B., Li, J., Bao, F., Gao, G., Li, H.: Teacher-student training for robust Tacotron-based TTS. In: ICASSP 2020–2020 IEEE International Conference on Acoustics, Speech and Signal Processing (ICASSP), pp. 6274–6278. IEEE (2020)
17. Wang, Y., et al.: Tacotron: towards end-to-end speech synthesis. arXiv preprint arXiv:1703.10135 (2017)
18. Bisani, M., Ney, H.: Joint-sequence models for grapheme-to-phoneme conversion. Speech Commun. **50**(5), 434–451 (2008)
19. Wang, Y., Bao, F., Zhang, H., Gao, G.: Joint alignment learning-attention based model for grapheme-to-phoneme conversion. In: ICASSP 2021–2021 IEEE International Conference on Acoustics, Speech and Signal Processing (ICASSP), pp. 7788–7792. IEEE (2021)
20. Yao, K., Zweig, G.: Sequence-to-sequence neural net models for grapheme-to-phoneme conversion. Comput. Sci. (2015)
21. Devlin, J., Chang, M.W., Lee, K., Toutanova, K.: BERT: pre-training of deep bidirectional transformers for language understanding, pp. 4171–4186. Association for Computational Linguistics, June 2019. https://doi.org/10.18653/v1/N19-1423, https://aclanthology.org/N19-1423
22. Yang, Z., Dai, Z., Yang, Y., Carbonell, J., Salakhutdinov, R.R., Le, Q.V.: Xlnet: generalized autoregressive pretraining for language understanding. In: Advances in Neural Information Processing Systems, vol. 32 (2019)
23. Cho, K., Van Merriënboer, B., Bahdanau, D., Bengio, Y.: On the properties of neural machine translation: encoder-decoder approaches. Comput. Sci. (2014)
24. Lu, M., Bao, F., Gao, G.: Language model for Mongolian polyphone proofreading. In: Sun, M., Wang, X., Chang, B., Xiong, D. (eds.) CCL/NLP-NABD -2017. LNCS (LNAI), vol. 10565, pp. 461–471. Springer, Cham (2017). https://doi.org/10.1007/978-3-319-69005-6_38
25. Uganbater, D.: Research on Cyrillic and Mongolian script's morphlolgy and conversion system. Ph.D. thesis, Inner Mongolia University (2014)
26. Dulamragchaa, U., Chadraabal, S., Ivanov, B., Baatarkhuu, M.: Mongolian language morphology and its database structure. In: 2017 International Conference on Green Informatics (ICGI), pp. 282–285. IEEE (2017)
27. Vaswani, A., et al.: Attention is all you need. In: Advances in Neural Information Processing Systems, vol. 30 (2017)

28. Zhang, Z.: New Mongolian Chinese Dictionary. Commercial Press, Beijing (2011)
29. Vaswani, A., et al.: Tensor2tensor for neural machine translation. arXiv preprint arXiv:1803.07416 (2018)
30. Popel, M., Bojar, O.: Training tips for the transformer model. arXiv preprint arXiv:1804.00247 (2018)

# Sequential Recommendation Based on Multi-View Graph Neural Networks

Hongshun Wang[1], Fangai Liu[1(✉)], Xuqiang Zhang[1], and Liying Wang[2]

[1] School of Information Science and Engineering, Shandong Normal University, Jinan 250358, China
lfa@sdnu.edn.cn

[2] College of Cloud Computing Technology and Application Industry, Shandong Institute of Commerce and Technology, Jinan 250103, China

**Abstract.** The problem of sequential recommendation aims to use a user's historical interaction sequence to predict and recommend the following item with which the user is most likely to interact. A range of outcomes has been attained in this field, from traditional methods to deep learning approaches. There are still two challenging problems with existing methods. First, users' interaction sequences will always include noisy interest preferences. Additionally, most methods ignore the potential collaborative information among different features and cannot fully explore users' true preferences. To solve these problems, we propose SR-MVG (Short for Sequential Recommendation based on Multi-View Graph Neural Networks) for sequential recommendation, which first transforms the user's behavioral sequence into an item-item graph so that similar items are closely connected to clearly distinguish the core interests of users. Second, the user's core preferences are retained adaptively by the designed multi-view attention network. In addition, we also designed a graph pooling strategy to reduce noise and extract more relevant user preferences. We carried out extensive tests on five public benchmarks, and the findings demonstrate that SR-MVG exhibits superior performance.

**Keywords:** Sequential Recommendation · Graph Neural Network · Graph Pooling · Deep Learning

## 1 Introduction

The user's behavior sequence contains a wealth of past behavioral data about them, indicating both their long-term, stable interest choices and their most recent shifts in interest. The interaction between users and items, which is serially correlated by nature, and the user's preference for items, which changes dynamically over time, are two key pieces of information that are ignored by traditional recommendation systems like collaborative filtering and content-based recommendation systems.


M. Tanveer et al. (Eds.): ICONIP 2022, CCIS 1793, pp. 626–637, 2023.
https://doi.org/10.1007/978-981-99-1645-0_52

Based on the above-mentioned issues, sequence recommendation has attracted a lot of attention from researchers due to its modeling of continuous user behavior, considering the dependencies in the sequence of user interactions. Existing works such as Markov-chain based approaches [1,2] apply k-order interactions to make suggestions; Recurrent Neural Networks (RNNs) based approaches [3,4] use gating units and long short-term memory modules to mine hidden associations in user sequences, and Convolutional Neural Networks (CNNs) based approaches [5,6] capture sequential patterns through convolutional filters.

In recent years, Graph Neural Networks (GNNs) have received a great deal of attention from researchers due to their high interpretability and performing end-to-end learning on graph data. Most of the data in recommendation systems are graph-structured, making gnn a powerful tool for solving recommendation tasks [7–12]. SR-GNN [7] first models conversation sequences as graph data structures for modeling complex item transformation relationships. Mengqi et al. [8] applied dynamic graph networks to sequential recommendation to mine dynamic collaborative signals from sequences of different users. Although the existing methods rely on their strengths to achieve satisfactory results, we believe that there are still several factors to be considered in sequence recommendations. First, Inevitably, there are interest preferences containing noise in the sequence of users' historical behaviors. Users may browse to follow items that are not of interest for a period of time but do not interact with them in subsequent behaviors, which can interfere with modeling the user's historical behavior and may persist throughout. Additionally, These approaches often ignore information about the potential collaboration between different item characteristics. Given a point in time, user preferences are influenced not only by stable interests but also by changes in interests in recent dynamics. Moreover, treating this information or features equally may lead to a bias towards the user's true intent. Therefore, it remains the focus of the recommendation model by considering the potential collaborative information between different item features in the history and clearly distinguishing the importance of the items.

We provide a multi-view graph neural networks-based method for sequential recommendation tasks to address the aforementioned issue. The architecture of SR-MVG is as follows: first, we transform the user's behavior sequence into an item-item graph such that similar items are connected to each other by an edge. Second, we design a multi-view graph attention network to adaptively retain users' core preferences by considering the importance of different item features. Third, to further reduce the interference of noisy preference information, we design a graph pooling strategy to extract more relevant user preferences. In summary, the main contributions of this paper are summarized as follows:

- We highlight the critical importance of explicitly modeling potential synergistic information between different item features in a sequential recommended approach.
- We propose a new multi-perspective attention-based graph neural network for sequential recommendation framework SR-MVG, which explicitly models three intuitive factors: 1) users' long-term and short-term interest preferences

and target perception information. 2) the synergy between the three types of information mentioned above. 3) sequential association pattern between items in the user history interactions.

- We have conducted extensive experiments on five real-world datasets. The experiments demonstrate that the performance of SR-MVG has been significantly improved.

## 2 Related Work

Sequential recommendation aims to use a user's historical interaction sequence to predict and recommend the next item with which the user is most likely to interact. Markov chain-based methods [1,2] construct correlations between user sequences of interactions play a pioneering influence in the early stages of sequential recommendation, e.g., FPMC [1] captures users' behavioral patterns with the help of MCs combined with matrix decomposition to recommend the next possible item for each user.

In recent years, the development of deep learning has provided a booming boost to sequence recommendation. For example, RNN-based methods can capture complex dependencies in sequences well and are widely used to solve sequential recommendation problems [3,4]. GRU4Rec [3] applied gated recursive units (GRUs) to session-based recommendation tasks for the first time. The CNN-based approach [5,6] models the overall information of sequences to perform abruptly. Caser [6] applied convolutional filters to capture sequential patterns. The superior mechanism of the attention networks also becomes a powerful tool for sequential recommendation. HGN [13] uses a hierarchical gating mechanism to identify potential features and important items and captures item relationships by item product display. Sun et al. [14] proposed BERT4Rec to improve the traditional Transformer to make recommendations for items in user history sequences considering contextual information. In recent work, the LSAN [15] proposed by uses a lightweight self-focused network that adaptively preserves global and local item dependencies. STOSA [16] designed a stochastic self-attention module based on a stochastic Gaussian distribution to model the uncertainty and collaborative transferability of continuous user behavior.

With the rapid development of Graph Neural Networks(GNN), the use of GNNs for sequential recommendation has received wide popularity [7–12]. SR-GNN [7] first modeled conversation sequences as graph data structures. Zhang et al. [9] proposed the A-PGNN model based on this joint personalized graph network and attention mechanism. In recent studies, the temporal enhancement model TASRec [10], is able to capture user interest and has efficient memory capability. GOTNet [11] combines k-Means with GNN to solve the over-smoothing problem brought by graph neural networks in remote dependency. MvDGAE [12] regards the cold start problem as a data missing problem and solves the cold start problem by randomly removing part of the user-item interaction graph and reconstructing the relationship graph, which effectively solves the cold start problem.

# 3 Method

The framework of our proposed SR-MVG is shown in Fig 1. The model consists of four parts: 1) Interest Graph Construction; 2) Multi-view Attention Network; 3) Interest Extraction Layer; and 4) Prediction Layer. We will introduce them one by one next.

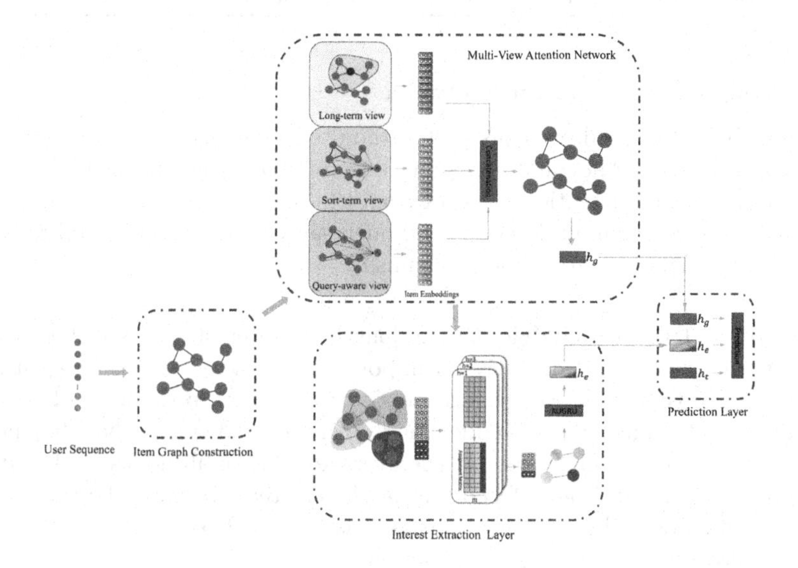


**Fig. 1.** The architecture of SR-MVG.

## 3.1 Interest Graph Construction

We first transform the user interaction sequence into an undirected graph $G$, where similar items are tightly connected and strong interests naturally form larger clusters. Specifically, we first encode the pairwise relationship of each node through a sparse attention mechanism:

$$M_{ij} = \sigma(\overrightarrow{a}(h_i \parallel h_j)) \tag{1}$$

where we employ a single layer neural network with weight vector $\overrightarrow{a} \in \mathbb{R}^{d \times d}$ as a parameter. $\sigma(\cdot)$ is an activation function similar to $ReLU(\cdot)$. The embedding representations of nodes $v_i$ and $v_j$ are, respectively, $h_i$ and $h_j$, $\parallel$ is the concatenation operator. Then, to guarantee the sparse distribution of the graph, we apply the sparsemax function to normalize them to easily identify comparable pairwise relational scores:

$$M_{ij}^{'} = sparsemax(M_{ij}) = [h_i - \tau h_j]_+ \tag{2}$$

where $[x]_+ = max\{0, x\}$, the $\tau(\cdot)$ is the threshold function, which returns a threshold value according to Algorithm 1. In light of the established threshold, $sparsemax(\cdot)$ adaptively activates the above-threshold score $M_{ij}$. A sparsely distributed graph that characterizes the user's core interests is constructed.

**Algorithm 1:** The calculation procedure of sparsemax()

**Require:** Input vector $z \in \mathbb{R}^n$
1: Sort $z$ into $u$: $u_1 \geq u_2 \geq ...u_n$.
2: Get $\rho = max\left\{1 \leq j \leq n : u_j + \frac{1}{j}(1 - \sum_{i=1}^{j} u_i > 0)\right\}$.
3: Define $\tau(z) = \frac{1}{\rho}(\sum_{i=1}^{\rho} u_i - 1)$
4: Return $x$ s.t. $x_i = max\left\{z_i - \tau(z), 0\right\}, i = 1, ..., n$

### 3.2 Multi-view Attention Network

As previously discussed, we constructed the user interest graph. In this section, we improve the accuracy and interpretability of recommendations by designing a multi-view graph attention network from the perspectives of users' long-term preferences, short-term preferences, and query-aware, respectively, and reassigning feature weights by attention mechanism.

**Long-Term Preference View.** The long-term interests of users best reflect their stable preferences and play an important role when interacting with the project [2]. In this work, we extract users' core interests in the form of clusters. Specifically, assuming that the node $v_i$ is the center, we define $v_i$ the k-hop neighbor nodes as their neighborhoods. The average value of all nodes in the cluster $h_{i_c}$ represents the average information of the cluster. To make the comparison between different nodes and their clusters easy, the softmax function is used for normalization, with the following equation:

$$\alpha_i = softmax(\frac{(W_1 h_i)^T (W_2 h_{i_c})}{\sqrt{d}}) \tag{3}$$

where $h_i$ and $h_{i_c}$ represent the nodes $v_i$ and its cluster $h_{i_c}$ of the embedding representation. $d$ is the dimensionality of the embedding vector, and the scale factor $\sqrt{d}$ is to avoid excessive dot product and to serve as a convergence accelerator. $W_1$ and $W_2$ are the learnable weight matrices.

**Short-Term Preference View.** The user's recent interests are dynamically reflected in the user's short-term information. Utilizing short-term information effectively has been shown in numerous works [13,17,18]. These works, however, disregard the reliance of current information on the past. As a result, we evaluate the consistency between the user's most recent two interactions and all available historical data:

$$\beta_i = softmax(\frac{(W_1 h_i)^T (W_2 h_l)}{\sqrt{d}}) \tag{4}$$

$$h_l = \sigma(FNN(h_n \parallel h_{n-1})) \tag{5}$$

where $h_{n-1}$ and $h_n$ represent the items that interacted at the n-th step and (n-1)-th step, respectively. The short-term information feature $h_l \in \mathbb{R}^d$ is represented by a two-layer feed-forward neural network with $Tanh(\cdot)$ as the activation function.

**Query-Aware View.** The correlation between the source node in the graph and the target item $h_t$ plays a key role in the final prediction, facilitating our understanding of the independent evolution of user interests across different target interests, whose importance is calculated by the following equation:

$$\gamma_i = softmax(\frac{(W_1 h_i)^T (W_2 h_t)}{\sqrt{d}}) \tag{6}$$

**Interest Integration.** The attention coefficient obtained $\alpha_i$, $\beta_i$ and $\gamma_i$ will be used to perform the aggregation of information about the project nodes with their neighbors, and after applying residual connectivity and nonlinear functions, the refined node embedding representations are obtained and they are aggregated to update each node's representation of the graph $G$:

$$h_i^L = \sigma(W_a(\sum_{j \in N_i} \alpha_i \times h_j) + h_i) \tag{7}$$

$$h_i^S = \sigma(W_b(\sum_{j \in N_i} \beta_i \times h_j) + h_i) \tag{8}$$

$$h_i^Q = \sigma(W_r(\sum_{j \in N_i} \gamma_i \times h_j) + h_i) \tag{9}$$

$$H^{'} = \left\{h_1^{'}, h_2^{'}, ..., h_n^{'}\right\}, h_i^{'} = \sigma(W_1 \left[h_i^L \parallel h_i^S \parallel h_i^Q\right]) \tag{10}$$

where $N_i$ is the set of neighboring nodes of $v_i$, and $h_j$ is the embedding vector of its neighbor nodes. $W_a$, $W_b$ and $W_r$ are the weight matrices of the corresponding linear transformations. $h_i^L$ ,$h_i^S$ and $h_i^Q$, respectively, represent the node embedding vector of each view. $H^{'}$ is the updated items embedding matrix.

### 3.3 Interest Extraction Layer

We add the CNNs pooling operator to the graph structure. In general, there shouldn't be information loss and the pooling operation should remain stable. Consequently, we construct the coarsened graph using the soft cluster assignment matrix approach [19]. First, we pass a learnable linear transformation matrix $W_t \in \mathbb{R}^{d^* \times d}$ to obtain a matrix that can characterize the content of the full graph: $T = H^{'} W_t$. T is the bridge between the interest $G^{'}$ and the coarsened graph $G^*$, which incorporates global properties into the pooling method. Then, we use the clustering-friendly distribution as a distance metric between nodes and clusters. Formally, we use the t-distribution (t-distribution) [20] as a kernel to compute the normalized similarity between nodes and clusters:

$$M_{ij} = \frac{(1 + \|T(i,:) - T(:,j)\|^2 / \tau)^{-\frac{\tau+1}{2}}}{\sum_{j'=0}^{m} (1 + \|T(i,:) - T(:,j')\|^2 / \tau)^{-\frac{\tau+1}{2}}} \tag{11}$$

where $m$ denotes the number of clusters, reflecting the degree of pooling, and $m < n$. Each row $T(i,:)$ corresponds to a node in graph $G^{'}$, and each column $T(:,j)$ represents a cluster node in the coarsened graph $G^{*}$. $\tau$ is the Gaussian variance centered on the data point $T(i,:)$ centered Gaussian variance, which represents the degrees of freedom of the $t$-distribution. We further employ a multi-headed attention mechanism [21] to stabilize the learning process of the cluster assignment matrix $M$:

$$M = softmax(\overset{\phi}{\underset{\delta=1}{\Vert}} M_{ij}^{\delta})W_o \tag{12}$$

where $M \in \mathbb{R}^{n\times m}$. $W_o \in \mathbb{R}^{\phi\times n\times m}$ is a representation of the linear transformation's weight matrix. $\phi$ is the number of attention heads. $M_{ij}^{\delta}$ is the distribution matrix obtained for the $\delta$-th attention head. With the work mentioned above, a new coarsened graph is generated:

$$H^{*} = M^{T}H^{'} = \{h_1^{*}, h_2^{*}, ..., h_m^{*}\} \tag{13}$$

### 3.4 Prediction Layer

**Graph-Level Readout.** We carry out a weighted readout in multi-view graph attention networks to emphasize the significance of various nodes:

$$h_g = Readout(\left\{\tilde{\gamma}_i \times h_i^{'} \mid i \in N_i\right\}) \tag{14}$$

where $\tilde{\gamma}_i$ is the normalized score of the attention coefficient $\gamma_i$. The Readout function utilizes a summing function to guarantee substitution invariance.

**Sequential Relevance.** Unlike the graph-level representation of $h_g$, we combine GRU plus attention update gate (i.e., AUGRU [22]) on the coarsened graph to model changes in user intense interest:

$$h_e = AUGRU(\{h_1^{*}, h_2^{*}, ..., h_m^{*}\}) \tag{15}$$

where $AUGRU$ uses $\tilde{r}_i$ to update all dimensions of the gate, reducing the shadow power of irrelevant interests and effectively avoiding interference in the process of interest change. Notably, we use relative position regularization to ensure that the pooled embedding matrix maintains the temporal order: $L_P = \|P_n M, P_m\|_2$. Both $P_n$ and $P_m$ are position encoding vectors$\{1, 2, ..., n\}$. Minimizing the $L_2$ norm keeps the cluster assignment matrix m to be orthogonal.

**Prediction.** We calculate the final score by $h_g$ (Eq. 14) , $h_e$ (Eq. 15) and $h_t$. For user u, given the interaction sequence $\{x_1, x_2, ..., x_n\}$, the final recommendation score $\hat{y}$ is calculated as follows:

$$\hat{y} = Predict(h_g \parallel h_e \parallel h_t \parallel h_g \odot h_t \parallel h_e \odot h_t) \tag{16}$$

Finally, we optimize the model using the negative log-likelihood function as the objective function. The model optimization is to minimize the objective function, defined as follows:

$$Loss = -\frac{1}{|O|}\sum_{o\in O}(y_o log\hat{y_o} + (1 - y_o log(1 - \hat{y_o})) + \lambda \left\|\Theta\right\|_2) \qquad (17)$$

where $O$ is the training set and $|O|$ is the number of instances in the training set. $\theta$ represents all model parameters. $||\cdot||_2$ is $L_2$ norm. $\lambda$ is to control regularization strength. $y_o = 1$ and $y_o = 0$ represent positive and negative instances, respectively.

# 4 Experiment

## 4.1 Datasets and Evaluation Metrics

We evaluated these methods on five real datasets, including Amazon-Games (Games), Amazon-CDs (CDs), and Amazon-Beauty (Beauty) from Amazon review, and MovieLens-1M (ML-1M) and MovieLens-20M (ML-20M) from MovieLens. These datasets are commonly used as benchmark datasets for evaluating recommendation models.

To be fair and efficient, for all datasets, we will filter out those users with less than 10 interactions and those with less than 5 occurrences. The dataset after data pre-processing is shown in Table 1. We take 90% of the dataset as the training set and the rest as the test set. For all models, we execute them three times separately and take the average value as the final result.

We use two widely used accuracy metrics, HR@10(Hit Rate) and NDCG@10 (Normalized Discounted cumulative gain), to evaluate all methods. HR@10 reflects whether the recommendation sequence contains items that users actually click on. A higher HR @10 indicates better performance. NDCG@10 is a position-aware metric that emphasizes that the target program should have a higher top position in the top-K recommendation ranking.

## 4.2 Parameter Setup

We implemented our model on the DGL Library. All parameters are initialized using a Gaussian distribution with mean 0 and standard deviation 0.1. Optimization is performed using Adam and the initial learning rate is set to 0.001. For all methods we set the batch size to 50 and the embedding size is fixed to 50. In the pooling layer, we configure three heads applied to the multi-head self-attention module for embedding learning. The pooling lengths for the Amazon and MovieLens datasets are at [10 ,20,30,40,50] and [60,70,80,90,100] were searched. The maximum sequence length is set to 50. The regularization rate $\lambda$ is adjusted between $\{0.1, 0.01, 0.001, 0.0001\}$ to alleviate the overfitting problem.

**Table 1.** The statistics of the datasets.

| Datasets | # of Users | # of Items | # of Interactions | Average length |
|---|---|---|---|---|
| Games | 31013 | 23715 | 287107 | 9.25 |
| CDs | 17052 | 35118 | 472265 | 27.69 |
| Beauty | 52024 | 57289 | 394908 | 7.59 |
| ML-1M | 5950 | 3125 | 573726 | 96.42 |
| ML-20M | 129780 | 13663 | 9926480 | 76.48 |

### 4.3 Performance Comparison

In the next recommended task, we compared SR-MVG with several advanced models, and Table 2 shows the experimental results for each method. We can obtain the following observations.

Our SR-MVG model achieves the best performance on all datasets, outperforming all baseline models in terms of accurate prediction and ranking quality. In particular, all models performed mediocrely on ML-1M, but SR-MVG got a significant performance improvement by explicitly propagating interest signals on the constructed interest graph through the attention mechanism to better encode user interest, specifically, SR-MVG scored about 5.3% and 3% higher on both HR@10 and NDCG@10 metrics. This validates that our method can better handle long sequences.

Effectiveness of joint modeling of long-term and short-term interests. HGN, SASRec and SliRec, which consider the user's long- and short-term interest signals, perform well , but the performance is still worse than SR-MVG. We believe the reason for this is that the GNN-based approach can better capture the complex transformation relationships of items, which is difficult to handle by previous traditional recommendation methods. Although SR-GNN is also a GNN-based approach, the SR-GNN approach has difficulty in forming user sequences into graph structures due to the lack of repetition in our data, resulting in its mediocre performance.

### 4.4 Ablation Experiments

To investigate the advantages of multi-view graph attention network and interest extraction layer in SR-MVG, we compared SR-MVG with different variants of SR-MVG on beauty, CDs and ML-20M datasets to actively explore the different effects and necessity of different modules on recommendation performance. In particular, we also replaced the multi-view graph attention network and graph pooling operations with a GCN with two layers and with a node discard strategy, respectively, and the experimental results are shown in Table 3, where we analytically obtained.

**Table 2.** Performance comparison of all methods at HR@10 and NDCG@10 metrics.

| Method | NDCG@10 | | | | | HR@10 | | | | |
|---|---|---|---|---|---|---|---|---|---|---|
| | Beauty | CDs | Games | ML-1M | ML-20M | Beauty | CDs | Games | ML-1M | ML-20M |
| FPMC [1] | 0.2791 | 0.3246 | 0.4668 | 0.5369 | 0.5163 | 0.4285 | 0.5023 | 0.6801 | 0.7594 | 0.7225 |
| Caser [6] | 0.2441 | 0.4307 | 0.3224 | 0.5528 | 0.5294 | 0.4342 | 0.6755 | 0.5296 | 0.7886 | 0.7491 |
| GRU4Rec+ [4] | 0.2524 | 0.4348 | 0.4756 | 0.5509 | 0.5185 | 0.4453 | 0.6681 | 0.6004 | 0.7496 | 0.7502 |
| Sli-Rec [18] | 0.2559 | 0.4352 | 0.4805 | 0.5869 | 0.5484 | 0.4456 | 0.6803 | 0.7154 | 0.8125 | 0.7865 |
| SR-GNN [7] | 0.3124 | 0.4798 | 0.4906 | 0.5864 | 0.5473 | 0.4698 | 0.6853 | 0.7368 | 0.8098 | 0.7894 |
| SASRec [17] | 0.3129 | 0.4885 | 0.5274 | 0.5902 | 0.5508 | 0.4696 | 0.7022 | 0.7411 | 0.8243 | 0.7915 |
| HGN [13] | 0.3133 | 0.4896 | 0.5298 | 0.6004 | 0.5633 | 0.4739 | 0.7043 | 0.7463 | 0.8296 | 0.7932 |
| SR-MVG | **0.3198** | **0.5033** | **0.5451** | **0.6184** | **0.5773** | **0.4829** | **0.7155** | **0.7612** | **0.8743** | **0.8312** |

**Table 3.** Comparison of the SR-MVG model and its variants.

| Variants | Beauty | | CDs | | ML-1M | |
|---|---|---|---|---|---|---|
| | NDCG@10 | HR@10 | NDCG@10 | HR@10 | NDCG@10 | HR@10 |
| w/o Long-term | 0.3171 | 0.4683 | 0.4924 | 0.7014 | 0.5965 | 0.8131 |
| w/o Short-term | 0.2688 | 0.4471 | 0.4881 | 0.6953 | 0.5922 | 0.8087 |
| w/o Query-aware | 0.3207 | 0.4692 | 0.4943 | 0.7042 | 0.5893 | 0.7649 |
| w/ GCN | 0.3161 | 0.4658 | 0.4852 | 0.6913 | 0.5903 | 0.8012 |
| w/o Graph Pooling | 0.3278 | 0.4760 | 0.4797 | 0.6840 | 0.5995 | 0.8155 |
| w/o Readout | 0.3258 | 0.4825 | 0.4824 | 0.6873 | 0.6034 | 0.8358 |
| w/ Node Drop | 0.3206 | 0.4788 | 0.4983 | 0.7084 | 0.6089 | 0.8361 |
| SR-MVG | 0.3324 | 0.4956 | 0.5033 | 0.7155 | 0.6136 | 0.8412 |

Each view in the multi-view graph attention network brought further performance improvement. We further found that the performance of the SR-MVG model without considering short-term interest decreased significantly in Beauty compared to the CDs and ML-1M datasets, suggesting that the shorter the user's interaction sequence, the more important the extraction of short-term interest in the sequence. In addition, replacing the multi-view graph attention network with a two-layer GCN achieves poor results, which is due to the fact that the GCN module treats all neighboring nodes as equally important, which introduces more noise in message propagation.

All variants with graph pooling exhibit better competition compared to those without graph pooling, due to the fact that the graph pooling feature filters out unnecessary noise interference and makes the model more focused on the most important parts of the user's interests. It is worth noting that the method using node-drop graph pooling gets good performance in CDs and ML-1M datasets, but underperforms in Beauty, indicating that although the node-drop pooling method can serve to filter noise, it leads to loss of information due to its working mechanism, which is particularly evident in sparse datasets.

## 5 Conclusion

In this work, we propose a graph neural network-based model SR-MVG. In SR-MVG, user sequences are first converted into graph structures, then collaborative information between different item features is considered from multiple perspectives to clearly distinguish the importance of different features, noise interference of irrelevant interests is further reduced by a designed graph pooling strategy, and finally a gated recurrent neural network is extracts the sequence signals to better capture the more relevant interests with the target items. Extensive experiments on five real datasets demonstrate the rationality and effectiveness of SR-MVG.

**Acknowledgements.** We are grateful for the support of the National Natural Science Foundation of Shandong ZR202011020044. We are grateful for the support of the National Natural Science Foundation of China 61772321. Natural Science Foundation of China 61772321.

## References

1. Rendle, S., Freudenthaler, C., Schmidt-Thieme, L.: Factorizing personalized Markov chains for next-basket recommendation. In: Proceedings of the 19th International Conference on World Wide Web, pp. 811–820 (2010)
2. He, R., McAuley, J.: Fusing similarity models with Markov chains for sparse sequential recommendation. In: 2016 IEEE 16th International Conference on Data Mining (ICDM), pp. 191–200. IEEE (2016)
3. Hidasi, B., Karatzoglou, A., Baltrunas, L., Tikk, D.: Session-based recommendations with recurrent neural networks. arXiv preprint arXiv:1511.06939 (2015)
4. Hidasi, B., Karatzoglou, A.: Recurrent neural networks with top-k gains for session-based recommendations. In: Proceedings of the 27th ACM International Conference on Information and Knowledge Management, pp. 843–852, 2018
5. Tang, J., Wang, K.: Personalized top-n sequential recommendation via convolutional sequence embedding. In: Proceedings of the Eleventh ACM International Conference on Web Search and Data Mining, pp. 565–573 (2018)
6. You, J., Wang, Y., Pal, A., Eksombatchai, P., Rosenburg, C., Leskovec, J.: Hierarchical temporal convolutional networks for dynamic recommender systems. In: The World Wide Web Conference, pp. 2236–2246 (2019)
7. Shu, W., Tang, Y., Zhu, Y., Wang, L., Xie, X., Tan, T.: Session-based recommendation with graph neural networks. In: Proceedings of the AAAI Conference on Artificial Intelligence, vol. 33, pp. 346–353 (2019)
8. Zhang, M., Wu, S., Yu, X., Liu, Q., Wang, L.: Dynamic graph neural networks for sequential recommendation. IEEE Trans. Knowl. Data Eng. (2022)
9. Zhang, M., Wu, S., Gao, M., Jiang, X., Xu, K., Wang, L.: Personalized graph neural networks with attention mechanism for session-aware recommendation. IEEE Trans. Knowl. Data Eng. (2020)
10. Zhou, H., Tan, Q., Huang, X., Zhou, K., Wang, X.: Temporal augmented graph neural networks for session-based recommendations. In: Proceedings of the 44th International ACM SIGIR Conference on Research and Development in Information Retrieval, pp. 1798–1802 (2021)

11. Chen, H., Yeh, C.-C.M., Wang, F., Yang, H.: Graph neural transport networks with non-local attentions for recommender systems. In: Proceedings of the ACM Web Conference 2022, pp. 1955–1964 (2022)
12. Zheng, J., Ma, Q., Gu, H., Zheng, Z.: Multi-view denoising graph auto-encoders on heterogeneous information networks for cold-start recommendation. In: Proceedings of the 27th ACM SIGKDD Conference on Knowledge Discovery and Data Mining, pp. 2338–2348 (2021)
13. Ma, C., Kang, P., Liu, X.: Hierarchical gating networks for sequential recommendation. In: Proceedings of the 25th ACM SIGKDD International Conference on Knowledge Discovery and Data Mining, pp. 825–833 (2019)
14. Sun, F., e tal.: BERT4Rec: sequential recommendation with bidirectional encoder representations from transformer. In: Proceedings of the 28th ACM International Conference on Information and Knowledge Management, pp. 1441–1450 (2019)
15. Li, Y., Chen, T., Zhang, P.-F., Yin, H.: Lightweight self-attentive sequential recommendation. In: Proceedings of the 30th ACM International Conference on Information and Knowledge Management, pp. 967–977 (2021)
16. Fan, Z., et al.: Sequential recommendation via stochastic self-attention. In: Proceedings of the ACM Web Conference 2022, pp. 2036–2047 (2022)
17. Kang, W.-C., McAuley, J.: Self-attentive sequential recommendation. In: 2018 IEEE International Conference on Data Mining (ICDM), pp. 197–206. IEEE (2018)
18. Yu, Z., Lian, J., Mahmoody, A., Liu, G., Xie, X.: Adaptive user modeling with long and short-term preferences for personalized recommendation. In: IJCAI, pp. 4213–4219 (2019)
19. Ying, Z., You, J., Morris, C., Ren, X., Hamilton, W., Leskovec, J.: Hierarchical graph representation learning with differentiable pooling. In: Advances in Neural Information Processing Systems, vol. 31 (2018)
20. Xie, J., Girshick, R., Farhadi, A.: Unsupervised deep embedding for clustering analysis. In: International Conference on Machine Learning, pp. 478–487. PMLR (2016)
21. Vaswani, A., et al.: Attention is all you need. In: Advances in Neural Information Processing Systems, vol. 30 (2017)
22. Zhou, G., et al.: Deep interest evolution network for click-through rate prediction. In: Proceedings of the AAAI Conference on Artificial Intelligence, vol. 33, pp. 5941–5948 (2019)

# Cross-Domain Reinforcement Learning for Sentiment Analysis

Hongye Cao, Qianru Wei(✉), and Jiangbin Zheng

School of Software, Northwestern Polytechnical University, Xi'an 710072, China
2020204278@mail.nwpu.edu.cn, {weiqianru,zhengjb}@nwpu.edu.cn

**Abstract.** By transferring knowledge from the source domain labeled data to the target domain, cross-domain sentiment analysis can predict sentiment polarity in the lacking of labeled data in the target domain. However, most existing cross-domain sentiment analysis methods establish the relationship of domains by extracting domain-invariant (pivot) features. These methods ignore domain-specific (non-pivot) features or introduce a lot of noisy features to make domain adaptability weak. Hence, we propose a cross-domain reinforcement learning framework for sentiment analysis. We extract pivot and non-pivot features separately to fully mine sentiment information. To avoid the Hughes phenomena caused by the feature redundancy, the proposed framework applies a multi-level policy to select appropriate features extracted from the data. And a sentiment predictor is applied to calculate delayed reward for policy improvement and predict the sentiment polarity. The decision-making capacity of reinforcement learning can effectively tackle the problem of noisy feature data and improve domain adaptability. Extensive experiments on the Amazon review datasets demonstrate that the proposed model for cross-domain sentiment analysis outperforms state-of-the-art methods.

**Keywords:** Reinforcement Learning · Cross-Domain · Sentiment Analysis

## 1 Introduction

With the rapid development of the Internet, sentiment analysis has become an important research topic. For example, in the e-commerce platform, sentiment analysis of consumer reviews can help make right choices.

The supervised learning method is widely used in the sentiment analysis [11]. This method requires sufficient labeled data to train the language model. However, in reality, no large amount of labeled data is found in different domains. Manual data labeling requires a lot of manpower. Cross-domain sentiment analysis realizes sentiment polarity prediction and reduces manual labor by transferring the source domain labeled data knowledge to the target domain. Hence, cross-domain sentiment analysis has received more attention.

Supported by organization x.


M. Tanveer et al. (Eds.): ICONIP 2022, CCIS 1793, pp. 638–649, 2023.
https://doi.org/10.1007/978-981-99-1645-0_53

Most existing cross-domain sentiment analysis methods are feature adaption [7], which extracts domain-invariant (pivot) features to establish connections between domains. With the development of deep learning, deep learning have been widely used in cross-domain sentiment analysis [4]. These methods can be applied in feature adaption to extract pivot features between the source and target domains automatically. However, the feature adaption method is highly dependent on the selection of pivots and does not make good use of domain-specific (non-pivot) features [4]. These methods are weak in capturing the sentiment features of domains and have weak domain adaptability. For example, "the voice recorder is light and small.", in this instance, light and small are often ignored as non-pivot features and judged as negative. How to better extract features and solve the noise problem is extremely important.

Hence, we propose a Cross-Domain Reinforcement Learning (CDRL) framework for sentiment analysis in response to the above problems. In order to fully mine the sentiment information, we extract pivot and non-pivot features separately. Then, CDRL applies a multi-level policy to select appropriate features extracted from the text. The extraction of the pivot and non-pivot feature fully captures the sentiment information in the domain. However, due to the extraction of large number of features, the introduction of many redundant features will cause the Hughes phenomenon [1]. To avoid the Hughes phenomenon caused by the feature redundancy, the feature selection policy is applied to reduce the interference of noisy features to balance pivot and non-pivot features. Finally, a sentiment predictor is proposed to calculate the delayed reward (DR) for policy improvement which improves domain adaptability. And it predicts the sentiment polarity of the data. The combination of the perception ability of deep learning and the decision-making ability of reinforcement learning can improve the performance of the model. The main contributions of this paper are as follows:

- We propose a cross-domain reinforcement learning framework for sentiment analysis. To the best of our knowledge, this is the first work to use reinforcement learning methods for cross-domain sentiment analysis.
- We extract pivot and non-pivot features to capture the sentiment information in the data fully. And a multi-level feature selection policy is applied to select appropriate features to tackle the problem of noisy feature data.
- The CDRL framework combines DRs for policy improvement. This mechanism can perform appropriate feature selection according to different domain which can effectively improve the domain adaptability.
- The experimental results on two Amazon review datasets demonstrate that the proposed method outperforms state-of-the-art (SOTA) methods. The ablation experiment further proves the contribution of each component to the improvement of model performance.

## 2 Related Work

The feature adaption method establishes the relationship between domains [7] by extracting pivot feature. The structure correspondence learning algorithm [2] is a multi-task learning algorithm with alternating structure optimization. However, these methods still require a lot of manpower to select feature information.

With the development of deep learning, more and more deep learning methods are applied to cross-domain sentiment analysis. The domain adversarial training of neural network is proposed to realize domain adaptation [4]. Du et al. [5] proposed a network based on Wasserstein distance to extract invariant information between domains. However, these methods are highly dependent on the selection of domain-invariant features. Li et al. [9] proposed a hierarchical attention transfer network which extracted the pivot and non-pivot features. However, due to the Hughes phenomenon caused by the feature redundancy, this model has weak domain adaptability. Hence, we propose a cross-domain reinforcement learning framework which can effectively solve the above problems.

In recent years, reinforcement learning has been successfully applied in natural language processing tasks. Language models based on reinforcement learning have been used in machine translation [16], abstract summarization [17], and text generation [13]. Wang et al. [14] proposed a hierarchical reinforcement learning framework to realize document-level aspect sentiment analysis. The reinforcement learning mechanism effectively tackled the noise problem in the model. To the best of our knowledge, this is the first work to use reinforcement learning methods for cross-domain sentiment analysis.

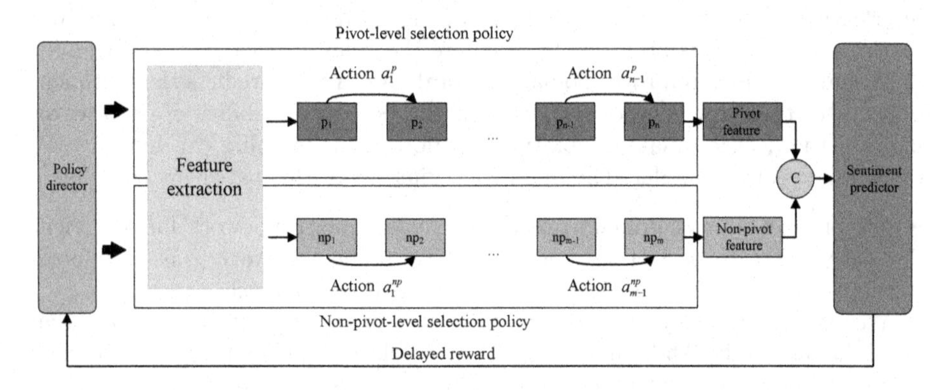


**Fig. 1.** The framework of cross-domain reinforcement learning.

# 3 Approach

## 3.1 Overview of the Framework

As shown in Fig. 1, we propose the CDRL framework. The policy director guides the feature selection in the policy layer and accepts DRs from the sentiment predictor to improve the policy. The feature selection policy layer is used to select the extracted features. It contains two parts. One is the pivot-level selection policy used to select pivot features. The other is the non-pivot-level selection policy. After selecting the pivot and non-pivot features, features are combined and input into the sentiment predictor. In the model training phase, the sentiment predictor is used to calculate the DR and feed it back to the policy director for policy improvement. In the sentiment polarity prediction phase, the sentiment predictor is used to predict the sentiment polarity. Cross-domain reinforcement learning is presented in Sect. 3.2, and the feature extraction in CDRL is introduced in Sect. 3.3.

### 3.2 Cross-Domain Reinforcement Learning

**Policy Director.** The policy director uses stochastic strategies and DRs to guide policy learning. We set the pivot feature sequence set in the input document as $p_1, \ldots, p_i, \ldots, p_n$. The action sequence is $a_1^p, \ldots, a_i^p, \ldots, a_n^p$, where $a_i^p = 0$ indicates $p_i$ is deleted at the i-th time-step state. When the feature is deleted, the attention [12] parameter will be updated. Otherwise, $a_i^p = 1$ indicates $p_i$ is retained at the i-th time-step state. The policy of the pivot-level at the i-th time-step state is shown as:

$$P_i, c_i^p = \begin{cases} P_{i-1}, c_{i-1}^p, a_i^p = 1 \\ f(P_{i-1}, c_{i-1}^p, P_i), a_i^p = 0 \end{cases} \tag{1}$$

where $f$ denotes the update function of features, $P_i$ is the pivot feature representation, $p_i$ is the single pivot feature, and $c_i^p$ is the attention coefficient at the i-th time-step state. The selection policy of non-pivot-level is shown as:

$$NP_i, c_i^{np} = \begin{cases} NP_{i-1}, c_{i-1}^{np}, a_i^{np} = 1 \\ f(NP_{i-1}, c_{i-1}^{np}, P_i), a_i^{np} = 0 \end{cases} \tag{2}$$

where $NP_j$ is the non-pivot feature representation and $c_j^{np}$ is the attention coefficient at the j-th time-step state. In each state, action is obtained, and the entire feature sequence is sampled for action.

**Feature Selection Policy.** The initial states of the pivot and non-pivot features contained in the input data are both retained. After the pivot selection is completed, the selection of non-pivot features is performed. As the training process progresses, feature selection takes turns. Finally, we choose state parameters that maximize the DR of the input.

State. The state $s_j$ of the j-th time step contains feature representation $F_{j-1}$ and the DR $r_{j-1}$ of the previous time step.

Action. We apply a binary action that contains retain and delete. Let $a_j$ denote the action at state $s_j$. And the policy is defined as follows:

$$\pi(a_j|s_j;\gamma) = \delta(C * s_j + m) \tag{3}$$

where $pi(a_j|s_j;\gamma)$ denotes the probability of choosing an action. $\delta$ denotes the sigmoid function and $\gamma = C, m$ denotes the parameters of policy director. In the training phase, action is sampled according to the probability calculated by Equation (3). In the testing phase, we choose the action with the highest probability.

The policy gradient is used to optimize parameters of the policy. Based on the method of REINFORCE [15], the optimization goal can be defined as the average reward at each time step:

$$J(\gamma) = \sum_{s} d_{\pi}(s) \sum_{a} \pi_{\gamma}(s,a)R \tag{4}$$

where $d_{\pi}(s)$ represents the steady-state distribution of Markov chain under strategy $\pi$. R represents the delayed reward of the policy $\pi$. The gradient of the final derivative used to update the policy network of $\gamma$ can be expressed as:

$$\bigtriangledown_{\gamma} J(\gamma) = E_{\pi_{\gamma}}[R \bigtriangledown_{\gamma} log\pi_{\gamma}(s,a) \tag{5}$$

**Sentiment Predictor.** The sentiment predictor aims to realize the calculation of the DR for policy improvement and the prediction of sentiment polarity. The DR is used to calculate the loss of the model under the current feature representation. We use the logarithm of the output probability of the sentiment predictor. In order to obtain more accurate domain features and prevent overfitting [14], the ratio of the number of deleted features to the length of the document is used as one of the reward indicators. The DR calculation is as follows:

$$R = \lambda_1 p_{\theta}(y|X) + \lambda_2 N'/N + \lambda_3 M'/M \tag{6}$$

where $y$ is the true label of the input sequence $X$; $N'$ is the number of deleted pivot features; $N$ is the number of pivot features in the initial input document; $M'$ is the number of deleted non-pivot features; $M$ is the number of non-pivot features in the initial input document. $\lambda_1$, $\lambda_2$, and $\lambda_3$ are weight parameters that balance indicators.

In the model training phase, the policy director improves the feature selection policy based on DR. In the model prediction phase, the sentiment predictor achieves sentiment polarity prediction:

$$P(y|X) = Softmax(CF + b) \tag{7}$$

where $F$ is the feature representation. $C$ is the weight matrix and $b$ is the bias of the connected layer. The sentiment label with the highest probability represents the predicted polarity value.

### 3.3 Feature Extraction

**Pivot Feature Extraction.** We propose the architecture of feature extraction in Fig. 2. It contains encoding module and feature extraction module. We first extract pivot features. Word2vec [10] is a pre-trained word vector as an

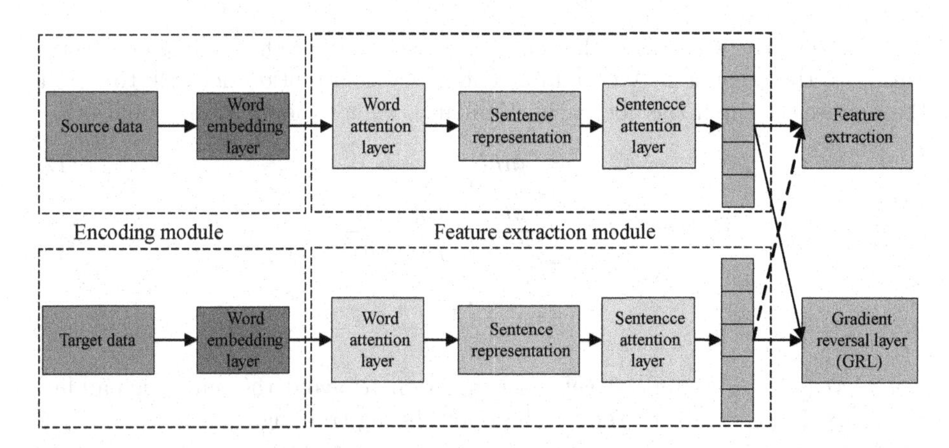


**Fig. 2.** The architecture of pivot and non-pivot feature extraction.

embedding. BERT [3] is used as a contextual embedding to learn richer semantic information.

We combine word2vec and BERT encoding to synthesize pre-training information and contextual semantic information [5].

Then, a word attention network is used to weight word vectors. The input vector $x_i$ of each word is represented by the output of word2vec and BERT:

$$x_i = [word2vec(w_i); BERTw_i] \tag{8}$$

where $w_i$ represents the i-th word. $word2vec(w_i)$ represents the pre-trained word vector. $BERTw_i$ represents the context embedding vector of word $w_i$.

The feature extraction module consists of a word attention layer and a sentence attention layer. The attention mechanism emphasizes the different effects of contextual vocabulary according to the weight coefficient. We input the encoded word into the attention layer to get the hidden layer representation of the word:

$$h_i = tanh(W_i x_i + b_i) \tag{9}$$

where $W_i$ represents the weight matrix, and $b_i$ represents the deviation learned by the network. Then, we use the SoftMax function to normalize the attention vector. It highlights the important words in the sentence as follows:

$$a_i = \frac{M_w(j)exp(h_i)}{\sum_{j=1}^{n} M_w(j)exp(h_i)} \tag{10}$$

where $M_w(j)$ is a word-level mask function to avoid the effect of padding words. When the word memory is occupied, $M_w(j)$ is equal to 1, otherwise $M_w(j)$ is equal to 0.

We combine the standardized attention score $a_i$ of each word with the corresponding hidden output hi to obtain the following final feature representation:

$$r = \{fea_1, ..., fea_i, ..., fea_n\} \tag{11}$$

where $fea_i = a_i * h_i$.

Subsequently, we analyze the sentences. Each sentence has a different contribution to the document. We build a sentence-level attention network to obtain the representation of the document. Similarly, we get the following equations.

$$h_i^s = tanh(W_s r_i + b_s) \tag{12}$$

$$\beta_i = \frac{M_s(j)exp(h_i^s)}{\sum_{j=1}^{m} M_s(j)exp(h_i^s)} \tag{13}$$

$$s = \sum_{i=1}^{m} h_i^s \beta_i \tag{14}$$

where $M_s(j)$ is a sentence-level mask function to avoid the effect of padding sentences. Next, we define the loss function during training.

$$L = L_{sen} + L_{dom} \tag{15}$$

Among them, the sentiment classification loss $L_{sen}$ is to minimize the cross entropy of the labeled data in the source domain. It is defined as follows:

$$L_{sen} = -\frac{1}{N_s}\sum_{i=1}^{N_s} \hat{y}_i ln y_i^p + (1-\hat{y}_i)ln(1-y_i^p) \tag{16}$$

where $\hat{y}_i, y_i^p \in \{0,1\}$, $\hat{y}_i$ and $y_i^p$ are the real sentiment label and the sentiment prediction label of the sample $x$ of the source data, respectively.

The domain adversarial loss is used to assist in the construction of the feature representation shared by the domain. Thereby, the domain-invariant features can be extracted. All hidden states of word embedding will go through the GRL.We define GRL as $Q_\lambda(x) = x$ with a reversal gradient $\partial Q_\lambda(x)/\partial x = -\lambda I$. The goal is to minimize the cross-entropy domain classification loss:

$$L_{dom} = -\frac{1}{N_s+N_t}\sum_{i=1}^{N_s+N_t} \hat{z}_i ln z_i^p + (1-\hat{z}_i)ln(1-z_i^p) \tag{17}$$

where $\hat{z}_i, z_i^p \in \{0,1\}$, $\hat{z}_i$ is the real sentiment label and $z_i^p$ is the sentiment prediction label of the sample $x$ of the source data. In the training process, we minimize the loss function to obtain the optimal model parameters. The word with the highest attention coefficient in the input sequences is set as a pivot feature. We integrate them into a set of pivot features.

**Non-pivot Feature Extraction.** Compared with pivot feature extraction, we first preprocess data and mask the pivot feature data. Subsequently, the feature extraction network is used to train the model on the masked data. Finally, the attention coefficient of the masked data is obtained. We set the features extracted from the data whose pivot features are predicted to be positive as positive non-pivot features. We arrange the training data in descending order of the attention coefficient. Finally, we select the words with the highest attention coefficient in the input sequences from the target domain dataset as non-pivot features.

**Table 1.** Dataset description.

| Dataset | D1 | | | | D2 | | | |
|---|---|---|---|---|---|---|---|---|
| | E | K | B | D | BK | E | BE | M |
| Train | 5600 | 5600 | 5600 | 5600 | 5400 | 5400 | 5400 | 5400 |
| Test | 6000 | 6000 | 6000 | 6000 | 6000 | 6000 | 6000 | 6000 |
| Unlabeled | 17009 | 13856 | 9750 | 11843 | 6000 | 6000 | 6000 | 6000 |

# 4 Experiment

## 4.1 Experimental Dataset and Settings

We select two Amazon review datasets for experiment, labeled as D1 and D2. Dataset D1 [2] has data in four domains: books (B), DVD (D), electronics (E), and kitchen (K). The description of the dataset is shown in Table 1. We create 12 source-target sentiment classification tasks to test the performance. We randomly select 2800 positive and 2800 negative reviews from the source domain for training. All unlabeled target domain data are used for training of the feature extraction, and all 6000 labeled reviews in the target domain are used for testing. D2 [6] has data in four domains: books (BK), electronics (E), Beauty (BE), and Music (M) and contains neutral sentiment polarity, so it is more realistic and convincing.

The basic BERT model includes 12 layers and 110 M parameters. The word vector is set to 300 dimensions. The batch size for training is 50. The learning rate for Adam Optimizer is 1e−4. The epsilon value for Adam Optimizer is 1e-8. The embedding size is set to 300. The hidden size is set to 300. The Pre-trained word vectors use the Google News-vectors. $\lambda_1$, $\lambda_2$, and $\lambda_3$ are 0.7,0.2 and 0.1. All experimental models are implemented in the TensorFlow framework. Accuracy and macro F1-score are used as performance metrics.

We compare proposed model with the following baseline methods.

- AMN [8]: AMN is based on the domain invariant representation of memory net-works and adversarial training.
- HATN [9]: HATN applies a hierarchical attention network with GRL to achieve fulcrum and predict the polarity.
- IATN [18]: IATN combines the sentence feature and aspect information with the interactive attention transfer network for cross-domain sentiment analysis.
- WTN [5]: WTN combines pre-trained model BERT and Wasserstein distance. Extract pivot features based on Wasserstein distance.
- DAAT [4]: DAAT combines pre-trained model BERT to extract pivot and non-pivot features.

## 4.2 Comparative Experiment

We first conduct comparative experiments with 5 methods on 12 cross-domain tasks of D1. Results in Table 2 show that CDRL outperforms other methods.

**Table 2.** Comparative experiment results in D1 of accuracy (%). Each result is averaged over 5 random seeds.

| Task | Comparative Methods | | | | | CDRL | | | |
|---|---|---|---|---|---|---|---|---|---|
| | AMN | HATN | IATN | WTN | DAAT | w/o pivot | w/o n-pivot | w/o RL | ours |
| E-K | 85.86 | 87.90 | 88.70 | 93.20 | 93.18 | 78.73 | 85.15 | 91.36 | **94.75** |
| E-B | 77.44 | 81.00 | 81.80 | 90.10 | 88.91 | 77.54 | 86.32 | 86.12 | **91.23** |
| E-D | 81.72 | 84.00 | 84.10 | 89.20 | 90.13 | 80.12 | 87.15 | 85.52 | **90.32** |
| K-E | 85.42 | 87.00 | 87.60 | 91.90 | 91.72 | 78.46 | 85.14 | 86.36 | **92.97** |
| K-B | 80.14 | 83.30 | 84.70 | 91.60 | 87.98 | 79.06 | 87.26 | 87.93 | **90.82** |
| K-D | 81.25 | 84.50 | 84.40 | 88.90 | 88.81 | 80.13 | 88.15 | 86.81 | **91.94** |
| B-E | 80.34 | 85.70 | 86.50 | 88.40 | 89.57 | 81.12 | 84.89 | 89.18 | **91.15** |
| B-K | 80.92 | 85.20 | 85.90 | 89.60 | 90.75 | 80.10 | 85.47 | 88.20 | **91.97** |
| B-D | 84.51 | 86.10 | 86.80 | 90.90 | 89.70 | 78.15 | 86.90 | 90.55 | **92.35** |
| D-E | 80.52 | 85.60 | 86.90 | 91.50 | 89.30 | 83.14 | 88.16 | 90.19 | **91.50** |
| D-K | 81.60 | 86.20 | 85.80 | 89.10 | 90.50 | 80.76 | 85.38 | 89.41 | **91.02** |
| D-B | 83.61 | 86.30 | 87.00 | 90.80 | 90.86 | 78.58 | 86.58 | 90.78 | **91.98** |
| AVE | 81.94 | 85.10 | 85.81 | 90.40 | 90.12 | 79.66 | 86.38 | 88.53 | **91.83** |

The AMN, HATN and IATN methods do not fully exploit the domain sentiment information and the average accuracy is less than 90%. The WTN and DAAT methods based on the pre-trained model achieve more accurate cross-domain sentiment analysis. CDRL outperforms the WTN and DAAT methods. This shows that combining the information perception ability of the pretrained model with the decision-making ability of reinforcement learning can further improve the model performance.

On tasks with large differences between the source and target domains, such as B-E and B-D, the accuracy of the other five models is low. However, CDRL can reach an accuracy rate of over 90%, showing that the proposed model can achieve accurate feature extraction in domains with large differences. The accuracy of CDRL on 12 tasks is higher than 90%, which indicates that the CDRL framework improves the domain adaptability. We choose the better performing WTN and DAAT methods for comparison in D2. Results of accuracy and macro-F1 in Table 3 show that CDRL achieves superior performance on all 12 tasks. Overall, CDRL outperforms existing SOTA methods.

### 4.3 Ablation Study

To verify the effectiveness of each component in CDRL, we set up ablation experiments for analysis. We set the model without (w/o) pivot feature, the model without non-pivot feature (n-pivot), and the model without reinforcement learning (RL) to compare with the proposed model. The experimental results are shown in Table 2.

The accuracy of the model without pivot feature extraction is highly reduced, indicating that the pivot features effectively mine the sentiment information between domains. The accuracy of the model without non-pivot features is

**Table 3.** Comparative experiment results in D2 of accuracy (%) and macro-F1. Each result is averaged over 5 random seeds.

| Task | WTN | | DAAT | | CDRL | |
|---|---|---|---|---|---|---|
| | Accuracy | Macro-F1 | Accuracy | Macro-F1 | Accuracy | Macro-F1 |
| BK-BT | 57.60 | 0.5524 | 58.41 | 0.5601 | **62.56** | **0.6012** |
| BK-E | 57.90 | 0.5637 | 60.32 | 0.5832 | **65.27** | **0.6428** |
| BK-M | 58.20 | 0.5728 | 56.20 | 0.5378 | **62.31** | **0.6002** |
| BT-BK | 64.00 | 0.6310 | 65.34 | 0.6310 | **70.14** | **0.6853** |
| BT-E | 63.10 | 0.6012 | 63.87 | 0.6019 | **66.37** | **0.6431** |
| BT-M | 57.60 | 0.5536 | 60.12 | 0.5901 | **64.25** | **0.6285** |
| E-BK | 58.80 | 0.5760 | 56.78 | 0.5375 | **60.17** | **0.5841** |
| E-BT | 59.00 | 0.5721 | 59.21 | 0.5784 | **62.83** | **0.6018** |
| E-M | 56.10 | 0.5417 | 55.75 | 0.5396 | **59.36** | **0.5702** |
| M-BK | 62.30 | 0.6014 | 64.20 | 0.6202 | **69.10** | **0.6730** |
| M-BT | 54.50 | 0.5201 | 60.12 | 0.5840 | **67.23** | **0.6531** |
| M-E | 54.40 | 0.5107 | 58.37 | 0.5671 | **64.17** | **0.6270** |
| AVE | 58.63 | 0.5664 | 59.89 | 0.5776 | **64.48** | **0.6259** |

reduced, which indicates that introducing the domain-specific features is an effective way to improve performance. The model without non-pivot outperforms HATN which extracts only pivot features. It shows that only extracting pivot features CDRL can also achieve good sentiment analysis. The accuracy of the model without pivot feature is lower than the model without non-pivot feature, which shows that the pivot feature contains more sentiment connotations than non-pivots. The average accuracy of the model without RL is 88.53%. For tasks with large domain differences, such as K-B and K-D, the classification accuracy is less than 88%, showing that the reinforcement learning mechanism can effectively improve the domain adaptability. Through the above analysis, we demonstrate the important contributions of each component of the CDRL.

### 4.4 Effect of Feature Selection Policy

To show the feature selection effect of CDRL framework intuitively, we select four sentences from the B-E and E-B tasks for the intuitive feature selection display. The results in Table 4 show that the selection of features by the feature selection policy in CDRL. In positive data, pivot features such as "hope" and "great" can be retained and non-pivot features such as "save", "light" and "small" are also retained. Noisy feature information such as "horrendous" and "only" are deleted through the cross-domain reinforcement learning framework. In negative data, pivot features such as "boring", "bad" and "little" are retained and non-pivot features such as "jumbling", "low" and "transitioning" are also retained. Noise feature information such as "interesting" and "like" are deleted. CDRL can tackle the problem of noisy data. The feature selection policy effectively selects important features in the domain and avoid the Hughes phenomena. The decision-making ability of reinforcement learning improves domain adaptability of cross-domain sentiment analysis.

**Table 4.** Visualization of feature selection in CDRL of the B-E and E-B tasks in D1. The deleted and retained pivot features are marked in red and green, respectively. Cyan and yellow mark the deleted and retained non-pivot features, respectively.

| Book domain | Electronic domain |
|---|---|
| Label: positive<br>This story gives me hope in mankind again. Despite the horrendous abuse, this woman gave herself – all of herself – to save a child from the torment she experienced | Label: positive<br>The only problem I had with it was that the voice recorder would play the recordings back fast in some parts and real slow in others but the disk told me how to fix that. I like it, I think it great, light and small |
| Label: negative<br>I found it interesting but somehow boring as the above story developed little and the focus was on the characters so that it seemed to me that it was slightly mumbling jumbling. Strategy, brilliance and adventure seemed to be low here | Label: negative<br>The quality of the recording was very bad. Little squares showed up all the time and in dramatic scene changes the whole screen was transitioning little squares, like it could just not keep up with the speed. I would definitely not buy it again |

## 5 Conclusion and Future Work

In this paper, a cross-domain reinforcement learning framework is proposed for sentiment analysis, which suffered from domain-specific sentiment information ignorance and weak ability to capture sentiment features in previous works. The proposed model separately extracts pivot and non-pivot features to capture the sentiment information fully. Through the feature selection policy and the guidance of delayed reward, CDRL effectively avoids the Hughes phenomena caused by the feature redundancy and improves the domain adaptability. Comparative experimental results demonstrate that CDRL outperforms the SOTA methods. Furthermore, the ablation experiment proves the effectiveness of the proposed framework components. In our future work, we will further improve the framework proposed to better adapt to more cross-domain issues.

**Acknowledgments.** The work is supported by the Social Science Foundation of Jiangsu Province (No. 20TQC003).

## References


1. Alonso, M.C., Malpica, J.A., De Agirre, A.M.: Consequences of the Hughes phenomenon on some classification techniques. In: Proceedings of the ASPRS 2001 Annual Conference, pp. 1–5 (2011)
2. Blitzer, J., Dredze, M., Pereira, F.: Biographies, bollywood, boom-boxes and blenders: domain adaptation for sentiment classification. In: ACL. Prague, Czech Republic, pp. 440–447 (2007)
3. Devlin, J., Chang, M.-W., Lee, K., Toutanova, K.: BERT: pre-training of deep bidirectional transformers for language understanding. arXiv preprint arXiv:1810.04805 (2018)
4. Du, C., Sun, H., Wang, J., Qi, Q., Liao, J.: Adversarial and domain-aware BERT for cross-domain sentiment analysis. In: ACL, pp. 4019–4028 (2020)
5. Du, Y., He, M., Wang, L., Zhang, H.: Wasserstein based transfer network for cross-domain sentiment classification. Knowl.-Based Syst. **204**, 106162–106178 (2020)
6. He, R., Lee, W.S., Ng, H.T., Dahlmeier, D.: Adaptive semi-supervised learning for cross-domain sentiment classification. arXiv preprint arXiv:1809.00530 (2018)
7. Jia, X., Jin, Y., Li, N., Su, X., Cardiff, B., Bhanu, B.: Words alignment based on association rules for cross-domain sentiment classification. Front. Inf. Technol. Electron. Eng. **19**(2), 260–272 (2018). https://doi.org/10.1631/FITEE.1601679
8. Li, Z., Li, X., Wei, Y., Bing, L., Zhang, Y., Yang, Q.: Transferable end-to-end aspect-based sentiment analysis with selective adversarial learning. arXiv preprint arXiv:1910.14192 (2019)
9. Li, Z., Wei, Y., Zhang, Y., Yang, Q.: Hierarchical attention transfer network for cross-domain sentiment classification. In: AAAI (2018)
10. Mikolov, T., Chen, K., Corrado, G., Dean, J.: Efficient estimation of word representations in vector space. arXiv preprint arXiv:1301.3781 (2013)
11. Ramshankar, N., Joe Prathap, P.M.: A novel recommendation system enabled by adaptive fuzzy aided sentiment classification for E-commerce sector using black hole-based grey wolf optimization. Sādhanā **46**(3), 1–24 (2021). https://doi.org/10.1007/s12046-021-01631-2
12. Vaswani, A., et al.: Attention is all you need. In: Advances in Neural Information Processing Systems, Long Beach, CA, USA, pp. 5998–6008 (2017)
13. Vijayaraghavan, P., Roy, D.: Generating black-box adversarial examples for text classifiers using a deep reinforced model. arXiv preprint arXiv:1909.07873 (2019)
14. Wang, J., et al.: Human-like decision making: document-level aspect sentiment classification via hierarchical reinforcement learning. arXiv preprint arXiv:1910.09260 (2019)
15. Williams, R.J.: Simple statistical gradient-following algorithms for connectionist reinforcement learning. Mach. Learn. **8**, 229–256 (1992)
16. Wu, L., Tian, F., Qin, T., Lai, J., Liu, T.-Y.: A study of reinforcement learning for neural machine translation. arXiv preprint arXiv:1808.08866 (2018)
17. Xiao, L., Wang, L., He, H., Jin, Y.: Copy or rewrite: hybrid summarization with hierarchical reinforcement learning. In: AAAI, pp. 9306–9313 (2020)
18. Zhang, K., Zhang, H., Liu, Q., Zhao, H., Zhu, H., Chen, E.: Interactive attention transfer network for cross-domain sentiment classification. In: AAAI, pp. 5773–5780 (2019)
19. Zhang, T., Huang, M., Zhao, L.: Learning structured representation for text classification via reinforcement learning. In: AAAI (2018)

# PPIR-Net: An Underwater Image Restoration Framework Using Physical Priors

Changhua Zhang[1(✉)], Xing Yang[1(✉)], Zhongshu Chen[1], Shimin Luo[1], Maolin Luo[1], Dapeng Yan[2], Hao Li[2], and Bingyan Wang[2]

[1] University of Electronic Science and Technology of China, Chengdu, China
zhangchanghua@uestc.edu.cn, 540952230@qq.com
[2] Nuclear Power Institute of China, Chengdu, China

**Abstract.** In recent years, underwater image processing has been a hot topic in machine vision, especially for underwater robots. A key part of underwater image processing is underwater image restoration. However, underwater image restoration is an essential but challenging task in the field of image processing. In this article, we propose an underwater image restoration framework based on physical priors, called PPIR-Net. The PPIR-Net combines prior knowledge with deep learning to greatly improve the structural texture and color information of underwater images. The framework estimates underwater transmission maps and underwater scattering maps through the structure restoration network (SRN). Moreover, the color correction network (CCN) is used to achieve image color correction. Extensive experimental results show that our method exceeds state-of-the-art methods on underwater image evaluation metrics.

**Keywords:** Underwater image restoration · Physical prior knowledge · Deep learning

## 1 Introduction

Underwater machine vision is widely used in underwater petroleum development, diving entertainment, underwater geomorphology survey, underwater data collection, and other fields. In contrast to common scenes, light is absorbed by water and scattered by suspended particles, resulting in degraded underwater image quality such as fogging, blurring, structural degradation, and color distortion, and this problem directly hinders the development of machine vision tasks [1–3]. For addressing image degradation, underwater image processing plays a vital role in underwater machine vision.

The existing traditional underwater image processing algorithms are mainly divided into two categories: non-physical methods and physical methods. Based on non-physical methods, some image enhancement methods adopt contrast enhancement to improve image quality, such as HE [4], AHE [5], CLAHE [6].

Supported by Sichuan Science and Technology Program (No. 2021YFG0201).


M. Tanveer et al. (Eds.): ICONIP 2022, CCIS 1793, pp. 650–659, 2023.
https://doi.org/10.1007/978-981-99-1645-0_54

Based on physical methods, underwater image restoration methods usually rely on physical prior knowledge or underwater physical models to restore underwater images, such as UDCP [7], RUIR [8]. However, these traditional methods exist some limitations, such as poor adaptability and difficulty in parameter tuning.

Recently, the rapid development of deep learning makes up for the shortcomings of traditional methods, in which deep convolutional neural networks (DCNN) have shown excellent performance in underwater image enhancement. Li et al. [9] proposed an unsupervised generative adversarial model Water-GAN for color correction of monocular underwater images, which can generate numerous underwater training data sets through adversarial learning. Wang et al. [10] proposed a DCNN-based underwater image enhancement network (UIE-Net), which enhances the underwater images by combining a color correction subnetwork and a dehazing sub-network. Li et al. [11] constructed an underwater image enhancement data set UIEBD and proposed an underwater image enhancement network Water-Net, greatly promoting the development of underwater images. Guo et al. [12] proposed a multi-scale dense GAN, and Li et al. [13] proposed an underwater image color correction network (CycleGAN) based on weakly-supervised color transfer, both of which enhance underwater images through adversarial learning. However, none of aforesaid networks consider the attenuation and scattering properties of the light in water, and the underwater image quality is not substantially improved.

In this paper, we propose an underwater image restoration framework using physical priors termed PPIR-Net. The end-to-end network contains two subnetworks, innovatively combining physical priors with deep learning to complete structure restoration and color correction tasks of degraded underwater images. The underwater image structure restoration network (SRN) is based on physical priors, and it contains two important branches: (1) accurately estimating underwater transmission map, and (2) estimating underwater scattering map; The underwater image color correction network (CCN) realizes the color correction of underwater image. At the same time, we design loss function strategies for underwater structure restoration network and underwater color correction network. Therefore, the PPIR-Net can effectively improve the structure texture and color information of underwater images. We used the UIEBD dataset to verify the effectiveness of the proposed method. The main contributions of this paper are summarized as follows:

- We propose an underwater image restoration framework using physical priors to improve structural texture and color information of underwater images; meanwhile, we also design loss function strategies.
- The SRN is proposed to recover underwater image structure texture by using physical priors; then, the CCN is used to correct its color information. We do extensive experiments to compare with state-of-the-art methods, which quantitatively and qualitatively show the superiority of our method.

## 2 Related Work

The essence of underwater image restoration is to construct a physical model and estimate the parameters of the model to obtain realistic underwater images.

**Prior-Based Methods:** To improve the image quality, the model parameters are estimated by prior knowledge, and then the physical model is constructed. He et al. [14] proposed a dark channel prior (DCP) model to estimate the transmission map and the atmospheric light, and achieved dehazing in the air. Drews et al. [7] developed an underwater dark channel prior algorithm (UDCP) to achieve underwater image restoration. Galdran et al. [8] proposed a red channel underwater image restoration algorithm (RUIR) which restores short-wavelength underwater colors and improves the color and contrast of underwater images. Inspired by these methods, we use prior knowledge to achieve very good results.

**Learning-Based Methods:** Deep learning is increasingly used in underwater image processing. Islam et al. [15] proposed the Deep SESR network model, which simultaneously achieved super-resolution and image enhancement of underwater images. Ronneberger et al. [16] developed the U-Net network to obtain diverse feature information. Woo et al. [17] proposed a convolutional attention module to highlight important regions of feature maps. Song et al. [18] used a multi-scale fusion module to improve underwater image quality, which improved the representation ability of the convolution neural network (CNN). The above methods are beneficial to the restoration and enhancement of underwater images.

# 3 Proposed Method

As shown in Fig. 1, the proposed underwater image restoration framework includes two modules: structure restoration network (SRN) and color correction network (CCN). The structure restoration network unifies physical priors to restore the structure of raw underwater images. The structure restoration network is the basis for the following color correction network (Sect. 3.1). The color correction network corrects the color cast of structurally restored images (Sect. 3.2). In addition, the residual connection is introduced to accelerate the training of the network. The detailed explanation of each module is as follows.

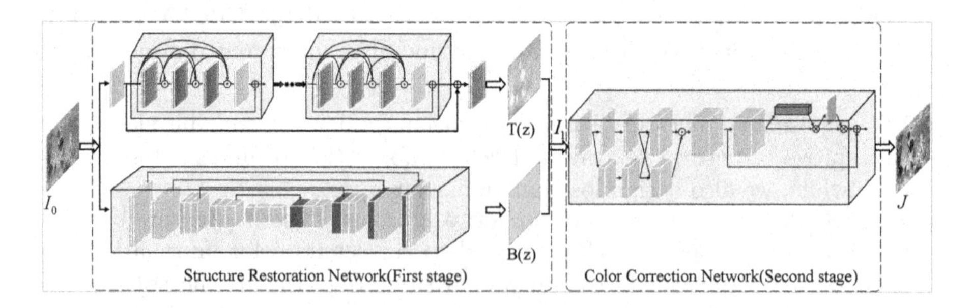


**Fig. 1.** The overall architecture of the proposed PPIR-Net for underwater image restoration. The first stage is the structure restoration network, and the second stage is the color correction network. $\otimes$ is element-wise multiplication; $\oplus$ denotes element-wise sum; $\odot$ refers to the dimension stacking of feature maps.

### 3.1 Structural Restoration Network

Since underwater images exist with structural distortions like blurred edges and fogging, the first stage is to perform structural restoration of raw images. In our model, inspired by prior information, we design a prior-based structure restoration network of underwater images. The network has two branches as follows:

The RD-Net [23] is used to accurately estimate underwater transmission maps. As shown in Fig. 2, the RD-Net consists of three parts: shallow features extraction layer, residual dense block, and dimensionality reduction layer. The first convolutional layer extracts a 64-dimensional feature map $F_1$ from a raw image $I_0$.

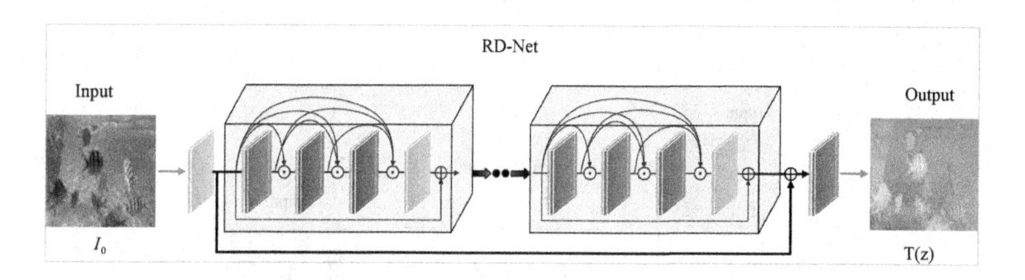


**Fig. 2.** The structure of the RD-Net model.

$$F_1 = f_1^{3\times 3}(I_0) \tag{1}$$

where $f_1^{3\times 3}(\bullet)$ denote convolution operation with convolution kernel size 3×3, $F_1$ will be used for residual learning. Then, we input $F_1$ into the residual dense network to extract a deeper feature map $F_d$.

$$F_d = M_d(M_{d-1}(\cdots(M_1(F_1))\cdots)) \tag{2}$$

where $M_d$ denotes the $d$-th residual dense block operation, and $d$ is set to 5. Finally, $F_d$ and $F_1$ perform the concatenation operation to attain $F_{d1}$. The dimensionality reduction operation of $F_{d1}$ can be computed as:

$$T(z) = f_2^{1\times 1}(F_{d1}) \tag{3}$$

where $f_2^{1\times 1}(\bullet)$ refers to the convolution operation with convolution kernel size 1×1, and $T$ is a 1D underwater transmission map.

The U-Net [16] is used to accurately estimate underwater scattering maps. As shown in Fig. 3, the U-Net structure learns different feature layers through an encoding-decoding manner. $B(z)$ is the underwater scattering map, which can be expressed as:

$$B(z) = H(I_0) \tag{4}$$

where $H(\bullet)$ denotes the U-Net network convolution operation.

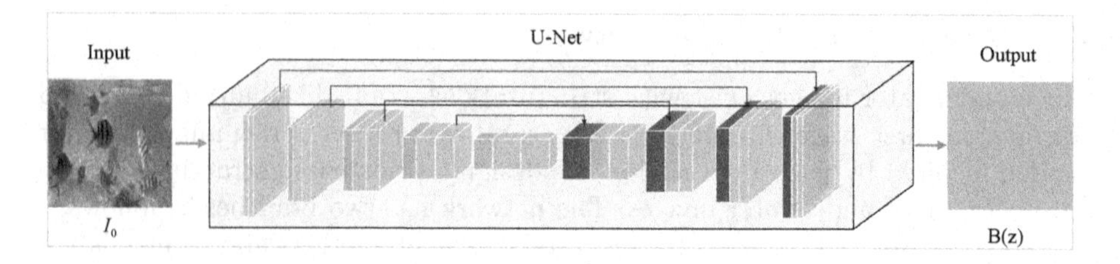


**Fig. 3.** The structure of the U-Net model.

As shown in Fig. 4, we put the generated underwater transmission map $T(z)$ and underwater scattering map $B(z)$ into the DCP model [14] to get clear images $I_1$.

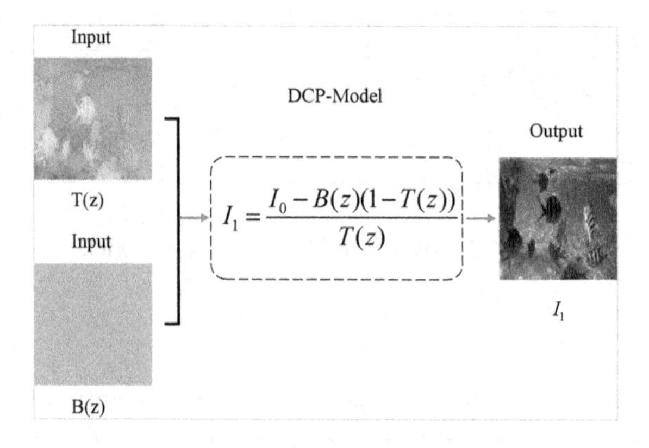


**Fig. 4.** The structure of the DCP model.

It can be clearly seen that the generated underwater image $I_1$ has clear edges, namely, achieves the dehazing. It means that the network realizes the restoration of the underwater image structure.

### 3.2 Color Correction Network

Since output underwater images $I_1$ exists with color deviation, the color of underwater images $I_1$ needs to be corrected. As shown in Fig. 5, the network is composed of a multi-scale fusion module and an attention module. Inspired by MFGS [18] and CBAM [17], the multi-scale fusion uses up-sampling and down-sampling operations to realize the cross-learning of the network and improve the characterization ability of the CNN. Moreover, the multi-scale fusion module is connected to the attention module to extract the area of interest of input images, so that the effective area of the image receives widespread attention. The mathematical expression of multi-scale fusion design is as follows: First, multiple convolution kernels extract features of the image $I_1$.

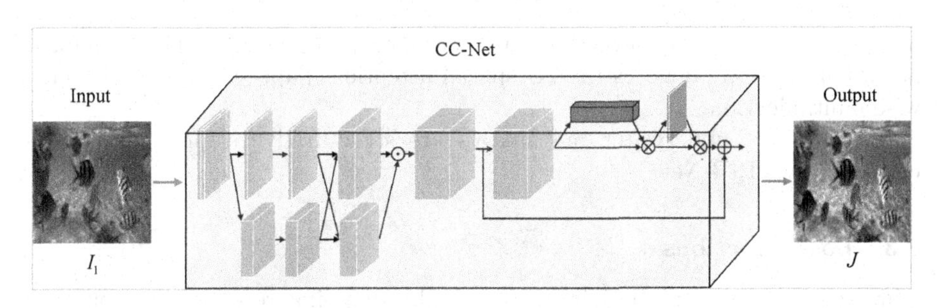


**Fig. 5.** The structure of color correction. The network structure is composed of a multi-scale fusion mechanism and an attention mechanism.

$$F = f^{3\times 3}(I_1) \tag{5}$$

where $f^{3\times3}(\bullet)$ denotes convolution operation with convolution kernel size 3×3, and $F$ denotes the extracted 64-dimensional feature map, which will be used for the down-sampling operation. Specifically, the mathematical expression for downsampling is described as:

$$\begin{aligned} f_1 &= \mathrm{conv}_{3\times 3}(\mathrm{conv}_{3\times 3}(F)) \\ f_3 &= \mathrm{conv}_{3\times 3}(f_1) \\ f_2 &= \mathrm{conv}_{3\times 3}(\text{ DownSample }(F)) \\ f_4 &= \mathrm{conv}_{3\times 3}(f_2) \end{aligned} \tag{6}$$

where $\mathrm{conv}_{3\times3}(\bullet)$ denotes the convolution operation with a convolution kernel size 3×3, and DownSample $(\bullet)$ refers to down-sampling operation; The sizes of $f_1$ and $f_3$ remain unchanged, and the sizes of $f_2$ and $f_4$ are reduced to half of the original size. Finally, the cross-fusion operation improves the diversity of the transmitted information, and its mathematical expression is as follows:

$$\begin{aligned} C_1 &= \mathrm{concat}(f_3, Up\,\mathrm{Sample}(f_2)) \\ C_2 &= \mathrm{Up\ Sample}(\mathrm{concat}(\text{ DownSample }(f_1), f_4)) \\ C &= \mathrm{concat}(C_1, C_2) \end{aligned} \tag{7}$$

where UpSample $(\bullet)$ denotes the upsampling operation, and concat$(\bullet)$ denotes the dimension stacking of feature maps; $C_1$ is the feature map fused with $f_3$ after upsampling; $C_2$ refers to the feature map after down-sampling, fusion and up-sampling; $C$ denotes the total output of multi-scale fusion, which will be used for the next step of residual learning.

Additionally, the feature map $C$ is sequentially input into the channel attention mechanism and the spatial attention mechanism, so that the various feature information is more concentrated and more obvious. Next, we will introduce the two attention mechanisms in detail. The design of the attention mechanism is as follows:

$$\begin{aligned} F' &= \mathrm{conv}_{3\times 3}(C) \\ F'' &= M_c(F') \otimes F' \\ F''' &= M_s(F'') \otimes F'' \end{aligned} \tag{8}$$

where $F'''$ is the output of the attention mechanism; $M_c$ denotes the 1D channel attention map; $M_s$ denotes the 2D spatial attention map; $\otimes$ refers to element-wise multiplication.

To sum up, underwater raw images are structurally restored and color-corrected by PPIR-Net.

### 3.3 Loss Functions

Our proposed network framework learns the mapping relationship from 800 image pairs $\left\{\left(X_i, \tilde{X}_i\right)\right\}_{i=1}^{800}$ through the loss function. $X_i$ denotes the raw underwater image, $\tilde{X}_i$ refers to the target label. The proposed network framework uses three loss functions for an end-to-end supervised training.

1) **MSE Loss.** During training, we apply $L_{MSE1}$ and $L_{MSE2}$ to the outputs of the underwater transmission map and the underwater scattering map, respectively, in the structural restoration network. $L_{MSE3}$ on the total output of the entire framework. $L_{MSE1}$, $L_{MSE2}$ and $L_{MSE3}$ are all mean square error, each of which can be expressed as follows:

$$L_{MSE} = \frac{1}{HWC}\sum_{h=1}^{H}\sum_{w=1}^{W}\sum_{c=1}^{C}\|\tilde{X} - \hat{X}\|^2 \tag{9}$$

where $H$, $W$, and $C$ are the height, width, and depth of the image, respectively, and $\hat{X}$ denotes the image predicted by our framework.

2) **Total loss.** The three loss functions are combined through a linearly weighted manner, and the ultimate loss is calculated by:

$$L_{\text{Total}} = \alpha L_{MSE1} + \beta L_{MSE2} + \gamma L_{MSE3} \tag{10}$$

where $\alpha$, $\beta$, and $\gamma$ are balance parameters which are set to 0.2, 0.2, and 0.4 in this study, respectively.

## 4 Experiments

### 4.1 Datasets

We use Underwater Image Enhancement Benchmark Dataset (UIEBD) to train the proposed network framework. UIEBD contains 800 pairs of training images, 90 pairs of test images, and 60 images without reference (challenge images). Before training, we employ the DCP model to estimate 800 underwater transmission maps and 800 underwater scattering maps, which are used as the target labels of the structural restoration network to assist in training the network. During the training process, we used 800 raw images and 800 ground truth (GT) images, 800 underwater transmission maps and 800 underwater scattering maps to train our network. Additionally, we uniformed the image size to $112px \times 112px$.

### 4.2 Evaluation Metrics

The Peak Signal Noise Ratio (PSNR) and Structural Similarity (SSIM) are utilized to evaluate the performance of the proposed model. The PSNR is an objective metric to measure the level of image distortion or noise; SSIM is an objective metric that measures the brightness, contrast, and structure of an image. High PSNR and SSIM are correspond to good image quality.

### 4.3 Implementation Details

In this experiment, our computing platform is equipped with Intel Core i7-9700K CPU and an NVIDIA RTX Titan GPU. Our proposed framework is implemented by PyTorch1.9.0 to realize underwater image restoration. We use Adam optimizer to optimize the network parameters, and the learning rate slowly decays from 0.0002 to 0.000001. Our proposed network framework is iterated 1000 times with a batch size of 5.

### 4.4 Comparison with State-of-the-Art Methods

In this section, our proposed method is compared with state-of-the-art image methods, including Deep SESR [15], Multi-Band-Enhancement [19], Underwater-Image-Fusion [20], Funie-GAN [21], RetinexNet [22], UDCP [7], and WaterNet [11]. To ensure the fairness of the experiment, each network model uses the UIEB test dataset and the same evaluation metrics.

**Visual Comparison.** Visual effects of all methods are shown in Fig. 6. As shown below, in terms of colors, our method is more realistic and closer to real images; in terms of structure, our method recovers abundant details of the edge information, enhances the contrast, and also achieves haze removal. In general, our proposed method is significantly better than state-of-the-art methods.

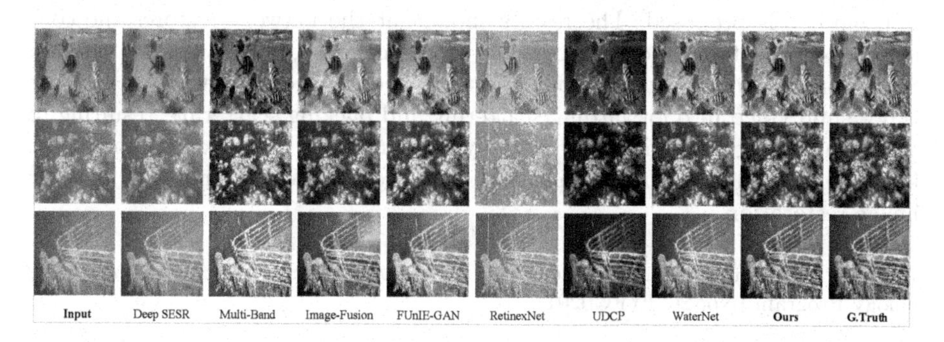


**Fig. 6.** Visual comparison results of paired images by different methods. From left to right are original images (Input), predicted images, and referenced images (G.Truth)

**Quantitative Evaluations.** To evaluate the performance of the proposed model accurately, we use PSNR and SSIM to quantitatively evaluate our model.

The experimental results are shown in Table 1, which indicates that our method is significantly better than other methods in the PSNR and SSIM metrics, and our PSNR and SSIM are 24.32 and 0.89, respectively.

**Table 1.** Metrics of our proposed framework and the state-of-the-art methods on UIEBD. Results for Rank-1 and Rank-2 are shown in red bold and blue bold, respectively.

| Model | PSNR | SSIM |
|---|---|---|
| Deep SESR [15] | $17.85 \pm 3.06$ | $0.62 \pm 0.12$ |
| Multi-Band-Enhancement [19] | $17.48 \pm 3.15$ | $0.69 \pm 0.12$ |
| Underwater-Image-Fusion [20] | $21.56 \pm 3.53$ | $0.83 \pm 0.10$ |
| FUnIE-GAN [21] | $20.82 \pm 2.86$ | $0.83 \pm 0.06$ |
| RetinexNet [22] | $15.28 \pm 2.94$ | $0.66 \pm 0.12$ |
| UDCP [7] | $11.86 \pm 3.31$ | $0.44 \pm 0.12$ |
| WaterNet [11] | $21.21 \pm 3.99$ | $0.76 \pm 0.09$ |
| (PPIR-Net)Ours | $24.32 \pm 4.32$ | $0.89 \pm 0.05$ |

## 5 Conclusion

In this work, we propose an underwater image restoration framework based on physical priors, which aims to achieve the enhancement and restoration of degraded underwater images. We use the relationship between the underwater transmission map and the underwater scattering map, combining with the deep learning method to achieve the restoration of underwater images. To effectively realize the color correction of underwater images, a multi-scale fusion and attention module is proposed. The results show that the network restores the contrast, texture details, and color of underwater images. Additionally, the network framework is superior to state-of-the-art methods. Since underwater scattering is not subdivided into forwarding scattering and backing scattering, resulting in a flaw in our proposed network. Notably, our proposed network combines prior knowledge and deep learning to provide a good research direction for underwater image restoration.

**Acknowledgments.** The presented work is supported by Sichuan Science and Technology Program (No. 2021YFG0201).

## References

1. Yeh, C., et al.: Lightweight deep neural network for joint learning of underwater object detection and color conversion. IEEE Trans. NNLS, 1–15 (2021)
2. Wu, Y., Ta, X., Xiao, R., Wei, Y., An, D., et al.: Survey of underwater robot positioning navigation. Appl. OR **90**, 101845 (2019)

3. Hu, X., Liu, Y., Zhao, Z., Liu, J., et al.: Real-time detection of uneaten feed pellets in underwater images for aquaculture using an improved YOLO-V4 network. Comput. Electron. Agric. **185**, 106135 (2021)
4. Hummel, R.: Real-time detection of uneaten feed pellets in underwater images for aquaculture using an improved YOLO-V4 network. GGIP **6**(2), 184–195 (1977)
5. Pizer, S., Amburn, E., Austin, J., et al.: Adaptive histogram equalization and its variations. CVGIP **39**(3), 355–368 (1987)
6. Zuiderveld, K.: Contrast limited adaptive histogram equalization. Academic, 474–485 (1994)
7. Drews, P., Nascimento, E., Moraes, F., et al.: Transmission estimation in underwater single images. In: ICCV Workshops, pp. 825–830 (2013)
8. Galdran, A., Pardo, D., Picón, A., et al.: Automatic red-channel underwater image restoration. JVCIR **26**, 132–145 (2015)
9. Li, J., Katherine, A., Ryan, M., et al.: WaterGAN: unsupervised generative network to enable real-time color correction of monocular underwater images. IEEE Robot. AL **3**(1), 387–394 (2018)
10. Wang, Y., Zhang, J., Cao, Y., Wang, Z.: A deep CNN method for underwater image enhancement. In: ICIP, pp. 1382–1386 (2017)
11. Li, C., Guo, C., Ren, W., Cong, R., et al.: An underwater image enhancement benchmark dataset and beyond. IEEE Trans. IP **29**, 4376–4389 (2020)
12. Guo, Y., Li, H., Zhuang, P.: Underwater image enhancement using a multiscale dense generative adversarial network. IEEE J. OE **45**(3), 862–870 (2020)
13. Li, C., Guo, J., Guo, C.: Emerging from water: underwater image color correction based on weakly supervised color transfer. IEEE Sig. Process. Lett. **25**(3), 323–327 (2018)
14. He, K., Sun, J., Tang, X.: Single image haze removal using dark channel prior. IEEE Trans. PAMI **33**(12), 2341–2353 (2011)
15. Islam, M., Luo, P., Sattar, J.: Simultaneous enhancement and super-resolution of underwater imagery for improved visual perception. cs. CV (2020). https://doi.org/10.48550/arXiv.2002.01155
16. Ronneberger, O., Fischer, P., Brox, T.: U-net: convolutional networks for biomedical image segmentation. In: Navab, N., Hornegger, J., Wells, W.M., Frangi, A.F. (eds.) MICCAI 2015. LNCS, vol. 9351, pp. 234–241. Springer, Cham (2015). https://doi.org/10.1007/978-3-319-24574-4_28
17. Woo, S., Park, J., Lee, J.-Y., Kweon, I.S.: CBAM: convolutional block attention module. In: Ferrari, V., Hebert, M., Sminchisescu, C., Weiss, Y. (eds.) ECCV 2018. LNCS, vol. 11211, pp. 3–19. Springer, Cham (2018). https://doi.org/10.1007/978-3-030-01234-2_1
18. Song, H., Wang, R.: Underwater image enhancement based on multi-scale fusion and global stretching of dual-model. MDPI **9**(6), 595 (2021)
19. Cho, Y., Jeong, J., Kim, A.: Model-assisted multiband fusion for single image enhancement and applications to robot vision. IEEE Robot. AL **3**(4), 2822–2829 (2018)
20. Cosmin, A., Codruta O., Tom, H., Philippe, B.: Enhancing underwater images and videos by fusion. In: CVPR, pp. 81–88 (2012)
21. Islam, M., Xia, Y., Sattar, J.: Fast underwater image enhancement for improved visual perception. IEEE Robot. AL **5**(2), 3227–3234 (2020)
22. Wei, C., Wang, W., Yang, W., Liu, J.: Deep retinex decomposition for low-light enhancement. In: BMVC (2018). https://doi.org/10.48550/arXiv.1808.04560
23. Xu, J., Chae, Y., et al.: Dense Bynet: residual dense network for image super resolution. In: ICIP, pp. 71–75 (2018)

# Denoising fMRI Message on Population Graph for Multi-site Disease Prediction

Yanyu Lin, Jing Yang(✉), and Wenxin Hu

East China Normal University, Shanghai, China
51205903085@stu.ecnu.edu.cn, jyang@cs.ecnu.edu.cn, wxhu@cc.ecnu.edu.cn

**Abstract.** In general, large-scale fMRI analysis helps to uncover functional biomarkers and diagnose neuropsychiatric disorders. However, the existence of multi-site problem caused by inter-site variation hinders the full exploitation of fMRI data from multiple sites. To address the heterogeneity across sites, we propose a novel end-to-end framework for multi-site disease prediction, which aims to build a robust population graph and denoise the message passing on it. Specifically, we decompose the fMRI feature into site-invariant and site-specific embeddings through representation disentanglement, and construct the edge of population graph through the site-specific embedding and represent each subject using its site-invariant embedding, followed by the feature propagation and transformation over the constructed population graph via graph convolutional networks. Compared to the state-of-the-art methods, we have demonstrated its superior performance of our framework on the challenging ABIDE dataset.

**Keywords:** Data heterogeneity · Disentangled representation · Graph neural network · Population-based disease prediction · Multi-site · Domain adaptation

## 1 Introduction

Resting-state functional Magnetic Resonance Imaging (rs-fMRI) has demonstrated its potential to capture interactions between brain regions. The resulting brain functional connectivity patterns could serve as diagnostic biomarkers for Alzheimer's disease, depression, ADHD, autism and others. In recent years, large-scale collaborative initiatives are collecting and sharing terabytes of fMRI data [16]. This makes it possible to utilize larger datasets, however, raises many new challenges. One of the most pressing is the multi-site problem [5]. Specifically, differences in acquisition protocol and scanner type can lead to uncontrolled inter-site variation, i.e., the sites might have different data distributions. Such inhomogeneity across the sites tends to distort diagnosis and biomarker extraction. And measuring and eliminating the potential heterogeneity is a very challenging task, which hinders some works [1,3] from fully exploring data from multiple sites. To verify that heterogeneity across the sites hinders the knowledge sharing and degrades the performance of jointly training, we conduct a


M. Tanveer et al. (Eds.): ICONIP 2022, CCIS 1793, pp. 660–671, 2023.
https://doi.org/10.1007/978-981-99-1645-0_55

comparative experiment, where in *Standalone Training* different sites train their models dependently on their own datasets and in *Jointly Training* the model is trained on the union set of all sites' dataset. Both training modes are under a 10-fold cross-validation. It can be observed from Fig. 1 that about one third of sites suffer from performance degeneration compared to standalone training on its private dataset, particularly for those with relatively large training dataset.

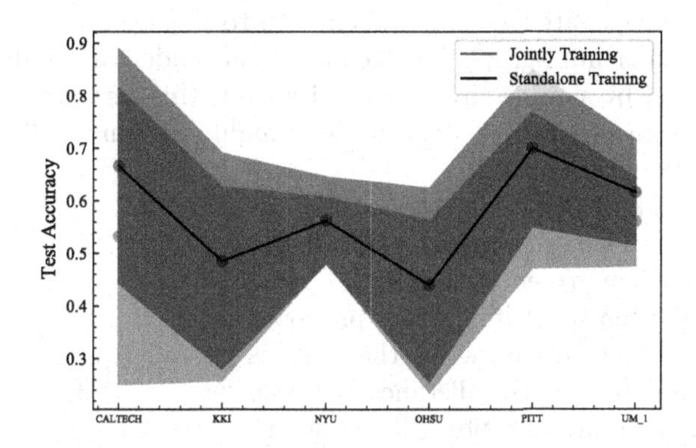


**Fig. 1.** Test accuracy and standard deviation of several large sites under Standalone Training and Jointly Training. There are a total of 20 different sites. Among them, NYU and UM_1 are the top two largest sites respectively.

There is also a line of works [5,7,8,12] that try to address the data heterogeneity across sites. Among all the previous works, the most common approach is to train a single model of machine learning or deep learning on the union set of all sites' dataset, based on the assumption that the model can implicitly learn site-invariant features [7,8]. Such approaches, however, are too simple to capture robust features without suffering from site-effect, especially for complex fMRI data. Another approaches are based on statistics. The most classic is Combat Harmonization method proposed by Fortin et al. [5], which eliminates site-effect using an empirical Bayes method to obtain post-processed data for training. Nevertheless, it is still not explicitly denoising and not adaptive, thus achieving limited boosts. Recently, Lee et al. [12] propose to calibrate the site-specific features into site-invariant features using meta-modulation network for multi-site disease prediction.

Additionally, besides the imaging information, large-scale datasets shares abundant phenotypic (i.e., demographic) information collected from multi sites, such as subject's age and sex. Population Graph (PG) was first introduced in [14], in which the nodes represent subjects' image-based features and the edges encode relationship between subjects. A number of existing works has confirmed the effectiveness of population graph for disease prediction [9,10]. Huang et al. [9] followed [14] and presented a edge-variational graph, in which the edge weight is adaptively learned from the specifically selected phenotypic features. Obviously, the success of PG relies heavily on the construction of the graph. How to represent a subject's features and how to calculate the similarity between subjects in the graph still remain as open problems. Due to multi-site problem, the site-invariant information and site-specific information of a given fMRI sample is entangled, where the site-invariant information is general and can be shared across all the sites and site-specific information represents the heterogeneity of the site that the sample is from. If we construct a "dirty" graph, where highly heterogeneous nodes are linked together with large weights, the site-specific information will be propagated as noises from one to its neighbors. In addition, if we just construct a mini-graph for each site, one node can easily avoid being adversely affected by samples from other sites, but this also prevents benefiting from samples from other sites, even if they might be similar. These two cases both give us sub-optimal results.

In practical, site-to-site heterogeneity is different, two independent sites might exhibit similar data distribution. To encourage such knowledge sharing between such sites, in our proposed method, the relationship between the sites is explicitly represented by the distance between their site-specific representations. It is worth noting that two different sites might exhibit similar data distribution when they utilize the same scanner type or adopt similar acquisition protocol. Therefore, it is preferred to characterize the similarity between two nodes in the population graph in terms of their data characteristic, which can extracted from the the raw imaging features. Simple demographic information, such as acquisition site, cannot adequately reflect the relationship between nodes. We illustrate our motivation in Fig. 2. It is a population graph, where there are three acquisition sites and different color denotes different data distribution. It can be observed that the bottom two sites with inter-site connections exhibit similar distributions. To make full use of data, we hope to explicitly characterize the similarity between nodes in terms of their data distribution and encourage similar nodes to share more.

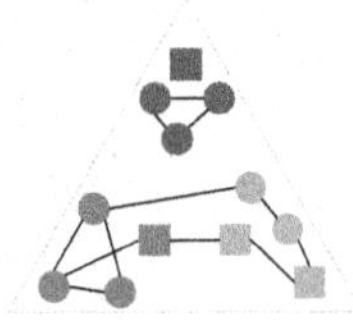

**Fig. 2.** Motivation illustration. There are three acquisition sites and different color denotes different data distribution. The circles denote the diseased and the square denote the healthy. (Color figure online)

Considering all the above, we present a novel end-to-end framework called **D**enoising **P**opulation **G**raph **C**onvolution (DPGC) for multi-site disease prediction. Specifically, we first perform disentanglement on the raw fMRI data to get the site-invariant and site-specific information. And we encode the edge weights using site-specific information together with some phenotypic features. Complementary to the novel edge construction, we propose to propagate and aggregate site-invariant information on the population graph. In case two heterogeneous nodes are connected, such strategy is able to avoid negative message passing between them.

The main contributions of our proposed DPGC fall into three folds.

1. To the best of our knowledge, we are the first to address data heterogeneity on the population graph for disease diagnosis.
2. We propose a novel and general framework DPGC, which manages to construct a robust population graph. We represent the individual features using site-invariant information and encode the edge weights using site-specific information together with some phenotypic features.
3. We conduct extensive experiments, and the results shows that our method outperforms existing competitive baselines on the challenging real-world ABIDE dataset.

## 2 Related Work

**Disentangled Representation Learning.** Learning representations that distinguish different independent ground truth is necessary, especially for heterogeneous input distribution [17]. There are two aspects should obey when learning disentangled representation. One is latent units account for data, the other is they change independently of each other. Among all, Ranzato et al. [15] is the first to use an encoder-decoder framework for implicit representation learning. We followed that framework, but we explicitly separated the encoded feature into two parts (site-invariant and site-specific feature) and designed three regularizers to learn disentangled representations.

**Graph Neural Networks.** Graph neural networks generate node representations by aggregating and transforming information over the graph structure. Aggregation and transformation processes can be different, which results in different graph neural networks. Among the various GNN variants, the vanilla Graph Convolutional Network (GCN) [11] motivated the convolutional architecture via a localized first-order approximation of spectral graph convolutions and suggests a renormalization trick on the adjacency matrix to prevent gradient explosion. The performance of GCN peaks at 2 layers and then declines due to over-smoothing problem. Xu et al. [18] proposed the jumping knowledge network (JK-Net) which could learn adaptive and structure-aware representations through "jumping" all the intermediate representations to the last layer to suppress over-smoothing. We will apply GCN on the population graph and take advantage of JK-Net to avoid suffering from over-smoothing. There is a line of works [9,14] that apply GNN on population graph for diesease prediction. Parisot et al. [14] first introduced population graph, in which the nodes represent subjects' image-based features and the edges encode relationship between subjects in terms of the phenotypic data. Huang et al. [9] followed [14] and presented a edge-variational graph, in which the edge weight is adaptively learned from the specifically selected phenotypic features. Despite this success, how to build a population graph and perform message aggregation is still in the air.

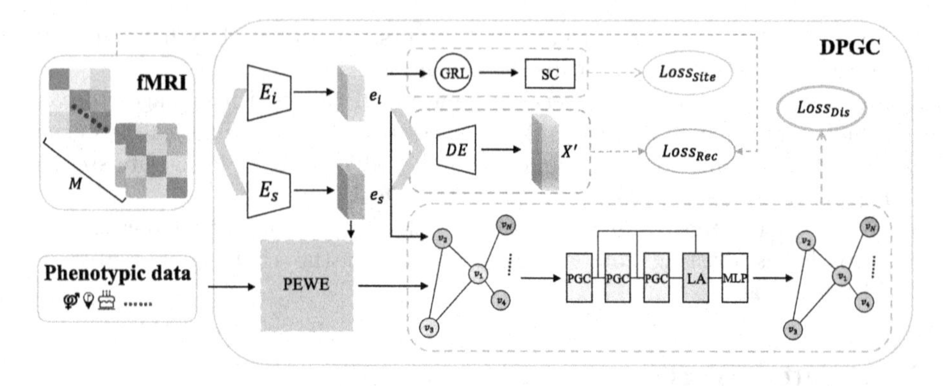


**Fig. 3.** Overview of DPGC. $E_i(E_s)$: encoder for capturing site-invariant (site-specific) representation. GRL: gradient reversal layer. SC: site classifier. DE: decoder for reconstruction. PEWE: population edge weight encoder. PGC: population graph convolution. LA: layer aggregation. Colours in the population graph: green, orange and grey denote the healthy, the diseased and the unlabeled, respectively.

## 3 Our Proposed Framework

### 3.1 Problem Statement

Assuming that there are $M$ subjects and $N$ sites denoted as $\mathcal{S}_1, \ldots, \mathcal{S}_N$, each site has a dataset $D_k = \{(x_i^k, y_i^k, p_i^k)\}_{i=1}^{M_k}$ sampled from data distribution $\mathcal{D}_k$, where $x_i^k$ is the $i^{th}$ rs-fMRI sample in site $S_k$, $y_i^k$ is the corresponding label (diseased or not) of the sample, $p_i^k$ is the corresponding non-imaging phenotypic data (e.g. age, gender, acquisition site, etc.) and $M_k$ is the number of samples of site $S_k$. Our objective is to predict the label of a subject and report the subject's disease state. However, data distribution varies considerably across the sites, i.e., $\mathcal{D}_u \neq \mathcal{D}_v$ if $u \neq v$, which hinders the construction of population graph and the message passing on it.

### 3.2 Proposed Framework

As in Fig. 3, our proposed framework DPGC consists of two key components, disentanglement of fMRI into site-invariant and site-specific representation and construction of population graph. We will detail the structure and training of DPGC in the following.

Through the dual-encoder, we explicitly capture the site-invariant representation $e_i$ and site-specific representation $e_s$ of the fMRI sample. To guide the training of disentangled representation learning, we design three losses, site classification loss, reconstructing loss and disease prediction (label prediction) loss. In addition, we specially present a Population Edge Weight Encoder (abbr. as PEWE), which takes two nodes' site-specific embedding and phenotypic information as input and outputs the edge weight linking the two nodes. Moreover, we represent each node in the graph using its site-invariant representation. A population graph is constructed, on which the site-invariant (i.e. denoised) message can be propagated and aggregated for large-scale multi-site disease prediction.

**Disentanglement Representation Learning.** Given an fMRI datapoint $x$ sampled from site $S_k$, to explicitly capture $e_i$ and $e_s$, we design a dual encoders $E_i$ and $E_s$ parameterized by $\theta_i$ and $\theta_s$, which are responsible for capturing site-invariant representation $e_i = E_i(x; \theta_i)$ and site-specific representation $e_s = E_s(x; \theta_s)$ respectively from the raw data. Inspired by classic unsupervised domain adaptation work [6], we adopt an adversary-based method to extract the site-invariant representation. Specifically, we feed the output of $E_i$, i.e. $e_i$, to the Gradient Reversal Layer (GRL) and encourage the site-classifier (SC, parameterized by $\theta_{sc}$) to recognize which site this datapoint is sampled from. We adopt the commonly used cross-entropy loss function, and the site classification loss $\mathcal{L}_{Site}$ is defined as:

$$\mathcal{L}_{Site} = \sum_{k=1}^{N} \sum_{j=1}^{M_k} CrossEntropy(SC(E_i(x_j^k; \theta_i); \theta_{sc}), k) \tag{1}$$

where $SC(E_i(x_j^k; \theta_i); \theta_{sc})$ and $k$ are the predicted and target classes of the site, respectively. While in the back-propagation process, we reverse the gradient of site classification loss w.r.t. the parameters of the site-invariant encoder. In this way, we force the site-invariant encoder to capture the site-invariant representation, which is indistinguishable for the site classifier.

Furthermore, we add a reconstruction loss to characterize the information loss incurred by disentanglement. It is expected to minimize the reconstruction loss to ensure the integrity of the representation after disentangling. Specifically, $e_i$ and $e_s$ are concatenated and fed into a decoder network DE parameterized by $\theta_d$, and a euclidean-norm based reconstruction loss is enforced on the decoded representation, as is formulated in Eq. (2).

$$\mathcal{L}_{Rec} = \sum_{k=1}^{N} \sum_{j=1}^{M_k} \|x_j^k - DE(Concat(E_i(x_j^k; \theta_i), E_s(x_j^k; \theta_s)); \theta_d)\|_2 \tag{2}$$

**Population Graph Construction**

*Edge Construction.* The site-specific representation extracted in the disentanglement representation learning can be regarded as complementary information to phenotypic features. To calculate the edge weight between two nodes $u$ and $v$ in population graph, we design a PEWE, which takes the site-specific embedding and normalized phenotypic features of two nodes as input and outputs a scalar $a_{u,v}$ ranged from 0 to 1 as the link weight. In detail, the phenotypic input is projected to a high-dimension representation $e_p$ via a MLP denoted as $\Omega$ with parameters $\theta_\omega$, i.e., $e_p = \Omega(p; \theta_\omega)$. And then we can calculate the edge weight as:

$$a_{u,v} = Cosine(concat(e_{u,s}, e_{u,p}), concat(e_{v,s}, e_{v,p}) + 1) * 0.5 \tag{3}$$

where $e_{u,s}$ and $e_{v,s}$ denote the site-specific embedding of node $u$ and $v$ respectively, $e_{u,p}$ and $e_{v,p}$ are the respective phenotypic embeddings. Following [9,14], we use the acquisition site and sex as phenotypic features. Finally, we can construct an adjacency matrix $A$ for population graph, where $a_{u,v}$ in the $u^{th}$ row and the $v^{th}$ column of $A$ reflects the similarity between node $u$ and $v$.

*Node Construction.* Different from existing population graph based methods for disease prediction [9,14], we represent each subject using its site-invariant embedding rather than the raw feature. Then we constrct the input node feature as $X = [e_{1i}, \ldots, e_{Mi}]^T$. Considering in case there are two heterogeneous nodes connected together, the heterogeneous information wouldn't be propagated from one to its neighbors through such strategy. Combined with the above strategy for constructing edges, our proposed node construction is capable for building a robust population graph, which can not only exploit the information of similar nodes from other sites, but also avoid negative information sharing.

*Disease Prediction.* The ultimate goal of our framework is for disease diagnosis. We apply graph convolution networks to propagate and transform $X$ on the constructed population graph, denoted as PGC(Population Graph Convolution). The PGC we use consists of three layers parameterized by $W$ and can be formulated as follow:

$$H = \hat{A}\sigma(\hat{A}\sigma(\hat{A}XW^{(0)})W^{(1)})W^{(2)} \tag{4}$$

Here, $\sigma(\cdot)$ is a non-linear activation function, e.g. ReLU, $W^{(0)} \in \mathbb{R}^{d_i \times d_h}$, $W^{(1,2)} \in \mathbb{R}^{d_h \times d_h}$ and $\hat{A}$ is a normalized adjacency matrix to perform the propagation and aggregation. We set $d_h = 16$ for all the experiments. Typically, we normalize the adjacency matrix in such fashion: $\hat{A} = \tilde{D}^{-\frac{1}{2}}\tilde{A}\tilde{D}^{-\frac{1}{2}}$, where $\tilde{D}_{ii} = \sum_j \tilde{A}_{ij}$ and $\tilde{A} = A + I$ is the adjacency matrix adding self-connections for nodes. Particularly, a layer aggregation module is deployed before the classifier which concatenates the output of each graph convolution layer to alleviate the over-smoothing problem. The classifier consists of a hidden fully connected layer with 256 units and a final softmax output layer. And a cross-entropy based loss function is employed for training. For simplicity, we use $\theta_{mlp}$ to denote the parameter of the disease label classifier. The disease prediction objective is to minimize the following loss function:

$$\mathcal{L}_{Dis} = CrossEntropy(MLP(H; \theta_{mlp}), y) \tag{5}$$

where $y$ is the label of all the training samples.

Putting all the components together, now we have the full framework as shown in Fig. 3. The overall optimization objective of DPGC is to minimize the following loss function:

$$\mathcal{L} = \mathcal{L}_{Dis} + \alpha\mathcal{L}_{Site} + \beta\mathcal{L}_{Rec} \tag{6}$$

where $\alpha$ and $\beta$ are balance coefficients for $\mathcal{L}_{Site}$ and $\mathcal{L}_{Rec}$, respectively.

# 4 Experiments

In this section, we conduct several experiments on ABIDE dataset to evaluate the performance of our proposed method and compare it with existing state-of-the-art methods on disease prediction. Furthermore, we perform an additional ablation study to verify the introduced components in DPGC.

## 4.1 Dataset and Preprocessing

The Autism Brain Imaging Data Exchange (ABIDE) [4] database involves $N = 20$ imaging sites[1], publicly sharing rs-fMRI and phenotypic datasets (e.g. site, gender and age). It includes 871 subjects, including 403 individuals suffering from Autism Spectrum Disorder (ASD) and 468 healthy controls (HC). For a fair comparison with state-of-the-art on the ABIDE, we chose the same 871 subjects and take the same Configurable Pipeline for the Analysis of Connectomes (C-PAC) [2] for data preprocessing. Then, we extract the Fisher $z$-transformed functional connectivity matrix from the Harvard Oxford (HO) atlas comprising $R = 110$ brain regions and take the upper triangular of it as model input, in which a single sample $x \in \mathbb{R}^{R(R+1)/2}$.

## 4.2 Baselines and Settings

To demonstrate the effectiveness of our proposed method, we consider following methods as our baselines.

**Ridge classifier** [1]: is a classic classification algorithm based on ridge regression, which only exploits the raw features extracted from medical imaging data.
**Population-GCN** [14]: first adopted GCN for fMRI analysis in populations, which combines imaging and non-imaging data.
**Combat-DNN** [5]: eliminated site-effect using an empirical Bayes method and then applied an MLP predictor for those post-processed data.
**GenM** [12]: proposed to calibrate the site-specific features into site-invariant features using meta-modulation network for multi-site disease prediction.
**EV-GCN** [9]: represented edge weights between subjects in terms of their phenotypical features and performed raw fMRI message propagation for disease prediction.

For evaluation, we follow previous works [9,14], perform a 10-fold cross-validation on the dataset and adopt overall accuracy (ACC) and area under the receiver operating characteristic curve (AUC) as the assessment criteria.

[1] Caltech, CMU, KKI, MAX_MUN, NYU, Olin, OHSU, SDSU, SBL, Stanford, Trinity, UCLA$_1$, UCLA$_2$, Leuven$_1$, Leuven$_2$, UM$_1$, UM$_2$, Pittsburgh, USM and Yale.

Particularly, we choose the model performing best on validation set for testing. For hyperparameters in our method, we set $\alpha = 0.1$, $\beta = 0.1$. All the methods are trained with an Adam optimizer and run with 600 epochs. And all the models are implemented in PyTorch and run on 1 T V100 GPU. The source codes will be made publicly available[2].

**Table 1.** Comparison of our method against the baselines on ABIDE dataset. For heterogeneity, ✗: no attempt is made to address the multi-site problem, and imaging data remains heterogeneous, ✓: taking the multi-site problem into account. For multimodal, ✗: only imaging data is used for prediction, ✓: both imaging and phenotypic data are used.

| Methods | Heterogeneity | Multimodal | ACC | AUC |
|---|---|---|---|---|
| Ridge classifier | ✗ | ✗ | 0.6333 | 0.6843 |
| Population-GCN | ✗ | ✓ | 0.6585 | 0.7193 |
| Combat-DNN | ✓ | ✗ | 0.6721 | 0.7418 |
| GenM | ✓ | ✗ | 0.7080 | 0.7117 |
| EV-GCN | ✗ | ✓ | 0.7743 | 0.8584 |
| DPGC | ✓ | ✓ | **0.8077** | **0.8830** |

### 4.3 Main Results

Table 1 shows the mean ACC and AUC of 10-fold cross-validation for five baselines and our method on ABIDE dataset. From the experiment result shown in Table 1, it can be observed that our method DPGC outperforms the competing baselines by a large margin. Ridge classifier performs the worst because it neither considers the multi-site problem nor exploits the non-imaging data. In contrast, both anti-heterogeneous methods (i.e. Combat-DNN, GenM and ours) and graph-based models (i.e. Population-GCN, EV-GCN and ours) perform better. DPGC takes into account the multi-site problem and non-imaging phenotypic data, which achieves substantial performance gains and performs the best over the baselines, exceeding ridge classifier by 17% and state-of-the-art method EV-GCN by more than 3% in terms of prediction accuracy.

### 4.4 Ablation Study

To show the effectiveness of the PEWE and the disentanglement mechanism, we conduct several ablation studies with several DPGC variants. Each variant removes or modifies one component in DPGC. The results is shown in Table 2. **(1) Edge Construction.** In DPGC, we calculate the similarity between two nodes based on their site-specific embeddings and phenotypic features. Existing works [9,14] have demonstrated the effectiveness of phenotypic features for

[2] https://github.com/missmissfreya/DPGC.

**Table 2.** Ablation study of each component in DPGC on ABIDE dataset.

| PG Construction | | | Disentanglement | | ACC | AUC |
|---|---|---|---|---|---|---|
| $e_s$ | $e_p$ | $e_i$ | site classification | reconstruction | | |
| ✗ | | | | | 0.7960 | 0.8604 |
| | ✗ | | | | 0.7650 | 0.8326 |
| | | ✗ | | | 0.6847 | 0.7263 |
| | | | ✗ | | 0.6432 | 0.6687 |
| | | | | ✗ | 0.7869 | 0.8598 |
| | | | | | **0.8077** | **0.8830** |

constructing population graph. We believe site-specific embeddings can provide more supplementary information about heterogeneity to phenotypic features. To verify this claim, we modify DPGC, in which we only use the phenotypic information or the site-specific embeddings for calculating edge weight. As in the first two rows of Table 2, without $e_s$ and $e_p$, DPGC decreases by 1% and 4% in ACC, respectively. It can be concluded that $e_s$ and $e_p$ both play important role for constructing population graph.

**(2) Node Representation.** To avoid negative sharing between two nodes in population graph, we represent each subject using its site-invariant representation. To examine whether it works for disease prediction, we still use our proposed edge construction method, but represent each subject using its raw features. As in the third row of Table 2, the modified version experiences drastic performance drop, decreasing from 80.77% to 68.47% in terms of ACC.

**(3) Disentanglement of fMRI.** Our proposed framework relies heavily on the success of disentanglement. The edge construction involves site-specific embedding $e_s$ and node representation involves site-invariant embedding $e_i$. So it is crucial to disentangle $e_i$ and $e_s$. Here, we further run DPGC without site classification loss and reconstruction loss on ABIDE with the same setting. As in the fourth and the fifth row in Table 2, both variants are negatively affected to varying degrees, with about 16% and 2% decrease in ACC, respectively.

## 4.5 Visualization

To further investigate the effectiveness of the learned site-specific embedding for constructing the population graph, we provide a visualization of the subjects' site representations through t-SNE [13]. Specifically, we pick datapoints from four sites $UM_1$, $UM_2$, $Leuven_1$ and $Leuven_2$ (represented by dark green, light green, orange and light orange dots in the figure), and it is worth noting that $UM_1$ and $UM_2$ are from the same site (i.e., similar data distribution), but their site ids are different, as are $Leuven_1$ and $Leuven_2$.

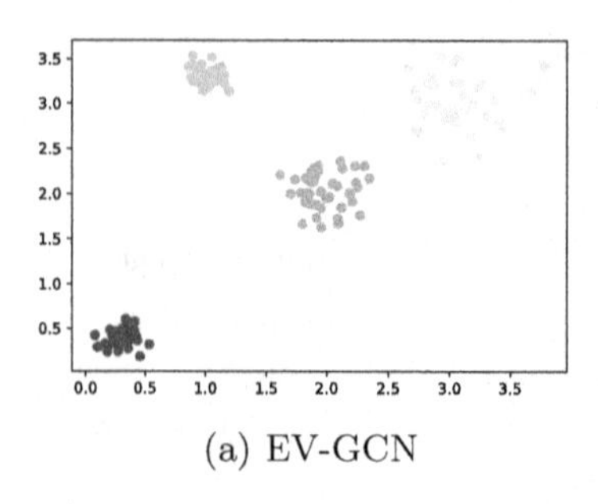


(a) EV-GCN

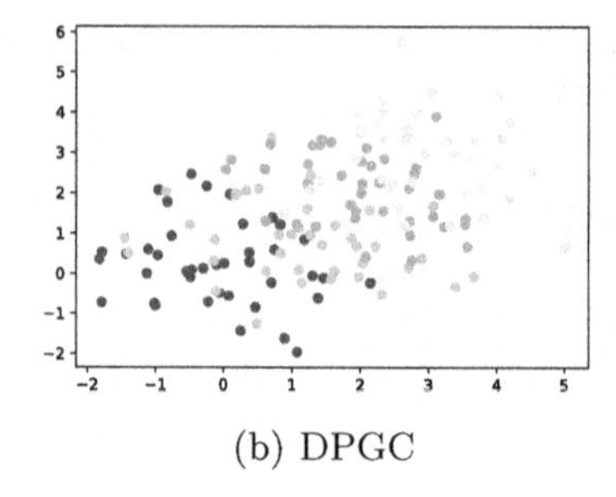


(b) DPGC

**Fig. 4.** Visualization of subjects' site representations extracted by EV-GCN and DPGC. Dots represent subjects, and different colors indicate that they belong to different sites. (Color figure online)

Figure 4 visualizes the site representations extracted by EV-GCN (the left panel) and DPGC (the right panel). EV-GCN only uses phenotypic features, resulting in similar site representations only for subjects belonging to the same site. In contrast, DPGC manages to learn data distribution similarities between sites from imaging data (specifically, site-specific embeddings), making the representations of subjects with similar data distributions close even if they belong to different sites. This enables data utilization from similar sites instead of being confined within one site, which echos our motivation illustrated in Fig. 2.

## 5 Conclusion

Nowadays, large-scale fMRI analysis helps to uncover functional biomarkers and diagnose neuropsychiatric disorders. However, data heterogeneity introduced by multi-site problem prevents us from fully exploit these datasets on a large scale. To address this, we propose a novel and end-to-end framework DPGC, which disentangles the site-specific and site-invariant embeddings for constructing edges and representing nodes in population graph. Extensive experiments and comparisons with existing state-of-the-art methods show DPGC's excellent performance. Additionally, we perform several ablation studies and the results verifies the effectiveness of our introduced components in DPGC. Furthermore, several extensions could be considered for this work. The phenotypic data we used is quite simple, so an interesting extension would be to exploiting richer non-imaging information. Constructing heterogeneous graph of population should also be taken into consideration (e.g. subject-site meta-path).

**Acknowledgements.** This research is funded by the Basic Research Project of Shanghai Science and Technology Commission (No.19JC1410101). The computation is supported by ECNU Multifunctional Platform for Innovation (001).

## References

1. Abraham, A., Milham, M.P., et al.: Deriving reproducible biomarkers from multi-site resting-state data: an autism-based example. NeuroImage **147**, 736–745 (2017)

2. Craddock, C., Sikka, S., Cheung, B., Khanuja, R., Ghosh, S.S., et al.: Towards automated analysis of connectomes: the configurable pipeline for the analysis of connectomes (C-PAC). Front. Neuroinform. **42**, 10–3389 (2013)
3. Dadi, K., et al.: Benchmarking functional connectome-based predictive models for resting-state FMRI. Neuroimage **192**, 115–134 (2019)
4. Di Martino, A., Yan, C.G., Li, Q., Denio, E., et al.: The autism brain imaging data exchange: towards a large-scale evaluation of the intrinsic brain architecture in autism. Mol. Psychiatry **19**(6), 659–667 (2014)
5. Fortin, J.P., Cullen, N., Sheline, Y.I., et al.: Harmonization of cortical thickness measurements across scanners and sites. Neuroimage **167**, 104–120 (2018)
6. Ganin, Y., Lempitsky, V.: Unsupervised domain adaptation by backpropagation. In: International Conference on Machine Learning, pp. 1180–1189. PMLR (2015)
7. Glocker, et al.: Machine learning with multi-site imaging data: an empirical study on the impact of scanner effects. arXiv preprint arXiv:1910.04597 (2019)
8. Heinsfeld, A.S., Franco, A.R., Craddock, R.C., Buchweitz, A., Meneguzzi, F.: Identification of autism spectrum disorder using deep learning and the abide dataset. NeuroImage: Clin. **17**, 16–23 (2018)
9. Huang, Y., Chung, A.C.S.: Edge-variational graph convolutional networks for uncertainty-aware disease prediction. In: Martel, A.L., et al. (eds.) MICCAI 2020. LNCS, vol. 12267, pp. 562–572. Springer, Cham (2020). https://doi.org/10.1007/978-3-030-59728-3_55
10. Kazi, A., et al.: InceptionGCN: receptive field aware graph convolutional network for disease prediction. In: Chung, A.C.S., Gee, J.C., Yushkevich, P.A., Bao, S. (eds.) IPMI 2019. LNCS, vol. 11492, pp. 73–85. Springer, Cham (2019). https://doi.org/10.1007/978-3-030-20351-1_6
11. Kipf, T.N., Welling, M.: Semi-supervised classification with graph convolutional networks. arXiv preprint arXiv:1609.02907 (2016)
12. Lee, J., Kang, E., Jeon, E., Suk, H.-I.: Meta-modulation network for domain generalization in multi-site fMRI classification. In: de Bruijne, M., et al. (eds.) MICCAI 2021. LNCS, vol. 12905, pp. 500–509. Springer, Cham (2021). https://doi.org/10.1007/978-3-030-87240-3_48
13. Van der Maaten, L., Hinton, G.: Visualizing data using t-SNE. J. Mach. Learn. Res. **9**(11), 2579–2605 (2008)
14. Parisot, S., et al.: Spectral graph convolutions for population-based disease prediction. In: Descoteaux, M., Maier-Hein, L., Franz, A., Jannin, P., Collins, D.L., Duchesne, S. (eds.) MICCAI 2017. LNCS, vol. 10435, pp. 177–185. Springer, Cham (2017). https://doi.org/10.1007/978-3-319-66179-7_21
15. Ranzato, M., Huang, F.J., Boureau, Y.L., LeCun, Y.: Unsupervised learning of invariant feature hierarchies with applications to object recognition. In: 2007 IEEE Conference on Computer Vision and Pattern Recognition, pp. 1–8. IEEE (2007)
16. Thompson, P.M., et al.: The enigma consortium: large-scale collaborative analyses of neuroimaging and genetic data. Brain Imaging Behav. **8**(2), 153–182 (2014)
17. Wang, G., Han, H., Shan, S., Chen, X.: Cross-domain face presentation attack detection via multi-domain disentangled representation learning. In: Proceedings of the IEEE/CVF Conference on Computer Vision and Pattern Recognition, pp. 6678–6687 (2020)
18. Xu, K., Li, C., Tian, Y., Sonobe, T., Kawarabayashi, K.I., Jegelka, S.: Representation learning on graphs with jumping knowledge networks. In: International Conference on Machine Learning, pp. 5453–5462. PMLR (2018)

# CATM: Candidate-Aware Temporal Multi-head Self-attention News Recommendation Model

Laiping Cui, Zhenyu Yang(✉), Yu Wang, Kaiyang Ma, and Yiwen Li

Qilu University of Technology (Shandong Academy of Sciences), Jinan, China
yzy@qlu.edu.cn

**Abstract.** User interests are diverse and change over time. Existing news recommendation models often do not consider the relationship and temporal changes between browsing news when modeling user characteristics. In addition, the wide range of user interests makes it difficult to match candidate news to users' interests precisely. This paper proposes a news recommendation model based on the candidate-aware time series self-attention mechanism(CATM). The method incorporates candidate news into user modeling based on considering the temporal relationship of news sequences browsed by users, effectively improving news recommendation performance. In addition, to obtain more rich semantic news information, we design a granular network to obtain more fine-grained segment features of news. Finally, we also designed a candidate-aware attention network to build candidate-aware user interest representations further to better match candidate news with user interests. Extensive experiments on the MIND dataset demonstrate that our method can effectively improve news recommendation performance.

**Keywords:** granular network · candidate-aware · MIND

## 1 Introduction

Personalized news recommendation can recommend interesting news for users [13]. Many existing techniques model the candidate news and the news sequences viewed by users separately to learn the candidate news representation and the user interest representation. For example, An et al. use ID information to select important words and news through a personalized attention mechanism to learn candidate news representations and user interest representations, respectively [12]. Wu et al. combine negative feedback information from users to learn user representations of positive and negative news [14]. Yang et al. implement hierarchical user interest modeling to capture users' multi-level interests [6]. These

This work was supported in part by Shandong Province Key R&D Program (Major Science and Technology Innovation Project) Project under Grants 2020CXGC010102 and the National Key Research and Development Plan under Grant No. 2019YFB1404701.


M. Tanveer et al. (Eds.): ICONIP 2022, CCIS 1793, pp. 672–684, 2023.
https://doi.org/10.1007/978-981-99-1645-0_56

methods are a great improvement in news recommendation. However, the temporal relationship between users' news browsing is not considered in modeling user interests (users clicking on the current news may be influenced by previous news clicks), making modeling user interests not optimal. In addition, candidate news is not involved in modeling while learning user interest representation, which makes it possible that candidate news may not be matched with user-specific interests.

| | **Historical Browsed News** |
|---|---|
| $D_1$ | **Dog** dies protecting Florida children from a deadly snake |
| | • • • • |
| $D_4$ | Is It Actually Harder to **Lose Weight** When You're Short? |
| | • • • • |
| $D_8$ | These Simple Diet Changes Helped This **Guy Lose 75 Pounds** in 9 Months |
| $D_9$ | Wolverines in the **NFL**: Top performances from Week 8 |
| $D_{10}$ | Should **NFL** be able to fine players for criticizing officiating? |
| $D_{11}$ | Bill Taylor spent years fighting corruption in Ukraine |
| | **Candidate News** |
| $C_1$ | **a Golden Retriever** hopped into a barber's chair for a much-needed haircut |
| $C_2$ | Setting Realistic Diet and Workout Goals Helped This Guy **Lose 100 Pounds** |
| $C_3$ | **NFL Films** signed off on airing of Sam Darnold's 'ghosts' comment |
| $C_4$ | **White House** is on verge of naming Chad Wolf acting **DHS secretary** |

**Fig. 1.** Example of a user's reading on MSN News. The text segments shown in yellow are key semantic clues. (Color figure online)

The sequential characteristics of news viewed by users' history can reflect their long-term and short-term interests. As shown in Fig. 1, we can infer from the news viewed by users that they are interested in news in the areas of animals, health, sports, and politics. We can also infer that users are interested in NFL sports from the correlation of their short-term behavior of the ninth and tenth clicks on the news. Also, the long-term behavioral correlations between the fourth and eighth news clicks give us enough information to infer users' long-term interest in health. Therefore, learning the representation of users' interests in combination with the sequential correlation information of users' historical browsing news can help improve the performance of news recommendation. In addition, candidate news may only match part of the user's interests. If the candidate news is not involved in the user modeling process, it may be difficult to find news related to the candidate news from the user's history of clicks. For example, the fourth candidate news only matches the user's political interests and has low relevance to the user's other interests (e.g., sports and fitness). In other words, it is difficult to identify the interests related to candidate news [5] from the many interests of users. Therefore, it is necessary to incorporate candidate news into user interest modeling to learn candidate-aware user interests.

In this paper, we propose a news recommendation model named CATM based on candidate-aware temporal multi-head self-attention. This approach incorporates candidate news into user interest modeling to improve news recommendation performance. The model first feeds the news representation vector browsed by the user into Bi-LSTM [17] to capture the temporal dependencies between news. Different from the previous use of Bi-LSTM to learn news representations, we use candidate news as the initial news features for users to browse news, in order to establish the dependencies between the user's historical click news and the candidate news. After that, we use the self-attention mechanism of multiple heads to capture the feature information of news from different subspaces, and realize the information selection of important news browsed by users. Finally, we use the candidate news to weight the important news that the user has viewed to learn the candidate-aware user interest representation, and further establish the dependencies between the user's historical clicked news and the candidate news. Furthermore, we design a gating structure to combine the local multi-granularity semantic representation of news with the global semantic representation of news to achieve news representation modeling. The experimental results in MIND [15] show that CATM can improve the user modeling performance for news recommendation. In short, the main contributions of this work are summarized as follows.

1) We design a multiple self-following news recommendation framework with candidate perceived temporal sequences. We incorporate the sequence features of candidate news and user browsing news into user interest modeling to achieve accurate matching of user interests.
2) We design a granular network module to achieve multi-granularity information extraction of news semantics. And apply a gated structure to fuse LSTM modules encoding news features to learn semantically richer news representations.
3) We conducted an ablation study on the real MIND dataset to validate the effectiveness of our approach.

## 2 Our Approach

Our CATM news recommendation method is shown in Fig. 2. It has three important components. They are the news encoder, the candidate-aware user encoder, and the click predictor. We will describe them in detail in the next few sections.

### 2.1 News Encoder

We first introduce our news encoder, which is used to learn news representations from news headlines. We use the pre-trained Glove word vectors to initialize the embedding matrix. We use M words to represent news headlines $D = [w_1, w_2, \ldots, w_M]$. Through the word vector search matrix $\mathbf{W}_e \in \mathcal{R}^{V \times D}$, the news title are mapped to the low-dimensional vector space to obtain

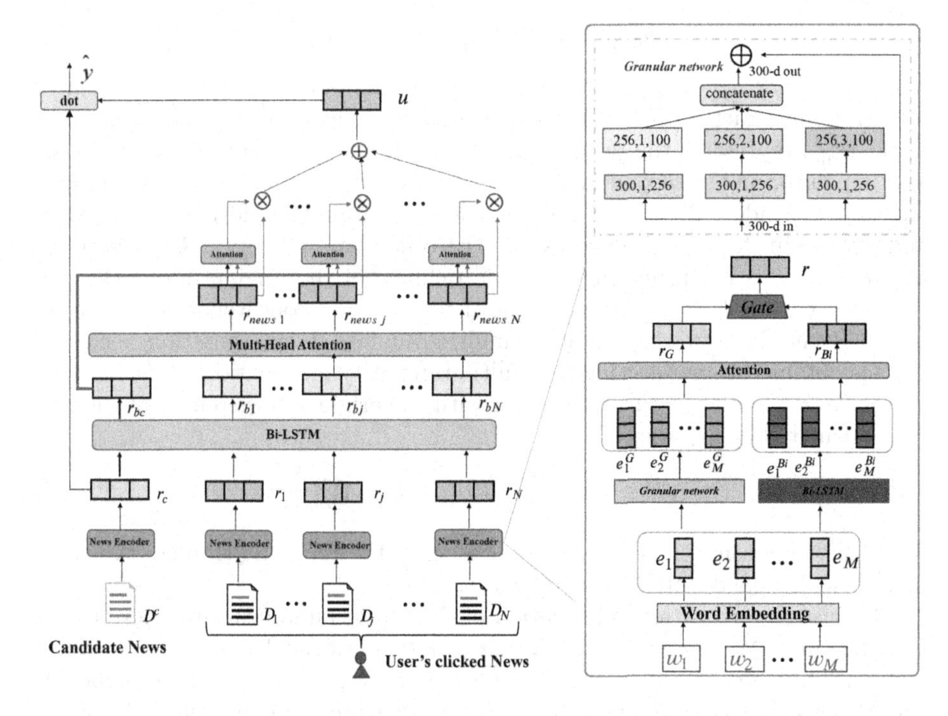


**Fig. 2.** The framework of our CATM news recommendation method.

$\mathbf{E} = [\mathbf{e}_1, \mathbf{e}_2, \ldots, \mathbf{e}_m]$, where $\mathbf{e}_\mathrm{m} \in \mathrm{W}_e$, $V$ and $D$ are the vocabulary size and word embedding dimension, respectively.

Next, we input the obtained vector representation $\mathbf{E}$ of the title into the granular network and Bi-LSTM to obtain the local multi-granularity semantic representation of news and the global semantic representation of news. Our granularity network is shown in Fig. 2. Borrowing from the idea of ResNeXt [16], the ResNeXt network plays an important role in image feature extraction. It can not only solve the degradation problem of deep neural networks but also help the convergence of the network. We introduce this residual network into the recommendation task to implement multi-granularity feature extraction of text information of news headlines to obtain deeper semantic representation. The three dimensions of each layer in the network represent the input feature dimension, the convolutional kernel size, and the output feature dimension. We replace the 2D convolution in the residual network with 1D convolution with different convolution windows for multi-granularity feature extraction for news texts. The network uses three groups of different convolution windows to convolve the news text, and each group of convolution windows extracts feature representations of different granularities so that the model can capture more semantic information in the news. The specific calculation formula of the multi-granularity network is as follows:

$$\mathbf{E}^{\mathbf{i}}_{\mathbf{reduct}} = \sigma\left(\mathbf{W}^{i}_{fr}\mathbf{E} + \mathbf{b}^{\mathbf{i}}\right) \tag{1}$$

$$\mathbf{E}^{\mathbf{i}}_{\mathbf{mul}} = \sigma\left(\mathbf{W}^{\mathbf{i}}_{\mathbf{mul}}\mathbf{E}_{\mathbf{reduct}} + \mathbf{b}^{\mathbf{i}}\right) \tag{2}$$

where $i \in \{1, 2, 3\}$; $\mathbf{E}^{\mathbf{i}}_{\mathbf{reduct}}$ represents the tensor representation of news after information extraction and dimension reduction in the first layer of granular network; $\mathbf{E}^{\mathbf{i}}_{\mathbf{mul}}$ represents the result of granular information extraction and dimension expansion in the second layer. $\mathbf{W}_{mul}$ represents a granular sliding window, and as the window slides, gradually extract the granular information of words and phrases in the news; $\sigma$ denotes the RELU activation function. $\mathbf{b}$ is the trainable parameter. To ensure the semantic richness of the original news, the last step of the granularity network employs the residual concatenation operation in the ResNeXt network. The local multi-granularity representation vector of news is obtained by connecting the multi-granularity information representation vector after granular feature extraction with the original news representation $\mathbf{E}$. It is calculated as follows:

$$\mathbf{E}^{G} = concat(\mathbf{E}, \mathbf{E}^{\mathbf{1}}_{\mathbf{mul}}, \mathbf{E}^{\mathbf{2}}_{\mathbf{mul}}, \mathbf{E}^{\mathbf{3}}_{\mathbf{mul}}) \tag{3}$$

where $\mathbf{E}^{G}$ represents the news representation of $\mathbf{E}$ after going through the granular network. We denote $\mathbf{E}^{G} = \left[\mathbf{e}^{G}_{1}, \mathbf{e}^{G}_{2}, \ldots, \mathbf{e}^{G}_{M}\right]$.

Because the important information of the news can appear anywhere in the news headline. We employ Bi-LSTM network to learn the dependencies between different characters in news. We input the news representation $\mathbf{E}$ into the Bi-LSTM network to obtain the global semantic feature representation $\mathbf{E}^{Bi}$ of the news. Calculated as follows:

$$\mathbf{E}^{Bi} = Bi - LSTM(\mathbf{E}) \tag{4}$$

where $\mathbf{E}^{Bi}$ represents the news representation of $\mathbf{E}$ after going through the Bi-LSTM network. We denote $\mathbf{E}^{Bi} = \left[\mathbf{e}^{Bi}_{1}, \mathbf{e}^{Bi}_{2}, \ldots, \mathbf{e}^{Bi}_{M}\right]$.

The importance of phrase information extracted by granular network and Bi-LSTM for news matching in the prediction stage is different. To further differentiate the importance of different words, we exploit additional attention to generate deep news representations. The formula is as follows:

$$a^{G}_{i} = \mathbf{q}^{\mathbf{T}}_{\mathbf{G}} tanh(\mathbf{W}_{\mathbf{G}} \times \mathbf{e}^{\mathbf{G}}_{\mathbf{i}} + \mathbf{b}_{\mathbf{G}}) \tag{5}$$

$$\alpha^{G}_{i} = \frac{\exp\left(a^{G}_{i}\right)}{\sum\limits_{j=1}^{M} \exp\left(a^{G}_{i}\right)} \tag{6}$$

$$a^{Bi}_{i} = \mathbf{q}^{\mathbf{T}}_{\mathbf{Bi}} tanh(\mathbf{W}_{\mathbf{Bi}} \times \mathbf{e}^{\mathbf{Bi}}_{\mathbf{i}} + \mathbf{b}_{\mathbf{Bi}}) \tag{7}$$

$$\alpha^{Bi}_{i} = \frac{\exp\left(a^{Bi}_{i}\right)}{\sum\limits_{j=1}^{M} \exp\left(a^{G}_{i}\right)} \tag{8}$$

where $\mathbf{W_G}$, $\mathbf{b_G}$, $\mathbf{W_{Bi}}$, $\mathbf{b_{Bi}}$ are trainable parameters, $\mathbf{q_G^T}$, $\mathbf{q_{Bi}^T}$ is the query vector. $\alpha_i$ is the attention weight of the ith word in the news title. The final representation of a news headline is the sum of its contextual word representations, weighted by their attention weights. The final representation of a news headline is the sum of its contextual word representations, weighted by their attention weights. i.e., $\mathbf{r_G} = \sum_{i=1}^{M} \alpha_i^G \mathbf{e}_i^G$, $\mathbf{r_{Bi}} = \sum_{i=1}^{M} \alpha_i^{Bl} \mathbf{e}_i^{Bi}$. Then we apply a gating network to aggregate news representations $\mathbf{r_G}$, $\mathbf{r_{Bi}}$ obtained by different encoders to enrich the semantic representation of news. We get the final news representation $\mathbf{r}$. The formula is as follows:

$$\mathbf{r} = \alpha \mathbf{r_G} + (1 - \alpha)\mathbf{r_{Bi}} \tag{9}$$

$$\alpha = softmax(MLP(\mathbf{r_G}, \mathbf{r_{Bi}})) \tag{10}$$

where $\alpha$ is the weight. MLP is a multilayer perceptron.

### 2.2 Candidate-Aware User Encoder

Because the news that the user clicks has a certain relationship before and after. The one-way recurrent neural network can predict the interest information of recent users by correlating the user's historical data. We adopt Bi-LSTM to capture data features between two directions to obtain long-term stable user interest information. We incorporate candidate news into user modeling. We first input the candidate news into the Bi-LSTM network as the initial news features of the user's historical click news in order to initially establish the connection between the candidate news and the user's historical click news. We denote the set of news vectors browsed by users as $\mathbf{R} = [\mathbf{r_1}, \mathbf{r_2}, \cdots, \mathbf{r_N}]$, where $\mathbf{r_N}$ represents the vector representation of the n-th news browsed by the user. The user browsing history news after inputting the Bi-LSTM network is denoted as $\hat{\mathbf{R}}$. The formula is as follows:

$$\hat{\mathbf{R}} = Bi - LSTM(\mathbf{R}) \tag{11}$$

where $\hat{\mathbf{R}}$ is represented as $\hat{\mathbf{R}} = [\mathbf{r_{b1}}, \mathbf{r_{b2}}, \cdots, \mathbf{r_{bN}}]$, $\mathbf{r_{bN}}$ represents the news representation of the n-th user's historically browsed news after passing through the Bi-LSTM network. The output candidate news vector from the Bi-LSTM network is $\mathbf{r_{bc}}$.

We further model the correlation between the news browsed by the user to obtain a deeper representation of the user. We apply multi-head attention to enhance news performance by capturing the interaction information of multiple news articles viewed by the same user. The multi-head attention mechanism is formed by stacking multiple scaled dot-product attention module base units. The input is the query matrix $\mathbf{Q}$, the keyword $\mathbf{K}$, and the eigenvalue $\mathbf{V}$ of the keyword. The formula is as follows:

$$\mathbf{Q} = \mathbf{W_q} * \hat{\mathrm{R}} \tag{12}$$

$$\mathbf{K} = \mathbf{W_k} * \hat{\mathrm{R}} \tag{13}$$

$$\mathbf{V} = \mathbf{W_v} * \hat{\mathrm{R}} \tag{14}$$

where $\mathbf{W_q}$, $\mathbf{W_k}$, $\mathbf{W_v}$ are trainable parameter matrices.

The calculation formula of the scaled dot product attention mechanism is:

$$Attention(\mathbf{Q}, \mathbf{K}, \mathbf{V}) = softmax(\frac{\mathbf{QK}^T}{\sqrt{d}})\mathbf{V} \tag{15}$$

where $d$ is the number of hidden units of the neural network.

In multi-head attention, $\mathbf{Q}$, $\mathbf{K}$, $\mathbf{V}$ first make a linear change and input into the scaled dot product attention. Here it is done $h$ times, and the linear transformation parameters $\mathbf{W^Q}$, $\mathbf{W^K}$, $\mathbf{W^V}$ of $\mathbf{Q}$, $\mathbf{K}$, $\mathbf{V}$ are different each time. Then the results of the $h$ scaled dot product attention are spliced together, and the value obtained by performing a linear transformation is used as the result of multi-head self-attention. The formula is as follows:

$$\mathbf{head_i} = Attention(\mathbf{QW_i^Q}, \mathbf{KW_i^{K}}, \mathbf{VW_i^V}) \tag{16}$$

$$\mathbf{R_{news}} = MultiHead(\mathbf{Q}, \mathbf{K}, \mathbf{V}) = concat(\mathbf{head_1}, \mathbf{head_2}, \cdots, \mathbf{head_h})\mathbf{W_q} \tag{17}$$

where we denote $\mathbf{R_{news}}$ as $\mathbf{R_{news}} = [\mathbf{r_{news1}}, \mathbf{r_{news2}}, \cdots, \mathbf{r_{newsN}}]$, $\mathbf{r_{newsN}}$ represents the n-th historical click news representation after multi-head attention.

We further strengthen the link between candidate news and click news. Since user interests modeled by different clicked news may have different importance to candidate news, we use a candidate-aware attention network to obtain the final user interest representation $\mathbf{u}$. i.e., $\mathbf{u} = \sum_{i=1}^{N} \alpha_i \mathbf{r_{newsi}}$, where $\alpha_i$ is the weight of the ith clicked news. The formula is as follows:

$$\alpha_i = \frac{\exp(MLP(\mathbf{r_{newsi}}, \mathbf{r_{bc}}))}{\sum_{j=1}^{N} \exp(MLP(\mathbf{r_{newsj}}, \mathbf{r_{bc}})} \tag{18}$$

## 2.3 Click Prediction and Model Training

The click prediction module is used to predict the probability of a user clicking on a candidate news item [15]. The dot product probability $\hat{y}$ is calculated from the inner product of the user representation vector and the candidate news representation vector. i.e., $\hat{y} = \mathbf{u^T r_c}$, where $\mathbf{r_c}$ is the vector representation of the output after the candidate news is input to the news encoder.

Motivated by Wu et al. [13]. The task of news recommendation can be viewed as a pseudo K+1-way classification problem. We employ a log-likelihood loss to train our model. The formula is as follows:

$$Loss = -\sum_{i \in S} \log \left( \frac{\exp(\hat{y}_i^+)}{\exp(\hat{y}_i^+) + \sum_{j=1}^{K} \exp(\hat{y}_j^-)} \right) \tag{19}$$

where $S$ is the set of positive training samples. $y^+$ represents the click probability score of the ith positive news, and $y_j^-$ represents the click probability score of the jth negative news in the same time period as the ith positive news.

# 3 Experiments

## 3.1 Experiment Setup

Our experiments are performed on a large public dataset named MIND [15]. Because the title is the decisive factor that affects the user's reading choice. In this article, we use news headlines as input. Our experiments are performed on the MIND-large dataset. Since no MIND-large test set label is provided, the performance on the test dataset is obtained by submitting to the MIND news recommendation competition[1]. The detailed statistics of the MIND-large train dataset are shown in Table 1.

**Table 1.** Statistical of the MIND-large training dataset.

| | | | |
|---|---|---|---|
| # Users | 711222 | # News items | 101527 |
| # Impressions | 2232748 | # Positive samples | 3383656 |
| Avg. # words per title | 11.52 | # Negative samples | 80123718 |

Next, we introduce the experimental setup and hyperparameters of CATM. In our experiments, the dimension of word embeddings is set to 300. We use our pretrained word embeddings using GloVe.840B.300d. The maximum number of news clicks represented by learning users is set to 50, and the maximum length of news headlines is set to 20. The batch size is set to 100. We applied a dropou ratio of 0.2 to mitigate overfitting. The negative sampling ratio K is 4. The number of filters in the first layer of the granular network CNN is set to 256 and the second layer is set to 100. The neurons of the Bi-LSTM in the news encoder are set to 150, and the neurons of the Bi-LSTM in the user encoder are set to 128. The self-attention network has 16 heads, and the output of each head is 16-dimensional. We train 3 ephocs with a learning rate of 0.001 via Adam [4]. We independently replicate each experiment 10 times and report the average results in terms of AUC, MRR, nDCG@5, and nDCG@10.

## 3.2 Performance Evaluation

We compare our model with following representative and state-of-the-art baselines, including: (1) DeepFM [1], It combines factorization machines with deep neural networks. (2) LibFM [7], Extract TF-IF features from news as input. (3) DKN [9], It leverages word embeddings and entity embeddings to learn news

[1] https://competitions.codalab.org/competitions/24122.

and user representations. (4) NPA [12], It personalizes the selection of important words and news articles. (5) NAML [11], It aggregates different components of news to learn user and news representations. (6) NRMS [13], Learning deep representations of news and users through self-attention. (7) FIM [8], The user's interest in candidate news is modeled from the semantic relevance of user clicked news and candidate news, which is then passed through a 3-D CNN network. (8) GnewRec [2], It uses GRU and GNN networks to learn user interest representation; (9) HieRec [6], It combines headlines and entities to better represent news and models multi-granularity interests to better represent users. (10) ANRS [10], It proposes a news aspect level encoder and a user aspect level encoder. (11) ATRN [3], It proposes a new adaptive transformer network to improve recommendation.

**Table 2.** The performance of different methods. CATM significantly outperforms all baselines (p <0.01)

| Methods | AUC | MRR | nDCG@5 | nDCG@10 |
|---|---|---|---|---|
| DeepFM [1] | 60.30 | 28.19 | 30.02 | 35.71 |
| LibFM [7] | 60.22 | 28.07 | 30.12 | 36.57 |
| DKN [9] | 65.11 | 31.76 | 33.53 | 39.35 |
| NPA [12] | 66.23 | 32.14 | 34.52 | 40.24 |
| NAML [11] | 66.35 | 32.85 | 35.45 | 41.67 |
| NRMS [13] | 68.12 | 33.36 | 36.42 | 42.21 |
| FIM [8] | 67.92 | 32.85 | 36.56 | 41.35 |
| GnewsRec [2] | 68.21 | 33.41 | 36.41 | 42.34 |
| HieRec [6] | 67.72 | 32.66 | 35.17 | 41.11 |
| ANRS [10] | 67.95 | 33.52 | 36.25 | 42.06 |
| ATRN [3] | 68.06 | 33.35 | 36.34 | 42.15 |
| CATM | **68.69** | **33.85** | **37.05** | **42.88** |

Since the experiments in MIND [15] are performed on a subset of published datasets, we reimplement all baselines on the full MIND dataset and report their results on the test set. Table 2 lists the average results, from which we have some observations. First, neural news recommendation methods (NPA, NAML, NRMS, etc.) are superior to traditional recommendation methods such as DeepFM. This may be because deep news and user feature representations can be learned more easily through neural networks than artificial feature engineering methods. Second, among deep learning based methods, we found that NRMS can outperform other methods such as DKN, NPA, etc. This may be because NRMS takes into account the relationship between users browsing news. It performs well by using multi-head self-attention to further capture news interactions of words and clicks. Third, we find that our model outperforms other baseline methods that model user interests with candidate news uninvolved. Such as (NRMS,

NAML). This may be because users usually have multiple interests, and the user interest modeling of candidate news participation helps to match candidate news with user interests accurately. Finally, our method can consistently outperform other baseline methods on all metrics. The significant improvement shows that the enhancement of text representation through granular network and Bi-LSTM, and then constructing a user feature model combining multi-head self-attention mechanism and candidate-aware additive attention mechanism can effectively exert the advantages of predicting the probability.

### 3.3 Ablation Studies

To highlight the individual contributions of each module of the user encoder, we run an ablation study on the MIND dataset using the following variant of CATM. (1) CATM-CAT: candidate perceptual attention after removing the multi-head attention mechanism module. (2) CATM-Bi-LSTM: removes the Bi-LSTM module in the user encoder. (3) CATM*: candidate news is not input into the Bi-LSTM module as the user's initial news features. The results are shown in Fig. 3. We can see that the CATM-CAT and CATM-Bi-LSTM variants respectively remove the candidate perceptual attention module and the temporal encoding module, which leads to the degradation of the model performance. This is reasonable, because CATM-Bi-LSTM can effectively obtain the long-distance dependencies between users' browsing news. Compared with CATM*, we find that using candidate news as the initial feature of historical click news can effectively improve the performance of the model, which may be because the user click news vector after the Bi-LSTM network establishes a connection with the candidate news, which provides clues for the user's accurate interest matching. In addition, CATM-CAT measures the importance of clicked news by the correlation between the user's historical clicked news and candidate news, which further improves the user's interest representation.

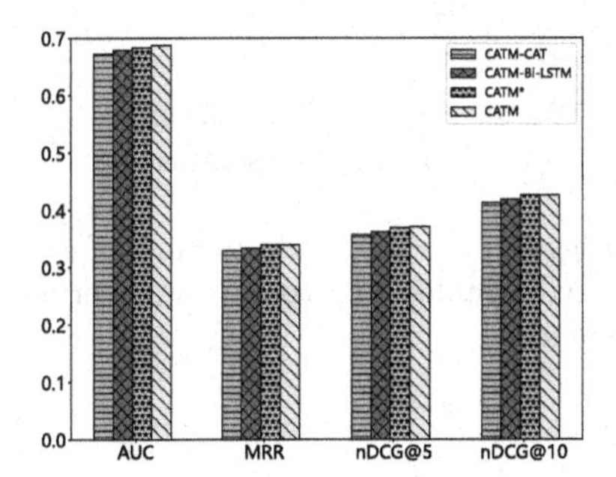


**Fig. 3.** Performance represented by different model variants

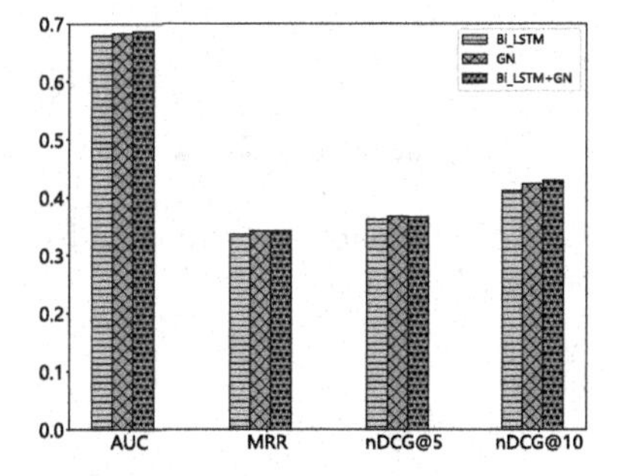


**Fig. 4.** Influence of different modules in news encoder

We study the effect of two encoding methods in the model, the granular network and Bi-LSTM in the news encoder, on the performance of our experiments. The experimental results are shown in Fig. 4, and we found that the encoding performance of the granular network alone is better than that of Bi-LSTM encoding. This may be because the texts of news headlines are mostly short texts, and it is more effective to grasp the global semantic information from the different granularities of local features than from the full text. The information of different granularity introduces more phrase information into the model, which helps to deepen the understanding of the deep semantics of news.

### 3.4 Parameter Analysis

We further investigate the effect of the number of groups G in granular networks. Figure 5 shows the results. As G increases from 1 to 5, it turns out that the depth of the representation layer does matter when it comes to recommendation accuracy. The deeper level makes the granularity of capturing semantic information finer and finer, so the recommendation performance of the model is improved. Nonetheless, as G continues to increase, model performance begins to degrade. This may be because the model has learned enough to be able to model a granular semantic feature representation of the news, and the deepening of the number of groups introduces noise that makes compromise the final news representation. Here we set G = 3 layers optimally.

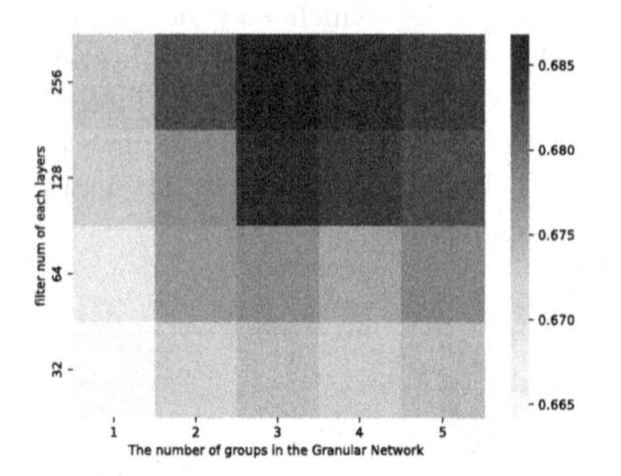


**Fig. 5.** Performance of Granular Networks on MIND Data

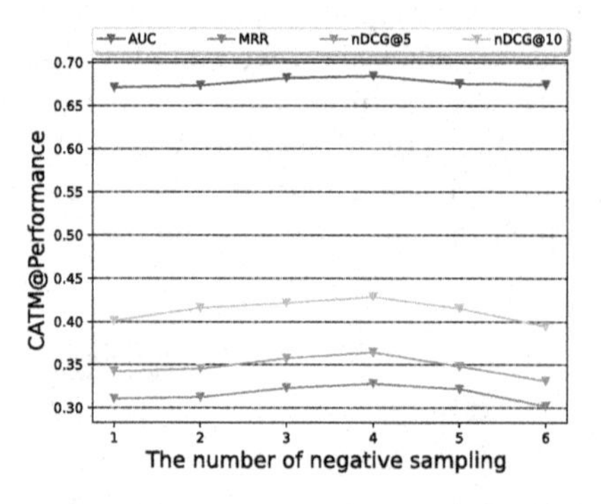


**Fig. 6.** The effect of the number of negative samples on experimental performance

We investigate the performance under different negative sampling rates k, and Fig. 6 shows the experimental results. We can find that when K is lower than 5, the performance keeps improving and then starts to decrease. The possible reason is that the value of K is too small, and the useful information mined from negative samples is limited. However, when too many negative samples are introduced, they may become dominant and the imbalance of the training data will increase. Therefore, it is more difficult for the model to accurately identify positive samples, which will also affect the recommendation performance. Overall, the optimal setting for K is moderate (e.g., K = 3)

## 4 Conclusion

In this paper, we propose a candidate-aware time-series multi-headed self-attentive mechanism based news recommendation model CATM. We learn the multi-granularity semantic representation of news efficiently through a gating structure. In addition we incorporate the sequence characteristics of candidate news and historical click news into user interest modeling to achieve accurate matching of candidate news with user interest. Extensive experiments on the MIND dataset validate the effectiveness of our approach.

## References

1. Guo, H., Tang, R., Ye, Y., Li, Z., He, X.: DeepFM: a factorization-machine based neural network for CTR prediction. In: IJCAI (2017)
2. Hu, L., Li, C., Shi, C., Yang, C., Shao, C.: Graph neural news recommendation with long-term and short-term interest modeling. Inf. Process. Manage. **57**(2), 102142 (2020)
3. Huang, J., Han, Z., Xu, H., Liu, H.: Adapted transformer network for news recommendation. Neurocomputing **469**, 119–129 (2022)
4. Kingma, D.P., Ba, J.: Adam: a method for stochastic optimization. In: ICLR (2014)
5. Qi, T., Wu, F., Wu, C., Huang, Y.: News recommendation with candidate-aware user modeling. In: SIGIR (2022)
6. Qi, T., et al.: HieRec: hierarchical user interest modeling for personalized news recommendation. In: ACL (2021)
7. Rendle, S.: Factorization machines with libFM. ACM Trans.Intell. Syst. Techno. (TIST) **3**(3), 1–22 (2012)
8. Wang, H., Wu, F., Liu, Z., Xie, X.: Fine-grained interest matching for neural news recommendation. In: Proceedings of the 58th Annual Meeting of the Association for Computational Linguistics, pp. 836–845 (2020)
9. Wang, H., Zhang, F., Xie, X., Guo, M.: DKN: deep knowledge-aware network for news recommendation. In: Proceedings of the 2018 World Wide Web Conference, pp. 1835–1844 (2018)
10. Wang, R., Lu, W., Wang, S., Peng, X., Wu, H., Zhang, Q.: Aspect-driven user preference and news representation learning for news recommendation. IEEE Trans. Intell. Transp. Syst. **23**, 25297–25307 (2022). https://doi.org/10.1109/TITS.2022.3182568
11. Wu, C., Wu, F., An, M., Huang, J., Huang, Y., Xie, X.: Neural news recommendation with attentive multi-view learning. In: IJCAI, pp. 3863–3869 (2019)
12. Wu, C., Wu, F., An, M., Huang, J., Huang, Y., Xie, X.: NPA: neural news recommendation with personalized attention. In: Proceedings of the 25th ACM SIGKDD International Conference on Knowledge Discovery & Data Mining, pp. 2576–2584 (2019)

13. Wu, C., Wu, F., Ge, S., Qi, T., Huang, Y., Xie, X.: Neural news recommendation with multi-head self-attention. In: Proceedings of the 2019 Conference on Empirical Methods in Natural Language Processing and the 9th International Joint Conference on Natural Language Processing (EMNLP-IJCNLP), pp. 6389–6394 (2019)
14. Wu, C., Wu, F., Huang, Y., Xie, X.: Neural news recommendation with negative feedback. CCF Trans. Pervasive Comput. Interaction **2**(3), 178–188 (2020). https://doi.org/10.1007/s42486-020-00044-0
15. Wu, F., et al.: Mind: a large-scale dataset for news recommendation. In: Proceedings of the 58th Annual Meeting of the Association for Computational Linguistics, pp. 3597–3606 (2020)
16. Xie, S., Girshick, R., Dollár, P., Tu, Z., He, K.: Aggregated residual transformations for deep neural networks. In: Proceedings of the IEEE Conference on Computer Vision and Pattern Recognition, pp. 1492–1500 (2017)
17. Zhou, P., et al.: Attention-based bidirectional long short-term memory networks for relation classification. In: Proceedings of the 54th Annual Meeting of the Association for Computational Linguistics (volume 2: Short papers), pp. 207–212 (2016)

# Variational Graph Embedding for Community Detection

Xu Sun, Weiyu Zhang(✉), Zhengkai Wang, and Wenpeng Lu

School of Computer Science and Technology, Qilu University of Technology (Shandong Academy of Sciences), Jinan, China
zwy@qlu.edu.cn

**Abstract.** Community detection aims to discover the community structure in the graph. In many systems, community detection plays an important role in its analysis, design, and majorization. However, previous community detection methods underutilize the information between nodes and their neighbors, and usually learn node representation separately from community detection while they are closely related. Therefore, we propose Variational Graph Embedding for Community Detection (VGECD), a new variational graph embedding generation model. VGECD introduces graph attention networks to do aggregation operations on neighbor nodes and jointly learns node representation and the embedding of community detection. We apply the inference model to generate node embedding and community assignment. Then we use the generative model to combine node embedding and community assignment to reconstruct the graph. Experiments on real-world datasets show that VGECD outperforms other comparison methods.

**Keywords:** Community detection · Joint learning · Variational embedding · Graph attention networks

## 1 Introduction

Graphs have flexible representation capabilities and can simulate complex relationships between entities. In graphs, nodes represent individuals in the system, and edges between nodes represent relationships between individuals. There are many examples of graphs in the real world, such as aviation networks, transportation networks, and social networks. Graph analysis is a crucial technique for learning the characteristics of networks, such as node classification [1], community detection [2] and link prediction [3] in social networks, and information diffusion forecast in citation networks. Community detection and node representation learning are two important tasks in graph analysis.

The research work is supported by National Key R&D Program of China under Grant No. 2018YFC0831704, National Nature Science Foundation of China under Grant No. 61806105 and Natural Science Foundation of Shandong Province under Grant No. ZR2022MF243.


M. Tanveer et al. (Eds.): ICONIP 2022, CCIS 1793, pp. 685–697, 2023.
https://doi.org/10.1007/978-981-99-1645-0_57

A community is a collection of nodes with similar functions or attributes in a graph. The connections between nodes within a community are closer than the connections between nodes that belong to different communities. The process of finding community structure in the network is called community detection. Early studies have proposed many community detection methods such as algorithm [4] and probabilistic model [5]. Spectral clustering is a classical method of community detection, and it solves the eigenvectors of Laplacian matrix to detect communities. Node representation learning can map the nodes in the graph to the vector space, and the learned node embeddings can be used for downstream tasks [1,6].

Node embedding can capture the local information of the graph, and community detection can capture the overall structure of the graph. These two tasks are usually processed separately. In fact, they are highly related. Studies by Cao et al. [1] and Kozdoba et al. [7] showed that adding node embeddings in the learning process can improve the effectiveness of community detection. A simple way to combine node embeddings and community assignments is first to obtain the node embeddings through DeepWalk [8], and then use k-means or Gaussian mixture models [2] to obtain the community assignments for each node. Recently, many approaches have been proposed to learn node embedding and community detection simultaneously in a unified framework, called joint learning methods [9,10]. However, these approaches underutilize the information between nodes and their neighbors and ignore the diversity of the graph. For example, in Amazon dataset, item A and item B have been purchased together 100 times and item A and item C have been purchased together 1000 times, then item A and item C are more likely to belong to the same community. Therefore different weights should be assigned to different neighbors when generating node embeddings.

To solve the above problems, we propose a new variational graph embedding generation model–VGECD. VGECD based on variational graph auto-encoders [11]. We suppose that each node is part of one or more communities, and connected nodes are more closely related than those without connections, and connected nodes are more likely to pertain to the same community. In our proposed model, firstly, the graph's adjacency matrix $A$ and feature matrix $X$ are input into the encoder to generate the embedding $z_v$ of node $v$. The corresponding community assignment $c_v$ is generated according to $z_v$. To generate an edge $(v, u)$, the encoder is also used to generate the embedding $z_u$ of node $u$, and the corresponding community assignment $c_u$ is generated according to $z_u$. Then the generated node embeddings $z_v$, $z_u$ and the corresponding community assignments $c_v$, $c_u$ are input to the decoder. The decoder reconstructs the graph to ensure that the node embeddings are highly similar to the community embeddings of the nodes connected to them. The learned embedding is useful in community detection tasks. In the generation process of node embedding, we introduce the Graph Attention Networks (GAT). GAT aggregates neighbor nodes by using the attention mechanism, and adaptively assigns different weights to neighbor nodes, thereby improving the performance of generating node embeddings and community assignments of model.

Compared with previous community detection methods, the contributions of this paper are summarized as follows:

1. We propose a new variational graph embedding model–VGECD, which jointly learns community detection and node representation to reconstruct the graph for community detection task.
2. In the process of learning node embedding, we design the encoder with two-layer GAT to better aggregate neighbor nodes.
3. We conduct extensive experiments on real-world datasets. The results show that VGECD has higher accuracy than other comparison methods.

## 2 Related Work

### 2.1 Community Detection and Node Representation Learning

Many community detection approaches are based on clustering algorithms [12], such as spectral clustering algorithm. In addition, there are many community detection algorithms based on matrix decomposition. These algorithms decompose the adjacency matrix, feature matrix, and other relationship matrices of the graph and restore the association matrix between nodes and communities [4]. For example, Yang et al. [2] regarded community detection as a non-negative matrix decomposition task to learn the latent factors of node community association. Yang et al. [13] extended the above model by extracting the relationship information between network structure and node attributes. However, because of the complexity and massive workload of matrix decomposition, the scalability of these methods is unsatisfactory. Compared with these models, our model is more scalable because it does not require matrix decomposition. Some generative models [14] are also used for community detection. These models transform the graph generation process and community detection task into inference problems.

Node representation learning can generate similar representations for nodes with similar connections. Many node representation approaches are based on random walks, which learn node representation by unsupervised learning. Such as DeepWalk, node2vec [15], and Large-scale information network embedding (LINE) [16]. However, these methods overemphasize the adjacent information and have a large dependence on hyperparameters. To solve the above problems, Gilmer et al. [17] proposed a graph convolutional encoder model, where graph convolution can force adjacent nodes to have similar representations.

### 2.2 Joint Learning of Community Detection and Node Representation

Many approaches have been proposed in joint learning of community detection and node representation, such as the model of alternating optimization between community assignment and node representation learning [10]. However, this method can't carry out node representation and community assignment tasks simultaneously. On this basis, sun et al. [14] proposed an extensible model and optimized the model end-to-end. Many recent studies have addressed node representation and community detection in the same framework. The proposed models in studies can learn node embeddings and improve community assignments [9,10], which are inspired by random walks but also inherit its shortcoming, i.e., have high complexity and need to learn two embeddings for each node. The generative model VGECD proposed in this paper can simultaneously learn node representation and community assignment. The model uses an inference model to generate node embeddings and then carries out community assignments under the condition of a single node. The decoder in VGECD is used to guarantee that the embeddings and communities of connected nodes have high similarity, so as to reconstruct the graph. In the process of node embedding generation, we introduce GAT to learn node embedding, which can improve the model's capacity in community detection tasks.

## 3 Proposed Method

### 3.1 Problem Definition

Given an undirected graph $G = (V, E)$, where $V=\{v_1, v_2, ..., v_n\}$ is a collection of vertexes with $N$ nodes, $E$ is a collection of edges. $A \in R^{N \times N}$ and $X \in R^{N \times F}$ are the adjacency matrix and node feature matrix of graph $G$, respectively. Given the number $K$ of communities, the purpose of our model is to conduct community detection task by jointly learning node embedding and community embedding. Firstly, node embeddings and community assignments are generated by the inference model, and then the generative model is used to combine node embeddings and community assignments to reconstruct the graph.

### 3.2 Graph Attention Networks

Assuming that the feature vector corresponding to any node $v_i$ in the $l$-th layer in the graph is $h_i$, when $l = 1$, the feature vector is $X$, $h_i \in R^{d^{(l)}}$, $d^{(l)}$ is the length of node feature. After passing through a graph attention layer, the output is the new eigenvector $h_i^{'}$ of each node, $h_i^{'} \in R^{d^{(l+1)}}$, $d^{(l+1)}$ is the length of the output eigenvector. Assume that the weight coefficient from node $v_i$ to neighbor node $v_j$ as

$$e_{ij} = \text{Leaky ReLU}\left(a^T \left[W h_i || W h_j\right]\right) \tag{1}$$

where $W \in R^{d^{(l+1)} \times d^{(l)}}$ represents the weight parameter, which is used to transform the node feature of this layer, and $a \in R^{2d^{(l+1)}}$ also represents the weight parameter. LeakyReLU is used for the activation function. In order to distribute weights preferably, softmax normalization is performed on the above formula i.e.

$$a_{ij} = \text{softmax}_j\left(e_{ij}\right) = \frac{\exp\left(e_{ij}\right)}{\sum_{v_k \in \tilde{N}(v_i)} \exp\left(e_{ik}\right)} \tag{2}$$

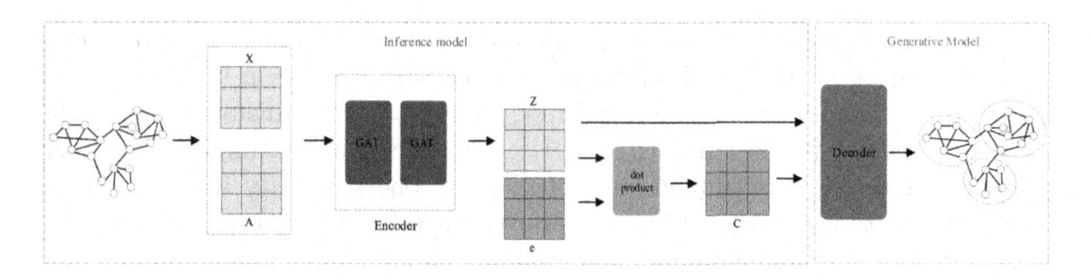


**Fig. 1.** The overall framework of VGECD.

where $a$ is the weight coefficient. (2) can ensure that the total weight coefficient of all neighbors is 1. The complete weight coefficient calculation formula is as follows

$$a_{ij} = \frac{\exp\left(\text{Leaky ReLU}\left(a^T\left[Wh_i||Wh_j\right]\right)\right)}{\sum_{v_k \in \tilde{N}(v_i)} \exp\left(\text{Leaky ReLU}\left(a^T\left[Wh_i||Wh_j\right]\right)\right)} \tag{3}$$

After calculating the weight coefficient, the new eigenvector of the node $v_i$ after weighted summation can be formulated as

$$h_i^{'} = \sigma\left(\sum_{v_j \in \tilde{N}(v_i)} a_{ij} W h_j\right) \tag{4}$$

### 3.3 Overall Framework

Figure 1 shows the overall framework of VGECD. It consists of inference model and generative model.

**Inference Model.** The inference model is essentially an approximate posterior, which makes the computation more tractable. Feed the feature matrix $X$ and adjacency matrix $A$ to it, the inference model uses the posterior probability to generate node embedding and community assignment

$$q_\phi\left(Z, c|X, A\right) = \prod_v q_\phi\left(z_v, c_v|X, A\right) \tag{5}$$

$$= \prod_v q_\phi\left(z_v|X, A\right) q_\phi\left(c_v|z_v, X, A\right) \tag{6}$$

where Random Variables $z_v \in R^d$ represents the embedding of node $v$, and $c_v$ represents the community assignment of node $v$, $c = [c_1, c_2, \cdots, c_N]$, matrix $Z = [z_1, z_2, \cdots, z_N]$. $q_\phi\left(z_v|X, A\right)$ in (6), as a node embedding encoder, is given by

$$q_\phi\left(z_v|X, A\right) = N\left(z_v|\mu_v, \text{diag}\left(\sigma_v^2\right)\right) \tag{7}$$

where $\mu = GAT_\mu\left(X, A\right)$ is the matrix of mean vector $\mu_v$, $\log\sigma = GAT_\sigma\left(X, A\right)$, we adopt a two-layer GAT to learn the parameters $\mu_v$ and $\sigma_v$ in $q_\phi\left(z_v|X, A\right)$.

GAT performs aggregation operation on the neighbor nodes of node $v$ and assigns different weights to the neighbors adaptively. To parameterize $q_\phi(c_v|z_v, X, A)$ in (6), we use a community embedding $\{e_1, \cdots, e_k\}$, $e_k \in R^d$ transformed from the adjacency matrix. We design $q_\phi(c_v = k|z_v, X, A)$ based on the similarity between $e_k$ and $z_v$ and neighbor embeddings of node $v$. The similarity is expressed as softmax operation on their dot product. We introduce a hyperparameter $\alpha$ to constrain the deviation between $z_v$ and the neighbor set $N_v$ of node $v$, i.e.

$$q_\phi(c_v = k|z_v, X, A) = \text{softmax}\left(\alpha\left(e_k^T z_v\right) + (1-\alpha)\frac{1}{|N_v|}\sum_{u\in N_v}\left(e_k^T z_u\right)\right) \tag{8}$$

We set the threshold of its weighted probability vector on the hyperparameter $\epsilon$ to obtain the community assignment of node $v$. The community assignment of node is given by

$$c_v = \left\{k \mid \frac{q_\phi(c_v = k|z_v, G)}{\max_\ell q_\phi(c_v = \ell|z_v, G)} \geq \epsilon\right\}, \epsilon\,[0,1] \tag{9}$$

**Generative Model.** Generative model is defined as

$$p(A) = \int \sum_c p(Z)\, p_\theta(c|Z)\, p_\theta(A|c, Z) dZ \tag{10}$$

where $\theta$ is the model parameter, we use $a_{vu}$ to represent the elements in $A$. According to methods [11,18], we regard $z_v$ as independently identically distribution random variables, and $c_v|z_v$ are further set as independently identically distribution random variables, then the joint distributions in (10) can be decomposed into

$$p(Z) = \prod_{v=1}^{N} p(z_v) \tag{11}$$

$$p_\theta(c|Z) = \prod_{v=1}^{N} p_\theta(c_v|z_v) \tag{12}$$

$$p_\theta(A|c, Z) = \prod_{v,u} p_\theta(a_{vu}|c_v, c_u, z_v, z_u) \tag{13}$$

where (13) assumes that $p_\theta(a_{vu}|c_v, c_u, z_v, z_u)$ depends only on $c_v, c_u, z_v, z_u$.

$p(z_v)$ in (11) is the standard Gaussian distribution of all nodes. The distribution $p_\theta(c_v|z_v)$ in (12) is modeled as a softmax operation of the dot product of $z_v$ and $e_k$, i.e.

$$p_\theta(c_v = k\,|z_v) = \text{softmax}\left(e_k^T z_v\right) \tag{14}$$

The decoder in (13) is based on the fact that connected nodes have a high probability of belonging to the same community, while unconnected nodes have only a small probability of pertaining to the same community. Thus, the decoder is defined as

$$p_\theta\left(a_{vu}=1|c_v=\ell,c_u=m,z_v,z_u\right)=\frac{1}{2}\left(\sigma\left(e_m^T z_v\right)+\sigma\left(e_\ell^T z_u\right)\right) \tag{15}$$

(15) uses both community embedding, node embedding, and community assignment information to reconstruct the edges between nodes.

**Optimize Variational Lower Bound.** Variational embedding aims to learn the model parameter $\theta$ to maximize $\log\left(p_\theta\left(A\right)\right)$. Therefore, we maximize the variational lower bound, i.e.

$$\begin{aligned}L=&\sum_{(v,u)\in\varepsilon}\mathbb{E}_{q_\phi(z_v,z_u,c_v,c_u|X,A)}\left\{\log\left(p_\theta\left(a_{vu}|c_v,c_u,z_v,z_u\right)\right)\right\}\\&-KL\left(q_\phi\left(Z|X,A\right)||p\left(Z\right)\right)-KL\left(q_\phi\left(c|Z,X,A\right)||p_\theta\left(c|Z\right)\right)\end{aligned} \tag{16}$$

where $KL\left(\cdot||\cdot\right)$ represents the Kullback-Leibler divergence (KL-divergence). The first term in (16) is the cross-entropy loss function. $q_\phi\left(z_v,z_u,c_v,c_u|X,A\right)$ can be decomposed into two conditionally independent distributions as

$$q_\phi\left(z_v,z_u,c_v,c_u|X,A\right)=q_\phi\left(z_v,c_v|X,A\right)q_\phi\left(z_u,c_u|X,A\right) \tag{17}$$

The overall process of VGECD is shown in algorithm 1. The complexity of VGECD is $O\left(N^2\right)$, where $N$ is the number of nodes.

**Algorithm 1** VGECD

**Input:** Adjacency matrix $A$, feature matrix $X$
**Output:** Reconstruct adjacency matrix $A'$
1: **for** $v=1,2,...,$N **do**
2: Obtain node embeding $Z_v$ for node $v$ from Eq.(7);
3: According to the node embedding $Z_v$, the corresponding community distribution $C_v$ is generated by Eq.(8);
4: **for** $u=1,2,...,$N **do**
5: Obtain node embeding $Z_u$ for node $u$ from Eq.(7);
6: According to the node embedding $Z_u$, the corresponding community distribution $C_u$ is generated by Eq.(8);
7: Feed node embeddings $Z_v$, $Z_u$ and the respective community assignments $C_v$, $C_u$ to decoder;
8: Decoder judges whether there is an edge between $v$ and $u$ to reconstruct the adjacency matrix $A$ by Eq.(15);
9: **end for**
10: **end for**

# 4 Experiments

In experiments, we use nine datasets to evaluate the performance of the proposed model, and Table 1 summarizes the statistics of these nine datasets.

**Table 1.** Statistics of datasets. $v$ represents the number of nodes, $\varepsilon$ represents the number of edges, $K$ represents the number of communities.

| Dataset | $v$ | $\varepsilon$ | $K$ |
|---|---|---|---|
| fb698 | 61 | 270 | 13 |
| fb414 | 150 | 1693 | 7 |
| fb686 | 168 | 1656 | 14 |
| fb348 | 224 | 3192 | 32 |
| fb3437 | 534 | 4813 | 46 |
| fb1912 | 747 | 30025 | 46 |
| amazon | 794 | 2109 | 5 |
| youtube | 5346 | 24121 | 5 |
| dblp | 24493 | 89063 | 5 |

### 4.1 Datasets

1. **Facebook:** Facebook datasets are obtained from Facebook users who participate in survey. The datasets include friend lists, node features, etc. fb698, fb414, fb686, fb348, fb3437, fb1912 represent six different Facebook social networks, from small to large in size.
2. **Amazon:** Amazon dataset is from the Amazon website. If two products are often purchased together, the graph contains an undirected edge between the two products. The Amazon dataset defines each product category as a real community.
3. **Youtube:** Youtube is a website where users can share videos, and it also contains a social network. Users can establish mutual friendships with others on the Youtube social network, create chat groups, and invite others to join. Such user-defined groups are called real communities.
4. **DBLP:** DBLP, as a complete list of research papers of computer science, is given by the Computer Science Index. A network of co-authors is built inside, i.e., if two authors publish one or more papers together, they are connected. Each published journal or conference defines a separate community, and the authors who publish in it compose that community.

### 4.2 Baselines

In order to verify the proposed algorithm well, we compare the VGECD algorithm with the following eight algorithms.

1. **SVI** [5]: SVI uses a Bayesian model of networks to represent the relationship between nodes in multiple communities.
2. **MNMF** [19]: MNMF adopts a joint NMF method based on modular regularization to learn the distribution of community members.

3. **vGraph** [14]: Vgraph represents nodes as members of a mixed community and learns node representation and community detection in the same framework.
4. **ComE** [10]: ComE uses the Gaussian mixture model formulation to jointly learn community detection and node embedding.
5. **J-ENC** [20]: J-ENC uses a generative model to learn node representation and community detection simultaneously in one node embedding.
6. **CNRL** [9]: CNRL jointly learns node embeddings and community detection by augmenting random walk sequences.
7. **ComGAN**: ComGAN, as a generative adversarial model, can learn node embeddings. The elements in vector of each node's embedding represent the node's relationship strength to different communities.
8. **DGI** [21]: DGI applies k-means to learn embeddings to obtain the community assignment.

### 4.3 Evaluation Metrics

In experiments, F1 score and jaccard similarity [13] are used to evaluate the accuracy of VGECD and baselines. F1 score is defined as the average of F1-score of the real community in graph that best matches each detected community and F1-score of the detected community that best matches each real community in graph. More formally, the F1 score is

$$\frac{1}{2}\left(\frac{1}{|C*|}\sum_{C_i\in C*} F1\left(C_i,\hat{C}_{g(i)}\right)+\frac{1}{\left|\hat{C}\right|}\sum_{\hat{C}_i\in\hat{C}} F1\left(C_{g'(i)},\hat{C}_i\right)\right) \tag{18}$$

where $C^*$ is a collection of real communities in graph, $\hat{C}$ is a collection of detected communities. Each real community $C_i \in C^*$ and each detected community $\hat{C}_i \in \hat{C}$ are composed by its collection of member nodes. $g(i) = \arg\max_j F1\left(C_i, \hat{C}_j\right)$ and $g'(i) = \arg\max_j F1\left(C_j, \hat{C}_i\right)$. $F1\left(C_i, \hat{C}_j\right)$ is the harmonic mean of precision and recall. Similar to F1 score, the jaccard similarity is defined as

$$\frac{1}{2}\left(\frac{1}{|C*|}\sum_{C_i\in C*} jac\left(C_i,\hat{C}_{g(i)}\right)+\frac{1}{\left|\hat{C}\right|}\sum_{\hat{C}_i\in\hat{C}} jac\left(C_{g'(i)},\hat{C}_i\right)\right) \tag{19}$$

### 4.4 Parameter Settings

We set hyperparameters $\alpha = 0.9, \epsilon = 0.3$. $K$, as the number of communities, is the real number of communities in each dataset. We learn the mean matrix $\mu$ and the log variance matrix $\sigma$ of 16-dimensional node embeddings. We train a variational graph auto-encoder first, then apply the Adam optimizer for gradient descent with the learning rate equals 0.01. (9) is used for community assignment. For the results of all baseline methods, we take them from their original papers.

**Table 2.** F1 score (%) of each dataset in the community detection task.

| Dataset | SVI | MNMF | vGraph | ComE | J-ENC | CNRL | ComGAN | DGI | VGECD |
|---|---|---|---|---|---|---|---|---|---|
| fb698 | 40.3 | 26.6 | 54.0 | 45.8 | 64.0 | 16.4 | 58.2 | 52.2 | **69.9** |
| fb414 | 38.9 | 22.1 | 64.7 | 55.3 | 69.6 | 25.3 | 43.9 | 56.9 | **72.1** |
| fb686 | 46.4 | NA | 47.8 | NA | NA | NA | NA | NA | **48.8** |
| fb348 | 46.1 | 20.0 | 55.4 | 46.2 | **58.2** | 34.1 | 55.8 | 54.7 | 58.0 |
| fb3437 | 15.4 | 13.7 | 20.9 | 21.3 | 50.2 | 3.9 | 39.3 | 19.7 | **52.1** |
| fb1912 | 28.0 | 14.9 | 25.8 | 28.7 | 45.8 | 8.0 | 35.6 | 32.6 | **48.5** |
| amazon | 47.3 | 38.2 | 53.3 | 50.1 | 58.1 | 53.5 | 51.4 | 44.7 | **58.3** |
| youtube | 41.4 | 59.9 | 50.7 | 65.5 | **67.3** | 51.4 | 43.6 | 47.8 | 60.6 |
| dblp | 33.7 | 21.8 | 39.3 | 47.1 | **53.9** | 46.8 | 34.9 | 44.0 | 44.6 |

**Table 3.** Jaccard similarity (%) of each dataset in the community detection task.

| Dataset | SVI | MNMF | vGraph | ComE | J-ENC | CNRL | ComGAN | DGI | VGECD |
|---|---|---|---|---|---|---|---|---|---|
| fb698 | 30.0 | 16.0 | 43.6 | 33.8 | 50.4 | 9.6 | 46.9 | 42.1 | **57.2** |
| fb414 | 29.3 | 12.8 | 51.8 | 42.2 | 58.4 | 15.4 | 53.6 | 46.4 | **63.1** |
| fb686 | **33.9** | NA | 32.7 | NA | NA | NA | NA | NA | 33.4 |
| fb348 | 33.6 | 11.3 | 41.0 | 34.4 | 43.5 | 21.7 | 23.2 | 41.8 | **44.1** |
| fb3437 | 9.0 | 7.7 | 12.0 | 12.5 | 36.2 | 2.0 | 33.4 | 11.6 | **37.0** |
| fb1912 | 20.1 | 8.4 | 18.6 | 18.5 | 37.3 | 4.6 | 13.5 | 22.5 | **39.7** |
| amazon | 36.4 | 25.2 | 36.9 | 34.6 | 41.9 | 38.7 | 38.0 | 29.0 | **42.2** |
| youtube | 28.7 | 46.7 | 34.3 | 52.5 | **53.3** | 35.5 | 44.0 | 32.7 | 44.3 |
| dblp | 20.9 | 20.9 | 25.0 | 27.9 | **37.3** | 32.8 | 25.0 | 29.2 | 28.9 |

### 4.5 Experimental Results

Table 2 and Table 3 summarize the comparison results of VGECD and each baseline in community detection task. VGECD is superior to the comparison baselines in six datasets in F1 score, especially in fb698, fb414, and fb1912. VGECD is 5.9% higher than the suboptimal J-ENC in fb698, 2.5% higher than J-ENC in fb414, and 2.7% higher than J-ENC in fb1912. The results of VGECD in the remaining three datasets are also at the upstream level. VGECD also outperforms the comparison baselines in six datasets in Jaccard similarity, with significant improvements in fb698, fb414, and fb1912. These results show that the performance of the model on the community detection task can be improved by combining the attention mechanism with joint learning. In addition, all methods that obtain suboptimal results are joint learning community detection task and node embedding task, which also confirms the view of this paper that joint learning is conducive to community detection task.

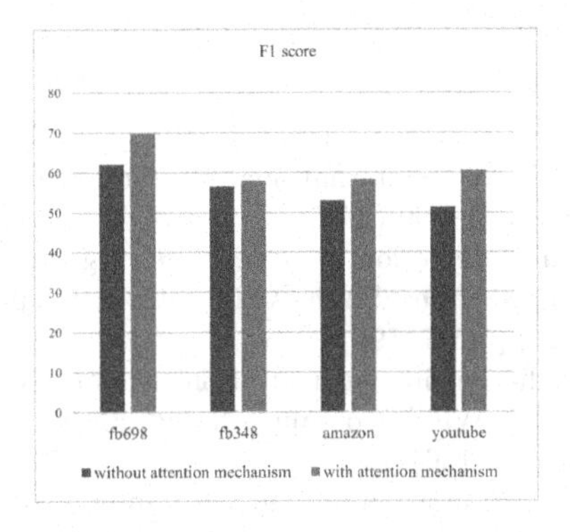


**Fig. 2.** F1 score comparison of datasets in two cases.

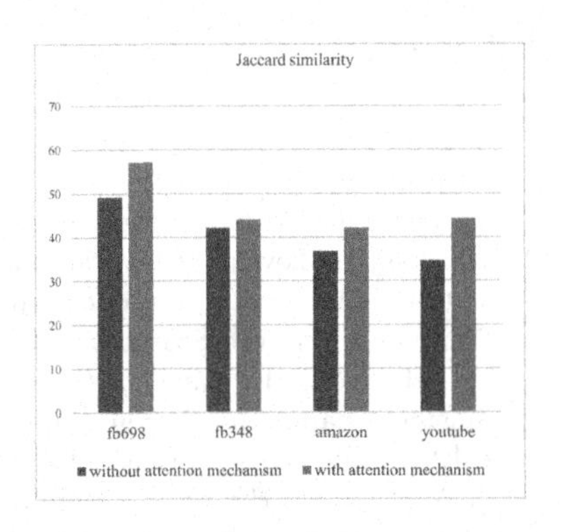


**Fig. 3.** Jaccard similarity comparison of datasets in two cases.

### 4.6 Ablation Study

We introduce GAT in our model, which is used to aggregate the neighbors of nodes, and adaptively allocate the weights of different neighbors to enhance the ability to generate node embeddings and community assignments. In order to verify its impact on VGECD, we conduct ablation experiments on four datasets, these datasets from small to large: fb698 ($v = 61$), fb348 ($v = 224$), amazon ($v = 794$), youtube ($v = 5346$). Figure 2 and Fig. 3 show the experimental results. Compared with the model without GAT, the model results after adding GAT are improved on the four datasets, especially in fb698 and youtube. The results prove that GAT has a positive effect on community detection task, and it improves the performance of VGECD.

## 5 Conclusion

In this paper, to improve the accuracy of the community detection task, we propose a new variational graph embedding generation model–VGECD. In our model, we use graph attention networks to assign weights to neighbor nodes adaptively. Graph attention networks fully exploit the neighborhood information of nodes and better generate node embeddings and community assignments. Furthermore, we use a joint learning approach, which learns community detection and node representation in the same framework to improve the model's capability in community detection task. Experimental results on nine datasets verify that VGECD outperforms the compared methods.

Community detection is widely used on homogeneous graphs but less on heterogeneous graphs. However, heterogeneous graphs contain rich information and are closer to real life. Therefore, it is an important research direction in future research to study community detection on heterogeneous graph networks.

## References

1. Cao, S., Lu, W., Xu, Q.: Grarep: learning graph representations with global structural information. In: Proceedings of the 24th ACM International Conference on Information and Knowledge Management, pp. 891–900 (2015)
2. Yang, J., Leskovec, J.: Overlapping community detection at scale: a nonnegative matrix factorization approach. In: Proceedings of the Sixth ACM International Conference on Web Search and Data Mining, pp. 587–596 (2013)
3. Liu, H., Kou, H., Yan, C., Qi, L.: Link prediction in paper citation network to construct paper correlation graph. EURASIP J. Wirel. Commun. Netw. **2019**(1), 1–12 (2019). https://doi.org/10.1186/s13638-019-1561-7
4. Li, Y., Sha, C., Huang, X.: Community detection in attributed graphs: an embedding approach. In: Thirty-Second AAAI Conference on Artificial Intelligence (2018)
5. Gopalan, P.K., Blei, D.M.: Efficient discovery of overlapping communities in massive networks. Proc. Natl. Acad. Sci. **110**(36), 14534–14539 (2013). https://doi.org/10.1073/pnas.1221839110
6. Tang, J., Aggarwal, C., Liu, H.: Node classification in signed social networks. In: Proceedings of the 2016 SIAM International Conference on Data Mining, pp. 54–62. SIAM (2016). https://doi.org/10.1137/1.9781611974348.7
7. Kozdoba, M., Mannor, S.: Community detection via measure space embedding. In: Advances in Neural Information Processing Systems, vol. 28 (2015)
8. Perozzi, B., Al-Rfou, R., Skiena, S.: Deepwalk: online learning of social representations. In: Proceedings of the 20th ACM SIGKDD International Conference on Knowledge Discovery and Data Mining, pp. 701–710 (2014)
9. Tu, C., et al.: A unified framework for community detection and network representation learning. IEEE Trans. Knowl. Data Eng. **31**(6), 1051–1065 (2018)
10. Cavallari, S., Zheng, V.W., Cai, H., Chang, K.C.C., Cambria, E.: Learning community embedding with community detection and node embedding on graphs. In: Proceedings of the 2017 ACM on Conference on Information and Knowledge Management, pp. 377–386 (2017). https://doi.org/10.1145/3132847.3132925
11. Kipf, T.N., Welling, M.: Variational graph auto-encoders. arXiv preprint arXiv:1611.07308 (2016). https://doi.org/10.48550/arXiv.1611.07308
12. Xie, J., Kelley, S., Szymanski, B.K.: Overlapping community detection in networks: the state-of-the-art and comparative study. ACM Comput. Surv. (CSUR) **45**(4), 1–35 (2013). https://doi.org/10.1145/2501654.2501657
13. Yang, J., McAuley, J., Leskovec, J.: Community detection in networks with node attributes. In: 2013 IEEE 13th International Conference on Data Mining, pp. 1151–1156. IEEE (2013). https://doi.org/10.1109/ICDM.2013.167
14. Sun, F.Y., Qu, M., Hoffmann, J., Huang, C.W., Tang, J.: vgraph: A generative model for joint community detection and node representation learning. In: Advances in Neural Information Processing Systems, vol. 32 (2019)
15. Grover, A., Leskovec, J.: node2vec: Scalable feature learning for networks. In: Proceedings of the 22nd ACM SIGKDD International Conference on Knowledge Discovery and Data Mining, pp. 855–864 (2016)
16. Tang, J., Qu, M., Wang, M., Zhang, M., Yan, J., Mei, Q.: Line: large-scale information network embedding. In: Proceedings of the 24th International Conference on World Wide Web, pp. 1067–1077 (2015)
17. Gilmer, J., Schoenholz, S.S., Riley, P.F., Vinyals, O., Dahl, G.E.: Neural message passing for quantum chemistry. In: International Conference on Machine Learning, pp. 1263–1272. PMLR (2017)

18. Khan, R.A., Anwaar, M.U., Kleinsteuber, M.: Epitomic variational graph autoencoder. In: 2020 25th International Conference on Pattern Recognition (ICPR), pp. 7203–7210. IEEE (2021). https://doi.org/10.1109/ICPR48806.2021.9412531
19. Wang, X., Cui, P., Wang, J., Pei, J., Zhu, W.: Community preserving network embedding. In: Thirty-First AAAI Conference on Artificial Intelligence (2017)
20. Khan, R.A., Anwaar, M.U., Kaddah, O., Han, Z., Kleinsteuber, M.: Unsupervised learning of joint embeddings for node representation and community detection. In: Oliver, N., Pérez-Cruz, F., Kramer, S., Read, J., Lozano, J.A. (eds.) ECML PKDD 2021. LNCS (LNAI), vol. 12976, pp. 19–35. Springer, Cham (2021). https://doi.org/10.1007/978-3-030-86520-7_2
21. Velickovic, P., Fedus, W., Hamilton, W.L., Liò, P., Bengio, Y., Hjelm, R.D.: Deep graph infomax. ICLR (Poster) **2**(3), 4 (2019)

# Counterfactual Causal Adversarial Networks for Domain Adaptation

Yan Jia, Xiang Zhang(✉), Long Lan, and Zhigang Luo

College of Computer Science and Technology,
National University of Defense Technology, Changsha, China
{jia.yan20,zhangxiang08,long.lan,zgluo}@nudt.edu.cn

**Abstract.** To eliminate domain shift in domain adaptation, most methods do so by encouraging the model to learn common features. However, the interpretability of these domain adaptation methods lacks in-depth research, and we note that the domain adaptation process can be regarded as a causal intervention, which can further form theoretical explanations in the causal relationship. Our proposed counterfactual causal adversarial network CCAN performs better on the domain adaptation task. Supported by causal theory, CCAN completes the adversarial learning of the network through counterfactual intervention, and uses the first proposed domain-adaptive causal effect to supervise the entire network. CCAN successfully validates the goal of evaluating the quality of domain adaptation through counterfactual intervention effects in causality to supervise the better completion of the entire task. The causal theory endows the whole CCAN with good interpretability. Experimental results on two challenging UDA benchmarks validate the superiority and effectiveness of CCAN for domain adaptation with counterfactual interventions based on causal theory, and analyze the role that domain adaptation causal effects play in the overall supervision.

**Keywords:** Unsupervised domain adaptation · Causal inference · Counterfactuals

## 1 Introduction

With the help of the interpretability and scalability of causal theory, the idea of combining deep learning and causal inference is gradually entering the field of vision of researchers. There have been successful cases of applying causal analysis tools in multiple domains, such as interpretable machine learning [18], natural language processing [27], and adversarial learning [11], to name a few. Deep learning brings good application and theoretical practice to causal theory. The causal theory can provide a scientific theoretical explanation for deep learning. Deep learning and causal theory complement each other. We can see that there have been many excellent works using causality as an effective tool to mitigate

This work was funded by Haihe Laboratory in Tianjin, Grants No. 22HHXCJC00007.


M. Tanveer et al. (Eds.): ICONIP 2022, CCIS 1793, pp. 698–709, 2023.
https://doi.org/10.1007/978-981-99-1645-0_58

the impact of dataset bias in vision tasks, including image classification [16], visual commonsense reasoning [26], etc.

Additionally, we note that domain adaptation tasks are deeply connected to causal inference. When applying a well-performing model learned from a source training set to a different but related target test set, it is usually assumed that it comes from the same distribution. But deep learning models have inherent generalization problems. A model trained on one dataset (source domain) performs poorly on another domain [21,24] due to domain shift. Domain adaptation is to solve the performance loss encountered by model transfer, and learn a good discriminative model precisely in the presence of a domain shift between the source and target domains. Unsupervised domain adaptation [13] attempts to transfer knowledge from a rich labeled source domain to an unlabeled target domain, making full use of unlabeled data in the target domain.

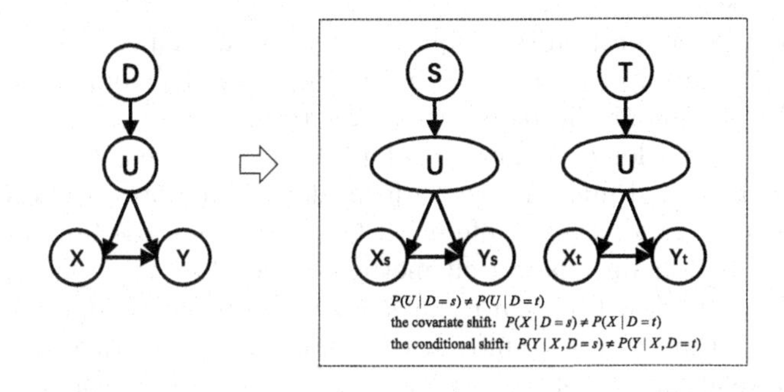


**Fig. 1.** Causal Map for a Single Domain in Domain Adaptation

When we approach the domain adaptation task from the perspective of causal inference, we can find that the goal of domain adaptation can be achieved as a causal intervention $P(Y|do(X), S)$ using the do-operator [20]. To understand this, we first abstract the DA problem turns into the causal map in Fig. 1. This figure shows the causal structure model when the network learns single-domain knowledge, D represents the instance in the domain, U represents the domain awareness, X represents the feature representation obtained by the network, and Y is the label. In this causal structure, D represents the input (usually the attributes of some things), U represents the confounding factor, Y represents the output (outcome), and X is the intervention (treatment). This directed acyclic graph (DAG) is similar to that in statistics, describing conditional dependencies and probabilities between different variables. In causal inference, a simple and effective way to determine causality is to change whether an intervention observation changes with it. If the outcome changes as the intervention changes, then it can be concluded that there is a causal relationship between the intervention and output Y. Obviously, this is consistent with what we generally expect, and the feature representation must directly affect the network's prediction of labels. After understanding the causal graph, the classification task X→Y is influenced by semantic features U (such as shape and background), where U→X corresponds to the generation of pixel-level image samples X, and U→Y corresponds to the definition from semantics to classes. This causal relationship has been verified in their respective parts [7]. For the whole domain adaptation, since $P(U|D = s) \neq P(U|D = t)$, the difference in domain variable D introduces the covariate shift $P(X|D = s) \neq P(X|D = t)$ [23] and the conditional shift $P(Y|X, D = s) \neq P(Y|X, D = t)$ [14], that is the domain shift. Domain-aware U is the confounding factor that prevents the model from learning the domain-invariant causal relationship X→Y. For this causal relationship, it has been theoretically demonstrated that statistical learning without causal intervention [19] cannot eliminate confounding effects. This shows that the causal intervention of the do-operator [20] is the task of domain adaptation methods. When we find the relationship between intervention and domain adaptation, counterfactual theory under further intervention will be applicable to domain adaptation tasks.

Many domain adaptation techniques have been developed in classical supervised machine learning to solve learning problems under domain transfer, but few have thought about the regularity and interpretability of model operation from the underlying logic of domain adaptation. We attempt to exploit their potential use in causal inference to perform domain adaptation tasks in a way that is interpretable for causal inference. Only the final loss function is supervised in current domain adaptation methods, and this direct likelihood-based method only supervises the final prediction. When we are supported by causal inference, we can introduce counterfactual interventions to take full advantage of the causal relationships of individual variables and predictions. Based on analyzing the deep connection between causal inference and domain adaptation, we propose the CCAN network and propose domain adaptation causal effects for the first time. Taking causality as the theoretical basis, fully excavate the causal relationship in domain adaptation, and complete the task of domain adaptation better.

The contributions of this paper are as follows: 1) Our proposed counterfactual causal adversarial network CCAN performs better on the domain adaptation task. Supported by causal theory, CCAN completes the adversarial learning of the network through counterfactual intervention, and uses the first proposed domain-adaptive causal effect to supervise the entire network. 2) We dissect the underlying theory of causality and domain adaptation task homology, enriching the application of causal theory and domain adaptation interpretability. And CCAN successfully validates that the goal of assessing the quality of domain adaptation through counterfactual intervention effects in causality is achievable to better supervise the completion of the entire task. 3) Experimental results on two challenging UDA benchmarks validate the superiority and effectiveness of CCAN for domain adaptation with counterfactual interventions based on causal theory, and analyze the role that domain adaptation causal effects play in the overall supervision.

## 2 Related Works

### 2.1 Unsupervised Domain Adaptation

Our work is mainly oriented towards unsupervised domain adaptation (UDA), where no labeled target images are available during training. Some methods are alignment-based domain adaptation. [4] improved adversarial feature adaptation to accomplish alignment. [28] adapt the feature specifications of these two domains to a wide range of values, making the learned features both task-discriminative and domain-invariant. Some methods are domain adaptation based on clustering. The clustering hypothesis states that the classification boundaries should not pass through high-density regions, but lie in low-density regions [3]. To strengthen the clustering assumption, conditional entropy minimization is widely used [5,28]. There are also methods based on potential domain discovery. The methods of latent domain discovery [17] assume that the data may actually contain multiple distinct distributions, and thus focus on capturing the underlying structure of source, target, or mixed data. However, our method will be based on causal inference. Compared with previous methods, it has outstanding interpretability, and there are few studies in this direction.

### 2.2 Causal Inference

Causal reasoning studies the effect on outcomes when some causes change variables, which is clearly different from traditional social reasoning. And the relational reasoning under this variable change makes it applicable in neural networks. Because most neural network models are now treated as black boxes, explicit cause-and-effect relationships can increase the transparency of deep models. Interest in combining deep learning and causal inference has grown rapidly in recent years. Typical counterfactuals [1] have recently gained interest in various fields of machine learning, especially applying insights from causal inference to augment training in reinforcement learning [2] and explanations [9]. Adversarial learning [8] has been found to be an excellent tool for using counterfactuals and has been shown to improve performance. However, most state-of-the-art techniques in this area have focused on the direct use of interventions, lacking a deeper integration of counterfactual causality, and not closely fitting the underlying link between the task and causal reasoning. Therefore, we propose CCAN. In addition to constructing counterfactual intervention instances, we additionally consider counterfactual intervention effects. For the first time, we point out the domain adaptation causal effect based on counterfactual causal inference on the task and use it to supervise the entire network. The successful application of the causal effect of domain adaptation further indicates that causal inference will have more theories that can be applied to the entire task to complete domain adaptation in a highly interpretable manner.

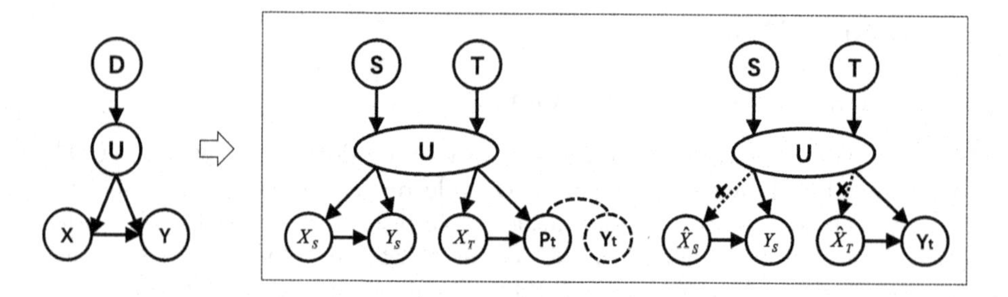


**Fig. 2.** Overview of our CCAN counterfactual causal inference.

# 3 Methodology

## 3.1 Counterfactual Causal Inference for Domain Adaptation

Combined with Fig. 1, in the introduction, we describe the general causality of domain adaptation in detail, and point out the corresponding part of domain shift in causality, which is applicable to all tasks in domain adaptation. The meanings of the variables shown in Fig. 2 are similar to those in Fig. 1. D represents the instance in the domain, U represents the domain perception, X represents the feature representation obtained by the network, and Y represents the label. In this causal structure, D represents the input (generally the attributes of some things), U represents the confounding factor, Y represents the output (effect), and X is the intervention (treatment). After obtaining a clear causal relationship, it is necessary to eliminate this domain shift in domain adaptation, that is, the confounding element. We thought of counterfactual causal inference, which has a huge impact in causality, and on the basis of Fig. 1, we found a new causal relationship after the introduction as shown in Fig. 3. When we use a network to implement domain adaptation end-to-end, both source and target domain data will be fed into the network simultaneously, which corresponds to the first-layer relationship in the graph.

And the domain-aware U for the same network will obtain the confounding factors of both the source and target domains. The semantic features that the network can perceive will be mixed, which will lead to the following results when the source and target domain semantic features are not similar: The source domain will always be able to achieve accurate prediction from X→Y due to its supervision information, while the weak target domain data will be under the influence of mixed U, the predicted P will be different from the actual label. The more complex the confounding element, in other words, the less similar the domains are, which will lead to the greater the difference between P and Y, and the weaker the target domain. For such causal relationships with confounders, statistical learning of causal intervention [39] has theoretically been demonstrated to eliminate confounding effects, using the do-operator [41] for

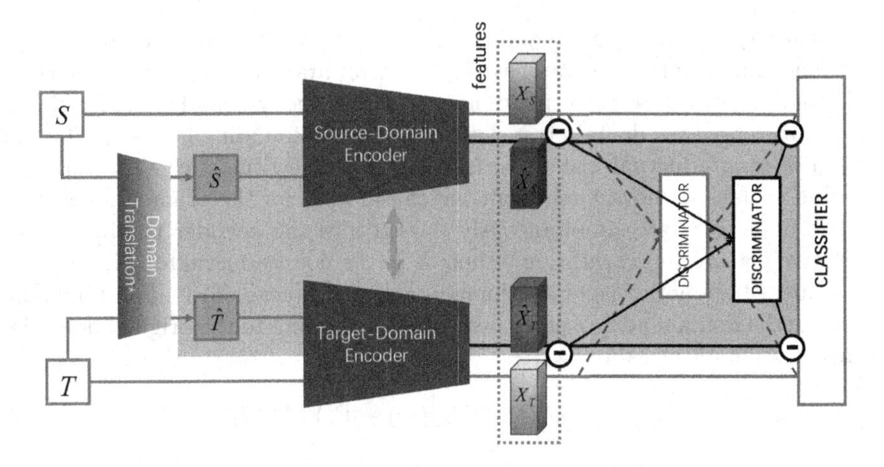


**Fig. 3. Outline of our CCAN Framework.** The blue part of the background in the structure corresponds to the counterfactual causal inference part. The source and target domain encoders share a set of parameters. Note that for the domain translator here, we choose CycleGan to generate the corresponding counterfactual instances, which is a generator with fixed parameters after pre-training. (Color figure online)

causal intervention $P(Y|do(X), S), P(Y|do(X), T)$to fulfill. In a network, we can directly intervene in X through the design of the network, and U is difficult to be accurately and comprehensively defined in all kinds of information hidden in the domain (background, foreground, light, color, style, etc.).

This brings us to the idea of using domain translation to generate counterfactual instances of each individual in the original domain, which can be understood as ideal samples (samples with certain confounding factors removed). We replace the original sample feature representation with these counterfactual sample feature representations, which can be regarded as intervention operations. Under this operation, the relationship between X and U will be cut off, and after eliminating the influence of the main contradiction U, accurate prediction of X-Y will be achieved. This is the effect reflected in the cause and effect diagram in the far right part of Fig. 2.

At the same time, when the counterfactual intervention mechanism is established, the theory used to evaluate the intervention effect in the cause-and-effect theory will also be able to be extended to domain adaptation. We elaborate on how to design a "domain-adaptation causal effect" in the discussion of the network in the next section.

### 3.2 Counterfactual Causal Adversarial Networks

Counterfactual causal intervention first requires constructing counterfactual conditions. When we employ an intervention on X, we need to get counterfactual instances. For domain adaptation tasks, counterfactual instances are specified individuals whose ties are removed from variable-fixing confounders. Therefore, we first use the source domain and target domain data to train a domain translator, and then use the generator with the fixed parameters to obtain the counterfactual instance $\hat{S}, \hat{T}$ corresponding to each instance S and T. The fixation of the parameters of the domain translator fixes the confounding factors in the generation of counterfactual instances. In CCAN, the new instance $\hat{S}, \hat{T}$ generated is a kind of data reinforcement from the causal angle of the original instance.

The data pairs are simultaneously passed into the encoder to obtain their respective feature representations, then $\hat{X}$ is the counterfactual of X, and they will be used together to generate domain-adaptive causal effects. Following [25], we define the domain adaptation causal effect according to the original definition of intervention effect as follows:

$$Y = \mathrm{E}_{\hat{X}} \left[ P(Y|do(\hat{X}), D) - P(Y|X, D) \right] \tag{1}$$

The effect of the entire domain adaptation network under intervention is expressed as the difference between the original representation's prediction $P(Y|X, D)$ and its counterfactual surrogate $P(Y|do(\hat{X}), D)$.

As shown in Fig. 3, the blue part of the background can be seen as the domain adaptation in the case of counterfactual intervention, and the outside of the blue part is the original domain adaptation operation. In the original part, we use the conditional adversarial loss to optimize the network, input the original feature representation and the prediction obtained by the classifier together into the domain discriminator, and calculate:

$$\mathcal{L}_{\mathrm{D}} = -\frac{1}{n_{\mathrm{s}}} \sum_{i=1}^{n_{\mathrm{s}}} \log \left[ D\left(f_i^{\mathrm{s}}, y_{\mathrm{c}i}^{\mathrm{s}}\right) \right] - \frac{1}{n_{\mathrm{t}}} \sum_{j=1}^{n_{\mathrm{t}}} \log \left[ 1 - D\left(f_j^{\mathrm{t}}, y_{\mathrm{c}j}^{\mathrm{t}}\right) \right] \tag{2}$$

To be able to utilize the domain adaptation causal effect in Eq. 1 to achieve supervision over the entire network, we find that we can cleverly design a new domain discriminator, but in particular by feeding the resulting feature representations of individual data pairs in the form of a difference operation. And the classifier prediction is also input to the discriminator in differential form, and the domain adaptation effect loss is calculated to supervise the current effect of the entire domain adaptation model:

$$\begin{aligned} \hat{\mathcal{L}}_{\mathrm{D-effect}} = &-\tfrac{1}{n_{\mathrm{s}}} \sum_{i=1}^{n_{\mathrm{s}}} \log \left[ D\left(\hat{f}_i^{\mathrm{s}} - f_i^{\mathrm{s}}, \hat{y}_{\mathrm{c}i}^{\mathrm{s}} - y_{\mathrm{c}i}^{\mathrm{s}}\right) \right] \\ &-\tfrac{1}{n_{\mathrm{t}}} \sum_{j=1}^{n_{\mathrm{t}}} \log \left[ 1 - D\left(\hat{f}_j^{\mathrm{t}} - f_j^{\mathrm{t}}, \hat{y}_{\mathrm{c}j}^{\mathrm{t}} - y_{\mathrm{c}j}^{\mathrm{t}}\right) \right] \end{aligned} \tag{3}$$

Domain discriminator has a similar principle as in [13]. Both domain discriminator structures in CCAN adopt a similar construction.

### 3.3 Objective Function

Outside the CCAN, the classifier is optimized by computing the standard cross-entropy loss $\mathcal{L}_{CE}$.

$$\mathcal{L}_{CE} = -\sum_{m=1}^{n} (y_m^{\mathrm{s}}) \log (f(x_{\mathrm{c}m}^{\mathrm{s}})) - \sum_{m=1}^{n} (\hat{y}_m^{\mathrm{s}}) \log (f(\hat{x}_{\mathrm{c}m}^{\mathrm{s}})) \tag{4}$$

Therefore, the overall training objective uses a combination of three losses, which can be formulated as follows:

$$\mathcal{L} = \mathcal{L}_{\mathrm{CE}} - \alpha * \mathcal{L}_D - \beta * \hat{\mathcal{L}}_{\mathrm{D-effect}} \tag{5}$$

## 4 Experiments

### 4.1 Datasets and Setups

We validate our method on three public benchmarks. Office-31 contains three distinct domains and consists of 4110 images belonging to 31 classes, namely Amazon, Webcam and DSLR. The dataset is imbalanced across domains. Amazon contains 2817 images from amazon.com; Webcam contains 795 images taken by a web camera; And DSLR contains 498 images taken by a digital SLR camera with different settings. Office-Home contains 15,588 images in 65 categories across 4 domains. Art represents an artistic depiction of images of multiple objects; Clipart describes picture collections of clipart; Product consists of images of objects with a clear background. Images of Real World are collected with regular cameras. Tt is challenging to perform domain adaptation on this dataset well.

The two encoders in CCAN share parameters and use ResNet-50 [10] pre-trained on ImageNet as our backbone network. The classifier consists of linear layer, BN layer and relu layer, and the whole network shares a classifier. We train the network using mini-batch stochastic gradient descent (SGD) with momentum 0.9. We follow the same learning rate schedule described in [15]. The initial learning rate, i.e. the convolutional layer, is 0.001.

### 4.2 Performance Analysis

We generate performance performance for different tasks on three datasets, Office-31, Office-Home. To get a better analysis of the performance of CCAN, we cite the performance in papers of multiple methods. We evaluate our CCAN and previous excellent UDA methods to illustrate the effectiveness of CCAN methods in solving domain adaptation problems. Detailed data can be seen in Table 1 and Table 2. Excellent works such as DAN [12], DANN [6], CDAN [13], SymNets [29] are compared using the performance reported in their paper. In addition, we also provide basic performance results of the backbone network Resnet50. From Table 1, we can see that CCAN has achieved an excellent 89.5% on average. Even on individual tasks, the performance has improved significantly. Table 2 shows

**Table 1.** Accuracy (%) on Office-31 dataset (based on ResNet50)

| Method | A→W | A→D | W→ A | W→D | D→A | D→W | Ave. |
|---|---|---|---|---|---|---|---|
| ResNet50 [10] | 68.4 | 68.9 | 60.7 | 99.3 | 62.5 | 96.7 | 76.1 |
| DAN [12] | 80.5 | 78.6 | 62.8 | 99.6 | 63.6 | 62.8 | 80.4 |
| DANN [6] | 82.0 | 79.7 | 67.4 | 99.1 | 68.2 | 96.9 | 82.2 |
| JAN [15] | 85.4 | 84.7 | 70.0 | 99.8 | 68.6 | 97.4 | 84.3 |
| GSM [30] | 85.9 | 84.1 | 75.5 | 97.2 | 73.6 | 97.1 | 85.3 |
| MCD [22] | 88.6 | 92.2 | 69.7 | 100 | 69.5 | 98.5 | 86.5 |
| CDAN+E [13] | 94.1 | 92.9 | 69.3 | 100 | 71.0 | 98.6 | 87.7 |
| SymNets [29] | 90.8 | 93.9 | 72.5 | 100 | 74.6 | 98.8 | 88.4 |
| MDD [31] | 94.5 | 93.5 | 72.2 | 100 | 74.6 | 98.4 | 88.9 |
| CCAN(ours) | **94.9** | 93.5 | **75.7** | 99.8 | **75.0** | 98.5 | **89.5** |

**Table 2.** Accuracy (%) on OfficeHome dataset (based on ResNet50)

| Method | A→C | A→P | A→R | C→A | C→P | C→R | P→A | P→C | P→R | R→A | R→C | R→P | Ave. |
|---|---|---|---|---|---|---|---|---|---|---|---|---|---|
| ResNet50 | 34.9 | 50.0 | 58.0 | 37.4 | 41.9 | 46.2 | 38.5 | 31.2 | 60.4 | 53.9 | 41.2 | 59.9 | 46.1 |
| DAN | 43.6 | 57.0 | 67.9 | 45.8 | 56.5 | 60.4 | 44.0 | 43.6 | 67.7 | 63.1 | 51.5 | 74.3 | 56.3 |
| DANN | 45.6 | 59.3 | 70.1 | 47.0 | 58.5 | 60.9 | 46.1 | 43.7 | 68.5 | 63.2 | 51.8 | 76.8 | 57.6 |
| JAN | 45.9 | 61.2 | 68.9 | 50.4 | 59.7 | 61.0 | 45.8 | 43.4 | 70.3 | 63.9 | 52.4 | 76.8 | 58.3 |
| MCD | 48.9 | 68.3 | 74.6 | 61.3 | 67.6 | 68.8 | 57.0 | 47.1 | 75.1 | 69.1 | 52.2 | 79.6 | 64.1 |
| CDAN+E | 50.7 | 70.6 | 76.0 | 57.6 | 70.0 | 70.0 | 57.4 | 50.9 | 77.3 | 70.9 | 56.7 | 81.6 | 65.8 |
| GSM | 49.4 | 75.5 | 80.2 | 62.9 | 70.6 | 70.3 | 65.6 | 50.0 | 80.8 | 72.4 | 50.4 | 81.6 | 67.5 |
| SymNets | 47.7 | 72.9 | 78.5 | 64.2 | 71.3 | 74.2 | 64.2 | 48.8 | 79.5 | 74.5 | 52.6 | 82.7 | 67.6 |
| MDD | 54.9 | 73.7 | 77.8 | 60.0 | 71.4 | 71.8 | 61.2 | 53.6 | 78.1 | 72.5 | 60.2 | 82.3 | 68.1 |
| CCAN | **56.8** | 72.6 | 77.7 | 62.9 | **74.0** | **74.8** | 62.6 | **54.6** | 76.7 | 74.0 | 58.7 | **83.8** | **69.1** |

the accuracy performance of ACCN on 12 tasks on the OfiiceHome dataset. On average, we also achieve best-in-class accuracy. The accuracy performance of individual tasks is similar to the advanced methods.

At the same time, in order to analyze the influence and effect of domain adaptation causal effect in the whole network model, we conduct ablation experiments on both datasets, and we list the data performance of some typical tasks in Table 3. From the data we can notice that the domain adaptation causal effect plays a significant role in the whole network. It achieves our original goal to supervise the optimization of the entire model which is effective. The successful application of domain adaptation causal effects will show that there can be many more causal theories that can be transferred to domain adaptation tasks, and based on our causal analysis of domain adaptation, more designs can be derived to promote better domain adaptation mission accomplished.

In order to visualize the domain adaptation effect of the entire model, we use the t-SNE technique to visualize the feature representation of the original instance in the network in Fig. 4. The three graphs reflect the performance of the

**Table 3.** The role of domain adaptation causal effects on Office-31 and OfficeHome

| Office-31 | A→D | D→A | A→W | W→A |
|---|---|---|---|---|
| w/o. $\hat{\mathcal{L}}_{\text{D-effect}}$ | 93.0 | 72.6 | 94.5 | 72.3 |
| increment | 0.5% | 2.4% | 0.4% | 3.4% |
| Office-Home | Pr→Ar | Pr→Cl | Ar→Pr | Cl→ Pr |
| w/o. $\hat{\mathcal{L}}_{\text{D-effect}}$ | 60.3 | 53.0 | 71.1 | 71.2 |
| increment | 2.3% | 1.7% | 1.5% | 2.8% |

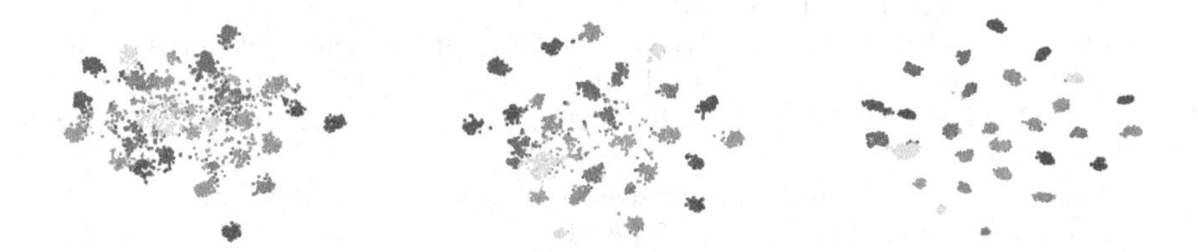

**Fig. 4.** Visualization of the CCAN on W→A with t-SNE.

model at the beginning of the optimization, during the optimization, and at the end of the final optimization. We can clearly observe that the boundary becomes clearer from the initial mixed state to the final superior state, which clearly meets our requirements for the domain adaptation task. This further validates the ideal state of our model design and the correctness of the application of causal theory.

## 5 Conclusion

Our proposed counterfactual causal adversarial network CCAN performs better on the domain adaptation task. Supported by causal theory, CCAN completes the adversarial learning of the network through counterfactual intervention, and uses the first proposed domain adaptation causal effect to supervise the entire network. We explain causality for general domain adaptation tasks and domain adaptation causality for inference under counterfactuals. CCAN successfully validates the goal of evaluating the quality of domain adaptation through counterfactual intervention effects in causality to supervise the better completion of the entire task. Experimental results on three challenging UDA benchmarks validate the superiority and effectiveness of CCAN for domain adaptation with counterfactual interventions based on causal theory, and analyze the role that domain adaptation causal effects play in the overall supervision.

## References

1. Bottou, L., et al.: Counterfactual reasoning and learning systems: the example of computational advertising. J. Mach. Learn. Res. **14**(11) (2013)

2. Buesing, L., et al.: Woulda, coulda, shoulda: counterfactually-guided policy search. arXiv preprint arXiv:1811.06272 (2018)
3. Chapelle, O., Zien, A.: Semi-supervised classification by low density separation. In: International Workshop on Artificial Intelligence and Statistics, pp. 57–64. PMLR (2005)
4. Chen, X., Wang, S., Long, M., Wang, J.: Transferability vs. discriminability: batch spectral penalization for adversarial domain adaptation. In: International Conference on Machine Learning, pp. 1081–1090. PMLR (2019)
5. Cicek, S., Soatto, S.: Unsupervised domain adaptation via regularized conditional alignment. In: Proceedings of the IEEE/CVF International Conference on Computer Vision, pp. 1416–1425 (2019)
6. Ghifary, M., Kleijn, W.B., Zhang, M.: Domain adaptive neural networks for object recognition. In: Pham, D.-N., Park, S.-B. (eds.) PRICAI 2014. LNCS (LNAI), vol. 8862, pp. 898–904. Springer, Cham (2014). https://doi.org/10.1007/978-3-319-13560-1_76
7. Goodfellow, I., et al.: Generative adversarial nets. In: Advances in Neural Information Processing Systems, vol. 27 (2014)
8. Goodfellow, I.J., Shlens, J., Szegedy, C.: Explaining and harnessing adversarial examples. arXiv preprint arXiv:1412.6572 (2014)
9. Goyal, Y., Wu, Z., Ernst, J., Batra, D., Parikh, D., Lee, S.: Counterfactual visual explanations. In: International Conference on Machine Learning, pp. 2376–2384. PMLR (2019)
10. He, K., Zhang, X., Ren, S., Sun, J.: Deep residual learning for image recognition. In: Proceedings of the IEEE Conference on Computer Vision and Pattern Recognition, pp. 770–778 (2016)
11. Kocaoglu, M., Snyder, C., Dimakis, A.G., Vishwanath, S.: CausalGAN: learning causal implicit generative models with adversarial training. arXiv preprint arXiv:1709.02023 (2017)
12. Long, M., Cao, Y., Wang, J., Jordan, M.: Learning transferable features with deep adaptation networks. In: International Conference on Machine Learning, pp. 97–105. PMLR (2015)
13. Long, M., Cao, Z., Wang, J., Jordan, M.I.: Conditional adversarial domain adaptation. In: Advances in Neural Information Processing Systems, vol. 31 (2018)
14. Long, M., Wang, J., Ding, G., Sun, J., Yu, P.S.: Transfer feature learning with joint distribution adaptation. In: Proceedings of the IEEE International Conference on Computer Vision, pp. 2200–2207 (2013)
15. Long, M., Zhu, H., Wang, J., Jordan, M.I.: Deep transfer learning with joint adaptation networks. In: International Conference on Machine Learning, pp. 2208–2217. PMLR (2017)
16. Lopez-Paz, D., Nishihara, R., Chintala, S., Scholkopf, B., Bottou, L.: Discovering causal signals in images. In: Proceedings of the IEEE Conference on Computer Vision and Pattern Recognition, pp. 6979–6987 (2017)
17. Mancini, M., Porzi, L., Bulo, S.R., Caputo, B., Ricci, E.: Boosting domain adaptation by discovering latent domains. In: Proceedings of the IEEE Conference on Computer Vision and Pattern Recognition, pp. 3771–3780 (2018)
18. Mothilal, R.K., Sharma, A., Tan, C.: Explaining machine learning classifiers through diverse counterfactual explanations. In: Proceedings of the 2020 Conference on Fairness, Accountability, and Transparency, pp. 607–617 (2020)
19. Parascandolo, G., Kilbertus, N., Rojas-Carulla, M., Schölkopf, B.: Learning independent causal mechanisms. In: International Conference on Machine Learning, pp. 4036–4044. PMLR (2018)

20. Pearl, J., Bareinboim, E.: External validity: from do-calculus to transportability across populations. In: Probabilistic and Causal Inference: The Works of Judea Pearl, pp. 451–482 (2022)
21. Pei, Z., Cao, Z., Long, M., Wang, J.: Multi-adversarial domain adaptation. In: Thirty-Second AAAI Conference on Artificial Intelligence (2018)
22. Saito, K., Watanabe, K., Ushiku, Y., Harada, T.: Maximum classifier discrepancy for unsupervised domain adaptation. In: Proceedings of the IEEE Conference on Computer Vision and Pattern Recognition, pp. 3723–3732 (2018)
23. Sugiyama, M., Krauledat, M., Müller, K.R.: Covariate shift adaptation by importance weighted cross validation. J. Mach. Learn. Res. **8**(5) (2007)
24. Tzeng, E., Hoffman, J., Saenko, K., Darrell, T.: Adversarial discriminative domain adaptation. In: Proceedings of the IEEE Conference on Computer Vision and Pattern Recognition, pp. 7167–7176 (2017)
25. VanderWeele, T.: Explanation in Causal Inference: Methods for Mediation and Interaction. Oxford University Press, Oxford (2015)
26. Wang, T., Huang, J., Zhang, H., Sun, Q.: Visual commonsense R-CNN. In: Proceedings of the IEEE/CVF Conference on Computer Vision and Pattern Recognition, pp. 10760–10770 (2020)
27. Wood-Doughty, Z., Shpitser, I., Dredze, M.: Challenges of using text classifiers for causal inference. In: Proceedings of the Conference on Empirical Methods in Natural Language Processing. Conference on Empirical Methods in Natural Language Processing, vol. 2018, p. 4586. NIH Public Access (2018)
28. Xu, R., Li, G., Yang, J., Lin, L.: Larger norm more transferable: an adaptive feature norm approach for unsupervised domain adaptation. In: Proceedings of the IEEE/CVF International Conference on Computer Vision, pp. 1426–1435 (2019)
29. Zhang, Y., Tang, H., Jia, K., Tan, M.: Domain-symmetric networks for adversarial domain adaptation. In: Proceedings of the IEEE/CVF Conference on Computer Vision and Pattern Recognition, pp. 5031–5040 (2019)
30. Zhang, Y., Xie, S., Davison, B.D.: Transductive learning via improved geodesic sampling. In: BMVC, p. 122 (2019)
31. Zhang, Y., Liu, T., Long, M., Jordan, M.: Bridging theory and algorithm for domain adaptation. In: International Conference on Machine Learning, pp. 7404–7413. PMLR (2019)

# Author Index




M. Tanveer et al. (Eds.): ICONIP 2022, CCIS 1793, pp. 711–714, 2023.
https://doi.org/10.1007/978-981-99-1645-0

Printed in the United States
by Baker & Taylor Publisher Services